Springer Finance

Springer Finance

Springer Finance is a programme of books addressing students, academics and practitioners working on increasingly technical approaches to the analysis of financial markets. It aims to cover a variety of topics, not only mathematical finance but foreign exchanges, term structure, risk management, portfolio theory, equity derivatives, and financial economics.

Ammann M., Credit Risk Valuation: Methods, Models, and Application (2001)
Back K., A Course in Derivative Securities: Introduction to Theory and Computation (2005)
Barucci E., Financial Markets Theory. Equilibrium, Efficiency and Information (2003)
Bielecki T.R. and Rutkowski M., Credit Risk: Modeling, Valuation and Hedging (2002)
Bingham N.H. and Kiesel R., Risk-Neutral Valuation: Pricing and Hedging of Financial Derivatives (1998, 2nd ed. 2004)
Brigo D. and Mercurio F., Interest Rate Models: Theory and Practice (2001, 2nd ed. 2006)
Buff R., Uncertain Volatility Models – Theory and Application (2002)
Carmona R.A. and Tehranchi M.R., Interest Rate Models: An Infinite Dimensional Stochastic Analysis Perspective (2006)
Dana R.-A. and Jeanblanc M., Financial Markets in Continuous Time (2003)
Deboeck G. and Kohonen T. (Editors), Visual Explorations in Finance with Self-Organizing Maps (1998)
Delbaen F. and Schachermayer W., The Mathematics of Arbitrage (2005)
Elliott R.J. and Kopp P.E., Mathematics of Financial Markets (1999, 2nd ed. 2005)
Fengler M.R., Semiparametric Modeling of Implied Volatility (2005)
Filipović D., Term-Structure Models (2009)
Fusai G. and Roncoroni A., Implementing Models in Quantitative Finance: Methods and Cases (2008)
Jeanblanc M., Yor M. and Chesney M., Mathematical Methods for Financial Markets (2009)
Geman H., Madan D., Pliska S.R. and Vorst T. (Editors), Mathematical Finance – Bachelier Congress 2000 (2001)
Gundlach M. and Lehrbass F. (Editors), CreditRisk$^+$ in the Banking Industry (2004)
Jondeau E., Financial Modeling Under Non-Gaussian Distributions (2007)
Kabanov Y.A. and Safarian M., Markets with Transaction Costs (2008 forthcoming)
Kellerhals B.P., Asset Pricing (2004)
Külpmann M., Irrational Exuberance Reconsidered (2004)
Kwok Y.-K., Mathematical Models of Financial Derivatives (1998, 2nd ed. 2008)
Malliavin P. and Thalmaier A., Stochastic Calculus of Variations in Mathematical Finance (2005)
Meucci A., Risk and Asset Allocation (2005, corr. 2nd printing 2007)
Pelsser A., Efficient Methods for Valuing Interest Rate Derivatives (2000)
Prigent J.-L., Weak Convergence of Financial Markets (2003)
Schmid B., Credit Risk Pricing Models (2004)
Shreve S.E., Stochastic Calculus for Finance I (2004)
Shreve S.E., Stochastic Calculus for Finance II (2004)
Yor M., Exponential Functionals of Brownian Motion and Related Processes (2001)
Zagst R., Interest-Rate Management (2002)
Zhu Y.-L., Wu X., Chern I.-L., Derivative Securities and Difference Methods (2004)
Ziegler A., Incomplete Information and Heterogeneous Beliefs in Continuous-time Finance (2003)
Ziegler A., A Game Theory Analysis of Options (2004)

Monique Jeanblanc · Marc Yor · Marc Chesney

Mathematical Methods for Financial Markets

Monique Jeanblanc
Université d'Evry
Dépt. Mathématiques
rue du Père Jarlan
91025 Evry CX
France
monique.jeanblanc@univ-evry.fr

Marc Chesney
Universität Zürich
Inst. Schweizerisches
Bankwesen (ISB)
Plattenstr. 14
8032 Zürich
Switzerland

Marc Yor
Université Paris VI
Labo. Probabilités et Modèles
Aléatoires
175 rue du Chevaleret
75013 Paris
France

ISBN 978-1-4471-2524-2 e-ISBN 978-1-84628-737-4
DOI 10.1007/978-1-84628-737-4
Springer Dordrecht Heidelberg London New York

British Library Cataloguing in Publication Data
A catalogue record for this book is available from the British Library

Mathematics Subject Classification (2000): 60-00; 60G51; 60H30; 91B28

Cover design: WMXDesign GmbH

Printed on acid-free paper

Springer is part of Springer Science+Business Media (www.springer.com)

Preface

We translate to the domain of mathematical finance what F. Knight wrote, in substance, in the preface of his *Essentials of Brownian Motion and Diffusion* (1981): "it takes some temerity for the prospective author to embark on yet another discussion of the concepts and main applications of mathematical finance". Yet, this is what we have tried to do in our own way, after considerable hesitation.

Indeed, we have attempted to fill the gap that exists in this domain between, on the one hand, mathematically oriented presentations which demand quite a bit of sophistication in, say, functional analysis, and are thus difficult for practitioners, and on the other hand, mainstream mathematical finance books which may be hard for mathematicians just entering into mathematical finance.

This has led us, quite naturally, to look for some compromise, which in the main consists of the gradual introduction, at the same time, of a financial concept, together with the relevant mathematical tools.

Interlacing: This program interlaces, on the one hand, the financial concepts, such as arbitrage opportunities, admissible strategies, contingent claims, option pricing, default risk and ruin problems, and on the other hand, Brownian motion, diffusion processes, Lévy processes, together with the basic properties of these processes. We have chosen to discuss essentially continuous-time processes, which in some sense correspond to the real-time efficiency of the markets, although it would also be interesting to study discrete-time models. We have not done so, and we refer the reader to some relevant bibliography in the Appendix at the end of this book. Another feature of our book is that in the first half we concentrate on continuous-path processes, whereas the second half deals with discontinuous processes.

Special features of the book: Intending that this book should be readable for both mathematicians and practitioners, we were led to a somewhat unusual organisation, in particular:

1. in a number of cases, when the discussion becomes too technical, in the Mathematics or the Finance direction, we give only the essence of the argument, and send the reader to the relevant references,
2. we sometimes wanted a given section, or paragraph, to contain most of the information available on the topic treated there. This led us to:
   a) some forward references to topics discussed further in the book, which we indicate throughout the book with an arrow ($\rightarrowtail$)
   b) some repetition or at least duplication of the same kind of topic in various degrees of generality. Let us give an important example: Itô's formula is presented successively for continuous path semi-martingales, Poisson processes, general semi-martingales, mixed processes and Lévy processes.

We understand that this way of writing breaks away with the academic tradition of book writing, but it may be more convenient to access an important result or method in a given context or model.

About the contents: At this point of the Preface, the reader may expect to find a detailed description of each chapter. In fact, such a description is found at the beginning of each chapter, and for the moment we simply refer the reader to the Contents and the user's guide, which follows the Contents.

Numbering: In the following, C,S,B,R are integers. The book consists of two parts, eleven chapters and two appendices. Each chapter C is divided into sections C.S., which in turn are divided into subsections C.S.B. A statement in Subsection C.S.B. is numbered as C.S.B.R. Although this system of numbering is a little heavy, it is the only way we could find of avoiding confusion between the numbering of statements and unrelated sections.

What is missing in this book? Besides discussing the content of this book, let us also indicate important topics that are not considered here: The term structure of interest rate (in particular Heath-Jarrow-Morton and Brace-Gatarek-Musiela models for zero-coupon bonds), optimization of wealth, transaction costs, control theory and optimal stopping, simulation and calibration, discrete time models (ARCH, GARCH), fractional Brownian motion, Malliavin Calculus, and so on.

History of mathematical finance: More than 100 years after the thesis of Bachelier [39, 41], mathematical finance has acquired a history that is only slightly evoked in our book, but by now many historical accounts and surveys are available. We recommend, among others, the book devoted to Bachelier by Courtault and Kabanov [199], the book of Bouleau [114] and the collective book [870], together with introductory papers of Broadie and Detemple [129], Davis [221], Embrechts [321], Girlich [392], Gobet [395, 396], Jarrow and Protter [480], Samuelson [758], Taqqu [819] and Rogers [738], as well as the seminal papers of Black and Scholes [105], Harrison and Kreps [421] and Harrison and Pliska [422, 423]. It is also interesting to read the talks given by the Nobel prize winners Merton [644] and Scholes [764] at the Royal Academy of Sciences in Stockholm.

A philosophical point: Mathematical finance raises a number of problems in probability theory. Some of the questions are deeply rooted in the developments of stochastic processes (let us mention Bachelier once again), while some other questions are new and necessitate the use of sophisticated probabilistic analysis, e.g., martingales, stochastic calculus, etc. These questions may also appear in apparently completely different fields, e.g., Bessel processes are at the core of the very recent Stochastic Loewner Evolutions (SLE) processes. We feel that, ultimately, mathematical finance contributes to the foundations of the stochastic world.

Any relation with the present financial crisis (2007-?)? The writing of this book began in February 2001, at a time when probabilists who had engaged in Mathematical Finance kept developing central topics, such as the no-arbitrage theory, resting implicitly on the "good health of the market", i.e.: its "natural" tendency towards efficiency. Nowadays, "the market" is in quite "bad health" as it suffers badly from illiquidity, lack of confidence, misappreciation of risks, to name a few points. Revisiting previous axioms in such a changed situation is a huge task, which undoubtedly shall be addressed in the future. However, for obvious reasons, our book does not deal with these new and essential questions.

Acknowledgements: We warmly thank Yann Le Cam, Olivier Le Courtois, Pierre Patie, Marek Rutkowski, Paavo Salminen and Michael Suchanecki, who carefully read different versions of this work and sent us many references and comments, and Vincent Torri for his advice on Tex language. We thank Ch. Bayer, B. Bergeron, B. Dengler, B. Forster, D. Florens, A. Hula, M. Keller-Ressel, Y. Miyahara, A. Nikeghbali, A. Royal, B. Rudloff, M. Siopacha, Th. Steiner and R. Warnung for their helpful suggestions. We also acknowledge help from Robert Elliott for his accurate remarks and his checking of the English throughout our text. All simulations were done by Yann Le Cam. Special thanks to John Preater and Hermann Makler from the Springer staff, who did a careful check of the language and spelling in the last version, and to Donatas Akmanavičius for editing work.

Drinking "sok z czarnych porzeczek" (thanks Marek!) was important while Monique was working on a first version. Marc Chesney greatly acknowledges support by both the University Research Priority Program "Finance and Financial Markets" and the National Center of Competence in Research

FINRISK. They are research instruments, respectively of the University of Zurich and of the Swiss National Science Foundation. He would also like to acknowledge the kind support received during the initial stages of this book project from group HEC (Paris), where he was a faculty member at the time.

All remaining errors are our sole responsibility. We would appreciate comments, suggestions and corrections from readers who may send e-mails to the corresponding author Monique Jeanblanc at monique.jeanblanc@univ-evry.fr.

Contents

Part I Continuous Path Processes

Part II Jump Processes

User's Guide

This book consists of two parts: the first part concerns continuous-path processes, and the second part concernes jump processes.

Part I:

Chapter 1 introduces the main results for continuous-path processes and presents many examples, including in particular Brownian motion.

Chapter 2 presents the main tools in finance: self-financing portfolios, valuation of contingent claims, hedging strategies.

Chapter 3 contains some useful information about hitting times and their laws. Closed form expressions are given in the case of (geometric) Brownian motion.

Chapter 4 discusses finer properties of Brownian motion, e.g., local times, bridges, excursions and meanders.

Chapter 5 is devoted mainly to the presentation of one-dimensional diffusions, thus extending the scope of Chapter 4. Filtration problems are also studied.

Chapter 6 focuses on Bessel processes and applications to finance.

Part II:

Chapter 7 is concerned with models of default risk, which involve stochastic processes with a single jump.

Chapter 8 introduces Poisson and compound Poisson processes, which are standard examples of jump processes.

Chapter 9 contains general theory of semi-martingales and aims at unifying results obtained in Chapters 1 and 8.

Chapter 10 presents some jump-diffusion processes and their applications to Finance.

Chapter 11 gives basic results about Lévy processes.

Chapter 12 consists of a list of useful formulae found throughout this book.

At the end of the book, the reader will find an extended bibliography, and a list of references, sorted by thema, followed by an index of authors, in which the page number where each author is quoted is specified.

In the text, some important words are in boldface. These words are also found in the subject index. Some notation can be found in the notation index.

To complete this guide, we emphasize some particular features of this book, already mentioned in the Preface:

- in some cases, proofs are sketched and/or omitted, but precise references are given;
- forward references to topics discussed further in the book are indicated with the arrow $\rightarrowtail$;
- we proceed by generalization: an important case/process is discussed, followed (a little later) by a general study.

Throughout this book, the symbol □ indicates the end of a proof, the symbol ◁ indicates the end of an exercise and the symbol ▶ is used to separate a long proof into different parts.

Section 2.1 refers to Chapter 2, Section 1, and Subsection 4.3.7 refers to Chapter 4, Section 3, Subsection 7. Theorem (Proposition, Lemma) 3.2.1.4 is the 4th in Chapter 3, Section 2, Subsection 1.

Begin at the beginning, and go on till you come to the end. Then, stop.

Lewis Carroll, Alice's Adventures in Wonderland.

Notation and Abbreviations

We shall use the standard notation and abbreviations.

We shall use **increasing** instead of nondecreasing and **positive** instead of non-negative.

u.i.	:	uniformly integrable (for a family of r.v.'s)
BM	:	Brownian motion
r.v.	:	random variable
e.m.m.	:	equivalent martingale measure
a.s.	:	almost surely
w.r.t.	:	with respect to
w.l.g.	:	without loss of generality
SDE	:	Stochastic Differential Equation
BSDE	:	Backward Stochastic Differential Equation
PRP	:	Predictable Representation Property
MCT	:	Monotone Class Theorem
$x \vee y$	=	$\sup(x, y)$
$x \wedge y$	=	$\inf(x, y)$
$x \bullet y$	:	scalar product of the vectors $x, y \in \mathbb{R}^d$
$H \star X$	:	stochastic integral of the process H with respect to the semi-martingale X
x^+	=	$x \vee 0$
x^-	=	$(-x) \vee 0$
$\partial_x f$	=	$\frac{\partial}{\partial x} f = f_x$
C_b^n	:	set of functions with continuous bounded derivatives up to n-th order
$\mu(t), \mu_t$	:	A function (or a process) evaluated at time t. If μ is a deterministic function, $\mu(t)$ is preferably used; if μ is a process, when the subscript is not too large, μ_t is prefered
$\int_a^b ds f(s)$	=	$\int_a^b f(s) ds$ when it seems convenient
$X \stackrel{\text{law}}{=} Y$	:	the random variables (or the processes) X and Y have the same law
$X \stackrel{\text{mart}}{=} Y$	:	the process $X - Y$ is a local martingale
$X \in \mathcal{F}$	:	X is a $\mathcal{F}$-measurable r.v., i.e., $X \in L^0(\mathcal{F})$
$X \in b\mathcal{F}$	:	X is a bounded $\mathcal{F}$-measurable random variable
$\mathcal{N}(x)$	=	$\frac{1}{\sqrt{2\pi}} \int_{-\infty}^x e^{-y^2/2} dy$, the cumulative function for a standard Gaussian law

Other notation can be found in the glossary at the end of the volume.

Part I

Continuous Path Processes

1

Continuous-Path Random Processes: Mathematical Prerequisites

Historically, in mathematical finance, continuous-time processes have been considered from the very beginning, e.g., Bachelier [39, 41] deals with Brownian motion, which has continuous paths. This may justify making our starting point in this book to deal with continuous-path random processes, for which, in this first chapter, we recall some well-known facts. We try to give all the definitions and to quote all the important facts for further use. In particular, we state, without proofs, results on stochastic calculus, change of probability and stochastic differential equations.

For proofs, the reader can refer to the books of Revuz and Yor [730], denoted hereafter [RY], Chung and Williams [186], Ikeda and Watanabe [456], Karatzas and Shreve [513], Lamberton and Lapeyre [559], Rogers and Williams [741, 742] and Williams, R. [845]. See also the reviews of Varadhan [826], Watanabe [836] and Rao [729]. The books of Øksendal [684] and Wong and Hajek [850] cover a large part of stochastic calculus.

1.1 Some Definitions

1.1.1 Measurability

Given a space Ω, a **σ-algebra** on Ω is a class $\mathcal{F}$ of subsets of Ω, such that $\mathcal{F}$ is closed under complements and countable intersection (hence under countable union) and $\emptyset \in \mathcal{F}$ (hence, $\Omega \in \mathcal{F}$). For a given class $\mathcal{C}$ of subsets of Ω, we denote by $\sigma(\mathcal{C})$ the smallest σ-algebra which contains $\mathcal{C}$ (i.e., the intersection of all the σ-algebras containing $\mathcal{G}$).

A **measurable space** $(\Omega, \mathcal{F})$ is a space Ω endowed with a σ-algebra $\mathcal{F}$. **A measurable map** X from $(\Omega, \mathcal{F})$ to another measurable space $(E, \mathcal{E})$ is a map from Ω to E such that, for any $B \in \mathcal{E}$, the set

$$X^{-1}(B) := \{\omega \in \Omega : X(\omega) \in B\}$$

belongs to $\mathcal{F}$.

M. Jeanblanc, M. Yor, M. Chesney, *Mathematical Methods for Financial Markets*, Springer Finance, DOI 10.1007/978-1-84628-737-4_1,

A real-valued random variable (r.v.) on $(\Omega, \mathcal{F})$ is a measurable map from $(\Omega, \mathcal{F})$ to $(\mathbb{R}, \mathcal{B})$ where $\mathcal{B}$ is the Borel σ-algebra, i.e., the smallest σ-algebra that contains the intervals.

Let X be a real-valued random variable on a measurable space $(\Omega, \mathcal{F})$. The σ-algebra generated by X, denoted $\sigma(X)$, is $\sigma(X) := \{X^{-1}(B)\,; B \in \mathcal{B}\}$. Doob's theorem asserts that any $\sigma(X)$-measurable real-valued r.v. can be written as $h(X)$ where h is a **Borel function**, i.e., a measurable map from $(\mathbb{R}, \mathcal{B})$ to $(\mathbb{R}, \mathcal{B})$ (a function such that $h^{-1}(B) := \{x \in \mathbb{R} : h(x) \in B\} \in \mathcal{B}$ for any $B \in \mathcal{B}$). The set of bounded Borel functions on a measurable space $(E, \mathcal{E})$ (i.e., the measurable maps from $(E, \mathcal{E})$ to $(\mathbb{R}, \mathcal{B})$) will be denoted by $b(\mathcal{E})$. If $\mathcal{H}$ is a σ-algebra on Ω, *we shall make the slight abuse of notation* by writing $X \in \mathcal{H}$ for: X is an $\mathcal{H}$-measurable r.v. and $X \in b\mathcal{H}$ for: X is a bounded r.v. in $\mathcal{H}$.

Let $(X_i, i \in I)$ be a set of random variables. There exists a unique r.v. with values in $\bar{\mathbb{R}}$, denoted $\text{esssup}_i X_i$ (**essential supremum** of the family $(X_i; i \in I)$) such that, for any r.v. Y,

$$X_i \leq Y a.s. \forall i \in I \iff \text{esssup}_i X_i \leq Y\,.$$

If the family is countable, $\text{esssup}_i X_i = \sup_i X_i$. In the case where the set I is not countable, the map $\sup_i X_i$ (pointwise supremum) may not be a random variable.

1.1.2 Monotone Class Theorem

We will frequently use the monotone class theorem which we state without proof (see Dellacherie and Meyer [242], Chapter 1). We give two different versions of that theorem, one dealing with sets, the other with functions.

Theorem 1.1.2.1 *Let $\mathcal{C}$ be a collection of subsets of Ω such that*

- $\Omega \in \mathcal{C}$,
- *if $A, B \in \mathcal{C}$ and $A \subset B$, then $B \backslash A = B \cap A^c \in \mathcal{C}$,*
- *if A_n is an increasing sequence of elements of $\mathcal{C}$, then $\cup_n A_n \in \mathcal{C}$.*

Then, if $\mathcal{F} \subset \mathcal{C}$ where $\mathcal{F}$ is closed under finite intersections, then $\sigma(\mathcal{F}) \subset \mathcal{C}$.

Theorem 1.1.2.2 *Let $\mathcal{V}$ be a vector space of bounded real-valued functions on Ω such that*

- *the constant functions are in $\mathcal{V}$,*
- *if h_n is an increasing sequence of positive elements of $\mathcal{V}$ such that $h = \sup h_n$ is bounded, then $h \in \mathcal{V}$.*

If $\mathcal{G}$ is a subset of $\mathcal{V}$ which is stable under pointwise multiplication, then $\mathcal{V}$ contains all the bounded $\sigma(\mathcal{G})$-measurable functions.

1.1.3 Probability Measures

A **probability measure** $\mathbb{P}$ on a measurable space $(\Omega, \mathcal{F})$ is a map from $\mathcal{F}$ to $[0,1]$ such that:

- $\mathbb{P}(\Omega) = 1$,
- $\mathbb{P}(\cup_{i=1}^{\infty} A_i) = \sum_{i=1}^{\infty} \mathbb{P}(A_i)$ for any countable family of disjoint sets $A_i \in \mathcal{F}$, i.e., such that $A_i \cap A_j = \emptyset$ for $i \neq j$.

Note that, for $A \in \mathcal{F}$, $\mathbb{P}(A) = 1 - \mathbb{P}(A^c)$ where A^c is the complement set of A, hence $P(\emptyset) = 0$.

We shall often write, for J a countable set, $\mathbb{P}(A_j, j \in J)$ for $\mathbb{P}(\cap_{j \in J} A_j)$.

Warning 1.1.3.1 The property $\mathbb{P}(\cup_{i=1}^{\infty} A_i) = \sum_{i=1}^{\infty} \mathbb{P}(A_i)$ does not extend to a non-countable family.

A measurable space $(\Omega, \mathcal{F})$ endowed with a probability measure $\mathbb{P}$ is called a **probability space**.

The "elementary" negligible sets are the sets $A \in \mathcal{F}$ such that $\mathbb{P}(A) = 0$. Sets $\Gamma \subset \Gamma'$ with $\Gamma' \in \mathcal{F}$ and $\mathbb{P}(\Gamma') = 0$ are said to be **$(\mathbb{P}, \mathcal{F})$-negligible**.

If $(\Omega, \mathcal{F})$ is a measurable space and $\mathbb{P}$ a probability measure on $\mathcal{F}$, the **completion** of $\mathcal{F}$ with respect to $\mathbb{P}$ is the σ-algebra of subsets A of Ω such that there exist A_1 and A_2 in $\mathcal{F}$ with $A_1 \subset A \subset A_2$ and $\mathbb{P}(A_1) = \mathbb{P}(A_2)$ (or, equivalently, $\mathbb{P}(A_2 \cap A_1^c) = 0$). In particular, the completion of $\mathcal{F}$ contains all the $\mathbb{P}$-negligible sets.

1.1.4 Filtration

A **filtration** $\mathbf{F} = (\mathcal{F}_t, t \geq 0)$ is a family of σ-algebras $\mathcal{F}_t$ on the same probability space $(\Omega, \mathcal{F}, \mathbb{P})$, which is increasing, i.e., such that $\mathcal{F}_s \subset \mathcal{F}_t$ for $s < t$ (that is: if $A \in \mathcal{F}_s$, then $A \in \mathcal{F}_t$ for $s < t$). We note $\mathcal{F}_\infty = \vee_{t \in \mathbb{R}} \mathcal{F}_t$.

It is generally assumed that the filtration satisfies the so-called "**usual hypotheses,**" that is,

(i) the filtration is right-continuous, i.e., $\mathcal{F}_t = \cap_{u>t} \mathcal{F}_u$,

(ii) the σ-algebra $\mathcal{F}_0$ contains the $(\mathbb{P}, \mathcal{F})$-negligible sets of $\mathcal{F}_\infty$.

Usually, (but not always) the σ-algebra $\mathcal{F}_0$ is the trivial σ-algebra, up to completion.

A probability space endowed with a filtration which satisfies the usual hypotheses is called a **filtered probability space**.

We shall say that a filtration $\mathbf{G}$ is larger than $\mathbf{F}$, and write $\mathbf{F} \subset \mathbf{G}$, if $\mathcal{F}_t \subset \mathcal{G}_t$, $\forall t$.

Comment 1.1.4.1 It is important that the *usual hypotheses* are satisfied in order to be able to apply general results on stochastic processes, especially when studying processes with jumps.

1.1.5 Law of a Random Variable, Expectation

The law of a real-valued r.v. X defined on the space $(\Omega, \mathcal{F}, \mathbb{P})$ is the probability measure $\mathbb{P}_X$ on $(\mathbb{R}, \mathcal{B})$ defined by

$$\forall A \in \mathcal{B},\ \mathbb{P}_X(A) = \mathbb{P}(X \in A)\,.$$

It is the image on $(\mathbb{R}, \mathcal{B})$ of $\mathbb{P}$ by the map $\omega \to X(\omega)$. This definition extends to an $\mathbb{R}^n$-valued random variable, and, more generally, to an E-valued random variable (a measurable map from $(\Omega, \mathcal{F})$ to $(E, \mathcal{E})$). If X and Y have the same law, we shall write $X \stackrel{\text{law}}{=} Y$.

The **cumulative distribution function** of a real valued r.v. X is the right-continuous function F defined as $F(x) = \mathbb{P}(X \leq x)$.

The expectation of a positive random variable Z is defined as

$$\mathbb{E}(Z) = \int Z d\mathbb{P} = \int_{\mathbb{R}^+} x\, d\mathbb{P}_Z(x)\,,$$

and, if $\mathbb{E}(|X|) < \infty$, then $\mathbb{E}(X) = \mathbb{E}(X^+) - \mathbb{E}(X^-)$. In case of ambiguity, we shall denote by $\mathbb{E}_{\mathbb{P}}$ the expectation with respect to the probability measure $\mathbb{P}$. The r.v. X is said to be $\mathbb{P}$-integrable (or integrable if there is no ambiguity) if $\mathbb{E}(|X|) < \infty$.

There are a few important transforms T of probabilities (on $\mathbb{R}$, say) which characterize a given probability μ, i.e., such that the map $\mu \to T(\mu)$ is one-to-one.

- The **Fourier transform** $F_\mu(t) = \int_{\mathbb{R}} e^{itx}\mu(dx)$ (where $t \in \mathbb{R}$).
- The **Laplace transform** $L_\mu(\lambda) = \int_{\mathbb{R}} e^{-\lambda x}\mu(dx)$ defined on the interval $\{\lambda \in \mathbb{R} : \mathbb{E}(e^{-\lambda X}) < \infty\}$. Note that the Laplace transform is well defined on $\mathbb{R}^+$ if X is positive. We shall also use, when it is defined, the Laplace transform $\mathbb{E}(e^{\lambda X})$, $\lambda \in \mathbb{R}$.

1.1.6 Independence

A family of random variables $(X_i, i \in I)$, defined on the space $(\Omega, \mathcal{F}, \mathbb{P})$, is said to be **independent** if, for any n distinct indices $(i_1, i_2, \dots, i_n)$ with $i_k \in I$ and for any $(A_1, \dots, A_n)$ where $A_k \in \mathcal{B}$,

$$\mathbb{P}\left(\cap_{k=1}^n (X_{i_k} \in A_k)\right) = \prod_{k=1}^n \mathbb{P}(X_{i_k} \in A_k)\,.$$

A classical application of the monotone class theorem is that, if the r.vs $(X_i, i \in I)$ are independent, then, with the same notation as above, for any bounded Borel functions f_k,

$$\mathbb{E}\left(\prod_{k=1}^{n} f_k(X_{i_k})\right) = \prod_{k=1}^{n} \mathbb{E}\left(f_k(X_{i_k})\right).$$

The converse holds true as well. In particular, two random variables X and Y are independent if and only if, for any pair of bounded Borel functions f and g, $\mathbb{E}(f(X)g(Y)) = \mathbb{E}(f(X))\mathbb{E}(g(Y))$. For the independence property to hold true, it suffices that this equality is satisfied for "enough" functions, for example:

- for f, g of the form $f = \mathbb{1}_{]-\infty,a]}, g = \mathbb{1}_{]-\infty,b]}$ for every pair of real numbers (a, b), i.e.,
$$\mathbb{P}(X \leq a, Y \leq b) = \mathbb{P}(X \leq a)\,\mathbb{P}(Y \leq b),$$
- for f, g of the form $f(x) = e^{i\lambda x}, g(x) = e^{i\mu x}$ for every pair of real numbers (λ, μ), i.e.,
$$\mathbb{E}(e^{i(\lambda X + \mu Y)}) = \mathbb{E}(e^{i\lambda X})\,\mathbb{E}(e^{i\mu Y}).$$
- in the case where X and Y are positive random variables, for f, g of the form $f(x) = e^{-\lambda x}, g(x) = e^{-\mu x}$ for every pair of positive real numbers (λ, μ), i.e.,
$$\mathbb{E}(e^{-\lambda X} e^{-\mu Y}) = \mathbb{E}(e^{-\lambda X})\mathbb{E}(e^{-\mu Y}).$$

It is important to note that if X and Y are independent r.vs, then for any bounded Borel function Φ defined on $\mathbb{R}^2$, $\mathbb{E}(\Phi(X,Y)) = \mathbb{E}(\varphi(X))$ where $\varphi(x) = \mathbb{E}(\Phi(x,Y))$. This result can be seen as a consequence of the monotone class theorem, or as an application of Fubini's theorem.

1.1.7 Equivalent Probabilities and Radon-Nikodým Densities

Let $\mathbb{P}$ and $\mathbb{Q}$ be two probabilities defined on the same measurable space $(\Omega, \mathcal{F})$. The probability $\mathbb{Q}$ is said to be **absolutely continuous** with respect to $\mathbb{P}$, (denoted $\mathbb{Q} \ll \mathbb{P}$) if $\mathbb{P}(A) = 0$ implies $\mathbb{Q}(A) = 0$, for any $A \in \mathcal{F}$. In that case, there exists a positive, $\mathcal{F}$-measurable random variable L, called the **Radon-Nikodým density** of $\mathbb{Q}$ with respect to $\mathbb{P}$, such that

$$\forall A \in \mathcal{F}, \mathbb{Q}(A) = \mathbb{E}_{\mathbb{P}}(L\mathbb{1}_A).$$

This random variable L satisfies $\mathbb{E}_{\mathbb{P}}(L) = 1$ and for any $\mathbb{Q}$-integrable random variable X, $\mathbb{E}_{\mathbb{Q}}(X) = \mathbb{E}_{\mathbb{P}}(XL)$. The notation $\frac{d\mathbb{Q}}{d\mathbb{P}} = L$ (or $\mathbb{Q}|_{\mathcal{F}} = L\,\mathbb{P}|_{\mathcal{F}}$) is in common use, in particular in the chain of equalities

$$\mathbb{E}_{\mathbb{Q}}(X) = \int X d\mathbb{Q} = \int X \frac{d\mathbb{Q}}{d\mathbb{P}}\, d\mathbb{P} = \int X L d\mathbb{P} = \mathbb{E}_{\mathbb{P}}(XL).$$

The probabilities $\mathbb{P}$ and $\mathbb{Q}$ are said to be **equivalent**, (this will be denoted $\mathbb{P} \sim \mathbb{Q}$), if they have the same negligible sets, i.e., if for any $A \in \mathcal{F}$,

$$\mathbb{Q}(A) = 0 \Leftrightarrow \mathbb{P}(A) = 0,$$

or equivalently, if $\mathbb{Q} \ll \mathbb{P}$ and $\mathbb{P} \ll \mathbb{Q}$. In that case, there exists a strictly positive, $\mathcal{F}$-measurable random variable L, such that $\mathbb{Q}(A) = \mathbb{E}_{\mathbb{P}}(L\mathbb{1}_A)$. Note that $\frac{d\mathbb{P}}{d\mathbb{Q}} = L^{-1}$ and $\mathbb{P}(A) = \mathbb{E}_{\mathbb{Q}}(L^{-1}\mathbb{1}_A)$.

Conversely, if L is a strictly positive $\mathcal{F}$-measurable r.v., with expectation 1 under $\mathbb{P}$, then $\mathbb{Q} = L \cdot \mathbb{P}$ defines a probability measure on $\mathcal{F}$, equivalent to $\mathbb{P}$. From the definition of equivalence, if a property holds almost surely (a.s.) with respect to $\mathbb{P}$, it also holds a.s. for any probability $\mathbb{Q}$ equivalent to $\mathbb{P}$. Two probabilities $\mathbb{P}$ and $\mathbb{Q}$ on the same filtered probability space $(\Omega, \mathbf{F})$ are said to be locally equivalent[1] if they have the same negligible sets on $\mathcal{F}_t$, for every $t \geq 0$, i.e., if $\mathbb{Q}|_{\mathcal{F}_t} \sim \mathbb{P}|_{\mathcal{F}_t}$. In that case, there exists a strictly positive $\mathbf{F}$-adapted process $(L_t, t \geq 0)$ such that $\mathbb{Q}|_{\mathcal{F}_t} = L_t \mathbb{P}|_{\mathcal{F}_t}$. (See ↣ Subsection 1.7.1 for more information.) Furthermore, if τ is a stopping time (see ↣ Subsection 1.2.3), then

$$\mathbb{Q}|_{\mathcal{F}_\tau \cap \{\tau < \infty\}} = L_\tau \cdot \mathbb{P}|_{\mathcal{F}_\tau \cap \{\tau < \infty\}} .$$

This will be important when dealing with Girsanov's theorem and explosion times (See ↣ Proposition 1.7.5.3).

Warning 1.1.7.1 If $\mathbb{P} \sim \mathbb{Q}$ and X is a $\mathbb{P}$-integrable random variable, it is not necessarily $\mathbb{Q}$-integrable.

1.1.8 Construction of Simple Probability Spaces

In order to construct a random variable with a given law, say a Gaussian law, the canonical approach is to take $\Omega = \mathbb{R}$, $X : \Omega \to \mathbb{R}; X(\omega) = \omega$ the identity map and $\mathbb{P}$ the law on $\Omega = \mathbb{R}$ with the Gaussian density with respect to the Lebesgue measure, i.e.,

$$\mathbb{P}(d\omega) = \frac{1}{\sqrt{2\pi}} \exp\left(-\frac{\omega^2}{2}\right) d\omega$$

(recall that here ω is a real number). Then the cumulative distribution function of the random variable X is

$$F_X(x) = \mathbb{P}(X \leq x) = \int_\Omega \mathbb{1}_{\{\omega \leq x\}} \mathbb{P}(d\omega) = \int_{-\infty}^{x} \frac{1}{\sqrt{2\pi}} \exp\left(-\frac{\omega^2}{2}\right) d\omega .$$

Hence, the map X is a Gaussian random variable. The construction of a real valued r.v. with any given law can be carried out using the same idea; for example, if one needs to construct a random variable with an exponential law, then, similarly, one may choose $\Omega = \mathbb{R}$ and the density $e^{-\omega}\mathbb{1}_{\{\omega \geq 0\}}$.

For two independent variables, we choose $\Omega = \Omega_1 \times \Omega_2$ where $\Omega_i, i = 1, 2$ are two copies of $\mathbb{R}$. On each Ω_i, one constructs a random variable as above,

[1] This commonly used terminology often refers to a sequence (T_n) of stopping times, with $T_n \uparrow \infty$ a.s.; here, it is preferable to restrict ourselves to the deterministic case $T_n = n$.

and defines $\mathbb{P} = \mathbb{P}_1 \otimes \mathbb{P}_2$ where the product probability $\mathbb{P}_1 \otimes \mathbb{P}_2$ is first defined on the sets $A_1 \times A_2$ for $A_i \in \mathcal{B}$, the Borel σ-field of $\mathbb{R}$, as

$$(\mathbb{P}_1 \otimes \mathbb{P}_2)(A_1 \times A_2) = \mathbb{P}_1(A_1)\mathbb{P}_2(A_2),$$

and then extended to $\mathcal{B} \times \mathcal{B}$.

1.1.9 Conditional Expectation

Let X be an integrable random variable on the space $(\Omega, \mathcal{F}, \mathbb{P})$ and $\mathcal{H}$ a σ-algebra contained in $\mathcal{F}$, i.e., $\mathcal{H} \subseteq \mathcal{F}$. The **conditional expectation** of X given $\mathcal{H}$ is the almost surely unique $\mathcal{H}$-measurable random variable Z such that, for any bounded $\mathcal{H}$-measurable random variable Y,

$$\mathbb{E}(ZY) = \mathbb{E}(XY).$$

The conditional expectation is denoted $\mathbb{E}(X|\mathcal{H})$ and the following properties hold (see, for example Breiman [123], Williams [842, 843]):

- If X is $\mathcal{H}$-measurable, $\mathbb{E}(X|\mathcal{H}) = X$, a.s.
- $\mathbb{E}(\mathbb{E}(X|\mathcal{H})) = \mathbb{E}(X)$.
- If $X \geq 0$, then $\mathbb{E}(X|\mathcal{H}) \geq 0$ a.s.
- Linearity: If Y is an integrable random variable and $a, b \in \mathbb{R}$,

$$\mathbb{E}(aX + bY|\mathcal{H}) = a\mathbb{E}(X|\mathcal{H}) + b\mathbb{E}(Y|\mathcal{H}), \quad \text{a.s.}$$

- If $\mathcal{G}$ is another σ-algebra and $\mathcal{G} \subseteq \mathcal{H}$, then

$$\mathbb{E}(\mathbb{E}(X|\mathcal{G})|\mathcal{H}) = \mathbb{E}(\mathbb{E}(X|\mathcal{H})|\mathcal{G}) = \mathbb{E}(X|\mathcal{G}), \quad \text{a.s.}$$

- If Y is $\mathcal{H}$-measurable and XY is integrable, $\mathbb{E}(XY|\mathcal{H}) = Y\mathbb{E}(X|\mathcal{H})$ a.s.
- Jensen's inequality: If f is a convex function such that $f(X)$ is integrable,

$$\mathbb{E}(f(X)|\mathcal{H}) \geq f(\mathbb{E}(X|\mathcal{H})), \quad \text{a.s.}$$

In the particular case where $\mathcal{H}$ is the σ-algebra generated by a r.v. Y, then $\mathbb{E}(X|\sigma(Y))$, which is usually denoted by $\mathbb{E}(X|Y)$, is $\sigma(Y)$-measurable, hence there exists a Borel function φ such that $\mathbb{E}(X|Y) = \varphi(Y)$. The function φ is uniquely defined up to a $\mathbb{P}_Y$-negligible set. The notation $\mathbb{E}(X|Y = y)$ is often used for $\varphi(y)$.

If X is an $\mathbb{R}^p$-valued random variable, and Y an $\mathbb{R}^n$-valued random variable, there exists a family of measures (conditional laws) $\mu(dx, y)$ such that, for any bounded Borel function h

$$\mathbb{E}(h(X)|Y = y) = \int h(x)\mu(dx, y).$$

If (X, Y) are independent random variables, and h is a bounded Borel function, then $\mathbb{E}(h(X, Y)|Y) = \Psi(Y)$, where $\Psi(y) = \mathbb{E}(h(X, y))$, i.e., the conditional law of X given $Y = y$ does not depend on y.

Note that, if X is square integrable, then $\mathbb{E}(X|\mathcal{H})$ may be defined as the projection of X on the space $L^2(\Omega, \mathcal{H})$ of $\mathcal{H}$-measurable square integrable random variables. The conditional variance of a square integrable random variable X is

$$\mathrm{var}(X|\mathcal{H}) = \mathbb{E}(X^2|\mathcal{H}) - (\mathbb{E}(X|\mathcal{H}))^2 .$$

Definition 1.1.9.1 *Two σ-algebras $\mathcal{G}_1$ and $\mathcal{G}_2$ are said to be conditionally independent with respect to the σ-algebra $\mathcal{H}$ if $\mathbb{E}(G_1 G_2|\mathcal{H}) = \mathbb{E}(G_1|\mathcal{H})\mathbb{E}(G_2|\mathcal{H})$ for any bounded random variables $G_i \in \mathcal{G}_i$. Two random variables X and Y are* ***conditionally independent*** *with respect to the σ-algebra $\mathcal{H}$ if $\sigma(X)$ and $\sigma(Y)$ are conditionally independent with respect to $\mathcal{H}$.*

This may be extended obviously to any finite family of r.v's. Two infinite families of random variables are conditionally independent if any finite subfamilies are conditionally independent.

1.1.10 Stochastic Processes

Definition 1.1.10.1 *A continuous time process X on $(\Omega, \mathcal{F}, \mathbb{P})$ is a family of random variables $(X_t, t \geq 0)$, such that the map $(\omega, t) \to X_t(\omega)$ is $\mathcal{F} \otimes \mathcal{B}(\mathbb{R}^+)$ measurable.*

We emphasize that when speaking of processes, we always mean a measurable process.

A process X is **continuous** if, for almost all ω, the map $t \to X_t(\omega)$ is continuous. The process is continuous on the right with limits on the left (in short **càdlàg** following the French acronym[2] if, for almost all ω, the map $t \to X_t(\omega)$ is càdlàg.

Definition 1.1.10.2 *A process X is* ***increasing*** *if $X_0 = 0$, X is right-continuous, and $X_s \leq X_t$, a.s. for $s \leq t$.*

Definition 1.1.10.3 *Let $(\Omega, \mathcal{F}, \mathbf{F}, \mathbb{P})$ be a filtered probability space. The process X is $\mathbf{F}$-adapted if for any $t \geq 0$, the random variable X_t is $\mathcal{F}_t$-measurable.*

The **natural filtration** $\mathbf{F}^X$ of a stochastic process X is the smallest filtration $\mathbf{F}$ which satisfies the usual hypotheses and such that X is $\mathbf{F}$-adapted. We shall write in short (with an abuse of notation) $\mathcal{F}_t^X = \sigma(X_s, s \leq t)$.

[2] In French, continuous on the right is continu à droite, and with limits on the left is admettant des limites à gauche. We shall also use càd for continuous on the right. The use of this acronym comes from P-A. Meyer.

Let $\mathbf{G} = (\mathcal{G}_t, t \geq 0)$ be another filtration on Ω. If $\mathbf{G}$ is larger than $\mathbf{F}$, and if X is an $\mathbf{F}$-adapted process, it is also $\mathbf{G}$-adapted.

Definition 1.1.10.4 *A real-valued process X is* ***progressively measurable*** *with respect to a given filtration $\mathbf{F} = (\mathcal{F}_t, t \geq 0)$, if, for every t, the map $(\omega, s) \to X_s(\omega)$ from $\Omega \times [0,t]$ into $\mathbb{R}$ is $\mathcal{F}_t \times \mathcal{B}([0,t])$-measurable.*

Any càd (or càg) $\mathbf{F}$-adapted process is progressively measurable. An $\mathbf{F}$-progressively measurable process is $\mathbf{F}$-adapted. If X is progressively measurable, then

$$\mathbb{E}\left(\int_0^\infty X_t dt\right) = \int_0^\infty \mathbb{E}(X_t)\, dt,$$

where the existence of one of these expressions implies the existence of the other.

Definition 1.1.10.5 *Two processes $(X_t, t \geq 0)$ and $(Y_t, t \geq 0)$ have the same law if, for any n and any $(t_1, t_2, \ldots, t_n)$*

$$(X_{t_1}, X_{t_2}, \ldots, X_{t_n}) \stackrel{\text{law}}{=} (Y_{t_1}, Y_{t_2}, \ldots, Y_{t_n}).$$

We shall write in short $X \stackrel{\text{law}}{=} Y$, or $X \stackrel{\text{law}}{=} \mu$ for a given probability law μ (on the canonical space).

The process X is a modification of Y if, for any t, $\mathbb{P}(X_t = Y_t) = 1$. The process X is **indistinguishable from** (or a **version** of) Y if $\{\omega : X_t(\omega) = Y_t(\omega), \forall t\}$ is a measurable set and $\mathbb{P}(X_t = Y_t, \forall t) = 1$. If X and Y are modifications of each other and are a.s. continuous, they are indistinguishable.

Let us state without proof a sufficient condition for the existence of a continuous version of a stochastic process.

Theorem 1.1.10.6 (Kolmogorov.) *If a collection $(X_t, t \geq 0)$ of random variables satisfies*

$$\mathbb{E}(|X_t - X_s|^p) \leq C|t-s|^{1+\epsilon}$$

for some $C > 0$, $p > 0$ and $\epsilon > 0$, then this collection admits a modification $(\widetilde{X}_t, t \geq 0)$ which is a.s. continuous, i.e., out of a negligible set, the map $t \to \widetilde{X}_t(\omega)$ is continuous.

PROOF: See, e.g., Ikeda and Watanabe [456], p. 20. □

Throughout the book, we shall see many applications of this theorem, in particular, for the existence of a.s. continuous Brownian paths (see ↣ Section 1.4).

Definition 1.1.10.7 *A process X has*
*- **independent increments** if for any pair $(s,t) \in \mathbb{R}_+^2$, the random variable $X_{t+s} - X_s$ is independent of $\mathcal{F}_s^X$,*
*- **stationary increments** if for any pair $(s,t) \in \mathbb{R}_+^2$,*

$$X_{t+s} - X_s \stackrel{\text{law}}{=} X_t .$$

A process is stationary if

$$\forall \text{ fixed } s > 0, \; (X_{t+s} - X_s, t \geq 0) \stackrel{\text{law}}{=} (X_t, t \geq 0) .$$

Definition 1.1.10.8 *A càd process A is of **finite variation** on $[0,t]$ if*

$$V_A(t,\omega) := \sup \sum_{i=1}^{n} |A_{t_i}(\omega) - A_{t_{i-1}}(\omega)| = \int_0^t |dA_s(\omega)|$$

is a.s. finite, where the supremum is taken over all finite partitions (t_i) of $[0,t]$.

A càd process A is of finite variation if it is of finite variation on any compact $[0,t]$. A càd finite variation process is the difference between two increasing processes. A càd finite variation process A is said to be integrable if $\mathbb{E}(\int_0^\infty |dA_s|) < \infty$. In the definition of finite variation processes, we do not restrict attention to adapted processes. Note that finite variation càd processes are càdlàg.

Exercise 1.1.10.9 One might naively think that a collection $(X_t, t \in \mathbb{R}^+)$ of independent r.v's may be chosen "measurably," i.e., with the map

$$(\mathbb{R}^+ \times \Omega, \mathcal{B}_{\mathbb{R}^+} \times \mathcal{F}) \to (\mathbb{R}, \mathcal{B}_{\mathbb{R}}) \; : (t,\omega) \to X_t(\omega)$$

being measurable, so that X is a "true" process. Prove that if the X_t's are centered and $\sup_t \mathbb{E}(X_t^2) < \infty$, then no measurable choice can be constructed, except $X = 0$.
`Hint:` $\mathbb{E}(\int_0^t X_s ds)^2 = \int_0^t \int_0^t \mathbb{E}(X_s X_u) ds\, du$ would be equal to 0, hence X would be null. ◁

1.1.11 Convergence

A sequence of processes Z^n converges in $L^2(\Omega \times [0,T])$ to a process Z if $\mathbb{E} \int_0^T |Z_s^n - Z_s|^2 ds$ converges to 0.

A sequence of processes Z^n converges uniformly on compacts in probability (ucp) to a process Z if, for any t, $\sup_{0 \leq s \leq t} |Z_s^n - Z_s|$ converges to 0 in probability.

1.1.12 Laplace Transform

The Laplace transform $\mathbb{E}(e^{\lambda X})$ of a r.v. X is well defined for $\lambda \geq 0$ when X is a negative r.v. (here, we use a slightly unorthodox definition of Laplace transform, with $\lambda \geq 0$). In some cases, the Laplace transform can be defined for every $\lambda \in \mathbb{R}$, as in the following important case, where we denote by $\mathcal{N}(\mu, \sigma^2)$ a Gaussian law with mean μ and variance σ^2:

Proposition 1.1.12.1 Laplace transform of a Gaussian variable. *The law of the random variable X is $\mathcal{N}(\mu, \sigma^2)$ if and only if, for any $\lambda \in \mathbb{R}$,*

$$\boxed{\mathbb{E}(\exp(\lambda X)) = \exp\left(\mu\lambda + \frac{1}{2}\lambda^2\sigma^2\right).}$$

This property extends to any $\lambda \in \mathbb{C}$, and to Gaussian random vectors: X is a d-dimensional Gaussian vector with mean μ and covariance matrix Σ if and only if for any $\lambda \in \mathbb{R}^d$,

$$\mathbb{E}(\exp(\lambda^* X)) = \exp\left(\lambda^*\mu + \frac{1}{2}\lambda^*\Sigma\lambda\right),$$

where the star stands for the transposition operator. If the matrix Σ is invertible, the random vector X admits the density

$$(2\pi)^{-d/2}(\det \Sigma)^{-1/2} \exp\left(-\frac{1}{2}(x-\mu)^*\Sigma^{-1}(x-\mu)\right).$$

Comment 1.1.12.2 Let $(X_t, t \geq 0)$ be a (measurable) process, $\lambda > 0$ and f a positive Borel function. Then, if Θ is a random variable, independent of X, with exponential law ($\mathbb{P}(\Theta \in dt) = \lambda e^{-\lambda t}\mathbb{1}_{\{t>0\}}dt$), one has

$$\mathbb{E}(f(X_\Theta)) = \lambda\mathbb{E}\left(\int_0^\infty e^{-\lambda t} f(X_t)dt\right) = \lambda\int_0^\infty e^{-\lambda t}\mathbb{E}\left(f(X_t)\right)dt.$$

Hence, if the process X is continuous, the value of $\mathbb{E}(f(X_\Theta))$ (for all λ and all bounded Borel functions f) characterizes the law of X_t, for any t, i.e., the law of the marginals of the process X. The law of the process assumed to be positive, may be characterized by $\mathbb{E}(\exp[-\int \mu(dt)X_t])$ for all positive measures μ on $(\mathbb{R}^+, \mathcal{B})$.

Exercise 1.1.12.3 Laplace Transforms for the Square of Gaussian Law. Let $X \stackrel{\text{law}}{=} \mathcal{N}(m, \sigma^2)$ and $\lambda > 0$. Prove that

$$\boxed{\mathbb{E}(e^{-\lambda X^2}) = \frac{1}{\sqrt{1+2\lambda\sigma^2}} \exp\left(-\frac{m^2\lambda}{1+2\lambda\sigma^2}\right)}$$

and more generally that

$$\mathbb{E}(\exp\{-\lambda X^2 + \mu X\}) = \frac{\widehat{\sigma}}{\sigma} \exp\left(\frac{\widehat{\sigma}^2}{2}\left(\mu + \frac{m}{\sigma^2}\right)^2 - \frac{m^2}{2\sigma^2}\right),$$

with $\widehat{\sigma}^2 = \dfrac{\sigma^2}{1+2\lambda\sigma^2}$. ◁

Exercise 1.1.12.4 Moments and Laplace Transform. If X is a positive random variable, prove that its negative moments are given by, for $r > 0$:

$$\text{(a)} \qquad \mathbb{E}(X^{-r}) = \frac{1}{\Gamma(r)} \int_0^\infty t^{r-1} \mathbb{E}(e^{-tX}) dt$$

where Γ is the Gamma function (see ↣ Subsection A.5.1 if needed) and its positive moments are, for $0 < r < 1$

$$\text{(b)} \qquad \mathbb{E}(X^r) = \frac{r}{\Gamma(1-r)} \int_0^\infty \frac{1 - \mathbb{E}(e^{-tX})}{t^{r+1}} dt$$

and for $n < r < n+1$, if $\phi(t) = \mathbb{E}(e^{-tX})$ belongs to C^n

$$\text{(c)} \qquad \mathbb{E}(X^r) = \frac{r-n}{\Gamma(n+1-r)} \int_0^\infty (-1)^n \frac{\phi^{(n)}(0) - \phi^{(n)}(t)}{t^{r+1-n}} dt\,.$$

`Hint:` For example, for (b), use Fubini's theorem and the fact that, for $0 < r < 1$,

$$s^r \Gamma(1-r) = r \int_0^\infty \frac{1 - e^{-st}}{t^{r+1}} dt\,.$$

For $r = n$, one has $\mathbb{E}(X^n) = (-1)^n \phi^{(n)}(0)$. See Schürger [774] for more results and applications. ◁

Exercise 1.1.12.5 Chi-squared Law. A noncentral chi-squared law $\chi^2(\delta, \alpha)$ with δ degrees of freedom and noncentrality parameter α has the density

$$\begin{aligned} f(x;\delta,\alpha) &= 2^{-\delta/2} \exp\left(-\frac{1}{2}(\alpha + x)\right) x^{\frac{\delta}{2}-1} \sum_{n=0}^{\infty} \left(\frac{\alpha}{4}\right)^n \frac{x^n}{n!\Gamma(n+\delta/2)} \mathbb{1}_{\{x>0\}} \\ &= \frac{e^{-\alpha/2}}{2\alpha^{\nu/2}} e^{-x/2} x^{\nu/2} I_\nu(\sqrt{x\alpha}) \mathbb{1}_{\{x>0\}}, \end{aligned}$$

where I_ν is the usual modified Bessel function (see ↣ Subsection A.5.2). Its cumulative distribution function is denoted $\chi^2(\delta, \alpha; \cdot)$.

Let $X_i, i = 1, \dots, n$ be independent random variables with $X_i \stackrel{\text{law}}{=} \mathcal{N}(m_i, 1)$. Check that $\sum_{i=1}^n X_i^2$ is a noncentral chi-squared variable with n degrees of freedom, and noncentrality parameter $\sum_{i=1}^n m_i^2$. $\triangleleft$

1.1.13 Gaussian Processes

A real-valued process $(X_t, t \geq 0)$ is a **Gaussian process** if any finite linear combination $\sum_{i=1}^n a_i X_{t_i}$ is a Gaussian variable. In particular, for each $t \geq 0$, the random variable X_t is a Gaussian variable. The law of a Gaussian process is characterized by its mean function $\varphi(t) = \mathbb{E}(X_t)$ and its covariance function $c(t, s) = \mathbb{E}(X_t X_s) - \varphi(t)\varphi(s)$ which satisfies

$$\sum_{i,j} \lambda_i \bar{\lambda}_j \, c(t_i, t_j) \geq 0, \; \forall \lambda \in \mathbb{C}^n .$$

Note that this property holds for every square integrable process, but that, conversely a Gaussian process may always be associated with a pair (φ, c) satisfying the previous conditions. See Janson [479] for many results on Gaussian processes.

1.1.14 Markov Processes

The $\mathbb{R}^d$-valued process X is said to be a **Markov process** if for any t, the past $\mathcal{F}_t^X = \sigma(X_s, s \leq t)$ and the future $\sigma(X_{t+u}, u \geq 0)$ are conditionally independent with respect to X_t, i.e., for any t, for any bounded random variable $Y \in \sigma(X_u, u \geq t)$:

$$\mathbb{E}(Y|\mathcal{F}_t^X) = \mathbb{E}(Y|X_t) .$$

This is equivalent to: for any bounded Borel function f, for any times $t > s \geq 0$

$$\mathbb{E}(f(X_t)|\mathcal{F}_s^X) = \mathbb{E}(f(X_t)|X_s) .$$

A **transition probability** is a family $(P_{s,t}, 0 \leq s < t)$ of probabilities such that the Chapman-Kolmogorov equation holds:

$$P_{s,t}(x, A) = \int P_{s,u}(x, dy) P_{u,t}(y, A) = \mathbb{P}(X_t \in A | X_s = x) .$$

A Markov process with transition probability $P_{s,t}$ satisfies

$$\mathbb{E}(f(X_t)|X_s) = P_{s,t} f(X_s) = \int f(y) P_{s,t}(X_s, dy) ,$$

for any $t > s \geq 0$, for every bounded Borel function f. If $P_{s,t}$ depends only on the difference $t - s$, the Markov process is said to be a **time-homogeneous Markov process** and we simply write P_t for $P_{0,t}$. Results for homogeneous Markov processes can be formally extended to inhomogeneous Markov processes by adding a time dimension to the space, i.e., by considering the process $((X_t, t), t \geq 0)$. For a time-homogeneous Markov process

$$\mathbb{P}_x(X_{t_1} \in A_1, \dots, X_{t_n} \in A_n) = \int_{A_1} P_{t_1}(x, dx_1) \cdots \int_{A_n} P_{t_n - t_{n-1}}(x_{n-1}, dx_n),$$

where $\mathbb{P}_x$ means that $X_0 = x$.

The (strong) **infinitesimal generator** of a time-homogeneous Markov process is the operator $\mathcal{L}$ defined as

$$\mathcal{L}(f)(x) = \lim_{t \to 0} \frac{\mathbb{E}_x(f(X_t)) - f(x)}{t},$$

where $\mathbb{E}_x$ denotes the expectation for the process starting from x at time 0. The domain of the generator is the set $\mathcal{D}(\mathcal{L})$ of bounded Borel functions f such that this limit exists in the norm $\|f\| = \sup |f(x)|$.

Let X be a time-homogeneous Markov process. The associated **semi-group** $P_t f(x) = \mathbb{E}_x(f(X_t))$ satisfies

$$\frac{d}{dt}(P_t f) = P_t \mathcal{L} f = \mathcal{L} P_t f, \; f \in \mathcal{D}(\mathcal{L}). \tag{1.1.1}$$

(See, for example, Kallenberg [505] or [RY], Chapter VII.)

A Markov process is said to be **conservative** if $P_t(x, \mathbb{R}^d) = 1$ for all t and $x \in \mathbb{R}^d$. A nonconservative process can be made conservative by adding an extra state ∂ (called the cemetery state) to $\mathbb{R}^d$. The conservative transition function P_t^∂ is defined by

$$\begin{aligned}
&P_t^\partial(x, A) := P_t(x, A), \quad x \in \mathbb{R}^d, A \in \mathcal{B}, \\
&P_t^\partial(x, \partial) := 1 - P_t(x, \mathbb{R}^d), \quad x \in \mathbb{R}^d, \\
&P_t^\partial(\partial, A) := \delta_{\{\partial\}}(A), \quad A \in \mathbb{R}^d \cup \partial.
\end{aligned}$$

Definition 1.1.14.1 *The **lifetime** of (the conservative process) X is the $\mathbf{F}^X$-stopping time*

$$\zeta(\omega) := \inf\{t \geq 0 : X_t(\omega) = \partial\}.$$

See ↣ Section 1.2.3 for the definition of stopping time.

Proposition 1.1.14.2 *Let X be a time-homogeneous Markov process with infinitesimal generator $\mathcal{L}$. Then, for any function f in the domain $\mathcal{D}(\mathcal{L})$ of the generator*

$$M_t^f := f(X_t) - f(X_0) - \int_0^t \mathcal{L} f(X_s) ds$$

is a martingale with respect to $\mathbb{P}_x, \forall x$. Moreover, if τ is a bounded stopping time

$$\mathbb{E}_x(f(X_\tau)) = f(x) + \mathbb{E}_x\left(\int_0^\tau \mathcal{L} f(X_s) ds\right).$$

PROOF: See $\rightarrowtail$ Section 1.2 for the definition of martingale. From

$$M^f_{t+s} - M^f_s = f(X_{t+s}) - f(X_s) - \int_s^{t+s} \mathcal{L}f(X_u)du$$

and the Markov property, one deduces

$$\mathbb{E}_x(M^f_{t+s} - M^f_s | \mathcal{F}_s) = \mathbb{E}_{X_s}(M^f_t) . \tag{1.1.2}$$

From (1.1.1),

$$\frac{d}{dt}\mathbb{E}_x[f(X_t)] = \mathbb{E}_x[\mathcal{L}f(X_t)],\ f \in \mathcal{D}(\mathcal{L})$$

hence, by integration

$$\mathbb{E}_x[f(X_t)] = f(x) + \int_0^t ds\, \mathbb{E}_x[\mathcal{L}f(X_s)] .$$

It follows that, for any x, $\mathbb{E}_x(M^f_t)$ equals 0, hence $\mathbb{E}_{X_s}(M^f_t) = 0$ and from (1.1.2), that M^f is a martingale. □

The family $(U_\alpha, \alpha > 0)$ of kernels defined by

$$U_\alpha f(x) = \int_0^\infty e^{-\alpha t}\mathbb{E}_x[f(X_t)]dt$$

is called the **resolvent** of the Markov process.(See also $\rightarrowtail$ Subsection 5.3.6.)

The **strong Markov property** holds if for any finite stopping time T and any $t \geq 0$, (see $\rightarrowtail$ Subsection 1.2.3 for the definition of a stopping time) and for any bounded Borel function f,

$$\mathbb{E}(f(X_{T+t})|\mathcal{F}^X_T) = \mathbb{E}(f(X_{T+t})|X_T) .$$

It follows that, for any pair of finite stopping times T and S, and any bounded Borel function f

$$\mathbb{1}_{\{S>T\}}\mathbb{E}(f(X_S)|\mathcal{F}^X_T) = \mathbb{1}_{\{S>T\}}\mathbb{E}(f(X_S)|X_T) .$$

Proposition 1.1.14.3 *Let X be a strong Markov process with continuous paths and b a continuous function. Define the first passage time of X over b as*

$$T_b = \inf\{t > 0 | X_t \geq b(t)\} .$$

Then, for $x \leq b(0)$ and $y > b(t)$

$$\mathbb{P}_x(X_t \in dy) = \int_0^t \mathbb{P}(X_t \in dy | X_s = b(s))F(ds)$$

where F is the law of T_b.

SKETCH OF THE PROOF: Let $B \subset [b(t), \infty[$.

$$\begin{aligned}\mathbb{P}_x(X_t \in B) = \mathbb{P}_x(X_t \in B, T_b \leq t) &= \mathbb{E}_x(\mathbb{1}_{\{T_b \leq t\}} \mathbb{E}_x(\mathbb{1}_{\{X_t \in B\}} | T_b)) \\ &= \int_0^t \mathbb{E}_x(\mathbb{1}_{\{X_t \in B\}} | T_b = s) \mathbb{P}_x(T_b \in ds) \\ &= \int_0^t \mathbb{P}(X_t \in B | X_s = b(s)) \mathbb{P}_x(T_b \in ds)\,.\end{aligned}$$

For a complete proof, see Peskir [707]. □

Definition 1.1.14.4 *Let X be a Markov process. A Borel set A is said to be* ***polar*** *if*

$$\mathbb{P}_x(T_A < \infty) = 0, \quad \textit{for every } x \in \mathbb{R}^d$$

where $T_A = \inf\{t > 0 \,:\, X_t \in A\}$.

This notion will be used (see ↣ Proposition 1.4.2.1) to study some particular cases.

Comment 1.1.14.5 See Blumenthal and Getoor [107], Chung [184], Dellacherie et al. [241], Dynkin [288], Ethier and Kurtz [336], Itô [462], Meyer [648], Rogers and Williams [741], Sharpe [785] and Stroock and Varadhan [812], for further results on Markov processes. Proposition 1.1.14.3 was obtained in Fortet [355] (see Peskir [707] for applications of this result to Brownian motion). Further examples of deterministic barriers will be given in ↣ Chapter 3.

Exercise 1.1.14.6 Let W be a Brownian motion (see ↣ Section 1.4 if needed), x, ν, σ real numbers, $X_t = x \exp(\nu t + \sigma W_t)$ and $M_t^X = \sup_{s \leq t} X_s$. Prove that the process $(Y_t = M_t^X / X_t, t \geq 0)$ is a Markov process. This fact (proved by Lévy) is used in particular in Shepp and Shiryaev [787] for the valuation of Russian options and in Guo and Shepp [412] for perpetual lookback American options. ◁

1.1.15 Uniform Integrability

A family of random variables $(X_i, i \in I)$, is **uniformly integrable** (u.i.) if $\sup_{i \in I} \int_{|X_i| \geq a} |X_i| d\mathbb{P}$ goes to 0 when a goes to infinity.

If $|X_i| \leq Y$ where Y is integrable, then $(X_i, i \in I)$ is u.i., but the converse does not hold.

Let $(\Omega, \mathcal{F}, \mathbf{F}, \mathbb{P})$ be a filtered probability space and X an $\mathcal{F}_\infty$-measurable integrable random variable. The family $(\mathbb{E}(X|\mathcal{F}_t), t \geq 0)$ is u.i.. More generally, if $(\Omega, \mathcal{F}, \mathbb{P})$ is a given probability space and X an integrable r.v., the family $\{\mathbb{E}(X|\mathcal{G}), \mathcal{G} \subseteq \mathcal{F}\}$ is u.i.

Very often, one uses the fact that if $(X_i, i \in I)$ is bounded in L^2, i.e., $\sup_i \mathbb{E}(X_i^2) < \infty$ then, it is a u.i. family.

Among the main uses of uniform integrability, the following is the most important: if $(X_n, n \geq 1)$ is u.i. and $X_n \xrightarrow{P} X$, then $X_n \xrightarrow{L^1} X$.

1.2 Martingales

Although our aim in this chapter is to discuss continuous path processes, there would be no advantage in this section of limiting ourselves to the scope of continuous martingales. We shall restrict our attention to continuous martingales in $\rightarrowtail$ Section 1.3.

1.2.1 Definition and Main Properties

Definition 1.2.1.1 *An* $\mathbf{F}$*-adapted process* $X = (X_t, t \geq 0)$*, is an* $\mathbf{F}$*-martingale (resp. sub-martingale, resp. super-martingale) if*

- $\mathbb{E}(|X_t|) < \infty$*, for every* $t \geq 0$*,*
- $\mathbb{E}(X_t|\mathcal{F}_s) = X_s$ *(resp.* $\mathbb{E}(X_t|\mathcal{F}_s) \geq X_s$*, resp.* $\mathbb{E}(X_t|\mathcal{F}_s) \leq X_s$*) a.s. for every pair* (s,t) *such that* $s < t$*.*

Roughly speaking, an $\mathbf{F}$-martingale is a process which is $\mathbf{F}$-conditionally constant, and a super-martingale is conditionally decreasing. Hence, one can ask the question: is a super-martingale the sum of a martingale and a decreasing process? Under some weak assumptions, the answer is positive (see the Doob-Meyer theorem quoted below as Theorem 1.2.1.6).

Example 1.2.1.2 The basic example of a martingale is the process X defined as $X_t := \mathbb{E}(X_\infty|\mathcal{F}_t)$, where X_∞ is a given $\mathcal{F}_\infty$-measurable integrable r.v.. In fact, X is a uniformly integrable martingale if and only if $X_t := \mathbb{E}(X_\infty|\mathcal{F}_t)$, for some $X_\infty \in L^1(\mathcal{F}_\infty)$.

Sometimes, we shall deal with processes indexed by $[0,T]$, which may be considered by a simple transformation as the above processes. If the filtration $\mathbf{F}$ is right-continuous, it is possible to show that any martingale has a càdlàg version.

If M is an $\mathbf{F}$-martingale and $\mathbf{H} \subseteq \mathbf{F}$, then $\mathbb{E}(M_t|\mathcal{H}_t)$ is an $\mathbf{H}$-martingale. In particular, if M is an $\mathbf{F}$-martingale, then it is an $\mathbf{F}^M$-martingale. A process is said to be a martingale if it is a martingale with respect to its natural filtration.

From now on, any martingale (super-martingale, sub-martingale) will be taken to be right-continuous with left-hand limits.

Warning 1.2.1.3 If M is an $\mathbf{F}$-martingale and $\mathbf{F} \subset \mathbf{G}$, it is *not* true in general that M is a $\mathbf{G}$-martingale (see $\rightarrowtail$ Section 5.9 on enlargement of filtrations for a discussion on that specific case).

Example 1.2.1.4 If X is a process with independent increments such that the r.v. X_t is integrable for any t, the process $(X_t - \mathbb{E}(X_t), t \geq 0)$ is a martingale. Sometimes, these processes are called self-similar processes (see $\rightarrowtail$ Chapter 11 for the particular case of Lévy processes).

Definition 1.2.1.5 *A process X is of the class (D), if the family of random variables $(X_\tau, \tau$ finite stopping time) is u.i..*

Theorem 1.2.1.6 (Doob-Meyer Decomposition Theorem) *The process $(X_t; t \geq 0)$ is a sub-martingale (resp. a super-martingale) of class (D) if and only if $X_t = M_t + A_t$ (resp. $X_t = M_t - A_t$) where M is a uniformly integrable martingale and A is an increasing predictable[3] process with $\mathbb{E}(A_\infty) < \infty$.*

PROOF: See Dellacherie and Meyer [244] Chapter VII, 12 or Protter [727] Chapter III. □

If M is a martingale such that $\sup_t \mathbb{E}(|M_t|) < \infty$ (i.e., M is L^1 bounded), there exists an integrable random variable M_∞ such that M_t converges almost surely to M_∞ when t goes to infinity (see [RY], Chapter I, Theorem 2.10). This holds, in particular, if M is uniformly integrable and in that case $M_t \rightarrow_{L^1} M_\infty$ and $M_t = \mathbb{E}(M_\infty|\mathcal{F}_t)$. However, an L^1-bounded martingale is not necessarily uniformly integrable as the following example shows:

Example 1.2.1.7 The martingale $M_t = \exp\left(\lambda W_t - \frac{\lambda^2}{2}t\right)$ where W is a Brownian motion (see $\rightarrowtail$ Section 1.4) is L^1 bounded (indeed $\forall t, \mathbb{E}(M_t) = 1$). From $\lim_{t\to\infty} \frac{W_t}{t} = 0$, a.s., we get that

$$\lim_{t\to\infty} M_t = \lim_{t\to\infty} \exp\left(t\left(\lambda\frac{W_t}{t} - \frac{\lambda^2}{2}\right)\right) = \lim_{t\to\infty} \exp\left(-t\frac{\lambda^2}{2}\right) = 0\,,$$

hence this martingale is not u.i. on $[0,\infty[$ (if it were, it would imply that M_t is null!).

Exercise 1.2.1.8 Let M be an $\mathbf{F}$-martingale and Z an adapted (bounded) continuous process. Prove that, for $0 < s < t$,

$$\mathbb{E}\left(M_t \int_s^t Z_u du|\mathcal{F}_s\right) = \mathbb{E}\left(\int_s^t M_u Z_u du|\mathcal{F}_s\right)\,. \qquad \triangleleft$$

Exercise 1.2.1.9 Consider the interval $[0,1]$ endowed with Lebesgue measure λ on the Borel σ-algebra $\mathcal{B}$. Define $\mathcal{F}_t = \sigma\{A : A \subset [0,t], A \in \mathcal{B}\}$. Let f be an integrable function defined on $[0,1]$, considered as a random variable.

[3] See Subsection 1.2.3 for the definition of predictable processes. In the particular case where X is continuous, then A is continuous.

Prove that

$$\mathbb{E}(f|\mathcal{F}_t)(u) = f(u)\mathbb{1}_{\{u\le t\}} + \mathbb{1}_{\{u>t\}}\frac{1}{1-t}\int_t^1 dx f(x)\,. \qquad \triangleleft$$

Exercise 1.2.1.10 Give another proof that $\lim_{t\to\infty} M_t = 0$ in the above Example 1.2.1.7 by using $T_{-a} = \inf\{t : W_t = -a\}$. $\triangleleft$

1.2.2 Spaces of Martingales

We denote by $\mathbf{H}^2$ (resp. $\mathbf{H}^2[0,T]$) the subset of **square integrable** martingales (resp. defined on [0,T]), i.e., martingales such that $\sup_t \mathbb{E}(M_t^2) < \infty$ (resp. $\sup_{t\le T} \mathbb{E}(M_t^2) < \infty$). From Jensen's inequality, if M is a square integrable martingale, M^2 is a sub-martingale. It follows that the martingale M is square integrable on $[0,T]$ if and only if $\mathbb{E}(M_T^2) < \infty$.

If $M \in \mathbf{H}^2$, the process M is u.i. and $M_t = \mathbb{E}(M_\infty|\mathcal{F}_t)$. From Fatou's lemma, the random variable M_∞ is square integrable and

$$\mathbb{E}(M_\infty^2) = \lim_{t\to\infty} \mathbb{E}(M_t^2) = \sup_t \mathbb{E}(M_t^2)\,.$$

From $M_t^2 \le \mathbb{E}(M_\infty^2|\mathcal{F}_t)$, it follows that $(M_t^2, t\ge 0)$ is uniformly integrable.

Doob's inequality states that, if $M \in \mathbf{H}^2$, then $\mathbb{E}(\sup_t M_t^2) \le 4\mathbb{E}(M_\infty^2)$. Hence, $\mathbb{E}(\sup_t M_t^2) < \infty$ is equivalent to $\sup_t \mathbb{E}(M_t^2) < \infty$. More generally, if M is a martingale or a positive sub-martingale, and $p > 1$,

$$\|\sup_{t\le T} |M_t|\|_p \le \frac{p}{p-1} \sup_{t\le T} \|M_t\|_p\,. \tag{1.2.1}$$

Obviously, the Brownian motion (see ↣ Section 1.4) does not belong to $\mathbf{H}^2$, however, it belongs to $\mathbf{H}^2([0,T])$ for any T.

We denote by $\mathbf{H}^1$ the set of martingales M such that $\mathbb{E}(\sup_t |M_t|) < \infty$. More generally, the space of martingales such that $M^* = \sup_t |M_t|$ is in L^p is denoted by $\mathbf{H}^p$. For $p > 1$, one has the equivalence

$$M^* \in L^p \Leftrightarrow M_\infty \in L^p\,.$$

Thus the space $\mathbf{H}^p$ for $p > 1$ may be identified with $L^p(\mathcal{F}_\infty)$. Note that $\sup_t \mathbb{E}(|M_t|) \le \mathbb{E}(\sup_t |M_t|)$, hence any element of $\mathbf{H}^1$ is L^1 bounded, but the converse if not true (see Azéma et al. [36]).

1.2.3 Stopping Times

Definitions

An $\mathbb{R}^+ \cup \{+\infty\}$-valued random variable τ is a **stopping time** with respect to a given filtration $\mathbf{F}$ (in short, an $\mathbf{F}$-stopping time), if $\{\tau \le t\} \in \mathcal{F}_t$, $\forall t \ge 0$.

If the filtration $\mathbf{F}$ is right-continuous, it is equivalent to demand that $\{\tau < t\}$ belongs to $\mathcal{F}_t$ for every t, or that the left-continuous process $\mathbb{1}_{]0,\tau]]}(t)$ is an $\mathbf{F}$-adapted process). If $\mathbf{F} \subset \mathbf{G}$, any $\mathbf{F}$-stopping time is a $\mathbf{G}$-stopping time.

If τ is an $\mathbf{F}$-stopping time, the σ-algebra of events prior to τ, $\mathcal{F}_\tau$ is defined as:

$$\mathcal{F}_\tau = \{A \in \mathcal{F}_\infty : A \cap \{\tau \le t\} \in \mathcal{F}_t, \forall t\}.$$

If X is $\mathbf{F}$-progressively measurable and τ a $\mathbf{F}$-stopping time, then the r.v. X_τ is $\mathcal{F}_\tau$-measurable on the set $\{\tau < \infty\}$.

The σ-algebra $\mathcal{F}_{\tau-}$ is the smallest σ-algebra which contains $\mathcal{F}_0$ and all the sets of the form $A \cap \{t < \tau\}, t > 0$ for $A \in \mathcal{F}_t$.

Definition 1.2.3.1 *A stopping time τ is* ***predictable*** *if there exists an increasing sequence (τ_n) of stopping times such that almost surely*
(i) $\lim_n \tau_n = \tau$,
(ii) $\tau_n < \tau$ *for every n on the set $\{\tau > 0\}$.*
A stopping time τ is ***totally inaccessible*** *if $\mathbb{P}(\tau = \vartheta < \infty) = 0$ for any predictable stopping time ϑ (or, equivalently, if for any increasing sequence of stopping times $(\tau_n, n \ge 0)$, $\mathbb{P}(\{\lim \tau_n = \tau\} \cap A) = 0$ where $A = \cap_n \{\tau_n < \tau\}$).*

If X is an $\mathbf{F}$-adapted process and τ a stopping time, the ($\mathbf{F}$-adapted) process X^τ where $X_t^\tau := X_{t\wedge\tau}$ is called the **process X stopped** at τ.

Example 1.2.3.2 If τ is a **random time**, (i.e., a positive random variable), the smallest filtration with respect to which τ is a stopping time is the filtration generated by the process $D_t = \mathbb{1}_{\{\tau \le t\}}$. The completed σ-algebra $\mathcal{D}_t$ is generated by the sets $\{\tau \le s\}, s \le t$ or, equivalently, by the random variable $\tau \wedge t$. This kind of times will be of great importance in ↣ Chapter 7 to model default risk events.

Example 1.2.3.3 If X is a continuous process, and a a real number, the first time T_a^+ (resp. T_a^-) when X is greater (resp. smaller) than a, is an $\mathbf{F}^X$-stopping time

$$T_a^+ = \inf\{t : X_t \ge a\}, \quad \text{resp. } T_a^- = \inf\{t : X_t \le a\}.$$

From the continuity of the process X, if the process starts below a (i.e., if $X_0 < a$), one has $T_a^+ = T_a$ where $T_a = \inf\{t : X_t = a\}$, and $X_{T_a} = a$ (resp. if $X_0 > a$, $T_a^- = T_a$). Note that if $X_0 \ge a$, then $T_a^+ = 0$, and $T_a > 0$.

More generally, if X is a continuous $\mathbb{R}^d$-valued processes, its **first entrance** time into a closed set F, i.e., $T_F = \inf\{t : X_t \in F\}$, is a stopping time (see [RY], Chapter I, Proposition 4.6.). If a real-valued process is progressive with respect to a standard filtration, the first entrance time of a Borel set is a stopping time.

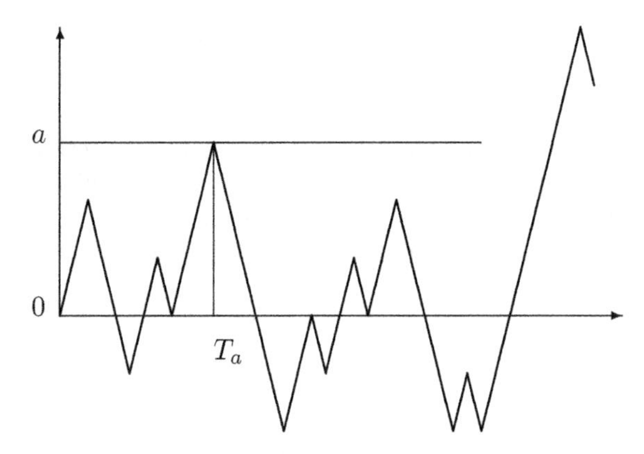

Fig. 1.1 First hitting time of a level a

Optional and Predictable Process

If τ and ϑ are two stopping times, the **stochastic interval** $]\!]\vartheta, \tau]\!]$ is the set $\{(t, \omega) : \vartheta(\omega) < t \leq \tau(\omega)\}$.

The **optional** σ**-algebra** $\mathcal{O}$ is the σ-algebra generated on $\mathcal{F} \times \mathcal{B}$ by the stochastic intervals $[\![\tau, \infty[\![$ where τ is an $\mathbf{F}$-stopping time.

The **predictable** σ**-algebra** $\mathcal{P}$ is the σ-algebra generated on $\mathcal{F} \times \mathcal{B}$ by the stochastic intervals $]\!]\vartheta, \tau]\!]$ where ϑ and τ are two $\mathbf{F}$-stopping times such that $\vartheta \leq \tau$.

A process X is said to be **F-predictable** (resp. $\mathbf{F}$-optional) if the map $(\omega, t) \rightarrow X_t(\omega)$ is $\mathcal{P}$-measurable (resp. $\mathcal{O}$-measurable).

Example 1.2.3.4 An adapted càg process is predictable.

Martingales and Stopping Times

If M is an $\mathbf{F}$-martingale and τ an $\mathbf{F}$-stopping time, the stopped process M^τ is an $\mathbf{F}$-martingale.

Theorem 1.2.3.5 (Doob's Optional Sampling Theorem.) *If M is a uniformly integrable martingale (e.g., bounded) and ϑ, τ are two stopping times with $\vartheta \leq \tau$, then*

$$M_\vartheta = \mathbb{E}(M_\tau | \mathcal{F}_\vartheta) = \mathbb{E}(M_\infty | \mathcal{F}_\vartheta), \ a.s.$$

If M is a positive super-martingale and ϑ, τ a pair of stopping times with $\vartheta \leq \tau$, then

$$\mathbb{E}(M_\tau | \mathcal{F}_\vartheta) \leq M_\vartheta \, .$$

Warning 1.2.3.6 This theorem often serves as a basic tool to determine quantities defined up to a first hitting time and laws of hitting times. However, in many cases, the u.i. hypothesis has to be checked carefully. For example, if W is a Brownian motion, (see the definition in ↣ Section 1.4), and T_a the first hitting time of a, then $\mathbb{E}(W_{T_a}) = a$, while a blind application of Doob's theorem would lead to equality between $\mathbb{E}(W_{T_a})$ and $W_0 = 0$. The process $(W_{t\wedge T_a}, t \geq 0)$ is not uniformly integrable, but $(W_{t\wedge T_a}, t \leq t_0)$ is, and obviously so is $(W_{t\wedge T_{-c}\wedge T_a}, t \geq 0)$ (here, $-c < 0 < a$).

The following proposition is an easy converse to Doob's optional sampling theorem:

Proposition 1.2.3.7 *If M is an adapted integrable process, and if for any two-valued stopping time τ, $\mathbb{E}(M_\tau) = \mathbb{E}(M_0)$, then M is a martingale.*

PROOF: Let $s < t$ and $\Gamma_s \in \mathcal{F}_s$. The random time

$$\tau = \begin{cases} s & \text{on } \Gamma_s^c \\ t & \text{on } \Gamma_s \end{cases}$$

is a stopping time, hence $\mathbb{E}(M_t \mathbb{1}_{\Gamma_s}) = \mathbb{E}(M_s \mathbb{1}_{\Gamma_s})$ and the result follows. □

The adapted integrable process M is a martingale if and only if the following property is satisfied ([RY], Chapter II, Sect. 3): if ϑ, τ are two bounded stopping times with $\vartheta \leq \tau$, then

$$M_\vartheta = \mathbb{E}(M_\tau | \mathcal{F}_\vartheta), \text{ a.s.}$$

Comments 1.2.3.8 (a) Knight and Maisonneuve [530] proved that a random time τ is an **F**-stopping time if and only if, for any bounded **F**-martingale M, $\mathbb{E}(M_\infty | \mathcal{F}_\tau) = M_\tau$. Here, $\mathcal{F}_\tau$ is the σ-algebra generated by the random variables Z_τ, where Z is any **F**-optional process. (See Dellacherie et al. [241], page 141, for more information.)
(b) Note that there exist some random times τ which are not stopping times, but nonetheless satisfy $\mathbb{E}(M_0) = \mathbb{E}(M_\tau)$ for any bounded **F**-martingale (see Williams [844]). Such times are called pseudo-stopping times. (See ↣ Subsection 5.9.4 and Comments 7.5.1.3.)

Definition 1.2.3.9 *A continuous uniformly integrable martingale M belongs to BMO space if there exists a constant m such that*

$$\mathbb{E}(\langle M \rangle_\infty - \langle M \rangle_\tau | \mathcal{F}_\tau) \leq m$$

for any stopping time τ.

See ↣ Subsection 1.3.1 for the definition of the bracket $\langle . \rangle$. It can be proved (see, e.g., Dellacherie and Meyer [244], Chapter VII,) that the space BMO is the dual of $\mathbf{H}^1$.

See Kazamaki [517] and Doléans-Dade and Meyer [257] for a study of Bounded Mean Oscillation (BMO) martingales.

Exercise 1.2.3.10 A Useful Lemma: Doob's Maximal Identity.
(1) Let M be a positive continuous martingale such that $M_0 = x$.
(i) Prove that if $\lim_{t\to\infty} M_t = 0$, then

$$\mathbb{P}(\sup M_t > a) = \left(\frac{x}{a}\right) \wedge 1 \tag{1.2.2}$$

and $\sup M_t \overset{\text{law}}{=} \frac{x}{U}$ where U is a random variable with a uniform law on $[0,1]$. (See [RY], Chapter 2, Exercise 3.12.)
(ii) Conversely, if $\sup M_t \overset{\text{law}}{=} \frac{x}{U}$, show that $M_\infty = 0$.
(2) Application: Find the law of $\sup_t(B_t - \mu t)$ for $\mu > 0$. (Use Example 1.2.1.7). For $T_a^{(-\mu)} = \inf\{t : B_t - \mu t \geq a\}$, compute $\mathbb{P}(T_a^{(-\mu)} < \infty)$.
`Hint:` Apply Doob's optional sampling theorem to $T_y \wedge t$ and prove, passing to the limit when t goes to infinity, that

$$a = \mathbb{E}(M_{T_y}) = y\mathbb{P}(T_y < \infty) = y\mathbb{P}(\sup M_t \geq y)\,.$$ ◁

1.2.4 Local Martingales

Definition 1.2.4.1 *An adapted, right-continuous process M is an* **F***-**local martingale** if there exists a sequence of stopping times (τ_n) such that:*

- *The sequence τ_n is increasing and $\lim_n \tau_n = \infty$, a.s.*
- *For every n, the stopped process $M^{\tau_n}\mathbb{1}_{\{\tau_n>0\}}$ is an* **F***-martingale.*

A sequence of stopping times such that the two previous conditions hold is called a localizing or reducing sequence. If M is a local martingale, it is always possible to choose the localizing sequence $(\tau_n, n \geq 1)$ such that each martingale $M^{\tau_n}\mathbb{1}_{\{\tau_n>0\}}$ is uniformly integrable.

Let us give some criteria that ensure that a local martingale is a martingale:

- Thanks to Fatou's lemma, a positive local martingale M is a supermartingale. Furthermore, it is a martingale if (and only if!) its expectation is constant ($\forall t$, $\mathbb{E}(M_t) = \mathbb{E}(M_0)$).
- A local martingale is a uniformly integrable martingale if and only if it is of the class (D) (see Definition 1.2.1.5).
- A local martingale is a martingale if and only if it is of the class (DL), that is, if for every $a > 0$ the family of random variables $(X_\tau, \tau \in \mathcal{T}_a)$ is uniformly integrable, where $\mathcal{T}_a$ is the set of stopping times smaller than a.
- If a local martingale M is in $\mathbf{H}^1$, i.e., if $\mathbb{E}(\sup_t |M_t|) < \infty$, then M is a uniformly integrable martingale (however, not every uniformly integrable martingale is in $\mathbf{H}^1$).

Later, we shall give explicit examples of local martingales which are not martingales. They are called strict local martingales (see, e.g., ↣ Example 6.1.2.6 and ↣ Subsection 6.4.1). Note that there exist strict local martingales with constant expectation (see ↣ Exercise 6.1.5.6).

Doob-Meyer decomposition can be extended to general sub-martingales:

Proposition 1.2.4.2 *A process X is a sub-martingale (resp. a super-martingale) if and only if $X_t = M_t + A_t$ (resp. $X_t = M_t - A_t$) where M is a local martingale and A an increasing predictable process.*

We also use the following definitions:
A local martingale M is locally square integrable if there exists a localizing sequence of stopping times (τ_n) such that $M^{\tau_n}\mathbb{1}_{\{\tau_n>0\}}$ is a square integrable martingale.
An increasing process A is locally integrable if there exists a localizing sequence of stopping times such that A^{τ_n} is integrable.
By similar localization, we may define locally bounded martingales, local super-martingales, and locally finite variation processes.

Let us state without proof (see [RY]) the following important result.

Proposition 1.2.4.3 *A continuous local martingale of locally finite variation is a constant.*

Warning 1.2.4.4 If X is a positive local super-martingale, then it is a super-martingale. If X is a positive local sub-martingale, it is not necessarily a sub-martingale (e.g., a positive strict local martingale is a positive local sub-martingale and a super-martingale).

Note that a locally integrable increasing process A does not necessarily satisfy $\mathbb{E}(A_t) < \infty$ for any t. As an example, if $A_t = \int_0^t ds/R_s^2$ where R is a 2-dimensional Bessel process (see ↣ Chapter 6) then A is locally integrable, however $\mathbb{E}(A_t) = \infty$, since, for any $s > 0$, $\mathbb{E}(1/R_s^2) = \infty$.

Comment 1.2.4.5 One can also define a continuous quasi-martingale as a continuous process X such that

$$\sup \sum_{i=1}^{p(n)} \mathbb{E}|\mathbb{E}(X_{t_{i+1}^n} - X_{t_i^n}|\mathcal{F}_{t_i^n})| < \infty$$

where the supremum is taken over the sequences $0 < t_i^n < t_{i+1}^n < T$. Super-martingales (sub-martingales) are quasi-martingales. In that case, the above condition reads

$$\mathbb{E}(|X_T - X_0|) < \infty .$$

1.3 Continuous Semi-martingales

A d-dimensional **continuous semi-martingale** is an $\mathbb{R}^d$-valued process X such that each component X^i admits a decomposition as $X^i = M^i + A^i$ where M^i is a continuous local martingale with $M^i_0 = 0$ and A^i is a continuous adapted process with locally finite variation. This decomposition is unique (see [RY]), and we shall say in short that M is the martingale part of the continuous semi-martingale X. This uniqueness property, which is not shared by general semi-martingales motivated us to restrict our study of semi-martingales at first to the continuous ones. Later ($\rightarrowtail$ Chapter 9) we shall consider general semi-martingales.

1.3.1 Brackets of Continuous Local Martingales

If M is a continuous local martingale, there exists a unique continuous increasing process $\langle M \rangle$, called the bracket (or predictable quadratic variation) of M such that $(M_t^2 - \langle M \rangle_t, t \geq 0)$ is a continuous local martingale (see [RY] Chap IV, Theorem 1.3 for the existence).

The process $\langle M \rangle$ is equal to the limit in probability of the quadratic variation $\sum_i (M_{t^n_{i+1}} - M_{t^n_i})^2$, where $0 = t^n_0 < t^n_1 < \cdots < t^n_{p(n)} = t$, when $\sup_{0 \leq i \leq p(n)-1} (t^n_{i+1} - t^n_i)$ goes to zero (see [RY], Chapter IV, Section 1). [4] Note that the limit of $\sum_i (M_{t^n_{i+1}} - M_{t^n_i})^2$ depends neither on the filtration nor on the probability measure on the space $(\Omega, \mathcal{F})$ (assuming that M remains a semi-martingale with respect to this filtration or to this probability) and the process $\langle M \rangle$ is $\mathbf{F}^M$-adapted.

Example 1.3.1.1 If W is a Brownian motion (defined in $\rightarrowtail$ Section 1.4),

$$\langle W \rangle_t = \lim \sum_{i=0}^{p(n)-1} (W_{t^n_{i+1}} - W_{t^n_i})^2 = t\,.$$

Here, the limit is in the L^2 sense (hence, in the probability sense). If $\sum_n \sup_i (t^n_{i+1} - t^n_i) < \infty$, the convergence holds also in the a.s. sense (see Kallenberg [505]). This is in particular the case for a dyadic sequence, where $t^n_i = \frac{i}{2^n} t$.

Definition 1.3.1.2 *If M and N are two continuous local martingales, the unique continuous process $(\langle M, N \rangle_t, t \geq 0)$ with locally finite variation such that $MN - \langle M, N \rangle$ is a continuous local martingale is called the* ***predictable bracket*** *(or the predictable covariation process) of M and N.*

[4] This is why the term *quadratic variation* is often used instead of bracket.

Let us remark that $\langle M \rangle = \langle M, M \rangle$ and

$$\langle M, N \rangle = \frac{1}{2} [\langle M + N \rangle - \langle M \rangle - \langle N \rangle] = \frac{1}{4} [\langle M + N \rangle - \langle M - N \rangle] .$$

These last identities are known as the polarization equalities.

In particular, if the bracket $\langle X, Y \rangle$ of two martingales X and Y is equal to zero, the product XY is a local martingale and X and Y are said to be **orthogonal**. Note that this is the case if X and Y are independent.

We present now some useful results, related to the predictable bracket. For the proofs, we refer to [RY], Chapter IV.

- A continuous local martingale M converges a.s. as t goes to infinity on the set $\{\langle M \rangle_\infty < \infty\}$.
- The **Kunita-Watanabe inequality** states that

$$|\langle M, N \rangle| \leq \langle M \rangle^{1/2} \langle N \rangle^{1/2} .$$

  More generally, for h, k positive measurable processes

$$\int_0^t h_s\, k_s |d\langle M, N \rangle_s| \leq \left(\int_0^t h_s^2 d\langle M \rangle_s \right)^{1/2} \left(\int_0^t k_s^2 d\langle N \rangle_s \right)^{1/2} .$$

- The **Burkholder-Davis-Gundy** (BDG) inequalities state that for $0 \leq p < \infty$, there exist two universal constants c_p and C_p such that if M is a continuous local martingale,

$$c_p\, \mathbb{E}[(\sup_t |M_t|)^p] \leq \mathbb{E}(\langle M \rangle_\infty^{p/2}) \leq C_p\, \mathbb{E}[(\sup_t |M_t|)^p] .$$

  (See Lenglart et al. [576] for a complete study.) It follows that, if a continuous local martingale M satisfies $\mathbb{E}(\langle M \rangle_\infty^{1/2}) < \infty$, then M is a martingale. Indeed, $\mathbb{E}(\sup_t |M_t|) < \infty$ (i.e., $M \in \mathbf{H}^1$) and, by dominated convergence, the martingale property follows.

We now introduce some spaces of processes, which will be useful for stochastic integration.

Definition 1.3.1.3 *For* $\mathbf{F}$ *a given filtration and* $M \in \mathbf{H}^{c,2}$, *the space of square integrable continuous* $\mathbf{F}$*-martingales, we denote by* $L^2(M, \mathbf{F})$ *the Hilbert space of equivalence classes of elements of* $\mathcal{L}^2(M)$, *the space of* $\mathbf{F}$*-progressively measurable processes* K *such that*

$$\mathbb{E}[\int_0^\infty K_s^2 d\langle M \rangle_s] < \infty .$$

We shall sometimes write only $L^2(M)$ when there is no ambiguity. If M is a continuous local martingale, we call $L^2_{loc}(M)$ the space of progressively measurable processes K such that there exists a sequence of stopping times (τ_n) increasing to infinity for which

$$\text{for every } n, \quad \mathbb{E}\left(\int_0^{\tau_n} K_s^2 d\langle M\rangle_s\right) < \infty\,.$$

The space $L^2_{loc}(M)$ consists of all progressively measurable processes K such that

$$\text{for every } t, \quad \int_0^t K_s^2 d\langle M\rangle_s < \infty \, a.s..$$

A continuous local martingale belongs to $\mathbf{H}^{c,2}$ (and is a martingale) if and only if $M_0 \in L^2$ and $\mathbb{E}(\langle M\rangle_\infty) < \infty$.

1.3.2 Brackets of Continuous Semi-martingales

Definition 1.3.2.1 *The bracket (or the predictable quadratic covariation) $\langle X, Y\rangle$ of two continuous semi-martingales X and Y is defined as the bracket of their local martingale parts M^X and M^Y.*

The bracket $\langle X, Y\rangle := \langle M^X, M^Y\rangle$ is also the limit in probability of the quadratic covariation of X and Y, i.e.,

$$\sum_{i=0}^{p(n)-1} (X_{t_{i+1}^n} - X_{t_i^n})(Y_{t_{i+1}^n} - Y_{t_i^n}) \tag{1.3.1}$$

for $0 = t_0^n \leq t_1^n \leq \cdots \leq t_{p(n)} = t$ when $\sup_{0\leq i\leq p(n)-1}(t_{i+1}^n - t_i^n)$ goes to 0. Indeed, the bounded variation parts A^X and A^Y do not contribute to the limit of the expression (1.3.1).

If τ is a stopping time, and X a semi-martingale, the stopped process X^τ is a semi-martingale and if Y is another semi-martingale, the bracket of the τ-stopped semi-martingales is the τ-stopped bracket:

$$\langle X^\tau, Y\rangle = \langle X^\tau, Y^\tau\rangle = \langle X, Y\rangle^\tau\,.$$

Remark 1.3.2.2 Let M be a continuous martingale of the form

$$M_t = \int_0^t \varphi_s dW_s$$

where φ is a continuous adapted process (such that $\int_0^t \varphi_s^2 ds < \infty$) and W a Brownian motion (see ↣ Sections 1.4 and 1.5.1 for definitions). The quadratic variation $\langle M\rangle$ is the process

$$\langle M\rangle_t = \int_0^t \varphi_s^2 ds = \mathbb{P} - \lim \sum_{i=1}^{p(n)} (M_{t_{i+1}^n} - M_{t_i}^n)^2\,,$$

hence, $\mathcal{F}_t^M$ contains $\sigma(\varphi_s^2, s \leq t)$.

Exercise 1.3.2.3 Let M be a Gaussian martingale with bracket $\langle M \rangle$. Prove that the process $\langle M \rangle$ is deterministic.
Hint: The Gaussian property implies that, for $t > s$, the r.v. $M_t - M_s$ is independent of $\mathcal{F}_s^M$, hence

$$\mathbb{E}((M_t - M_s)^2 | \mathcal{F}_s^M) = \mathbb{E}((M_t - M_s)^2) = A(t) - A(s)$$

with $A(t) = \mathbb{E}(M_t^2)$ which is deterministic. ◁

1.4 Brownian Motion

1.4.1 One-dimensional Brownian Motion

Let X be an $\mathbb{R}$-valued *continuous* process starting from 0 and $\mathbf{F}^X$ its natural filtration.

Definition 1.4.1.1 *The continuous process X is said to be a Brownian motion, (in short, a BM), if one of the following equivalent properties is satisfied:*

(i) *The process X has stationary and independent increments, and for any $t > 0$, the r.v. X_t follows the $\mathcal{N}(0, t)$ law.*
(ii) *The process X is a Gaussian process, with mean value equal to 0 and covariance $t \wedge s$.*
(iii) *The processes $(X_t, t \geq 0)$ and $(X_t^2 - t, t \geq 0)$ are $\mathbf{F}^X$-local martingales.*
(iii′) *The process X is an $\mathbf{F}^X$-local martingale with bracket t.*
(iv) *For every real number λ, the process $\left(\exp\left(\lambda X_t - \frac{\lambda^2}{2}t\right), t \geq 0\right)$ is an $\mathbf{F}^X$-local martingale.*
(v) *For every real number λ, the process $\left(\exp\left(i\lambda X_t + \frac{\lambda^2}{2}t\right), t \geq 0\right)$ is an $\mathbf{F}^X$-local martingale.*

To establish the existence of Brownian motion, one starts with the canonical space $\Omega = C(\mathbb{R}^+, \mathbb{R})$ of continuous functions. The canonical process $X_t : \omega \rightarrow \omega(t)$ (ω is now a generic continuous function) is defined on Ω. There exists a unique probability measure on this space Ω such that the law of X satisfies the above properties. This probability measure is called Wiener measure and is often denoted by $\mathbf{W}$ in deference to Wiener (1923) who proved its existence. We refer to [RY] Chapter I, for the proofs.

It can be proved, as a consequence of Kolmogorov's continuity criterion 1.1.10.6 that a process (not assumed to be continuous) which satisfies (i) or (ii) admits in fact a continuous modification. There exist discontinuous processes that satisfy (iii) (e.g., the martingale associated with a Poisson process, see ↣ Chapter 8).

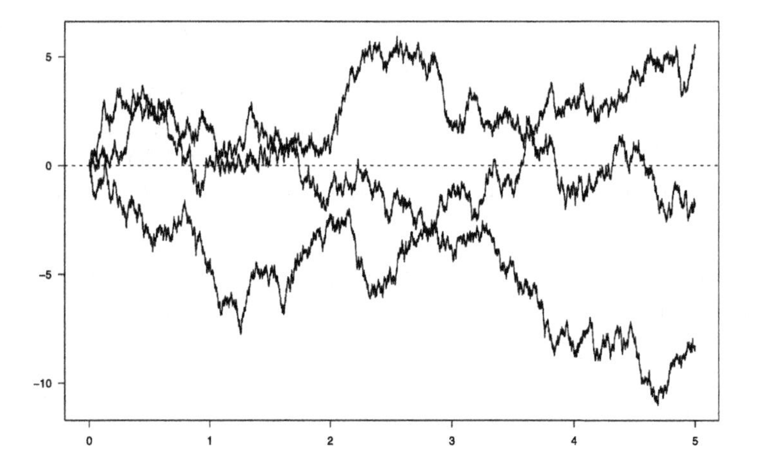

Fig. 1.2 Simulation of Brownian paths

Extending Definition 1.4.1.1, a continuous process X is said to be a BM with respect to a filtration $\mathbf{F}$ larger than $\mathbf{F}^X$ if for any (t, s), the random variable $X_{t+s} - X_t$ is independent of $\mathcal{F}_t$ and is $\mathcal{N}(0, s)$ distributed.

The **transition probability** of the Brownian motion starting from x (i.e., such that $\mathbb{P}_x(W_0 = x) = 1$) is $p_t(x, y)$ defined as

$$p_t(x, y)dy = \mathbb{P}_x(W_t \in dy) = \mathbb{P}_0(x + W_t \in dy)$$

and

$$\boxed{p_t(x, y) = \frac{1}{\sqrt{2\pi t}} \exp\left(-\frac{1}{2t}(x - y)^2\right).} \tag{1.4.1}$$

We shall also use the notation $p_t(x)$ for $p_t(0, x) = p_t(x, 0)$, hence

$$p_t(x, y) = p_t(x - y).$$

We shall prove in $\rightarrowtail$ Exercise 1.5.3.3 Lévy's characterization of Brownian motion, which is a generalization of (iii) above.

Theorem 1.4.1.2 (Lévy's Characterization of Brownian Motion.) *The process X is an $\mathbf{F}$-Brownian motion if and only if the processes $(X_t, t \geq 0)$ and $(X_t^2 - t, t \geq 0)$ are continuous $\mathbf{F}$-local martingales.*

In this case, the processes are $\mathbf{F}^X$-local martingales, and in fact $\mathbf{F}^X$-martingales. If X is a Brownian motion, the local martingales in (iv) and (v) Definition 1.4.1.1 are martingales. See also [RY], Chapter IV, Theorem 3.6.

An important fact is that in a Brownian filtration, i.e., in a filtration generated by a BM, every stopping time is predictable ([RY], Chapter IV, Corollary 5.7) which is equivalent to the property that all martingales are continuous.

Comment 1.4.1.3 In order to prove property (a), it must be established that $\lim_{t \to 0} tW_{1/t} = 0$, which follows from $(W_t, t > 0) \overset{\text{law}}{=} (tW_{1/t}, t > 0)$.

Definition 1.4.1.4 *A process $X_t = \mu t + \sigma B_t$ where B is a Brownian motion is called a drifted Brownian motion, with drift μ.*

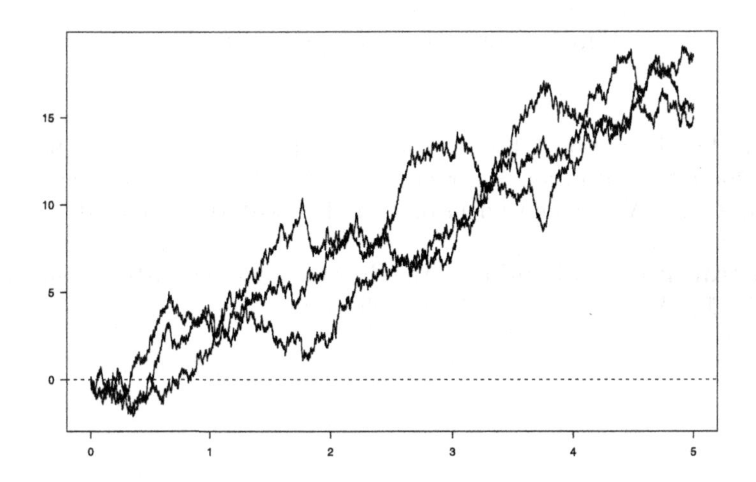

Fig. 1.3 Simulation of drifted Brownian paths $X_t = 3(t + B_t)$

Example 1.4.1.5 Let W be a Brownian motion. Then,

(a) The processes $(-W_t, t \geq 0)$ and $(tW_{1/t}, t \geq 0)$ are BMs. The second result is called the **time inversion** property of the BM.
(b) For any $c \in \mathbb{R}^+$, the process $(\frac{1}{c} W_{c^2 t}, t \geq 0)$ is a BM (scaling property).
(c) The process $B_t = \int_0^t \text{sgn}(W_s) dW_s$ is a Brownian motion with respect to $\mathbf{F}^W$ (and to $\mathbf{F}^B$): indeed the processes B and $(B_t^2 - t, t \geq 0)$ are $\mathbf{F}^W$-martingales. (See $\rightarrowtail$ 1.5.1 for the definition of the stochastic integral and the proofs of the martingale properties). It can be proved that the natural filtration of B is strictly smaller than the filtration of W (see $\rightarrowtail$ Section 5.8).

(d) The process $\widehat{B}_t = W_t - \int_0^t W_s \frac{ds}{s}$ is a Brownian motion with respect to $\mathbf{F}^B$) (but not w.r.t. $\mathbf{F}^W$): indeed, the process $\widehat{B}$ is a Gaussian process and an easy computation establishes that its mean is 0 and its covariance is $s \wedge t$. It can be noted that the process $\widehat{B}$ is not an $\mathbf{F}^W$-martingale and that its natural filtration is strictly smaller than the filtration of W (see $\rightarrowtail$ Section 5.8).

Comment 1.4.1.6 A Brownian filtration is large enough to contain a strictly smaller Brownian filtration (see Examples 1.4.1.5, (c) and (d)). On the other hand, if the processes $W^{(i)}, i = 1, 2$ are independent real-valued Brownian motions, it is not possible to find a real-valued Brownian motion B such that $\sigma(B_s, s \leq t) = \sigma(W_s^{(1)}, W_s^{(2)}, s \leq t)$. This will be proved using the predictable representation theorem. (See $\rightarrowtail$ Subsection 1.6.1.)

Exercise 1.4.1.7 Prove that, for $\lambda > 0$, one has

$$\boxed{\int_0^\infty e^{-\lambda t} p_t(x, y) dt = \frac{1}{\sqrt{2\lambda}} e^{-|x-y|\sqrt{2\lambda}}.}$$

Prove that if f is a bounded Borel function, and $\lambda > 0$,

$$\mathbb{E}_x\left(\int_0^\infty e^{-\lambda^2 t/2} f(W_t) dt\right) = \frac{1}{\lambda} \int_{-\infty}^\infty e^{-\lambda|y-x|} f(y) dy\,. \qquad \triangleleft$$

Exercise 1.4.1.8 Prove that (v) of Definition 1.4.1.1 characterizes a BM, i.e., if the process $(Z_t = \exp(i\lambda X_t + \frac{\lambda^2}{2} t), t \geq 0)$ is a $\mathbf{F}^X$-local martingale for any λ, then X is a BM.
`Hint:` Establish that Z is a martingale, then prove that, for $t > s$,

$$\forall A \in \mathcal{F}_s,\ \mathbb{E}[\mathbb{1}_A \exp(i\lambda(X_t - X_s))] = \mathbb{P}(A) \exp\left(-\tfrac{1}{2}\lambda^2(t-s)\right). \qquad \triangleleft$$

Exercise 1.4.1.9 Prove that, for any $\lambda \in \mathbb{C}$, $(e^{-\lambda^2 t/2} \cosh(\lambda W_t), t \geq 0)$ is a martingale. $\triangleleft$

Exercise 1.4.1.10 Let W be a BM and φ be an adapted process.
(a) Prove that $\int_0^t \varphi_s dW_s$ is a BM if and only if $|\varphi_s| = 1$, ds a.s.
(b) Assume now that φ is deterministic. Prove that $W_t - \int_0^t ds\, \varphi_s\, W_s$ is a BM if and only if $\varphi \equiv 0$ or $\varphi \equiv \frac{1}{s}$, ds a.s..
`Hint:` The function φ satisfies, for $t > s$,

$$\mathbb{E}\left((W_t - \int_0^t du\, \varphi_u\, W_u)\,(W_s - \int_0^s du\, \varphi_u\, W_u)\right) = s$$

if and only if $s\varphi_s = \varphi_s \int_0^s du\, u\, \varphi_u$. $\triangleleft$

1.4.2 d-dimensional Brownian Motion

A continuous process $X = (X^1, \dots, X^d)$, taking values in $\mathbb{R}^d$ is a d-dimensional Brownian motion if one of the following equivalent properties is satisfied:

- all its components X^i are independent Brownian motions.
- The processes X^i and $(X_t^i X_t^j - \delta_{i,j} t, t \geq 0)$, where $\delta_{i,j}$ is the Kronecker symbol ($\delta_{i,j} = 1$ if $i = j$ and $\delta_{i,j} = 0$ otherwise) are continuous local $\mathbf{F}^X$-martingales.
- For any $\lambda \in \mathbb{R}^d$, the process $\left(\exp\left(i\lambda \centerdot X_t + \frac{\|\lambda\|^2}{2} t\right), t \geq 0\right)$ is a continuous $\mathbf{F}^X$-local martingale, where the notation $\lambda \centerdot x$ indicates the Euclidian scalar product between λ and x.

Proposition 1.4.2.1 *Let B be a $\mathbb{R}^d$-valued Brownian motion, and T_x the first hitting time of x, defined as $T_x = \inf\{t > 0 : B_t = x\}$.*

- *If $d = 1$, $\mathbb{P}(T_x < \infty) = 1$, for every $x \in \mathbb{R}$,*
- *If $d \geq 2$, $\mathbb{P}(T_x < \infty) = 0$, for every $x \in \mathbb{R}^d$, i.e., the one-point sets are polar.*
- *If $d \leq 2$, the BM is recurrent, i.e., almost surely, the set $\{t : B_t \in A\}$ is unbounded for all open subsets $A \in \mathbb{R}^d$.*
- *If $d \geq 3$, the BM is transient, more precisely, $\lim_{t \to \infty} |B_t| = +\infty$ almost surely.*

PROOF: We refer to [RY], Chapter V, Section 2. □

1.4.3 Correlated Brownian Motions

If W^1 and W^2 are two independent BMs and ρ a constant satisfying $|\rho| \leq 1$, the process

$$W^3 = \varrho W^1 + \sqrt{1 - \varrho^2}\, W^2$$

is a BM, and $\langle W^1, W^3 \rangle_t = \varrho t$. This leads to the following definition.

Definition 1.4.3.1 *Two $\mathbf{F}$-Brownian motions B and W are said to be $\mathbf{F}$-correlated with correlation ρ if $\langle B, W \rangle_t = \rho t$.*

Proposition 1.4.3.2 *The components of the 2-dimensional correlated BM (B, W) are independent if and only if $\rho = 0$.*

PROOF: If the Brownian motions are independent, their product is a martingale, hence $\rho = 0$. Note that this can also be proved using the integration by parts formula (see ↣ Subsection 1.5.2).

If the bracket is null, then the product BW is a martingale, and it follows that for $t > s$,

$$\mathbb{E}(B_s W_t) = \mathbb{E}(B_s \mathbb{E}(W_t|\mathcal{F}_s)) = \mathbb{E}(B_s W_s) = 0 .$$

Therefore, the Gaussian processes W and B are uncorrelated, hence they are independent. □

If B and W are correlated BMs, the process $(B_t W_t - \rho t, t \geq 0)$ is a martingale and $\mathbb{E}(B_t W_t) = \rho t$. From the Cauchy-Schwarz inequality, it follows that $|\rho| \leq 1$. In the case $|\rho| < 1$, the process X defined by the equation

$$W_t = \rho B_t + \sqrt{1-\rho^2} X_t$$

is a Brownian motion independent of B. Indeed, it is a continuous martingale, and it is easy to check that its bracket is t. Moreover $\langle X, B \rangle = 0$.

Note that, for any pair $(a,b) \in \mathbb{R}^2$ the process $Z_t = aB_t + bW_t$ is, up to a multiplicative factor, a Brownian motion. Indeed, setting $c = \sqrt{a^2 + b^2 + 2ab\rho}$ the two processes $\left(\widetilde{Z}_t := \frac{1}{c} Z_t, t \geq 0\right)$ and $(\widetilde{Z}_t^2 - t, t \geq 0)$ are continuous martingales, hence $\widetilde{Z}$ is a Brownian motion.

Proposition 1.4.3.3 *Let $B_t = \Gamma W_t$ where W is a d-dimensional Brownian motion and $\Gamma = (\gamma_{i,j})$ is a $d \times d$ matrix with $\sum_{j=1}^d \gamma_{i,j}^2 = 1$. The process B is a vector of correlated Brownian motions, with correlation matrix $\rho = \Gamma \Gamma^*$.*

Exercise 1.4.3.4 Prove Proposition 1.4.3.3. ◁

Exercise 1.4.3.5 Let B be a Brownian motion and let $\widehat{B}_t = B_t - \int_0^t ds \frac{B_s}{s}$. Prove that for every t, the r.v's B_t and $\widehat{B}_t$ are not correlated, hence are independent. However, clearly, the two Brownian motions B and $\widehat{B}$ are not independent. There is no contradiction with our previous discussion, as $\widehat{B}$ is not an $\mathbf{F}^B$-Brownian motion. ◁

Remark 1.4.3.6 It is possible to construct two Brownian motions W and B such that the pair (W, B) is not a Gaussian process. For example, let W be a Brownian motion and set $B_t = \int_0^t \text{sgn}(W_s) dW_s$ where the stochastic integral is defined in ↣ Subsection 1.5.1. The pair (W, B) is not Gaussian, since $aW_t + B_t = \int_0^t (a + \text{sgn}(W_s)) dW_s$ is not a Gaussian process. Indeed, its bracket is not deterministic, whereas the bracket of a Gaussian martingale is deterministic (see Exercise 1.3.2.3). Note that $\langle B, W \rangle_t = \int_0^t \text{sgn}(W_s) ds$, hence the bracket is not of the form as in Definition 1.4.3.1. Nonetheless, there is some "correlation" between these two Brownian motions.

1.5 Stochastic Calculus

Let $(\Omega, \mathcal{F}, \mathbf{F}, \mathbb{P})$ be a filtered probability space. We recall very briefly the definition of a stochastic integral with respect to a square integrable martingale. We refer the reader to [RY] for details.

1.5.1 Stochastic Integration

An elementary **F**-predictable process is a process K which can be written

$$K_t := K_0 \mathbb{1}_{\{0\}}(t) + \sum_i K_i \mathbb{1}_{]T_i, T_{i+1}]}(t),$$

with

$$0 = T_0 < T_1 < \cdots < T_i < \cdots \text{ and } \lim_i T_i = +\infty.$$

Here, the T_i's are **F**-stopping times and the r.v's K_i are $\mathcal{F}_{T_i}$-measurable and uniformly bounded, i.e., there exists a constant C such that $\forall i$, $|K_i| \leq C$ a.s..

Let M be a continuous local martingale.

▶ For any elementary predictable process K, the stochastic integral $\int_0^t K_s dM_s$ is defined path-by-path as

$$\int_0^t K_s dM_s := \sum_{i=0}^{\infty} K_i (M_{t \wedge T_{i+1}} - M_{t \wedge T_i}).$$

▶ The stochastic integral $\int_0^t K_s dM_s$ can be defined for any continuous process $K \in L^2(M)$ as follows. For any $p \in \mathbb{N}$, one defines the sequence of stopping times

$$\begin{aligned} T_0 &:= 0 \\ T_1^p &:= \inf \left\{ t : |K_t - K_0| > \frac{1}{p} \right\} \\ T_n^p &:= \inf \left\{ t > T_{n-1}^p : |K_t - K_{T_{n-1}^p}| > \frac{1}{p} \right\}. \end{aligned}$$

Set $K_s^{(p)} = \sum_i K_{T_i^p} \mathbb{1}_{]T_i^p, T_{i+1}^p]}(s)$. The sequence $\int_0^t K_s^{(p)} dM_s$ converges in L^2 to a continuous local martingale denoted by $(K{\star}M)_t := \int_0^t K_s dM_s$.

▶ Then, by density arguments, one can define the stochastic integral for any process $K \in L^2(M)$, and by localization for $K \in L^2_{loc}(M)$.

If $M \in \mathbf{H}^{c,2}$, there is an isometry between $L^2(M)$ and the space of stochastic integrals, i.e.,

$$\mathbb{E}\left(\int_0^t K_s^2 \, d\langle M \rangle_s \right) = \mathbb{E}\left(\int_0^t K_s dM_s \right)^2.$$

(See [RY], Chapter IV for details.)

Let M and N belong to $\mathbf{H}^{c,2}$ and $\phi \in L^2(M), \psi \in L^2(N)$. For the martingales X and Y, where $X_t = (\phi\star M)_t$ and $Y_t = (\psi\star N)_t$, we have $\langle X\rangle_t = \int_0^t \phi_s^2 d\langle M\rangle_s$ and $\langle X, Y\rangle_t = \int_0^t \psi_s\phi_s d\langle M, N\rangle_s$. In particular, for any fixed T, the process $(X_t, t \leq T)$ is a square integrable martingale.

If X is a semi-martingale, the integral of a predictable process K, where $K \in L^2_{loc}(M) \cap L^1_{loc}(|dA|)$ with respect to X is defined to be

$$\int_0^t K_s dX_s = \int_0^t K_s dM_s + \int_0^t K_s dA_s$$

where $\int_0^t K_s dA_s$ is defined path-by-path as a Stieltjes integral (we have required that $\int_0^t |K_s(\omega)|\,|dA_s(\omega)| < \infty$).

For a Brownian motion, we obtain in particular the following proposition:

Proposition 1.5.1.1 *Let W be a Brownian motion, τ a stopping time and θ an adapted process such that $\mathbb{E}\left(\int_0^\tau \theta_s^2 ds\right) < \infty$. Then $\mathbb{E}\left(\int_0^\tau \theta_s dW_s\right) = 0$ and $\mathbb{E}\left(\int_0^\tau \theta_s dW_s\right)^2 = \mathbb{E}\left(\int_0^\tau \theta_s^2 ds\right)$.*

PROOF: We apply the previous results with $\widetilde{\theta} = \theta \mathbb{1}_{\{]0,\tau]\}}$.

Comment 1.5.1.2 In the previous proposition, the integrability condition $\mathbb{E}\left(\int_0^\tau \theta_s^2 ds\right) < \infty$ is important (the case where $\tau = \inf\{t : W_t = a\}$ and $\theta = 1$ is an example where the condition does not hold).

In general, there is the inequality

$$\mathbb{E}\left(\int_0^\tau K_s dM_s\right)^2 \leq \mathbb{E}\left(\int_0^\tau K_s^2\, d\langle M\rangle_s\right) \tag{1.5.1}$$

and it may happen that

$$\mathbb{E}\left(\int_0^\tau K_s^2\, d\langle M\rangle_s\right) = \infty, \text{ and } \mathbb{E}\left(\int_0^\tau K_s dM_s\right)^2 < \infty\,.$$

This is the case if $K_t = 1/R_t^2$ for $t \geq 1$ and $K_t = 0$ for $t < 1$ where R is a Bessel process of dimension 3 and M the driving Brownian motion for R (see ↣ Section 6.1).

Comment 1.5.1.3 In the case where K is continuous, the stochastic integral $\int K_s dM_s$ is the limit of the "Riemann sums" $\sum_i K_{u_i}(M_{t_{i+1}} - M_{t_i})$ where $u_i \in [t_i, t_{i+1}[$. But these sums do not converge pathwise because the paths of M are a.s. not of bounded variation. This is why we use L^2 convergence. It can be proved that the Riemann sums converge uniformly on compacts in probability to the stochastic integral.

Exercise 1.5.1.4 Let b and θ be continuous deterministic functions. Prove that the process $Y_t = \int_0^t b(u)du + \int_0^t \theta(u)dW_u$ is a Gaussian process, with mean $\mathbb{E}(Y_t) = \int_0^t b(u)du$ and covariance $\int_0^{s\wedge t} \theta^2(u)du$. ◁

Exercise 1.5.1.5 Prove that, if W is a Brownian motion, from the definition of the stochastic integral as an L^2 limit, $\int_0^t W_s dW_s = \frac{1}{2}(W_t^2 - t)$. ◁

1.5.2 Integration by Parts

The integration by parts formula follows directly from the definition of the bracket. If (X, Y) are two continuous semi-martingales, then

$$d(XY) = XdY + YdX + d\langle X, Y \rangle$$

or, in an integrated form

$$X_t Y_t = X_0 Y_0 + \int_0^t X_s dY_s + \int_0^t Y_s dX_s + \langle X, Y \rangle_t \,.$$

Definition 1.5.2.1 *Two square integrable continuous martingales are orthogonal if their product is a martingale.*

Exercise 1.5.2.2 If two martingales are independent, they are orthogonal. Check that the converse does not hold.
`Hint:` Let B and W be two independent Brownian motions. The martingales W and M where $M_t = \int_0^t W_s dB_s$ are orthogonal and not independent. Indeed, the martingales W and M satisfy $\langle W, M \rangle = 0$. However, the bracket of M, that is $\langle M \rangle_t = \int_0^t W_s^2 ds$ is $\mathbf{F}^W$-adapted. One can also note that

$$\mathbb{E}\left(\exp\left(i\lambda \int_0^t W_s dB_s\right) |\mathcal{F}_\infty^W\right) = \exp\left(-\frac{\lambda^2}{2}\int_0^t W_s^2 ds\right),$$

and the right-hand side is not a constant as it would be if the independence property held. The martingales M and N where $N_t = \int_0^t B_s dW_s$ (or M and $\widetilde{N}_t := \int_0^t W_s dW_s$) are also orthogonal and not independent. ◁

Exercise 1.5.2.3 Prove that the two martingales N and $\widetilde{N}$, defined in Exercise 1.5.2.2 are not orthogonal although as r.v's, for fixed t, N_t and $\widetilde{N}_t$ are orthogonal in L^2. ◁

1.5.3 Itô's Formula: The Fundamental Formula of Stochastic Calculus

The vector space of semi-martingales is invariant under "smooth" transformations, as established by Itô (see [RY] Chapter IV, for a proof):

Theorem 1.5.3.1 (Itô's formula.) *Let F belong to $C^{1,2}(\mathbb{R}^+ \times \mathbb{R}^d, \mathbb{R})$ and let $X = M + A$ be a continuous d-dimensional semi-martingale. Then the process $(F(t, X_t), t \geq 0)$ is a continuous semi-martingale and*

$$F(t, X_t) = F(0, X_0) + \int_0^t \frac{\partial F}{\partial t}(s, X_s) ds + \sum_{i=1}^d \int_0^t \frac{\partial F}{\partial x_i}(s, X_s) dX_s^i$$
$$+ \quad \frac{1}{2} \sum_{i,j} \int_0^t \frac{\partial^2 F}{\partial x_j \, \partial x_i}(s, X_s) d\langle X^i, X^j \rangle_s .$$

Hence, the bounded variation part of $F(t, X_t)$ is

$$\int_0^t \frac{\partial F}{\partial t}(s, X_s) ds + \sum_{i=1}^d \int_0^t \frac{\partial F}{\partial x_i}(s, X_s) dA_s^i \qquad (1.5.2)$$
$$+ \frac{1}{2} \sum_{i,j} \int_0^t \frac{\partial^2 F}{\partial x_j \partial x_i}(s, X_s) d\langle X^i, X^j \rangle_s .$$

An important consequence is the following: in the one-dimensional case, if X is a martingale ($X = M$) and $d\langle M \rangle_t = h(t)dt$ with h deterministic (i.e., X is a Gaussian martingale), and if F is a $C^{1,2}$ function such that $\partial_t F + h(t) \frac{1}{2} \partial_{xx} F = 0$, then the process $F(t, X_t)$ is a local martingale. A similar result holds in the d-dimensional case.

Note that the application of Itô's formula does not depend on whether or not the processes (A_t^i) or $\langle M^i, M^j \rangle_t$ are absolutely continuous with respect to Lebesgue measure. In particular, if $F \in C^{1,1,2}(\mathbb{R}^+ \times \mathbb{R} \times \mathbb{R}^d, \mathbb{R})$ and V is a continuous bounded variation process, then

$$dF(t, V_t, X_t) = \frac{\partial F}{\partial t}(t, V_t, X_t) dt + \frac{\partial F}{\partial v}(t, V_t, X_t) dV_t + \sum_i \frac{\partial F}{\partial x_i}(t, V_t, X_t) dX_t^i$$
$$+ \frac{1}{2} \sum_{i,j} \frac{\partial^2 F}{\partial x_j \, \partial x_i}(t, V_t, X_t) d\langle X^i, X^j \rangle_t .$$

We now present an extension of Itô's formula, which is useful in the study of stochastic flows and in some cases in finance, when dealing with factor models (see Douady and Jeanblanc [264]) or with credit derivatives dynamics in a multi-default setting (see Bielecki et al. [96]).

Theorem 1.5.3.2 (Itô-Kunita-Ventzel's formula.) *Let $F_t(x)$ be a family of stochastic processes, continuous in $(t, x) \in (\mathbb{R}^+ \times \mathbb{R}^d)$ a.s. satisfying:*

(i) *For each $t > 0$, $x \to F_t(x)$ is C^2 from $\mathbb{R}^d$ to $\mathbb{R}$.*
(ii) *For each x, $(F_t(x), t \geq 0)$ is a continuous semi-martingale*

$$dF_t(x) = \sum_{j=1}^n f_t^j(x) dM_t^j$$

where M^j are continuous semi-martingales, and $f^j(x)$ are stochastic processes continuous in (t,x), such that $\forall s > 0$, $x \to f^j_s(x)$ are C^1 maps, and $\forall x$, $f^j(x)$ are adapted processes.

Let $X = (X^1, \ldots, X^d)$ be a continuous semi-martingale. Then

$$F_t(X_t) = F_0(X_0) + \sum_{j=1}^{n} \int_0^t f^j_s(X_s) dM^j_s + \sum_{i=1}^{d} \int_0^t \frac{\partial F_s}{\partial x_i}(X_s) dX^i_s$$

$$+ \sum_{i=1}^{d} \sum_{j=1}^{n} \int_0^t \frac{\partial f_s}{\partial x_i}(X_s) d\langle M^j, X^i \rangle_s + \frac{1}{2} \sum_{i,k=1}^{d} \int_0^t \frac{\partial^2 F_s}{\partial x_i \partial x_k} d\langle X^k, X^i \rangle_s .$$

PROOF: We refer to Kunita [546] and Ventzel [828]. □

Exercise 1.5.3.3 Prove Theorem 1.4.1.2, i.e., if X is continuous, X_t and $X_t^2 - t$ are martingales, then X is a BM.
Hint: Apply Itô's formula to the complex valued martingale $\exp(i\lambda X_t + \frac{1}{2}\lambda^2 t)$ and use Exercise 1.4.1.8. ◁

Exercise 1.5.3.4 Let $f \in C^{1,2}([0,T] \times \mathbb{R}^d, \mathbb{R})$. We write $\partial_x f(t,x)$ for the row vector $\left[\frac{\partial f}{\partial x_i}(t,x)\right]_{i=1,\ldots,d}$; $\partial_{xx} f(t,x)$ for the matrix $\left[\frac{\partial^2 f}{\partial x_i \, \partial x_j}(t,x)\right]_{i,j}$, and $\partial_t f(t,x)$ for $\frac{\partial f}{\partial t}(t,x)$. Let $B = (B^1, \ldots, B^n)$ be an n-dimensional Brownian motion and $Y_t = f(t, X_t)$, where X_t satisfies $dX^i_t = \mu^i_t dt + \sum_{j=1}^n \eta^{i,j}_t dB^j_t$. Prove that

$$dY_t = \left\{ \partial_t f(t,X_t) + \partial_x f(t,X_t)\mu_t + \frac{1}{2}\left[\eta_t \partial_{xx} f(t,X_t) \eta^T_t\right] \right\} dt + \partial_x f(t,X_t) \eta_t \, dB_t .$$

◁

Exercise 1.5.3.5 Let B be a d-dimensional Brownian motion, with $d \geq 2$ and β defined as

$$d\beta_t = \frac{1}{\|B_t\|} B_t \cdot dB_t = \frac{1}{\|B_t\|} \sum_{i=1}^{d} B^i_t dB^i_t, \quad \beta_0 = 0.$$

Prove that β is a Brownian motion. This will be the starting point of the study of Bessel processes (see ↣ Chapter 6). ◁

Exercise 1.5.3.6 Let $dX_t = b_t dt + dB_t$ where B is a Brownian motion and b a given bounded $\mathbf{F}^B$-adapted process. Let

$$L_t = \exp\left(-\int_0^t b_s dB_s - \frac{1}{2}\int_0^t b_s^2 ds\right) .$$

Show that L and LX are local martingales. (This will be used while dealing with Girsanov's theorem in ↣ Section 1.7.) ◁

Exercise 1.5.3.7 Let X and Y be continuous semi-martingales. The Stratonovich integral of X w.r.t. Y may be defined as

$$\int_0^t X_s \circ dY_s = \int_0^t X_s dY_s + \frac{1}{2}\langle X, Y\rangle_t \,.$$

Prove that

$$\int_0^t X_s \circ dY_s = (ucp) \lim_{n\to\infty} \sum_{i=0}^{p(n)-1} \left(\frac{X_{t_i^n} + X_{t_{i+1}^n}}{2}\right)(Y_{t_{i+1}^n} - Y_{t_i^n}),$$

where $0 = t_0 < t_1^n < \cdots < t_{p(n)}^n = t$ is a subdivision of $[0,T]$ such that $\sup_i(t_{i+1}^n - t_i^n)$ goes to 0 when n goes to infinity. Prove that for $f \in C^3$, we have

$$f(X_t) = f(X_0) + \int_0^t f'(X_s) \circ dX_s \,.$$

For a Brownian motion, the Stratonovich integral may also be approximated as

$$\int_0^t \varphi(B_s) \circ dB_s = \lim_{n\to\infty} \sum_{i=0}^{p(n)-1} \varphi(B_{(t_i+t_{i+1})/2})(B_{t_{i+1}} - B_{t_i}),$$

where the limit is in probability; however, such an approximation does not hold in general for continuous semi-martingales (see Yor [859]). See Stroock [811], page 226, for a discussion on the C^3 assumption on f in the integral form of $f(X_t)$. The Stratonovich integral can be extended to general semi-martingales (not necessarily continuous): see Protter [727], Chapter 5. ◁

Exercise 1.5.3.8 Let B be a BM and $M_t^B := \sup_{s\le t} B_s$. Let $f(t,x,y)$ be a $C^{1,2,1}(\mathbb{R}^+ \times \mathbb{R} \times \mathbb{R}^+)$ function such that

$$\frac{1}{2}f_{xx} + f_t = 0$$
$$f_x(t,0,y) + f_y(t,0,y) = 0 \,.$$

Prove that $f(t, M_t^B - B_t, M_t^B)$ is a local martingale. In particular, for $h \in C^1$

$$h(M_t^B) - h'(M_t^B)(M_t^B - B_t)$$

is a local martingale. See Carraro et al. [157] and El Karoui and Meziou [304] for application to finance. ◁

Exercise 1.5.3.9 (Kennedy Martingales.) Let $B_t^{(\mu)} := B_t + \mu t$ be a BM with drift μ and $M^{(\mu)}$ its running maximum, i.e., $M_t^{(\mu)} = \sup_{s\le t} B_s^{(\mu)}$. Let $R_t = M_t^{(\mu)} - B_t^{(\mu)}$ and $T_a = T_a(R) = \inf\{t : R_t \ge a\}$.

1. Set $\mu = 0$. Prove that, for any (α, β) the process

$$e^{-\alpha M_t - \frac{1}{2}\beta^2 t}\left(\cosh(\beta(M_t - B_t)) + \frac{\alpha}{\beta}\sinh(\beta(M_t - B_t))\right)$$

is a martingale. Deduce that

$$\mathbb{E}\left(\exp\left(-\alpha M_{T_a} - \frac{1}{2}\beta^2 T_a\right)\right) = \beta\left(\beta\cosh\beta a + \alpha\sinh\beta a\right)^{-1} := \varphi(\alpha, \beta; a).$$

2. For any μ, prove that

$$\mathbb{E}\left(\exp\left(-\alpha M^{(\mu)}_{T_a} - \frac{1}{2}\beta^2 T_a\right)\right) = e^{-\mu a}\varphi(\alpha_\mu, \beta_\mu; a)$$

where $\alpha_\mu = \alpha - \mu, \beta_\mu = \sqrt{\beta^2 + \mu^2}$.

◁

Exercise 1.5.3.10 Let $(B^{(\mu)}_t, t \geq 0)$ be a Brownian motion with drift μ, and let b, c be real numbers. Define $X_t = \exp(-cB^{(\mu)}_t)\left(x + \int_0^t \exp(bB^{(\mu)}_s)ds\right)$. Prove that

$$X_t = x - c\int_0^t X_s dB^{(\mu)}_s + \frac{c^2}{2}\int_0^t X_s ds + \int_0^t e^{(b-c)B^{(\mu)}_s} ds.$$

In particular, for $b = c$, X is a diffusion (see ↣ Section 5.3) with infinitesimal generator

$$\frac{c^2}{2}x^2\partial_{xx} + \left[\left(\frac{c^2}{2} - c\mu\right)x + 1\right]\partial_x.$$

(See Donati-Martin et al. [258].) ◁

Exercise 1.5.3.11 Let $B^{(\mu)}$ be as defined in Exercice 1.5.3.9 and let $M^{(\mu)}$ be its running maximum. Prove that, for $t < T$,

$$\mathbb{E}(M^{(\mu)}_T | \mathcal{F}_t) = M^{(\mu)}_t + \int_{M^{(\mu)}_t - B^{(\mu)}_t}^{\infty} G(T-t, u)\, du$$

where $G(T-t, u) = \mathbb{P}(M^{(\mu)}_{T-t} > u)$. ◁

Exercise 1.5.3.12 Let $M_t = \int_0^t (X_s dY_s - Y_s dX_s)$ where X and Y are two real-valued independent Brownian motions. Prove that

$$M_t = \int_0^t \sqrt{X_s^2 + Y_s^2}\, dB_s$$

where B is a BM. Prove that

$$X_t^2 + Y_t^2 = 2 \int_0^t (X_u dY_u + Y_u dX_u) + 2t$$
$$= 2 \int_0^t \sqrt{X_u^2 + Y_u^2}\, d\beta_u + 2t$$

where β is a Brownian motion, with $d\langle B, \beta\rangle_t = 0$. ◁

1.5.4 Stochastic Differential Equations

We start with a general result ([RY], Chapter IX). Let $\mathcal{W} = C(\mathbb{R}^+, \mathbb{R}^d)$ be the space of continuous functions from $\mathbb{R}^+$ into $\mathbb{R}^d$, $w(s)$ the coordinate mappings and $\mathcal{B}_t = \sigma(w(s), s \le t)$. A function f defined on $\mathbb{R}^+ \times \mathcal{W}$ is said to be **predictable** if it is predictable as a process defined on $\mathcal{W}$ with respect to the filtration $(\mathcal{B}_t)$. If X is a continuous process defined on a probability space $(\Omega, \mathbf{F}, \mathbb{P})$, we write $f(t, X_\bullet)$ for the value of f at time t on the path $t \to X_t(\omega)$. We emphasize that we write $X_\bullet$ because $f(t, X_\bullet)$ may depend on the path of X up to time t.

Definition 1.5.4.1 *Let g and f be two predictable functions on $\mathcal{W}$ taking values in the sets of $d \times n$ matrices and n-dimensional vectors, respectively. A solution of the stochastic differential equation* **e***(f,g) is a pair (X, B) of adapted processes on a probability space $(\Omega, \mathcal{F}, \mathbb{P})$ with filtration* **F** *such that:*

- *The n-dimensional process B is a standard* **F***-Brownian motion.*
- *For $i = 1, \ldots, d$ and for any $t \in \mathbb{R}^+$*

$$X_t^i = X_0^i + \int_0^t f_i(s, X_\bullet)ds + \sum_{j=0}^n \int_0^t g_{i,j}(s, X_\bullet)dB_s^j . \qquad \mathbf{e}(f,g)$$

We shall also write this equation as

$$dX_t^i = f_i(t, X_\bullet)dt + \sum_{j=0}^n g_{i,j}(t, X_\bullet)dB_t^j .$$

Definition 1.5.4.2 *(1) There is* ***pathwise uniqueness*** *for* **e***(f,g) if whenever two pairs (X, B) and $(\widehat{X}, \widehat{B})$ are solutions defined on the same probability space with $B = \widehat{B}$ and $X_0 = \widehat{X}_0$, then X and $\widehat{X}$ are indistinguishable.*

(2) There is ***uniqueness in law*** *for* **e***(f,g) if whenever (X, B) and $(\widehat{X}, \widehat{B})$ are two pairs of solutions possibly defined on different probability spaces with $X_0 \stackrel{\text{law}}{=} \widehat{X}_0$, then $X \stackrel{\text{law}}{=} \widehat{X}$.*

(3) A solution (X, B) is said to be ***strong*** *if X is adapted to the filtration $\mathbf{F}^B$. A general solution is often called a* ***weak*** *solution, and if not strong, a* ***strictly weak*** *solution.*

Theorem 1.5.4.3 *Assume that f and g satisfy the Lipschitz condition, for a constant $K > 0$, which does not depend on t,*

$$\|f(t,w) - f(t,w')\| + \|g(t,w) - g(t,w')\| \leq K \sup_{s \leq t} \|w(s) - w'(s)\| .$$

Then, $\mathbf{e}(f,g)$ admits a unique strong solution, up to indistinguishability.

See [RY], Chapter IX for a proof. The following theorem, due to Yamada and Watanabe (see also [RY] Chapter IX, Theorem 1.7) establishes a hierarchy between different uniqueness properties.

Theorem 1.5.4.4 *If pathwise uniqueness holds for $\mathbf{e}(f,g)$, then uniqueness in law holds and the solution is strong.*

Example 1.5.4.5 Pathwise uniqueness is strictly stronger than uniqueness in law. For example, in the one-dimensional case, let $\sigma(x) = \text{sgn}(x)$, with $\text{sgn}(0) = -1$. Any solution (X,B) of $\mathbf{e}(0,\sigma)$ (meaning that $g(t,X_\cdot) = \sigma(X_t)$) starting from 0 is a standard BM, thus uniqueness in law holds. On the other hand, if β is a BM, and $B_t = \int_0^t \text{sgn}(\beta_s) d\beta_s$, then (β, B) and $(-\beta, B)$ are two solutions of $\mathbf{e}(0,\sigma)$ (indeed, $dB_t = \sigma(\beta_t) d\beta_t$ is equivalent to $d\beta_t = \sigma(\beta_t) dB_t$), and pathwise uniqueness does not hold. If (X,B) is any solution of $\mathbf{e}(0,\sigma)$, then $B_t = \int_0^t \text{sgn}(X_s) dX_s$, and $\mathbf{F}^B = \mathbf{F}^{|X|}$ which establishes that any solution is strictly weak (see $\rightarrowtail$ Comments 4.1.7.9 and $\rightarrowtail$ Subsection 5.8.2 for the study of the filtrations).

A simple case is the following:

Theorem 1.5.4.6 *Let $b : [0,T] \times \mathbb{R}^d \to \mathbb{R}^d$ and $\sigma : [0,T] \times \mathbb{R}^d \to \mathbb{R}^{d \times n}$ be Borel functions satisfying*

$$\|b(t,x)\| + \|\sigma(t,x)\| \leq C(1 + \|x\|) \,, x \in \mathbb{R}^d, t \in [0,T],$$
$$\|b(t,x) - b(t,y)\| + \|\sigma(t,x) - \sigma(t,y)\| \leq C\|x - y\| \,, x,y \in \mathbb{R}^d, t \in [0,T]$$

and let X_0 be a square integrable r.v. independent of the n-dimensional Brownian motion B. Then, the stochastic differential equation (SDE)

$$dX_t = b(t,X_t)dt + \sigma(t,X_t)dB_t, \; t \leq T, \; X_0 = x$$

has a unique continuous strong solution, up to indistinguishability. Moreover, this process is a strong (inhomogeneous) Markov process.

SKETCH OF THE PROOF: The proof relies on Picard's iteration procedure.
▶ In a first step, one considers the mapping $Z \to K(Z)$ where

$$(K(Z))_t = x + \int_0^t b(s,Z_s)ds + \int_0^t \sigma(s,Z_s)dB_s \,,$$

and one defines a sequence $(X^n)_{n=0}^\infty$ of processes by setting $X^0 = x$, and $X^n = K(X^{n-1})$. Then, one proves that

$$\mathbb{E}\left(\sup_{s\le t}(X_s^n - X_s^{n-1})^2\right) \le kc^n \frac{t^n}{n!}$$

where k, c are constants. This proves the existence of a solution.
▶ In a second step, one establishes the uniqueness by use of Gronwall's lemma. See [RY], Chapter IV for details. □

The solution depends continuously on the initial value.

Example 1.5.4.7 Geometric Brownian Motion. If B is a Brownian motion and μ, σ are two real numbers, the solution S of

$$dS_t = S_t(\mu dt + \sigma dB_t)$$

is called a geometric Brownian motion with parameters μ and σ. The process S will often be written in this book as

$$S_t = S_0 \exp(\mu t + \sigma B_t - \sigma^2 t/2) = S_0 \exp(\sigma X_t) \tag{1.5.3}$$

where

$$X_t = \nu t + B_t,\ \nu = \frac{\mu}{\sigma} - \frac{\sigma}{2}. \tag{1.5.4}$$

The process $(S_t e^{-\mu t}, t \ge 0)$ is a martingale. The Markov property of S may be seen from the equality

$$S_t = S_s \exp(\mu(t-s) + \sigma(B_t - B_s) - \sigma^2(t-s)/2),\ t > s\,.$$

Let s be fixed. The process $Y_u = \exp(\mu u + \sigma\widehat{B}_u - \sigma^2 u/2), u \ge 0)$ where $\widehat{B}_u = B_{s+u} - B_s$ is independent of $\mathcal{F}_s^S$ and has the same law as S_u/S_0. Moreover, the decomposition $S_t = S_s Y_{t-s}$, for $t > s$ where Y is independent of $\mathcal{F}_s^S$ and has the same law as S/S_0 will be of frequent use.

Example 1.5.4.8 Affine Coefficients: Method of Variation of Constants. The solution of

$$dX_t = (a(t)X_t + b(t))dt + (c(t)X_t + f(t))dB_t,\ X_0 = x$$

where a, b, c, f are (bounded) Borel functions is $X = YZ$ where Y is the solution of

$$dY_t = Y_t[a(t)dt + c(t)dB_t],\ Y_0 = 1$$

and

$$Z_t = x + \int_0^t Y_s^{-1}[b(s) - c(s)f(s)]ds + \int_0^t Y_s^{-1} f(s)dB_s\,.$$

Note that one can write Y in a closed form as

$$Y_t = \exp\left(\int_0^t a(s)ds + \int_0^t c(s)dB_s - \frac{1}{2}\int_0^t c^2(s)ds\right)$$

Remark 1.5.4.9 Under Lipschitz conditions on the coefficients, the solution of

$$dX_t = b(X_t)dt + \sigma(X_t)dB_t,\ t \leq T,\ X_0 = x \in \mathbb{R}$$

is a homogeneous Markov process. More generally, under the conditions of Theorem 1.5.4.6, the solution of

$$dX_t = b(t, X_t)dt + \sigma(t, X_t)dB_t,\ t \leq T,\ X_0 = x \in \mathbb{R}$$

is an inhomogeneous Markov process. The pair (X_t, t) is a homogeneous Markov process.

Definition 1.5.4.10 (Explosion Time.) *Suppose that X is a solution of an SDE with locally Lipschitz coefficients. Then, a localisation argument allows to define unambiguously, for every n, $(X_t, t \leq \tau_n)$, when τ_n is the first exit time from $[-n, n]$. Let $\tau = \sup \tau_n$. When $\tau < \infty$, we say that X explodes at time τ.*

If the functions $b : \mathbb{R}^d \to \mathbb{R}^d$ and $\sigma : \mathbb{R}^d \to \mathbb{R}^d \times \mathbb{R}^n$ are continuous, the SDE

$$dX_t = b(X_t)dt + \sigma(X_t)dB_t \tag{1.5.5}$$

admits a weak solution up to its explosion time.

Under the regularity assumptions

$$\|\sigma(x) - \sigma(y)\|^2 \leq C|x-y|^2\, r(|x-y|^2),\ \text{for } |x-y| < 1$$
$$|b(x) - b(y)| \leq C|x-y|\, r(|x-y|^2),\ \text{for } |x-y| < 1,$$

where $r :]0, 1[\to \mathbb{R}^+$ is a C^1 function satisfying
(i) $\lim_{x\to 0} r(x) = +\infty$,
(ii) $\lim_{x\to 0} \dfrac{xr'(x)}{r(x)} = 0$,
(iii) $\displaystyle\int_0^a \frac{ds}{sr(s)} = +\infty$, for any $a > 0$,
Fang and Zhang [340, 341] have established the pathwise uniqueness of the solution of the equation (1.5.5).

If, for $|x| \geq 1$,

$$\|\sigma(x)\|^2 \leq C\,(|x|^2\,\rho(|x|^2) + 1)$$
$$|b(x)| \leq C\,(|x|\,\rho(|x|^2) + 1)$$

for a function ρ of class C^1 satisfying
(i) $\lim_{x\to\infty} \rho(x) = +\infty$,
(ii) $\lim_{x\to\infty} \dfrac{x\rho'(x)}{\rho(x)} = 0$,
(iii) $\displaystyle\int_1^\infty \frac{ds}{s\rho(s) + 1} = +\infty$,
then, the solution of the equation (1.5.5) does not explode.

1.5.5 Stochastic Differential Equations: The One-dimensional Case

In the case of dimension one, the following result requires less regularity for the existence of a solution of the equation

$$X_t = X_0 + \int_0^t b(s, X_s)ds + \int_0^t \sigma(s, X_s)dB_s\,. \tag{1.5.6}$$

Theorem 1.5.5.1 *Suppose* $\varphi\ :]0,\infty[\to]0,\infty[$ *is a Borel function such that* $\int_{0+} da/\varphi(a) = +\infty$.
Under any of the following conditions:

(i) *the Borel function* b *is bounded, the function* σ *does not depend on the time variable and satisfies*

$$|\sigma(x) - \sigma(y)|^2 \le \varphi(|x-y|)$$

and $|\sigma| \ge \epsilon > 0\,,$

(ii) $|\sigma(s,x) - \sigma(s,y)|^2 \le \varphi(|x-y|)$ *and* b *is Lipschitz continuous,*

(iii) *the function* σ *does not depend on the time variable and satisfies*

$$|\sigma(x) - \sigma(y)|^2 \le |f(x) - f(y)|$$

where f *is a bounded increasing function,* $\sigma \ge \epsilon > 0$ *and* b *is bounded,*

the equation (1.5.6) admits a unique solution which is strong, and the solution X *is a Markov process.*

See [RY], Chapter IV, Section 3 for a proof. Let us remark that condition (iii) on σ holds in particular if σ is bounded below and has bounded variation: indeed

$$|\sigma(x) - \sigma(y)|^2 \le V|\sigma(x) - \sigma(y)| \le V|f(x) - f(y)|$$

with $V = \int |d\sigma|$ and $f(x) = \int_{-\infty}^x |d\sigma(y)|$.

The existence of a solution for $\sigma(x) = \sqrt{|x|}$ and more generally for the case $\sigma(x) = |x|^\alpha$ with $\alpha \ge 1/2$ can be proved using $\varphi(a) = ca$. For $\alpha \in [0, 1/2[$, pathwise uniqueness does not hold, see Girsanov [394], McKean [637], Jacod and Yor [472].

This criterion does not extend to higher dimensions. As an example, let Z be a complex valued Brownian motion. It satisfies

$$Z_t^2 = 2\int_0^t Z_s dZ_s = 2\int_0^t |Z_s| d\gamma_s$$

where $\gamma_t = \displaystyle\int_0^t \frac{Z_s dZ_s}{|Z_s|}$ is a $\mathbb{C}$-valued Brownian motion (see also $\rightarrowtail$ Subsection 5.1.3). Now, the equation $\zeta_t = 2\int_0^t \sqrt{|\zeta_s|} d\gamma_s$ where γ is a Brownian motion admits at least two solutions: the constant process 0 and the process Z.

Comment 1.5.5.2 The proof of (iii) was given in the homogeneous case, using time change and Cameron-Martin's theorem, by Nakao [666] and was improved by LeGall [566]. Other interesting results are proved in Barlow and Perkins [49], Barlow [46], Brossard [132] and Le Gall [566].

The reader will find in ↣ Subsection 5.5.2 other results about existence and uniqueness of stochastic differential equations.

It is useful (and sometimes unavoidable!) to allow solutions to explode. We introduce an absorbing state δ so that the processes are $\mathbb{R}^d \cup \delta$-valued. Let τ be the explosion time (see Definition 1.5.4.10) and set $X_t = \delta$ for $t > \tau$.

Proposition 1.5.5.3 *Equation* $\mathbf{e}(f,g)$ *has no exploding solution if*

$$\sup_{s \leq t} |f(s, x_{\bullet})| + \sup_{s \leq t} |g(s, x_{\bullet})| \leq c(1 + \sup_{s \leq t} |x_{\bullet}|) .$$

PROOF: See Kallenberg [505] and Stroock and Varadhan [812]. □

Example 1.5.5.4 Zvonkin's Argument. The equation

$$dX_t = dB_t + b(X_t)dt$$

where b is a bounded Borel function has a solution. Indeed, assume that there is a solution and let $Y_t = h(X_t)$ where h satisfies $\frac{1}{2}h''(x) + b(x)h'(x) = 0$ (so h is of the form

$$h(x) = C \int_0^x dy \, \exp(-2\widehat{b}(y)) + D$$

where $\widehat{b}$ is an antiderivative of b, hence h is strictly monotone). Then

$$Y_t = h(x) + \int_0^t h'(h^{-1}(Y_s))dB_s .$$

Since $h' \circ h^{-1}$ is Lipschitz, Y exists, hence X exists. The law of X is

$$\mathbb{P}_x^{(b)}|_{\mathcal{F}_t} = \exp\left(\int_0^t b(X_s)dX_s - \frac{1}{2}\int_0^t b^2(X_s)ds\right) \mathbf{W}_x|_{\mathcal{F}_t} .$$

In a series of papers, Engelbert and Schmidt [331, 332, 333] prove results concerning existence and uniqueness of solutions of

$$X_t = x + \int_0^t \sigma(X_s)dB_s$$

that we recall now (see Cherny and Engelbert [168], Karatzas and Shreve [513] p. 332, or Kallenberg [505]). Let

$$N_\sigma = \{x \in \mathbb{R} : \sigma(x) = 0\}$$
$$I_\sigma = \{x \in \mathbb{R} : \int_{-a}^{a} \sigma^{-2}(x+y)dy = +\infty, \forall a > 0\}.$$

The condition $I_\sigma \subset N_\sigma$ is necessary and sufficient for the existence of a solution for arbitrary initial value, and $N_\sigma \subset I_\sigma$ is sufficient for uniqueness in law of solutions. These results are generalized to the case of SDE with drift by Rutkowski [751].

Example 1.5.5.5 The equation

$$dX_t = \frac{1}{2} X_t dt + \sqrt{1 + X_t^2} dB_t, \quad X_0 = 0$$

admits the unique solution $X_t = \sinh(B_t)$. Indeed, it suffices to note that, setting $\varphi(x) = \sinh(x)$, one has $d\varphi(B_t) = b(X_t)dt + \sigma(X_t)dW_t$ where

$$\sigma(x) = \varphi'(\varphi^{-1}(x)) = \sqrt{1+x^2},\; b(x) = \frac{1}{2}\varphi''(\varphi^{-1}(x)) = \frac{x}{2}\,. \tag{1.5.7}$$

More generally, if φ is a strictly increasing, C^2 function, which satisfies $\varphi(-\infty) = -\infty, \varphi(\infty) = \infty$, the process $Z_t = \varphi(B_t)$ is a solution of

$$Z_t = Z_0 + \int_0^t \varphi' \circ \varphi^{-1}(Z_s)dB_s + \frac{1}{2}\int_0^t \varphi'' \circ \varphi^{-1}(Z_s)ds\,.$$

One can characterize more explicitly SDEs of this form. Indeed, we can check that

$$dZ_t = b(Z_t)dt + \sigma(Z_t)dB_t$$

where

$$b(z) = \frac{1}{2}\sigma(z)\sigma'(z)\,. \tag{1.5.8}$$

Example 1.5.5.6 Tsirel'son's Example. Let us give Tsirel'son's example [822] of an equation with diffusion coefficient equal to one, for which there is no strong solution, as an SDE of the form $dX_t = f(t, X_\cdot)dt + dB_t$. Introduce the bounded function T on path space as follows: let $(t_i, i \in -\mathbb{N})$ be a sequence of positive reals which decrease to 0 as i decreases to $-\infty$. Let

$$T(s, X_\cdot) = \sum_{k \in -\mathbb{N}^*} \left[\left[\frac{X_{t_k} - X_{t_{k-1}}}{t_k - t_{k-1}}\right]\right] \mathbb{1}_{]t_k, t_{k+1}]}(s)\,.$$

Here, $[[x]]$ is the fractional part of x. Then, the equation $e(T, 1)$ has no strong solution because, for each fixed k, $\left[\left[\frac{X_{t_k} - X_{t_{k-1}}}{t_k - t_{k-1}}\right]\right]$ is independent of B, and uniformly distributed on $[0, 1]$. Thus Zvonkin's result does not extend to the case where the coefficients depend on the past of the process. See Le Gall and Yor [568] for further examples.

Example 1.5.5.7 Some stochastic differential equations of the form

$$dX_t = b(t, X_t)dt + \sigma(t, X_t)dW_t$$

can be reduced to an SDE with affine coefficients (see Example 1.5.4.8) of the form

$$dY_t = (a(t)Y_t + b(t))dt + (c(t)Y_t + f(t))dW_t\,,$$

by a change of variable $Y_t = U(t, X_t)$. Many examples are provided in Kloeden and Platen [524]. For example, the SDE

$$dX_t = -\frac{1}{2}\exp(-2X_t)dt + \exp(-X_t)dW_t$$

can be transformed (with $U(x) = e^x$) to $dY_t = dW_t$. Hence, the solution is $X_t = \ln(W_t + e^{X_0})$ up to the explosion time $\inf\{t : W_t + e^{X_0} = 0\}$.

Flows of SDE

Here, we present some results on the important topic of the stochastic flow associated with the initial condition.

Proposition 1.5.5.8 *Let*

$$X_t^x = x + \int_0^t b(s, X_s^x)ds + \int_0^t \sigma(s, X_s^x)dW_s$$

and assume that the functions b and σ are globally Lipschitz and have locally Lipschitz first partial derivatives. Then, the explosion time is equal to ∞. Furthermore, the solution is continuously differentiable w.r.t. the initial value, and the process $Y_t = \partial_x X_t$ satisfies

$$Y_t = 1 + \int_0^t Y_s\,\partial_x b(s, X_s^x)ds + \int_0^t Y_s \partial_x \sigma(s, X_s^x)dW_s\,.$$

We refer to Kunita [547, 548] or Protter, Chapter V [727] for a proof.

SDE with Coefficients Depending of a Parameter

We assume that $b(t, x, a)$ and $\sigma(t, x, a)$, defined on $\mathbb{R}^+ \times \mathbb{R} \times \mathbb{R}$, are C^2 with respect to the two last variables x, a, with bounded derivatives of first and second order.

Let

$$X_t = x + \int_0^t b(s, X_s, a)ds + \int_0^t \sigma(s, X_s, a)dW_s$$

and $Z_t = \partial_a X_t$. Then,

$$\begin{aligned} Z_t = &\int_0^t (\partial_a b(s, X_s, a) + Z_s \partial_x b(s, X_s, a))\,ds \\ &+ \int_0^t (\partial_a \sigma(s, X_s, a) + Z_s \partial_x \sigma(s, X_s, a))\,dW_s\,. \end{aligned}$$

See Métivier [645].

Comparison Theorem

We conclude this paragraph with a comparison theorem.

Theorem 1.5.5.9 (Comparison Theorem.) *Let*

$$dX_i(t) = b_i(t, X_i(t))dt + \sigma(t, X_i(t))dW_t\,,\; i = 1, 2$$

where $b_i, i = 1, 2$ are bounded Borel functions and at least one of them is Lipschitz and σ satisfies (ii) or (iii) of Theorem 1.5.5.1. Suppose also that $X_1(0) \geq X_2(0)$ and $b_1(x) \geq b_2(x)$. Then $X_1(t) \geq X_2(t)\,, \forall t, a.s.$

PROOF: See [RY], Chapter IX, Section 3. □

Exercise 1.5.5.10 Consider the equation $dX_t = \mathbb{1}_{\{X_t \geq 0\}} dB_t$. Prove (in a direct way) that this equation has no solution starting from 0. Prove that the equation $dX_t = \mathbb{1}_{\{X_t > 0\}} dB_t$ has a solution.
Hint: For the first part, one can consider a smooth function f vanishing on $\mathbb{R}^+$. From Itô's formula, it follows that X remains positive, and the contradiction is obtained from the remark that X is a martingale. ◁

Comment 1.5.5.11 Doss and Süssmann Method. Let σ be a C^2-function with bounded derivatives of the first two orders, and let b be Lipschitz continuous. Let h be the solution of the ODE

$$\frac{\partial h}{\partial t}(x, t) = \sigma(h(x, t)),\; h(x, 0) = x\,.$$

Let X be a continuous semi-martingale which vanishes at time 0 and let D be the solution of the ODE

$$\frac{dD_t}{dt} = b(h(D_t, X_t(\omega))) \exp\left\{ - \int_0^{X_t(\omega)} \sigma'(h(D_s, s))ds \right\},\; D_0 = y\,.$$

Then, $Y_t = h(D_t, X_t)$ is the unique solution of

$$Y_t = y + \int_0^t \sigma(Y_s) \circ dX_s + \int_0^t b(Y_s)ds$$

where $\circ$ stands for the Stratonovich integral (see Exercise 1.5.3.7). See Doss [261] and Süssmann [815].

1.5.6 Partial Differential Equations

We now give an important relation between two problems: to compute the (conditional) expectation of a function of the terminal value of the solution of an SDE and to solve a second-order PDE with boundary conditions.

Proposition 1.5.6.1 *Let $\mathcal{A}$ be the second-order operator defined on $C^{1,2}$ functions by*

$$\mathcal{A}(\varphi)(t,x) = \frac{\partial \varphi}{\partial t}(t,x) + b(t,x)\frac{\partial \varphi}{\partial x}(t,x) + \frac{1}{2}\sigma^2(t,x)\frac{\partial^2 \varphi}{\partial x^2}(t,x).$$

Let X be the diffusion (see $\rightarrowtail$ Section 5.3)

$$dX_t = b(t,X_t)dt + \sigma(t,X_t)dW_t.$$

We assume that this equation admits a unique solution. Then, for $f \in C_b(\mathbb{R})$ the bounded solution to the Cauchy problem

$$\mathcal{A}\varphi = 0,\ \varphi(T,x) = f(x), \tag{1.5.9}$$

is given by

$$\varphi(t,x) = \mathbb{E}(f(X_T)|X_t = x).$$

Conversely, if $\varphi(t,x) = \mathbb{E}(f(X_T)|X_t = x)$ is $C^{1,2}$, then it solves (1.5.9).

PROOF: From the Markov property of X, the process

$$\varphi(t,X_t) = \mathbb{E}(f(X_T)|X_t) = \mathbb{E}(f(X_T)|\mathcal{F}_t),$$

is a martingale. Hence, its bounded variation part is equal to 0. From (1.5.2), assuming that $\varphi \in C^{1,2}$,

$$\partial_t \varphi + b(t,x)\partial_x \varphi + \frac{1}{2}\sigma^2(t,x)\partial_{xx}\varphi = 0.$$

The smoothness of φ is established from general results on diffusions under suitable conditions on b and σ (see Kallenberg [505], Theorem 17-6 and Durrett [286]). □

Exercise 1.5.6.2 Let $dX_t = rX_t dt + \sigma(X_t)dW_t$, Ψ a bounded continuous function and $\psi(t,x) = \mathbb{E}(e^{-r(T-t)}\Psi(X_T)|X_t = x)$. Assuming that ψ is $C^{1,2}$, prove that

$$\partial_t \psi + rx\partial_x \psi + \frac{1}{2}\sigma^2(x)\partial_{xx}\psi = r\psi,\ \psi(T,x) = \Psi(x).$$

◁

1.5.7 Doléans-Dade Exponential

Let M be a continuous local martingale. For any $\lambda \in \mathbb{R}$, the process

$$\mathcal{E}(\lambda M)_t := \exp\left(\lambda M_t - \frac{\lambda^2}{2}\langle M\rangle_t\right)$$

is a positive local martingale (hence, a super-martingale), called the **Doléans-Dade exponential** of λM (or, sometimes, the stochastic exponential of λM). It is a martingale if and only if $\forall t,\ \mathbb{E}(\mathcal{E}(\lambda M)_t) = 1$.

If $\lambda \in L^2(M)$, the process $\mathcal{E}(\lambda M)$ is the unique solution of the stochastic differential equation

$$dY_t = Y_t \lambda_t dM_t,\ Y_0 = 1\,.$$

This definition admits an extension to semi-martingales as follows. If X is a continuous semi-martingale vanishing at 0, the **Doléans-Dade exponential** of X is the unique solution of the equation

$$Z_t = 1 + \int_0^t Z_s dX_s\,.$$

It is given by

$$\mathcal{E}(X)_t := \exp\left(X_t - \frac{1}{2}\langle X\rangle_t\right)\,.$$

Let us remark that in general $\mathcal{E}(\lambda M)\,\mathcal{E}(\mu M)$ is not equal to $\mathcal{E}((\lambda+\mu)M)$. In fact, the general formula

$$\mathcal{E}(X)_t\,\mathcal{E}(Y)_t = \mathcal{E}(X + Y + \langle X, Y\rangle)_t \tag{1.5.10}$$

leads to

$$\mathcal{E}(\lambda M)_t \mathcal{E}(\mu M)_t = \mathcal{E}((\lambda+\mu)M + \lambda\mu\langle M\rangle)_t\,,$$

hence, the product of the exponential local martingales $\mathcal{E}(M)\mathcal{E}(N)$ is a local martingale if and only if the local martingales M and N are orthogonal.

Example 1.5.7.1 For later use (see ↣ Proposition 2.6.4.1) we present the following computation. Let f and g be two continuous functions and W a Brownian motion starting from x at time 0. The process

$$Z_t = \exp\left(\int_0^t [f(s)W_s + g(s)]dW_s - \frac{1}{2}\int_0^t [f(s)W_s + g(s)]^2 ds\right)$$

is a local martingale. Using ↣ Proposition 1.7.6.4, it can be proved that it is a martingale, therefore its expectation is equal to 1. It follows that

$$\mathbb{E}\left(\exp\left[\int_0^t [f(s)W_s + g(s)]dW_s - \frac{1}{2}\int_0^t [f^2(s)W_s^2 + 2W_s f(s)g(s)]ds\right]\right)$$
$$= \exp\left(\frac{1}{2}\int_0^t g^2(s)ds\right)\,.$$

If moreover f and g are C^1, integration by parts yields

$$\int_0^t g(s)dW_s = g(t)W_t - g(0)W_0 - \int_0^t g'(s)W_s ds$$
$$\int_0^t f(s)W_s dW_s = \frac{1}{2}\left(W_t^2 f(t) - W_0^2 f(0) - \int_0^t f(s)ds - \int_0^t f'(s)W_s^2 ds\right)\,,$$

therefore,

$$\mathbb{E}\left[\exp\left(g(t)W_t + \frac{1}{2}f(t)W_t^2 - \frac{1}{2}\int_0^t \left([f^2(s) + f'(s)]W_s^2 + 2W_s(f(s)g(s) + g'(s))\right) ds\right)\right]$$
$$= \exp\left(g(0)W_0 + \frac{1}{2}\left(f(0)W_0^2 + \int_0^t f(s)ds + \int_0^t g^2(s)ds\right)\right).$$

Exercise 1.5.7.2 Check formula (1.5.10), by showing, e.g., that both sides satisfy the same linear SDE. ◁

Exercise 1.5.7.3 Let H and Z be continuous semi-martingales. Check that the solution of the equation $X_t = H_t + \int_0^t X_s dZ_s$, is

$$X_t = \mathcal{E}(Z)_t \left(H_0 + \int_0^t \frac{1}{\mathcal{E}(Z)_s}(dH_s - d\langle H, Z\rangle_s)\right).$$

See Protter [727], Chapter V, Section 9, for the case where H, Z are general semi-martingales. ◁

Exercise 1.5.7.4 Prove that if θ is a bounded function, then the process $(\mathcal{E}(\theta\star W)_t, t \leq T)$ is a u.i. martingale.
`Hint:`

$$\exp\left(\int_0^t \theta_s dW_s - \frac{1}{2}\int_0^t \theta_s^2 ds\right) \leq \exp\left(\sup_{t\leq T}\int_0^t \theta_s dW_s\right) = \exp\widehat{\beta}_{\int_0^T \theta_s^2 ds}$$

with $\widehat{\beta}_t = \sup_{u\leq t}\beta_u$ where β is a BM. ◁

Exercise 1.5.7.5 Multiplicative Decomposition of Positive Sub-martingales. Let $X = M + A$ be the Doob-Meyer decomposition of a strictly positive continuous sub-martingale. Let Y be the solution of

$$dY_t = Y_t \frac{1}{X_t} dM_t, Y_0 = X_0$$

and let Z be the solution of $dZ_t = -Z_t \frac{1}{X_t} dA_t, Z_0 = 1$. Prove that $U = Y/Z$ satisfies $dU_t = U_t \frac{1}{X_t} dX_t$ and deduce that $U = X$.
`Hint:` Use that the solution of $dU_t = U_t \frac{1}{X_t} dX_t$ is unique. See Meyer and Yoeurp [649] and Meyer [647] for a generalization to discontinuous sub-martingales. Note that this decomposition states that a strictly positive continuous sub-martingale is the product of a martingale and an increasing process. ◁

1.6 Predictable Representation Property

1.6.1 Brownian Motion Case

Let W be a real-valued Brownian motion and $\mathbf{F}^W$ its natural filtration. We recall that the space $L^2(W)$ was presented in Definition 1.3.1.3.

Theorem 1.6.1.1 *Let $(M_t, t \geq 0)$ be a square integrable $\mathbf{F}^W$-martingale (i.e., $\sup_t \mathbb{E}(M_t^2) < \infty$). There exists a constant μ and a unique predictable process m in $L^2(W)$ such that*

$$\forall t,\ \ M_t = \mu + \int_0^t m_s \, dW_s \, .$$

If M is an $\mathbf{F}^W$-local martingale, there exists a unique predictable process m in $L^2_{loc}(W)$ such that

$$\forall t,\ \ M_t = \mu + \int_0^t m_s \, dW_s \, .$$

PROOF: The first step is to prove that for any square integrable $\mathcal{F}^W_\infty$-measurable random variable F, there exists a unique predictable process H such that

$$F = \mathbb{E}(F) + \int_0^\infty H_s \, dW_s \, , \tag{1.6.1}$$

and $\mathbb{E}[\int_0^\infty H_s^2 ds] < \infty$. Indeed, the space of random variables F of the form (1.6.1) is closed in L^2. Moreover, it contains any random variable of the form

$$F = \exp\left(\int_0^\infty f(s) dW_s - \frac{1}{2}\int_0^\infty f(s)^2 ds\right)$$

with $f = \sum_i \lambda_i \mathbb{1}_{]t_{i-1},t_i]}$, $\lambda_i \in \mathbb{R}^d$, and this space is total in L^2. Then density arguments complete the proof. See [RY], Chapter V, for details. □

Example 1.6.1.2 A special case of Theorem 1.6.1.1 is when $M_t = f(t, W_t)$ where f is a smooth function (hence, f is space-time harmonic, i.e., it satisfies $\frac{\partial f}{\partial t} + \frac{1}{2}\frac{\partial^2 f}{\partial x^2} = 0$). In that case, Itô's formula leads to $m_s = \partial_x f(s, W_s)$.

This theorem holds in the multidimensional Brownian setting. Let W be a n-dimensional BM and M be a square integrable $\mathbf{F}^W$-martingale. There exists a constant μ and a unique n-dimensional predictable process m in $L^2(W)$ such that

$$\forall t,\ \ M_t = \mu + \sum_{i=1}^n \int_0^t m_s^i \, dW_s^i \, .$$

Corollary 1.6.1.3 *Every $\mathbf{F}^W$-local martingale admits a continuous version.*

As a consequence, every optional process in a Brownian filtration is predictable.
From now on, we shall abuse language and say that every $\mathbf{F}^W$-local martingale is continuous.

Corollary 1.6.1.4 *Let W be a $\mathbf{G}$-Brownian motion with natural filtration $\mathbf{F}$. Then, for every square integrable $\mathbf{G}$-adapted process φ,*

$$\mathbb{E}\left(\int_0^t \varphi_s dW_s | \mathcal{F}_t\right) = \int_0^t \mathbb{E}(\varphi_s|\mathcal{F}_s)dW_s\,,$$

where $\mathbb{E}(\varphi_s|\mathcal{F}_s)$ denotes the predictable version of the conditional expectation.

PROOF: Since the r.v. $\int_0^t \mathbb{E}(\varphi_s|\mathcal{F}_s)dW_s$ is $\mathcal{F}_t$-measurable, it suffices to check that, for any bounded r.v. $F_t \in \mathcal{F}_t$

$$\mathbb{E}\left(F_t \int_0^t \varphi_s dW_s\right) = \mathbb{E}\left(F_t \int_0^t \mathbb{E}(\varphi_s|\mathcal{F}_s)dW_s\right).$$

The predictable representation theorem implies that $F_t = \mathbb{E}(F_t) + \int_0^t f_s dW_s$, for some $\mathbf{F}$-predictable process $f \in L^2(W)$, hence

$$\begin{aligned}\mathbb{E}\left(F_t \int_0^t \varphi_s dW_s\right) &= \mathbb{E}\left(\int_0^t f_s\varphi_s ds\right) = \int_0^t \mathbb{E}(f_s\varphi_s)ds \\ &= \int_0^t \mathbb{E}(f_s\mathbb{E}(\varphi_s|\mathcal{F}_s))ds = \mathbb{E}\left(\int_0^t f_s\mathbb{E}(\varphi_s|\mathcal{F}_s)ds\right) \\ &= \mathbb{E}\left(\left\{\mathbb{E}(F_t) + \int_0^t f_s dW_s\right\}\int_0^t \mathbb{E}(\varphi_s|\mathcal{F}_s)dW_s\right),\end{aligned}$$

which ends the proof. □

Example 1.6.1.5 If $F = \int_0^\infty ds\, h(s, W_s)$ where $\int_0^\infty ds\, \mathbb{E}(|h(s, W_s)|) < \infty$, then from the Markov property, $M_t = \mathbb{E}(F|\mathcal{F}_t) = \int_0^t ds\, h(s, W_s) + \varphi(t, W_t)$, for some function φ. Assuming that φ is smooth, the martingale property of M and Itô's formula lead to

$$h(t, W_t) + \partial_t\varphi(t, W_t) + \frac{1}{2}\partial_{xx}\varphi(t, W_t) = 0$$

and $M_t = \varphi(0,0) + \int_0^t \partial_x\varphi(s, W_s)dW_s$. See the papers of Graversen et al. [405] and Shiryaev and Yor [793] for some examples of functionals of the Brownian motion which are explicitly written as stochastic integrals.

Proposition 1.6.1.6 *Let $M_t = \mathbb{E}(f(W_T)|\mathcal{F}_t)$, for $t \le T$ where f is a C_b^1 function. Then,*

$$M_t = \mathbb{E}(f(W_T)) + \int_0^t \mathbb{E}(f'(W_T)|\mathcal{F}_s)dW_s = \mathbb{E}(f(W_T)) + \int_0^t P_{T-s}(f')(W_s)dW_s\,.$$

PROOF: From the independence and stationarity of the increments of the Brownian motion,

$$\mathbb{E}(f(W_T)|\mathcal{F}_t) = \psi(t, W_t)$$

where $\psi(t,x) = \mathbb{E}(f(x + W_{T-t}))$. Itô's formula and the martingale property of $\psi(t, W_t)$ lead to

$$\partial_x \psi(t,x) = \mathbb{E}(f'(x + W_{T-t})) = \mathbb{E}(f'(W_T)|W_t = x).$$

□

Comment 1.6.1.7 In a more general setting, one can use Malliavin's derivative. For T fixed, and $h \in L^2([0,T])$, we define $W(h) = \int_0^T h(s)dW_s$. Let $F = f(W(h_1), \dots, W(h_n))$ where f is a smooth function. The derivative of F is defined as the process $(D_t F, t \le T)$ by

$$D_t F = \sum_{i=1}^{n} \frac{\partial f}{\partial x_i}(W(h_1), \dots, W(h_n))h_i(t).$$

The Clark-Ocone representation formula states that for random variables which satisfy some suitable integrability conditions,

$$F = \mathbb{E}(F) + \int_0^T \mathbb{E}(D_t F|\mathcal{F}_t)dW_t.$$

We refer the reader to the books of Nualart [681] for a study of Malliavin calculus and of Malliavin and Thalmaier [616] for applications in finance. See also the issue [560] of *Mathematical Finance* devoted to applications to finance of Malliavin calculus.

Exercise 1.6.1.8 Let $W = (W^1, \dots, W^d)$ be a d-dimensional BM. Is the space of martingales $\sum_{i=1}^d \int_0^t H_i(W^i_\cdot)_s dW^i_s$ dense in the space of square integrable martingales?
Hint: The answer is negative. Look for $Y \in L^2(\mathcal{W}_\infty)$ such that Y is orthogonal to all these variables. ◁

1.6.2 Towards a General Definition of the Predictable Representation Property

Besides the Predictable Representation Property (PRP) of Brownian motion, let us recall the Kunita-Watanabe orthogonal decomposition of a martingale M with respect to another one X:

Lemma 1.6.2.1 ***(Kunita-Watanabe Decomposition.)*** *Let X be a given continuous local $\mathbf{F}$-martingale. Then, every continuous $\mathbf{F}$-local martingale M vanishing at 0 may be uniquely written*

$$M = H \star X + N \tag{1.6.2}$$

where H is predictable and N is a local martingale orthogonal to X.

Referring to the Brownian motion case (previous subsection), one may wonder for which local martingales X it is true that every N in (1.6.2) is a constant. This leads us to the following definition.

Definition 1.6.2.2 *A continuous local martingale X enjoys the* ***predictable representation property*** *(PRP) if for any $\mathbf{F}^X$-local martingale $(M_t, t \geq 0)$, there is a constant m and an $\mathbf{F}^X$-predictable process $(m_s, s \geq 0)$ such that*

$$M_t = m + \int_0^t m_s dX_s, \, t \geq 0.$$

Exercise 1.6.2.3 Prove that $(m_s, s \geq 0)$ is unique in $L^2_{loc}(X)$. ◁

More generally, a continuous $\mathbf{F}$-local martingale X enjoys the $\mathbf{F}$-predictable representation property if any $\mathbf{F}$-adapted martingale M can be written as $M_t = m + \int_0^t m_s \, dX_s$, with $\int_0^t m_s^2 d\langle X \rangle_s < \infty$. We do not require in that last definition that $\mathbf{F}$ is the natural filtration of X. (See an important example in ↣ Subsection 1.7.7.)

We now look for a characterization of martingales that enjoy the PRP. Given a continuous $\mathbf{F}$-adapted process Y, we denote by $\mathcal{M}(Y)$ the subset of probability measures $\mathbb{Q}$ on $(\Omega, \mathbf{F})$, for which the process Y is a $(\mathbb{Q}, \mathbf{F})$-local martingale. This set is convex. A probability measure $\mathbb{P}$ is called extremal in $\mathcal{M}(Y)$ if whenever $\mathbb{P} = \lambda\mathbb{P}_1 + (1-\lambda)\mathbb{P}_2$ with $\lambda \in]0,1[$ and $\mathbb{P}_1, \mathbb{P}_2 \in \mathcal{M}(Y)$, then $\mathbb{P} = \mathbb{P}_1 = \mathbb{P}_2$.

Note that if $\mathbb{P} = \lambda\mathbb{P}_1 + (1-\lambda)\mathbb{P}_2$, then $\mathbb{P}_1$ and $\mathbb{P}_2$ are absolutely continuous with respect to $\mathbb{P}$. However, the $\mathbb{P}_i$'s are not necessarily equivalent. The following theorem relates the PRP for Y under $\mathbb{P} \in \mathcal{M}(Y)$ and the extremal points of $\mathcal{M}(Y)$.

Theorem 1.6.2.4 *The process Y enjoys the PRP with respect to $\mathbf{F}^Y$ and $\mathbb{P}$ if and only if $\mathbb{P}$ is an extremal point of $\mathcal{M}(Y)$.*

PROOF: See Jacod [468], Yor [861] and Jacod and Yor [472]. □

Comments 1.6.2.5 (a) The PRP is essential in finance and is deeply linked with Delta hedging and completeness of the market. If the price process enjoys the PRP under an equivalent probability measure, the market is complete. It is worthwhile noting that the key process is the price process itself, rather than the processes that may drive the price process. See ↣ Subsection 2.3.6 for more details.

(b) We compare Theorems 1.6.1.1 and 1.6.2.4. It turns out that the Wiener measure is an extremal point in $\mathcal{M}$, the set of martingale laws on $C(\mathbb{R}^+, \mathbb{R})$ where $Y_t(\omega) = \omega(t)$. This extremality property follows from Lévy's characterization of Brownian motion.

(c) Let us give an example of a martingale which does not enjoy the PRP. Let $M_t = \int_0^t e^{aB_s - a^2 s/2} d\beta_s = \int_0^t \mathcal{E}(aB)_s d\beta_s$, where B, β are two independent one-dimensional Brownian motions. We note that $d\langle M\rangle_t = (\mathcal{E}(aB)_t)^2 dt$, so that $(\mathcal{E}_t := \mathcal{E}(aB)_t, t \geq 0)$ is $\mathbf{F}^M$-adapted and hence is an $\mathbf{F}^M$-martingale. Since $\mathcal{E}_t = 1 + a\int_0^t \mathcal{E}_s dB_s$, the martingale $\mathcal{E}$ cannot be obtained as a stochastic integral w.r.t. β or equivalently w.r.t. M. In fact, every $\mathbf{F}^M$-martingale can be written as the sum of a stochastic integral with respect to M (or equivalently to β) and a stochastic integral with respect to B.

(d) It is often asked what is the minimal number of orthogonal martingales needed to obtain a representation formula in a given filtration. We refer the reader to Davis and Varaiya [224] who defined the notion of multiplicity of a filtration. See also Davis and Obłój [223] and Barlow et al. [50].

Example 1.6.2.6 We give some examples of martingales that enjoy the PRP.

(a) Let W be a BM and $\mathbf{F}$ its natural filtration. Set $X_t = x + \int_0^t x_s\, dW_s$ where $(x_s, s \geq 0)$ is continuous and does not vanish. Then X enjoys the PRP.

(b) A continuous martingale is a time-changed Brownian motion. Let X be a martingale, then $X_t = \beta_{\langle X\rangle_t}$ where β is a Brownian motion. If $\langle X\rangle$ is measurable with respect to β, then X is said to be pure, and $\mathbb{P}_X$ is extremal. However, the converse does not hold. See Yor [862].

Exercise 1.6.2.7 Let $\mathcal{M}_{\mathbb{P}}(X) = \{\mathbb{Q} \ll \mathbb{P} : X \text{ is a } \mathbb{Q}\text{-martingale}\}$. For any convex set $\mathcal{K}$, we denote by $\text{ext}(\mathcal{K})$ the set of extremal points of $\mathcal{K}$. Prove that

$$\text{ext}(\mathcal{M}_{\mathbb{P}}(X)) = \text{ext}(\mathcal{M}(X)) \cap \mathcal{M}_{\mathbb{P}}(X)\,.$$

An open question is: does the equality

$$\text{ext}(\mathcal{M}_{\mathbb{P}}^{eq}(X)) = \text{ext}\mathcal{M}(X) \cap \mathcal{M}_{\mathbb{P}}^{eq}(X)$$

where $\mathcal{M}_{\mathbb{P}}^{eq}(X) = \{\mathbb{Q} \sim \mathbb{P} : X \text{ is a } \mathbb{Q}\text{-martingale}\}$, hold? $\triangleleft$

Exercise 1.6.2.8 We present an example where the representation of a bounded r.v. considered as the terminal variable of a martingale can be explicitly computed. Let B be a Brownian motion and $T_a = \inf\{t \geq 0 : B_t = a\}$ where $a > 0$.

1. Using the Doléans-Dade exponential of λB, prove that, for $\lambda > 0$

$$\mathbb{E}(e^{-\lambda^2 T_a/2}|\mathcal{F}_t) = e^{-\lambda a} + \lambda \int_0^{T_a \wedge t} e^{-\lambda(a-B_u)-\lambda^2 u/2} dB_u \tag{1.6.3}$$

and that

$$e^{-\lambda^2 T_a/2} = e^{-\lambda a} + \lambda \int_0^{T_a} e^{-\lambda(a-B_u)-\lambda^2 u/2} dB_u\,.$$

Check that $\mathbb{E}(\int_0^{T_a} (e^{-\lambda(a-B_u)-\lambda^2 u/2})^2 du) < \infty$. Prove that (1.6.3) is not true for $\lambda < 0$, i.e., that, in the case $\mu := -\lambda > 0$ the quantities

$\mathbb{E}(e^{-\mu^2 T_a/2}|\mathcal{F}_t)$ and $e^{\mu a} - \mu \int_0^{T_a \wedge t} e^{\mu(a-B_u)-\mu^2 u/2}\, dB_u$ are not equal. Prove that, nonetheless,

$$e^{-\lambda^2 T_a/2} = e^{\lambda a} - \lambda \int_0^{T_a} e^{\lambda(a-B_u)-\lambda^2 u/2} dB_u$$

but $\mathbb{E}(\int_0^{T_a} (e^{\lambda(a-B_u)-\lambda^2 u/2})^2\, du) = \infty$. Deduce, from the previous results, that

$$\sinh(\lambda a) = \lambda \int_0^{T_a} e^{-\lambda^2 u/2} \cosh((a-B_u)\lambda)\, dB_u\,.$$

2. By differentiating the Laplace transform of T_a, and using the fact that φ satisfies the Kolmogorov equation $\partial_t \varphi(t,x) = \frac{1}{2}\partial_{xx}\varphi(t,x)$, (see $\rightarrowtail$ Subsection 5.4.1), prove that

$$\lambda e^{-\lambda c} = 2 \int_0^\infty e^{-\lambda^2 t/2} \partial_t \varphi(t,c)\, dt$$

where $\varphi(t,x) = \frac{1}{\sqrt{2\pi t}} e^{-x^2/(2t)}$.
3. Prove that, for any bounded Borel function f

$$\mathbb{E}(f(T_a)|\mathcal{F}_t) = \mathbb{E}(f(T_a)) + 2 \int_0^{T_a \wedge t} dB_s \int_0^\infty f(u+s) \frac{\partial}{\partial u} \varphi(u, B_s - a) du\,.$$

4. Deduce that, for fixed T,

$$\mathbb{1}_{\{T_a<T\}} = \mathbb{P}(T_a < T) + 2 \int_0^{T_a \wedge T} \varphi(T-s, B_s - a)\, dB_s\,.$$

See Shiryaev and Yor [793], Graversen et al. [405] for other examples. ◁

1.6.3 Dudley's Theorem

In the previous exercise, we were careful to check the integrability of the stochastic integrals. This may be contrasted with Dudley's result [269], which states that every $\mathcal{F}_T^W$-random variable can be represented as an Itô stochastic integral $\int_0^T \theta_s dW_s$ where θ is predictable and satisfies $\int_0^T \theta_s^2 ds < \infty$, a.s. where W is a Brownian motion. In fact, this result has no relation with the predictable representation property, as shown by Émery et al. [330]. Indeed, the authors proved that, in a filtration where any martingale is continuous, if τ is a stopping time and X is an $\mathcal{F}_\tau$-measurable random variable, there exists a local martingale M, null at 0, such that $M_\tau = X$.

Comment 1.6.3.1 In mathematical finance, Dudley's result is related to arbitrage opportunities (see $\rightarrowtail$ Chapter 2 for the definition of financial terms if needed). Let us study the simple case where $dS_t = S_t \sigma dW_t$, $S_0 = x > 0$

is the price of the risky asset and where the interest rate is null. Consider a process θ such that $\int_0^T \theta_s^2 ds < \infty$, a.s.. and $\int_0^T \theta_s dW_s = 1$ (the existence is a consequence of Dudley's theorem). Had we chosen $\pi_s = \theta_s/(S_s\sigma)$ as the risky part of a self-financing strategy with a zero initial wealth, then we would obtain an arbitrage opportunity. However, the wealth X associated with this strategy, i.e., $X_t = \int_0^t \theta_s dW_s$ is not bounded below (otherwise, X would be a super-martingale with initial value equal to 0, hence $\mathbb{E}(X_T) \leq 0$). These strategies are linked with the well-known doubling strategy of coin tossing (see Harrison and Pliska [422]).

1.6.4 Backward Stochastic Differential Equations

In deterministic case studies, it is easy to solve an ODE with a terminal condition just by time reversal. In a stochastic setting, if one insists that the solution is adapted w.r.t. a given filtration, it is not possible in general to use time reversal.

A probability space $(\Omega, \mathcal{F}, \mathbb{P})$, an n-dimensional Brownian motion W and its natural filtration $\mathbf{F}$, an $\mathcal{F}_T$-measurable square integrable random variable ζ and a family of $\mathbf{F}$-adapted, $\mathbb{R}^d$-valued processes $f(t, \centerdot, x, y), x, y \in \mathbb{R}^d \times \mathbb{R}^{d\times n}$ are given (we shall, as usual, forget the dependence in ω and write only $f(t, x, y)$). The problem we now consider is to solve a stochastic differential equation where the terminal condition ζ as well as the form of the drift term f (called the generator) are given, however, the diffusion term is left unspecified.

The **Backward Stochastic Differential Equation** (BSDE) (f, ζ) has the form

$$\begin{aligned} -dX_t &= f(t, X_t, Y_t)\, dt - Y_t \centerdot dW_t \\ X_T &= \zeta\,. \end{aligned}$$

Here, we have used the usual convention of signs which is in force while studying BSDEs. The solution of a BSDE is *a pair* (X, Y) *of adapted processes* which satisfy

$$X_t = \zeta + \int_t^T f(s, X_s, Y_s)\, ds - \int_t^T Y_s \centerdot dW_s\,, \tag{1.6.4}$$

where X is $\mathbb{R}^d$-valued and Y is $d \times n$-matrix valued.

We emphasize that the diffusion coefficient Y is a part of the solution, as it is clear from the obvious case when f is null: in that case, we are looking for a martingale with given terminal value. Hence, the quantity Y is the predictable process arising in the representation of the martingale X in terms of the Brownian motion.

Example 1.6.4.1 Let us study the easy case where f is a deterministic function of time (or a given process such that $\int_0^T f_s ds$ is square integrable) and

$d = n = 1$. If there exists a solution to $X_t = \zeta + \int_t^T f(s)\, ds - \int_t^T Y_s\, dW_s$, then the $\mathbf{F}$-adapted process $X_t + \int_0^t f(s)\, ds$ is equal to $\zeta + \int_0^T f(s)\, ds - \int_t^T Y_s\, dW_s$. Taking conditional expectation w.r.t. $\mathcal{F}_t$ of the two sides, and assuming that Y is square integrable, we get

$$X_t + \int_0^t f(s)\, ds = \mathbb{E}(\zeta + \int_0^T f(s)\, ds | \mathcal{F}_t) \tag{1.6.5}$$

therefore, the process $X_t + \int_0^t f(s)\, ds$ is an $\mathbf{F}$-martingale with terminal value $\zeta + \int_0^T f(s)\, ds$. (A more direct proof is to write $dX_t + f(t)dt = Y_t dW_t$.) The predictable representation theorem asserts that there exists an adapted square integrable process Y such that $X_t + \int_0^t f(s)\, ds = X_0 + \int_0^t Y_s dW_s$ and the pair (X, Y) is the solution of the BSDE. The process X can be written in terms of the generator f and the terminal condition as $X_t = \mathbb{E}(\zeta + \int_t^T f(s) ds | \mathcal{F}_t)$. In particular, if $\zeta^1 \geq \zeta^2$ and $f_1 \geq f_2$, and if X^i is the solution of (f_i, ζ^i) for $i = 1, 2$, then, for $t \in [0, T]$, $X_t^1 \geq X_t^2$.

Definition 1.6.4.2 *Let $L^2([0, T] \times \Omega; \mathbb{R}^d)$ be the set of $\mathbb{R}^d$-valued square integrable $\mathbf{F}$-progressively measurable processes, i.e., processes Z such that*

$$\mathbb{E}\left[\int_0^T \|Z_s\|^2 ds\right] < \infty .$$

Theorem 1.6.4.3 *Let us assume that for any $(x, y) \in \mathbb{R}^n \times \mathbb{R}^{d \times n}$, the process $f(\centerdot, x, y)$ is progressively measurable, with $f(\centerdot, 0, 0) \in L^2([0, T] \times \Omega; \mathbb{R}^d)$ and that the function $f(t, \centerdot, \centerdot)$ is uniformly Lipschitz, i.e., there exists a constant K such that*

$$\|f(t, x_1, y_1) - f(t, x_2, y_2)\| \leq K[\, \|x_1 - x_2\| + \|y_1 - y_2\|\,], \quad \forall t, \mathbb{P}, a.s.$$

Then there exists a unique pair (X, Y) of adapted processes belonging to $L^2([0, T] \times \Omega; \mathbb{R}^n) \times L^2([0, T] \times \Omega, \mathbb{R}^{d \times n})$ which satisfies (1.6.4).

SKETCH OF THE PROOF: Example (1.6.4.1) provides the proof when f does not depend on (x, y). The general case is established using Picard's iteration: let Φ be the map $\Phi(x, y) = (X, Y)$ where (x, y) is a pair of adapted processes and (X, Y) is the solution of

$$-dX_t = f(t, x_t, y_t)\, dt - Y_t \centerdot dW_t,\ X_T = \zeta\, .$$

The map Φ is proved to be a contraction.
The uniqueness is proved by introducing the norm $\|\Phi\|_\beta^2 = \mathbb{E}(\int_0^T e^{\beta s} |\phi_s| ds)$ and giving a priori estimates of the norm $\|Y_1 - Y_2\|_\beta$ for two solutions of the BSDE. See Pardoux and Peng [694] and El Karoui et al. [309] for details. □

An important result is the following comparison theorem for BSDE

Theorem 1.6.4.4 *Let $f^i, i = 1, 2$ be two real-valued processes satisfying the previous hypotheses and $f^1(t,x,y) \leq f^2(t,x,y)$. Let ζ^i be two $\mathcal{F}_T$-measurable, square integrable real-valued random variables such that $\zeta^1 \leq \zeta^2$ a.s.. Let (X^i, Y^i) be the solution of*

$$-dX_t^i = f^i(t, X_t^i, Y_t^i)\, dt - Y_t^i \centerdot dW_t\,, X_T^i = \zeta.$$

Then $X_t^1 \leq X_t^2, \forall t \leq T$.

Linear Case. Let us consider the particular case of a linear generator $f : \mathbb{R}^+ \times \mathbb{R} \times \mathbb{R}^d \to \mathbb{R}$ defined as $f(t,x,y) = a_t x + b_t \centerdot y + c_t$ where a, b, c are bounded adapted processes. We define the adjoint process Γ as the solution of the SDE

$$\begin{cases} d\Gamma_t = \Gamma_t[a_t dt + b_t \centerdot dW_t] \\ \Gamma_0 \ = 1 \end{cases} . \qquad (1.6.6)$$

Theorem 1.6.4.5 *Let $\zeta \in \mathcal{F}_T$, square integrable. The solution of the linear BSDE*

$$-dX_t = (a_t X_t + b_t \centerdot Y_t + c_t)dt - Y_t \centerdot dW_t,\ X_T = \zeta$$

is given by

$$X_t = (\Gamma_t)^{-1} \mathbb{E}\left(\Gamma_T \zeta + \int_t^T \Gamma_s c_s ds | \mathcal{F}_t \right).$$

PROOF: If (X, Y) is a solution of

$$-dX_t = (a_t X_t + b_t \centerdot Y_t + c_t)dt - Y_t \centerdot dW_t$$

with the terminal condition $X_T = \zeta$, then

$$-d\widehat{X}_t = \widehat{c}_t dt - Y_t \centerdot (dW_t - b_t dt),\ \widehat{X}_T = \zeta \exp\left(\int_0^T a_s ds \right)$$

where $\widehat{X}_t = X_t \exp\left(\int_0^t a_s ds\right)$ and $\widehat{c}_t = c_t \exp\left(\int_0^t a_s ds\right)$. We use Girsanov's theorem (see $\rightarrowtail$ Section 1.7) to eliminate the term $Y \centerdot b$. Let $\mathbb{Q}|_{\mathcal{F}_t} = L_t \mathbb{P}|_{\mathcal{F}_t}$ where $dL_t = L_t b_t \centerdot dW_t$. Then,

$$-d\widehat{X}_t = \widehat{c}_t dt - Y_t \centerdot d\widetilde{W}_t$$

where $\widetilde{W}$ is a $\mathbb{Q}$-Brownian motion and the process $\widehat{X}_t + \int_0^t \widehat{c}_s ds$ is a $\mathbb{Q}$-martingale with terminal value $\zeta + \int_0^T \widehat{c}_s ds$. Hence, $\widehat{X}_t = \mathbb{E}_{\mathbb{Q}}(\zeta + \int_t^T \widehat{c}_s ds | \mathcal{F}_t)$. The result follows by application of Exercise 1.2.1.8. □

Backward stochastic differential equations are of frequent use in finance. Suppose, for example, that an agent would like to obtain a terminal wealth

X_T while consuming at a given rate c (an adapted positive process). The financial market consists of d securities

$$dS_t^i = S_t^i(b_i(t)dt + \sum_{j=1}^{d} \sigma_{i,j}(t)dW_t^{(j)})$$

and a riskless bond with interest rate denoted by r. We assume that the market is complete and arbitrage free (see $\rightarrowtail$ Chapter 2 if needed). The wealth associated with a portfolio $(\pi_i, i = 0, \ldots, d)$ is the sum of the wealth invested in each asset, i.e., $X_t = \pi_0(t)S_t^0 + \sum_{i=1}^{d} \pi_i(t)S_t^i$. The self-financing condition for a portfolio with a given consumption c, i.e.,

$$dX_t = \pi_0(t)dS_t^0 + \sum_{i=1}^{d} \pi_i(t)dS_t^i - c_t dt$$

allows us to write

$$dX_t = X_t r dt + \pi_t \centerdot (b_t - r\mathbf{1})dt - c_t dt + \pi_t \centerdot \sigma_t dW_t \,,$$

where $\mathbf{1}$ is the d-dimensional vector with all components equal to 1. Therefore, the pair (wealth process, portfolio) is obtained via the solution of the BSDE

$$dX_t = f(t, X_t, Y_t)dt + Y_t \centerdot dW_t, \; X_T \text{ given}$$

with $f(t, \cdot, x, y) = rx + y \centerdot \sigma_t^{-1}(b_t - r\mathbf{1}) - c_t$ and the portfolio $(\pi_i, i = 1, \ldots, d)$ is given by $\pi_t = Y_t \centerdot \sigma_t^{-1}$. This is a particular case of a linear BSDE. Then, the process Γ introduced in (1.6.6) satisfies

$$d\Gamma_t = \Gamma_t(rdt + \sigma_t^{-1}(b_t - r\mathbf{1})dW_t), \; \Gamma_0 = 1$$

and Γ_t is the product of the discounted factor e^{-rt} and the strictly positive martingale L, which satisfies

$$dL_t = L_t \sigma_t^{-1}(b_t - r\mathbf{1})dW_t, L_0 = 1 \,,$$

i.e., $\Gamma_t = e^{-rt}L_t$. If $\mathbb{Q}$ is defined as $\mathbb{Q}|_{\mathcal{F}_t} = L_t \mathbb{P}|_{\mathcal{F}_t}$, denoting $R_t = e^{-rt}$, the process $R_t X_t + \int_0^t c_s R_s ds$ is a local martingale under the e.m.m. $\mathbb{Q}$ (see $\rightarrowtail$ Chapter 2 if needed). Therefore,

$$\Gamma_t X_t = \mathbb{E}_{\mathbb{P}} \left(X_T \Gamma_T + \int_t^T c_s \Gamma_s ds | \mathcal{F}_t \right) .$$

In particular, the value of wealth at time t needed to hedge a positive terminal wealth X_T and a positive consumption is always positive. Moreover, from the comparison theorem, if $X_T^1 \leq X_T^2$ and $c^1 \leq c^2$, then $X_t^1 \leq X_t^2$. This can be explained using the arbitrage principle. If a contingent claim ζ_1 is greater than a contingent claim ζ_2, and if there is no consumption, then the initial wealth is the price of ζ_1 and is greater than the price of ζ_2.

Exercise 1.6.4.6 Quadratic BSDE: an example. This exercise provides an example where there exists a solution although the Lipschitz condition is not satisfied.
Let a and b be two constants and ζ a bounded $\mathcal{F}_T$-measurable r.v.. Prove that the solution of $-dX_t = (aY_t^2 + bY_t)dt - Y_t dW_t, X_T = \zeta$ is

$$X_t = \frac{1}{2a}\left(\frac{1}{2}b^2(t-T) - bW_t + \ln \mathbb{E}\left(e^{bW_T + 2a\zeta}|\mathcal{F}_t\right)\right).$$

`Hint:` First, prove that the solution of the BSDE

$$-dX_t = aY_t^2 dt - Y_t dW_t,\ X_T = \zeta$$

is $X_t = \frac{1}{2a}\ln \mathbb{E}(e^{2a\zeta}|\mathcal{F}_t)$. Then, using Girsanov's theorem, the solution of

$$-dX_t = (aY_t^2 + bY_t)dt - Y_t dW_t, X_T = \zeta$$

is given by

$$X_t = \frac{1}{2a}\ln \widehat{\mathbb{E}}(e^{2a\zeta}|\mathcal{F}_t)$$

where $\widehat{\mathbb{Q}}_{|\mathcal{F}_t} = e^{bW_t - \frac{1}{2}b^2 t}\mathbb{P}_{|\mathcal{F}_t}$. Therefore,

$$\begin{aligned} X_t &= \frac{1}{2a}\ln\left(\mathbb{E}(e^{bW_T - \frac{1}{2}b^2 T}e^{2a\zeta}|\mathcal{F}_t)e^{-bW_t + \frac{1}{2}b^2 t}\right) \\ &= \frac{1}{2a}\left(\ln \mathbb{E}\left(e^{bW_T - \frac{1}{2}b^2 T}e^{2a\zeta}|\mathcal{F}_t\right) - bW_t + \frac{1}{2}b^2 t\right). \quad \triangleleft \end{aligned}$$

Comments 1.6.4.7 (a) The main references on this subject are the collective book [303], the book of Ma and Yong [607], the El Karoui and Quenez lecture in [308], El Karoui et al. [309] and Buckdhan's lecture in [134]. See also the seminal papers of Lepeltier and San Martin [578, 579] where general existence theorems for continuous generators with linear growth are established.

(b) In El Karoui and Rouge [310], the indifference price is characterized as a solution of a BSDE with a quadratic generator.

(c) BSDEs are used to solve control problems in Bielecki et al. [98], Hamadène [419], Hu and Zhou [448] and Mania and Tevzadze [619].

(d) Backward stochastic differential equations are also studied in the case where the driving martingale is a process with jumps. The reader can refer to Barles et al. [43], Royer [744], Nualart and Schoutens [683] and Rong [743].

(e) Reflected BSDE are studied by El Karoui and Quenez [308] in order to give the price of an American option, without using the notion of a Snell envelope.

(f) One of the main applications of BSDE is the notion of non-linear expectation (or G-expectation), and the link between this notion and risk measures (see Peng [705, 706]).

1.7 Change of Probability and Girsanov's Theorem

1.7.1 Change of Probability

We start with a general filtered probability space $(\Omega, \mathcal{F}, \mathbf{F}, \mathbb{P})$ where, as usual $\mathcal{F}_0$ is trivial.

Proposition 1.7.1.1 *Let $\mathbb{P}$ and $\mathbb{Q}$ be two equivalent probabilities on $(\Omega, \mathcal{F}_T)$. Then, there exists a strictly positive $(\mathbb{P}, \mathbf{F})$-martingale $(L_t, t \leq T)$, such that $\mathbb{Q}|_{\mathcal{F}_t} = L_t \mathbb{P}|_{\mathcal{F}_t}$, that is $\mathbb{E}_{\mathbb{Q}}(X) = \mathbb{E}_{\mathbb{P}}(L_t X)$ for any $\mathcal{F}_t$-measurable positive random variable X with $t \leq T$. Moreover, $L_0 = 1$ and $\mathbb{E}_{\mathbb{P}}(L_t) = 1$, $\forall t \leq T$.*

PROOF: If $\mathbb{P}$ and $\mathbb{Q}$ are equivalent on $(\Omega, \mathcal{F}_T)$, from the Radon-Nikodým theorem there exists a strictly positive $\mathcal{F}_T$-measurable random variable L_T such that $\mathbb{Q} = L_T \mathbb{P}$ on $\mathcal{F}_T$. From the definition of $\mathbb{Q}$, the expectation under $\mathbb{Q}$ of any $\mathcal{F}_T$-measurable $\mathbb{Q}$-integrable r.v. X is defined as $\mathbb{E}_{\mathbb{Q}}(X) = \mathbb{E}_{\mathbb{P}}(L_T X)$. In particular, $\mathbb{E}_{\mathbb{P}}(L_T) = 1$.

The process $L = (L_t = \mathbb{E}_{\mathbb{P}}(L_T|\mathcal{F}_t), t \leq T)$ is a $(\mathbb{P}, \mathbf{F})$-martingale and is the Radon-Nikodým density of $\mathbb{Q}$ with respect to $\mathbb{P}$ on $\mathcal{F}_t$. Indeed, if X is $\mathcal{F}_t$-measurable (hence $\mathcal{F}_T$-measurable) and $\mathbb{Q}$-integrable

$$\mathbb{E}_{\mathbb{Q}}(X) = \mathbb{E}_{\mathbb{P}}(L_T X) = \mathbb{E}_{\mathbb{P}}[\mathbb{E}_{\mathbb{P}}(X L_T|\mathcal{F}_t)] = \mathbb{E}_{\mathbb{P}}[X \mathbb{E}_{\mathbb{P}}(L_T|\mathcal{F}_t)] = \mathbb{E}_{\mathbb{P}}(X L_t).$$

□

Note that $\mathbb{P}|_{\mathcal{F}_T} = (L_T)^{-1}\mathbb{Q}|_{\mathcal{F}_T}$ so that, for any positive r.v. $Y \in \mathcal{F}_T$, $\mathbb{E}_{\mathbb{P}}(Y) = \mathbb{E}_{\mathbb{Q}}(L_T^{-1} Y)$ and L^{-1} is a $\mathbb{Q}$-martingale.

We shall speak of the law of a random variable (or of a process) under $\mathbb{P}$ or under $\mathbb{Q}$ to make precise the choice of the probability measure on the space Ω. From the equivalence between the measures, a property which holds $\mathbb{P}$-a.s. holds also $\mathbb{Q}$-a.s. However, a $\mathbb{P}$-integrable random variable is not necessarily $\mathbb{Q}$-integrable.

Definition 1.7.1.2 *A probability $\mathbb{Q}$ on a filtered probability space $(\Omega, \mathcal{F}, \mathbf{F}, \mathbb{P})$ is said to be locally equivalent to $\mathbb{P}$ if there exists a strictly positive $\mathbf{F}$-martingale L such that $\mathbb{Q}|_{\mathcal{F}_t} = L_t \mathbb{P}|_{\mathcal{F}_t}$, $\forall t$. The martingale L is called the Radon-Nikodým density of $\mathbb{Q}$ w.r.t. $\mathbb{P}$.*

Warning 1.7.1.3 This definition, which is standard in mathematical finance, is different from the more general one used by the Strasbourg school, where locally refers to a sequence of $\mathbf{F}$-stopping times, increasing to infinity.

Proposition 1.7.1.4 *Let $\mathbb{P}$ and $\mathbb{Q}$ be locally equivalent, with Radon-Nikodým density L. Then, for any stopping time τ,*

$$\mathbb{Q}|_{\mathcal{F}_\tau \cap (\tau < \infty)} = L_\tau \, \mathbb{P}|_{\mathcal{F}_\tau \cap (\tau < \infty)} \, .$$

PROOF: Let $A \in \mathcal{F}_\tau$. Then,

$$\mathbb{Q}(\mathbb{1}_A \mathbb{1}_{\{\tau \le t\}}) = \mathbb{E}_\mathbb{P}(L_t \mathbb{1}_A \mathbb{1}_{\{\tau \le t\}}) = \mathbb{E}_\mathbb{P}(L_\tau \mathbb{1}_A \mathbb{1}_{\{\tau \le t\}}) .$$

The result follows by letting $t \to \infty$. □

Proposition 1.7.1.4 may be quite useful to shift computations under $\mathbb{Q}$ into computations under $\mathbb{P}$ when L_τ has a simple expression. See ↣ Subsection 3.2.3 and ↣ Exercice 4.3.5.7.

Proposition 1.7.1.5 (Bayes Formula.) *Suppose that $\mathbb{Q}$ and $\mathbb{P}$ are equivalent on $\mathcal{F}_T$ with Radon-Nikodým density L. Let X be a $\mathbb{Q}$-integrable $\mathcal{F}_T$-measurable random variable, then, for $t < T$*

$$\mathbb{E}_\mathbb{Q}(X|\mathcal{F}_t) = \frac{\mathbb{E}_\mathbb{P}(L_T X|\mathcal{F}_t)}{L_t} .$$

PROOF: The proof follows immediately from the definition of conditional expectation. To check that the $\mathcal{F}_t$-measurable r.v. $Z = \dfrac{\mathbb{E}_\mathbb{P}(L_T X|\mathcal{F}_t)}{L_t}$ is the $\mathbb{Q}$-conditional expectation of X, we prove that $\mathbb{E}_\mathbb{Q}(F_t X) = \mathbb{E}_\mathbb{Q}(F_t Z_t)$ for any bounded $\mathcal{F}_t$-measurable random variable F_t. This follows from the equalities

$$\begin{aligned}\mathbb{E}_\mathbb{Q}(F_t X) &= \mathbb{E}_\mathbb{P}(L_T F_t X) = \mathbb{E}_\mathbb{P}(F_t \mathbb{E}_\mathbb{P}(X L_T|\mathcal{F}_t)) \\ &= \mathbb{E}_\mathbb{Q}(F_t L_t^{-1} \mathbb{E}_\mathbb{P}(X L_T|\mathcal{F}_t)) = \mathbb{E}_\mathbb{Q}(F_t Z) .\end{aligned}$$

□

Proposition 1.7.1.6 *Let $\mathbb{P}$ and $\mathbb{Q}$ be two locally equivalent probability measures with Radon-Nikodým density L. A process M is a $\mathbb{Q}$-martingale if and only if the process LM is a $\mathbb{P}$-martingale. By localization, this result remains true for local martingales.*

PROOF: Let M be a $\mathbb{Q}$-martingale. From the Bayes formula, we obtain, for $s \le t$,

$$M_s = \mathbb{E}_\mathbb{Q}(M_t|\mathcal{F}_s) = \frac{\mathbb{E}_\mathbb{P}(L_t M_t|\mathcal{F}_s)}{L_s}$$

and the result follows. The converse part is now obvious. □

Exercise 1.7.1.7 Let $(\Omega, \mathcal{F}, \mathbf{F}, \mathbb{P})$ be a filtered probability space and denote by $(L_t, t \ge 0)$ the Radon-Nikodým density of $\mathbb{Q}$ with respect to $\mathbb{P}$. Then, if $\widetilde{\mathbf{F}}$ is a subfiltration of $\mathbf{F}$, prove that $\mathbb{Q}|_{\widetilde{\mathcal{F}}_t} = \widetilde{L}_t \mathbb{P}|_{\widetilde{\mathcal{F}}_t}$, where $\widetilde{L}_t = \mathbb{E}_\mathbb{P}(L_t|\widetilde{\mathcal{F}}_t)$. ◁

Exercise 1.7.1.8 Give conditions on the function h so that the measure $\mathbb{Q}$ defined on $\mathcal{F}_T$ as $\mathbb{Q} = h(W_T)\mathbb{P}$ is a probability equivalent to $\mathbb{P}$. Prove that, for $t < T$ $\mathbb{Q}|_{\mathcal{F}_t} = L_t \mathbb{P}|_{\mathcal{F}_t}$ where

$$L_t = \int_{-\infty}^{\infty} dy\, h(y) \frac{e^{-(y-W_t)^2/(2(T-t))}}{\sqrt{2\pi(T-t)}} .$$

Prove that

$$L_t = 1 + \int_0^t dW_s \int_{-\infty}^{\infty} dy \frac{h(y)e^{-(y-W_s)^2/(2(T-s))}}{\sqrt{2\pi(T-s)}} \frac{y-W_s}{T-s} .$$

For $h \in C^1$ with compact support, prove that

$$L_t = 1 + \int_0^t dW_s \int_{-\infty}^{\infty} dy \frac{h'(y)e^{-(y-W_s)^2/(2(T-s))}}{\sqrt{2\pi(T-s)}} .$$

See Baudoin [60] for applications. ◁

Exercise 1.7.1.9 (1) Let f a Borel function satisfying $0 < \int_0^\infty f^2(u)du < \infty$. Compute, for any t, $\mathbb{P}\left(\int_0^\infty f(s)dW_s > 0 \,|\mathcal{F}_t\right) =: Z_t^f$. Prove that, as a consequence $Z_t^f > 0$ a.s., but $\mathbb{P}(Z_\infty^f = 0) = 1/2$.

(2) Prove that there exist pairs $(\mathbb{Q}, \mathbb{P})$ of probabilities that are locally equivalent, but $\mathbb{Q}$ is not equivalent to $\mathbb{P}$ on $\mathcal{F}_\infty$. ◁

1.7.2 Decomposition of $\mathbb{P}$-Martingales as $\mathbb{Q}$-semi-martingales

Theorem 1.7.2.1 *Let $\mathbb{P}$ and $\mathbb{Q}$ be locally equivalent, with Radon-Nikodým density L. We assume that the process L is continuous.*

If M is a continuous $\mathbb{P}$-local martingale, then the process $\widetilde{M}$ defined by

$$d\widetilde{M} = dM - \frac{1}{L} d\langle M, L\rangle$$

is a continuous $\mathbb{Q}$-local martingale. If N is another continuous $\mathbb{P}$-local martingale,

$$\langle M, N\rangle = \langle \widetilde{M}, \widetilde{N}\rangle = \langle M, \widetilde{N}\rangle .$$

PROOF: From Proposition 1.7.1.6, it is enough to check that $\widetilde{M}L$ is a $\mathbb{P}$-local martingale, which follows easily from Itô's calculus. □

Corollary 1.7.2.2 *Under the hypotheses of Theorem 1.7.2.1, we may write the process L as a Doléans-Dade martingale: $L_t = \mathcal{E}(\zeta)_t$, where ζ is an $\mathbf{F}$-local martingale. The process $\widetilde{M} = M - \langle M, \zeta\rangle$ is a $\mathbb{Q}$-local martingale.*

1.7.3 Girsanov's Theorem: The One-dimensional Brownian Motion Case

If the filtration $\mathbf{F}$ is generated by a Brownian motion W, and $\mathbb{P}$ and $\mathbb{Q}$ are locally equivalent, with Radon-Nikodým density L, the martingale L admits a representation of the form $dL_t = \psi_t dW_t$. Since L is strictly positive, this equality takes the form $dL_t = -\theta_t L_t dW_t$, where $\theta = -\psi/L$. (The minus sign will be convenient for further use in finance (see $\rightarrowtail$ Subsection 2.2.2), to obtain the usual risk premium). It follows that

$$L_t = \exp\left(-\int_0^t \theta_s dW_s - \frac{1}{2}\int_0^t \theta_s^2 ds\right) = \mathcal{E}(\zeta)_t\,,$$

where $\zeta_t = -\int_0^t \theta_s dW_s$.

Proposition 1.7.3.1 (Girsanov's Theorem) *Let W be a $(\mathbb{P}, \mathbf{F})$-Brownian motion and let θ be an $\mathbf{F}$-adapted process such that the solution of the SDE*

$$dL_t = -L_t\theta_t dW_t,\ L_0 = 1$$

is a martingale. We set $\mathbb{Q}|_{\mathcal{F}_t} = L_t\,\mathbb{P}|_{\mathcal{F}_t}$. Then the process W admits a $\mathbb{Q}$-semi-martingale decomposition $\widetilde{W}$ as $W_t = \widetilde{W}_t - \int_0^t \theta_s ds$ where $\widetilde{W}$ is a $\mathbb{Q}$-Brownian motion.

PROOF: From $dL_t = -L_t\theta_t dW_t$, using Girsanov's theorem 1.7.2.1, we obtain that the decomposition of W under $\mathbb{Q}$ is $\widetilde{W}_t - \int_0^t \theta_s ds$. The process W is a $\mathbb{Q}$-semi-martingale and its martingale part $\widetilde{W}$ is a BM. This last fact follows from Lévy's theorem, since the bracket of W does not depend on the (equivalent) probability. □

Warning 1.7.3.2 Using a real-valued, or complex-valued martingale density L, with respect to Wiener measure, induces a real-valued or complex-valued measure on path space. The extension of the Girsanov theorem in this framework is tricky; see Dellacherie et al. [241], paragraph (39), page 349, as well as Ruiz de Chavez [748] and Begdhdadi-Sakrani [66].

When the coefficient θ is deterministic, we shall refer to this result as **Cameron-Martin's** theorem due to the origin of this formula [137], which was extended by Maruyama [626], Girsanov [393], and later by Van Schuppen and Wong [825].

Example 1.7.3.3 Let S be a geometric Brownian motion

$$dS_t = S_t(\mu dt + \sigma dW_t)\,.$$

Here, W is a Brownian motion under a probability $\mathbb{P}$. Let $\theta = (\mu - r)/\sigma$ and $dL_t = -\theta L_t dW_t$. Then, $B_t = W_t + \theta t$ is a Brownian motion under $\mathbb{Q}$, where $\mathbb{Q}|_{\mathcal{F}_t} = L_t\,\mathbb{P}|_{\mathcal{F}_t}$ and

$$dS_t = S_t(rdt + \sigma dB_t)\,.$$

Comment 1.7.3.4 In the previous example, the equality

$$S_t(\mu dt + \sigma dW_t) = S_t(r dt + \sigma dB_t)$$

holds under both $\mathbb{P}$ and $\mathbb{Q}$. The rôle of the probabilities $\mathbb{P}$ and $\mathbb{Q}$ makes precise the dynamics of the driving process W (or B). Therefore, the equation can be computed in an "algebraic" way, by setting $dB_t = dW_t + \theta dt$. This leads to

$$\mu dt + \sigma dW_t = r dt + \sigma[dW_t + \theta dt] = r dt + \sigma dB_t \,.$$

The explicit computation of S can be made with W or B

$$\begin{aligned} S_t &= S_0 \exp\left(\mu t + \sigma W_t - \frac{1}{2}\sigma^2 t\right) \\ &= S_0 \exp\left(r t + \sigma B_t - \frac{1}{2}\sigma^2 t\right) . \end{aligned}$$

As a consequence, the importance of the probability appears when we compute the expectations

$$\mathbb{E}_{\mathbb{P}}(S_t) = S_0 e^{\mu t}, \; \mathbb{E}_{\mathbb{Q}}(S_t) = S_0 e^{rt} \,,$$

with the help of the above formulae. Note that $(S_t e^{-\mu t}, t \geq 0)$ is a $\mathbb{P}$-martingale and that $(S_t e^{-rt}, t \geq 0)$ is a $\mathbb{Q}$-martingale.

Example 1.7.3.5 Let

$$dX_t = a\, dt + 2\sqrt{X_t} dW_t \tag{1.7.1}$$

where we choose $a \geq 0$ so that there exists a positive solution $X_t \geq 0$. (See $\rightarrowtail$ Chapter 6 for more information.) Let F be a C^1 function. The continuity of F implies that the local martingale

$$L_t = \exp\left(\int_0^t F(s)\sqrt{X_s} dW_s - \frac{1}{2}\int_0^t F^2(s) X_s ds\right)$$

is in fact a martingale, therefore $\mathbb{E}(L_t) = 1$. From the definition of X and the integration by parts formula,

$$\begin{aligned} \int_0^t F(s)\sqrt{X_s} dW_s &= \frac{1}{2}\int_0^t F(s)(dX_s - a ds) \qquad (1.7.2) \\ &= \frac{1}{2}\left(F(t)X_t - F(0)X_0 - \int_0^t F'(s) X_s ds - a\int_0^t F(s) ds\right) . \end{aligned}$$

Therefore, one obtains the general formula

$$\begin{aligned} &\mathbb{E}\left[\exp\left(\frac{1}{2}\left\{F(t)X_t - \int_0^t [F'(s) + F^2(s)] X_s ds\right\}\right)\right] \\ &= \exp\left(\frac{1}{2}\left[F(0)X_0 + a\int_0^t F(s) ds\right]\right) . \end{aligned}$$

In the particular case $F(s) = -k/2$, setting

$$\mathbb{Q}|_{\mathcal{F}_t} = L_t \mathbb{P}|_{\mathcal{F}_t},$$

we obtain

$$dX_t = k(\theta - X_t)dt + 2\sqrt{X_t}dB_t = (a - kX_t)dt + 2\sqrt{X_t}dB_t \tag{1.7.3}$$

where B is a $\mathbb{Q}$-Brownian motion. Hence, if $\mathbb{Q}^a$ is the law of the process (1.7.1) and ${}^k\mathbb{Q}^a$ the law of the process defined in (1.7.3) with $a = k\theta$, we get from (1.7.2) the absolute continuity relationship

$${}^k\mathbb{Q}^a|_{\mathcal{F}_t} = \exp\left(\frac{k}{4}(at - X_t + x) - \frac{k^2}{8}\int_0^t X_s ds\right)\mathbb{Q}^a|_{\mathcal{F}_t}.$$

See Donati-Martin et al. [258] for more information.

Exercise 1.7.3.6 See Exercise 1.7.1.8 for the notation. Prove that B defined by

$$dB_t = dW_t - \frac{\int_{-\infty}^{\infty} dy\, h'(y)e^{-(y-W_t)^2/(2(T-t))}}{\int_{-\infty}^{\infty} dy\, h(y)e^{-(y-W_t)^2/(2(T-t))}}dt$$

is a $\mathbb{Q}$-Brownian motion. See Baudoin [61] for an application to finance. ◁

Exercise 1.7.3.7 (1) Let $dS_t = S_t\sigma dW_t, S_0 = x$. Prove that for any bounded function f,

$$\mathbb{E}(f(S_T)) = \mathbb{E}\left(\frac{S_T}{x}f\left(\frac{x^2}{S_T}\right)\right).$$

(2) Prove that, if $dS_t = S_t(\mu dt + \sigma dW_t)$, there exists γ such that S^γ is a martingale. Prove that for any bounded function f,

$$\mathbb{E}(f(S_T)) = \mathbb{E}\left(\left(\frac{S_T}{x}\right)^\gamma f\left(\frac{x^2}{S_T}\right)\right).$$

Prove that, for bounded function f,

$$\mathbb{E}(S_T^\alpha f(S_T)) = x^\alpha e^{\mu(\alpha)T}\mathbb{E}\left(f(e^{\alpha\sigma^2 T}S_T)\right),$$

where $\mu(\alpha) = \alpha(\mu + \frac{1}{2}\sigma^2(\alpha - 1))$. See ↣ Lemma 3.6.6.1 for application to finance. ◁

Exercise 1.7.3.8 Let W be a $\mathbb{P}$-Brownian motion, and $B_t = W_t + \nu t$ be a $\mathbb{Q}$-Brownian motion, under a suitable change of probability. Check that, in the case $\nu > 0$, the process e^{W_t} tends towards 0 under $\mathbb{Q}$ when t goes to infinity, whereas this is not the case under $\mathbb{P}$. ◁

Comment 1.7.3.9 The relation obtained in question (1) in Exercise 1.7.3.7 can be written as

$$\mathbb{E}(\varphi(W_T - \sigma T/2)) = \mathbb{E}(e^{-\sigma(W_T+\sigma T/2)}\varphi(W_T + \sigma T/2))$$

which is an "h-process" relationship between a Brownian motion with drift $\sigma/2$ and a Brownian motion with drift $-\sigma/2$.

Exercise 1.7.3.10 Examples of a martingale with respect to two different probabilities:
Let W be a $\mathbb{P}$-BM, and set $d\mathbb{Q}|_{\mathcal{F}_t} = L_t d\mathbb{P}|_{\mathcal{F}_t}$ where $L_t = \exp(\lambda W_t - \frac{1}{2}\lambda^2 t)$. Prove that the process X, where

$$X_t = W_t - \int_0^t \frac{W_s}{s} ds$$

is a Brownian motion with respect to its natural filtration under both $\mathbb{P}$ and $\mathbb{Q}$.
`Hint:` (a) Under $\mathbb{P}$, for any t, $(X_u, u \leq t)$ is independent of W_t and is a Brownian motion.

(b) Replacing W_u by $(W_u + \lambda u)$ in the definition of X does not change the value of X. (See Atlan et al. [26].) See also ↣ Example 5.8.2.3. ◁

1.7.4 Multidimensional Case

Let W be an n-dimensional Brownian motion and θ an n-dimensional adapted process such that $\int_0^t ||\theta_s||^2 ds < \infty, a.s.$. Define the local martingale L as the solution of $dL_t = L_t \theta_t \centerdot dW_t = L_t(\sum_{i=1}^n \theta_t^i dW_t^i)$, so that

$$L_t = \exp\left(\int_0^t \theta_s \centerdot dW_s - \frac{1}{2}\int_0^t ||\theta_s||^2 ds\right).$$

If L is a martingale, the n-dimensional process $(\widetilde{W}_t = W_t - \int_0^t \theta_s ds, t \geq 0)$ is a $\mathbb{Q}$-martingale, where $\mathbb{Q}$ is defined by $\mathbb{Q}|_{\mathcal{F}_t} = L_t \mathbb{P}|_{\mathcal{F}_t}$. Then, $\widetilde{W}$ is an n-dimensional Brownian motion (and in particular its components are independent).

If W is a Brownian motion with correlation matrix Λ, then, since the brackets do not depend on the probability, under $\mathbb{Q}$, the process

$$\widetilde{W}_t = W_t - \int_0^t \theta_s \centerdot \Lambda ds$$

is a correlated Brownian motion with the same correlation matrix Λ.

1.7.5 Absolute Continuity

In this section, we describe Girsanov's transformation in terms of absolute continuity. We start with elementary remarks. In what follows, $\mathbf{W}_x$ denotes the Wiener measure such that $\mathbf{W}_x(X_0 = x) = 1$ and $\mathbf{W}$ stands for $\mathbf{W}_0$. The notation $\mathbf{W}^{(\nu)}$ for the law of a BM with drift ν on the canonical space will be used:

$$\mathbf{W}^{(\nu)}[F(X_t, t \leq T)] = \mathbb{E}[F(\nu t + W_t, t \leq T)] .$$

On the left-hand side the process X is the canonical process, whose law is that of a Brownian motion with drift ν, on the right-hand side, W stands for a standard Brownian motion.

The right-hand side could be written as $\mathbf{W}^{(0)}[F(\nu t + X_t, t \leq T)]$. We also use the notation $\mathbf{W}^{(f)}$ for the law of the solution of $dX_t = f(X_t)dt + dW_t$.

Comment 1.7.5.1 Throughout our book, $(X_t, t \geq 0)$ may denote a particular stochastic process, often defined in terms of BM, or $(X_t, t \geq 0)$ may be the canonical process on $C(\mathbb{R}^+, \mathbb{R}^d)$. Each time, the context should not bring any ambiguity.

Proposition 1.7.5.2 (Cameron-Martin's Theorem.)
The Cameron-Martin theorem reads:

$$\mathbf{W}^{(\nu)}[F(X_t, t \leq T)] = \mathbf{W}^{(0)}[e^{\nu X_T - \nu^2 T/2} F(X_t, t \leq T)] .$$

More generally:

Proposition 1.7.5.3 (Girsanov's Theorem.) *Assume that the solution of $dX_t = f(X_t)dt + dW_t$ does not explode. Then, Girsanov's theorem reads: for any T,*

$$\mathbf{W}^{(f)}[F(X_t, t \leq T)]$$
$$= \mathbf{W}^{(0)}\left[\exp\left(\int_0^T f(X_s)dX_s - \frac{1}{2}\int_0^T f^2(X_s)ds\right) F(X_t, t \leq T)\right] .$$

This result admits a useful extension to stopping times (in particular to explosion times):

Proposition 1.7.5.4 *Let ζ be the explosion time of the solution of the SDE $dX_t = f(X_t)dt + dW_t$. Then, for any stopping time $\tau \leq \zeta$,*

$$\mathbf{W}^{(f)}[F(X_t, t \leq \tau)]$$
$$= \mathbf{W}^{(0)}\left[\exp\left(\int_0^\tau f(X_s)dX_s - \frac{1}{2}\int_0^\tau f^2(X_s)ds\right) F(X_t, t \leq \tau)\right] .$$

Example 1.7.5.5 From Cameron-Martin's theorem applied to the particular random variable $F(X_t, t \leq \tau) = h(e^{\sigma X_\tau})$, we deduce

$$\begin{aligned}\mathbf{W}^{(\nu)}(h(e^{\sigma X_\tau})) = \mathbb{E}(h(e^{\sigma(W_\tau + \nu\tau)})) &= \mathbf{W}^{(0)}(e^{-\nu^2\tau/2 + \nu X_\tau} h(e^{\sigma X_\tau})) \\ &= \mathbb{E}(e^{-\nu^2\tau/2} e^{\nu W_\tau} h(e^{\sigma W_\tau})) .\end{aligned}$$

Example 1.7.5.6 If $T_a(S)$ is the first hitting time of a for the geometric Brownian motion $S = xe^{\sigma X}$ defined in (1.5.3), with $a > x$ and $\sigma > 0$, and $T_\alpha(X)$ is the first hitting time of $\alpha = \frac{1}{\sigma}\ln(a/x)$ for the drifted Brownian motion X defined in (1.5.4), then

$$\begin{aligned}\mathbb{E}(F(S_t, t \leq T_a(S))) &= \mathbf{W}^{(\nu)}\left[F(xe^{\sigma X_t}, t \leq T_\alpha(X))\right] \\ &= \mathbf{W}^{(0)}\left[e^{\nu\alpha - \frac{\nu^2}{2}T_\alpha(X)} F(xe^{\sigma X_t},\, t \leq T_\alpha(X))\right] \\ &= \mathbb{E}\left(e^{\nu\alpha - \frac{\nu^2}{2}T_\alpha(W)} F(xe^{\sigma W_t},\, t \leq T_\alpha(W))\right) . \qquad (1.7.4)\end{aligned}$$

Exercise 1.7.5.7 Let W be a standard Brownian motion, $a > 1$, and τ the stopping time $\tau = \inf\{t : e^{W_t - t/2} > a\}$. Prove that, $\forall \lambda \geq 1/2$,

$$\mathbb{E}\left(\mathbb{1}_{\{\tau<\infty\}} \exp(\lambda W_\tau - \frac{1}{2}\lambda^2 \tau\right) = 1 .$$

`Hint:`

$$\mathbb{E}\left(\mathbb{1}_{\tau<\infty} \exp\left(\lambda W_\tau - \frac{1}{2}\lambda^2 \tau\right)\right) = \mathbf{W}^{(\lambda)}(\tau < \infty) .$$

The process $(W_t - \frac{1}{2}t, t \geq 0)$ is, under $\mathbf{W}^{(\lambda)}$, a BM with drift $\lambda - \frac{1}{2}$. ◁

Exercise 1.7.5.8 Let W be a $\mathbb{P}$-Brownian motion and $d\mathbb{Q}|_{\mathcal{F}_t} = e^{W_t - t/2} d\mathbb{P}|_{\mathcal{F}_t}$. Let $\tau = \inf\{t : W_t = -m\}$ for $m > 0$. Compute $\mathbb{P}(\tau < \infty)$ and $\mathbb{Q}(\tau < \infty)$.
`Hint:` $\mathbb{P}(\tau < \infty) = 1$, and using results on hitting times of BM (see ↣ Proposition 3.1.6.1) $\mathbb{Q}(\tau < \infty) = e^{-m}\mathbb{E}_{\mathbb{P}}(e^{-\tau/2}) = e^{-2m}$. ◁

1.7.6 Condition for Martingale Property of Exponential Local Martingales

As noted previously, if $\mathbb{Q}$ is a probability measure equivalent to $\mathbb{P}$, then the Radon-Nikodým density is a martingale: A strict local martingale cannot be a density between two probabilities.

In many cases we have to solve a problem of the following form: let W be a Brownian motion and

$$X_t^{\Phi} := W_t - \int_0^t ds\, \Phi_s \qquad (1.7.5)$$

where Φ is an $\mathbf{F}^W$-predictable process such that $\int_0^1 ds\, |\Phi_s| < \infty$; find a probability measure $\mathbb{Q}$ equivalent to $\mathbb{P}$, such that $(X_t^{\Phi}, t \leq 1)$ is a $(\mathbb{Q}, \mathbf{F})$-martingale.

Suppose that $\mathbb{Q}$ exists. Then $\mathbb{Q}|_{\mathcal{F}_t} = L_t \mathbb{P}|_{\mathcal{F}_t}$ and $d\langle L, W\rangle_t = \Phi_t L_t dt$. Hence $L_t = 1 + \int_0^t L_s \Phi_s dW_s$ and $\int_0^t ds\, \Phi_s^2 < \infty, a.s.$. It remains to check that the local martingale L is a martingale. The positive local martingale L is a supermartingale and is a martingale when $\mathbb{E}(L_t) = 1$. We give below criteria which can be more widely applied. A first condition is due to Novikov.

Proposition 1.7.6.1 (Novikov's Condition.) *If the continuous martingale ζ satisfies:*

$$\mathbb{E}\left(\exp\left(\frac{1}{2}\langle\zeta\rangle_\infty\right)\right) < \infty \tag{1.7.6}$$

then ζ belongs to $\mathbf{H}^p$ for every $p \in [1, \infty[$ and $L = \mathcal{E}(\zeta)$ is a uniformly integrable martingale.

PROOF: See [RY], Chapter VIII, Proposition 1.15. □

The constant 1/2 in (1.7.6) is the best possible (see Kazamaki [517], Chapter 1, Example 1.5).

In the case where $\zeta_t = \int_0^t \theta_s dW_s$, Novikov's condition reads

$$\mathbb{E}\left(\exp\left(\frac{1}{2}\int_0^\infty \theta_s^2 ds\right)\right) < \infty .$$

Obviously, if we restrict our attention to the time interval $[0, T]$, Novikov's condition

$$\mathbb{E}\left(\exp\left(\frac{1}{2}\int_0^T \theta_s^2 ds\right)\right) < \infty \tag{1.7.7}$$

implies that $(L_t; 0 \leq t \leq T)$ is a martingale where $dL_t = \theta_t L_t dW_t$. Note that, Novikov's condition (1.7.7) is satisfied whenever θ is bounded.

It should be noted that if the local martingale $\mathcal{E}(\zeta)$ is uniformly integrable, i.e., if the family of r.v. $(\mathcal{E}(\zeta)_t, t \geq 0)$ is u.i., it is not necessarily a martingale (see Kazamaki [517], Chapter 1, Example 1.1. for a counter-example). If the local martingale $\mathcal{E}(\zeta)$ belongs to class D, i.e., if the family of r.v. $(\mathcal{E}(\zeta)_\tau, \tau$ stopping time) is u.i., then $\mathcal{E}(\zeta)$ is a martingale. The process $\mathcal{E}(\zeta)$ can be a martingale which is not uniformly integrable: take $\zeta = B$ where B is a Brownian motion.

Let us give two theorems (see Kazamaki [517]).

Theorem 1.7.6.2 (Kazamaki's Criterion.) *If ζ is a continuous local martingale such that the process $\exp(\frac{1}{2}\zeta)$ is a uniformly integrable submartingale, then the process $L = \mathcal{E}(\zeta)$ is a uniformly integrable martingale.*

Theorem 1.7.6.3 (BMO Criterion.) *Let ζ be a continuous martingale in BMO, then the process $L = \mathcal{E}(\zeta)$ is a uniformly integrable martingale.*

These conditions are often difficult to check and the following proposition is a useful tool. In a Markovian case, an easy condition is the following:

Proposition 1.7.6.4 (Non-explosion Criteria.) *Let* $\zeta_t = \int_0^t b(s, W_s)dW_s$ *where* b *satisfies*

$$\begin{cases} |b(s,x) - b(s,y)| \leq C|x-y|\,, \\ \sup_{s\leq t} |b(s,0)| \quad \leq C\,. \end{cases}$$

Then, the process $Z_t = \exp(\zeta_t - \frac{1}{2}\langle\zeta\rangle_t)\,; t \geq 0$ *is a martingale. More generally,* Z *is a martingale as soon as the stochastic equation*

$$dX_t = b(t, X_t)dt + dW_t,\ X_0 = 0$$

has a unique solution in law, without explosion.

PROOF: If the stochastic differential equation $X_t = W_t + \int_0^t b(s, X_s)ds$ has a unique solution, its law is locally equivalent to the Wiener measure (here, locally refers to the existence of a localizing sequence of stopping times). Let $T_n = \inf\{t : |X_t| = n\}$. We define an equivalent probability measure $\mathbf{W}^b$ via:

$$\mathbf{W}^b|_{\mathcal{F}_{t\wedge T_n}} = \exp\left[\int_0^{t\wedge T_n} b(s, X_s)dX_s - \frac{1}{2}\int_0^{t\wedge T_n} b^2(s, X_s)ds\right] \mathbf{W}|_{\mathcal{F}_{t\wedge T_n}}\,.$$

Then, for any $F_t \in \mathcal{F}_t$

$$\mathbf{W}^b(F_t \mathbb{1}_{\{t<T_n\}}) = \mathbf{W}\left(F_t \mathbb{1}_{\{t<T_n\}} \exp\left[\int_0^t b(s, X_s)dX_s - \frac{1}{2}\int_0^t b^2(s, X_s)ds\right]\right)$$

Letting n go to infinity, and using the fact that $T_n \to \infty$ both under $\mathbf{W}^b$ and $\mathbf{W}$, we obtain:

$$\mathbf{W}^b|_{\mathcal{F}_t} = \exp\left[\int_0^t b(s, X_s)dX_s - \frac{1}{2}\int_0^t b^2(s, X_s)ds\right] \mathbf{W}|_{\mathcal{F}_t}\,,$$

hence, the process

$$\exp\left(\int_0^t b(s, X_s)dX_s - \frac{1}{2}\int_0^t b^2(s, X_s)ds\right),\ t \geq 0$$

is a martingale. □

In the particular case $b(x) = \lambda x$ of the OU process, we deduce that the process

$$\exp\left(\lambda\frac{B_t^2 - t}{2} - \frac{\lambda^2}{2}\int_0^t dsB_s^2\right), \quad t \geq 0$$

is a martingale, for any $\lambda \in \mathbb{R}$.

Example 1.7.6.5 If $dX_t = dB_t + f(X_t)dt$, where $f : \mathbb{R} \to \mathbb{R}$ is a C^1-function, the Feller criterion (see McKean [636] or $\rightarrowtail$ Proposition 5.3.3.4) gives a sufficient condition for no explosion. Note that if $\mathbf{W}_x^{(f)}$ is the law of the solution, and τ the explosion time, then

$$\begin{aligned}\mathbf{W}_x^{(f)}|_{\mathcal{F}_t \cap \{t<\tau\}} &= \exp\left(\int_0^t f(X_s)dX_s - \frac{1}{2}\int_0^t f^2(X_s)ds\right)\mathbf{W}_x|_{\mathcal{F}_t} \\ &= \exp\left(F(X_t) - F(x) - \frac{1}{2}\int_0^t (f^2 + f')(X_s)ds\right)\mathbf{W}_x|_{\mathcal{F}_t}\end{aligned}$$

where F is an antiderivative of f. If $f(x) = |x|^\gamma$ with $\gamma > 1$, then there is explosion. In the case $f(x) = cx^{2n}$, one gets

$$\begin{aligned}\mathbb{P}^{(c,n)}(\tau > t) &= \mathbb{E}\left(\exp\left(c\int_0^t B_s^{2n}dB_s - \frac{c^2}{2}\int_0^t B_s^{4n}ds\right)\right) \\ &= \mathbb{E}\left(\exp\left(ct^{n+1/2}\int_0^1 B_s^{2n}dB_s - \frac{c^2}{2}t^{2n+1}\int_0^1 B_s^{4n}ds\right)\right),\end{aligned}$$

which gives an implicit description of the law of τ in terms of the joint law of $\left(\int_0^1 B_s^{2n}dB_s, \int_0^1 B_s^{4n}ds\right)$.

Example 1.7.6.6 Let us give one example of a local martingale which is not a martingale (we say that the local martingale is a strict local martingale). Let α be a positive real number and

$$dX_t = X_t Y_t^\alpha \sigma dB_t; \quad dY_t = Y_t a\, dB_t\,.$$

Using the fact that the process Z defined by $dZ_t = Z_t a dW_t + Z_t^{\alpha+1}\mu dt$ with $\mu > 0$ has a finite explosion time, Sin [800] proves that the process X is a strict local martingale.

Comment 1.7.6.7 There is an extensive literature on uniformly integrable exponential martingales. Let us mention Cherny and Shiryaev [169], Choulli et al. [181], Kazamaki [517] and Lépingle and Mémin [580].

1.7.7 Predictable Representation Property under a Change of Probability

Let $\mathbf{F}$ be the filtration of a Brownian motion W and θ an $\mathbf{F}$-adapted process such that the local martingale $L_t := \exp(\int_0^t \theta_s dW_s - \frac{1}{2}\int_0^t \theta_s^2\, ds)$ is a martingale. Let $\mathbb{Q}$ be the probability law, equivalent to $\mathbb{P}$ on $\mathcal{F}_t$ for any t, defined as $\mathbb{Q}|_{\mathcal{F}_t} = L_t\,\mathbb{P}|_{\mathcal{F}_t}$. Girsanov's theorem implies that $\widetilde{W}_t := W_t - \int_0^t \theta_s ds$ is an $(\mathbf{F}, \mathbb{Q})$-Brownian motion. Since, obviously, the process $\widetilde{W}$ is $\mathbf{F}$-adapted, the inclusion $\widetilde{\mathcal{F}}_t = \sigma(\widetilde{W}_s, s \le t) \subseteq \mathcal{F}_t$ holds. If θ is deterministic, then both filtrations are equal, but this is not the case in general (see Tsirel'son's example 1.5.5.6 or [822]). However, the representation theorem (see Section 1.6.1) extends to this framework.

Proposition 1.7.7.1 *Let W be a Brownian motion under $\mathbb{P}$, $\mathbf{F}$ its natural filtration, and $\mathbb{Q}$ a probability measure locally equivalent to $\mathbb{P}$. Let $\widetilde{W}$ be the martingale part of the $\mathbb{Q}$-semimartingale W. If M is a $(\mathbf{F},\mathbb{Q})$-local martingale, there exists an $\mathbf{F}$-predictable process H such that*

$$\forall t, \ \ M_t = M_0 + \int_0^t H_s \, d\widetilde{W}_s \, .$$

PROOF: It is enough to write the predictable representation of the $\mathbb{P}$-martingale ML as $M_t L_t = M_0 + \int_0^t \psi_s dW_s$. From Itô's formula and the obvious relation $M = (ML)L^{-1}$, the process M can be written as a stochastic integral w.r.t. $\widetilde{W}$. □

We have here an example of a "weakly Brownian filtration." We shall give other examples in ↣ Chapter 5.

Exercise 1.7.7.2 Prove the result recalled in Comment 1.4.1.6.
Hint: If $W_T^{(i)}$ could be written as $\int_0^T \phi_s^{(i)} dB_s$ for $i = 1, 2$, the properties of $\phi^{(i)}$ would lead to a contradiction. ◁

1.7.8 An Example of Invariance of BM under Change of Measure

Let $\mathbb{P}$ and $\mathbb{Q}$ be two equivalent probabilities on $(\Omega, \mathcal{F})$ and X a r.v. (or a process). We present a simple condition under which the law of X is the same under $\mathbb{P}$ and $\mathbb{Q}$, as well as an example (see also ↣ Example 1.7.3.10).

Proposition 1.7.8.1 *Let X be a real-valued $\mathbf{F}$-Brownian motion under $\mathbb{P}$ and L be the Radon-Nikodým density of $\mathbb{Q}$ w.r.t. $\mathbb{P}$. Then X is a $\mathbb{Q}$-Brownian motion if and only if X and L are $(\mathbf{F},\mathbb{P})$-orthogonal martingales*

PROOF: Note that

$$\widetilde{X}_t = X_t - \int_0^t \frac{d\langle X, L\rangle_s}{L_s}$$

is a $(\mathbf{F},\mathbb{Q})$- Brownian motion. □

This result admits an extension to the multidimensional case: Let W be an n-dimensional Brownian motion and $X_t = x + \int_0^t x_s \centerdot dW_s$ where $(x_t, t \geq 0)$ is an n-dimensional predictable process. The process X is a BM if and only if $|x_t|^2 = 1, ds \times d\mathbb{P} a.s.$ Let L be a Radon-Nikodým density. The process L admits the representation $L_t = 1 + \int_0^t \ell_s \centerdot dW_s$. The process X is a $(\mathbf{F},\mathbb{Q})$-Brownian motion if and only if $x_t \centerdot \ell_t = 0$, $dt \times d\mathbb{P}$ a.s.

Example 1.7.8.2 If $W = (X, Y)$ is a 2-dimensional Brownian motion starting from (a, b), the pair (x, ℓ) where $x_t = W_t/|W_t|$ (stopped at the first time $|W|$ vanishes) and $\ell_t = (Y_t, -X_t)$ satisfies the previous condition.

2

Basic Concepts and Examples in Finance

In this chapter, we present briefly the main concepts in mathematical finance as well as some straightforward applications of stochastic calculus for continuous-path processes. We study in particular the general principle for valuation of contingent claims, the Feynman-Kac approach, the Ornstein-Uhlenbeck and Vasicek processes, and, finally, the pricing of European options.

Derivatives are products whose payoffs depend on the prices of the traded underlying assets. In order for the model to be arbitrage free, the link between derivatives and underlying prices has to be made precise. We shall present the mathematical setting of this problem, and give some examples. In this area we recommend Portait and Poncet [723], Lipton [596], Overhaus et al. [689], and Brockhaus et al. [131].

Important assets are the zero-coupon bonds which deliver one monetary unit at a terminal date. The price of this asset depends on the interest rate. We shall present some basic models of the dynamics of the interest rate (Vasicek and CIR) and the dynamics of associated zero-coupon bonds. We refer the reader to Martellini et al. [624] and Musiela and Rutkowski [661] for a study of modelling of zero-coupon prices and pricing derivatives.

2.1 A Semi-martingale Framework

In a first part, we present in a general setting the modelling of the stock market and the hypotheses in force in mathematical finance. The dynamics of prices are semi-martingales, which is justified from the hypothesis of no-arbitrage (see the precise definition in $\rightarrowtail$ Subsection 2.1.2). Roughly speaking, this hypothesis excludes the possibility of starting with a null amount of money and investing in the market in such a way that the value of the portfolio at some fixed date T is positive (and not null) with probability 1. We shall comment upon this hypothesis later.

We present the definition of self-financing strategies and the concept of hedging portfolios in a case where the tradeable asset prices are given as

M. Jeanblanc, M. Yor, M. Chesney, *Mathematical Methods for Financial Markets*, Springer Finance, DOI 10.1007/978-1-84628-737-4_2,

semi-martingales. We give the definition of an arbitrage opportunity and we state the fundamental theorem which links the non-arbitrage hypothesis with the notion of equivalent martingale measure. We define a complete market and we show how this definition is related to the predictable representation property.

In this first section, we do not require path-continuity of asset prices.

An important precision: Concerning all financial quantities presented in that chapter, these will be defined up to a finite horizon T, called the maturity. On the other hand, when dealing with semi-martingales, these will be implicitly defined on $\mathbb{R}^+$.

2.1.1 The Financial Market

We study a financial market where assets (stocks) are traded in continuous time. We assume that there are d assets, and that the prices $S^i, i = 1, \dots, d$ of these assets are modelled as semi-martingales with respect to a reference filtration $\mathbf{F}$. We shall refer to these assets as **risky assets** or as **securities.** We shall also assume that there is a **riskless asset** (also called the savings account) with dynamics

$$dS_t^0 = S_t^0 r_t dt, \; S_0^0 = 1$$

where r is the (positive) **interest rate**, assumed to be $\mathbf{F}$-adapted. One monetary unit invested at time 0 in the riskless asset will give a payoff of $\exp\left(\int_0^t r_s ds\right)$ at time t. If r is deterministic, the price at time 0 of one monetary unit delivered at time t is

$$\boxed{R_t := \exp\left(-\int_0^t r_s ds\right).}$$

The quantity $R_t = (S_t^0)^{-1}$ is called the **discount factor**, whether or not it is deterministic. The discounted value of S_t^i is $S_t^i R_t$; in the case where r and S_t^i are deterministic, this is the monetary value at time 0 of S_t^i monetary units delivered at time t. More generally, if a process $(V_t, t \geq 0)$ describes the value of a financial product at any time t, its discounted value process is $(V_t R_t, t \geq 0)$. The asset that delivers one monetary unit at time T is called a **zero-coupon bond** (ZC) of maturity T. If r is deterministic, its price at time t is given by

$$P(t,T) = \exp\left(-\int_t^T r(s) ds\right),$$

the dynamics of the ZC's price is then $d_t P(t,T) = r_t P(t,T) dt$ with the terminal condition $P(T,T) = 1$. If r is a stochastic process, the problem of giving the price of a zero-coupon bond is more complex; we shall study this case later. In that setting, the previous formula for $P(t,T)$ would be absurd, since $P(t,T)$ is known at time t (i.e., is $\mathcal{F}_t$-measurable), whereas the quantity $\exp\left(-\int_t^T r(s)ds\right)$ is not. Zero-coupon bonds are traded and are at the core of trading in financial markets.

Comment 2.1.1.1 In this book, we assume, as is usual in mathematical finance, that borrowing and lending interest rates are equal to $(r_s, s \geq 0)$: one monetary unit borrowed at time 0 has to be reimbursed by $S_t^0 = \exp\left(\int_0^t r_s ds\right)$ monetary units at time t. One monetary unit invested in the riskless asset at time 0 produces $S_t^0 = \exp\left(\int_0^t r_s ds\right)$ monetary units at time t. In reality, borrowing and lending interest rates are not the same, and this equality hypothesis, which is assumed in mathematical finance, oversimplifies the "real-world" situation. Pricing derivatives with different interest rates is very similar to pricing under constraints. If, for example, there are two interest rates with $r_1 < r_2$, one has to assume that it is impossible to borrow money at rate r_1 (see also ↣ Example 2.1.2.1). We refer the reader to the papers of El Karoui et al. [307] for a study of pricing with constraints.

A **portfolio** (or a strategy) is a $(d+1)$-dimensional $\mathbf{F}$-predictable process $(\widehat{\pi}_t = (\pi_t^i, i = 0, \ldots, d) = (\pi_t^0, \pi_t); t \geq 0)$ where π_t^i represents the number of shares of asset i held at time t. Its time-t value is

$$V_t(\widehat{\pi}) := \sum_{i=0}^{d} \pi_t^i S_t^i = \pi_t^0 S_t^0 + \sum_{i=1}^{d} \pi_t^i S_t^i .$$

We assume that the integrals $\int_0^t \pi_s^i dS_s^i$ are well defined; moreover, we shall often place more integrability conditions on the portfolio $\widehat{\pi}$ to avoid arbitrage opportunities (see ↣ Subsection 2.1.2).

We shall assume that the market is liquid: there is no transaction cost (the buying price of an asset is equal to its selling price), the number of shares of the asset available in the market is not bounded, and **short-selling** of securities is allowed (i.e., $\pi^i, i \geq 1$ can take negative values) as well as borrowing money $(\pi^0 < 0)$.

We introduce a constraint on the portfolio, to make precise the idea that instantaneous changes to the value of the portfolio are due to changes in prices, not to instantaneous rebalancing. This **self-financing** condition is an extension of the discrete-time case and we impose it as a constraint in continuous time. *We emphasize that this constraint is not a consequence of Itô's lemma* and that, if a portfolio $(\widehat{\pi}_t = (\pi_t^i, i = 0, \ldots, d) = (\pi_t^0, \pi_t); t \geq 0)$ is given, this condition has to be satisfied.

Definition 2.1.1.2 *A portfolio* $\widehat{\pi}$ *is said to be* ***self-financing*** *if*

$$dV_t(\widehat{\pi}) = \sum_{i=0}^{d} \pi_t^i dS_t^i ,$$

or, in an integrated form, $V_t(\widehat{\pi}) = V_0(\widehat{\pi}) + \sum_{i=0}^{d} \int_0^t \pi_s^i dS_s^i$.

If $\widehat{\pi} = (\pi^0, \pi)$ is a self-financing portfolio, then some algebraic computation establishes that

$$\begin{aligned} dV_t(\widehat{\pi}) &= \pi_t^0 S_t^0 r_t dt + \sum_{i=1}^{d} \pi_t^i dS_t^i = r_t V_t(\widehat{\pi}) dt + \sum_{i=1}^{d} \pi_t^i (dS_t^i - r_t S_t^i dt) \\ &= r_t V_t(\widehat{\pi}) dt + \pi_t (dS_t - r_t S_t dt) \end{aligned}$$

where the vector $\pi = (\pi^i; i = 1, \ldots, d)$ is written as a $(1, d)$ matrix. We prove now that the self-financing condition holds for discounted processes, i.e., if all the processes V and S^i are discounted (note that the discounted value of S_t^0 is 1):

Proposition 2.1.1.3 *If* $\widehat{\pi}$ *is a self-financing portfolio, then*

$$R_t V_t(\widehat{\pi}) = V_0(\widehat{\pi}) + \sum_{i=1}^{d} \int_0^t \pi_s^i d(R_s S_s^i) . \tag{2.1.1}$$

Conversely, if x *is a given positive real number, if* $\pi = (\pi^1, \ldots, \pi^d)$ *is a vector of predictable processes, and if* V^π *denotes the solution of*

$$\boxed{dV_t^\pi = r_t V_t^\pi dt + \pi_t (dS_t - r_t S_t dt) , \; V_0^\pi = x ,} \tag{2.1.2}$$

then the $\mathbb{R}^{d+1}$*-valued process* $(\widehat{\pi}_t = (V_t^\pi - \pi_t S_t, \pi_t); t \geq 0)$ *is a self-financing strategy, and* $V_t^\pi = V_t(\widehat{\pi})$.

PROOF: Equality (2.1.1) follows from the integration by parts formula:

$$d(R_t V_t) = R_t dV_t - V_t r_t R_t dt = R_t \pi_t (dS_t - r_t S_t dt) = \pi_t d(R_t S_t) .$$

Conversely, if $(x, \pi = (\pi^1, \ldots, \pi^d))$ are given, then one deduces from (2.1.2) that the value V_t^π of the portfolio at time t is given by

$$V_t^\pi R_t = x + \int_0^t \pi_s d(R_s S_s)$$

and the wealth invested in the riskless asset is

$$\pi_t^0 S_t^0 = V_t^\pi - \sum_{i=1}^{d} \pi_t^i S_t^i = V_t^\pi - \pi_t S_t .$$

The portfolio $\widehat{\pi} = (\pi^0, \pi)$ is obviously self-financing since

$$dV_t = r_t V_t dt + \pi_t (dS_t - r_t S_t dt) = \pi_t^0 S_t^0 r_t dt + \pi_t dS_t \,.$$

The process $(\sum_{i=1}^d \int_0^t \pi_s^i d(R_s S_s^i),\ t \geq 0)$ is the discounted gain process. □

This important result proves that a self-financing portfolio is characterized by its initial value $V_0(\widehat{\pi})$ and the strategy $\pi = (\pi^i, i = 1, \ldots, d)$ which represents the investment in the risky assets. The equality (2.1.1) can be written in terms of the savings account S^0 as

$$\boxed{\frac{V_t(\widehat{\pi})}{S_t^0} = V_0(\widehat{\pi}) + \sum_{i=1}^d \int_0^t \pi_s^i \, d\left(\frac{S_s^i}{S_s^0}\right)} \tag{2.1.3}$$

or as

$$dV_t^0 = \sum_{i=1}^d \pi_t^i dS_t^{i,0}$$

where

$$V_t^0 = V_t R_t = V_t / S_t^0, \quad S_t^{i,0} = S_t^i R_t = S_t^i / S_t^0$$

are the prices in terms of time-0 monetary units. We shall extend this property in ↣ Section 2.4 by proving that the self-financing condition does not depend on the choice of the numéraire.

By abuse of language, we shall also call $\pi = (\pi^1, \ldots, \pi^d)$ a self-financing portfolio.

The investor is said to have a **long position** at time t on the asset S if $\pi_t \geq 0$. In the case $\pi_t < 0$, the investor is **short**.

Exercise 2.1.1.4 Let $dS_t = (\mu dt + \sigma dB_t)$ and $r = 0$. Is the portfolio $\widehat{\pi}(t, 1)$ self-financing? If not, find π^0 such that $(\pi_t^0, 1)$ is self-financing. ◁

2.1.2 Arbitrage Opportunities

Roughly speaking, an **arbitrage opportunity** is a self-financing strategy π with zero initial value and with terminal value $V_T^\pi \geq 0$, such that $\mathbb{E}(V_T^\pi) > 0$.

From Dudley's result (see Subsection 1.6.3), it is obvious that we have to impose conditions on the strategies to exclude arbitrage opportunities. Indeed, if B is a BM, for any constant A, it is possible to find an adapted process φ such that $\int_0^T \varphi_s dB_s = A$. Hence, in the simple case $dS_s = \sigma S_s dB_s$ and null interest rate, it is possible to find π such that $\int_0^T \pi_s dS_s = A > 0$. The process $V_t = \int_0^t \pi_s dS_s$ would be the value of a self-financing strategy, with null initial wealth and strictly positive terminal value, therefore, π would be an arbitrage opportunity. These strategies are often called doubling strategies, by extension to an infinite horizon of a tossing game: a player with an initial wealth 0 playing such a game will have, with probability 1, at some time a wealth equal to 10^{63} monetary units: he only has to wait long enough (and to agree to lose a large amount of money before that). It "suffices" to play in continuous time to win with a BM.

Example 2.1.2.1 If there are two riskless assets in the market with interest rates r_1 and r_2, then in order to exclude arbitrage opportunities, we must have $r_1 = r_2$: otherwise, if $r_1 < r_2$, an investor might borrow an amount k of money at rate r_1, and invest the same amount at rate r_2. The initial wealth is 0 and the wealth at time T would be $ke^{r_2 T} - ke^{r_1 T} > 0$. So, in the case of different interest rates with $r_1 < r_2$, one has to restrict the strategies to those for which the investor can only borrow money at rate r_2 and invest at rate r_1. One has to add one dimension to the portfolio; the quantity of shares of the savings account, denoted by π^0 is now a pair of processes $\pi^{0,1}, \pi^{0,2}$ with $\pi^{0,1} \geq 0, \pi^{0,2} \leq 0$ where the wealth in the bank account is $\pi_t^{0,1} S_t^{0,1} + \pi_t^{0,2} S_t^{0,2}$ with $dS_t^{0,j} = r_j S_t^{0,j} dt$.

Exercise 2.1.2.2 There are many examples of relations between prices which are obtained from the absence of arbitrage opportunities in a financial market. As an exercise, we give some examples for which we use call and put options (see ↣ Subsection 2.3.2 for the definition). The reader can refer to Cox and Rubinstein [204] for proofs. We work in a market with constant interest rate r. We emphasize that these relations are model-independent, i.e., they are valid whatever the dynamics of the risky asset.

- Let C (resp. P) be the value of a European call (resp. a put) on a stock with current value S, and with strike K and maturity T. Prove the put-call parity relationship

$$\boxed{C = P + S - Ke^{-rT} .}$$

- Prove that $S \geq C \geq \max(0, S - K)$.
- Prove that the value of a call is decreasing w.r.t. the strike.
- Prove that the call price is concave w.r.t. the strike.
- Prove that, for $K_2 > K_1$,

$$K_2 - K_1 \geq C(K_2) - C(K_1) ,$$

where $C(K)$ is the value of the call with strike K.

◁

2.1.3 Equivalent Martingale Measure

We now introduce the key definition of equivalent martingale measure (or risk-neutral probability). It is a major tool in giving the prices of derivative products as an expectation of the (discounted) terminal payoff, and the existence of such a probability is related to the non-existence of arbitrage opportunities.

Definition 2.1.3.1 *An* **equivalent martingale measure** *(e.m.m.) is a probability measure $\mathbb{Q}$, equivalent to $\mathbb{P}$ on $\mathcal{F}_T$, such that the discounted prices $(R_t S^i_t, t \le T)$ are $\mathbb{Q}$-local martingales.*

It is proved in the seminal paper of Harrison and Kreps [421] in a discrete setting and in a series of papers by Delbaen and Schachermayer [233] in a general framework, that the existence of e.m.m. is more or less equivalent to the absence of arbitrage opportunities. One of the difficulties is to make precise the choice of "admissible" portfolios. We borrow from Protter [726] the name of Folk theorem for what follows:

Folk Theorem: *Let S be the stock price process. There is absence of arbitrage essentially if and only if there exists a probability $\mathbb{Q}$ equivalent to $\mathbb{P}$ such that the discounted price process is a $\mathbb{Q}$-local martingale.*

From (2.1.3), we deduce that not only the discounted prices of securities are local-martingales, but that more generally, any price, and in particular prices of derivatives, are local martingales:

Proposition 2.1.3.2 *Under any e.m.m. the discounted value of a self-financing strategy is a local martingale.*

Comment 2.1.3.3 Of course, it can happen that discounted prices are strict local martingales. We refer to Pal and Protter [692] for an interesting discussion.

2.1.4 Admissible Strategies

As mentioned above, one has to add some regularity conditions on the portfolio to exclude arbitrage opportunities. The most common such condition is the following admissibility criterion.

Definition 2.1.4.1 *A self-financing strategy π is said to be admissible if there exists a constant A such that $V_t(\pi) \ge -A$, a.s. for every $t \le T$.*

Definition 2.1.4.2 *An arbitrage opportunity on the time interval $[0,T]$ is an admissible self-financing strategy π such that $V^\pi_0 = 0$ and $V^\pi_T \ge 0, \mathbb{E}(V^\pi_T) > 0$.*

In order to give a precise meaning to the fundamental theorem of asset pricing, we need some definitions (we refer to Delbaen and Schachermayer [233]). In the following, we assume that the interest rate is equal to 0. Let us define the sets

$$\begin{aligned}
\mathcal{K} &= \left\{ \int_0^T \pi_s dS_s \,:\, \pi \text{ is admissible} \right\}, \\
\mathcal{A}_0 &= \mathcal{K} - L^0_+ = \left\{ X = \int_0^T \pi_s dS_s - f \,:\, \pi \text{ is admissible}, f \geq 0, f \text{ finite} \right\}, \\
\mathcal{A} &= \mathcal{A}_0 \cap L^\infty, \\
\bar{\mathcal{A}} &= \text{closure of } \mathcal{A} \text{ in } L^\infty.
\end{aligned}$$

Note that $\mathcal{K}$ is the set of terminal values of admissible self-financing strategies with zero initial value. Let L^∞_+ be the set of positive random variables in L^∞.

Definition 2.1.4.3 *A semi-martingale S satisfies the no-arbitrage condition if $\mathcal{K} \cap L^\infty_+ = \{0\}$. A semi-martingale S satisfies the* **No-Free Lunch with Vanishing Risk** *(NFLVR) condition if $\bar{\mathcal{A}} \cap L^\infty_+ = \{0\}$.*

Obviously, if S satisfies the no-arbitrage condition, then it satisfies the NFLVR condition.

Theorem 2.1.4.4 (Fundamental Theorem.) *Let S be a locally bounded semi-martingale. There exists an equivalent martingale measure $\mathbb{Q}$ for S if and only if S satisfies NFLVR.*

PROOF: The proof relies on the Hahn-Banach theorem, and goes back to Harrison and Kreps [421], Harrison and Pliska [423] and Kreps [545] and was extended by Ansel and Stricker [20], Delbaen and Schachermayer [233], Stricker [809]. We refer to the book of Delbaen and Schachermayer [236], Theorem 9.1.1. □

The following result (see Delbaen and Schachermayer [236], Theorem 9.7.2.) establishes that the dynamics of asset prices have to be semi-martingales:

Theorem 2.1.4.5 *Let S be an adapted càdlàg process. If S is locally bounded and satisfies the no free lunch with vanishing risk property for simple integrands, then S is a semi-martingale.*

Comments 2.1.4.6 (a) The study of the absence of arbitrage opportunities and its connection with the existence of e.m.m. has led to an extensive literature and is fully presented in the book of Delbaen and Schachermayer [236]. The survey paper of Kabanov [500] is an excellent presentation of arbitrage theory. See also the important paper of Ansel and Stricker [20] and Cherny [167] for a slightly different definition of arbitrage.

(b) Some authors (e.g., Karatzas [510], Levental and Skorokhod [583]) give the name of tame strategies to admissible strategies.

(c) It should be noted that the condition for a strategy to be admissible is restrictive from a financial point of view. Indeed, in the case $d = 1$, it excludes short position on the stock. Moreover, the condition depends on the choice of numéraire. These remarks have led Sin [799] and Xia and Yan [851, 852] to introduce allowable portfolios, i.e., by definition there exists $a \geq 0$ such that $V_t^\pi \geq -a \sum_i S_t^i$. The authors develop the fundamental theory of asset pricing in that setting.

(d) Frittelli [364] links the existence of e.m.m. and NFLVR with results on optimization theory, and with the choice of a class of utility functions.

(e) The condition $\mathcal{K} \cap L_+^\infty = \{0\}$ is too restrictive to imply the existence of an e.m.m.

2.1.5 Complete Market

Roughly speaking, a market is complete if any derivative product can be perfectly hedged, i.e., is the terminal value of a self-financing portfolio.

Assume that there are d risky assets S^i which are $\mathbf{F}$-semi-martingales and a riskless asset S^0. A **contingent claim** H is defined as a square integrable $\mathcal{F}_T$-random variable, where T is a fixed horizon.

Definition 2.1.5.1 *A contingent claim H is said to be* **hedgeable** *if there exists a predictable process $\pi = (\pi^1, \ldots, \pi^d)$ such that $V_T^\pi = H$. The self-financing strategy $\hat{\pi} = (V^\pi - \pi S, \pi)$ is called the* ***replicating strategy*** *(or the* ***hedging strategy****) of H, and $V_0^\pi = h$ is the initial price. The process V^π is the price process of H.*

In some sense, this initial value is an equilibrium price: the seller of the claim agrees to sell the claim at an initial price p if he can construct a portfolio with initial value p and terminal value greater than the claim he has to deliver. The buyer of the claim agrees to buy the claim if he is unable to produce the same (or a greater) amount of money while investing the price of the claim in the financial market.

It is also easy to prove that, if the price of the claim is not the initial value of the replicating portfolio, there would be an arbitrage in the market: assume that the claim H is traded at v with $v > V_0$, where V_0 is the initial value of the replicating portfolio. At time 0, one could

- ▶ invest V_0 in the financial market using the replicating strategy
- ▶ sell the claim at price v
- ▶ invest the amount $v - V_0$ in the riskless asset.

The terminal wealth would be (if the interest rate is a constant r)

- ▶ the value of the replicating portfolio, i.e., H
- ▶ minus the value of the claim to deliver, i.e., H
- ▶ plus the amount of money in the savings account, that is $(v - V_0)e^{rT}$

and that quantity is strictly positive. If the claim H is traded at price v with $v < V_0$, we invert the positions, buying the claim at price v and selling the replicating portfolio.

Using the characterization of a self-financing strategy obtained in Proposition 2.1.1.3, we see that the contingent claim H is hedgeable if there exists a pair (h, π) where h is a real number and π a d-dimensional predictable process such that

$$H/S_T^0 = h + \sum_{i=1}^{d} \int_0^T \pi_s^i d(S_s^i/S_s^0) .$$

From (2.1.3) the discounted value at time t of this strategy is given by

$$V_t^\pi / S_t^0 = h + \sum_{i=1}^{d} \int_0^t \pi_s^i dS_s^{i,0} .$$

We shall say that V_0^π is the initial value of H, and that π is the **hedging portfolio**. Note that the discounted price process $V^{\pi,0}$ is a $\mathbb{Q}$-local martingale under any e.m.m. $\mathbb{Q}$.

To give precise meaning the notion of market completeness, one needs to take care with the measurability conditions. The filtration to take into account is, in the case of a deterministic interest rate, the filtration generated by the traded assets.

Definition 2.1.5.2 *Assume that r is deterministic and let $\mathbf{F}^S$ be the natural filtration of the prices. The market is said to be* ***complete*** *if any contingent claim $H \in L^2(\mathcal{F}_T^S)$ is the value at time T of some self-financing strategy π.*

If r is stochastic, the standard attitude is to work with the filtration generated by the discounted prices.

Comments 2.1.5.3 (a) We emphasize that the definition of market completeness depends strongly on the choice of measurability of the contingent claims (see ↣ Subsection 2.3.6) and on the regularity conditions on strategies (see below).

(b) It may be that the market is complete, but there exists no e.m.m. As an example, let us assume that a riskless asset S^0 and two risky assets with dynamics

$$dS_t^i = S_t^i(b_i dt + \sigma dB_t), \; i = 1, 2$$

are traded. Here, B is a one-dimensional Brownian motion, and $b_1 \neq b_2$. Obviously, there does not exist an e.m.m., so arbitrage opportunities exist, however, the market is complete. Indeed, any contingent claim H can be written as a stochastic integral with respect to S^1/S^0 (the market with the two assets S^0, S^1 is complete).

(c) In a model where $dS_t = S_t(b_t dt + \sigma dB_t)$, where b is $\mathbf{F}^B$-adapted, the value of the trend b has no influence on the valuation of hedgeable contingent claims. However, if b is a process adapted to a filtration bigger than the filtration $\mathbf{F}^B$, there may exist many e.m.m.. In that case, one has to write the dynamics of S in its natural filtration, using filtering results (see ↣ Section 5.10). See, for example, Pham and Quenez [711].

Theorem 2.1.5.4 *Let $\tilde{S}$ be a process which represents the discounted prices. If there exists a unique e.m.m. $\mathbb{Q}$ such that $\tilde{S}$ is a $\mathbb{Q}$-local martingale, then the market is complete and arbitrage free.*

PROOF: This result is obtained from the fact that if there is a unique probability measure such that $\tilde{S}$ is a local martingale, then the process $\tilde{S}$ has the representation property. See Jacod and Yor [472] for a proof or $\rightarrowtail$ Subsection 9.5.3. □

Theorem 2.1.5.5 *In an arbitrage free and complete market, the time-t price of a (bounded) contingent claim H is*

$$\boxed{V_t^H = R_t^{-1}\mathbb{E}_{\mathbb{Q}}(R_T H|\mathcal{F}_t)} \tag{2.1.4}$$

where $\mathbb{Q}$ is the unique e.m.m. and R the discount factor.

PROOF: In a complete market, using the predictable representation theorem, there exists π such that $HR_T = h + \sum_{i=1}^d \int_0^T \pi_s dS_s^{i,0}$, and $S^{i,0}$ is a $\mathbb{Q}$-martingale. Hence, the result follows. □

Working with the historical probability yields that the process Z defined by $Z_t = L_t R_t V_t^H$, where L is the Radon-Nikodým density, is a $\mathbb{P}$-martingale; therefore we also obtain the price V_t^H of the contingent claim H as

$$V_t^H R_t L_t = \mathbb{E}_{\mathbb{P}}(L_T R_T H|\mathcal{F}_t)\,. \tag{2.1.5}$$

Remark 2.1.5.6 Note that, in an incomplete market, if H is hedgeable, then the time-t value of the replicating portfolio is $V_t^H = R_t^{-1}\mathbb{E}_{\mathbb{Q}}(R_T H|\mathcal{F}_t)$, for any e.m.m. $\mathbb{Q}$.

2.2 A Diffusion Model

In this section, we make precise the dynamics of the assets as Itô processes, we study the market completeness and, in a Markovian setting, we present the PDE approach.

Let $(\Omega, \mathcal{F}, \mathbb{P})$ be a probability space. We assume that an n-dimensional Brownian motion B is constructed on this space and we denote by $\mathbf{F}$ its natural filtration. We assume that the dynamics of the assets of the financial market are as follows: the dynamics of the savings account are

$$dS_t^0 = r_t S_t^0 dt,\ S_0^0 = 1\,, \tag{2.2.1}$$

and the vector valued process $(S^i, 1 \leq i \leq d)$ consisting of the prices of d risky assets is a d-dimensional diffusion which follows the dynamics

$$dS_t^i = S_t^i(b_t^i dt + \sum_{j=1}^n \sigma_t^{i,j} dB_t^j), \tag{2.2.2}$$

where r, b^i, and the volatility coefficients $\sigma^{i,j}$ are supposed to be given $\mathbf{F}$-predictable processes, and satisfy for any t, almost surely,

$$r_t > 0, \int_0^t r_s ds < \infty, \quad \int_0^t |b_s^i| ds < \infty, \quad \int_0^t (\sigma_s^{i,j})^2 ds < \infty.$$

The solution of (2.2.2) is

$$S_t^i = S_0^i \exp\left(\int_0^t b_s^i ds + \sum_{j=1}^n \int_0^t \sigma_s^{i,j} dB_s^j - \frac{1}{2}\sum_{j=1}^n \int_0^t (\sigma_s^{i,j})^2 ds\right).$$

In particular, the prices of the assets are strictly positive. As usual, we denote by

$$R_t = \exp\left(-\int_0^t r_s ds\right) = 1/S_t^0$$

the discount factor. We also denote by $S^{i,0} = S^i/S^0$ the discounted prices and $V^0 = V/S^0$ the discounted value of V.

2.2.1 Absence of Arbitrage

Proposition 2.2.1.1 *In the model* (2.2.1–2.2.2), *the existence of an e.m.m. implies absence of arbitrage.*

PROOF: Let π be an admissible self-financing strategy, and assume that $\mathbb{Q}$ is an e.m.m. Then,

$$dS_t^{i,0} = R_t\,(dS_t^i - r_t S_t^i dt) = S_t^{i,0} \sum_{j=1}^n \sigma_t^{i,j} dW_t^j$$

where W is a $\mathbb{Q}$-Brownian motion. Then, the process $V^{\pi,0}$ is a $\mathbb{Q}$-local martingale which is bounded below (admissibility assumption), and therefore, it is a supermartingale, and $V_0^{\pi,0} \geq \mathbb{E}_{\mathbb{Q}}(V_T^{\pi,0})$. Therefore, $V_T^{\pi,0} \geq 0$ implies that the terminal value is null: there are no arbitrage opportunities. □

2.2.2 Completeness of the Market

In the model (2.2.1, 2.2.2) when $d = n$ (i.e., the number of risky assets equals the number of driving BM), and when σ is invertible, the e.m.m. exists and is unique as long as some regularity is imposed on the coefficients. More precisely, we require that we can apply Girsanov's transformation in such a way that the d-dimensional process W where

$$dW_t = dB_t + \sigma_t^{-1}(b_t - r_t\mathbf{1})dt = dB_t + \theta_t dt\,,$$

is a $\mathbb{Q}$-Brownian motion. In other words, we assume that the solution L of

$$dL_t = -L_t\sigma_t^{-1}(b_t - r_t\mathbf{1})dB_t = -L_t\theta_t dB_t, \quad L_0 = 1$$

is a martingale (this is the case if θ is bounded). The process

$$\boxed{\theta_t = \sigma_t^{-1}(b_t - r_t\mathbf{1})}$$

is called the **risk premium**[1]. Then, we obtain

$$dS_t^{i,0} = S_t^{i,0}\sum_{j=1}^{d}\sigma_t^{i,j}dW_t^j\,.$$

We can apply the predictable representation property under the probability $\mathbb{Q}$ and find for any $H \in L^2(\mathcal{F}_T)$ a d-dimensional predictable process $(h_t, t \leq T)$ with $\mathbb{E}_{\mathbb{Q}}(\int_0^T |h_s|^2 ds) < \infty$ and

$$HR_T = \mathbb{E}_{\mathbb{Q}}(HR_T) + \int_0^T h_s\, dW_s\,.$$

Therefore,

$$HR_T = \mathbb{E}_{\mathbb{Q}}(HR_T) + \sum_{i=1}^{d}\int_0^T \pi_s^i\, dS_s^{i,0}$$

where π satisfies $\sum_{i=1}^d \pi_s^i S_s^{i,0}\sigma_s^{i,j} = h_s^j$. Hence, the market is complete, the price of H is $\mathbb{E}_{\mathbb{Q}}(HR_T)$, and the hedging portfolio is $(V_t - \pi_t S_t, \pi_t)$ where the time-t discounted value of the portfolio is given by

$$V_t^0 = R_t^{-1}\mathbb{E}_{\mathbb{Q}}(HR_T|\mathcal{F}_t) = \mathbb{E}_{\mathbb{Q}}(HR_T) + \int_0^t R_s\pi_s(dS_s - r_sS_sds)\,.$$

Remark 2.2.2.1 In the case $d < n$, the market is generally incomplete and does not present arbitrage opportunities. In some specific cases, it can be reduced to a complete market as in the $\rightarrowtail$ Example 2.3.6.1.

In the case $n < d$, the market generally presents arbitrage opportunities, as shown in Comments 2.1.5.3, but is complete.

[1] In the one-dimensional case, σ is, in finance, a positive process. Roughly speaking, the investor is willing to invest in the risky asset only if $b > r$, i.e., if he will get a positive "premium."

2.2.3 PDE Evaluation of Contingent Claims in a Complete Market

In the particular case where $H = h(S_T)$, r is deterministic, h is bounded, and S is an inhomogeneous diffusion

$$dS_t = DS_t(b(t, S_t)dt + \Sigma(t, S_t)dB_t)\,,$$

where DS is the diagonal matrix with S^i on the diagonal, we deduce from the Markov property of S under $\mathbb{Q}$ that there exists a function $V(t, x)$ such that

$$\mathbb{E}_{\mathbb{Q}}(R(T)h(S_T)|\mathcal{F}_t) = R(t)V(t, S_t) = V^0(t, S_t)\,.$$

The process $(V^0(t, S_t),\ t \geq 0)$ is a martingale, hence its bounded variation part is equal to 0. Therefore, as soon as V is smooth enough (see Karatzas and Shreve [513] for conditions which ensure this regularity), Itô's formula leads to

$$\begin{aligned} V^0(t, S_t) &= V^0(0, S_0) + \sum_{i=1}^{d} \int_0^t \partial_{x_i} V^0(s, S_s)(dS_s^i - r(s)S_s^i ds) \\ &= V(0, S_0) + \sum_{i=1}^{d} \int_0^t \partial_{x_i} V(s, S_s) dS_s^{i,0}\,, \end{aligned}$$

where we have used the fact that

$$\partial_{x_i} V^0(t, x) = R(t)\,\partial_{x_i} V(t, x)\,.$$

We now compare with (2.1.1)

$$V^0(t, S_t) = \mathbb{E}_{\mathbb{Q}}(HR(T)) + \sum_{i=1}^{d} \int_0^t \pi_s^i dS_s^{i,0}$$

and we obtain that $\pi_s^i = \partial_{x_i} V(s, S_s)$.

Proposition 2.2.3.1 *Let*

$$dS_t^i = S_t^i\,(r(t)dt + \sum_{j=1}^{d} \sigma_{i,j}(t, S_t)dB_t^j)\,,$$

be the risk-neutral dynamics of the d risky assets where the interest rate is deterministic. Assume that V solves the PDE, for $t < T$ and $x_i > 0, \forall i$,

$$\boxed{\partial_t V + r(t) \textstyle\sum_{i=1}^d x_i \partial_{x_i} V + \dfrac{1}{2} \displaystyle\sum_{i,j} x_i x_j \partial_{x_i x_j} V \sum_{k=1}^{d} \sigma_{i,k}\sigma_{j,k} = r(t)V} \tag{2.2.3}$$

with terminal condition $V(T, x) = h(x)$. Then, the value at time t of the contingent claim $H = h(S_T)$ is equal to $V(t, S_t)$.

The hedging portfolio is $\pi_t^i = \partial_{x_i} V(t, S_t), i = 1, \ldots, d$.

In the one-dimensional case, when $dS_t = S_t(b(t, S_t)dt + \sigma(t, S_t)dB_t)$, the PDE reads, for $x > 0, t \in [0, T[$,

$$\boxed{\partial_t V(t,x) + r(t)x\partial_x V(t,x) + \frac{1}{2}\sigma^2(t,x)x^2\partial_{xx}V(t,x) = r(t)V(t,x)} \tag{2.2.4}$$

with the terminal condition $V(T, x) = h(x)$.

Definition 2.2.3.2 *Solving the equation (2.2.4) with the terminal condition is called the Partial Derivative Equation (PDE) evaluation procedure.*

In the case when the contingent claim H is path-dependent (i.e., when the payoff $H = h(S_t, t \leq T)$ depends on the past of the price process, and not only on the terminal value), it is not always possible to associate a PDE to the pricing problem (see, e.g., Parisian options (see ↣ Section 4.4) and Asian options (see ↣ Section 6.6)).

Thus, we have two ways of computing the price of a contingent claim of the form $h(S_T)$, either we solve the PDE, or we compute the conditional expectation (2.1.5). The quantity RL is often called **the state-price density** or the pricing kernel. Therefore, in a complete market, we can characterize the processes which represent the value of a self-financing strategy.

Proposition 2.2.3.3 *If a given process V is such that VR is a $\mathbb{Q}$-martingale (or VRL is a $\mathbb{P}$-martingale), it defines the value of a self-financing strategy.*

In particular, the process $(N_t = 1/(R_t L_t), t \geq 0)$ is the value of a portfolio (NRL is a $\mathbb{P}$-martingale), called the **numéraire portfolio** or the growth optimal portfolio. It satisfies

$$dN_t = N_t((r(t) + \theta_t^2)dt + \theta_t dB_t)\,.$$

(See Becherer [63], Long [603], Karatzas and Kardaras [511] and the book of Heath and Platen [429] for a study of the numéraire portfolio.) It is a main tool for consumption-investment optimization theory, for which we refer the reader to the books of Karatzas [510], Karatzas and Shreve [514], and Korn [538].

2.3 The Black and Scholes Model

We now focus on the well-known Black and Scholes model, which is a very particular and important case of the diffusion model.

2.3.1 The Model

The **Black and Scholes model** [105] (see also Merton [641]) assumes that there is a riskless asset with interest rate r and that the dynamics of the price of the underlying asset are

$$dS_t = S_t(bdt + \sigma dB_t)$$

under the historical probability $\mathbb{P}$. Here, the risk-free rate r, the trend b and the volatility σ are supposed to be constant (note that, for valuation purposes, b may be an $\mathbf{F}$-adapted process). In other words, the value at time t of the risky asset is

$$S_t = S_0 \exp\left(bt + \sigma B_t - \frac{\sigma^2}{2}t\right).$$

From now on, we fix a finite horizon T and our processes are only indexed by $[0, T]$.

Notation 2.3.1.1 In the sequel, for two semi-martingales X and Y, we shall use the notation $X \overset{\text{mart}}{=} Y$ (or $dX_t \overset{\text{mart}}{=} dY_t$) to mean that $X - Y$ is a local martingale.

Proposition 2.3.1.2 *In the Black and Scholes model, there exists a unique e.m.m.* $\mathbb{Q}$, *precisely* $\mathbb{Q}|_{\mathcal{F}_t} = \exp(-\theta B_t - \frac{1}{2}\theta^2 t)\mathbb{P}|_{\mathcal{F}_t}$ *where* $\theta = \frac{b-r}{\sigma}$ *is the risk-premium. The risk-neutral dynamics of the asset are*

$$dS_t = S_t(rdt + \sigma dW_t)$$

where W *is a* $\mathbb{Q}$*-Brownian motion.*

PROOF: If $\mathbb{Q}$ is equivalent to $\mathbb{P}$, there exists a strictly positive martingale L such that $\mathbb{Q}|_{\mathcal{F}_t} = L_t \mathbb{P}|_{\mathcal{F}_t}$. From the predictable representation property under $\mathbb{P}$, there exists a predictable ψ such that

$$dL_t = \psi_t dB_t = L_t \phi_t dB_t$$

where $\phi_t L_t = \psi_t$. It follows that

$$d(LRS)_t \overset{\text{mart}}{=} (LRS)_t(b - r + \phi_t \sigma)dt\,.$$

Hence, in order for $\mathbb{Q}$ to be an e.m.m., or equivalently for LRS to be a $\mathbb{P}$-local martingale, there is one and only one process ϕ such that the bounded variation part of LRS is null, that is

$$\phi_t = \frac{r - b}{\sigma} = -\theta\,,$$

where θ is the risk premium. Therefore, the unique e.m.m. has a Radon-Nikodým density L which satisfies $dL_t = -L_t\theta dB_t, L_0 = 1$ and is given by $L_t = \exp(-\theta B_t - \frac{1}{2}\theta^2 t)$.

Hence, from Girsanov's theorem, $W_t = B_t + \theta t$ is a $\mathbb{Q}$-Brownian motion, and

$$dS_t = S_t(bdt + \sigma dB_t) = S_t(rdt + \sigma(dB_t + \theta dt)) = S_t(rdt + \sigma dW_t).$$

□

In a closed form, we have

$$S_t = S_0 \exp\left(bt + \sigma B_t - \frac{\sigma^2}{2}t\right) = S_0 e^{rt} \exp\left(\sigma W_t - \frac{\sigma^2}{2}t\right) = S_0 e^{\sigma X_t}$$

with $X_t = \nu t + W_t$, and $\nu = \frac{r}{\sigma} - \frac{\sigma}{2}$.

In order to price a contingent claim $h(S_T)$, we compute the expectation of its discounted value under the e.m.m.. This can be done easily, since $\mathbb{E}_\mathbb{Q}(h(S_T)e^{-rT}) = e^{-rT}\,\mathbb{E}_\mathbb{Q}(h(S_T))$ and

$$\mathbb{E}_\mathbb{Q}(h(S_T)) = \mathbb{E}\left[h(S_0 e^{rT - \frac{\sigma^2}{2}T} \exp(\sigma\sqrt{T}\,G))\right]$$

where G is a standard Gaussian variable.

We can also think about the expression $\mathbb{E}_\mathbb{Q}(h(S_T)) = \mathbb{E}_\mathbb{Q}(h(xe^{\sigma X_T}))$ as a computation for the drifted Brownian motion $X_t = \nu t + W_t$. As an exercise on Girsanov's transformation, let us show how we can reduce the computation to the case of a standard Brownian motion. The process X is a Brownian motion under $\mathbb{Q}^*$, defined on $\mathcal{F}_T$ as

$$\mathbb{Q}^* = \exp\left(-\nu W_T - \frac{1}{2}\nu^2 T\right)\mathbb{Q} = \zeta_T \mathbb{Q}.$$

Therefore,

$$\mathbb{E}_\mathbb{Q}(h(xe^{\sigma X_T})) = \mathbb{E}_{\mathbb{Q}^*}(\zeta_T^{(-1)} h(xe^{\sigma X_T})).$$

From

$$\zeta_T^{(-1)} = \exp\left(\nu W_T + \frac{1}{2}\nu^2 T\right) = \exp\left(\nu X_T - \frac{1}{2}\nu^2 T\right),$$

we obtain

$$\mathbb{E}_\mathbb{Q}\left(h(xe^{\sigma X_T})\right) = \exp\left(-\frac{1}{2}\nu^2 T\right)\mathbb{E}_{\mathbb{Q}^*}(\exp(\nu X_T) h(xe^{\sigma X_T})), \tag{2.3.1}$$

where on the left-hand side, X is a $\mathbb{Q}$-Brownian motion with drift ν and on the right-hand side, X is a $\mathbb{Q}^*$-Brownian motion. We can and do write the quantity on the right-hand side as $\exp(-\frac{1}{2}\nu^2 T)\,\mathbb{E}(\exp(\nu W_T)h(xe^{\sigma W_T}))$, where W is a generic Brownian motion.

We can proceed in a more powerful way using Cameron-Martin's theorem, i.e., the absolute continuity relationship between a Brownian motion with drift and a Brownian motion. Indeed, as in Exercise 1.7.5.5

$$\mathbb{E}_{\mathbb{Q}}(h(xe^{\sigma X_T})) = \mathbf{W}^{(\nu)}(h(xe^{\sigma X_T})) = \mathbb{E}\left(e^{\nu W_T - \frac{\nu^2}{2}T} h(xe^{\sigma W_T})\right) \tag{2.3.2}$$

which is exactly (2.3.1).

Proposition 2.3.1.3 *Let us consider the Black and Scholes framework*

$$dS_t = S_t(rdt + \sigma dW_t),\ S_0 = x$$

where W is a $\mathbb{Q}$-Brownian motion and $\mathbb{Q}$ is the e.m.m. or risk-neutral probability. In that setting, the value of the contingent claim $h(S_T)$ is

$$\mathbb{E}_{\mathbb{Q}}(e^{-rT}h(S_T)) = e^{-(r+\frac{\nu^2}{2})T}\mathbf{W}\left(e^{\nu X_T}h(xe^{\sigma X_T})\right)$$

where $\nu = \frac{r}{\sigma} - \frac{\sigma}{2}$ and X is a Brownian motion under $\mathbf{W}$.

The time-t value of the contingent claim $h(S_T)$ is

$$\mathbb{E}_{\mathbb{Q}}(e^{-r(T-t)}h(S_T)|\mathcal{F}_t) = e^{-(r+\frac{\nu^2}{2})(T-t)}\mathbf{W}\left(e^{\nu X_{T-t}}h(ze^{\sigma X_{T-t}})\right)|_{z=S_t}\,.$$

The value of a path-dependent contingent claim $\Phi(S_t, t \le T)$ is

$$\mathbb{E}_{\mathbb{Q}}(e^{-rT}\Phi(S_t, t \le T)) = e^{-(r+\frac{\nu^2}{2})T}\mathbf{W}\left(e^{\nu X_T}\Phi(xe^{\sigma X_t}, t \le T)\right).$$

PROOF: It remains to establish the formula for the time-t value. From

$$\mathbb{E}_{\mathbb{Q}}(e^{-r(T-t)}h(S_T)|\mathcal{F}_t) = \mathbb{E}_{\mathbb{Q}}(e^{-r(T-t)}h(S_t S_T^t)|\mathcal{F}_t)$$

where $S_T^t = S_T/S_t$, using the independence between S_T^t and $\mathcal{F}_t$ and the equality $S_T^t \stackrel{\text{law}}{=} S_{T-t}^1$, where S^1 has the same dynamics as S, with initial value 1, we get

$$\mathbb{E}_{\mathbb{Q}}(e^{-r(T-t)}h(S_T)|\mathcal{F}_t) = \Psi(S_t)$$

where

$$\Psi(x) = \mathbb{E}_{\mathbb{Q}}(e^{-r(T-t)}h(xS_T^t)) = \mathbb{E}_{\mathbb{Q}}(e^{-r(T-t)}h(xS_{T-t}))\,.$$

This last quantity can be computed from the properties of BM. Indeed,

$$\mathbb{E}_{\mathbb{Q}}\left(h(S_T)|\mathcal{F}_t\right) = \frac{1}{\sqrt{2\pi}}\int_{\mathbb{R}} dy\, h\left(S_t e^{r(T-t)+\sigma\sqrt{T-t}y-\sigma^2(T-t)/2}\right)e^{-y^2/2}\,.$$

(See Example 1.5.4.7 if needed.) □

Notation 2.3.1.4 In the sequel, when working in the Black and Scholes framework, we shall use systematically the notation $\nu = \frac{r}{\sigma} - \frac{\sigma}{2}$ and the fact that for $t \geq s$ the r.v. $S_t^s = S_t/S_s$ is independent of S_s.

Exercise 2.3.1.5 The payoff of a power option is $h(S_T)$, where the function h is given by $h(x) = x^\beta(x-K)^+$. Prove that the payoff can be written as the difference of European payoffs on the underlying assets $S^{\beta+1}$ and S^β with strikes depending on K and β . ◁

Exercise 2.3.1.6 We consider a contingent claim with a terminal payoff $h(S_T)$ and a continuous payoff $(x_s, s \leq T)$, where x_s is paid at time s. Prove that the price of this claim is

$$V_t = \mathbb{E}_{\mathbb{Q}}(e^{-r(T-t)}h(S_T) + \int_t^T e^{-r(s-t)}x_s ds|\mathcal{F}_t)\,.$$

◁

Exercise 2.3.1.7 In a Black and Scholes framework, prove that the price at time t of the contingent claim $h(S_T)$ is

$$C_h(x, T-t) = e^{-r(T-t)}\mathbb{E}_{\mathbb{Q}}(h(S_T)|S_t = x) = e^{-r(T-t)}\mathbb{E}_{\mathbb{Q}}(h(S_T^{t,x}))$$

where $S_s^{t,x}$ is the solution of the SDE

$$dS_s^{t,x} = S_s^{t,x}(rds + \sigma dW_s),\; S_t^{t,x} = x$$

and the hedging strategy consists of holding $\partial_x C_h(S_t, T-t)$ shares of the underlying asset.

Assuming some regularity on h, and using the fact that $S_T^{t,x} \stackrel{\text{law}}{=} xe^{\sigma X_{T-t}}$, where X_{T-t} is a Gaussian r.v., prove that

$$\partial_x C_h(x, T-t) = \frac{1}{x}\mathbb{E}_{\mathbb{Q}}\left(h'(S_T^{t,x})S_T^{t,x}\right)e^{-r(T-t)}\,.$$

◁

2.3.2 European Call and Put Options

Among the various derivative products, the most popular are the European Call and Put Options, also called vanilla[2] options.

A **European call** is associated with some underlying asset, with price $(S_t, t \geq 0)$. At maturity (a given date T), the holder of a call receives $(S_T - K)^+$ where K is a fixed number, called the strike. The price of a call is the amount of money that the buyer of the call will pay at time 0 to the seller. The time-t price is the price of the call at time t, equal to $\mathbb{E}_{\mathbb{Q}}(e^{-r(T-t)}(S_T - K)^+|\mathcal{F}_t)$, or, due to the Markov property, $\mathbb{E}_{\mathbb{Q}}(e^{-r(T-t)}(S_T - K)^+|S_t)$. At maturity (a given date T), the holder of a European put receives $(K - S_T)^+$.

[2] To the best of our knowledge, the name "vanilla" (or "plain vanilla") was given to emphasize the standard form of these products, by reference to vanilla, a standard flavor for ice cream, or to plain vanilla, a standard font in printing.

Theorem 2.3.2.1 Black and Scholes formula.
Let $dS_t = S_t(bdt + \sigma dB_t)$ be the dynamics of the price of a risky asset and assume that the interest rate is a constant r. The value at time t of a European call with maturity T and strike K is $\mathcal{BS}(S_t, \sigma, t)$ where

$$\boxed{\begin{aligned}\mathcal{BS}(x,\sigma,t) := {} & x\mathcal{N}\left[d_1\left(\frac{x}{Ke^{-r(T-t)}}, T-t\right)\right] \\ & - Ke^{-r(T-t)}\mathcal{N}\left[d_2\left(\frac{x}{Ke^{-r(T-t)}}, T-t\right)\right]\end{aligned}} \tag{2.3.3}$$

where

$$d_1(y,u) = \frac{1}{\sqrt{\sigma^2 u}}\ln(y) + \frac{1}{2}\sqrt{\sigma^2 u}, \quad d_2(y,u) = d_1(y,u) - \sqrt{\sigma^2 u}\,,$$

where we have written $\sqrt{\sigma^2}$ so that the formula does not depend on the sign of σ.

PROOF: It suffices to solve the evaluation PDE (2.2.4) with terminal condition $C(x,T) = (x-K)^+$. Another method is to compute the conditional expectation, under the e.m.m., of the discounted terminal payoff, i.e., $\mathbb{E}_{\mathbb{Q}}(e^{-rT}(S_T-K)^+|\mathcal{F}_t)$. For $t=0$,

$$\mathbb{E}_{\mathbb{Q}}(e^{-rT}(S_T-K)^+) = \mathbb{E}_{\mathbb{Q}}(e^{-rT}S_T\mathbb{1}_{\{S_T\geq K\}}) - Ke^{-rT}\mathbb{Q}(S_T\geq K)\,.$$

Under $\mathbb{Q}$, $dS_t = S_t(rdt+\sigma dW_t)$ hence, $S_T \stackrel{\text{law}}{=} S_0 e^{rT-\sigma^2T/2}e^{\sigma\sqrt{T}G}$, where G is a standard Gaussian law, hence

$$\mathbb{Q}(S_T\geq K) = \mathcal{N}\left[d_2\left(\frac{x}{Ke^{-rT)}}, T\right)\right]\,.$$

The equality

$$\mathbb{E}_{\mathbb{Q}}(e^{-rT}S_T\mathbb{1}_{\{S_T\geq K\}}) = x\mathcal{N}\left(d_1\left(\frac{S_0}{Ke^{-rT}}, T\right)\right)$$

can be proved using the law of S_T, however, we shall give in $\rightarrowtail$ Subsection 2.4.1 a more pleasant method.

The computation of the price at time t is carried out using the Markov property. □

Let us emphasize that a pricing formula appears in Bachelier [39, 41] in the case where S is a drifted Brownian motion. The central idea in Black and Scholes' paper is the hedging strategy. Here, the hedging strategy for a call is to keep a long position of $\Delta(t,S_t) = \frac{\partial C}{\partial x}(S_t, T-t)$ in the underlying asset (and to have $C - \Delta S_t$ shares in the savings account). It is well known that this quantity is equal to $\mathcal{N}(d_1)$. This can be checked by a tedious differentiation of (2.3.3). One can also proceed as follows: as we shall see in $\rightarrowtail$ Comments 2.3.2.2

$$C(x,T-t) = \mathbb{E}_{\mathbb{Q}}(e^{-r(T-t)}(S_T - K)^+ | S_t = x) = \mathbb{E}_{\mathbb{Q}}(R_T^t(xS_T^t - K)^+)\,,$$

where $S_T^t = S_T/S_t$, so that $\Delta(t,x)$ can be obtained by a differentiation with respect to x under the expectation sign. Hence,

$$\Delta(t,x) = \mathbb{E}(R_T^t S_T^t \mathbb{1}_{\{xS_T^t \geq K\}}) = \mathcal{N}\left(d_1(S_t/(Ke^{-r(T-t)}), T-t)\right).$$

This quantity, called the "Delta" (see $\rightarrowtail$ Subsection 2.3.3) is positive and bounded by 1. The second derivative with respect to x (the "Gamma") is $\frac{1}{\sigma x \sqrt{T-t}} \mathcal{N}'(d_1)$, hence $C(x,T-t)$ is convex w.r.t. x.

Comment 2.3.2.2 It is remarkable that the PDE evaluation was obtained in the seminal paper of Black and Scholes [105] without the use of any e.m.m.. Let us give here the main arguments. In this paper, the objective is to replicate the risk-free asset with simultaneous positions in the contingent claim and in the underlying asset. Let (α, β) be a replicating portfolio and

$$V_t = \alpha_t C_t + \beta_t S_t$$

the value of this portfolio assumed to satisfy the self-financing condition, i.e.,

$$dV_t = \alpha_t dC_t + \beta_t dS_t$$

Then, assuming that C_t is a smooth function of time and underlying value, i.e., $C_t = C(S_t, t)$, by relying on Itô's lemma the differential of V is obtained:

$$dV_t = \alpha_t(\partial_x C dS_t + \partial_t C dt + \frac{1}{2}\sigma^2 S_t^2 \partial_{xx} C dt) + \beta_t dS_t\,,$$

where $\partial_t C$ (resp. $\partial_x C$) is the derivative of C with respect to the second variable (resp. the first variable) and where all the functions $C, \partial_x C, \ldots$ are evaluated at (S_t, t). From $\alpha_t = (V_t - \beta_t S_t)/C_t$, we obtain

$$\begin{aligned} dV_t &= ((V_t - \beta_t S_t)(C_t)^{-1} \partial_x C + \beta_t)\sigma S_t dB_t \qquad (2.3.4)\\ &+ \left(\frac{V_t - \beta_t S_t}{C_t}\left(\partial_t C + \frac{1}{2}\sigma^2 S_t^2 \partial_{xx} C + bS_t \partial_x C\right) + \beta_t S_t b\right) dt\,. \end{aligned}$$

If this replicating portfolio is risk-free, one has $dV_t = V_t r dt$: the martingale part on the right-hand side vanishes, which implies

$$\beta_t = (S_t \partial_x C - C_t)^{-1} V_t \partial_x C$$

and

$$\frac{V_t - \beta_t S_t}{C_t}\left(\partial_t C + \frac{1}{2}\sigma^2 S_t^2 \partial_{xx} C + S_t b \partial_x C\right) + \beta_t S_t b = rV_t\,. \tag{2.3.5}$$

Using the fact that

$$(V_t - \beta_t S_t)(C_t)^{-1}\partial_x C + \beta_t = 0$$

we obtain that the term which contains b, i.e.,

$$bS_t\left(\frac{V_t - \beta S_t}{C_t}\partial_x C + \beta_t\right)$$

vanishes. After simplifications, we obtain

$$\begin{aligned} rC &= \left(1 + \frac{S\partial_x C}{C - S\partial_x C}\right)\left(\partial_t C + \frac{1}{2}\sigma^2 x^2 \partial_{xx} C\right) \\ &= \frac{C}{C - S\partial_x C}\left(\partial_t C + \frac{1}{2}\sigma^2 x^2 \partial_{xx} C\right) \end{aligned}$$

and therefore the PDE evaluation

$$\begin{aligned} \partial_t C(x,t) + rx\partial_x C(x,t) + \frac{1}{2}\sigma^2 x^2 \partial_{xx} C(x,t) \\ = rC(x,t),\ x > 0, t \in [0,T[\end{aligned} \tag{2.3.6}$$

is obtained. Now,

$$\beta_t = V_t \partial_x C(S\partial_x C - C)^{-1} = V_0 \frac{\mathcal{N}(d_1)}{Ke^{-rT}\mathcal{N}(d_2)}\,.$$

Note that the hedging ratio is

$$\frac{\beta_t}{\alpha_t} = -\partial_x C(t, S_t)\,.$$

Reading carefully [105], it seems that the authors assume that there exists a self-financing strategy $(-1, \beta_t)$ such that $dV_t = rV_t dt$, which is not true; in particular, the portfolio $(-1, \mathcal{N}(d_1))$ is not self-financing and its value, equal to $-C_t + S_t\mathcal{N}(d_1) = Ke^{-r(T-t)}\mathcal{N}(d_2)$, is not the value of a risk-free portfolio.

Exercise 2.3.2.3 Robustness of the Black and Scholes formula. Let

$$dS_t = S_t(bdt + \sigma_t dB_t)$$

where $(\sigma_t, t \geq 0)$ is an adapted process such that for any t, $0 < a \leq \sigma_t \leq b$. Prove that

$$\forall t,\ \mathcal{BS}(S_t, a, t) \leq \mathbb{E}_{\mathbb{Q}}(e^{-r(T-t)}(S_T - K)^+ | \mathcal{F}_t) \leq \mathcal{BS}(S_t, b, t)\,.$$

Hint: This result is obtained by using the fact that the $\mathcal{BS}$ function is convex with respect to x. ◁

Comment 2.3.2.4 The result of the last exercise admits generalizations to other forms of payoffs as soon as the convexity property is preserved, and to the case where the volatility is a given process, not necessarily **F**-adapted. See El Karoui et al. [301], Avellaneda et al. [29] and Martini [625]. This convexity property holds for a d-dimensional price process only in the geometric Brownian motion case, see Ekström et al. [296]. See Mordecki [413] and Bergenthum and Rüschendorf [74], for bounds on option prices.

Exercise 2.3.2.5 Suppose that the dynamics of the risky asset are given by $dS_t = S_t(bdt + \sigma(t)dB_t)$, where σ is a deterministic function. Characterize the law of S_T under the risk-neutral probability $\mathbb{Q}$ and prove that the price of a European option on the underlying S, with maturity T and strike K, is $\mathcal{BS}(x, \Sigma(t), t)$ where $(\Sigma(t))^2 = \frac{1}{T-t}\int_t^T \sigma^2(s)ds$. $\triangleleft$

Exercise 2.3.2.6 Assume that, under $\mathbb{Q}$, S follows a Black and Scholes dynamics with $\sigma = 1, r = 0, S_0 = 1$. Prove that the function $t \to C(1,t;1) := \mathbb{E}_{\mathbb{Q}}((S_t - 1)^+)$ is a cumulative distribution function of some r.v. X; identify the law of X.
`Hint:` $\mathbb{E}_{\mathbb{Q}}((S_t - 1)^+) = \mathbb{Q}(4B_1^2 \leq t)$ where B is a $\mathbb{Q}$-BM. See Bentata and Yor [72] for more comments. $\triangleleft$

2.3.3 The Greeks

It is important for practitioners to have a good knowledge of the sensitivity of the price of an option with respect to the parameters of the model.

The **Delta** is the derivative of the price of a call with respect to the underlying asset price (the spot). In the BS model, the Delta of a call is $\mathcal{N}(d_1)$. The Delta of a portfolio is the derivative of the value of the portfolio with respect to the underlying price. A portfolio with zero Delta is said to be delta neutral. Delta hedging requires continuous monitoring and rebalancing of the hedge ratio.

The **Gamma** is the derivative of the Delta w.r.t. the underlying price. In the BS model, the Gamma of a call is $\mathcal{N}'(d_1)/S\sigma\sqrt{T-t}$. It follows that the BS price of a call option is a convex function of the spot. The Gamma is important because it makes precise how much hedging will cost in a small interval of time.

The **Vega** is the derivative of the option price w.r.t. the volatility. In the BS model, the Vega of a call is $\mathcal{N}'(d_1)S\sqrt{T-t}$.

2.3.4 General Case

Let us study the case where

$$dS_t = S_t(\alpha_t dt + \sigma_t dB_t)\,.$$

Here, B is a Brownian motion with natural filtration $\mathbf{F}$ and α and σ are bounded $\mathbf{F}$-predictable processes. Then,

$$S_t = S_0 \exp\left(\int_0^t \left(\alpha_s - \frac{\sigma_s^2}{2}\right) ds + \int_0^t \sigma_s dB_s\right)$$

and $\mathcal{F}_t^S \subset \mathcal{F}_t$. We assume that r is the constant risk-free interest rate and that $\sigma_t \geq \epsilon > 0$, hence the risk premium $\theta_t = \dfrac{\alpha_t - r}{\sigma_t}$ is bounded. It follows that the process

$$L_t = \exp\left(-\int_0^t \theta_s dB_s - \frac{1}{2}\int_0^t \theta_s^2 ds\right),\ t \leq T$$

is a uniformly integrable martingale. We denote by $\mathbb{Q}$ the probability measure satisfying $\mathbb{Q}|_{\mathcal{F}_t} = L_t \mathbb{P}|_{\mathcal{F}_t}$ and by W the Brownian part of the decomposition of the $\mathbb{Q}$-semi-martingale B, i.e., $W_t = B_t + \int_0^t \theta_s ds$. Hence, from integration by parts formula, $d(RS)_t = R_t S_t \sigma_t dW_t$.

Then, from the predictable representation property (see Section 1.6), for any square integrable $\mathcal{F}_T$-measurable random variable H, there exists an $\mathbf{F}$-predictable process ϕ such that $HR_T = \mathbb{E}_{\mathbb{Q}}(HR_T) + \int_0^T \phi_s dW_s$ and $\mathbb{E}(\int_0^T \phi_s^2 ds) < \infty$; therefore

$$HR_T = \mathbb{E}_{\mathbb{Q}}(HR_T) + \int_0^T \psi_s d(RS)_s$$

where $\psi_t = \phi_t/(R_t S_t \sigma_t)$. It follows that H is hedgeable with the self-financing portfolio $(V_t - \psi_t S_t, \psi_t)$ where

$$V_t = R_t^{-1}\mathbb{E}_{\mathbb{Q}}(HR_T|\mathcal{F}_t) = H_t^{-1}\mathbb{E}_{\mathbb{P}}(HH_T|\mathcal{F}_t)$$

with $H_t = R_t L_t$. The process H is called the deflator or the pricing kernel.

2.3.5 Dividend Paying Assets

In this section, we suppose that the owner of one share of the stock receives a dividend. Let S be the stock process. Assume in a first step that the stock pays dividends Δ_i at fixed increasing dates $T_i, i \leq n$ with $T_n \leq T$. The price of the stock at time 0 is the expectation under the risk-neutral probability $\mathbb{Q}$ of the discounted future payoffs, that is

$$S_0 = \mathbb{E}_{\mathbb{Q}}(S_T R_T + \sum_{i=1}^{n} \Delta_i R_{T_i}) \,.$$

We now assume that the dividends are paid in continuous time, and let D be the cumulative dividend process (that is D_t is the amount of dividends received between 0 and t). The discounted price of the stock is the risk-neutral expectation (one often speaks of risk-adjusted probability in the case of dividends) of the future dividends, that is

$$S_t R_t = \mathbb{E}_{\mathbb{Q}}\left(S_T R_T + \int_t^T R_s dD_s | \mathcal{F}_t \right) \,.$$

Note that the discounted price $R_t S_t$ is no longer a $\mathbb{Q}$-martingale. On the other hand, the discounted cum-dividend price[3]

$$S_t^{cum} R_t := S_t R_t + \int_0^t R_s dD_s$$

is a $\mathbb{Q}$-martingale. Note that $S_t^{cum} = S_t + \frac{1}{R_t}\int_0^t R_s dD_s$. If we assume that the reference filtration is a Brownian filtration, there exists σ such that

$$d(S_t^{cum} R_t) = \sigma_t S_t R_t dW_t,$$

and we obtain

$$d(S_t R_t) = -R_t dD_t + S_t R_t \sigma_t dW_t.$$

Suppose now that the asset S pays a proportional dividend, that is, the holder of one share of the asset receives $\delta S_t dt$ in the time interval $[t, t+dt]$. In that case, under the risk-adjusted probability $\mathbb{Q}$, the discounted value of an asset equals the expectation on the discounted future payoffs, i.e.,

$$R_t S_t = \mathbb{E}_{\mathbb{Q}}(R_T S_T + \delta \int_t^T R_s S_s ds | \mathcal{F}_t) \,.$$

Hence, the discounted cum-dividend process

$$R_t S_t + \int_0^t \delta R_s S_s ds$$

is a $\mathbb{Q}$-martingale so that the risk-neutral dynamics of the underlying asset are given by

$$dS_t = S_t \left((r - \delta)dt + \sigma dW_t \right) \,. \tag{2.3.7}$$

One can also notice that the process $(S_t R_t e^{\delta t}, t \geq 0)$ is a $\mathbb{Q}$-martingale.

---

[3] Nothing to do with scum!

If the underlying asset pays a proportional dividend, the self-financing condition takes the following form. Let

$$dS_t = S_t(b_t dt + \sigma_t dB_t)$$

be the historical dynamics of the asset which pays a dividend at rate δ. A trading strategy π is self-financing if the wealth process $V_t = \pi_t^0 S_t^0 + \pi_t^1 S_t$ satisfies

$$dV_t = \pi_t^0 dS_t^0 + \pi_t^1(dS_t + \delta S_t dt) = rV_t dt + \pi_t^1(dS_t + (\delta - r)S_t dt).$$

The term $\delta\pi_t^1 S_t$ makes precise the fact that the gain from the dividends is reinvested in the market. The process VR satisfies

$$d(V_t R_t) = R_t \pi_t^1(dS_t + (\delta - r)S_t dt) = R_t \pi_t^1 S_t \sigma dW_t$$

hence, it is a (local) $\mathbb{Q}$-martingale.

2.3.6 Rôle of Information

When dealing with completeness the choice of the filtration is very important; this is now discussed in the following examples:

Example 2.3.6.1 Toy Example. Assume that the riskless interest rate is a constant r and that the historical dynamics of the risky asset are given by

$$dS_t = S_t(bdt + \sigma_1 dB_t^1 + \sigma_2 dB_t^2)$$

where $(B^i, i = 1, 2)$ are two independent BMs and b a constant[4]. It is not possible to hedge every $\mathcal{F}_T^{B^1,B^2}$-measurable contingent claim with strategies involving only the riskless and the risky assets, hence the market consisting of the $\mathcal{F}_T^{B^1,B^2}$-measurable contingent claims is incomplete.

The set $\mathcal{Q}$ of e.m.m's is obtained via the family of Radon-Nikodým densities $dL_t = L_t(\psi_t dB_t^1 + \gamma_t dB_t^2)$ where the predictable processes ψ, γ satisfy $b + \psi_t\sigma_1 + \gamma_t\sigma_2 = r$. Thus, the set $\mathcal{Q}$ is infinite.

However, writing the dynamics of S as a semi-martingale in its own filtration leads to $dS_t = S_t(bdt + \sigma dB_t^3)$ where B^3 is a Brownian motion and $\sigma^2 = \sigma_1^2 + \sigma_2^2$. Note that $\mathcal{F}_t^{B^3} = \mathcal{F}_t^S$. It is now clear that any $\mathcal{F}_T^S$-measurable contingent claim can be hedged, and the market is $\mathbf{F}^S$-complete.

Example 2.3.6.2 More generally, a market where the riskless asset has a price given by (2.2.1) and where the d risky assets' prices follow

$$dS_t^i = S_t^i(b^i(t, S_t)dt + \sum_{j=1}^{n} \sigma^{i,j}(t, S_t)dB_t^j),\ S_0^i = x_i\,, \qquad (2.3.8)$$

[4] Of course, the superscript 2 is not a power!

where B is a n-dimensional BM, with $n > d$, can often be reduced to the case of an $\mathcal{F}_T^S$-complete market. Indeed, it may be possible, under some regularity assumptions on the matrix σ, to write the equation (2.3.8) as

$$dS_t^i = S_t^i(b^i(t, S_t)dt + \sum_{j=1}^{d} \widetilde{\sigma}^{i,j}(t, S_t)d\widetilde{B}_t^j),\ S_0^i = x_i\,,$$

where $\widetilde{B}$ is a d-dimensional Brownian motion. The concept of a strong solution for an SDE is useful here. See the book of Kallianpur and Karandikar [506] and the paper of Kallianpur and Xiong [507].

When

$$dS_t^i = S_t^i \left(b_t^i dt + \sum_{j=1}^{n} \sigma_t^{i,j} dB_t^j \right),\ S_0^i = x_i,\ i = 1, \dots, d, \tag{2.3.9}$$

and $n > d$, if the coefficients are adapted with respect to the Brownian filtration $\mathbf{F}^B$, then the market is generally incomplete, as was shown in Exercice 2.3.6.1 (for a general study, see Karatzas [510]). Roughly speaking, a market with a riskless asset and risky assets is complete if the number of sources of noise is equal to the number of risky assets.

An important case of an incomplete market (the stochastic volatility model) is when the coefficient σ is adapted to a filtration different from $\mathbf{F}^B$. (See ↣ Section 6.7 for a presentation of some stochastic volatility models.)

Let us briefly discuss the case $dS_t = S_t\sigma_t dW_t$. The square of the volatility can be written in terms of S and its bracket as $\sigma_t^2 = \frac{d\langle S\rangle_t}{S_t^2 dt}$ and is obviously $\mathbf{F}^S$-adapted. However, except in the particular case of regular local volatility, where $\sigma_t = \sigma(t, S_t)$, the filtration generated by S is not the filtration generated by a one-dimensional BM. For example, when $dS_t = S_t e^{B_t} dW_t$, where B is a BM independent of W, it is easy to prove that $\mathcal{F}_t^S = \mathcal{F}_t^W \vee \mathcal{F}_t^B$, and in the warning (1.4.1.6) we have established that the filtration generated by S is not generated by a one-dimensional Brownian motion and that S does not possess the predictable representation property.

2.4 Change of Numéraire

The value of a portfolio is expressed in terms of a monetary unit. In order to compare two numerical values of two different portfolios, one has to express these values in terms of the same **numéraire**. In the previous models, the numéraire was the savings account. We study some cases where a different choice of numéraire is helpful.

2.4.1 Change of Numéraire and Black-Scholes Formula

Definition 2.4.1.1 *A numéraire is any strictly positive price process. In particular, it is a semi-martingale.*

As we have seen, in a Black and Scholes model, the price of a European option is given by:

$$\begin{aligned}C(S_0,T) &= \mathbb{E}_{\mathbb{Q}}(e^{-rT}(S_T-K)\mathbb{1}_{\{S_T\geq K\}})\\ &= \mathbb{E}_{\mathbb{Q}}(e^{-rT}S_T\mathbb{1}_{\{S_T\geq K\}}) - e^{-rT}K\mathbb{Q}(S_T\geq K)\,.\end{aligned}$$

Hence, if

$$k = \frac{1}{\sigma}\left(\ln(K/x) - (r-\frac{1}{2}\sigma^2)T\right),$$

using the symmetry of the Gaussian law, one obtains

$$\mathbb{Q}(S_T\geq K) = \mathbb{Q}(W_T\geq k) = \mathbb{Q}(W_T\leq -k) = \mathcal{N}\left(d_2\left(\frac{x}{Ke^{-rT}},T\right)\right)$$

where the function d_2 is given in Theorem 2.3.2.1.

From the dynamics of S, one can write:

$$e^{-rT}\mathbb{E}_{\mathbb{Q}}(S_T\mathbb{1}_{\{S_T\geq K\}}) = S_0\mathbb{E}_{\mathbb{Q}}\left(\mathbb{1}_{\{W_T\geq k\}}\exp\left(-\frac{\sigma^2}{2}T+\sigma W_T\right)\right).$$

The process $(\exp(-\frac{\sigma^2}{2}t+\sigma W_t), t\geq 0)$ is a positive $\mathbb{Q}$-martingale with expectation equal to 1. Let us define the probability $\mathbb{Q}^*$ by its Radon-Nikodým derivative with respect to $\mathbb{Q}$:

$$\mathbb{Q}^*|_{\mathcal{F}_t} = \exp\left(-\frac{\sigma^2}{2}t+\sigma W_t\right)\mathbb{Q}|_{\mathcal{F}_t}\,.$$

Hence,

$$e^{-rT}\mathbb{E}_{\mathbb{Q}}(S_T\mathbb{1}_{\{S_T\geq K\}}) = S_0\mathbb{Q}^*(W_T\geq k)\,.$$

Girsanov's theorem implies that the process $(\widehat{W}_t = W_t - \sigma t, t\geq 0)$ is a $\mathbb{Q}^*$-Brownian motion. Therefore,

$$\begin{aligned}e^{-rT}\mathbb{E}_{\mathbb{Q}}(S_T\mathbb{1}_{\{S_T\geq K\}}) &= S_0\mathbb{Q}^*\left(W_T-\sigma T\geq k-\sigma T\right)\\ &= S_0\mathbb{Q}^*\left(\widehat{W}_T\leq -k+\sigma T\right),\end{aligned}$$

i.e.,

$$e^{-rT}\mathbb{E}_{\mathbb{Q}}(S_T\mathbb{1}_{\{S_T\geq K\}}) = S_0\mathcal{N}\left(d_1\left(\frac{x}{Ke^{-rT}},T\right)\right).$$

Note that this change of probability measure corresponds to the choice of $(S_t, t\geq 0)$ as numéraire (see $\rightarrowtail$ Subsection 2.4.3).

2.4.2 Self-financing Strategy and Change of Numéraire

If N is a numéraire (e.g., the price of a zero-coupon bond), we can evaluate any portfolio in terms of this numéraire. If V_t is the value of a portfolio, its value in the numéraire N is V_t/N_t. The choice of the numéraire does not change the fundamental properties of the market. We prove below that the set of self-financing portfolios does not depend on the choice of numéraire.

Proposition 2.4.2.1 *Let us assume that there are d assets in the market, with prices $(S_t^i;\, i = 1,\dots,d,\, t \geq 0)$ which are continuous semi-martingales with S^1 there to be strictly positive.(We do not require that there is a riskless asset.) We denote by $V_t^\pi = \sum_{i=1}^d \pi_t^i S_t^i$ the value at time t of the portfolio $\pi_t = (\pi_t^i, i = 1,\dots,d)$. If the portfolio $(\pi_t,\, t \geq 0)$ is self-financing, i.e., if $dV_t^\pi = \sum_{i=1}^d \pi_t^i dS_t^i$, then,choosing S_t^1 as a numéraire, and*

$$dV_t^{\pi,1} = \sum_{i=2}^d \pi_t^i dS_t^{i,1}$$

where $V_t^{\pi,1} = V_t^\pi / S_t^1$, $S_t^{i,1} = S_t^i / S_t^1$.

PROOF: We give the proof in the case $d = 2$ (for two assets). We note simply V (instead of V^π) the value of a self-financing portfolio $\pi = (\pi^1, \pi^2)$ in a market where the two assets $S^i, i = 1, 2$ (there is no savings account here) are traded. Then

$$\begin{aligned} dV_t &= \pi_t^1 dS_t^1 + \pi_t^2 dS_t^2 = (V_t - \pi_t^2 S_t^2) dS_t^1 / S_t^1 + \pi_t^2 dS_t^2 \\ &= (V_t^1 - \pi_t^2 S_t^{2,1}) dS_t^1 + \pi_t^2 dS_t^2 \,. \end{aligned} \tag{2.4.1}$$

On the other hand, from $V_t^1 S_t^1 = V_t$ one obtains

$$dV_t = V_t^1 dS_t^1 + S_t^1 dV_t^1 + d\langle S^1, V^1 \rangle_t \,, \tag{2.4.2}$$

hence,

$$\begin{aligned} dV_t^1 &= \frac{1}{S_t^1} \left(dV_t - V_t^1 dS_t^1 - d\langle S^1, V^1 \rangle_t \right) \\ &= \frac{1}{S_t^1} \left(\pi_t^2 dS_t^2 - \pi_t^2 S_t^{2,1} dS_t^1 - d\langle S^1, V^1 \rangle_t \right) \end{aligned}$$

where we have used (2.4.1) for the last equality. The equality $S_t^{2,1} S_t^1 = S_t^2$ implies

$$dS_t^2 - S_t^{2,1} dS_t^1 = S_t^1 dS_t^{2,1} + d\langle S^1, S^{2,1} \rangle_t$$

hence,

$$dV_t^1 = \pi_t^2 dS_t^{2,1} + \frac{\pi_t^2}{S_t^1} d\langle S^1, S^{2,1} \rangle_t - \frac{1}{S_t^1} d\langle S^1, V^1 \rangle_t \,.$$

This last equality implies that

$$\left(1+\frac{1}{S_t^1}\right)d\langle V^1, S^1\rangle_t = \pi_t^2\left(1+\frac{1}{S_t^1}\right)d\langle S^1, S^{2,1}\rangle_t$$

hence, $d\langle S^1, V^1\rangle_t = \pi_t^2 d\langle S^1, S^{2,1}\rangle_t$, hence it follows that $dV_t^1 = \pi_t^2 dS_t^{2,1}$. □

Comment 2.4.2.2 We refer to Benninga et al. [71], Duffie [270], El Karoui et al. [299], Jamshidian [478], and Schroder [773] for details and applications of the change of numéraire method. Change of numéraire has strong links with optimization theory, see Becherer [63] and Gourieroux et al. [401]. See also an application to hedgeable claims in a default risk setting in Bielecki et al. [89]. We shall present applications of change of numéraire in ↣ Subsection 2.7.1 and in the proof of symmetry relations (e.g., ↣ formula (3.6.1.1)).

2.4.3 Change of Numéraire and Change of Probability

We define a change of probability associated with any numéraire Z. The numéraire is a price process, hence the process $(Z_t R_t, t \geq 0)$ is a strictly positive $\mathbb{Q}$-martingale. Define $\mathbb{Q}^Z$ as $\mathbb{Q}^Z|_{\mathcal{F}_t} := (Z_t R_t)\mathbb{Q}|_{\mathcal{F}_t}$.

Proposition 2.4.3.1 *Let $(X_t, t \geq 0)$ be the dynamics of a price and Z a new numéraire. The price of X, in the numéraire Z: $(X_t/Z_t, 0 \leq t \leq T)$, is a $\mathbb{Q}^Z$-martingale.*

PROOF: If X is a price process, the discounted process $\widetilde{X}_t := X_t R_t$ is a $\mathbb{Q}$-martingale. Furthermore, from Proposition 1.7.1.1, it follows that X_t/Z_t is a $\mathbb{Q}^Z$-martingale if and only if $(X_t/Z_t)Z_t R_t = R_t X_t$ is a $\mathbb{Q}$-martingale. □

In particular, if the market is arbitrage-free, and if a riskless asset S^0 is traded, choosing this asset as a numéraire leads to the risk-neutral probability, under which X_t/S_t^0 is a martingale.

Comments 2.4.3.2 (a) If the numéraire is the numéraire portfolio, defined at the end of Subsection 2.2.3, i.e., $N_t = 1/R_t S_t$, then the risky assets are $\mathbb{Q}^N$-martingales.

(b) See ↣ Subsection 2.7.2 for another application of change of numéraire.

2.4.4 Forward Measure

A particular choice of numéraire is the zero-coupon bond of maturity T. Let $P(t,T)$ be the price at time t of a zero-coupon bond with maturity T. If the interest rate is deterministic, $P(t,T) = R_T/R_t$ and the computation of the value of a contingent claim X reduces to the computation of $P(t,T)\mathbb{E}_{\mathbb{Q}}(X|\mathcal{F}_t)$ where $\mathbb{Q}$ is the risk-neutral probability measure.

When the spot rate r is a stochastic process, $P(t,T) = (R_t)^{-1}\mathbb{E}_{\mathbb{Q}}(R_T|\mathcal{F}_t)$ where $\mathbb{Q}$ is the risk-neutral probability measure and the price of a contingent claim H is $(R_t)^{-1}\mathbb{E}_{\mathbb{Q}}(HR_T|\mathcal{F}_t)$. The computation of $\mathbb{E}_{\mathbb{Q}}(HR_T|\mathcal{F}_t)$ may be difficult and a change of numéraire may give some useful information. Obviously, the process

$$\zeta_t := \frac{1}{P(0,T)}\mathbb{E}_{\mathbb{Q}}(R_T|\mathcal{F}_t) = \frac{P(t,T)}{P(0,T)}R_t$$

is a strictly positive $\mathbb{Q}$-martingale with expectation equal to 1. Let us define the **forward measure** $\mathbb{Q}^T$ as the probability associated with the choice of the zero-coupon bond as a numéraire:

Definition 2.4.4.1 *Let $P(t,T)$ be the price at time t of a zero-coupon with maturity T. The T-forward measure is the probability $\mathbb{Q}^T$ defined on $\mathcal{F}_t$, for $t \leq T$, as*

$$\mathbb{Q}^T|_{\mathcal{F}_t} = \zeta_t \, \mathbb{Q}|_{\mathcal{F}_t}$$

where $\zeta_t = \dfrac{P(t,T)}{P(0,T)}R_t$.

Proposition 2.4.4.2 *Let $(X_t, t \geq 0)$ be the dynamics of a price. Then the forward price $(X_t/P(t,T), 0 \leq t \leq T)$ is a $\mathbb{Q}^T$-martingale.*
The price of a contingent claim H is

$$V_t^H = \mathbb{E}_{\mathbb{Q}}\left(H \exp\left(-\int_t^T r_s ds\right)|\mathcal{F}_t\right) = P(t,T)\mathbb{E}_{\mathbb{Q}^T}(H|\mathcal{F}_t)\,.$$

Remark 2.4.4.3 Obviously, if the spot rate r is deterministic, $\mathbb{Q}^T = \mathbb{Q}$ and the forward price is equal to the spot price.

Comment 2.4.4.4 A **forward contract** on H, made at time 0, is a contract that stipulates that its holder pays the deterministic amount K at the delivery date T and receives the stochastic amount H. Nothing is paid at time 0. The forward price of H is K, determined at time 0 as $K = \mathbb{E}_{\mathbb{Q}_T}(H)$. See Björk [102], Martellini et al. [624] and Musiela and Rutkowski [661] for various applications.

2.4.5 Self-financing Strategies: Constrained Strategies

We present a very particular case of hedging with strategies subject to a constraint. The change of numéraire technique is of great importance in characterizing such strategies. This result is useful when dealing with default risk (see Bielecki et al. [93]).

We assume that the $k \geq 3$ assets S^i traded in the market are continuous semi-martingales, and we assume that S^1 and S^k are strictly positive processes. We do not assume that there is a riskless asset (we can consider this case if we specify that $dS_t^1 = r_t S_t^1 dt$).

Let $\pi = (\pi^1, \pi^2, \dots, \pi^k)$ be a self-financing trading strategy satisfying the following constraint:

$$\sum_{i=\ell+1}^{k} \pi_t^i S_t^i = Z_t, \quad \forall\, t \in [0,T], \tag{2.4.3}$$

for some $1 \leq \ell \leq k-1$ and a predetermined, **F**-predictable process Z. Let $\Phi_\ell(Z)$ be the class of all self-financing trading strategies satisfying the condition (2.4.3). We denote by $S^{i,1} = S^i/S^1$ and $Z^1 = Z/S^1$ the prices and the value of the constraint in the numéraire S^1.

Proposition 2.4.5.1 *The relative time-t wealth $V_t^{\pi,1} = V_t^\pi (S_t^1)^{-1}$ of a strategy $\pi \in \Phi_\ell(Z)$ satisfies*

$$\begin{aligned} V_t^{\pi,1} = V_0^{\pi,1} &+ \sum_{i=2}^{\ell} \int_0^t \pi_u^i \, dS_u^{i,1} + \sum_{i=\ell+1}^{k-1} \int_0^t \pi_u^i \left(dS_u^{i,1} - \frac{S_u^{i,1}}{S_u^{k,1}} \, dS_u^{k,1} \right) \\ &+ \int_0^t \frac{Z_u^1}{S_u^{k,1}} \, dS_u^{k,1}. \end{aligned}$$

PROOF: Let us consider discounted values of price processes $S^1, S^2, \dots, S^k$, with S^1 taken as a numéraire asset. In the proof, for simplicity, we do not indicate the portfolio π as a superscript for the wealth. We have the numéraire invariance

$$V_t^1 = V_0^1 + \sum_{i=2}^{k} \int_0^t \pi_u^i \, dS_u^{i,1}. \tag{2.4.4}$$

The condition (2.4.3) implies that

$$\sum_{i=\ell+1}^{k} \pi_t^i S_t^{i,1} = Z_t^1,$$

and thus

$$\pi_t^k = (S_t^{k,1})^{-1} \Big(Z_t^1 - \sum_{i=\ell+1}^{k-1} \pi_t^i S_t^{i,1} \Big). \tag{2.4.5}$$

By inserting (2.4.5) into (2.4.4), we arrive at the desired formula. □

Let us take $Z = 0$, so that $\pi \in \Phi_\ell(0)$. Then the constraint condition becomes $\sum_{i=\ell+1}^{k} \pi_t^i S_t^i = 0$, and (2.4.4) reduces to

$$V_t^{\pi,1} = \sum_{i=2}^{\ell} \int_0^t \pi_s^i \, dS_s^{i,1} + \sum_{i=\ell+1}^{k-1} \int_0^t \pi_s^i \left(dS_s^{i,1} - \frac{S_s^{i,1}}{S_s^{k,1}} \, dS_s^{k,1} \right). \tag{2.4.6}$$

The following result provides a different representation for the (relative) wealth process in terms of correlations (see Bielecki et al. [92] for the case where Z is not null).

Lemma 2.4.5.2 *Let $\pi = (\pi^1, \pi^2, \dots, \pi^k)$ be a self-financing strategy in $\Phi_\ell(0)$. Assume that the processes S^1, S^k are strictly positive. Then the relative wealth process $V_t^{\pi,1} = V_t^\pi (S_t^1)^{-1}$ satisfies*

$$V_t^{\pi,1} = V_0^{\pi,1} + \sum_{i=2}^{\ell} \int_0^t \pi_u^i \, dS_u^{i,1} + \sum_{i=\ell+1}^{k-1} \int_0^t \widehat{\pi}_u^{i,k,1} \, d\widehat{S}_u^{i,k,1}, \quad \forall t \in [0,T],$$

where we denote

$$\widehat{\pi}_t^{i,k,1} = \pi_t^i (S_t^{1,k})^{-1} e^{\alpha_t^{i,k,1}}, \quad \widehat{S}_t^{i,k,1} = S_t^{i,k} e^{-\alpha_t^{i,k,1}}, \tag{2.4.7}$$

with $S_t^{i,k} = S_t^i (S_t^k)^{-1}$ and

$$\alpha_t^{i,k,1} = \langle \ln S^{i,k}, \ln S^{1,k} \rangle_t = \int_0^t (S_u^{i,k})^{-1} (S_u^{1,k})^{-1} \, d\langle S^{i,k}, S^{1,k} \rangle_u. \tag{2.4.8}$$

PROOF: Let us consider the relative values of all processes, with the price S^k chosen as a numéraire, and $V_t^k := V_t (S_t^k)^{-1} = \sum_{i=1}^k \pi_t^i S_t^{i,k}$ (we do not indicate the superscript π in the wealth). In view of the constraint we have that $V_t^k = \sum_{i=1}^\ell \pi_t^i S_t^{i,k}$. In addition, as in Proposition 2.4.2.1 we get

$$dV_t^k = \sum_{i=1}^{k-1} \pi_t^i \, dS_t^{i,k}.$$

Since $S_t^{i,k} (S_t^{1,k})^{-1} = S_t^{i,1}$ and $V_t^1 = V_t^k (S_t^{1,k})^{-1}$, using an argument analogous to that of the proof of Proposition 2.4.2.1, we obtain

$$V_t^1 = V_0^1 + \sum_{i=2}^{\ell} \int_0^t \pi_u^i \, dS_u^{i,1} + \sum_{i=\ell+1}^{k-1} \int_0^t \widehat{\pi}_u^{i,k,1} \, d\widehat{S}_u^{i,k,1}, \quad \forall t \in [0,T],$$

where the processes $\widehat{\pi}_t^{i,k,1}$, $\widehat{S}_t^{i,k,1}$ and $\alpha_t^{i,k,1}$ are given by (2.4.7)–(2.4.8). □

The result of Proposition 2.4.5.1 admits a converse.

Proposition 2.4.5.3 *Let an $\mathcal{F}_T$-measurable random variable H represent a contingent claim that settles at time T. Assume that there exist $\mathbf{F}$-predictable processes π^i, $i = 2, 3, \dots, k-1$ such that*

$$\begin{aligned} \frac{H}{S_T^1} &= x + \sum_{i=2}^{l} \int_0^T \pi_t^i \, dS_t^{i,1} \\ &+ \sum_{i=l+1}^{k-1} \int_0^T \pi_t^i \left(dS_t^{i,1} - \frac{S_t^{i,1}}{S_t^{k,1}} \, dS_t^{k,1} \right) + \int_0^T \frac{Z_t^1}{S_t^{k,1}} \, dS_t^{k,1}. \end{aligned}$$

Then there exist two $\mathbf{F}$*-predictable processes* π^1 *and* π^k *such that the strategy* $\pi = (\pi^1, \pi^2, \dots, \pi^k)$ *belongs to* $\Phi_\ell(Z)$ *and replicates* H*. The wealth process of* π *equals, for every* $t \in [0, T]$,

$$\frac{V_t^{\pi})}{S_t^1} = x + \sum_{i=2}^{l} \int_0^t \pi_u^i \, dS_u^{i,1}$$
$$+ \sum_{i=l+1}^{k-1} \int_0^t \pi_u^i \left(dS_u^{i,1} - \frac{S_u^{i,1}}{S_u^{k,1}} dS_u^{k,1} \right) + \int_0^t \frac{Z_u^1}{S_u^{k,1}} dS_u^{k,1}.$$

PROOF: The proof is left as an exercise. □

2.5 Feynman-Kac

In what follows, $\mathbb{E}_x$ is the expectation corresponding to the probability distribution of a Brownian motion W starting from x.

2.5.1 Feynman-Kac Formula

Theorem 2.5.1.1 *Let* $\alpha \in \mathbb{R}^+$ *and let* $k : \mathbb{R} \to \mathbb{R}^+$ *and* $g : \mathbb{R} \to \mathbb{R}$ *be continuous functions with* g *bounded. Then the function*

$$f(x) = \mathbb{E}_x \left[\int_0^\infty dt \, g(W_t) \exp\left(-\alpha t - \int_0^t k(W_s) ds \right) \right] \tag{2.5.1}$$

is piecewise C^2 *and satisfies*

$$(\alpha + k) f = \frac{1}{2} f'' + g \,. \tag{2.5.2}$$

PROOF: We refer to Karatzas and Shreve [513] p.271. □

Let us assume that f is a bounded solution of (2.5.2). Then, one can check that equality (2.5.1) is satisfied.

We give a few hints for this verification. Let us consider the increasing process Z defined by:

$$Z_t = \alpha t + \int_0^t k(W_s) ds \,.$$

By applying Itô's lemma to the process

$$U_t^\varphi := \varphi(W_t) e^{-Z_t} + \int_0^t g(W_s) e^{-Z_s} ds \,,$$

where φ is C^2, we obtain

$$dU_t^{\varphi} = \varphi'(W_t)e^{-Z_t}dW_t + \left(\frac{1}{2}\varphi''(W_t) - (\alpha + k(W_t))\,\varphi(W_t) + g(W_t)\right) e^{-Z_t}dt$$

Now let $\varphi = f$ where f is a bounded solution of (2.5.2). The process U^f is a local martingale:

$$dU_t^f = f'(W_t)e^{-Z_t}dW_t\,.$$

Since U^f is bounded, U^f is a uniformly integrable martingale, and

$$\mathbb{E}_x(U_\infty^f) = \mathbb{E}_x\left(\int_0^\infty g(W_s)e^{-Z_s}ds\right) = U_0^f = f(x)\,.$$

□

2.5.2 Occupation Time for a Brownian Motion

We now give Kac's proof of Lévy's arcsine law as an application of the Feynman-Kac formula:

Proposition 2.5.2.1 *The random variable* $A_t^+ := \int_0^t \mathbb{1}_{[0,\infty[}(W_s)ds$ *follows the arcsine law with parameter* t*:*

$$\mathbb{P}(A_t^+ \in ds) = \frac{ds}{\pi\sqrt{s(t-s)}}\mathbb{1}\{0 \le s < t\}\,.$$

PROOF: By applying Theorem 2.5.1.1 to $k(x) = \beta\mathbb{1}_{\{x\ge 0\}}$ and $g(x) = 1$, we obtain that for any $\alpha > 0$ and $\beta > 0$, the function f defined by:

$$f(x) := \mathbb{E}_x\left[\int_0^\infty dt \exp\left(-\alpha t - \beta\int_0^t \mathbb{1}_{[0,\infty[}(W_s)ds\right)\right] \tag{2.5.3}$$

solves the following differential equation:

$$\begin{cases} \alpha f(x) = \frac{1}{2}f''(x) - \beta f(x) + 1, & x \ge 0 \\ \alpha f(x) = \frac{1}{2}f''(x) + 1, & x \le 0 \end{cases}. \tag{2.5.4}$$

Bounded and continuous solutions of this differential equation are given by:

$$f(x) = \begin{cases} Ae^{-x\sqrt{2(\alpha+\beta)}} + \frac{1}{\alpha+\beta}, & x \ge 0 \\ Be^{x\sqrt{2\alpha}} + \frac{1}{\alpha}, & x \le 0 \end{cases}.$$

Relying on the continuity of f and f' at zero, we obtain the unique bounded C^2 solution of (2.5.4):

$$A = \frac{\sqrt{\alpha+\beta} - \sqrt{\alpha}}{(\alpha+\beta)\sqrt{\alpha}}\,,\; B = \frac{\sqrt{\alpha} - \sqrt{\alpha+\beta}}{\alpha\sqrt{\alpha+\beta}}\,.$$

The following equality holds:

$$f(0) = \int_0^\infty dt e^{-\alpha t} \mathbb{E}_0 \left[e^{-\beta A_t^+} \right] = \frac{1}{\sqrt{\alpha(\alpha+\beta)}} .$$

We can invert the Laplace transform using the identity

$$\int_0^\infty dt e^{-\alpha t} \left(\int_0^t du \frac{e^{-\beta u}}{\pi \sqrt{u(t-u)}} \right) = \frac{1}{\sqrt{\alpha(\alpha+\beta)}} ,$$

and the density of A_t^+ is obtained:

$$\mathbb{P}(A_t^+ \in ds) = \frac{ds}{\pi \sqrt{s(t-s)}} \mathbb{1}_{\{s<t\}} .$$

Therefore, the law of A_t^+ is the arcsine law on $[0,t]$, and its distribution function is, for $s \in [0,t]$:

$$\mathbb{P}(A_t^+ \leq s) = \frac{2}{\pi} \arcsin \sqrt{\frac{s}{t}} .$$

Note that, by scaling, $A_t^+ \stackrel{\text{law}}{=} t A_1^+$. □

Comment 2.5.2.2 This result is due to Lévy [584] and a different proof was given by Kac [502]. Intensive studies for the more general case $\int_0^t f(W_s) ds$ have been made in the literature. Biane and Yor [86] and Jeanblanc et al. [483] present a study of the laws of these random variables for particular functions f, using excursion theory and the Ray-Knight theorem for Brownian local times at an exponential time.

2.5.3 Occupation Time for a Drifted Brownian Motion

The same method can be applied in order to compute the density of the occupation times above and below a level $L > 0$ up to time t for a Brownian motion with drift ν, i.e.,

$$A_t^{+,L,\nu} = \int_0^t ds \mathbb{1}_{\{X_s > L\}}, \quad A_t^{-,L,\nu} = \int_0^t ds \mathbb{1}_{\{X_s < L\}}$$

where $X_t = \nu t + W_t$. We start with the computation of Ψ where

$$\Psi(\alpha, \beta) := \mathbf{W}_0^{(\nu)} \left(\int_0^\infty dt \exp \left(-\alpha t - \beta \int_0^t ds \mathbb{1}_{\{X_s < 0\}} \right) \right) .$$

From the Feynman-Kac result, Ψ is the unique bounded solution of the equation (see Akahori [1] for details)

$$-\frac{1}{2}f'' - \nu f' + \alpha f + \beta \mathbb{1}_{\{x<0\}} f = 1 .$$

Hence,

$$\begin{aligned}\Psi(\alpha,\beta) &= \frac{\nu}{2\alpha}\frac{\sqrt{\nu^2+2(\alpha+\beta)}}{\alpha+\beta} - \frac{\nu}{2(\alpha+\beta)}\frac{\sqrt{\nu^2+2\alpha}}{\alpha} \\ &+\frac{1}{2}\frac{\sqrt{\nu^2+2(\alpha+\beta)}}{\alpha+\beta}\frac{\sqrt{\nu^2+2\alpha}}{\alpha} - \frac{\nu^2}{2}\frac{1}{\alpha(\alpha+\beta)} .\end{aligned}$$

Inverting the Laplace transform, we get

$$\begin{aligned}\mathbb{P}(A_t^{-,0,\nu} \in du)/du &= \left[\sqrt{\frac{2}{\pi u}} \exp\left(-\frac{\nu^2}{2}u\right) - 2\nu\, \Theta(\nu\sqrt{u})\right] \\ \times &\left[\nu + \frac{1}{\sqrt{2\pi(t-u)}} \exp\left(-\frac{\nu^2}{2}(t-u)\right) - \nu\, \Theta(\nu\sqrt{t-u})\right] \quad (2.5.5)\end{aligned}$$

where $\Theta(x) = \frac{1}{\sqrt{2\pi}}\int_x^\infty \exp(-\frac{y^2}{2})dy$. More generally, the law of $A_t^{-,L,\nu}$ for $L > 0$ is obtained from

$$\mathbb{P}(A_t^{-,L,\nu} \le u) = \int_0^u ds\, \varphi(s,L;\nu)\mathbb{P}(A_{t-s}^{-,0,\nu} < u-s)$$

where $\varphi(s,L;\nu)$ is the density $\mathbb{P}(T_L(X) \in ds)/ds$ (see ↣ (3.2.3) for its closed form). The law of $A_t^{+,L,\nu}$ follows from $A_t^{+,L,\nu} + A_t^{-,L,\nu} = t$.

The law of $A_t^{+,L,0}$ can also be obtained in a more direct way. It is easy to compute the double Laplace transform

$$\Psi(\alpha,\beta;L) := \int_0^\infty dt\, e^{-\alpha t}\mathbb{E}_0\left(e^{-\beta A_t^{+,L,0}}\right)$$

as follows: let, for $L > 0$, $T_L = \inf\{t : X_t = L\}$. Then,

$$\begin{aligned}\Psi(\alpha,\beta;L) &= \mathbb{E}_0\left(\int_0^{T_L} dt\, e^{-\alpha t} + \int_{T_L}^\infty dt\, e^{-\alpha t}\exp\left(-\beta\int_{T_L}^\infty ds\, \mathbb{1}_{\{W_s>L\}}\right)\right) \\ &= \frac{1}{\alpha}\,\mathbb{E}_0(1-e^{-\alpha T_L}) + \frac{1}{\sqrt{\alpha(\alpha+\beta)}}\,\mathbb{E}_0(e^{-\alpha T_L}) \\ &= \frac{1}{\alpha}\,(1-e^{-L\sqrt{2\alpha}}) + \frac{1}{\sqrt{\alpha(\alpha+\beta)}}\,e^{-L\sqrt{2\alpha}} .\end{aligned}$$

This quantity is the double Laplace transform of

$$f(t,u)du := \mathbb{P}(T_L > t)\,\delta_0(du) + \frac{1}{\sqrt{u(t-u)}}\,e^{-L^2/(2(t-u))}\mathbb{1}_{\{u<t\}}du ,$$

i.e.,

$$\Psi(\alpha,\beta;L) = \int_0^\infty\int_0^\infty e^{-\alpha t}e^{-\beta u}f(t,u)dtdu .$$

Comment 2.5.3.1 For a general presentation of Feynman-Kac formula, we refer to Durrett [286] and Karatzas and Shreve [513]. For extensions, see Chung [185], Evans [337], Fusai and Tagliani [371] and Pitman and Yor [718, 719]. Occupation time densities for CEV processes (see ↣ Section 6.4) are presented in Leung and Kwok [582].

2.5.4 Cumulative Options

Let S be a given process. The occupation time of S above (resp. below) a level L up to time t is the random variable $A_t^{+,L} := \int_0^t ds \mathbb{1}_{\{S_s \geq L\}}$ (resp. $A_t^{-,L} = \int_0^t ds \mathbb{1}_{\{S_s \leq L\}}$). An occupation time derivative is a contingent claim whose payoff depends on the terminal value of the underlying asset and on an occupation time. We are mainly interested in terminal payoff of the form $f(S_T, A_T^{-,L})$, (or $f(S_T, A_T^{+,L})$). In a **Black and Scholes model**, as given in Proposition 2.3.1.3, the price of such a claim is

$$\mathbb{E}_{\mathbb{Q}}(e^{-rT} f(S_T, A_T^{-,L})) = e^{-rT-\nu^2 T/2}\mathbb{E}\left(e^{\nu W_T} f(xe^{\sigma W_T}, A_T^{-,\ell}(W))\right)$$

where $\ell = \sigma^{-1}\ln(L/x)$.

We study the particular case $f(x,a) = (x-K)^+ e^{-\rho a}$, called a step option by Linetsky [590]. Let

$$C_{step}(x) = e^{-(r+\nu^2/2)T}\mathbb{E}\left(e^{\nu W_T}(xe^{\sigma W_T} - K)^+ e^{-\rho A_T^{-,\ell}}\right)$$

where W starts from 0. Setting $\gamma = r + \nu^2/2$, we obtain

$$\begin{aligned} C_{step}(x) &= e^{-\gamma T}\mathbb{E}_{-\ell}\left(e^{\nu(W_T+\ell)}(xe^{\sigma(W_T+\ell)} - K)^+ e^{-\rho A_T^{-,0}}\right) \\ &= e^{-\gamma T + \nu\ell}(xe^{\sigma\ell}\Psi(-\ell, \nu+\sigma) - K\Psi(-\ell,\nu)) \end{aligned}$$

where $\Psi(x,a) = \mathbb{E}_x\left(e^{aW_T}\mathbb{1}_{\{W_T \geq \frac{1}{\sigma}\ln(K/L)\}} e^{-\rho A_T^{-,0}}\right)$. The function Ψ can be computed from the joint law of $(A_T^{-,0}, W_T)$.

Proposition 2.5.4.1 *The density of the pair* $(A_t^{-,0}, W_t)$ *is*

$$\mathbb{P}(A_t^{-,0} \in du, W_t \in dx) = du\, dx \frac{|x|}{\sqrt{2\pi}} \int_u^t \frac{1}{\sqrt{s^3(t-s)^3}} e^{-x^2/(2(t-s))} ds \, \mathbb{1}_{\{u<t\}}.$$

PROOF: Let, for $a > 0, \rho > 0$,

$$f(t,x) = \mathbb{E}\left(\mathbb{1}_{[a,\infty[}(x+W_t)\exp\left(-\rho\int_0^t \mathbb{1}_{]-\infty,0]}(x+W_s)ds\right)\right).$$

From the Feynman-Kac theorem, the function f satisfies the PDE

$$\partial_t f = \frac{1}{2}\partial_{xx} f - \rho \mathbb{1}_{]-\infty,0]}(s) f, \quad f(0,x) = \mathbb{1}_{[a,\infty[}(x) .$$

Letting $\widehat{f}$ be the Laplace transform in time of f, i.e.,

$$\widehat{f}(\lambda, x) = \int_0^\infty e^{-\lambda t} f(x,t) dt ,$$

we obtain

$$-\mathbb{1}_{[a,\infty[}(x) + \lambda \widehat{f} = \frac{1}{2}\partial_{xx}\widehat{f} - \rho \mathbb{1}_{]-\infty,0]}(x)\widehat{f} .$$

Solving this ODE with the boundary conditions at 0 and a leads to

$$\widehat{f}(\lambda, 0) = \frac{\exp(-a\sqrt{2\lambda})}{\sqrt{\lambda}\left(\sqrt{\lambda} + \sqrt{\lambda+\rho}\right)} = \widehat{f_1}(\lambda)\widehat{f_2}(\lambda) , \tag{2.5.6}$$

with

$$\widehat{f_1}(\lambda) = \frac{1}{\sqrt{\lambda}\left(\sqrt{\lambda} + \sqrt{\lambda+\rho}\right)}, \quad \widehat{f_2}(\lambda) = \exp(-a\sqrt{2\lambda}) .$$

Then, one gets

$$-\partial_a \widehat{f}(\lambda, 0) = \sqrt{2}\frac{\exp(-a\sqrt{2\lambda})}{\sqrt{\lambda} + \sqrt{\lambda+\rho}} .$$

The right-hand side of (2.5.6) may be recognized as the product of the Laplace transforms of the functions

$$f_1(t) = \frac{1 - e^{-\rho t}}{\rho\sqrt{2\pi t^3}}, \quad \text{and} \quad f_2(t) = \frac{a}{\sqrt{2\pi t^3}} e^{-a^2/2t} ,$$

hence, it is the Laplace transform of the convolution of these two functions. The result follows. □

Comment 2.5.4.2 Cumulative options are studied in Chesney et al. [175, 196], Dassios [211], Detemple [251], Fusai [370], Hugonnier [451] and Moraux [657]. In [370], Fusai determines the Fourier transform of the density of the occupation time $\tau = \int_0^T \mathbb{1}_{\{a<\nu s+W_s<b\}} ds$, in order to compute the price of a corridor option, i.e., an option with payoff $(\tau - K)^+$. The joint law of W_T and A_T^+ can be found in Fujita and Miura [366] where the authors present, among other results, options which are knocked-out at the time

$$\tau = \inf\left\{ t \,:\, \int_{T_a}^t \mathbb{1}_{\{S_u \le a\}} \, du \ge \alpha(T - T_a) \right\} .$$

Exercise 2.5.4.3 (1) Deduce from Proposition 2.5.4.1 $\mathbb{P}(A_t^{-,0} \in du | W_t = x)$.
(2) Recover the formula (2.5.5) for $\mathbb{P}(A_t^{-,0} \in du)$. ◁

2.5.5 Quantiles

Proposition 2.5.5.1 *Let $X_t = \mu t + \sigma W_t$ and $M_t^X = \sup_{s\leq t} X_s$. We assume $\sigma > 0$. Define, for a fixed t, $\theta_t^X = \sup\{s \leq t : X_s = M_s^X\}$. Then*

$$\theta_t^X \stackrel{\text{law}}{=} \int_0^t \mathbb{1}_{\{X_s>0\}} ds \,.$$

PROOF: We shall prove the result in the case $\sigma = 1, \mu = 0$ in $\rightarrowtail$ Exercise 4.1.7.5. The drifted Brownian motion case follows from an application of Girsanov's theorem. □

Proposition 2.5.5.2 *Let $X_t = \mu t + \sigma W_t$ with $\sigma > 0$, and*

$$q^X(\alpha, t) = \inf\left\{x : \int_0^t \mathbb{1}_{\{X_s\leq x\}} ds > \alpha t\right\}.$$

Let $X^i, i = 1, 2$ be two independent copies of X. Then

$$q^X(\alpha, t) \stackrel{\text{law}}{=} \sup_{0\leq s\leq \alpha t} X_s^1 + \inf_{0\leq s\leq (1-\alpha t)} X_s^2 \,.$$

PROOF: We give the proof for $t = 1$. We note that

$$A^X(x) = \int_0^1 \mathbb{1}_{\{X_s>x\}} ds = \int_{T_x}^1 \mathbb{1}_{\{X_s>x\}} ds = 1 - \int_0^{1-T_x} \mathbb{1}_{\{X_{s+T_x}\leq x\}} ds$$

where $T_x = \inf\{t : X_t = x\}$. Then, denoting $q(\alpha) = q^X(\alpha, 1)$, one has

$$\mathbb{P}(q(\alpha) > x) = \mathbb{P}(A^X(x) > 1 - \alpha) = \mathbb{P}\left(\int_0^{1-T_x} \mathbb{1}_{\{X_{s+T_x}-x>0\}} ds > 1 - \alpha\right).$$

The process $(X_s^1 = X_{s+T_x} - x,\ s \geq 0)$ is independent of $(X_s, s \leq T_x; T_x)$ and has the same law as X. Hence,

$$\mathbb{P}(q(\alpha) > x) = \int_0^\alpha \mathbb{P}(T_x \in du)\mathbb{P}\left(\int_0^{1-u} \mathbb{1}_{\{X_s^1>0\}} ds > 1 - \alpha\right).$$

Then, from Proposition 2.5.5.1,

$$\mathbb{P}\left(\int_0^{1-u} \mathbb{1}_{\{X_s^1>0\}} ds > 1 - \alpha\right) = \mathbb{P}(\theta_{1-u}^{X^1} > 1 - \alpha).$$

From the definition of θ_s^1, for $s > a$,

$$\mathbb{P}(\theta_s^{X^1} > a) = \mathbb{P}\left(\sup_{u\leq a}(X_u^1 - X_a^1) < \sup_{a\leq v\leq s}(X_v^1 - X_a^1)\right).$$

It is easy to check that

$$\left(\sup_{u\leq a}(X^1_u - X^1_a),\ \sup_{a\leq v\leq s}(X^1_v - X^1_a)\right) \stackrel{\text{law}}{=} \left(-\inf_{u\leq a} X^2_u,\ \sup_{0<v\leq s-a} X^3_v\right)$$

where X^2 and X^3 are two independent copies of X. The result follows. □

Exercise 2.5.5.3 Prove that, in the case $\nu = 0$, setting $\beta = ((1-\alpha)/\alpha)^{1/2}$, and $\Phi^*(x) = \sqrt{2/\pi}\int_x^\infty e^{-y^2/2}dy$

$$\mathbb{P}(q(\alpha) \in dx) = \begin{cases} \sqrt{2/\pi}\, e^{-x^2/2}\Phi^*(\beta x)dx \text{ for } x \geq 0 \\ \sqrt{2/\pi}\, e^{-x^2/2}\Phi^*(-x\beta^{-1})dx \text{ for } x \leq 0 \end{cases}.$$

◁

Comment 2.5.5.4 See Akahori [1], Dassios [211, 212], Detemple [252], Embrechts et al. [324], Fujita and Yor [368], Fusai [370], Miura [653] and Yor [866] for results on quantiles and pricing of quantile options.

2.6 Ornstein-Uhlenbeck Processes and Related Processes

In this section, we present a particular SDE, the solution of which was used to model interest rates. Even if this kind of model is nowadays not so often used by practitioners for interest rates, it can be useful for modelling underlying values in a real options framework.

2.6.1 Definition and Properties

Proposition 2.6.1.1 *Let k, θ and σ be bounded Borel functions, and W a Brownian motion. The solution of*

$$dr_t = k(t)(\theta(t) - r_t)dt + \sigma(t)dW_t \tag{2.6.1}$$

is

$$r_t = e^{-K(t)}\left(r_0 + \int_0^t e^{K(s)}k(s)\theta(s)ds + \int_0^t e^{K(s)}\sigma(s)dW_s\right)$$

where $K(t) = \int_0^t k(s)ds$. The process $(r_t, t \geq 0)$ is a Gaussian process with mean

$$\mathbb{E}(r_t) = e^{-K(t)}\left(r_0 + \int_0^t e^{K(s)}k(s)\theta(s)ds\right)$$

and covariance

$$e^{-K(t)-K(s)}\int_0^{t\wedge s} e^{2K(u)}\sigma^2(u)du\,.$$

PROOF: The solution of (2.6.1) is a particular case of Example 1.5.4.8. The values of the mean and of the covariance follow from Exercise 1.5.1.4. □

The **Hull and White model** corresponds to the dynamics (2.6.1) where k is a positive function. In the particular case where k, θ and σ are constant, we obtain

Corollary 2.6.1.2 *The solution of*

$$\boxed{dr_t = k(\theta - r_t)dt + \sigma dW_t} \tag{2.6.2}$$

is

$$r_t = (r_0 - \theta)e^{-kt} + \theta + \sigma \int_0^t e^{-k(t-u)} dW_u .$$

The process $(r_t, t \geq 0)$ *is a Gaussian process with mean* $(r_0 - \theta)e^{-kt} + \theta$ *and covariance*

$$\text{Cov}(r_s, r_t) = \frac{\sigma^2}{2k} e^{-k(s+t)}(e^{2ks} - 1) = \frac{\sigma^2}{k} e^{-kt} \sinh(ks)$$

for $s \leq t$.

In finance, the solution of (2.6.2) is called a **Vasicek** process. In general, k is chosen to be positive, so that $\mathbb{E}(r_t) \to \theta$ as $t \to \infty$ (this is why this process is said to enjoy the **mean reverting** property). The process (2.6.1) is called a **Generalized Vasicek process** (GV). Because r is a Gaussian process, it takes negative values. This is one of the reasons why this process is no longer used to model interest rates. When $\theta = 0$, the process r is called an **Ornstein-Uhlenbeck** (OU) process. Note that, if r is a Vasicek process, the process $r - \theta$ is an OU process with parameter k. More formally,

Definition 2.6.1.3 *An Ornstein-Uhlenbeck (OU) process driven by a BM follows the dynamics* $dr_t = -kr_t dt + \sigma dW_t$.

An OU process can be constructed in terms of time-changed BM (see also ↣ Section 5.1):

Proposition 2.6.1.4 (i) *If* W *is a BM starting from* x *and* $a(t) = \sigma^2 \frac{e^{2kt}-1}{2k}$, *the process* $Z_t = e^{-kt} W_{a(t)}$ *is an OU process starting from* x.

(ii) *Conversely, if* U *is an OU process starting from* x, *then there exists a BM* W *starting from* x *such that* $U_t = e^{-kt} W_{a(t)}$.

PROOF: Indeed, the process Z is a Gaussian process, with mean xe^{-kt} and covariance $e^{-k(t+s)}(a(t) \wedge a(s))$. □

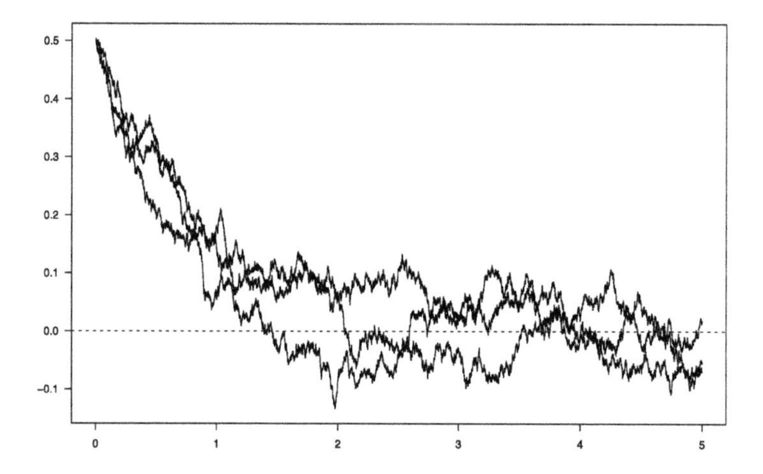

Fig. 2.1 Simulation of Ornstein-Uhlenbeck paths $\theta = 0, k = 3/2, \sigma = 0.1$

More generally, one can define an Ornstein-Uhlenbeck process driven by a Lévy process (see ↣ Chapter 11). Here, we note that the Vasicek process defined in (2.6.2) is an OU process, driven by the Brownian motion with drift $\sigma W_t + k\theta t$.

From the Markov and Gaussian properties of a Vasicek process r we deduce:

Proposition 2.6.1.5 *Let r be a Vasicek process, the solution of* (2.6.2) *and let* $\mathbf{F}$ *be its natural filtration. For $s < t$, the conditional expectation and the conditional variance of r_t, with respect to $\mathcal{F}_s$ (denoted as* $\mathrm{Var}_s(r_t)$*) are given by*

$$\mathbb{E}(r_t|\mathcal{F}_s) = \mathbb{E}(r_t|r_s) = (r_s - \theta)e^{-k(t-s)} + \theta$$

$$\mathrm{Var}_s\ (r_t) = \frac{\sigma^2}{2k}(1 - e^{-2k(t-s)})\,.$$

Note that the filtration generated by the process r is equal to that of the driving Brownian motion. Owing to the Gaussian property of the process r, the law of the integrated process $\int_0^t r_s ds$ can be characterized as follows:

Proposition 2.6.1.6 *Let r be a solution of* (2.6.2).
The process $\left(\int_0^t r_s ds,\ t \geq 0\right)$ *is Gaussian with mean and variance given by*

$$\mathbb{E}\left(\int_0^t r_s ds\right) = \theta t + (r_0 - \theta)\frac{1-e^{-kt}}{k},$$

$$\text{Var}\left(\int_0^t r_s ds\right) = -\frac{\sigma^2}{2k^3}(1-e^{-kt})^2 + \frac{\sigma^2}{k^2}\left(t - \frac{1-e^{-kt}}{k}\right)$$

and covariance (for $s < t$)

$$\frac{\sigma^2}{k^2}\left(s - e^{-kt}\frac{e^{ks}-1}{k} - \frac{1-e^{-ks}}{k} + e^{-k(t+s)}\frac{e^{2ks}-1}{2k}\right).$$

PROOF: From the definition, $r_t = r_0 + k\theta t - k\int_0^t r_s ds + \sigma W_t$, hence

$$\begin{aligned}\int_0^t r_s ds &= \frac{1}{k}[-r_t + r_0 + k\theta t + \sigma W_t] \\ &= \frac{1}{k}[k\theta t + (r_0 - \theta)(1-e^{-kt}) - \sigma e^{-kt}\int_0^t e^{ku} dW_u + \sigma W_t].\end{aligned}$$

Obviously, from the properties of the Wiener integral, the right-hand side defines a Gaussian process. It remains to compute the expectation and the variance of the Gaussian variable on the right-hand side, which is easy, since the variance of a Wiener integral is well known. □

Note that one can also justify directly the Gaussian property of an integral process $(\int_0^t y_s ds, t \geq 0)$ where y is a Gaussian process.

More generally, for $t \geq s$,

$$\mathbb{E}\left(\int_s^t r_u\, du|\mathcal{F}_s\right) = \theta(t-s) + (r_s - \theta)\frac{1-e^{-k(t-s)}}{k} := M(s,t)\,, \quad (2.6.3)$$

$$\begin{aligned}\text{Var}_s\left(\int_s^t r_u\, du\right) &= -\frac{\sigma^2}{2k^3}(1-e^{-k(t-s)})^2 \\ &\quad + \frac{\sigma^2}{k^2}\left(t - s - \frac{1-e^{-k(t-s)}}{k}\right) := V(s,t)\,. \quad (2.6.4)\end{aligned}$$

Exercise 2.6.1.7 Compute the transition probability for an OU process. ◁

Exercise 2.6.1.8 (1) Let B be a Brownian motion, and define the probability P^b via

$$\mathbb{P}^b|_{\mathcal{F}_T} := \exp\left\{-b\int_0^T B_s\, dB_s - \frac{b^2}{2}\int_0^T B_s^2\, ds\right\}\mathbb{P}|_{\mathcal{F}_T}\,.$$

Prove that the process $(B_t, t \geq 0)$ is a $\mathbb{P}^b$-Ornstein-Uhlenbeck process and that

$$\mathbb{E}\left(\exp\left(-\alpha B_t^2 - \frac{b^2}{2}\int_0^t B_s^2\, ds\right)\right) = \mathbb{E}^b\left(\exp\left(-\alpha B_t^2 + \frac{b}{2}(B_t^2 - t)\right)\right),$$

where $\mathbb{E}^b$ is the expectation w.r.t. the probability measure $\mathbb{P}^b$. One can also prove that if B is an n-dimensional BM starting from a

$$\mathbb{E}_a\left(\exp(-\alpha|B_t|^2 - \frac{b^2}{2}\int_0^t |B_s|^2 ds)\right)$$

$$= \left(\cosh bt + \frac{2\alpha}{b}\sinh bt\right)^{-n/2} \exp\left(-\frac{|a|^2 b}{2}\frac{1+\frac{2\alpha}{b}\coth bt}{\coth bt + 2\alpha/b}\right),$$

where $\mathbb{E}_a$ is the expectation for a BM starting from a. (See Yor [864].)

(2) Use the Gaussian property of the variable B_t to obtain that

$$\mathbb{E}_{\mathbb{P}}\left(\exp\left(-\alpha B_t^2 - \frac{b^2}{2}\int_0^t B_s^2\, ds\right)\right) = \left(\cosh\, bt + 2\frac{\alpha}{b}\sinh\, bt\right)^{-\frac{1}{2}}.$$

If B and C are two independent Brownian motions starting form 0, prove that

$$\mathbb{E}_{\mathbb{P}}\left(\exp(-\alpha(B_t^2 + C_t^2) - \frac{b^2}{2}\int_0^t (B_s^2 + C_s^2)\, ds)\right) = \left(\cosh\, bt + 2\frac{\alpha}{b}\sinh\, bt\right)^{-1}.$$

(3) Deduce Lévy's area formula:

$$\mathbb{E}(\exp i\lambda\mathcal{A}_t \big| \, |Z_t|^2 = r^2) = \mathbb{E}\left(\exp -\frac{\lambda^2}{8}\int_0^t |Z_s|^2 ds \big| \, |Z_t|^2 = r^2\right)$$

$$= \frac{t\lambda/2}{\sinh(t\lambda/2)} \exp -\frac{r^2}{2}(\lambda t \coth \lambda t - 1),$$

where

$$\mathcal{A}_t := \frac{1}{2}\int_0^t (B_s dC_s - C_s dB_s) = \frac{1}{2}\gamma\left(\int_0^t (B_s^2 + C_s^2) ds\right)$$

where γ is a Brownian motion independent of $|Z|^2 := B^2 + C^2$ (see ⤚ Exercise 5.1.3.9)

Hint: Note that $\int_0^t B_s dB_s = \frac{1}{2}(B_t^2 - t)$. ◁

2.6.2 Zero-coupon Bond

Suppose that the dynamics of the interest rate under the risk-neutral probability are given by (2.6.2). The value $P(t,T)$ of a zero-coupon bond maturing at date T is given as the conditional expectation under the e.m.m. of the discounted payoff. Using the Laplace transform of a Gaussian law (see Proposition 1.1.12.1), and Proposition 2.6.1.6, we obtain

$$P(t,T) = \mathbb{E}\left(\exp\left(-\int_t^T r_u\, du\right) \big| \mathcal{F}_t\right) = \exp\left(-M(t,T) + \frac{1}{2}V(t,T)\right),$$

where M and V are defined in (2.6.3) and (2.6.4).

Proposition 2.6.2.1 *In a Vasicek model, the price of a zero-coupon with maturity T is*

$$\begin{aligned} P(t,T) &= \exp\left[-\theta(T-t) - (r_t - \theta)\frac{1-e^{-k(T-t)}}{k} - \frac{\sigma^2}{4k^3}(1-e^{-k(T-t)})^2 \right. \\ &\qquad \left. + \frac{\sigma^2}{2k^2}\left(T - t - \frac{1-e^{-k(T-t)}}{k}\right)\right] \\ &= \exp(a(t,T) - b(t,T)r_t)\,, \end{aligned}$$

with $b(t,T) = \frac{1-e^{-k(T-t)}}{k}$.

Without any computation, we know that

$$d_tP(t,T) = P(t,T)(r_t dt - \sigma_t dW_t)\,,$$

since the discounted value of the zero-coupon bond is a martingale. It suffices to identify the volatility term. It is not difficult, using Itô's formula, to check that the risk-neutral dynamics of the zero-coupon bond are

$$d_tP(t,T) = P(t,T)(r_t dt - b(t,T)dW_t)\,.$$

2.6.3 Absolute Continuity Relationship for Generalized Vasicek Processes

Let W be a $\mathbb{P}$-Brownian motion starting from x, θ a bounded Borel function and L the solution of $dL_t = kL_t(\theta(t) - W_t)dW_t,\ L_0 = 1$, that is,

$$L_t = \exp\left(\int_0^t k(\theta(s) - W_s)dW_s - \frac{1}{2}\int_0^t k^2(\theta(s) - W_s)^2 ds\right)\,. \tag{2.6.5}$$

This process is a martingale, from the non-explosion criteria. We define

$$\mathbb{P}^{k,\theta}|_{\mathcal{F}_t} = L_t\,\mathbb{P}|_{\mathcal{F}_t}\,.$$

Then,

$$W_t = x + \beta_t + \int_0^t k(\theta(s) - W_s)ds$$

where, thanks to Girsanov's theorem, β is a $\mathbb{P}^{k,\theta}$-Brownian motion starting from 0. Hence, we have proved that the $\mathbb{P}$-Brownian motion W is a GV process under $\mathbb{P}^{k,\theta}$ (and thus we generalize Exercise 2.6.1.8).

Proposition 2.6.3.1 *Let θ be a differentiable function and let $\mathbb{P}^{k,\theta}_x$ be the law of the GV process*

$$dr_t = dW_t + k(\theta(t) - r_t)dt,\ r_0 = x\,.$$

We denote by $\mathbf{W}_x$ the law of a Brownian motion starting from x. Then the following absolute continuity relationship holds

$$\mathbb{P}_x^{k,\theta}|_{\mathcal{F}_t} = \exp\left[\frac{k}{2}\left(t + x^2 - k\int_0^t \theta^2(s)ds - 2x\theta(0)\right)\right]$$
$$\times \exp\left[k\theta(t)X_t - \frac{k}{2}X_t^2 + \int_0^t (k^2\theta(s) - k\theta'(s))X_s ds - \frac{k^2}{2}\int_0^t X_s^2 ds\right] \mathbf{W}_x|_{\mathcal{F}_t}.$$

PROOF: We have seen that $\mathbb{P}_x^{k,\theta}|_{\mathcal{F}_t} = L_t \mathbf{W}_x|_{\mathcal{F}_t}$ where L is given in (2.6.5). Since θ is differentiable, an integration by parts under $\mathbf{W}_x$ leads to

$$\int_0^t (\theta(s) - X_s)dX_s = \theta(t)X_t - x\theta(0) - \int_0^t \theta'(s)X_s ds - \frac{1}{2}(X_t^2 - x^2 - t).$$

□

Corollary 2.6.3.2 *Let r be a Vasicek process*

$$dr_t = k(\theta - r_t)dt + \sigma dW_t,\ r_0 = x.$$

Then

$$\mathbb{P}_x^{k,\theta}|_{\mathcal{F}_t} = \exp\left(\frac{k}{2}\left(t + x^2 - k\theta^2 t - 2x\theta\right)\right) \tag{2.6.6}$$
$$\times \exp\left(-\frac{k}{2}X_t^2 + k\theta X_t + k^2\theta\int_0^t X_s ds - \frac{k^2}{2}\int_0^t X_s^2 ds\right)\mathbf{W}_x|_{\mathcal{F}_t}.$$

PROOF: The absolute continuity relation (2.6.6) follows from Proposition 2.6.3.1. □

Example 2.6.3.3 As an exercise, we present the computation of

$$A = \mathbb{E}_x^{k,\theta}\left(\exp\left(-\alpha X_t - \lambda X_t^2 - \beta\int_0^t X_s ds - \frac{\gamma^2}{2}\int_0^t X_s^2 ds\right)\right),$$

where $(\alpha, \beta, \theta, \lambda, \gamma)$ are real numbers with $\lambda > 0$. From (2.6.6)

$$A = \exp\left(\frac{k}{2}\left(t + x^2 - k\theta^2 t - 2x\theta\right)\right)$$
$$\times \mathbf{W}_x\left(\exp\left(-\lambda_1 X_t^2 + \alpha_1 X_t + (k^2\theta - \beta)\int_0^t X_s ds - \frac{\gamma_1^2}{2}\int_0^t X_s^2 ds\right)\right)$$

where $\lambda_1 = \lambda + \frac{k}{2}$, $\alpha_1 = k\theta - \alpha$, $\gamma_1^2 = \gamma^2 + k^2$. From

$$(k^2\theta - \beta)\int_0^t X_s ds - \frac{\gamma_1^2}{2}\int_0^t X_s^2 ds = -\frac{\gamma_1^2}{2}\int_0^t (X_s + \beta_1)^2 ds + \frac{\beta_1^2\gamma_1^2}{2}t$$

with $\beta_1 = \frac{\beta - k^2\theta}{\gamma_1^2}$ and setting $Z_s = X_s + \beta_1$, one gets

$$A = \exp\left(\frac{k}{2}\left(t + x^2 - k\theta^2 t - 2x\theta\right) + \frac{\beta_1^2\gamma_1^2}{2}t\right) \times \mathbf{W}_{x+\beta_1}\left(\exp\left(-\lambda_1(Z_t - \beta_1)^2 + \alpha_1(Z_t - \beta_1) - \frac{\gamma_1^2}{2}\int_0^t Z_s^2 ds\right)\right).$$

Now,

$$-\lambda_1(Z_t - \beta_1)^2 + \alpha_1(Z_t - \beta_1) = -\lambda_1 Z_t^2 + (\alpha_1 + 2\lambda_1\beta_1)Z_t - \beta_1(\lambda_1\beta_1 + \alpha_1).$$

Hence,

$$\mathbf{W}_{x+\beta_1}\left(\exp\left(-\lambda_1(Z_t - \beta_1)^2 + \alpha_1(Z_t - \beta_1) - \frac{\gamma_1^2}{2}\int_0^t Z_s^2 ds\right)\right)$$
$$= e^{-\beta_1(\lambda_1\beta_1+\alpha_1)}\mathbf{W}_{x+\beta_1}\left(\exp\left(-\lambda_1 Z_t^2 + (\alpha_1 + 2\lambda_1\beta_1)Z_t - \frac{\gamma_1^2}{2}\int_0^t Z_s^2 ds\right)\right).$$

From (2.6.6) again

$$\mathbf{W}_{x+\beta_1}\left(\exp\left(-\lambda_1 Z_t^2 + (\alpha_1 + 2\lambda_1\beta_1)Z_t - \frac{\gamma_1^2}{2}\int_0^t Z_s^2 ds\right)\right)$$
$$= \exp\left(-\frac{\gamma_1}{2}\left(t + (x+\beta_1)^2\right)\right) \times \mathbb{E}_{x+\beta_1}^{\gamma_1,0}\left(\exp\left((-\lambda_1 + \frac{\gamma_1}{2})X_t^2 + (\alpha_1 + 2\lambda_1\beta_1)X_t\right)\right).$$

Finally

$$A = e^C\, \mathbb{E}_{x+\beta_1}^{\gamma_1,0}\left(\exp\left((-\lambda_1 + \frac{\gamma_1}{2})X_t^2 + (\alpha_1 + 2\lambda_1\beta_1)X_t\right)\right)$$

where

$$C = \frac{k}{2}\left(t + x^2 - k\theta^2 t - 2x\theta\right) + \frac{\beta_1^2\gamma_1^2}{2}t - \beta_1(\lambda_1\beta_1 + \alpha_1) - \frac{\gamma_1}{2}\left(t + (x+\beta_1)^2\right).$$

One can then finish the computation since, under $\mathbb{P}_{x+\beta_1}^{\gamma_1,0}$ the r.v. X_t is a Gaussian variable with mean $m = (x+\beta_1)e^{-\gamma_1 t}$ and variance $\frac{\sigma^2}{2\gamma_1}(1 - e^{-2\gamma_1 t})$. Furthermore, from Exercise 1.1.12.3, if $U \stackrel{\text{law}}{=} \mathcal{N}(m, \sigma^2)$

$$\mathbb{E}(\exp\{\lambda U^2 + \mu U\}) = \frac{\Sigma}{\sigma}\exp\left(\frac{\Sigma^2}{2}(\mu + \frac{m}{\sigma^2})^2 - \frac{m^2}{2\sigma^2}\right).$$

with $\Sigma^2 = \dfrac{\sigma^2}{1 - 2\lambda\sigma^2}$, for $2\lambda\sigma^2 < 1$.

2.6.4 Square of a Generalized Vasicek Process

Let r be a GV process with dynamics

$$dr_t = k(\theta(t) - r_t)dt + dW_t$$

and $\rho_t = r_t^2$. Hence

$$d\rho_t = (1 - 2k\rho_t + 2k\theta(t)\sqrt{\rho_t})dt + 2\sqrt{\rho_t}dW_t\,.$$

By construction, the process ρ takes positive values, and can represent a spot interest rate. Then, the value of the corresponding zero-coupon bond can be computed as an application of the absolute continuity relationship between a GV and a BM, as we present now.

Proposition 2.6.4.1 *Let*

$$d\rho_t = (1 - 2k\rho_t + 2k\theta(t)\sqrt{\rho_t})dt + 2\sqrt{\rho_t}dW_t,\ \rho_0 = x^2.$$

Then

$$\mathbb{E}\left[\exp\left(-\int_0^T \rho_s ds\right)\right] = A(T)\exp\left(\frac{k}{2}(T + x^2 - k\int_0^T \theta^2(s)ds - 2\theta(0)x)\right)$$

where

$$A(T) = \exp\left(\frac{1}{2}\left(\int_0^T f(s)ds + \int_0^T g^2(s)ds\right)\right)\,.$$

Here,

$$f(s) = K\frac{\alpha e^{Ks} + e^{-Ks}}{\alpha e^{Ks} - e^{-Ks}},$$

$$g(s) = k\frac{\theta(T)v(T) - \int_s^T(\theta'(u) - k\theta(u))v(u)du}{v(s)}$$

with $v(s) = \alpha e^{Ks} - e^{-Ks}$, $K = \sqrt{k^2 + 2}$ *and* $\alpha = \frac{k - K}{k + K}e^{-2TK}$.

PROOF: From Proposition 2.6.3.1,

$$\mathbb{E}\left[\exp\left(-\int_0^T \rho_s ds\right)\right] = A(T)\exp\left(\frac{k}{2}(T + x^2 - k\int_0^T \theta^2(s)ds - 2\theta(0)x)\right)$$

where $A(T)$ is equal to the expectation, under $\mathbf{W}$, of

$$\exp\left(k\theta(T)X_T - \frac{k}{2}X_T^2 + \int_0^T (k^2\theta(s) - k\theta'(s))X_s ds - \frac{k^2+2}{2}\int_0^T X_s^2 ds\right)\,.$$

The computation of $A(T)$ follows from Example 1.5.7.1 which requires the solution of

$$f^2(s) + f'(s) = k^2 + 2, \; s \leq T,$$
$$f(s)g(s) + g'(s) = k\theta'(s) - k^2\theta(s),$$

with the terminal condition at time T

$$f(T) = -k, \quad g(T) = k\theta(T).$$

Let us set $K^2 = k^2 + 2$. The solution follows by solving the classical Ricatti equation $f^2(s) + f'(s) = K^2$ whose solution is

$$f(s) = K\frac{\alpha e^{Ks} + e^{-Ks}}{\alpha e^{Ks} - e^{-Ks}}.$$

The terminal condition yields $\alpha = \frac{k-K}{k+K}e^{-2TK}$. A straightforward computation leads to the expression of g given in the proposition. □

2.6.5 Powers of δ-Dimensional Radial OU Processes, Alias CIR Processes

In the case $\theta = 0$, the process

$$d\rho_t = (1 - 2k\rho_t)dt + 2\sqrt{\rho_t}dW_t$$

is called a one-dimensional square OU process which is justified by the computation at the beginning of this subsection. Let U be a δ-dimensional OU process, i.e., the solution of

$$U_t = u + B_t - k\int_0^t U_s ds$$

where B is a δ-dimensional Brownian motion and k a real number, and set $V_t = ||U_t||^2$. From Itô's formula,

$$\boxed{dV_t = (\delta - 2kV_t)dt + 2\sqrt{V_t}dW_t}$$

where W is a one-dimensional Brownian motion. The process V is called either a squared δ-dimensional radial Ornstein-Uhlenbeck process or more commonly in mathematical finance a Cox-Ingersoll-Ross (CIR) process with dimension δ and linear coefficient k, and, for $\delta \geq 2$, does not reach 0 (see $\rightarrowtail$ Subsection 6.3.1).

Let $\gamma \neq 0$ be a real number, and $Z_t = V_t^\gamma$. Then,

$$Z_t = z + 2\gamma \int_0^t Z_s^{1-1/(2\gamma)} dW_s - 2k\gamma \int_0^t Z_s ds + \gamma(2(\gamma-1)+\delta) \int_0^t Z_s^{1-1/\gamma} ds .$$

In the particular case $\gamma = 1 - \delta/2$,

$$Z_t = z + 2\gamma \int_0^t Z_s^{1-1/(2\gamma)} dW_s - 2k\gamma \int_0^t Z_s ds ,$$

or in differential notation

$$dZ_t = Z_t(\mu dt + \sigma Z_t^\beta dW_t) ,$$

with

$$\mu = -2k\gamma, \beta = -1/(2\gamma) = 1/(\delta - 2), \sigma = 2\gamma .$$

The process Z is called a CEV process.

Comment 2.6.5.1 We shall study CIR processes in more details in $\rightarrowtail$ Section 6.3. See also Pitman and Yor [716, 717]. See $\rightarrowtail$ Section 6.4, where squares of OU processes are of major interest in constructing CEV processes.

2.7 Valuation of European Options

In this section, we give a few applications of Itô's lemma, changes of probabilities and Girsanov's theorem to the valuation of options.

2.7.1 The Garman and Kohlhagen Model for Currency Options

In this section, European currency options will be considered. It will be shown that the Black and Scholes formula corresponds to a specific case of the Garman and Kohlhagen [373] model in which the foreign interest rate is equal to zero. As in the Black and Scholes model, let us assume that trading is continuous and that the historical dynamics of the underlying (the currency) S are given by

$$dS_t = S_t(\alpha dt + \sigma dB_t) .$$

whereas the risk-neutral dynamics satisfy the Garman-Kohlhagen dynamics

$$dS_t = S_t((r - \delta)dt + \sigma dW_t) . \tag{2.7.1}$$

Here, $(W_t, t \geq 0)$ is a $\mathbb{Q}$-Brownian motion and $\mathbb{Q}$ is the risk-neutral probability defined by its Radon-Nikodým derivative with respect to $\mathbb{P}$ as $\mathbb{Q}|_{\mathcal{F}_t} = \exp(-\theta B_t - \frac{1}{2}\theta^2 t)\, \mathbb{P}|_{\mathcal{F}_t}$ with $\theta = \dfrac{\alpha - (r-\delta)}{\sigma}$. It follows that

$$S_t = S_0 e^{(r-\delta)t} e^{\sigma W_t - \frac{\sigma^2}{2}t}.$$

The domestic (resp. foreign) interest rate r (resp. δ) and the volatility σ are constant. The term δ corresponds to a dividend yield for options (see Subsection 2.3.5).

The method used in the Black and Scholes model will give us the PDE evaluation for a European call. We give the details for the reader's convenience.

In that setting, the PDE evaluation for a contingent claim $H = h(S_T)$ takes the form

$$-\partial_u V(x,T-t) + (r-\delta)x\partial_x V(x,T-t) + \frac{1}{2}\sigma^2 x^2 \partial_{xx} V(x,T-t) = rV(x,T-t) \tag{2.7.2}$$

with the initial condition $V(x,0) = h(x)$. Indeed, the process $e^{-rt}V(S_t,t)$ is a $\mathbb{Q}$-martingale, and an application of Itô's formula leads to the previous equality. Let us now consider the case of a European call option:

Proposition 2.7.1.1 *The time-t value of the European call on an underlying with risk-neutral dynamics* (2.7.1) *is $C_E(S_t, T-t)$. The function C_E satisfies the following PDE:*

$$\begin{aligned} -\frac{\partial C_E}{\partial u}(x,T-t) &+ \frac{1}{2}\sigma^2 x^2 \frac{\partial^2 C_E}{\partial x^2}(x,T-t) \\ &+ (r-\delta)x\frac{\partial C_E}{\partial x}(x,T-t) = rC_E(x,T-t) \end{aligned} \tag{2.7.3}$$

with initial condition $C_E(x,0) = (x-K)^+$, and is given by

$$C_E(x,u) = xe^{-\delta u}\mathcal{N}\left[d_1\left(\frac{xe^{-\delta u}}{Ke^{-ru}}, u\right)\right] - Ke^{-ru}\mathcal{N}\left[d_2\left(\frac{xe^{-\delta u}}{Ke^{-ru}}, u\right)\right], \tag{2.7.4}$$

where the d_i's are given in Theorem 2.3.2.1.

PROOF: The evaluation PDE (2.7.3) is obtained from (2.7.2). Formula (2.7.4) is obtained by a direct computation of $\mathbb{E}_{\mathbb{Q}}(e^{-rT}(S_T-K)^+)$, or by solving (2.7.3). □

2.7.2 Evaluation of an Exchange Option

An exchange option is an option to exchange one asset for another. In this domain, the original reference is Margrabe [623]. The model corresponds to an extension of the Black and Scholes model with a stochastic strike price, (see Fischer [345]) in a risk-adjusted setting. Let us assume that under the risk-adjusted neutral probability $\mathbb{Q}$ the stock prices' (respectively, S^1 and S^2) dynamics[5] are given by:

[5] Of course, 1 and 2 are only superscripts, not powers.

$$dS_t^1 = S_t^1\left((r-\nu)dt + \sigma_1 dW_t\right),\ dS_t^2 = S_t^2\left((r-\delta)dt + \sigma_2 dB_t\right)$$

where r is the risk-free interest rate and ν and δ are, respectively, the stock 1 and 2 dividend yields and σ_1 and σ_2 are the stock prices' volatilities. The correlation coefficient between the two Brownian motions W and B is denoted by ρ. It is assumed that all of these parameters are constant. The payoff at maturity of the **exchange call option** is $(S_T^1 - S_T^2)^+$. The option price is therefore given by:

$$\begin{aligned} C_{EX}(S_0^1, S_0^2, T) &= \mathbb{E}_{\mathbb{Q}}(e^{-rT}(S_T^1 - S_T^2)^+) = \mathbb{E}_{\mathbb{Q}}(e^{-rT} S_T^2 (X_T - 1)^+) \\ &= S_0^2 \mathbb{E}_{\mathbb{Q}^*}(e^{-\delta T}(X_T - 1)^+)\,. \end{aligned} \tag{2.7.5}$$

Here, $X_t = S_t^1/S_t^2$, and the probability measure $\mathbb{Q}^*$ is defined by its Radon-Nikodým derivative with respect to $\mathbb{Q}$

$$\frac{d\mathbb{Q}^*}{d\mathbb{Q}}\Big|_{\mathcal{F}_t} = e^{-(r-\delta)t}\frac{S_t^2}{S_0^2} = \exp\left(\sigma_2 B_t - \frac{\sigma_2^2}{2}t\right). \tag{2.7.6}$$

Note that this change of probability is associated with a change of numéraire, the new numéraire being the asset S^2. Using Itô's lemma, the dynamics of X are

$$dX_t = X_t[(\delta - \nu + \sigma_2^2 - \rho\sigma_1\sigma_2)dt + \sigma_1 dW_t - \sigma_2 dB_t]\,.$$

Girsanov's theorem for correlated Brownian motions (see Subsection 1.7.4) implies that the processes $\widetilde{W}$ and $\widetilde{B}$ defined as

$$\widetilde{W}_t = W_t - \rho\sigma_2 t,\ \widetilde{B}_t = B_t - \sigma_2 t\,,$$

are $\mathbb{Q}^*$-Brownian motions with correlation ρ. Hence, the dynamics of X are

$$dX_t = X_t[(\delta - \nu)dt + \sigma_1 d\widetilde{W}_t - \sigma_2 d\widetilde{B}_t] = X_t[(\delta - \nu)dt + \sigma dZ_t]$$

where Z is a $\mathbb{Q}^*$-Brownian motion defined as

$$dZ_t = \sigma^{-1}(\sigma_1 d\widetilde{W}_t - \sigma_2 d\widetilde{B}_t)$$

and where

$$\sigma = \sqrt{\sigma_1^2 + \sigma_2^2 - 2\rho\sigma_1\sigma_2}\,.$$

As shown in equation (2.7.5), δ plays the rôle of a discount rate. Therefore, by relying on the Garman and Kohlhagen formula (2.7.4), the exchange option price is given by:

$$C_{EX}(S_0^1, S_0^2, T) = S_0^1 e^{-\nu T}\mathcal{N}(b_1) - S_0^2 e^{-\delta T}\mathcal{N}(b_2)$$

with

$$b_1 = \frac{\ln(S_0^1/S_0^2) + (\delta - \nu)T}{\Sigma\sqrt{T}} + \frac{1}{2}\Sigma\sqrt{T}\,,\ b_2 = b_1 - \Sigma\sqrt{T}\,.$$

This value is independent of the domestic risk-free rate r. Indeed, since the second asset is the numéraire, its dividend yield, δ, plays the rôle of the domestic risk-free rate. The first asset dividend yield ν, plays the rôle of the foreign interest rate in the foreign currency option model developed by Garman and Kohlhagen [373]. When the second asset plays the rôle of the numéraire, in the risk-neutral economy the risk-adjusted trend of the process $(S_t^1/S_t^2, t \geq 0)$ is the dividend yield differential $\delta - \nu$.

2.7.3 Quanto Options

In the context of the international diversification of portfolios, **quanto options** can be useful. Indeed with these options, the problems of currency risk and stock market movements can be managed simultaneously. Using the model established in El Karoui and Cherif [298], the valuation of these products can be obtained.

Let us assume that under the domestic risk-neutral probability $\mathbb{Q}$, the dynamics of the stock price S, in foreign currency units and of the currency price X, in domestic units, are respectively given by:

$$\begin{aligned} dS_t &= S_t \left((\delta - \nu - \rho\sigma_1\sigma_2)dt + \sigma_1 dW_t \right) \\ dX_t &= X_t \left((r - \delta)dt + \sigma_2 dB_t \right) \end{aligned} \tag{2.7.7}$$

where r, δ and ν are respectively the domestic, foreign risk-free interest rate and the dividend yield and σ_1 and σ_2 are, respectively, the stock price and currency volatilities. Again, the correlation coefficient between the two Brownian motions is denoted by ρ. It is assumed that the parameters are constant.

The trend in equation (2.7.7) is equal to $\mu_1 = \delta - \nu - \rho\sigma_1\sigma_2$ because, in the domestic risk-neutral economy, we want the trend of the stock price (in domestic units: XS) dynamics to be equal to $r - \nu$.

We now present four types of quanto options:

Foreign Stock Option with a Strike in a Foreign Currency

In this case, the payoff at maturity is $X_T(S_T - K)^+$, i.e., the value in the domestic currency of the standard Black and Scholes payoff in the foreign currency $(S_T - K)^+$. The call price is therefore given by:

$$C_{qt1}(S_0, X_0, T) := \mathbb{E}_{\mathbb{Q}}(e^{-rT}(X_T S_T - K X_T)^+)\,.$$

This quanto option is an exchange option, an option to exchange at maturity T, an asset of value $X_T S_T$ for another of value $K X_T$. By relying on the previous Subsection 2.7.2

$$C_{qt1}(S_0, X_0, T) = X_0 \mathbb{E}_{\mathbb{Q}^*}(e^{-\delta T}(S_T - K)^+)$$

where the probability measure $\mathbb{Q}^*$ is defined by its Radon-Nikodým derivative with respect to $\mathbb{Q}$, in equation (2.7.6).

Cameron-Martin's theorem implies that the two processes $(B_t - \sigma_2 t, t \geq 0)$ and $(W_t - \rho\sigma_2 t, t \geq 0)$ are $\mathbb{Q}^*$ -Brownian motions. Now, by relying on equation (2.7.7)

$$dS_t = S_t \left((\delta - \nu)dt + \sigma_1 d(W_t - \rho\sigma_2 t) \right) .$$

Therefore, under the $\mathbb{Q}^*$ measure, the trend of the process $(S_t, t \geq 0)$ is equal to $\delta - \nu$ and the volatility of this process is σ_1. Therefore, using the Garman and Kohlhagen formula (2.7.4), the exchange option price is given by

$$C_{qt1}(S_0, X_0, T) = X_0(S_0 e^{-\nu T}\mathcal{N}(b_1) - Ke^{-\delta T}\mathcal{N}(b_2))$$

with

$$b_1 = \frac{\ln(S_0/K) + (\delta - \nu)T}{\sigma_1\sqrt{T}} + \frac{1}{2}\sigma_1\sqrt{T}\,,\; b_2 = b_1 - \sigma_1\sqrt{T}\,.$$

This price could also be obtained by a straightforward arbitrage argument. If a stock call option (with payoff $(S_T - K)^+$) is bought in the domestic country, its payoff at maturity is the quanto payoff $X_T(S_T - K)^+$and its price at time zero is known. It is the Garman and Kohlhagen price (in the foreign risk-neutral economy where the trend is $\delta - \nu$ and the positive dividend yield is ν), times the exchange rate at time zero.

Foreign Stock Option with a Strike in the Domestic Currency

In this case, the payoff at maturity is $(X_T S_T - K)^+$. The call price is therefore given by

$$C_{qt2}(S_0, X_0, T) := \mathbb{E}_{\mathbb{Q}}(e^{-rT}(X_T S_T - K)^+)\,.$$

This quanto option is a standard European option, with a new underlying process XS, with volatility given by

$$\sigma_{XS} = \sqrt{\sigma_1^2 + \sigma_2^2 + 2\rho\sigma_1\sigma_2}$$

and trend equal to $r - \nu$ in the risk-neutral domestic economy. The risk-free discount rate and the dividend rate are respectively r and ν. Its price is therefore given by

$$C_{qt2}(S_0, X_0, T) = X_0 S_0 e^{-\nu T}\mathcal{N}(b_1) - Ke^{-rT}\mathcal{N}(b_2)$$

with

$$b_1 = \frac{\ln(X_0 S_0/K) + (r - \nu)T}{\sigma_{XS}\sqrt{T}} + \frac{1}{2}\sigma_{XS}\sqrt{T}\,,\; b_2 = b_1 - \sigma_1\sqrt{T}\,.$$

Quanto Option with a Given Exchange Rate

In this case, the payoff at maturity is $\bar{X}(S_T - K)^+$, where $\bar{X}$ is a given exchange rate ($\bar{X}$ is fixed at time zero). The call price is therefore given by:

$$C_{qt3}(S_0, X_0, T) := \mathbb{E}_{\mathbb{Q}}(e^{-rT}\bar{X}(S_T - K)^+)$$

i.e.,

$$C_{qt3}(S_0, X_0, T) = \bar{X}e^{-(r-\delta)T}\mathbb{E}_{\mathbb{Q}}(e^{-\delta T}(S_T - K)^+) .$$

We obtain the expectation, in the risk-neutral domestic economy, of the standard foreign stock option payoff discounted with the foreign risk-free interest rate. Now, under the domestic risk-neutral probability $\mathbb{Q}$, the foreign asset trend is given by $\delta - \nu - \rho\sigma_1\sigma_2$ (see equation (2.7.7)).

Therefore, the price of this quanto option is given by

$$C_{qt3}(S_0, X_0, T) = \bar{X}e^{-(r-\delta)T}\left[S_0 e^{-(\nu+\rho\sigma_1\sigma_2)T}\mathcal{N}(b_1) - Ke^{-\delta T}\mathcal{N}(b_2)\right]$$

with

$$b_1 = \frac{\ln(S_0/K) + (\delta - \nu - \rho\sigma_1\sigma_2)T}{\sigma_1\sqrt{T}} + \frac{1}{2}\sigma_1\sqrt{T},\ b_2 = b_1 - \sigma_1\sqrt{T} .$$

Foreign Currency Quanto Option

In this case, the payoff at maturity is $S_T(X_T - K)^+$. The call price is therefore given by

$$C_{qt4}(S_0, X_0, T) := \mathbb{E}_{\mathbb{Q}}(e^{-rT}S_T(X_T - K)^+) .$$

Now, the price can be obtained by relying on the first quanto option. Indeed, the stock price now plays the rôle of the currency price and vice-versa. Therefore, μ_1 and σ_1 can be used respectively instead of $r - \delta$ and σ_2, and vice versa. Thus

$$C_{qt4}(S_0, X_0, T) = S_0(X_0 e^{(r-\delta+\rho\sigma_1\sigma_2-(r-\mu_1))T}\mathcal{N}(b_1) - Ke^{-(r-\mu_1)T}\mathcal{N}(b_2))$$

or, in a closed form

$$C_{qt4}(S_0, X_0, T) = S_0(X_0 e^{-\nu T}\mathcal{N}(b_1) - Ke^{-(r-\delta+\nu+\rho\sigma_1\sigma_2)\,T}\mathcal{N}(b_2))$$

with

$$b_1 = \frac{\ln(X_0/K) + (r - \delta + \rho\sigma_1\sigma_2)T}{\sigma_2\sqrt{T}} + \frac{1}{2}\sigma_2\sqrt{T},\ b_2 = b_1 - \sigma_2\sqrt{T} .$$

Indeed, $r - \delta + \rho\sigma_1\sigma_2$ is the trend of the currency price under the probability measure $\mathbb{Q}^*$, defined by its Radon-Nikodým derivative with respect to $\mathbb{Q}$ as

$$\mathbb{Q}^*|_{\mathcal{F}_t} = \exp\left(\sigma_1 W_t - \frac{1}{2}\sigma_1^2 t\right)\mathbb{Q}|_{\mathcal{F}_t} .$$

3

Hitting Times: A Mix of Mathematics and Finance

In this chapter, a Brownian motion $(W_t, t \geq 0)$ starting from 0 is given on a probability space $(\Omega, \mathcal{F}, \mathbb{P})$, and $\mathbf{F} = (\mathcal{F}_t, t \geq 0)$ is its natural filtration. As before, the function $\mathcal{N}(x) = \frac{1}{\sqrt{2\pi}} \int_{-\infty}^{x} e^{-u^2/2} du$ is the cumulative function of a standard Gaussian law $\mathcal{N}(0, 1)$. We establish well known results on first hitting times of levels for BM, BM with drift and geometric Brownian motion, and we study barrier and lookback options. However, we emphasize that the main results on barrier option valuation are obtained below without any knowledge of hitting time laws but using only the strong Markov property. In the last part of the chapter, we present applications to the structural approach of default risk and real options theory and we give a short presentation of American options.

For a continuous path process X, we denote by $T_a(X)$ (or, if there is no ambiguity, T_a) the first hitting time of the level a for the process X defined as

$$T_a(X) = \inf\{t \geq 0 \,:\, X_t = a\}\,.$$

The first time when X is above (resp. below) the level a is

$$T_a^+ = \inf\{t \geq 0 \,:\, X_t \geq a\}, \quad \text{resp.} \quad T_a^- = \inf\{t \geq 0 \,:\, X_t \leq a\}\,.$$

For $X_0 = x$ and $a > x$, we have $T_a^+ = T_a$, and $T_a^- = 0$ whereas for $a < x$, we have $T_a^- = T_a$, and $T_a^+ = 0$. In what follows, we shall write hitting time for first hitting time. We denote by M_t^X (resp. m_t^X) the running maximum (resp. minimum)

$$M_t^X = \sup_{s \leq t} X_s, \; m_t^X = \inf_{s \leq t} X_s\,.$$

In case X is a BM, we shall frequently omit the superscript and denote by M_t the running maximum of the BM. In this chapter, no martingale will be denoted M_t!

M. Jeanblanc, M. Yor, M. Chesney, *Mathematical Methods for Financial Markets*, Springer Finance, DOI 10.1007/978-1-84628-737-4_3,

3.1 Hitting Times and the Law of the Maximum for Brownian Motion

We first study the law of the pair of random variables (W_t, M_t) where M is the maximum process of the Brownian motion W, i.e., $M_t := \sup_{s \le t} W_s$. In a similar way, we define the minimum process m as $m_t := \inf_{s \le t} W_s$. Let us remark that the process M is an increasing process, with positive values, and that $M \stackrel{\text{law}}{=} (-m)$. Then, we deduce the law of the hitting time of a given level by the Brownian motion.

3.1.1 The Law of the Pair of Random Variables (W_t, M_t)

Let us prove the reflection principle.

Proposition 3.1.1.1 (Reflection principle.) *For $y \ge 0$, $x \le y$, one has:*

$$\boxed{\mathbb{P}(W_t \le x\,, M_t \ge y) = \mathbb{P}(W_t \ge 2y - x)\,.} \tag{3.1.1}$$

PROOF: Let $T_y^+ = \inf\{t : W_t \ge y\}$ be the first time that the BM W is greater than y. This is an **F**-stopping time and $\{T_y^+ \le t\} = \{M_t \ge y\}$ for $y \ge 0$. Furthermore, for $y \ge 0$ and by relying on the continuity of Brownian motion paths, $T_y^+ = T_y$ and $W_{T_y} = y$. Therefore

$$\mathbb{P}(W_t \le x\,, M_t \ge y) = \mathbb{P}(W_t \le x\,, T_y \le t) = \mathbb{P}(W_t - W_{T_y} \le x - y\,, T_y \le t)\,.$$

For the sake of simplicity, we denote $\mathbb{E}_{\mathbb{P}}(\mathbb{1}_A | T_y) = \mathbb{P}(A|T_y)$. By relying on the strong Markov property, we obtain

$$\begin{aligned}\mathbb{P}(W_t - W_{T_y} \le x - y\,, T_y \le t) &= \mathbb{E}(\mathbb{1}_{\{T_y \le t\}}\, \mathbb{P}(W_t - W_{T_y} \le x - y\,|T_y)) \\ &= \mathbb{E}(\mathbb{1}_{\{T_y \le t\}}\, \Phi(T_y))\end{aligned}$$

with $\Phi(u) = \mathbb{P}(\widetilde{W}_{t-u} \le x - y)$ where $(\widetilde{W}_u := W_{T_y+u} - W_{T_y}, u \ge 0)$ is a Brownian motion independent of $(W_t, t \le T_y)$. The process $\widetilde{W}$ has the same law as $-\widetilde{W}$. Therefore $\Phi(u) = \mathbb{P}(\widetilde{W}_{t-u} \ge y - x)$ and by proceeding backward

$$\begin{aligned}\mathbb{E}(\mathbb{1}_{\{T_y \le t\}}\, \Phi(T_y)) &= \mathbb{E}[\mathbb{1}_{\{T_y \le t\}} \mathbb{P}(W_t - W_{T_y} \ge y - x\,|T_y)] \\ &= \mathbb{P}(W_t \ge 2y - x\,, T_y \le t)\,.\end{aligned}$$

Hence,

$$\mathbb{P}(W_t \le x, M_t \ge y) = \mathbb{P}(W_t \ge 2y - x, M_t \ge y)\,. \tag{3.1.2}$$

The right-hand side of (3.1.2) is equal to $\mathbb{P}(W_t \ge 2y - x)$ since, from $x \le y$ we have $2y - x \ge y$ which implies that, on the set $\{W_t \ge 2y - x\}$, one has

$M_t \geq y$ (i.e., the hitting time T_y is smaller than t). □

From the symmetry of the normal law, it follows that

$$\mathbb{P}(W_t \leq x, M_t \geq y) = \mathbb{P}(W_t \geq 2y - x) = \mathcal{N}\left(\frac{x - 2y}{\sqrt{t}}\right).$$

We now give the joint law of the pair of r.v.'s (W_t, M_t) for fixed t.

Theorem 3.1.1.2 *Let W be a BM starting from 0 and $M_t = \sup_{s \leq t} W_s$. Then,*

$$\text{for} \quad y \geq 0, x \leq y, \quad \mathbb{P}(W_t \leq x, M_t \leq y) = \mathcal{N}\left(\frac{x}{\sqrt{t}}\right) - \mathcal{N}\left(\frac{x - 2y}{\sqrt{t}}\right) \tag{3.1.3}$$

$$\begin{aligned} \text{for} \quad y \geq 0, x \geq y, \quad \mathbb{P}(W_t \leq x, M_t \leq y) &= \mathbb{P}(M_t \leq y) \\ &= \mathcal{N}\left(\frac{y}{\sqrt{t}}\right) - \mathcal{N}\left(\frac{-y}{\sqrt{t}}\right), \end{aligned} \tag{3.1.4}$$

$$\text{for} \quad y \leq 0, \quad \mathbb{P}(W_t \leq x, M_t \leq y) = 0.$$

The distribution of the pair of r.v.'s (W_t, M_t) is

$$\boxed{\mathbb{P}(W_t \in dx, M_t \in dy) = \mathbb{1}_{\{y \geq 0\}} \mathbb{1}_{\{x \leq y\}} \frac{2(2y - x)}{\sqrt{2\pi t^3}} \exp\left(-\frac{(2y - x)^2}{2t}\right) dx\, dy} \tag{3.1.5}$$

PROOF: From the reflection principle it follows that, for $y \geq 0, x \leq y$,

$$\begin{aligned} \mathbb{P}(W_t \leq x\,, M_t \leq y) &= \mathbb{P}(W_t \leq x) - \mathbb{P}(W_t \leq x\,, M_t \geq y) \\ &= \mathbb{P}(W_t \leq x) - \mathbb{P}(W_t \geq 2y - x)\,, \end{aligned}$$

hence the equality (3.1.3) is obtained.

For $0 \leq y \leq x$, since $M_t \geq W_t$ we get:

$$\mathbb{P}(W_t \leq x, M_t \leq y) = \mathbb{P}(W_t \leq y, M_t \leq y) = \mathbb{P}(M_t \leq y).$$

Furthermore, by setting $x = y$ in (3.1.3)

$$\mathbb{P}(W_t \leq y, M_t \leq y) = \mathcal{N}\left(\frac{y}{\sqrt{t}}\right) - \mathcal{N}\left(\frac{-y}{\sqrt{t}}\right),$$

hence the equality (3.1.5) is obtained. Finally, for $y \leq 0$,

$$\mathbb{P}(W_t \leq x, M_t \leq y) = 0$$

since $M_t \geq M_0 = 0$. □

Note that we have also proved that the process B defined for $y > 0$ as

$$B_t = W_t \mathbb{1}_{\{t<T_y\}} + (2y - W_t)\mathbb{1}_{\{T_y \leq t\}}$$

is a Brownian motion.

Comment 3.1.1.3 It is remarkable that Bachelier [39, 40] obtained the reflection principle for Brownian motion, extending the result of Désiré André for random walks. See Taqqu [819] for a presentation of Bachelier's work.

Remark 3.1.1.4 Let $T_0 = \inf\{t > 0 : W_t = 0\}$. Then $\mathbb{P}(T_0 = 0) = 1$.

Exercise 3.1.1.5 We have proved that

$$\mathbb{P}(W_t \in dx, M_t \in dy) = \mathbb{1}_{\{y \geq 0\}}\mathbb{1}_{\{x \leq y\}} \frac{1}{\sqrt{t}} g(\frac{x}{\sqrt{t}}, \frac{y}{\sqrt{t}})\, dx\, dy$$

where

$$g(x,y) = \frac{2(2y-x)}{\sqrt{2\pi}} \exp\left(-\frac{(2y-x)^2}{2}\right).$$

Prove that $(M_t, W_t, t \geq 0)$ is a Markov process and give its semi-group in terms of g. ◁

3.1.2 Hitting Times Process

Proposition 3.1.2.1 *Let W be a Brownian motion and, for any $y > 0$, define $T_y = \inf\{t : W_t = y\}$. The increasing process $(T_y, y \geq 0)$ has independent and stationary increments. It enjoys the scaling property*

$$(T_{\lambda y}, y \geq 0) \overset{\text{law}}{=} (\lambda^2 T_y, y \geq 0).$$

PROOF: The increasing property follows from the continuity of paths of the Brownian motion. For $z > y$,

$$T_z - T_y = \inf\{t \geq 0 : W_{T_y + t} - W_{T_y} = z - y\}.$$

Hence, the independence and the stationarity properties follow from the strong Markov property. From the scaling property of BM, for $\lambda > 0$,

$$T_{\lambda y} = \inf\left\{t : \frac{1}{\lambda} W_t = y\right\} \overset{\text{law}}{=} \lambda^2 \inf\{t : \widehat{W}_t = y\}$$

where $\widehat{W}$ is the BM defined by $\widehat{W}_t = \frac{1}{\lambda} W_{\lambda^2 t}$. □

The process $(T_y, y \geq 0)$ is a particular stable subordinator (with index $1/2$) (see ↣ Section 11.6). Note that this process is not continuous but admits a right-continuous left-limited version. The non-continuity property may seem surprising at first, but can easily be understood by looking at the following case. Let W be a BM and $T_1 = \inf\{t : W_t = 1\}$. Define two random times g and θ as

$$g = \sup\{t \leq T_1 : W_t = 0\}, \; \theta = \inf\left\{t \leq g : W_t = \sup_{s \leq g} W_s\right\}$$

and $\Sigma = W_\theta$. Obviously

$$\theta = T_\Sigma < g < T_{\Sigma+} := \inf\{t : W_t > \Sigma\}.$$

See Karatzas and Shreve [513] Chapter 6, Theorem 2.1. for more comments and ↣ Example 11.2.3.5 for a different explanation.

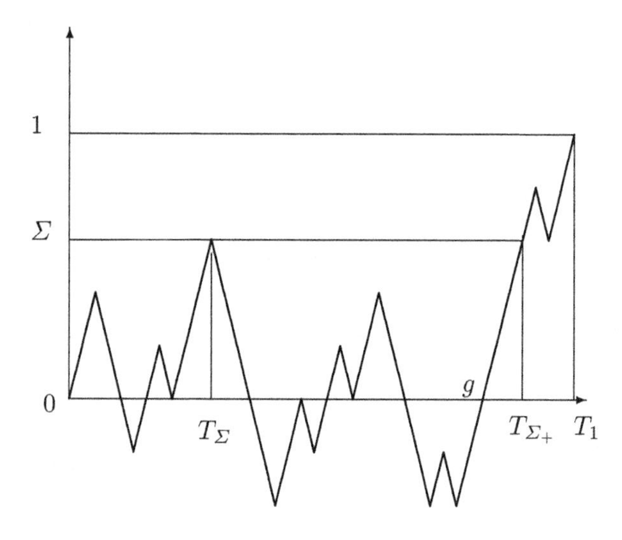

Fig. 3.1 Non continuity of T_y

3.1.3 Law of the Maximum of a Brownian Motion over $[0, t]$

Proposition 3.1.3.1 *For fixed t, the random variable M_t has the same law as $|W_t|$.*

PROOF: This follows from the equality (3.1.4) which states that

$$\mathbb{P}(M_t \leq y) = \mathbb{P}(W_t \leq y) - \mathbb{P}(W_t \leq -y).$$

□

Comments 3.1.3.2 (a) Obviously, the process M does not have the same law as the process $|W|$. Indeed, the process M is an increasing process, whereas this is not the case for the process $|W|$. Nevertheless, there are some further equalities in law, e.g., $M - W \stackrel{\text{law}}{=} |W|$, this identity in law taking place between processes (see Lévy's equivalence Theorem 4.1.7.2 in ↣ Subsection 4.1.7).

(b) Seshadri's result states that the two random variables $M_t(M_t - W_t)$ and W_t are independent and that $M_t(M_t - W_t)$ has an exponential law (see Yor [867, 869]).

Exercise 3.1.3.3 Prove that as a consequence of the reflection principle (formula (3.1.1)), for any fixed t:
(i) $2M_t - W_t$ is distributed as $\|B_t^{(3)}\|$ where $B^{(3)}$ is a 3-dimensional BM, starting from 0,
(ii) given $2M_t - W_t = r$, both M_t and $M_t - W_t$ are uniformly distributed on $[0, r]$.
This result is a small part of Pitman's theorem (see ↣ Comments 4.1.7.3 and ↣ Section 5.7). ◁

3.1.4 Laws of Hitting Times

For $x > 0$, the law of the hitting time T_x of the level x is now easily deduced from

$$\begin{aligned} \mathbb{P}(T_x \leq t) &= \mathbb{P}(x \leq M_t) = \mathbb{P}(x \leq |W_t|) \\ &= \mathbb{P}(x \leq |G|\sqrt{t}) = \mathbb{P}\left(\frac{x^2}{G^2} \leq t\right), \end{aligned} \tag{3.1.6}$$

where, as usual, G stands for a Gaussian random variable, with zero expectation and unit variance. Hence,

$$\boxed{T_x \stackrel{\text{law}}{=} \frac{x^2}{G^2}} \tag{3.1.7}$$

and the density of the r.v. T_x is given by:

$$\mathbb{P}(T_x \in dt) = \frac{x}{\sqrt{2\pi t^3}} \exp\left(-\frac{x^2}{2t}\right) \mathbb{1}_{\{t \geq 0\}} \, dt \, .$$

For $x < 0$, we have, using the symmetry of the law of BM

$$T_x = \inf\{t : W_t = x\} = \inf\{t : -W_t = -x\} \stackrel{\text{law}}{=} T_{-x}$$

and it follows that, for any $x \neq 0$,

$$\boxed{\mathbb{P}(T_x \in dt) = \frac{|x|}{\sqrt{2\pi t^3}} \exp\left(-\frac{x^2}{2t}\right) \mathbb{1}_{\{t\geq 0\}}\, dt\,.} \tag{3.1.8}$$

In particular, for $x \neq 0$, $\mathbb{P}(T_x < \infty) = 1$ and $\mathbb{E}(T_x) = \infty$. More precisely, $\mathbb{E}((T_x)^\alpha) < \infty$ if and only if $\alpha < 1/2$, which is immediate from (3.1.7).

Remark 3.1.4.1 Note that, for $x > 0$, from the explicit form of the density of T_x given in (3.1.8), we have

$$t\mathbb{P}(T_x \in dt) = x\mathbb{P}(W_t \in dx)\,.$$

This relation, known as Kendall's identity (see Borovkov and Burq [110]) will be generalized in ↣ Subsection 11.5.3.

Exercise 3.1.4.2 Prove that, for $0 \leq a < b$,

$$\mathbb{P}(W_s \neq 0, \forall t \in [a,b]) = \frac{2}{\pi} \arcsin\sqrt{\frac{a}{b}}\,.$$

Hint: From elementary properties of Brownian motion, we have

$$\begin{aligned}\mathbb{P}(W_s \neq 0, \forall s \in [a,b]) &= \mathbb{P}(\forall s \in [a,b],\, W_s - W_a \neq -W_a)\\ &= \mathbb{P}(\forall s \in [a,b],\, W_s - W_a \neq W_a) = \mathbb{P}(\widehat{T}_{W_a} > b-a)\,,\end{aligned}$$

where $\widehat{T}$ is associated with the BM $(\widehat{W}_t = W_{t+a} - W_a, t \geq 0)$. Using the scaling property, we compute the right-hand side of this equality

$$\begin{aligned}\mathbb{P}(W_s \neq 0, \forall s \in [a,b]) &= \mathbb{P}(aW_1^2 \widehat{T}_1 > b-a) = \mathbb{P}\left(\frac{G^2}{\widehat{G}^2} > \frac{b}{a} - 1\right)\\ &= \mathbb{P}\left(\frac{1}{1+C^2} < \frac{a}{b}\right) = \frac{2}{\pi}\arcsin\left(\sqrt{\frac{a}{b}}\right),\end{aligned}$$

where G and $\widehat{G}$ are two independent standard Gaussian variables and C a standard Cauchy variable (see ↣ A.4.2 for the required properties of Gaussian variables). ◁

Exercise 3.1.4.3 Prove that $\sigma(M_s - W_s, s \leq t) = \sigma(W_s, s \leq t)$.
Hint: This equality follows from $\int_0^t \mathbb{1}_{\{M_s - W_s = 0\}} d(M_s - W_s) = M_t$. Use the fact that dM_s is carried by $\{s : M_s = B_s\}$. ◁

Exercise 3.1.4.4 The right-hand side of formula (3.1.5) reads, on the set $y \geq 0, y - x \geq 0$,

$$\frac{\mathbb{P}(T_{y-x} \in dt)}{dt} dx dy = \frac{2y-x}{t} p_t(2y-x) dx dy$$

Check simply that this probability has total mass equal to 1! ◁

3.1.5 Law of the Infimum

The law of the infimum of a Brownian motion may be obtained by relying on the same procedure as the one used for the maximum. It can also be deduced by observing that

$$m_t := \inf_{s\leq t} W_s = -\sup_{s\leq t}(-W_s) = -\sup_{s\leq t}(B_s)$$

where $B = -W$ is a Brownian motion. Hence

$$\begin{aligned}
&\text{for} \quad y \leq 0, x \geq y \qquad && \mathbb{P}(W_t \geq x, m_t \geq y) = \mathcal{N}\left(\frac{-x}{\sqrt{t}}\right) - \mathcal{N}\left(\frac{2y-x}{\sqrt{t}}\right), \\
&\text{for} \quad y \leq 0, \, x \leq y && \mathbb{P}(W_t \geq x, m_t \geq y) = \mathcal{N}\left(\frac{-y}{\sqrt{t}}\right) - \mathcal{N}\left(\frac{y}{\sqrt{t}}\right) \\
& && \qquad\qquad\qquad\quad = 1 - 2\mathcal{N}\left(\frac{y}{\sqrt{t}}\right), \\
&\text{for} \quad y \geq 0 && \mathbb{P}(W_t \geq x, m_t \geq y) = 0 \, .
\end{aligned}$$

In particular, for $y \leq 0$, the second equality reduces to

$$\mathbb{P}(m_t \geq y) = \mathcal{N}\left(\frac{-y}{\sqrt{t}}\right) - \mathcal{N}\left(\frac{y}{\sqrt{t}}\right) .$$

If the Brownian motion W starts from z at time 0 and if T_0 is the first hitting time of 0, i.e., $T_0 = \inf\{t : W_t = 0\}$, then, for $z > 0, x > 0$, we obtain

$$\mathbb{P}_z(W_t \in dx, T_0 \geq t) = \mathbb{P}_0(W_t + z \in dx, T_{-z} \geq t) = \mathbb{P}_0(W_t + z \in dx, m_t \geq -z) .$$

The right-hand side of this equality can be obtained by differentiating w.r.t. x the following equality, valid for $x \geq 0, z \geq 0$ (hence $x - z \geq -z, -z \leq 0$)

$$\mathbb{P}(W_t \geq x - z, m_t \geq -z) = \mathcal{N}\left(-\frac{x-z}{\sqrt{t}}\right) - \mathcal{N}\left(-\frac{x+z}{\sqrt{t}}\right) .$$

Thus, we obtain, using the notation (1.4.2)

$$\boxed{\begin{aligned}
\mathbb{P}_z(W_t \in dx, T_0 \geq t) &= \frac{\mathbb{1}_{\{x\geq 0\}}}{\sqrt{2\pi t}}\left[\exp\left(-\frac{(z-x)^2}{2t}\right) - \exp\left(-\frac{(z+x)^2}{2t}\right)\right] dx \, , \\
&= \mathbb{1}_{\{x\geq 0\}}(p_t(z-x) - p_t(z+x))dx \, .
\end{aligned}}$$

(3.1.9)

3.1.6 Laplace Transforms of Hitting Times

The law of first hitting time of a level y is characterized by its Laplace transforms, which is given in the next proposition.

Proposition 3.1.6.1 *Let T_y be the first hitting time of $y \in \mathbb{R}$ for a standard Brownian motion. Then, for $\lambda > 0$*

$$\mathbb{E}\left[\exp\left(-\frac{\lambda^2}{2}T_y\right)\right] = \exp(-|y|\lambda)\,.$$

PROOF: Recall that, for any $\lambda \in \mathbb{R}$, the process $(\exp(\lambda W_t - \frac{1}{2}\lambda^2 t),\, t \geq 0)$ is a martingale. Now, for $y \geq 0$, $\lambda \geq 0$ the martingale

$$(\exp(\lambda W_{t\wedge T_y} - \frac{1}{2}\lambda^2(t \wedge T_y)),\, t \geq 0)$$

is bounded by $e^{\lambda y}$, hence it is u.i.. Using $\mathbb{P}(T_y < \infty) = 1$, Doob's optional sampling theorem yields

$$\mathbb{E}\left[\exp\left(\lambda W_{T_y} - \frac{1}{2}\lambda^2 T_y\right)\right] = 1\,.$$

Since $W_{T_y} = y$, we obtain the Laplace transform of T_y. The case where $y < 0$ follows since $W \stackrel{\text{law}}{=} -W$. □

Warning 3.1.6.2 In order to apply Doob's optional sampling theorem, we had to check carefully that the martingale $\exp(\lambda W_{t\wedge T_y} - \frac{1}{2}\lambda^2(t \wedge T_y))$ is uniformly integrable. In the case $\lambda > 0$ and $y < 0$, a wrong use of this theorem would lead to the equality between 1 and

$$\mathbb{E}[\exp(\lambda W_{T_y} - \frac{1}{2}\lambda^2 T_y)] = e^{\lambda y}\mathbb{E}\left[\exp\left(-\frac{1}{2}\lambda^2 T_y\right)\right],$$

that is the two quantities $\mathbb{E}[\exp(-\frac{1}{2}\lambda^2 T_y)]$ and $\exp(-y\lambda)$ would be the same. This is obviously false since the quantity $\mathbb{E}[\exp(-\frac{1}{2}\lambda^2 T_y)]$ is smaller than 1 whereas $\exp(-y\lambda)$ is strictly greater than 1.

Remark 3.1.6.3 From the equality (3.1.8) and Proposition 3.1.6.1, we check that for $\lambda > 0$

$$\exp(-|y|\lambda) = \int_0^\infty dt\,\frac{|y|}{\sqrt{2\pi t^3}}\exp\left(-\frac{y^2}{2t}\right)\exp\left(-\frac{\lambda^2 t}{2}\right). \tag{3.1.10}$$

This equality may be directly obtained, in the case $y > 0$, by checking that the function

$$H(\mu) = \int_0^\infty dt\,\frac{1}{\sqrt{t^3}}\,e^{-\mu t}\exp\left(-\frac{1}{t}\right)$$

satisfies $\mu H'' + \frac{1}{2}H' - H = 0$. A change of function $G(\sqrt{\mu}) = H(\mu)$ leads to $\frac{1}{4}G'' - G = 0$, and the form of H follows. Let us remark that, for $y > 0$, one can write the equality (3.1.10) in the form

$$1 = \int_0^\infty dt \, \frac{y}{\sqrt{2\pi t^3}} \exp\left(-\frac{1}{2}\left(\frac{y}{\sqrt{t}} - \lambda\sqrt{t}\right)^2\right) . \qquad (3.1.11)$$

Note that the quantity

$$\frac{y}{\sqrt{2\pi t^3}} \exp\left(-\frac{1}{2}\left(\frac{y}{\sqrt{t}} - \lambda\sqrt{t}\right)^2\right)$$

in the right-hand member is the density of the hitting time of the level y by a drifted Brownian motion (see ↣ formula (3.2.3)). Another proof relies on the knowledge of the resolvent of the Brownian motion: the result can be obtained via a differentiation w.r.t. y of the equality obtained in Exercise 1.4.1.7

$$\int_0^\infty e^{-\lambda^2 t/2} p_t(0,y)dt = \int_0^\infty e^{-\lambda^2 t/2} \frac{1}{\sqrt{2\pi t}} e^{-\frac{y^2}{2t}} dt = \frac{1}{\lambda} e^{-|y|\lambda}$$

Comment 3.1.6.4 We refer the reader to Lévy's equivalence ↣ Theorem 4.1.7.2 which allows translation of all preceding results to the running maximum involving results on the Brownian motion local time.

Exercise 3.1.6.5 Let $T_a^* = \inf\{t \geq 0 : |W_t| = a\}$. Using the fact that the process $(e^{-\lambda^2 t/2} \cosh(\lambda W_t), t \geq 0)$ is a martingale, prove that

$$\mathbb{E}(\exp(-\lambda^2 T_a^*/2)) = [\cosh(a\lambda)]^{-1} .$$

See ↣ Subsection 3.5.1 for the density of T_a^*. ◁

Exercise 3.1.6.6 Let $\tau = \inf\{t : M_t - W_t > a\}$. Prove that M_τ follows the exponential law with parameter a^{-1}.
`Hint:` The exponential law stems from

$$\mathbb{P}(M_\tau > x + y | M_\tau > y) = \mathbb{P}(\tau > T_{x+y} | \tau > T_y) = \mathbb{P}(M_\tau > x) .$$

The value of the mean of M_τ is obtained by passing to the limit in the equality $\mathbb{E}(M_{\tau\wedge n}) = \mathbb{E}(M_{\tau\wedge n} - W_{\tau\wedge n})$. ◁

Exercise 3.1.6.7 Let W be a Brownian motion, $\mathbf{F}$ its natural filtration and $M_t = \sup_{s \leq t} W_s$. Prove that, for $t < 1$,

$$\mathbb{E}(f(M_1)|\mathcal{F}_t) = F(1-t, W_t, M_t)$$

with

$$F(s,a,b) = \sqrt{\frac{2}{\pi s}} \left(f(b) \int_0^{b-a} e^{-u^2/(2s)} du + \int_b^\infty f(u) \exp\left(-\frac{(u-a)^2}{2s}\right) du \right) .$$

Hint: Note that

$$\sup_{s\leq 1} W_s = \sup_{s\leq t} W_s \vee \sup_{t\leq s\leq 1} W_s = \sup_{s\leq t} W_s \vee (\widehat{M}_{1-t} + W_t)$$

where $\widehat{M}_s = \sup_{u\leq s} \widehat{W}_u$ for $\widehat{W}_u = W_{u+t} - W_t$.

Another method consists in an application of $\rightarrowtail$ Theorem 4.1.7.8. Apply Doob's Theorem to the martingale $h(M_t)(M_t - W_t) + \int_{M_t}^{\infty} du\, h(u)$. ◁

Exercise 3.1.6.8 Let a and σ be continuous deterministic functions, B a BM and X the solution of $dX_t = a(t)X_t dt + \sigma(t)dB_t$, $X_0 = x$.

Let $T_0 = \inf\{t \geq 0,\, X_t \leq 0\}$. Prove that, for $x > 0, y > 0$,

$$\mathbb{P}(X_t \geq y, T_0 \leq t) = \mathbb{P}(X_t \leq -y)\,.$$

Hint: Use the fact that $X_t e^{-A_t} = W^{(x)}_{\alpha(t)}$ where $A_t = \int_0^t a(s)ds$ and $W^{(x)}$ is a Brownian motion starting from x. Here α denotes the increasing function $\alpha(t) = \int_0^t e^{-2A(s)}\sigma^2(s)ds$. Then, use the reflection principle to obtain $\mathbb{P}(W^{(x)}_u \geq z, T_0 \leq u) = \mathbb{P}(W^{(x)}_u \leq -z)$. We refer the reader to $\rightarrowtail$ Theorem 4.1.7.2 which allows computations relative to the maximum M to be couched in terms of Brownian local time.

◁

Exercise 3.1.6.9 Let f be a (bounded) function. Prove that

$$\lim_{t\to\infty} \sqrt{t}\,\mathbb{E}(f(M_t)|\mathcal{F}_s) = c(f(M_s)(M_s - W_s) + F(M_s))$$

where c is a constant and $F(x) = \int_x^{\infty} duf(u)$.

Hint: Write $M_t = M_s \vee (W_s + \widehat{M}_{t-s})$ where $\widehat{M}$ is the supremum of a Brownian motion $\widehat{W}$, independent of $W_u, u \leq s$. ◁

3.2 Hitting Times for a Drifted Brownian Motion

We now study the first hitting times for the process $X_t = \nu t + W_t$, where W is a Brownian motion and ν a constant. Let $M^X_t = \sup(X_s, s \leq t)$, $m^X_t = \inf(X_s, s \leq t)$ and $T_y(X) = \inf\{t \geq 0\,|\, X_t = y\}$. We recall that $\mathbf{W}^{(\nu)}$ denotes the law of the Brownian motion with drift ν, i.e., $\mathbf{W}^{(\nu)}(X_t \in A)$ is the probability that a Brownian motion with drift ν belongs to A at time t.

3.2.1 Joint Laws of (M^X, X) and (m^X, X) at Time t

Proposition 3.2.1.1 *For $y \geq 0, y \geq x$*

$$\mathbf{W}^{(\nu)}(X_t \leq x, M^X_t \leq y) = \mathcal{N}\left(\frac{x - \nu t}{\sqrt{t}}\right) - e^{2\nu y}\mathcal{N}\left(\frac{x - 2y - \nu t}{\sqrt{t}}\right)$$

and for $y \leq 0, y \leq x$

$$\mathbf{W}^{(\nu)}(X_t \geq x, m_t^X \geq y) = \mathcal{N}\left(\frac{-x+\nu t}{\sqrt{t}}\right) - e^{2\nu y}\mathcal{N}\left(\frac{-x+2y+\nu t}{\sqrt{t}}\right).$$

PROOF: From Cameron-Martin's theorem (see Proposition 1.7.5.2)

$$\mathbf{W}^{(\nu)}(X_t \leq x, M_t^X \geq y) = \mathbb{E}\left[\exp\left(\nu W_t - \frac{\nu^2}{2}t\right) \mathbb{1}_{\{W_t \leq x, M_t^W \geq y\}}\right].$$

From the reflection principle (3.1.2) for $y \geq 0, x \leq y$, it holds that

$$\mathbb{P}(W_t \leq x, M_t^W \geq y) = \mathbb{P}(x \geq 2y - W_t, M_t^W \geq y),$$

hence, on the set $y \geq 0, x \leq y$, one has

$$\mathbb{P}(W_t \in dx, M_t^W \in dy) = \mathbb{P}(2y - W_t \in dx, M_t^W \in dy).$$

It follows that

$$\begin{aligned}
&\mathbb{E}\left[\exp\left(\nu W_t - \frac{\nu^2}{2}t\right) \mathbb{1}_{\{W_t \leq x, M_t^W \geq y\}}\right] \\
&= \mathbb{E}\left[\exp\left(\nu(2y - W_t) - \frac{\nu^2}{2}t\right) \mathbb{1}_{\{2y - W_t \leq x, M_t^W \geq y\}}\right] \\
&= e^{2\nu y}\mathbb{E}\left[\exp\left(-\nu W_t - \frac{\nu^2}{2}t\right) \mathbb{1}_{\{W_t \geq 2y - x\}}\right].
\end{aligned}$$

Applying Cameron-Martin's theorem again we obtain

$$\mathbb{E}\left[\exp\left(-\nu W_t - \frac{\nu^2}{2}t\right) \mathbb{1}_{\{W_t \geq 2y - x\}}\right] = \mathbf{W}^{(-\nu)}(X_t \geq 2y - x).$$

It follows that for $y \geq 0, y \geq x$,

$$\begin{aligned}
\mathbf{W}^{(\nu)}(X_t \leq x, M_t^X \geq y) &= e^{2\nu y}\mathbb{P}(W_t \geq 2y - x + \nu t) \\
&= e^{2\nu y}\mathcal{N}\left(\frac{-2y + x - \nu t}{\sqrt{t}}\right).
\end{aligned}$$

Therefore, for $y \geq 0$ and $y \geq x$,

$$\begin{aligned}
\mathbf{W}^{(\nu)}(X_t \leq x, M_t^X \leq y) &= \mathbf{W}^{(\nu)}(X_t \leq x) - \mathbf{W}^{(\nu)}(X_t \leq x, M_t^X \geq y) \\
&= \mathcal{N}\left(\frac{x - \nu t}{\sqrt{t}}\right) - e^{2\nu y}\mathcal{N}\left(\frac{x - 2y - \nu t}{\sqrt{t}}\right),
\end{aligned}$$

and for $y \leq 0, y \leq x$,

$$\begin{aligned}\mathbf{W}^{(\nu)}(X_t \geq x, m_t^X \leq y) &= \mathbb{P}(W_t + \nu t \geq x, \inf_{s \leq t}(W_s + \nu s) \leq y) \\ &= \mathbb{P}(-W_t - \nu t \leq -x, \sup_{s \leq t}(-W_s - \nu s) \geq -y) \\ &= \mathbb{P}(W_t - \nu t \leq -x, \sup_{s \leq t}(W_s - \nu s) \geq -y) \\ &= e^{2\nu y}\mathcal{N}\left(\frac{2y - x + \nu t}{\sqrt{t}}\right). \end{aligned} \tag{3.2.1}$$

The result of the proposition follows. □

Corollary 3.2.1.2 *Let $X_t = \nu t + W_t$ and $M_t^X = \sup_{s \leq t} X_s$. The joint density of the pair X_t, M_t^X is*

$$\mathbf{W}^{(\nu)}(X_t \in dx, M_t^X \in dy) = \mathbb{1}_{x<y}\mathbb{1}_{0<y}\frac{2(2y - x)}{\sqrt{2\pi t^3}}e^{\nu x - \frac{1}{2}\nu^2 t - \frac{1}{2t}(2y-x)^2}dxdy$$

Exercise 3.2.1.3 Prove that for $y \geq 0$ and $y \geq x$

$$\mathbf{W}^{(\nu)}(X_t \leq x, M_t^X \geq y) = e^{2\nu y}\mathbb{P}(W_t + \nu t \leq x - 2y)$$

and that for $y \leq 0$ and $y \leq x$

$$\mathbf{W}^{(\nu)}(X_t \geq x, m_t^X \leq y) = e^{2\nu y}\mathbb{P}(W_t + \nu t \geq x - 2y).$$

◁

3.2.2 Laws of Maximum, Minimum, and Hitting Times

The laws of the maximum and the minimum of a drifted Brownian motion are deduced from the obvious equalities

$$\mathbf{W}^{(\nu)}(M_t^X \leq y) = \mathbf{W}^{(\nu)}(X_t \leq y, M_t^X \leq y)$$

and $\mathbf{W}^{(\nu)}(m_t^X \geq y) = \mathbf{W}^{(\nu)}(X_t \geq y, m_t^X \geq y)$. The right-hand sides of these equalities are computed from Proposition 3.2.1.1. In a closed form, we obtain

$$\begin{aligned}\mathbf{W}^{(\nu)}(M_t^X \leq y) &= \mathcal{N}\left(\frac{y - \nu t}{\sqrt{t}}\right) - e^{2\nu y}\mathcal{N}\left(\frac{-y - \nu t}{\sqrt{t}}\right), \quad y \geq 0 \\ \mathbf{W}^{(\nu)}(M_t^X \geq y) &= \mathcal{N}\left(\frac{-y + \nu t}{\sqrt{t}}\right) + e^{2\nu y}\mathcal{N}\left(\frac{-y - \nu t}{\sqrt{t}}\right), \quad y \geq 0 \\ \mathbf{W}^{(\nu)}(m_t^X \geq y) &= \mathcal{N}\left(\frac{-y + \nu t}{\sqrt{t}}\right) - e^{2\nu y}\mathcal{N}\left(\frac{y + \nu t}{\sqrt{t}}\right), \quad y \leq 0 \\ \mathbf{W}^{(\nu)}(m_t^X \leq y) &= \mathcal{N}\left(\frac{y - \nu t}{\sqrt{t}}\right) + e^{2\nu y}\mathcal{N}\left(\frac{y + \nu t}{\sqrt{t}}\right), \quad y \leq 0. \end{aligned}$$

For $y > 0$, from the equality $\mathbf{W}^{(\nu)}(T_y(X) \geq t) = \mathbf{W}^{(\nu)}(M_t^X \leq y)$, we deduce that the law of the random variable $T_y(X)$ is

$$\mathbf{W}^{(\nu)}(T_y(X) \in dt) = \frac{y}{\sqrt{2\pi t^3}} e^{\nu y} \exp\left(-\frac{1}{2}\left(\frac{y^2}{t} + \nu^2 t\right)\right) \mathbb{1}_{\{t \geq 0\}} dt \qquad (3.2.2)$$

or, in a more pleasant form

$$\boxed{\mathbf{W}^{(\nu)}(T_y(X) \in dt) = \frac{y}{\sqrt{2\pi t^3}} \exp\left(-\frac{1}{2t}(y - \nu t)^2\right) \mathbb{1}_{\{t \geq 0\}}\, dt\,.} \qquad (3.2.3)$$

This law is the inverse Gaussian law with parameters (y, ν). (See $\rightarrowtail$ Appendix A.4.4.)

Note that, for $\nu < 0$ and $y > 0$, when $t \to \infty$ in $\mathbf{W}^{(\nu)}(T_y \geq t)$, we obtain $\mathbf{W}^{(\nu)}(T_y = \infty) = 1 - e^{2\nu y}$. In this case, the density of T_y under $\mathbf{W}^{(\nu)}$ is defective. For $\nu > 0$ and $y > 0$, we obtain $\mathbf{W}^{(\nu)}(T_y = \infty) = 1$, which can also be obtained from (3.1.11). See also Exercise 1.2.3.10.

Let us point out the simple (Cameron-Martin) absolute continuity relationship between the Brownian motion with drift ν and the Brownian motion with drift $-\nu$: from both formulae

$$\begin{cases} \mathbf{W}^{(\nu)}|_{\mathcal{F}_t} = \exp\left(\nu X_t - \frac{1}{2}\nu^2 t\right) \mathbf{W}|_{\mathcal{F}_t} \\ \mathbf{W}^{(-\nu)}|_{\mathcal{F}_t} = \exp\left(-\nu X_t - \frac{1}{2}\nu^2 t\right) \mathbf{W}|_{\mathcal{F}_t} \end{cases} \qquad (3.2.4)$$

we deduce

$$\mathbf{W}^{(\nu)}|_{\mathcal{F}_t} = \exp(2\nu X_t) \mathbf{W}^{(-\nu)}|_{\mathcal{F}_t}\,. \qquad (3.2.5)$$

(See $\rightarrowtail$ Exercise 3.6.6.4 for an application of this relation.) In particular, we obtain again, using Proposition 1.7.1.4,

$$\mathbf{W}^{(\nu)}(T_y < \infty) = e^{2\nu y}, \text{ for } \nu y < 0\,.$$

Exercise 3.2.2.1 Let $X_t = W_t + \nu t$ and $m_t^X = \inf_{s \leq t} X_s$. Prove that, for $y < 0, y < x$

$$\mathbb{P}(m_t^X \leq y | X_t = x) = \exp\left(-\frac{2y(y-x)}{t}\right).$$

`Hint:` Note that, from Cameron-Martin's theorem, the left-hand side does not depend on ν. ◁

3.2.3 Laplace Transforms

From Cameron-Martin's relationship (3.2.4),

$$\mathbf{W}^{(\nu)}\left(\exp\left(-\frac{\lambda^2}{2} T_y(X)\right)\right) = \mathbb{E}\left(\exp\left(\nu W_{T_y} - \frac{\nu^2 + \lambda^2}{2} T_y(W)\right)\right),$$

where $\mathbf{W}^{(\nu)}(\cdot)$ is the expectation under $\mathbf{W}^{(\nu)}$. From Proposition 3.1.6.1, the right-hand side equals

$$e^{\nu y}\mathbb{E}\left[\exp\left(-\frac{1}{2}(\nu^2+\lambda^2)T_y(W)\right)\right] = e^{\nu y}\exp\left(-|y|\sqrt{\nu^2+\lambda^2}\right).$$

Therefore

$$\mathbf{W}^{(\nu)}\left(\exp\left[-\frac{\lambda^2}{2}T_y(X)\right]\right) = e^{\nu y}\exp\left(-|y|\sqrt{\nu^2+\lambda^2}\right). \qquad (3.2.6)$$

In particular, letting λ go to 0 in (3.2.6), in the case $\nu y < 0$

$$\mathbf{W}^{(\nu)}(T_y < \infty) = e^{2\nu y},$$

which proves again that the probability that a Brownian motion with strictly positive drift hits a negative level is not equal to 1. In the case $\nu y \geq 0$, obviously $\mathbf{W}^{(\nu)}(T_y < \infty) = 1$. This is explained by the fact that $(W_t + \nu t)/t$ goes to ν when t goes to infinity, hence the drift drives the process to infinity. In the case $\nu y > 0$, taking the derivative (w.r.t. $\lambda^2/2$) of (3.2.6) for $\lambda = 0$, we obtain $\mathbf{W}^{(\nu)}(T_y(X)) = y/\nu$. When $\nu y < 0$, the expectation of the stopping time is equal to infinity.

3.2.4 Computation of $\mathbf{W}^{(\nu)}(\mathbb{1}_{\{T_y(X)<t\}}\, e^{-\lambda T_y(X)})$

We present the computation of $\mathbf{W}^{(\nu)}[\mathbb{1}_{\{T_y(X)<t\}}\exp(-\lambda T_y(X))]$. This will be useful for finance purposes, for example while studying Boost options in ↣ Subsection 3.9.2 and last passage times (↣ Subsections 4.3.9 and 5.6.4). Obviously, the computation could be done using the density of T_y, however this direct method is rather tedious.

For any γ, Cameron-Martin's theorem leads to

$$\begin{aligned}&\mathbf{W}^{(\nu)}(e^{-\lambda T_y(X)}\mathbb{1}_{\{T_y(X)<t\}})\\ &= \mathbf{W}^{(\gamma)}\left(e^{-\lambda T_y(X)}\exp\left[(\nu-\gamma)X_{T_y} - \frac{\nu^2-\gamma^2}{2}T_y\right]\mathbb{1}_{\{T_y(X)<t\}}\right).\end{aligned}$$

Choosing γ such that $2\lambda = \gamma^2 - \nu^2$, we obtain

$$\mathbf{W}^{(\nu)}(e^{-\lambda T_y(X)}\mathbb{1}_{\{T_y(X)<t\}}) = \exp[(\nu-\gamma)y]\,\mathbf{W}^{(\gamma)}(T_y(X) < t).$$

Hence, using the results on the law of the hitting time established in Subsection 3.2.2 for $y > 0$,

$$\mathbf{W}^{(\nu)}(e^{-\lambda T_y}\mathbb{1}_{\{T_y<t\}}) = e^{(\nu-\gamma)y}\mathcal{N}\left(\frac{\gamma t - y}{\sqrt{t}}\right) + e^{(\nu+\gamma)y}\mathcal{N}\left(\frac{-\gamma t - y}{\sqrt{t}}\right)$$

and, for $y < 0$

$$\mathbf{W}^{(\nu)}(e^{-\lambda T_y}\mathbb{1}_{\{T_y<t\}}) = e^{(\nu-\gamma)y}\mathcal{N}\left(\frac{-\gamma t + y}{\sqrt{t}}\right) + e^{(\nu+\gamma)y}\mathcal{N}\left(\frac{\gamma t + y}{\sqrt{t}}\right).$$

Setting

$$H(a, y, t) := e^{-ay}\mathcal{N}\left(\frac{at - y}{\sqrt{t}}\right) + e^{ay}\mathcal{N}\left(\frac{-at - y}{\sqrt{t}}\right), \tag{3.2.7}$$

we get

$$\begin{aligned}\mathbf{W}^{(\nu)}(e^{-\lambda T_y}\mathbb{1}_{\{T_y<t\}}) &= e^{\nu y}H(\gamma, |y|, t) \\ &= e^{\nu y}H(\sqrt{2\lambda + \nu^2}, |y|, t).\end{aligned}$$

In particular, for $\nu = 0$,

$$\mathbb{E}(e^{-\lambda T_y(W)}\mathbb{1}_{\{T_y(W)<t\}}) = H(\sqrt{2\lambda}, |y|, t).$$

3.2.5 Normal Inverse Gaussian Law

Let (W, B) be a two-dimensional Brownian motion, $X_t = x + \nu t + W_t$, and $T_y^{(\mu)} = \inf\{t : \mu t + B_t = y\}$. Then, the density of $X_{T_y^{(\mu)}}$ is the **Normal Inverse Gaussian law** $NIG(\alpha, \nu, x, y)$ where $\alpha = \sqrt{\nu^2 + \mu^2}$. (If needed, see ↣ Appendix A.4.5 for the expression of the density.) This can be checked from

$$\mathbb{P}(X_{T_y^{(\mu)}} \in A) = \int_0^\infty \mathbb{P}(X_u \in A)\mathbb{P}(T_y^{(\mu)} \in du)$$

and the integral representation of the Bessel function K_ν.

Another method of finding the law of $X_{T_y^{(\mu)}}$ is to compute its characteristic function as follows:

$$\begin{aligned}\mathbb{E}\left(\exp(i\zeta(x + \nu T_y^{(\mu)} + W_{T_y^{(\mu)}}))\right) &= \mathbb{E}\left(\exp(i\zeta(x + \nu T_y^{(\mu)}) - \frac{\zeta^2}{2}T_y^{(\mu)})\right) \\ &= \exp(i\zeta x)\mathbb{E}\left(\exp\left[(i\zeta\nu - \frac{\zeta^2}{2})T_y^{(\mu)}\right]\right) \\ &= \exp(i\zeta x)e^{\mu y}\mathbb{E}\left(\exp\left[-\frac{1}{2}(\zeta^2 + \mu^2 - 2i\zeta\nu)T_y^{(0)}\right]\right) \\ &= \exp(i\zeta x)e^{\mu y}e^{-y\sqrt{(\zeta - i\nu)^2 + \mu^2 + \nu^2}}.\end{aligned}$$

Comment 3.2.5.1 See Barndorff-Nielsen [51], Eberlein [289] and Barndorff-Nielsen et al. [53] for applications of these laws in finance.

3.3 Hitting Times for Geometric Brownian Motion

Let us assume that

$$dS_t = S_t(\mu dt + \sigma dW_t)\,, S_0 = x > 0 \tag{3.3.1}$$

with $\sigma > 0$, i.e.,

$$S_t = x \exp\left((\mu - \sigma^2/2)t + \sigma W_t\right) = x e^{\sigma X_t}\,,$$

where $X_t = \nu t + W_t$, $\nu = (\mu - \sigma^2/2)\,\sigma^{-1}$. We denote by $T_a(S)$ the first hitting time of a by the process S and $T_\alpha(X)$ the first hitting time of α by the process X. From

$$T_a(S) = \inf\{t \geq 0 \,:\, S_t = a\} = \inf\{t \geq 0 \,:\, X_t = \frac{1}{\sigma}\ln(a/x)\}$$

we obtain $T_a(S) = T_\alpha(X)$ where

$$\alpha = \frac{1}{\sigma}\ln(a/x)\,.$$

When another level b is considered for the geometric Brownian motion S, we shall denote

$$\beta = \frac{1}{\sigma}\ln(b/x)\,.$$

Using the previous results, we give below the law of the hitting time, as well as the law of the maximum M_t^S (resp. minimum m_t^S) of S over the time interval $[0,t]$.

3.3.1 Laws of the Pairs (M_t^S, S_t) and (m_t^S, S_t)

We deduce, from the results obtained for drifted Brownian motion in Proposition 3.2.1.1, that for $a > b, a > x$

$$\begin{aligned}\mathbb{P}_x(S_t \leq b, M_t^S \leq a) &= \mathbf{W}^{(\nu)}(X_t \leq \beta, M_t^X \leq \alpha)\\ &= \mathcal{N}\left(\frac{\beta - \nu t}{\sqrt{t}}\right) - e^{2\nu\alpha}\mathcal{N}\left(\frac{\beta - 2\alpha - \nu t}{\sqrt{t}}\right)\end{aligned}$$

whereas, for $b > a, a < x$

$$\begin{aligned}\mathbb{P}_x(S_t \geq b, m_t^S \geq a) &= \mathbf{W}^{(\nu)}(X_t \geq \beta, m_t^X \geq \alpha)\\ &= \mathcal{N}\left(\frac{-\beta + \nu t}{\sqrt{t}}\right) - e^{2\nu\alpha}\mathcal{N}\left(\frac{-\beta + 2\alpha + \nu t}{\sqrt{t}}\right).\end{aligned}$$

Proposition 3.3.1.1 *Let $S_t = xe^{\mu t + \sigma W_t}$ and $M_t^S = \sup_{s \leq t} S_s$. The joint density of the pair S_t, M_t^S is*

$$\begin{aligned}&\mathbb{P}(S_t \in dz, M_t^S \in dy)\\&= \frac{2}{\sigma^3\sqrt{2\pi t^3}} \frac{\ln(y^2/(xz))}{zy} \exp\left(-\frac{\ln^2(y^2/(xz))}{2\sigma^2 t} + \frac{\rho}{\sigma}\ln(z/x) - \frac{\rho^2 t}{2}\right) dzdy\end{aligned}$$

where $\rho = \mu/\sigma + \sigma/2$.

It follows that, for $a > x$ (or $\alpha > 0$)

$$\begin{aligned}\mathbb{P}_x(T_a(S) < t) &= \mathbf{W}^{(\nu)}(T_\alpha(X) < t)\\&= \mathcal{N}\left(\frac{-\alpha + \nu t}{\sqrt{t}}\right) + e^{2\nu\alpha}\mathcal{N}\left(\frac{-\nu t - \alpha}{\sqrt{t}}\right)\end{aligned} \tag{3.3.2}$$

and, for $a < x$ (or $\alpha < 0$)

$$\mathbb{P}_x(T_a(S) < t) = \mathcal{N}\left(\frac{\alpha - \nu t}{\sqrt{t}}\right) + e^{2\nu\alpha}\mathcal{N}\left(\frac{\nu t + \alpha}{\sqrt{t}}\right). \tag{3.3.3}$$

The density of the hitting time $T_a(S)$ is obtained by differentiation, or more directly, from (3.2.3) and the equality $T_a(S) = T_\alpha(X)$:

$$\mathbb{P}_x(T_a(S) \in dt) = \frac{dt}{\sqrt{2\pi t^3}}\alpha \exp\left(-\frac{1}{2t}(\alpha - \nu t)^2\right)\mathbb{1}_{\{t \geq 0\}}. \tag{3.3.4}$$

Exercise 3.3.1.2 Prove that, for $a > S_0$, and $t \leq T$

$$\mathbb{P}(T_a(S) > T|\mathcal{F}_t) = \mathbb{1}_{\{\max_{s \leq t} S_s < a\}}\left(\mathcal{N}(d_1) - \left(\frac{a}{S_t}\right)^{2(r-\delta-\sigma^2/2)\sigma^{-2}}\mathcal{N}(d_2)\right)$$

with

$$\begin{aligned}d_1 &= \frac{1}{\sigma\sqrt{T-t}}\left(\ln\left(\frac{a}{S_t}\right) - \left(r - \delta - \frac{\sigma^2}{2}\right)\right)\\d_2 &= \frac{1}{\sigma\sqrt{T-t}}\left(\ln\left(\frac{S_t}{a}\right) - \left(r - \delta - \frac{\sigma^2}{2}\right)\right).\end{aligned}$$

◁

3.3.2 Laplace Transforms

From the equality $T_a(S) = T_\alpha(X)$,

$$\mathbb{E}_x\left(\exp\left[-\frac{\lambda^2}{2}T_a(S)\right]\right) = \mathbf{W}^{(\nu)}\left(\exp\left[-\frac{\lambda^2}{2}T_\alpha(X)\right]\right).$$

Therefore, from (3.2.6)

$$\mathbb{E}_x\left(\exp\left[-\frac{\lambda^2}{2}T_a(S)\right]\right) = \exp\left(\nu\alpha - |\alpha|\sqrt{\nu^2 + \lambda^2}\right). \tag{3.3.5}$$

3.3.3 Computation of $\mathbb{E}(e^{-\lambda T_a(S)}\mathbb{1}_{\{T_a(S)<t\}})$

For $a > x$ (or $\alpha > 0$) we obtain, using the results of Subsection 3.2.4 about drifted Brownian motion, and choosing γ such that $2\lambda = \gamma^2 - \nu^2$,

$$\mathbb{E}_x(e^{-\lambda T_a(S)}\mathbb{1}_{\{T_a(S)<t\}}) = e^{(\nu-\gamma)\alpha}\mathcal{N}\left(\frac{\gamma t - \alpha}{\sqrt{t}}\right) + e^{(\gamma+\nu)\alpha}\mathcal{N}\left(\frac{-\gamma t - \alpha}{\sqrt{t}}\right).$$

In the case $\lambda = \mu$, $2\lambda + \nu^2 = (\mu\sigma^{-1} + \sigma/2)^2$, we choose $\gamma = -(\mu\sigma^{-1} + \sigma/2)$ so that $\gamma + \nu = -\sigma, \nu - \gamma = 2\mu/\sigma$. Then, for $a > x$

$$\mathbb{E}_x(e^{-\mu T_a(S)}\mathbb{1}_{\{T_a(S)<t\}}) = e^{2\mu\alpha/\sigma}\mathcal{N}\left(\frac{\gamma t - \alpha}{\sqrt{t}}\right) + e^{-\alpha\sigma}\mathcal{N}\left(\frac{-\gamma t - \alpha}{\sqrt{t}}\right).$$

In the case where $a < x$, we obtain

$$\mathbb{E}_x(e^{-\mu T_a(S)}\mathbb{1}_{\{T_a(S)<t\}}) = e^{2\mu\alpha/\sigma}\mathcal{N}\left(\frac{\alpha - \gamma t}{\sqrt{t}}\right) + e^{-\alpha\sigma}\mathcal{N}\left(\frac{\gamma t + \alpha}{\sqrt{t}}\right).$$

3.4 Hitting Times in Other Cases

3.4.1 Ornstein-Uhlenbeck Processes

Proposition 3.4.1.1 *Let $(X_t, t \geq 0)$ be an OU process defined as*

$$dX_t = -kX_t\,dt + dW_t, \quad X_0 = x,$$

and $T_0 = \inf\{t \geq 0 : X_t = 0\}$. For any $x > 0$, the density function of T_0 equals

$$f(t) = \frac{x}{\sqrt{2\pi}}\exp\left(\frac{kx^2}{2}\right)\exp\left(\frac{k}{2}(t - x^2\coth(kt))\right)\left(\frac{k}{\sinh(kt)}\right)^{3/2}.$$

PROOF: We present here the proof of Alili et al. [10]. As proved in Corollary 2.6.1.2, the OU process can be written $X_t = e^{-kt}(x + \int_0^t e^{ks}dW_s)$. Hence

$$\begin{aligned}T_0 &= \inf\{t \geq 0 : X_t = 0\} = \inf\{t : x + \int_0^t e^{ks}dW_s = 0\}\\ &= \inf\{t : \widehat{W}_{A(t)} = -x\}\end{aligned}$$

where we have written the martingale $\int_0^t e^{ks}dW_s$ as a Brownian motion $\widehat{W}$, time changed by $A(t) = \int_0^t e^{2ks}ds$ (see ↣ Section 5.1 for comments). It follows that $A(T_0) = T_{-x}(\widehat{W})$, hence

$$\begin{aligned}\mathbb{P}_x(T_0 \in dt) &= A'(t)\mathbb{P}_0(T_{-x}(\widehat{W}) \in dA(t))\\ &= e^{2kt}\exp\left(-\frac{x^2}{2A(t)}\right)\frac{|x|}{\sqrt{2\pi A^3(t)}}\,dt\,.\end{aligned}$$

Some easy computations, based on $A(t) = \frac{\sinh(kt)}{k}e^{kt}$ lead to the result. □

Comments 3.4.1.2 (a) We shall present a different proof in ↣ Subsection 6.5.2. See also ↣ Subsection 5.3.7.

(b) Ricciardi and Sato [732] obtained, for $x > a$, that the density of the hitting time of a is

$$-ke^{k(x^2-a^2)/2}\sum_{n=1}^{\infty}\frac{D_{\nu_{n,a}}(x\sqrt{2k})}{D'_{\nu_{n,a}}(a\sqrt{2k})}e^{-k\nu_{n,a}t}$$

where $0 < \nu_{1,a} < \cdots < \nu_{n,a} < \cdots$ are the zeros of $\nu \to D_\nu(-a)$. Here D_ν is the parabolic cylinder function with index ν (see ↣ Appendix A.5.4). The expression $D'_{\nu_{n,a}}$ denotes the derivative of $D_\nu(a)$ with respect to ν, evaluated at point $\nu = \nu_{n,a}$. Note that the formula in Leblanc et al. [573] for the law of the hitting time of a is only valid for $a = 0$. See also the discussion in Subsection 3.4.1.

(c) See other related results in Borodin and Salminen [109], Alili et al. [10], Göing-Jaeschke and Yor [398, 397], Novikov [679, 678], Patie [697], Pitman and Yor [719], Salminen [752], Salminen et al. [755] and Shepp [786].

Exercise 3.4.1.3 Prove that the Ricciardi and Sato result given in Comments 3.4.1.2 (b) allows us to express the density of

$$\tau := \inf\{t : x + W_t = \sqrt{1+2kt}\}.$$

Hint: The hitting time of a for an OU process is

$$\inf\{t\,:\, e^{-kt}(x+\widehat{W}_{A(t)}) = a\} = \inf\{u\,:\, x+\widehat{W}_u = ae^{kA^{-1}(u)}\}.$$

◁

3.4.2 Deterministic Volatility and Nonconstant Barrier

Valuing barrier options has some interest in two different frameworks:

(i) in a Black and Scholes model with deterministic volatility and a constant barrier
(ii) in a Black and Scholes model with a barrier which is a deterministic function of time.

As we discuss now, these two problems are linked. Let us study the case where the process S is a geometric BM with deterministic volatility $\sigma(t)$:

$$dS_t = S_t(rdt + \sigma(t)dW_t), \quad S_0 = x,$$

and let $T_a(S)$ be the first hitting time of a constant barrier a:

$$T_a(S) = \inf\{t\,:\, S_t = a\} = \inf\left\{t\,:\, rt - \frac{1}{2}\int_0^t \sigma^2(s)ds + \int_0^t \sigma(s)dW_s = \alpha\right\},$$

where $\alpha = \ln(a/x)$. The process $U_t = \int_0^t \sigma(s)dW_s$ is a Gaussian martingale and can be written as $Z_{A(t)}$ where Z is a BM and $A(t) = \int_0^t \sigma^2(s)ds$ (see ↣ Section 5.1 for a general presentation of time change). Let C be the inverse of the function A. Then,

$$T_a(S) = \inf\{t : rt - \frac{1}{2}A(t) + Z_{A(t)} = \alpha\} = \inf\left\{C(u) : rC(u) - \frac{1}{2}u + Z_u = \alpha\right\}$$

hence, the computation of the law of $T_a(S)$ reduces to the study of the hitting time of the non-constant boundary $\alpha - rC(u)$ by the drifted Brownian motion $(Z_u - \frac{1}{2}u, u \geq 0)$. This is a difficult and as yet unsolved problem (see references and comments below).

Comments 3.4.2.1 Deterministic Barriers and Brownian Motion. Groeneboom [409] studies the case

$$T = \inf\{t : x + W_t = \alpha t^2\} = \inf\{t : X_t = -x\}$$

where $X_t = W_t - \alpha t^2$. He shows that the densities of the first passage times for the process X can be written as functionals of a Bessel process of dimension 3, by means of the Cameron-Martin formula. For any $x > 0$ and $\alpha < 0$,

$$\mathbb{P}_x(T \in dt) = 2(\alpha c)^2 \sum_{n=0}^{\infty} \exp\left(\lambda_n/c - \frac{2}{3}\alpha^2 t^3\right) \frac{\mathrm{Ai}(\lambda_n - 2\alpha c x)}{\mathrm{Ai}'(\lambda_n)},$$

where λ_n are the zeros on the negative half-line of the Airy function Ai, the unique bounded solution of $u'' - xu = 0$, $u(0) = 1$, and $c = (1/2\alpha^2)^{1/3}$. (See ↣ Appendix A.5.5 for a closed form.) This last expression was obtained by Salminen [753].

Breiman [122] studies the case of a square root boundary when the stopping time T is $T = \inf\{t : W_t = \sqrt{\alpha + \beta t}\}$ and relates this study to that of the first hitting times of an OU process.

The hitting time of a nonlinear boundary by a Brownian motion is studied in a general framework in Alili's thesis [6], Alili and Patie [9], Daniels [210], Durbin [285], Ferebee [344], Hobson et al. [443], Jennen and Lerche [491, 492], Kahalé [503], Lerche [581], Park and Paranjape [695], Park and Schuurmann [696], Patie's thesis [697], Peskir and Shiryaev [708], Robbins and Siegmund [734], Salminen [753] and Siegmund and Yuh [798].

Deterministic Barriers and Diffusion Processes. We shall study hitting times for Bessel processes in ↣ Chapter 6 and for diffusions in Subsection 5.3.6. See Borodin and Salminen [109], Delong [245], Kent [519] or Pitman and Yor [715] for more results on first hitting time distributions for diffusions. See also Barndorff-Nielsen et al. [52], Kent [520, 521], Ricciardi et al. [732, 731], and Yamazato [854]. We shall present in ↣ Subsection 5.4.3 a method based on the Fokker-Planck equation in the case of general diffusions.

3.5 Hitting Time of a Two-sided Barrier for BM and GBM

3.5.1 Brownian Case

For $a < 0 < b$ let T_a, T_b be the two hitting times of a and b, where

$$T_y = \inf\{t \geq 0 \, : \, W_t = y\} \, ,$$

and let $T^* = T_a \wedge T_b$ be the exit time from the interval $[a, b]$. As before M_t denotes the maximum of the Brownian motion over the interval $[0, t]$ and m_t the minimum.

Proposition 3.5.1.1 *Let W be a BM starting from x and let $T^* = T_a \wedge T_b$. Then, for any a, b, x with $a < x < b$*

$$\mathbb{P}_x(T^* = T_a) = \mathbb{P}_x(T_a < T_b) = \frac{b - x}{b - a}$$

and $\mathbb{E}_x(T^*) = (x - a)(b - x)$.

PROOF: We apply Doob's optional sampling theorem to the bounded martingale $(W_{t \wedge T_a \wedge T_b}, t \geq 0)$, so that

$$x = \mathbb{E}_x(W_{T_a \wedge T_b}) = a\mathbb{P}_x(T_a < T_b) + b\mathbb{P}_x(T_b < T_a) \, ,$$

and using the obvious equality

$$\mathbb{P}_x(T_a < T_b) + \mathbb{P}_x(T_b < T_a) = 1 \, ,$$

one gets $\mathbb{P}_x(T_a < T_b) = \dfrac{b - x}{b - a}$.

The process $\{W^2_{t \wedge T_a \wedge T_b} - (t \wedge T_a \wedge T_b), t \geq 0\}$ is a bounded martingale, hence applying Doob's optional sampling theorem again, we get

$$x^2 = \mathbb{E}_x(W^2_{t \wedge T_a \wedge T_b}) - \mathbb{E}_x(t \wedge T_a \wedge T_b) \, .$$

Passing to the limit when t goes to infinity, we obtain

$$x^2 = a^2\mathbb{P}_x(T_a < T_b) + b^2\mathbb{P}_x(T_b < T_a) - \mathbb{E}_x(T_a \wedge T_b) \, ,$$

hence $\mathbb{E}_x(T_a \wedge T_b) = x(b + a) - ab - x^2 = (x - a)(b - x)$. □

Comment 3.5.1.2 The formula established in Proposition 3.5.1.1 will be very useful in giving a definition for the scale function of a diffusion (see Subsection 5.3.2).

Proposition 3.5.1.3 *Let W be a BM starting from 0, and let $a < 0 < b$. The Laplace transform of $T^* = T_a \wedge T_b$ is*

$$\mathbb{E}_0\left[\exp\left(-\frac{\lambda^2}{2}T^*\right)\right] = \frac{\cosh[\lambda(a+b)/2]}{\cosh[\lambda(b-a)/2]}.$$

The joint law of (M_t, m_t, W_t) is given by

$$\mathbb{P}_0(a \leq m_t < M_t \leq b, W_t \in E) = \int_E \varphi(t,y)\,dy \tag{3.5.1}$$

where, for $y \in [a,b]$,

$$\begin{aligned}\varphi(t,y) &= \mathbb{P}_0(W_t \in dy\,,\, T^* > t)\,/dy \\ &= \sum_{n=-\infty}^{\infty} p_t(y+2n(b-a)) - p_t(2b-y+2n(b-a)) \end{aligned}\tag{3.5.2}$$

and p_t is the Brownian density

$$p_t(y) = \frac{1}{\sqrt{2\pi t}}\exp\left(-\frac{y^2}{2t}\right).$$

PROOF: We only give the proof of the form of the Laplace transform. We refer the reader to formula 5.7 in Chapter X of Feller [343], and Freedman [357], for the form of the joint law. The Laplace transform of T^* is obtained by Doob's optional sampling theorem. Indeed, the martingale

$$\exp\left(\lambda\left(W_{t\wedge T^*} - \frac{a+b}{2}\right) - \frac{\lambda^2\,(t\wedge T^*)}{2}\right)$$

is bounded and T^* is finite, hence

$$\begin{aligned}\exp\left[-\lambda\left(\frac{a+b}{2}\right)\right] &= \mathbb{E}\left[\exp\left(\lambda\left(W_{T^*} - \frac{a+b}{2}\right) - \frac{\lambda^2\,T^*}{2}\right)\right] \\ &= \exp\left(\lambda\frac{b-a}{2}\right)\mathbb{E}\left[\exp\left(-\frac{\lambda^2\,T^*}{2}\right)\mathbb{1}_{\{T^*=T_b\}}\right] \\ &\quad + \exp\left(\lambda\frac{a-b}{2}\right)\mathbb{E}\left[\exp\left(-\frac{\lambda^2\,T^*}{2}\right)\mathbb{1}_{\{T^*=T_a\}}\right]\end{aligned}$$

and using $-W$ leads to

$$\begin{aligned}\exp\left[-\lambda\left(\frac{a+b}{2}\right)\right] &= \mathbb{E}\left[\exp\left(\lambda\left(-W_{T^*} - \frac{a+b}{2}\right) - \frac{\lambda^2\,T^*}{2}\right)\right] \\ &= \exp\left(\lambda\frac{-3b-a}{2}\right)\mathbb{E}\left[\exp\left(-\frac{\lambda^2\,T^*}{2}\right)\mathbb{1}_{\{T^*=T_b\}}\right] \\ &\quad + \exp\left(\lambda\frac{-b-3a}{2}\right)\mathbb{E}\left[\exp\left(-\frac{\lambda^2\,T^*}{2}\right)\mathbb{1}_{\{T^*=T_a\}}\right].\end{aligned}$$

By solving a linear system of two equations, the following result is obtained:

$$\begin{cases} \mathbb{E}\left[\exp\left(-\frac{\lambda^2 T^*}{2}\right)\mathbb{1}_{\{T^*=T_b\}}\right] = \frac{\sinh(-\lambda a)}{\sinh(\lambda(b-a))} \\ \mathbb{E}\left[\exp\left(-\frac{\lambda^2 T^*}{2}\right)\mathbb{1}_{\{T^*=T_a\}}\right] = \frac{\sinh(\lambda b)}{\sinh(\lambda(b-a))} \end{cases} . \tag{3.5.3}$$

The proposition is finally derived from

$$\mathbb{E}\left[e^{-\lambda^2 T^*/2}\right] = \mathbb{E}\left[e^{-\lambda^2 T^*/2)}\mathbb{1}_{\{T^*=T_b\}}\right] + \mathbb{E}\left[e^{-\lambda^2 T^*/2}\mathbb{1}_{\{T^*=T_a\}}\right].$$

□

By inverting this Laplace transform using series expansions, written in terms of $e^{-\lambda c}$ (for various c) which is the Laplace transform in $\lambda^2/2$ of T_c, the density of the exit time T^* of $[a,b]$ for a BM starting from $x \in [a,b]$ follows: for $y \in [a,b]$,

$$\mathbb{P}_x(B_t \in dy, T^* > t) = dy \sum_{n\in\mathbb{Z}} p_t(y - x + 2n(b-a)) - p_t(2b - y - x + 2n(b-a))$$

and the density of T^* is

$$\mathbb{P}_x(T^* \in dt) = (\mathrm{ss}_t(b-x, b-a) + \mathrm{ss}_t(x-a, b-a))\, dt$$

where, using the notation of Borodin and Salminen [109],

$$\mathrm{ss}_t(u,v) = \frac{1}{\sqrt{2\pi t^3}} \sum_{k=-\infty}^{\infty} (v - u + 2kv) e^{-(v-u+2kv)^2/2t} . \tag{3.5.4}$$

In particular,

$$\mathbb{P}_x(T^* \in dt, B_{T^*} = a) = \mathrm{ss}_t(x-a, b-a)dt .$$

In the case $-a = b$ and $x = 0$, we get the formula obtained in Exercise 3.1.6.5 for $T_b^* = \inf\{t : |B_t| = b\}$:

$$\mathbb{E}_0\left[\exp\left(-\frac{\lambda^2}{2}T_b^*\right)\right] = (\cosh(b\lambda))^{-1}$$

and inverting the Laplace transform leads to the density

$$\mathbb{P}_0(T_b^* \in dt) = \frac{1}{b^2} \sum_{n=-\infty}^{\infty} \left(n + \frac{1}{2}\right) e^{-(1/2)(n+1/2)^2\pi^2 t/b^2} dt .$$

□

Comments 3.5.1.4 (a) Let $M_1^* = \sup_{s\leq 1}|B_s|$ where B is a d-dimensional Brownian motion. As a consequence of Brownian scaling, $M_1^* \stackrel{\text{law}}{=} (T_1^*)^{-1/2}$ where $T_1^* = \inf\{t : |B_t| = 1\}$. In [774], Schürger computes the moments of the random variable M_1^* using the formula established in Exercise 1.1.12.4. See also Biane and Yor [86] and Pitman and Yor [720].

(b) Proposition 3.5.1.1 can be generalized to diffusions by using the corresponding scale functions. See $\rightarrowtail$ Subsection 5.3.2.

(c) The law of the hitting time of a two-sided barrier was studied in Bachelier [40], Borodin and Salminen [109], Cox and Miller [204], Freedman [357], Geman and Yor [384], Harrison [420], Karatzas and Shreve [513], Kunitomo and Ikeda [551], Knight [528], Itô and McKean [465] (Chapter I) and Linetsky [593]. See also Biane, Pitman and Yor [85].

(d) Another approach, following Freedman [357] and Knight [528] is given in [RY], Chap. III, Exercise 3.15.

(e) The law of T^* and generalizations can be obtained using spider-martingales (see Yor [868], p. 107).

3.5.2 Drifted Brownian Motion

Let $X_t = \nu t + W_t$ be a drifted Brownian motion and $T^*(X) = T_a(X) \wedge T_b(X)$ with $a < 0 < b$. From Cameron-Martin's theorem, writing T^* for $T^*(X)$,

$$\begin{aligned}
\mathbf{W}^{(\nu)}\left(\exp\left(-\frac{\lambda^2}{2}T^*\right)\right) &= \mathbb{E}\left(\exp\left(\nu W_{T^*} - \frac{\nu^2}{2}T^*\right)\exp\left(-\frac{\lambda^2}{2}T^*\right)\right)\\
&= \mathbb{E}(\mathbb{1}_{\{T^*=T_a\}}\, e^{\nu W_{T^*}-(\nu^2+\lambda^2)T^*/2}) + \mathbb{E}(\mathbb{1}_{\{T^*=T_b\}}\, e^{\nu W_{T^*}-(\nu^2+\lambda^2)T^*/2})\\
&= e^{\nu a}\mathbb{E}(\mathbb{1}_{\{T^*=T_a\}}\, e^{-(\nu^2+\lambda^2)T^*/2}) + e^{\nu b}\mathbb{E}(\mathbb{1}_{\{T^*=T_b\}}\, e^{-(\nu^2+\lambda^2)T^*/2}).
\end{aligned}$$

From the result (3.5.3) obtained in the case of a standard BM, it follows that

$$\mathbf{W}^{(\nu)}\left(\exp\left(-\frac{\lambda^2}{2}T^*\right)\right) = \exp(\nu a)\frac{\sinh(\mu b)}{\sinh(\mu(b-a))} + \exp(\nu b)\frac{\sinh(-\mu a)}{\sinh(\mu(b-a))}$$

where $\mu^2 = \nu^2 + \lambda^2$. Inverting the Laplace transform,

$$\begin{aligned}
&\mathbb{P}_x(T^* \in dt)\\
&\quad = e^{-\nu^2 t/2}\left(e^{\nu(a-x)}\mathrm{ss}_t(b-x, b-a) + e^{\nu(b-x)}\mathrm{ss}_t(x-a, b-a)\right)dt\,,
\end{aligned}$$

where the function ss is defined in (3.5.4). In the particular case $a = -b$, the Laplace transform is

$$\mathbf{W}^{(\nu)}\left(\exp\left(-\frac{\lambda^2}{2}T^*\right)\right) = \frac{\cosh(\nu b)}{\cosh(b\sqrt{\nu^2+\lambda^2})}\,.$$

The formula (3.5.1) can also be extended to drifted Brownian motion thanks to the Cameron-Martin relationship.

3.6 Barrier Options

In this section, we study the price of barrier options in the case where the underlying asset S follows the Garman-Kohlhagen risk-neutral dynamics

$$dS_t = S_t((r-\delta)dt + \sigma dW_t), \tag{3.6.1}$$

where r is the risk-free interest rate, δ the dividend yield generated by the asset and W a BM. If needed, we shall denote by $(S_t^x, t \geq 0)$ the solution of (3.6.1) with initial condition x. In a closed form,

$$S_t^x = xe^{(r-\delta)t}e^{\sigma W_t - \sigma^2 t/2}.$$

We follow closely El Karoui [297] and El Karoui and Jeanblanc [300]. In a first step, we recall some properties of standard Call and Put options. We also recall that an option is **out-of-the-money** (resp. in-the-money) if its intrinsic value $(S_t - K)^+$ is equal to 0 (resp. strictly positive).

3.6.1 Put-Call Symmetry

In the particular case where $r = \delta = 0$, Garman and Kohlhagen's formulae (2.7.4) for the time-t price of a European call C_E^* and a put option P_E^* with strike price K and maturity T on the underlying asset S reduce to

$$C_E^*(x,K,T-t) = x\mathcal{N}\left[d_1\left(\frac{x}{K},T-t\right)\right] - K\mathcal{N}\left[d_2\left(\frac{x}{K},T-t\right)\right] \tag{3.6.2}$$

$$P_E^*(x,K,T-t) = K\mathcal{N}\left[d_1\left(\frac{K}{x},T-t\right)\right] - x\mathcal{N}\left[d_2\left(\frac{K}{x},T-t\right)\right] \tag{3.6.3}$$

The functions d_i are defined on $\mathbb{R}^+ \times [0,T]$ as:

$$\begin{aligned} d_1(y,u) &:= \frac{1}{\sqrt{\sigma^2 u}}\ln(y) + \frac{1}{2}\sqrt{\sigma^2 u} \\ d_2(y,u) &:= d_1(y,u) - \sqrt{\sigma^2 u}, \end{aligned} \tag{3.6.4}$$

and x is the value of the underlying at time t. Note that these formulae do not depend on the sign of σ and $d_1(y,u) = -d_2(1/y,\, u)$.

In the general case, the time-t prices of a European call C_E and a put option P_E with strike price K and maturity T on the underlying currency S are

$$\begin{aligned} C_E(x,K;r,\delta;T-t) &= C_E^*(xe^{-\delta(T-t)}, Ke^{-r(T-t)}, T-t) \\ P_E(x,K;r,\delta;T-t) &= P_E^*(xe^{-\delta(T-t)}, Ke^{-r(T-t)}, T-t) \end{aligned}$$

or, in closed form

$$C_E\,(x,K;r,\delta;T-t) = xe^{-\delta(T-t)}\mathcal{N}\left[d_1\left(\frac{xe^{-\delta(T-t)}}{Ke^{-r(T-t)}},T-t\right)\right] - Ke^{-r(T-t)}\mathcal{N}\left[d_2\left(\frac{xe^{-\delta(T-t)}}{Ke^{-r(T-t)}},T-t\right)\right] \quad (3.6.5)$$

$$P_E\,(x,K;r,\delta;T-t) = Ke^{-r(T-t)}\mathcal{N}\left[d_1\left(\frac{Ke^{-r(T-t)}}{xe^{-\delta(T-t)}},T-t\right)\right] - xe^{-\delta(T-t)}\mathcal{N}\left[d_2\left(\frac{Ke^{-r(T-t)}}{xe^{-\delta(T-t)}},T-t\right)\right]. \quad (3.6.6)$$

Notation: The quantity $C_E^*(\alpha,\beta;u)$ depends on three arguments: the first one, α, is the value of the underlying, the second one β is the value of the strike, and the third one, u, is the time to maturity. For example, $C_E^*\,(K,x;T-t)$ is the time-t value of a call on an underlying with time-t value equal to K and strike x. We shall use the same kind of convention for the function $C_E(x,K;r,\delta;u)$ which depends on 5 arguments.

As usual, $\mathcal{N}$ represents the cumulative distribution function of a standard Gaussian variable.

If σ is a deterministic function of time, $d_i(y,T-t)$ has to be changed into $d_i(y,T,t)$, where

$$\begin{aligned} d_1(y;T,t) &= \frac{1}{\Sigma_{t,T}}\ln(y) + \frac{1}{2}\Sigma_{t,T} \\ d_2(y;T,t) &= d_1(y;T,t) - \Sigma_{t,T} \end{aligned} \quad (3.6.7)$$

with $\Sigma_{t,T}^2 = \int_t^T \sigma^2(s)ds$.

Note that, from the definition and the fact that the geometric Brownian motion (solution of (3.6.1)) satisfies $S_t^{\lambda x} = \lambda S_t^x$, the call (resp. the put) is a homogeneous function of degree 1 with respect to the first two arguments, the spot and the strike:

$$\begin{aligned} \lambda C_E\,(x,K;r,\delta;T-t) &= C_E\,(\lambda x,\lambda K;r,\delta;T-t) \\ \lambda P_E\,(x,K;r,\delta;T-t) &= P_E\,(\lambda x,\lambda K;r,\delta;T-t)\,. \end{aligned} \quad (3.6.8)$$

This can also be checked from the formula (3.6.5). The **Deltas**, i.e., the first derivatives of the option price with respect to the underlying, are given by

$$\text{DeltaC}(x,K;r,\delta;T-t) = e^{-\delta(T-t)}\mathcal{N}\left[d_1\left(\frac{xe^{-\delta(T-t)}}{Ke^{-r(T-t)}},T-t\right)\right]$$

$$\text{DeltaP}(x,K;r,\delta;T-t) = -e^{-\delta(T-t)}\mathcal{N}\left[d_2\left(\frac{Ke^{-r(T-t)}}{xe^{-\delta(T-t)}},T-t\right)\right].$$

The Deltas are homogeneous of degree 0 in the first two arguments, the spot and the strike:

$$\text{DeltaC}\ (x, K; r, \delta; T-t) = \text{DeltaC}\ (\lambda x, \lambda K; r, \delta; T-t), \qquad (3.6.9)$$
$$\text{DeltaP}\ (x, K; r, \delta; T-t) = \text{DeltaP}\ (\lambda x, \lambda K; r, \delta; T-t).$$

Using the explicit formulae (3.6.5, 3.6.6), the following result is obtained.

Proposition 3.6.1.1 *The put-call symmetry is given by the following expressions*

$$C_E^*(K, x; T-t) = P_E^*(x, K; T-t)$$
$$P_E(x, K; r, \delta; T-t) = C_E(K, x; \delta, r; T-t).$$

PROOF: The formula is straightforward from the expressions (3.6.2, 3.6.3) of C_E^* and P_E^*. Hence, the general case for C_E and P_E follows. This formula is in fact obvious when dealing with exchange rates: the seller of US dollars is the buyer of Euros. From the homogeneity property, this can also be written

$$P_E(x, K; r, \delta; T-t) = xKC_E(1/x, 1/K; \delta, r; T-t). \qquad \square$$

Remark 3.6.1.2 A different proof of the put-call symmetry which does not use the closed form formulae (3.6.2, 3.6.3) relies on Cameron-Martin's formula and a change of numéraire. Indeed

$$C_E(x, K, r, \delta, T) = \mathbb{E}_{\mathbb{Q}}(e^{-rT}(S_T - K)^+) = \mathbb{E}_{\mathbb{Q}}(e^{-rT}(S_T/x)(x - KxS_T^{-1})^+).$$

The process $Z_t = e^{-(r-\delta)t}S_t/x$ is a strictly positive martingale with expectation 1. Set $\widehat{\mathbb{Q}}|_{\mathcal{F}_t} = Z_t\mathbb{Q}|_{\mathcal{F}_t}$. Under $\widehat{\mathbb{Q}}$, the process $Y_t = xK(S_t)^{-1}$ follows dynamics $dY_t = Y_t((\delta - r)dt - \sigma dB_t)$ where B is a $\widehat{\mathbb{Q}}$-Brownian motion, and $Y_0 = K$. Hence,

$$C_E(x, K, r, \delta, T) = \mathbb{E}_{\mathbb{Q}}(e^{-\delta T}Z_T(x - Y_T)^+) = \widehat{\mathbb{E}}(e^{-\delta T}(x - Y_T)^+),$$

and the right-hand side represents the price of a put option on the underlying Y, when δ is the interest rate, r the dividend, K the initial value of the underlying asset, $-\sigma$ the volatility and x the strike. It remains to note that the value of a put option is the same for σ and $-\sigma$.

Comments 3.6.1.3 (a) This symmetry relation extends to American options (see Carr and Chesney [147], McDonald and Schroder [633] and Detemple [251]). See ⤚ Subsections 10.4.2 and 11.7.3 for an extension to mixed diffusion processes and Lévy processes.

(b) The homogeneity property does not extend to more general dynamics.

Exercise 3.6.1.4 Prove that

$$C_E(x, K; r, \delta; T-t) = P_E^*(Ke^{-\mu(T-t)}, xe^{\mu(T-t)}; T-t)$$
$$= e^{-\mu(T-t)}P_E^*(K, xe^{2\mu(T-t)}; T-t),$$

where $\mu = r - \delta$ is called the cost of carry. ◁

3.6.2 Binary Options and Δ's

Among the exotic options traded on the market, binary options are the simplest ones. Their valuation is straightforward, but hedging is more difficult. Indeed, the hedging ratio is discontinuous in the neighborhood of the strike price.

A **binary call** (in short BinC) (resp. binary put, BinP) is an option that generates one monetary unit if the underlying value is higher (resp. lower) than the strike, and 0 otherwise. In other words, the payoff is $\mathbb{1}_{\{S_T \geq K\}}$ (resp. $\mathbb{1}_{\{S_T \leq K\}}$). Binary options are also called digital options.

Since $\frac{1}{h}\left((x-k)^+ - (x-(k+h))^+\right) \to \mathbb{1}_{\{x \geq k\}}$ as $h \to 0$, the value of a binary call is the limit, as $h \to 0$ of the call-spread

$$\frac{1}{h}[C(x,K,T) - C(x,K+h,T)],$$

i.e., is equal to the negative of the derivative of the call with respect to the strike. Along the same lines, a binary put is the derivative of the put with respect to the strike.

By differentiating the formula obtained in Exercise 3.6.1.4 with respect to the variable K, we obtain the following formula:

Proposition 3.6.2.1 *In the Garman-Kohlhagen framework, with carrying cost $\mu = r - \delta$ the following results are obtained:*

$$\begin{aligned} \mathrm{BinC}(x,K;r,\delta;T-t) &= -e^{-\mu(T-t)}\mathrm{DeltaP}_E^*(K, xe^{2\mu(T-t)}; T-t) \\ &= e^{-r(T-t)}\mathcal{N}\left[d_2\left(\frac{xe^{\mu(T-t)}}{K}, T-t\right)\right] \end{aligned} \tag{3.6.10}$$

$$\begin{aligned} \mathrm{BinP}(x,K;r,\delta;T-t) &= e^{-\mu(T-t)}\mathrm{DeltaC}_E^*(K, xe^{2\mu(T-t)}; T-t) \\ &= e^{-r(T-t)}\mathcal{N}\left[d_1\left(\frac{K}{xe^{\mu(T-t)}}, T-t\right)\right], \end{aligned} \tag{3.6.11}$$

where d_1, d_2 are defined in (3.6.4).

Exercise 3.6.2.2 Prove that

$$\begin{cases} \mathrm{DeltaC}\,(x,K;r,\delta) = \dfrac{1}{x}\left[C_E\,(x,K;r,\delta) + K\mathrm{BinC}\,(x,K;r,\delta)\right] \\ \\ \mathrm{DeltaP}\,(x,K;r,\delta) = \dfrac{1}{x}\left[P_E\,(x,K;r,\delta) - K\mathrm{BinP}\,(x,K;r,\delta)\right] \end{cases} \tag{3.6.12}$$

where the quantities are evaluated at time $T - t$. ◁

Comments 3.6.2.3 The price of a BinC can also be computed via a PDE approach, by solving

$$\begin{cases} \partial_t u + \frac{1}{2}\sigma^2 x^2 \partial_{xx} u + \mu \, \partial_x u = ru \\ \qquad\qquad u(x,T) = \mathbb{1}_{\{K<x\}} \,. \end{cases} \tag{3.6.13}$$

See Ingersoll [459] and Rubinstein and Reiner [747] for a discussion on binary options. Navatte and Quittard-Pinon [667] have studied binary options in a stochastic interest case (one factor Gaussian model); their results are extended to a Lévy model in Eberlein and Kluge [291].

3.6.3 Barrier Options: General Characteristics

Practitioners give the name *barrier options* to options with a payoff that depends on whether or not the underlying value has reached a given level (the barrier) before maturity. They are particular types of path-dependent options, because the final payoff depends on the asset price trajectory and they are classified into two categories:

- *Knock-out options:* The option ceases to exist at the first passage time of the underlying value at the barrier.
- *Knock-in options:* The option is activated as soon as the barrier is reached.

Let us consider for instance:

- A DOC (**down-and-out call**) with strike K, barrier L and maturity T is the option to buy the underlying at price K (at maturity T) if the underlying value never falls below the (low) barrier L before time T. The value of a DOC is therefore null for $S_0 < L$ and, for $S_0 \geq L$,
$$\mathrm{DOC}(S_0, K, L, T) := \mathbb{E}_{\mathbb{Q}}(e^{-rT}(S_T - K)^+ \mathbb{1}_{\{T<T_L\}})$$
where:
$$T_L := \inf\{t \,|\, S_t \leq L\} = \inf\{t \,|\, S_t = L\} \,.$$
In what follows, we consider DOC options only in the case $S_0 \geq L$.
- An UOC (**up-and-out call**) has the same characteristics but the (high) barrier H is above the initial underlying value, $S_0 \leq H$. Its price is
$$\mathrm{UOC}(S_0, K, H, T) := \mathbb{E}_{\mathbb{Q}}(e^{-rT}(S_T - K)^+ \mathbb{1}_{\{T<T_H\}})$$
where $T_H := \inf\{t \,|\, S_t \geq H\} = \inf\{t \,|\, S_t = H\}$.
- A DIC (**down-and-in call**) is activated if the underlying value falls below the barrier L before time T. Its price is, for $S_0 > L$,
$$\mathrm{DIC}(S_0, K, L, T) := \mathbb{E}_{\mathbb{Q}}(e^{-rT}(S_T - K)^+ \mathbb{1}_{\{T>T_L\}}) \,.$$

- An UIC (**up-and-in call**) is activated as soon as the underlying value hits the barrier H from below. Its price is

$$\mathrm{UIC}(S_0, K, H, T) := \mathbb{E}_{\mathbb{Q}}(e^{-rT}(S_T - K)^+ \mathbb{1}_{\{T > T_H\}}) .$$

The same definitions apply to puts, binary options and bonds. For example

- A DIP is a **down-and-in put**.
- A **binary down-and-in call** (BinDIC) is a binary call, activated only if the underlying value falls below the barrier, before maturity. The payoff is $\mathbb{1}_{\{S_T > K\}} \mathbb{1}_{\{T_L < T\}}$.
- A DIB (**down-and-in bond**) is a product which generates one monetary unit at maturity if the barrier L has been reached beforehand by the underlying. Its value is $\mathbb{E}_{\mathbb{Q}}(e^{-rT} \mathbb{1}_{\{T_L < T\}}) = e^{-rT} \mathbb{Q}(T_L < T)$.

Barrier options are often used on currency markets. Their prices are smaller than the corresponding standard European prices. This provides an advantage for the marketing of these products. However, they are more difficult to hedge.

Depending on the "at the barrier" intrinsic value, these exotic options can be classified further :

- A barrier option that is out of the money when the barrier L is reached is called a **regular** option. As an example, note that the time-t intrinsic value $(x - K)^+ \mathbb{1}_{\{T_L \leq t\}}$ of a DIC such that $K \geq L$ is equal to 0 for $x = L$.
- A barrier option which is in the money when the barrier is reached is called a **reverse** option.
- Some barrier options generate a *rebate* received in cash when the barrier L is reached. The value of the rebate corresponds to the payoff of a binary option. In particular, the rebate is often chosen in such a way that the payoff continuity is kept at the barrier, e.g., if the payoff is $f(S_T)$ at time T, the rebate is $f(L)$.

Let us remark that by relying on the absence of arbitrage opportunities, to being long on one in-option and on one out-option is equivalent to be long on a plain-vanilla option. Therefore, we restrict our attention to in-options.

Comments 3.6.3.1 (a) Barrier options are studied in a discrete time setting in Wilmott et al. [847], Chesney et al. [176], Musiela and Rutkowski [661], Zhang [872], Pliska [721] and Wilmott [846].

(b) In continuous time, the main papers are Andersen et al. [16], Rubinstein and Reiner [746], Bowie and Carr [116], Rich [733], Heynen and Kat [434], Douady [262], Carr and Chou [148], Baldi et al. [42], Linetsky [593] and Suchanecki [814]. Broadie et al. [130] present some correction terms between discrete and continuous time barrier options.

(c) Roberts and Shortland [735] study a case where the underlying has time dependent coefficients. The books of Kat [516], Musiela and Rutkowski [661], Zhang [872] and Wilmott [846] contain more information. Taleb [818] studies hedging strategies.

3.6.4 Valuation and Hedging of a Regular Down-and-In Call Option When the Underlying is a Martingale

In this section, we suppose that the barrier option is written on an underlying S *without carrying costs* – hence a martingale – (i.e., $\mu = 0$ or $r = \delta$) with dynamics having *deterministic* volatility:

$$dS_t = S_t \sigma(t) dW_t .$$

Furthermore, when there is no ambiguity, the instantaneous time t, the maturity time T and the volatility will not appear as arguments in the formulae. The value of the underlying at time t is denoted by x.

Let L be the barrier. We denote by $\mathrm{DIC}^M(x, K, L)$ the DIC option price and by $C_E^M(x, K)$ (resp. $P_E^M(x, K)$) the standard European call (resp. put) price (where the first variable is the underlying and the second variable is the strike), when the underlying is a martingale (hence the superscript M). Relying on the assumption that the carrying cost is zero, the symmetry formula established in Proposition 3.6.1.1 is

$$C_E^M(x, K) = P_E^M(K, x) . \tag{3.6.14}$$

In particular $\partial_K C_E^M(x, K) = \mathrm{Delta} P_E^M(K, x)$.

We now follow closely Carr et al. [149]. We recall that for a regular DIC option, the barrier L is lower than the strike ($K \geq L$).

Proposition 3.6.4.1 *Consider a regular DIC option on an underlying without carrying costs.*

(a) *Its price is given by:*

$$\textit{for } x \leq L, \quad \mathrm{DIC}^M(x, K, L) = C_E^M(x, K) , \tag{3.6.15}$$

$$\textit{for } x \geq L, \quad \mathrm{DIC}^M(x, K, L) = \frac{K}{L} P_E^M\left(x, \frac{L^2}{K}\right) = C_E^M\left(L, K\frac{x}{L}\right) , \tag{3.6.16}$$

(b) *The static hedging consists of:*

(i) *a long call for $x \leq L$,*

(ii) *for $x \geq L$, a long position of K/L puts of strike L^2/K.*

PROOF: We shall give a proof "without mathematics."

▶ If the value x of the underlying (at date t) is smaller than the barrier L, the option is already activated, therefore it is a plain vanilla option and the equality (3.6.15) is satisfied.

▶ If the value of the underlying (at date t) is higher than the barrier, we proceed as follows. Let t be fixed and denote by

$$T_L = \inf\{s \geq t ; S_s \leq L\} \tag{3.6.17}$$

the first passage time after t of the underlying value below the barrier.

In order to price the option at time t, by relying on the absence of arbitrage opportunities, we compute the option value at date T_L, and we denote by V this value. In a second step, we compute the value at time t of the claim V, to be received at time T_L.

At the barrier, the level of the underlying is known and only the remaining maturity $T - T_L$ is unknown. The DIC^M option is equivalent to a call of maturity $T - T_L$ on an underlying with value L, i.e., $C_E^M(L, K, T - T_L)$. The underlying is a martingale, and the volatility is deterministic. Therefore, the underlying dynamics with starting time T_L and starting point L is, conditionally with respect to the past before T_L, log-normally distributed. The symmetry formula (3.6.14) and the homogeneity of the put price yield

$$C_E^M(L, K, T - T_L) = \frac{K}{L} P_E^M\left(L, \frac{L^2}{K}, T - T_L\right). \tag{3.6.18}$$

Now, since the underlying is equal to the barrier, the down-and-in option values are equal to standard option values, in particular, for any strike k, one has $\text{DIP}^M(L, k, L) = P_E^M(L, k)$. Therefore, formula (3.6.18) implies that the option $\text{DIC}^M(x, K, L)$ is equivalent to K/L options $\text{DIP}^M\left(x, L^2K^{-1}, L\right)$.

At maturity, the terminal payoff of the DIP is strictly positive only if the underlying value is below L^2K^{-1}. Since $L \leq K$, the quantity L^2K^{-1} is smaller than L. Hence, if the DIP is in the money at maturity, the barrier L was reached with probability 1; therefore, the barrier is no longer relevant in pricing the option. The $\text{DIP}^M\left(x, L^2K^{-1}, L\right)$ barrier option is thus equal to the plain vanilla $P_E^M\left(x, L^2K^{-1}\right)$ for $L \leq K$ and the result is obtained.

In order to conclude, the symmetry formula is applied again. □

Corollary 3.6.4.2 *In an explicit form,*

$$\text{for } x \leq L,\ \text{DIC}^M = e^{-r(T-t)}\left\{x\mathcal{N}\left(d_1\left(\frac{x}{K}, T-t\right)\right) - K\mathcal{N}\left(d_2\left(\frac{x}{K}, T-t\right)\right)\right\},$$

$$\begin{aligned}\text{for } x \geq L,\ \text{DIC}^M = e^{-r(T-t)} &\left\{L\mathcal{N}\left(d_1\left(\frac{L^2}{Kx}, T-t\right)\right)\right.\\ &\left. -\frac{Kx}{L}\mathcal{N}\left(d_2\left(\frac{L^2}{Kx}, T-t\right)\right)\right\},\end{aligned}$$

where d_1, d_2 are defined in (3.6.4).

Proposition 3.6.4.3 *The price of a regular up-and-in put ($H \geq K$) on an underlying without carrying cost is given by:*

(i) *for $x \geq L$,* $\text{UIP}^M(x, K, H) = P_E^M(x, K)$,

(ii) *for $x \leq H$,* $\text{UIP}^M(x, K, H) = \dfrac{K}{H} C_E^M\left(x, \dfrac{H^2}{K}\right)$.

Proposition 3.6.4.4 *Let* $x \geq L$. *The regular binary option* BinDIC^M *satisfies*

(i)

$$\mathrm{BinDIC}^M(x,K,L) = \frac{x}{L}\,\mathrm{BinC}^M\left(L, \frac{Kx}{L}\right), \tag{3.6.19}$$

(ii)

$$\begin{aligned}\mathrm{DeltaDIC}^M(x,K,L) &= -\frac{K}{L}\mathrm{BinC}^M\left(L, \frac{Kx}{L}\right) \\ &= -\frac{Ke^{-rT}}{L}\mathcal{N}\left(d_2\left(\frac{L^2}{xK}, T\right)\right).\end{aligned} \tag{3.6.20}$$

The price of the DIB *option is given by*

$$\mathrm{DIB}^M(x,L) = e^{-rT}\left[\frac{x}{L}\mathcal{N}\left(d_2(L/x, T)\right) + \mathcal{N}\left(d_1(L/x, T)\right)\right]. \tag{3.6.21}$$

The binary put value is obtained by proceeding along the same lines.

PROOF: By definition, $\mathrm{BinDIC}^M(x,K,L) = -\partial_K \mathrm{DIC}^M(x,K,L)$. We differentiate the first and third terms of (3.6.16) with respect to K. We get

$$\mathrm{BinDIC}^M(x,K,L) = -\frac{x}{L}\partial_K C_E^M\left(L, \frac{Kx}{L}\right) = -\frac{x}{L}\mathrm{DeltaP}^{\mathrm{M}}\left(\frac{Kx}{L}, L\right)$$

where we have used the symmetry formula for the second equality. It remains to apply Proposition 3.6.2.1 to obtain (i). By differentiating the two sides of the first and second terms of equality (3.6.16) w.r.t. x, one gets

$$\mathrm{DeltaDIC}^{\mathrm{M}}(x,K,L) = \frac{K}{L}\mathrm{DeltaP}^{\mathrm{M}}\left(x, \frac{L^2}{K}\right) = \frac{K}{L}\mathrm{DeltaP}^{\mathrm{M}}\left(\frac{Kx}{L}, L\right)$$

where we have used the homogeneity property of degree 0 for the last equality, hence (ii) is obtained using Proposition 3.6.2.1 again. The payoff of the DIB option is equal to 1 if the barrier is reached before time T, and, using

$$\{T_L \leq T\} = \{T_L \leq T,\, S_T > L\} \cup \{S_T \leq L\},$$

we obtain

$$\begin{aligned}\mathrm{DIB}^M(x,L) &= \mathrm{BinDIC}^M(x,L,L) + \mathrm{BinP}^M(x,L) \\ &= \frac{x}{L}\mathrm{BinC}^M(L,x) + \mathrm{BinP}^M(x,L) \\ &= e^{-rT}\left\{\frac{x}{L}\mathcal{N}\left(d_2\left(\frac{L}{x}, T\right)\right) + \mathcal{N}\left(d_1\left(\frac{L}{x}, T\right)\right)\right\}.\end{aligned}$$

One can check that the value of the DIB is smaller than 1. □

Hedge of a Regular Down-and-In Call Option

In this section, we do not write the time argument in $d_i(x,T)$. A static hedge for a DIC regular option consists in holding K/L puts as long as the underlying value remains above the barrier, and a standard call after the barrier is crossed. At the barrier, the put-call symmetry implies the continuity of the price. This is not the case for the hedge ratio, which admits a right limit given from (3.6.20) by

$$\Delta_+ \mathrm{DIC}^M(L,K,L) = -\frac{Ke^{-rT}}{L}\mathcal{N}\left(d_2(LK^{-1})\right)$$

whereas, from (3.6.15) the left limit is

$$\Delta_- \mathrm{DIC}^M(L,K,L) = \mathrm{DeltaC}^M(L,K) = e^{-rT}\mathcal{N}\left(d_1(LK^{-1})\right) .$$

Hence, the Delta is not continuous at the barrier and admits a negative jump equal to minus the discounted probability that the underlying with starting point K reaches the barrier before T: indeed from (3.6.21)

$$\begin{aligned}[\Delta_+ - \Delta_-]\mathrm{DIC}^M(L,K,L) &= -\frac{Ke^{-rT}}{L}\mathcal{N}\left(d_2(LK^{-1})\right) - e^{-rT}\mathcal{N}\left(d_1(LK^{-1})\right) \\ &= -\mathrm{DIB}^M(K,L) .\end{aligned}$$

The absolute value of the jump is smaller than 1.

3.6.5 Mathematical Results Deduced from the Previous Approach

In this section, we do not write the time argument T in $d_i(x,T)$. We consider a *martingale* $(S_t, t \geq 0)$ with *deterministic* volatility $\sigma = (\sigma(t), t \geq 0)$ which represents the price of an asset without carrying costs under the risk neutral probability $\mathbb{Q}$, that is

$$S_t = x \exp\left(\int_0^t \sigma(s)dW_s - \frac{1}{2}\int_0^t \sigma^2(s)ds\right) . \tag{3.6.22}$$

A Result on Change of Probability

In a first step, we translate the symmetry formula in terms of a change of probability: equality (3.6.14) reads for any K,

$$\mathbb{E}_{\mathbb{Q}}((S_T - K)^+) = \mathbb{E}_{\mathbb{Q}}((x - KS_T/x)^+) .$$

We note that if X and Y are positive random variables with density, satisfying $\mathbb{E}((X-K)^+) = \mathbb{E}((Y-K)^+)$ for any $K \geq 0$, then $X \overset{\text{law}}{=} Y$. Therefore, from

$$\mathbb{E}_{\mathbb{Q}}((S_T - K)^+) = \mathbb{E}_{\mathbb{Q}}((x - KS_T/x)^+) = \mathbb{E}_{\mathbb{Q}}\left(\frac{S_T}{x}\left(\frac{x^2}{S_T} - K\right)^+\right)$$

it follows that the law of S_T under $\mathbb{Q}$ is equal to the law of x^2/S_T under $\widehat{\mathbb{Q}}$, where $\widehat{\mathbb{Q}}|_{\mathcal{F}_T} = \frac{S_T}{x}\mathbb{Q}|_{\mathcal{F}_T}$.

One can also obtain the same result using Cameron-Martin's relationship.

Exercise 3.6.5.1 Let X be a integrable random variable with density φ such that $\mathbb{E}(f(X)) = \mathbb{E}(Xf(1/X))$ for any bounded function f.

Prove that $\varphi(x) = \frac{1}{x^2}\varphi(\frac{1}{x})$. Check that the density of $X = e^{B_T - T/2}$ satisfies this equality.

Hint: Consider $\xi(s) := \mathbb{E}(X^s)$ for $s \in \mathbb{C}$, which satisfies $\xi(s) = \xi(1-s)$. $\triangleleft$

Joint Law of (m_T^S, S_T)

Here, we assume that $x \geq L$. Let us introduce the first passage time below the barrier:

$$T_L = \inf\{t : S_t \leq L\}$$

where we set $\inf(\emptyset) = +\infty$ and note that

$$\{T_L \leq T\} = \left\{\inf_{0 \leq t \leq T} S_t \leq L\} = \{m_T^S \leq L\right\},$$

where $m_t^S = \inf_{s \leq t} S_s$. The prices at time 0 for barrier and binary options are given as:

$$\begin{aligned} \mathrm{DIC}^M(x, K, L) &= e^{-rT}\mathbb{E}_Q[\mathbb{1}_{\{T_L \leq T\}}(S_T - K)^+], \\ \mathrm{BinDIC}^M(x, K, L) &= e^{-rT}\mathbb{Q}[\{T_L \leq T\} \cap \{S_T \geq K\}] \\ &= e^{-rT}\mathbb{Q}[\{m_T^S \leq L\} \cap \{S_T \geq K\}]. \end{aligned}$$

Proposition 3.6.5.2 *Let $(S_t, t \geq 0)$ be a martingale with the following dynamics*

$$dS_t = S_t\sigma(t)dW_t\,, \; S_0 = x$$

where W is a $\mathbb{Q}$-Brownian motion, with initial value x with $x \geq L$.

For any $K \geq L$, the law of the pair (m_T^S, S_T) is given by

$$\mathbb{Q}(m_T^S \leq L, S_T \geq K) = \frac{x}{L}\mathbb{Q}\left(S_T \geq \frac{Kx^2}{L^2}\right) = \frac{x}{L}\mathcal{N}\left(d_2\left(\frac{L^2}{Kx}\right)\right)$$

and the law of the minimum $m_T^S = \inf_{t \leq T} S_t$:

$$\mathbb{Q}(m_T^S \leq L) = \frac{x}{L}\mathcal{N}\left(d_2(Lx^{-1})\right) + \mathcal{N}\left(d_1(Lx^{-1})\right),$$

where d_1, d_2 are given by (3.6.7).

PROOF: Formula (3.6.19) leads to

$$\mathbb{Q}(m_T^S \leq L, S_T \geq K) = \frac{x}{L}\mathbb{Q}\left(L\frac{S_T}{x} \geq \frac{K}{L}x\right) = \frac{x}{L}\mathbb{Q}\left(S_T \geq \frac{Kx^2}{L^2}\right).$$

The law of the minimum follows, taking $K = L$. $\square$

The equality

$$\mathbb{Q}(m_T^S \leq L, S_T \geq K) = \frac{x}{L}\mathbb{Q}\left(S_T \geq \frac{Kx^2}{L^2}\right)$$

corresponds to the reflection principle obtained for Brownian motion. Indeed, writing, for $x = 1$, $S_t = e^{\sigma X_t}$ where $X_t = W_t - \nu t$ and $\nu = \sigma/2$ and taking the logarithm, when σ is constant, one obtains the formula given in Exercise 3.2.1.3 for the drifted Brownian motion:

$$\mathbb{P}(W_T - \nu T \geq \alpha, \inf_{0\leq t\leq T}(W_t - \nu t) \leq \beta) = e^{2\nu\beta}\mathbb{P}(W_T - \nu T \geq \alpha - 2\beta).$$

By considering current prices, we shall obtain the conditional distribution (with respect to the information at time t) of the underlying value at time T and its minimum on the time interval (t, T). Let $S_t = y$ and let $m_t^S = \inf_{s\leq t} S_s = m$ (with $m \leq y$) be the minimum over the time interval $[0, t]$. In the case $m \leq L$, the barrier has been reached during the time interval $[0, t]$, whereas the barrier has not been reached when $m > L$. In the second case, the two events $(\inf_{0\leq u\leq T} S_u \leq L)$ and $(\inf_{t\leq u\leq T} S_u \leq L)$ are identical.

The equality (3.6.19) concerning barrier options

$$\text{BinDIC}^M(S_t, K, L, T-t) = \frac{S_t}{L}\text{BinC}^M\left(L, \frac{KS_t}{L}, T-t\right)$$

can be written, on the set $\{T_L \geq t\}$, as follows:

$$\begin{aligned}\mathbb{Q}(\{T_L \leq T\} \cap \{S_T \geq K\}|\mathcal{F}_t) &= \mathbb{Q}(\{\inf_{t\leq u\leq T} S_u \leq L\} \cap \{S_T \geq K\}|\mathcal{F}_t)\\ &= \frac{S_t}{L}\mathbb{Q}\left(S_T\frac{L}{S_t} \geq \frac{KS_t}{L}|\mathcal{F}_t\right) = \frac{S_t}{L}\mathcal{N}\left(d_2\left(\frac{L^2}{KS_t}, T-t\right)\right). \qquad (3.6.23)\end{aligned}$$

The equality (3.6.23) gives the conditional distribution function of the pair $(m^S[t, T], S_T)$ where $m^S[t, T] = \min_{t\leq s\leq T} S_s$, on the set $\{T_L \geq t\}$, as a differentiable function.

Hence, the conditional law of the pair $(m^S[t, T], S_T)$ with respect to $\mathcal{F}_t$ admits a density $f(h, k)$ on the set $0 < h < k$ which can be computed from the density p of a log-normal random variable with expectation 1 and with variance $\Sigma_{t,T}^2 = \int_t^T \sigma^2(s)ds$,

$$p(y) = \frac{1}{y\Sigma_{t,T}\sqrt{2\pi}}\exp\left(-\frac{1}{2\Sigma_{t,T}^2}\left(\ln(y) - \frac{1}{2}\Sigma_{t,T}^2\right)^2\right). \qquad (3.6.24)$$

Indeed,

$$\mathbb{Q}(m^S[t,T] \leq L, S_T \geq K | S_t = x) = \frac{x}{L} \mathcal{N} \left(d_2 \left(\frac{L^2}{Kx}, T - t \right) \right)$$

$$= \frac{x}{L} \frac{1}{\sqrt{2\pi}} \int_{-\infty}^{d_2(L^2/(Kx),,T-t)} e^{-u^2/2} du = \frac{x}{L} \int_{Kx/L^2}^{+\infty} p(y) dy \,.$$

Hence, we obtain the following proposition:

Proposition 3.6.5.3 *Let $dS_t = \sigma(t) S_t dB_t$. The conditional density f of the pair $(\inf_{t \leq u \leq T} S_u, S_T)$ is given, on the set $\{0 < h < k\}$, by*

$$\mathbb{Q}(\inf_{t \leq u \leq T} S_u \in dh, S_T \in dk |\, S_t = x)$$

$$= \left(-\frac{3x^2}{h^4} p(kxh^{-2}) - \frac{2kx^3}{h^6} p'(kxh^{-2}) \right) dh\, dk \,.$$

where p is defined in (3.6.24).

Comment 3.6.5.4 In the case $dS_t = \sigma(t) S_t dB_t$, the law of $(S_T, \sup_{s \leq T} S_s)$ can also be obtained from results on BM. Indeed,

$$S_s = S_0 \exp \left(\int_0^s \sigma(u) dB_u - \frac{1}{2} \int_0^s \sigma^2(u) du \right)$$

can be written using a change of time as

$$S_t = S_0 \exp \left(B_{\Sigma_t} - \frac{1}{2} \Sigma_t \right)$$

where B is a BM and $\Sigma(t) = \int_0^t \sigma^2(u) du$. The law of $(S_T, \sup_{s \leq T} S_s)$ is deduced from the law of $(B_u, \sup_{s \leq u} B_s)$ where $u = \Sigma_T$.

3.6.6 Valuation and Hedging of Regular Down-and-In Call Options: The General Case

Valuation

We shall keep the same notation for options. However under the risk neutral probability, the dynamics of the underlying are now:

$$dS_t = S_t \left((r - \delta) dt + \sigma dW_t \right) , S_0 = x \,. \tag{3.6.25}$$

A standard method exploiting the martingale framework consists of studying the associated forward price $S_t^F = S_t e^{(r-\delta)(T-t)}$. This is a martingale under the risk-neutral forward probability measure. In this case, it is necessary to discount the barrier.

We can avoid this problem by noticing that any log-normally distributed asset is the power of a martingale asset. In what follows, we shall denote by DIC^S the price of a DIC option on the underlying S with dynamics given by equation (3.6.25).

Lemma 3.6.6.1 *Let S be an underlying whose dynamics are given by (3.6.25) under the risk-neutral probability $\mathbb{Q}$. Then, setting*

$$\gamma = 1 - \frac{2(r-\delta)}{\sigma^2}, \tag{3.6.26}$$

(i) *the process $S^\gamma = (S_t^\gamma, t \geq 0)$ is a martingale with dynamics*

$$dS_t^\gamma = S_t^\gamma \, \widehat{\sigma} \, dW_t$$

where $\widehat{\sigma} = \gamma\sigma$.

(ii) *for any positive Borel function f*

$$\mathbb{E}_\mathbb{Q}(f(S_T)) = \mathbb{E}_\mathbb{Q}\left(\left(\frac{S_T}{x}\right)^\gamma f\left(\frac{x^2}{S_T}\right)\right).$$

PROOF: The proof of (i) is obvious. The proof of (ii) was the subject of Exercise 1.7.3.7 (see also Exercise 3.6.5.1). □

The important fact is that the process $S_t^\gamma = \exp(\widehat{\sigma} W_t - \frac{1}{2}\widehat{\sigma}^2 t)$ is a martingale, hence we can apply the results of Subsection 3.6.4.

The valuation and the instantaneous replication of the BinDIC^S on an underlying S with dynamics (3.6.25), and more generally of a DIC option, are possible by relying on Lemma 3.6.6.1.

Theorem 3.6.6.2 *The price of a regular down-and-in binary option on an underlying with dynamics (3.6.25) is, for $x \geq L$,*

$$\mathrm{BinDIC}^S(x, K, L) = \left(\frac{x}{L}\right)^\gamma \mathrm{BinC}^S\left(L, \frac{Kx}{L}\right). \tag{3.6.27}$$

The price of a regular DIC option is, for $x \geq L$,

$$\mathrm{DIC}^S(x, K, L) = \left(\frac{x}{L}\right)^{\gamma-1} C_E^S\left(L, \frac{Kx}{L}\right). \tag{3.6.28}$$

PROOF: In the first part of the proof, we assume that γ is positive, so that the underlying with carrying cost is an increasing function of the underlying martingale.

It is therefore straightforward to value the binary options:

$$\begin{aligned}
\mathrm{BinC}^S(x, K; \sigma) &= \mathrm{BinC}^M(x^\gamma, K^\gamma; \widehat{\sigma}) \\
\mathrm{BinDIC}^S(x, K, L; \sigma) &= \mathrm{BinDIC}^M(x^\gamma, K^\gamma, L^\gamma; \widehat{\sigma}),
\end{aligned}$$

where we indicate (when it seems important) the value of the volatility, which is σ for S and $\widehat{\sigma}$ for S^γ. The right-hand sides of the last two equations are known from equation (3.6.19):

$$\begin{aligned}\mathrm{BinDIC}^M(x^\gamma, K^\gamma, L^\gamma; \widehat{\sigma}) &= \left(\frac{x}{L}\right)^\gamma \mathrm{BinC}^M\left(L^\gamma, \left(\frac{Kx}{L}\right)^\gamma; \widehat{\sigma}\right) \\ &= \left(\frac{x}{L}\right)^\gamma \mathrm{BinC}^S\left(L, \frac{Kx}{L}; \sigma\right). \end{aligned} \tag{3.6.29}$$

Hence, we obtain the equality (3.6.27). Note that, from formulae (3.6.10) and (3.6.9) (we drop the dependence w.r.t. σ)

$$\begin{aligned}\mathrm{BinC}^S\left(L, \frac{Kx}{L}\right) &= -e^{-\mu T}\mathrm{DeltaP}^S\left(\frac{Kx}{L}, Le^{2\mu T}\right) \\ &= -e^{-\mu T}\mathrm{DeltaP}^S\left(x, \frac{(Le^{\mu T})^2}{K}\right).\end{aligned}$$

By taking the integral of this option's value between K and $+\infty$, the price DIC^S is obtained

$$\begin{aligned}\mathrm{DIC}^S(x, K, L) &= \int_K^\infty \mathrm{BinDIC}^S(x, k, L)dk = \left(\frac{x}{L}\right)^\gamma \int_K^\infty \mathrm{BinC}^S\left(L, k\frac{x}{L}\right) dk \\ &= \left(\frac{x}{L}\right)^{\gamma - 1} C_E^S\left(L, \frac{Kx}{L}\right).\end{aligned}$$

By relying on the put-call symmetry relationship of Proposition 3.6.1.1, and on the homogeneity property (3.6.8), the equality

$$\mathrm{DIC}^S(x, L, K) = \left(\frac{x}{L}\right)^{\gamma - 1} \frac{K}{L} P_E^S\left(x, \frac{L^2}{K}\right)$$

is obtained.

When γ is negative, a DIC binary option on the underlying becomes a UIP binary option on an underlying which is a martingale. In particular,

$$\mathrm{BinDIC}^S(x, K, L; \sigma) = \mathrm{BinUIP}^M\left(x^\gamma, K^\gamma, L^\gamma; \widehat{\sigma}\right),$$

and

$$\mathrm{BinP}^M\left(L^\gamma, \left(\frac{Kx}{L}\right)^\gamma; \widehat{\sigma}\right) = \mathrm{BinC}^S\left(L, \frac{Kx}{L}; \sigma\right)$$

because the payoffs of the two options are the same. From Proposition 3.6.4.3 corresponding to UIP options, we obtain

$$\begin{aligned}\mathrm{BinUIP}^M(x^\gamma, K^\gamma, H^\gamma \widehat{\sigma}) &= \left(\frac{x}{H}\right)^\gamma \mathrm{BinP}^M\left(H^\gamma, \left(\frac{Kx}{H}\right)^\gamma; \widehat{\sigma}\right) \\ &= \left(\frac{x}{H}\right)^\gamma \mathrm{BinC}^S\left(H, \frac{Kx}{H}; \sigma\right).\end{aligned}$$

□

Remark 3.6.6.3 Let us remark that, when $\mu = 0$ (i.e., $\gamma = 1$) the equality (3.6.28) is formula (3.6.16). The presence of carrying costs induces us to consider a forward boundary, already introduced by Carr and Chou [148], in order to give two-sided bounds for the option's price. Indeed, if μ is positive and $(x/L)^{\gamma-1} \leq 1$, the right-hand side gives Carr's upper bound, while if μ is negative, the lower bound is obtained.

Therefore, the smaller $\frac{2\mu}{\sigma^2}$, the more accurate is Carr's approximation. This is also the case when x is close to L, because at the boundary, the two formulae are the same.

Hedging of the Regular Down-and-In Call Option in the General Case

As for the case of a regular DIC option without carrying costs, the Delta is discontinuous at the boundary. By relying on the above developments and on equation (3.6.29), the following equation is obtained

$$\begin{aligned} \Delta_+ \mathrm{DIC}^S(L,K,L) &= \frac{\gamma-1}{L} C_E^S(L,K) - \frac{K}{L} \mathrm{BinC}^S(L,K) \\ &= \frac{\gamma}{L} C_E^S(L,K) - \mathrm{DeltaC}^S(L,K)\,. \end{aligned}$$

Thus,

$$(\Delta_+ - \Delta_-)\mathrm{DIC}^S(L,K,L) = \frac{\gamma}{L} C_E^S(L,K) - 2\,\mathrm{DeltaC}^S(L,K)\,.$$

However, the absolute value of this quantity is not always smaller than 1, as it was in the case without carrying costs. Therefore, depending on the level of the carrying costs, the discontinuity can be either positive or negative.

Exercise 3.6.6.4 Recover (ii) with the help of formula (3.2.4) which expresses a simple absolute continuity relationship between Brownian motions with opposite drifts $\triangleleft$

Exercise 3.6.6.5 A power put option (see Exercise 2.3.1.5) is an option with payoff $S_T^\alpha (K - S_T)^+$, its price is denoted $\mathrm{PowP}^\alpha(x,K)$. Prove that there exists γ such that

$$\mathrm{DIC}^S(x,K,L) = \frac{1}{L^\gamma} \mathrm{PowP}^{\gamma-1}(Kx, L^2)\,.$$

`Hint:` From (ii) in Lemma 3.6.6.1, $\mathrm{DIC}^S(x,K,L) = \frac{1}{L^\gamma} \mathbb{E}(S_T^\gamma (\frac{L^2}{S_T} - K)^+)$. $\triangleleft$

3.6.7 Valuation and Hedging of Reverse Barrier Options

Valuation of the Down-and-In Bond

The payoff of a down-and-in bond (DIB) is one monetary unit at maturity, if the barrier is reached before maturity. It is straightforward to obtain these prices by relying on $\text{BinDIC}(x, L, L)$ prices and on a standard binary put. Indeed, the payoff of the BinDIC option is one monetary unit if the underlying value is greater than L and if the barrier is hit. The payoff of the standard binary put is also 1 if the underlying value is below the barrier at maturity. Being long on these two options generates a payoff of 1 if the barrier was reached before maturity. Hence,

$$\begin{aligned}
&\text{for } x \geq L, \text{DIB}(x, L) = \text{BinP}(x, L) + \text{BinDIC}(x, L, L) \\
&\text{for } x \leq L, \text{DIB}(x, L) = B(0, T)\,.
\end{aligned}$$

By relying on equations (3.6.10, 3.6.11, 3.6.28) and on Black and Scholes' formula, we obtain, for $x \geq L$,

$$\begin{aligned}
\text{DIB}(x, L) &= \text{BinP}^S(x, L) + \left(\frac{x}{L}\right)^{\gamma} \text{BinC}^S(L, x) \\
&= e^{-rT}\left[\mathcal{N}\left(d_1\left(\frac{L}{xe^{\mu T}}\right)\right) + \frac{x^{\gamma}}{L^{\gamma}}\mathcal{N}\left(d_2\left(\frac{Le^{\mu T}}{x}\right)\right)\right]. \quad (3.6.30)
\end{aligned}$$

Example 3.6.7.1 Prove the following relationships:

$$\begin{aligned}
&DIC^S(x, L, L) + L\,\text{BinDIC}^S(x, L, L) \\
&= \left(\frac{x}{L}\right)^{\gamma-1} e^{-\mu T}\left[P_E^S(x, Le^{2\mu T}) - L\frac{x}{L}\text{Delta}P_E^S(x, Le^{2\mu T})\right] \\
&= \left(\frac{x}{L}\right)^{\gamma-1} e^{\mu T} L\,\text{BinP}^S(x, Le^{2\mu T}), \\
\text{DIB}(x, L) &= \text{BinP}^S(x, L) + \left(\frac{x}{L}\right)^{\gamma-1} e^{\mu T}\text{BinP}^S(x, Le^{2\mu T}) \\
&\quad - \frac{1}{L}\text{DIC}^S(x, L, L)\,. \quad (3.6.31)
\end{aligned}$$

`Hint:` Use formulae (3.6.12) and (3.6.28).

Valuation of a Reverse DIC, Case $K < L$

Let us study the reverse DIC option, with strike smaller than the barrier, that is $K \leq L$. Depending on the value of the underlying with respect to the barrier at maturity, the payoff of such an option can be decomposed. Let us consider the case where $x \geq L$.

- The option with a payoff $(S_T - K)^+$ if the underlying value is higher than L at maturity and if the barrier was reached can be hedged with a $\text{DIC}(x, L, L)$ with payoff $(S_T - L)$ at maturity if the barrier was reached and by $(L - K)$ $\text{BinDIC}(x, L, L)$ options, with a payoff $L - K$ if the barrier was reached.
- The option with a payoff $(S_T - K)^+$ if the underlying value is between K and L at maturity (which means that the barrier was reached) can be hedged by the following portfolio:

$$-P_E(x,L) + P_E(x,K) + (L-K)\mathrm{DIB}(x,L)\,.$$

Indeed the corresponding payoff is

$$\begin{aligned}(S_T - K)^+ \mathbb{1}_{\{K \le S_T \le L\}} &= (S_T - L - K + L)\mathbb{1}_{\{K \le S_T \le L\}} \\ &= (S_T - L)\mathbb{1}_{\{K \le S_T \le L\}} + (L-K)\mathbb{1}_{\{K \le S_T \le L\}} \\ &= (S_T - L)\mathbb{1}_{\{S_T \le L\}} - (S_T - L)\mathbb{1}_{\{S_T \le K\}} \\ &\qquad + (L-K)\mathbb{1}_{\{S_T \le L\}} - (L-K)\mathbb{1}_{\{S_T \le K\}} \\ &= -(L - S_T)^+ + (K - S_T)^+ + (L-K)\mathbb{1}_{\{S_T \le L\}}\,.\end{aligned}$$

This very general formula is a simple consequence of the no arbitrage principle and can be obtained without specific assumptions concerning the underlying dynamics, unlike the DIB valuation formula.

The hedging of such an option requires plain vanilla options, regular DIC options with the barrier equal to the strike, and $\mathrm{DIB}(x,L)$ options, and is not straightforward. The difficulty corresponds to the hedging of the standard binary option.

In the particular case of a deterministic volatility, by relying on (3.6.31),

$$\begin{aligned}\mathrm{DIC}_{rev}(x,K,L) = & \left(\frac{K}{L} - 1\right)\mathrm{DIC}(x,L,L) - P_E(x,L) + P_E(x,K) \\ & + (L-K)\mathrm{BinP}(x,L) \\ & + (L-K)\left(\frac{x}{L}\right)^{\gamma-1} e^{\mu T}\mathrm{BinP}(x, Le^{2\mu T})\,.\end{aligned}$$

3.6.8 The Emerging Calls Method

Another way to understand barrier options is the study of the first passage time of the underlying at the barrier, and of the prices of the calls at this first passage time. This corresponds to integration of the calls with respect to the hitting time distribution.

Let us assume that the initial underlying value x is higher than the barrier, i.e., $x > L$. We denote, as usual,

$$T_L = \inf\{t \,:\, S_t \le L\}$$

the hitting time of the barrier L.

The term $e^{rT}\mathrm{DIB}(x,L,T)$ is equal to the probability that the underlying reaches the barrier before maturity T. Hence, its derivative, i.e., the quantity $f_L(x,t) = \partial_T[e^{rT}\mathrm{DIB}(x,L,T)]_{T=t}$ is the density $\mathbb{Q}(T_L \in dt)/dt$, and the following decomposition of the barrier option is obtained:

$$\mathrm{DIC}(x,K,L,T) = \int_0^T C_E(L,K,T-\tau)e^{-r\tau} f_L(x,\tau)d\tau\,. \qquad (3.6.32)$$

The density f_L is obtained by differentiating $e^{rT}\text{DIB}$ with respect to T in (3.6.30). Hence

$$f_L(x,t) = \frac{h}{\sqrt{2\pi t^3}} \exp(-\frac{1}{2t}(h-\nu t)^2),$$

where

$$h = \frac{1}{\sigma} \ln\left(\frac{x}{L}\right), \nu = \frac{\mu}{\sigma} - \frac{\sigma}{2}.$$

(See Subsection 3.2.2 for a different proof.)

3.6.9 Closed Form Expressions

Here, we give the previous results in a closed form.

▶ For $K \leq L$,

$$\begin{aligned}\text{DIC}^S(L,K) &= S_0\Big(\mathcal{N}(z_1) - \mathcal{N}(z_2) + \left(\frac{L}{x}\right)^{\frac{2r}{\sigma^2}+1} \mathcal{N}(z_3)\Big)\\ &\quad - Ke^{-rT}\Big(\mathcal{N}(z_4) - \mathcal{N}(z_5) + \left(\frac{L}{x}\right)^{\frac{2r}{\sigma^2}-1} \mathcal{N}(z_6)\Big)\end{aligned}$$

where

$$\begin{aligned}
z_1 &= \frac{1}{\sigma\sqrt{T}}\left(\left(r+\frac{1}{2}\sigma^2\right)T + \ln\left(\frac{x}{K}\right)\right), & z_4 &= z_1 - \sigma\sqrt{T}\\
z_2 &= \frac{1}{\sigma\sqrt{T}}\left(\left(r+\frac{1}{2}\sigma^2\right)T + \ln\left(\frac{x}{L}\right)\right), & z_5 &= z_2 - \sigma\sqrt{T}\\
z_3 &= \frac{1}{\sigma\sqrt{T}}\left(\left(r+\frac{1}{2}\sigma^2\right)T - \ln\left(\frac{x}{L}\right)\right), & z_6 &= z_3 - \sigma\sqrt{T}.
\end{aligned}$$

▶ In the case $K \geq L$, we find that

$$\text{DIC}^S(L,K) = x\left(\frac{L}{x}\right)^{\frac{2r}{\sigma^2}+1} \mathcal{N}(z_7) - Ke^{-rT}\left(\frac{L}{x}\right)^{\frac{2r}{\sigma^2}-1} \mathcal{N}(z_8)$$

where

$$\begin{aligned}
z_7 &= \frac{1}{\sigma\sqrt{T}}\left(\ln(L^2/xK) + \left(r+\frac{1}{2}\sigma^2\right)T\right)\\
z_8 &= z_7 - \sigma\sqrt{T}.
\end{aligned}$$

3.7 Lookback Options

A **lookback** option on the minimum is an option to buy at maturity T the underlying S at a price equal to K times the minimum value m_T^S of the underlying during the maturity period (here, $m_T^S = \min_{0\leq u\leq T} S_u$). The terminal payoff is $(S_T - Km_T^S)^+$. We assume in this section that the dynamics of the underlying asset value under the risk-adjusted probability is given in a Garman-Kohlhagen model by equation (3.6.25).

3.7.1 Using Binary Options

The BinDIC^S price formula can be used in order to value and hedge options on a minimum. Let $\mathrm{MinC}^S(x,K)$ be the price of the lookback option. The terminal payoff can be written

$$(S_T - Km_T^S)^+ = \int_0^{+\infty} \mathbb{1}_{\{S_T\geq k\geq Km_T^S\}}\,dk\,.$$

The expectation of this quantity can be expressed in terms of barrier options:

$$\begin{aligned}\mathrm{MinC}^S(x,K) &= e^{-rT}\mathbb{E}_{\mathbb{Q}}((S_T-Km_T^S)^+) = \int_0^{+\infty}\mathrm{BinDIC}^S\left(x,k,\frac{k}{K}\right)dk\\ &= \int_0^{xK}\mathrm{BinDIC}^S\left(x,k,\frac{k}{K}\right)dk + \int_{xK}^{\infty}\mathrm{BinDIC}^S\left(x,k,\frac{k}{K}\right)dk\\ &= I_1 + I_2\,.\end{aligned}$$

In the second integral I_2, since $x < k/K$, the BinDIC is activated at time 0 and $\mathrm{BinDIC}^S(x,k,\frac{k}{K}) = \mathrm{BinC}^S(x,k)$, hence

$$I_2 = e^{-rT}\int_{xK}^{\infty}\mathbb{E}_{\mathbb{Q}}(\mathbb{1}_{\{S_T\geq k\}})dk = e^{-rT}\mathbb{E}_{\mathbb{Q}}((S_T-xK)^+) = C_E^S(x,xK)\,.$$

The first term I_1 is more difficult to compute than I_2. From Theorem 3.6.6.2, we obtain, for $k < Kx$,

$$\mathrm{BinDIC}^S\left(x,k,\frac{k}{K}\right) = \left(\frac{xK}{k}\right)^{\gamma}\mathrm{BinC}^S\left(\frac{k}{K},xK\right),$$

where γ is the real number such that $(S_t^\gamma, t\geq 0)$ is a martingale, i.e., $S_t = xM_t^{1/\gamma}$ where M is a martingale with initial value 1. From the identity $\mathrm{BinC}^S(x,K) = e^{-rT}\mathbb{Q}(xM_T^{1/\gamma} > K)$, we get:

$$\begin{aligned}\int_0^{xK}\mathrm{BinDIC}^S\left(x,k,\frac{k}{K}\right)dk &= \int_0^{xK}\left(\frac{xK}{k}\right)^{\gamma}\mathrm{BinC}^S\left(\frac{k}{K},xK\right)dk\\ &= e^{-rT}\mathbb{E}_{\mathbb{Q}}\left(\int_0^{xK}\left(\frac{xK}{k}\right)^{\gamma}\mathbb{1}_{\{kM_T^{1/\gamma}>xK^2\}}dk\right)\end{aligned}$$

$$= e^{-rT}(xK)^{\gamma}\mathbb{E}_{\mathbb{Q}}\left(\int_0^{\infty} k^{-\gamma}\mathbb{1}_{\{xK>k>xK^2M_T^{-1/\gamma}\}}dk\right).$$

For $\gamma \neq 1$, the integral can be computed as follows:

$$\begin{aligned}
&\int_0^{xK} \mathrm{BinDIC}^S\left(x, k, \frac{k}{K}\right) dk \\
&= e^{-rT}\frac{(xK)^{\gamma}}{1-\gamma}\mathbb{E}_{\mathbb{Q}}\left[\left((xK)^{1-\gamma} - (xK^2M_T^{-1/\gamma})^{1-\gamma}\right)^+\right] \\
&= e^{-rT}\frac{xK}{1-\gamma}\mathbb{E}_{\mathbb{Q}}\left[\left(1 - K^{1-\gamma}M_T^{-(1-\gamma)/\gamma}\right)^+\right] \\
&= e^{-rT}\frac{xK}{1-\gamma}\mathbb{E}_{\mathbb{Q}}\left[\left(1 - \frac{K^{1-\gamma}S_T^{\gamma-1}}{x^{\gamma-1}}\right)^+\right].
\end{aligned}$$

Using Itô's formula and recalling that $1-\gamma = \frac{2\mu}{\sigma^2}$, we have

$$d(S_t^{\gamma-1}) = S_t^{\gamma-1}\left(\mu dt - \frac{2\mu}{\sigma}dW_t\right)$$

hence the following formula is derived

$$\mathrm{MinC}^S(x,K) = x\left[C_E^S(1,K;\sigma) + \frac{K\sigma^2}{2\mu}P_E^S\left(K^{1-\gamma}, 1; \frac{2\mu}{\sigma}\right)\right]$$

where $C_E^S(x,K;\sigma)$ (resp. $P_E^S(x,K;\sigma)$) is the call (resp. put) value on an underlying with carrying cost μ and volatility σ with strike K. The price at date t is $\mathrm{MinC}^S(S_t, Km_t^S; T-t)$ where $m_t^S = \min_{s\le t} S_s$.

For $\gamma = 1$ we obtain

$$\mathrm{MinC}^S(x,K) = C_E^S(x, xK) + xK\mathbb{E}_{\mathbb{Q}}\left[\left(\ln\frac{S_T}{xK}\right)^+\right].$$

Let $C_{\ln}^S(x,K)$ be the price of an option with payoff $(\ln(S_T/x) - \ln K)^+$, then

$$\mathrm{MinC}^S(x,K) = C_E^S(x,xK) + xKC_{\ln}^S(x,xK).$$

3.7.2 Traditional Approach

The payoff for a **standard lookback** call option is $S_T - m_T^S$. Let us remark that the quantity $S_T - m_T^S$ is positive. The price of such an option is

$$\mathrm{MinC}^S(x,1;T) = e^{-rT}\mathbb{E}_{\mathbb{Q}}(S_T - m_T^S)$$

whereas $\mathrm{MinC}^S(x,1;T-t)$, the price at time t, is given by

$$\mathrm{MinC}^S(x,1;T-t) = e^{-r(T-t)}\mathbb{E}_{\mathbb{Q}}(S_T - m_T^S|\mathcal{F}_t)\,.$$

We now forget the superscript S in order to simplify the notation. The relation $m_T = m_t \wedge m_{t,T}$, with $m_{t,T} = \inf\{S_u, u \in [t,T]\}$ leads to

$$e^{-rt}\mathrm{MinC}^S(x,1;T-t) = e^{-rT}\mathbb{E}_{\mathbb{Q}}(S_T|\mathcal{F}_t) - e^{-rT}\mathbb{E}_{\mathbb{Q}}(m_t \wedge m_{t,T}|\mathcal{F}_t)\,.$$

Using the $\mathbb{Q}$-martingale property of the process $(e^{-\mu t}S_t, t \geq 0)$, the first term is $e^{-rt}e^{-\delta(T-t)}S_t$. As far as the second term is concerned, the expectation is decomposed as follows:

$$\mathbb{E}_{\mathbb{Q}}(m_t \wedge m_{t,T}|\mathcal{F}_t) = \mathbb{E}_{\mathbb{Q}}(m_t \mathbb{1}_{\{m_t < m_{t,T}\}}|\mathcal{F}_t) + \mathbb{E}_{\mathbb{Q}}(m_{t,T}\mathbb{1}_{\{m_{t,T} < m_t\}}|\mathcal{F}_t)\,.$$

Using measurability and independence arguments, we obtain

$$\mathbb{E}_{\mathbb{Q}}(m_t \mathbb{1}_{\{m_t < m_{t,T}\}}|\mathcal{F}_t) = m_t\,\Phi(T-t, m_t, S_t)$$

where $\Phi(u,m,x) = \mathbb{Q}(m < x m_u^Y)$, with $Y \stackrel{\text{law}}{=} (S/S_0)$. An explicit expression for Φ is obtained from the results concerning the law of the minimum of the drifted Brownian motion or by relying on barrier options results:

$$\Phi(u,m,x) = \mathcal{N}(d - \sigma\sqrt{u}) - \left(\frac{x}{m}\right)^{1-2\mu/\sigma^2}\mathcal{N}\left(-d + \frac{2\mu}{\sigma}\sqrt{u}\right)$$

where

$$d = d_1\left(\frac{xe^{ru}}{m}\right) = \frac{\ln\left(\frac{x}{m}\right) + (\mu + \sigma^2/2)u}{\sigma\sqrt{u}}\,.$$

The quantity

$$\mathbb{E}_{\mathbb{Q}}(m_{t,T}\mathbb{1}_{\{m_{t,T} < m_t\}}|\mathcal{F}_t)$$

can be written $\Psi(T-t, m_t, S_t)$ with $\Psi(u,m,x) = \mathbb{E}_{\mathbb{Q}}(x m_u \mathbb{1}_{\{x m_u < m\}})$ which can be computed from the law of m_u. The following proposition (obtained also in the previous section, setting $K=1$) is derived:

Proposition 3.7.2.1 *The lookback option price is*

$$\begin{aligned}\mathrm{Min}^S(S_t,1;T-t) &= S_t e^{-\delta(T-t)}\mathcal{N}(d_t) - e^{-r(T-t)} m_t \mathcal{N}\left(d_t - \sigma\sqrt{T-t}\right)\\ &\quad + e^{-r(T-t)}\frac{S_t\sigma^2}{2\mu}\left[\left(\frac{m_t}{S_t}\right)^{\frac{2\mu}{\sigma^2}}\mathcal{N}\left(-d_t + \frac{2\mu\sqrt{T-t}}{\sigma}\right) - e^{r(T-t)}\mathcal{N}(-d_t)\right]\end{aligned}$$

with $d_t = \dfrac{1}{\sigma\sqrt{T-t}}\ln\left(\dfrac{S_t}{m_t} + \left(\mu + \dfrac{1}{2}\sigma^2\right)(T-t)\right)$ *and* $m_t = \inf_{s \leq t} S_s$.

Comment 3.7.2.2 Other results on lookback options are presented in Conze and Viswanathan [193] and He et al. [426]. A PDE approach for European options whose terminal payoff involves path-dependent lookback variables is presented in Xu and Kwok [853]. See also Elliott and Kopp [317] p. 182–183 for the case $\delta = 0$ and Musiela and Rutkowski [661] p. 214–218 and Shreve [795], p. 314–320.

3.8 Double-barrier Options

The payoff of a double-barrier option is $(S_T - K)^+$ if the underlying asset has remained in the range $[L, H]$ for all times between 0 and maturity, otherwise, the payoff is null. Its price is

$$C_{db}(x, K, L, H, T) := \mathbb{E}_{\mathbb{Q}}(e^{-rT}(S_T - K)^+ \mathbb{1}_{\{T^* > T\}})$$

where $T^* := T_H(S) \wedge T_L(S)$. We give the computation of

$$\mathbb{E}_{\mathbb{Q}}(e^{-rT}(S_T - K)^+ \mathbb{1}_{\{T^* < T\}}) = \mathbb{E}_{\mathbb{Q}}(e^{-rT}(S_T - K)^+) - C_{db}(x, K, L, H, T),$$

in the case where the risk-neutral dynamics of S are

$$dS_t = S_t(rdt + \sigma dW_t);$$

the price of the double barrier will follow. With a change of probability the quantity $\mathbb{E}_{\mathbb{Q}}(e^{-rT}(S_T - K)^+ \mathbb{1}_{\{T^* < T\}})$ can be written as

$$e^{-(r+\frac{1}{2}\nu^2)T} \mathbb{E}_{\mathbb{Q}}((xe^{\sigma B_T} - K)^+ e^{\nu B_T} \mathbb{1}_{\{T^* < T\}}),$$

where B is a generic BM. The explicit computation can be performed using the law of the pair (B_T, T^*) which may be obtained from the two-sided series (3.5.2).

Another approach is to proceed as in Geman and Yor [384] where the Laplace transform Φ of $\varphi(t) = \mathbb{E}_{\mathbb{Q}}[e^{\nu B_t}(xe^{\sigma B_t} - K)^+ \mathbb{1}_{\{T^* < t\}}]$ is computed. From Markov's property

$$\begin{aligned}\Phi(\lambda) &= \int_0^\infty \exp\left(-\frac{\lambda^2 t}{2}\right) \varphi(t)\, dt = \mathbb{E}_{\mathbb{Q}}\left(\int_{T^*}^\infty \exp\left(-\frac{\lambda^2 t}{2}\right) \psi(B_t)\, dt\right) \\ &= \mathbb{E}_{\mathbb{Q}}\left(\exp\left(-\frac{\lambda^2 T^*}{2}\right) \int_0^\infty \exp\left(-\frac{\lambda^2 t}{2}\right) \psi(\widetilde{B}_t + B_{T^*})\, dt\right)\end{aligned}$$

where $\psi(y) = e^{\nu y}(xe^{\sigma y} - K)^+$ and $\widetilde{B} = (\widetilde{B}_t = B_{t+T^*} - B_{T^*}; t \geq 0)$ is a Brownian motion independent of $(B_s, s \leq T^*)$. The computation of the expectation can be simplified by splitting the expression into two parts depending on the stopping time values:

$$\Phi(\lambda) = \Psi(h)\mathbb{E}_{\mathbb{Q}}\left[e^{-\lambda^2 T^*/2} \mathbb{1}_{\{T^* = T_h\}}\right] + \Psi(\ell)\mathbb{E}_{\mathbb{Q}}\left[e^{-\lambda^2 T^*/2} \mathbb{1}_{\{T^* = T_\ell\}}\right],$$

where $h = \ln(H/x)\sigma^{-1}, \ell = \ln(L/x)\sigma^{-1}$ and, from Exercise 1.4.1.7

$$\Psi(z) = \mathbb{E}\int_0^\infty e^{-\lambda^2 t/2} \psi(\widetilde{B}_t + z) dt = \frac{1}{\lambda}\int_{-\infty}^\infty e^{-\lambda|z-y|}\psi(y) dy\,.$$

We have obtained an explicit form for

$$\mathbb{E}_{\mathbb{Q}}\left[\exp\left(-\frac{\lambda^2 T^*}{2}\right)\mathbb{1}_{\{T^*=T_h\}}\right]$$

in the proof of Proposition 3.5.1.3; we now present the computation of $\Psi(x)$.

▶ Let $K \in [L, H]$ and let $k = \ln(K/x)\sigma^{-1}, \ell = \ln(L/x)\sigma^{-1}$. For values of λ such that $\nu + \sigma - \lambda < 0$, and by relying on the resolvent:

$$\begin{aligned}\Psi(h) &= g(h,\lambda)[Kg(h,\nu-\lambda) - xg(h,\sigma+\nu-\lambda)] \\ &\quad + g(-h,\lambda)\big[x(g(h,\sigma+\nu+\lambda) - g(k,\sigma+\nu+\lambda)) \\ &\qquad - K(g(h,\nu+\lambda) - g(k,\nu+\lambda))\big]\end{aligned}$$

with $g(u,\alpha) = \frac{1}{u}e^{u\alpha}$ and

$$\Psi(\ell) = \frac{e^{\lambda\ell}}{\lambda}[Kg(k,\nu-\lambda) - xg(k,\sigma+\nu-\lambda)] .$$

▶ For $K < L$, and $z = \ell$ or $z = k$, we find

$$\begin{aligned}\Psi(z) &= g(h,\lambda)\big(Kg(h,\nu-\lambda) - xg(z,\sigma+\nu-\lambda)\big) \\ &\quad + g(-h,\lambda)\Big(x\left(g(h,\sigma+m+\lambda) - g(z,\sigma+\nu+\lambda)\right) \\ &\qquad - K\left(g(h,\nu+\lambda) - g(z,\nu+\lambda)\right)\Big) .\end{aligned}$$

The Laplace transform must now be inverted.

The main papers concerning double-barrier options are those of Kunitomo and Ikeda [551], Geman and Yor [384], Goldman et al. [399], Pelsser [704], Hui et al. [600], Schröder [768] and Davydov and Linetsky [226].

3.9 Other Options

We give a few examples of other traded options. We assume as previously that

$$dS_t = S_t((r-\delta)dt + \sigma dW_t),\ S_0 = x$$

under the risk-neutral probability $\mathbb{Q}$ and we denote by $T_a = T_a(S)$ the first time when level a is reached by the process S.

3.9.1 Options Involving a Hitting Time

Digital Options

The *asset-or-nothing* options depend on an exercise price K. The terminal payoff is equal to the value of the underlying, if it is in the money at maturity and 0 otherwise, i.e., $S_T\mathbb{1}_{\{S_T \geq K\}}$. The strike price plays the rôle of a barrier.

The value of such an option is $e^{-rT}\mathbb{E}_{\mathbb{Q}}(S_T \mathbb{1}_{\{S_T \geq K\}})$ and is straightforward to evaluate. Indeed, this is the first term in the Black and Scholes formula (2.3.3).

These options can also have an up-and-in feature which depends on a barrier. The price is $e^{-rT}\mathbb{E}_{\mathbb{Q}}(S_T \mathbb{1}_{\{S_T \geq K\}} \mathbb{1}_{\{T_L > T\}})$. They are used for hedging barrier options.

Barrier Forward-start or Early-ending Options

In this case, the barrier is activated at time T', with $T' < T$ where T is the maturity. In the case of an up-and-out forward-start call option, the payoff is $(S_T - K)^+ \mathbb{1}_{\{T_H^{T'} \geq T\}}$ with $T_H^{T'} = \inf\{u \geq T' : S_u \geq H\}$. For *early-ending* options, the barrier is active only until T'.

3.9.2 Boost Options

The BOOST (Banking On Overall Stability) options were introduced in the market by Société Générale in 1994. They are characterized by two levels, a and b, with $a \leq b$. When the boundary of a given range $[a, b]$ is reached for the first time, the BOOST option terminates, and its owner receives a payoff equal to a daily amount multiplied by the number of days during which the underlying asset remained in the range before the first exit. A BOOST option is, most of the time, a strictly decreasing function of the volatility; therefore it enables its owner to bet on a decrease in the volatility.

One-level

The one-level BOOST pays, at maturity, an amount equal to the time that the underlying asset remains continuously above a level a. Therefore, its price is

$$\mathbb{E}_{\mathbb{Q}}[e^{-rT}(T \wedge T_a)] = e^{-rT} T\, \mathbb{Q}(T < T_a) + e^{-rT}\mathbb{E}_{\mathbb{Q}}(T_a \mathbb{1}_{\{T_a < T\}}) .$$

Assume that $a < x$ and let us introduce, as in Subsection 3.2.4,

$$\Psi(\lambda) := \mathbb{E}(e^{-\lambda T_a(S)} \mathbb{1}_{\{T_a(S) < T\}}) = e^{(\nu-\gamma)\alpha} \mathcal{N}\left(\frac{\alpha - \gamma T}{\sqrt{T}}\right) + e^{(\nu+\gamma)\alpha} \mathcal{N}\left(\frac{\alpha + \gamma T}{\sqrt{T}}\right)$$

with $\nu = (r - \delta)(\sigma)^{-1} - \sigma/2$, $\gamma^2 = 2\lambda + \nu^2$, $\alpha = \sigma^{-1}\ln(a/x)$.

Then $\mathbb{E}(T_a \mathbb{1}_{\{T_a < T\}}) = -\Psi'(0)$, i.e.,

$$\begin{aligned} \mathbb{E}(T_a \mathbb{1}_{\{T_a < T\}}) &= -\frac{\alpha}{\nu}\left[\mathcal{N}\left(-\frac{\nu T + \alpha}{\sqrt{T}}\right) - e^{2\nu\alpha}\mathcal{N}\left(\frac{\nu T + \alpha}{\sqrt{T}}\right)\right] \\ &+ \frac{\sqrt{T}}{\nu\sqrt{2\pi}}\left[\exp\left[-\frac{1}{2T}(\nu T - \alpha)^2\right] - e^{2\nu\alpha}\exp\left[-\frac{1}{2T}(\nu T + \alpha)^2\right]\right] \end{aligned}$$

and

$$\mathbb{Q}(T < T_a) = 1 - \Psi(0) = \mathcal{N}\left(\frac{-\alpha + \nu T}{\sqrt{T}}\right) - e^{2\nu\alpha}\mathcal{N}\left(\frac{\alpha + \nu T}{\sqrt{T}}\right).$$

Corridor

The BOOST option value $B_{cor}(S_0, T)$ is given by the expected discounted payoff,

$$B_{cor}(S_0, T) := \mathbb{E}_{\mathbb{Q}}(e^{-rT^*}T^*\mathbb{1}_{\{T^*<T\}} + e^{-rT}T\mathbb{1}_{\{T^*\geq T\}}) \tag{3.9.1}$$

with

$$T^* = T_a(S) \wedge T_b(S).$$

We suppose that $a < S_0 < b$. The valuation problem reduces to the knowledge of the law of T^*.

Let us consider a **perpetual corridor BOOST with payment at hit.** Its price is given by (3.9.1) with $T = \infty$, i.e.,

$$B_{cor}(S_0, \infty) := \mathbb{E}_{\mathbb{Q}}(e^{-rT^*}T^*).$$

The problem reduces to the computation of $\Psi(\lambda) = \mathbb{E}_{\mathbb{Q}}(\exp(-\frac{1}{2}\lambda^2 T^*))$. Indeed, the computation of $\mathbb{E}_{\mathbb{Q}}(e^{-rT^*}T^*)$ will follow after differentiation with respect to λ: $\mathbb{E}_{\mathbb{Q}}(e^{-rT^*}T^*) = -\dfrac{\Psi'(\sqrt{2r})}{\sqrt{2r}}$. Let us remark that

$$T^* = \inf\{t | X_t \leq \alpha \text{ or } X_t \geq \beta\} := T^*(X)$$

where $X_t = \nu t + B_t = (\frac{r-\delta}{\sigma} - \frac{\sigma}{2})t + B_t$. Using the results obtained in Subsection 3.5.2, we get, in the case $\frac{a}{x} = \frac{x}{b}$,

$$\mathbb{E}_{\mathbb{Q}}(e^{-rT^*}) = \frac{b}{x}\frac{x^\theta + b^\theta}{x^{\theta-2} + b^{\theta-2}}$$

with

$$\theta = -\frac{2\nu}{\sigma} = -\frac{2(r-\delta)}{\sigma^2} + 1.$$

It follows that

$$\mathbb{E}(T^* e^{-rT^*}) = \frac{2b\,(bx)^{\theta-2}}{x\sigma^2\,[x^{\theta-2} + b^{\theta-2}]^2}(x^2 - b^2)\,\ln\frac{x}{b}.$$

Comments 3.9.2.1 (a) Many other examples are presented in Haug [425], Kat [516], Pechtl [703], and Zhang [872].

(b) Crucial hedging problems are not considered here: we refer to Bhansali [84] and Taleb [818].

(c) BOOST options have been studied by Douady [263] and Leblanc [572]. Path-dependent options with payoff of the form

$$\left(\int_0^T \mathbb{1}_{\{S_s \geq a\}} ds - K\right)^+ \mathbb{1}_{\{M_T^S \leq b\}}$$

are studied in Fujita et al. [367].

3.9.3 Exponential Down Barrier Option

We apply the results given in Subsections 3.3.1 and 3.2.2 to obtain the price of an option with a deterministic exponential barrier. As usual, we work in the Black and Scholes model where the dynamics of the underlying stock value in the risk-neutral economy are:

$$dS_t = S_t(rdt + \sigma dB_t),\ S_0 = x$$

where the risk-free rate r and the volatility σ are constant and where B is a Brownian motion under the risk-neutral probability $\mathbb{Q}$. The barrier $b(t)$ is a deterministic function of time

$$b(t) = z \exp(\eta t),$$

where $z < x$, $\eta > 0$ and $ze^{\eta T} < K$. The first hitting time of the barrier is the time τ

$$\tau = \inf\{t \geq 0,\ S_t \leq b(t)\} = \inf\{t \geq 0,\ \widehat{S}_t \leq z\}$$

where $\widehat{S}_t = S_t e^{-\eta t}$. The dynamics of $\widehat{S}$ are:

$$d\widehat{S}_t = \widehat{S}_t((r-\eta)dt + \sigma dB_t),\ \widehat{S}_0 = x\,.$$

We assume that the payoff $(K - S_\tau)^+ = (K - S_\tau)$ is paid at hit, i.e., at time τ in the case $\tau < T$ and that, if $T \leq \tau$, the payoff is $(K - S_T)^+$, paid at T, where K is the strike price. Therefore, the value of this down-paid at hit option with exponential barrier is given by:

$$\begin{aligned}
&P^{\eta,z}_{\text{expbar}}(S_0, T) \\
&= \mathbb{E}_{\mathbb{Q}}((K - S_\tau)^+ e^{-r\tau}\mathbb{1}_{\{\tau<T\}}) + e^{-rT}\mathbb{E}_{\mathbb{Q}}((K - S_T)^+\mathbb{1}_{\{\tau \geq T\}}) \\
&= \mathbb{E}_{\mathbb{Q}}((K - S_\tau)e^{-r\tau}\mathbb{1}_{\{\tau<T\}}) + e^{(\eta-r)T}\mathbb{E}_{\mathbb{Q}}((e^{-\eta T}K - \widehat{S}_T)^+\mathbb{1}_{\{\tau \geq T\}}) \\
&= \int_0^T (K - b(t))e^{-rt}\mathbb{Q}(\tau \in dt) \\
&\qquad + e^{(\eta-r)T}\int_z^{Ke^{-\eta T}} (Ke^{-\eta T} - y)\,\mathbb{Q}(\widehat{S}_T \in dy, \widehat{m}_T > z)
\end{aligned}$$

where $\widehat{m}_T$ is the minimum

$$\widehat{m}_T = \inf_{u \in [0,T]} \widehat{S}_u \,.$$

By relying on the dynamics of the process $\widehat{S}$ and on Subsections 3.2.2 and 3.3.1 the two densities are known: setting

$$\alpha = \frac{\ln(z/x)}{\sigma}, \quad \text{and} \quad \nu = \frac{r-\eta}{\sigma} - \frac{\sigma}{2},$$

we obtain

$$\mathbb{Q}(\tau \in dt) = |\alpha| \frac{1}{\sqrt{2\pi t^3}} \exp\left(-\frac{1}{2t}(\alpha - \nu t)^2\right) dt$$

and, for $y > z, x > z$,

$$\mathbb{Q}(\widehat{S}_T \in dy, m_T > z) = -\frac{d}{dy}\mathbb{Q}(\widehat{S}_T \geq y, m_T > z)\,.$$

Hence, setting $\beta(y) = \frac{\ln(y/x)}{\sigma}$

$$\mathbb{Q}(\widehat{S}_T \in dy, m_T > z)/dy =$$

$$\frac{1}{\sigma y \sqrt{2\pi T}} \left(\exp\left(-\frac{(-\beta(y) + \nu T)^2}{2T}\right) - e^{2\nu\alpha} \exp\left(-\frac{(-\beta(y) + 2\alpha + \nu T)^2}{2T}\right)\right)$$

The value of the option follows.

Comments 3.9.3.1 By assuming that the exercise boundary of the American put (see ↣ Section 3.11) written on a non-dividend-paying stock is an exponential function of time to expiration, Omberg [686] obtains an approximation of the put price P_A. The author makes the assumption that the exercise boundary for an American put can be approximated by

$$b_{p,z,\eta}(t) = z^* \exp(\eta^* t), \; t \in]0, T]$$

where the two unknowns $z^* = b_p(0)$ and η^* are positive and constant. Each function of this form corresponds to a possible exercise policy which is defined as follows: to exercise the put as soon as the underlying process S reaches $b_{p,z,\eta}$ before maturity, that is to say at time τ if $\tau < T$, or at maturity if the put is in the money and if $\tau \geqslant T$. In this context, the put option value is given by means of the previous computation:

$$P_A(S_0, T) = \sup_{z,\eta} P^{\eta,z}_{\text{expbar}}(S_0, T)$$

and z^*, η^* are the values of (z, η) which maximise this expression. By simplifying further the option value, Omberg [686] obtains a weighted sum of cumulative functions of the standard Gaussian law.

It is worthwhile mentioning that the above approximation is in reality a lower bound for the put value, since an exponential exercise boundary is in general suboptimal. Indeed, for example, at maturity, it is known that the exercise boundary is a non-differentiable function of time (the slope is infinite). As shown in equation (3.11.7), the approximation of the exercise boundary near to maturity is different from an exponential function of time. However, as shown by Omberg, the level of accuracy obtained with this approximation formula is high.

3.10 A Structural Approach to Default Risk

Credit risk, or default risk, concerns the case where a promised payoff is not delivered if some event (the default) happens before the delivery date. The default occurs at time τ where τ is a random variable.

In the structural approach, a default event is specified in terms of the evolution of the firm's assets. Given the value of the assets of the firm, the aim is to deduce the value of corporate debt.

3.10.1 Merton's Model

In this approach – pioneered by Merton [642] – the default occurs if the assets of the firm are insufficient to meet payments on debt *at maturity*. The firm is financed by the issue of bonds, and the face value L of the bonds must be paid at time T. At time T, the bondholders will receive $\min(V_T, L)$ where L is the debt value and V_T the value of the firm. Thus, writing

$$\min(V_T, L) = L - (L - V_T)^+$$

we are essentially dealing with an option pricing problem. Merton assumes that the risk-neutral dynamics of the value of the firm are

$$dV_t = V_t(rdt + \sigma dB_t),\, V_0 = v > L\,,$$

where r is the (constant) risk-free interest rate, and σ is the constant volatility. In that context, the contingent claim pricing methodology can be used: the market where $(V_t, t \geq 0)$ is a tradeable asset is complete and arbitrage free, the equivalent martingale measure is the historical one, hence the value of the corporate bonds at time t is

$$\mathbb{E}(e^{-r(T-t)} \min(V_T, L)|\mathcal{F}_t) = Le^{-r(T-t)} - P_E(t, V_t, L)$$

where $P_E(t, x, L)$ is the value at time t of a put option on the underlying V with strike L and maturity T.

We denote by $P(t,T) = e^{-r(T-t)}$ the value of a default-free zero-coupon and by $D(t,T)$ the value of the defaultable zero-coupon of maturity T, with payment $L = 1$, i.e.,

$$D(t,T) = e^{-r(T-t)}\mathbb{E}(\mathbb{1}_{\{V_T>1\}} + V_T\mathbb{1}_{\{V_T<1\}}|\mathcal{F}_t)\,.$$

Then, from the valuation formula for the European put option

$$D(t,T) = V_t \mathcal{N}(-d_1(V_t, T-t)) + P(t,T)\mathcal{N}(d_2(V_t, T-t)),$$

where

$$d_1(V_t, T-t) = \frac{\log(V_t) + \left(r + \frac{1}{2}\sigma^2\right)(T-t)}{\sigma\sqrt{T-t}}$$

$$d_2(V_t, T-t) = \frac{\log(V_t) + \left(r - \frac{1}{2}\sigma^2\right)(T-t)}{\sigma\sqrt{T-t}}.$$

We denote by

$$Y(t,T) = -\frac{\ln P(t,T)}{T-t},$$

and

$$Y_d(t,T) = -\frac{\ln D(t,T)}{T-t},$$

the yield to maturity. The spread on corporate debt, i.e.,

$$S(t,T) = Y_d(t,T) - Y(t,T)$$

is

$$S(t,T) = -\frac{1}{T-t} \ln\left(\mathcal{N}(d_2(V_t,t)) + \frac{V_t}{P(t,T)}\mathcal{N}(-d_1(V_t,t))\right).$$

We can specify the probability of default given the information at date t: if the dynamics of the firm are

$$dV_t = V_t(\mu dt + \sigma dB_t)$$

under the historical probability,

$$\mathbb{P}(V_T \leq L|\mathcal{F}_t) = \mathcal{N}(-d_t)$$

where now

$$d_t = \frac{1}{\sigma\sqrt{T-t}}\left(\ln(V_t/L) + (\mu - \sigma^2/2)(T-t)\right)$$

is the so-called *distance-to-default.*

Comment 3.10.1.1 Computation in the case where L is not assumed to be equal to 1 can be found, e.g., in Bielecki and Rutkowski [99]. If the default barrier is an exponential function, computations can be done using the previous subsection. Results are given in Bielecki and Rutkowski [99].

3.10.2 First Passage Time Models

Merton's model does not allow for a premature default; Black and Cox [104] extend Merton's model to the case where safety covenants provide the firm's bondholders with the right to force the firm into bankruptcy and obtain the ownership of the assets. They postulate that as soon as the firm's asset cross a lower threshold, the bondholders take over the firm. The safety convenant takes the form of an exponential. In this subsection, the model is simplified. We assume that the firm defaults when its value falls below a pre-specified level, i.e.,

$$\tau = T_L(V) = \inf\{t \,:\, V_t \le L\},$$

where $V_0 \ge L$. In this case, the default time τ is a stopping time in the asset's filtration. The valuation of a defaultable claim X reduces to the problem of pricing the claim $X\mathbb{1}_{\{T<\tau\}}$. The valuation of the defaultable claim within the structural approach is a standard problem which needs the knowledge of the law of the pair (τ, X).

Let us assume that

$$dV_t = V_t((r-\delta)dt + \sigma dB_t)\,,$$

where δ stands for the dividend yield. The value of a defaultable T-maturity bond with face value 1 and $L \le 1$ is $D(t,T) = P(t,T)\mathbb{E}(\mathbb{1}_{\{T<\tau\}}|\mathcal{F}_t)$, i.e., using the results on hitting time of a barrier for a geometric BM (see Exercise 3.3.1.2):

$$D(t,T) = P(t,T)\left(\mathcal{N}(b_1(V_t,T-t)) - \left[\frac{L}{V_t}\right]^{2\nu\sigma^{-2}} \mathcal{N}(b_2(V_t,T-t))\right)$$

where

$$b_1(x,T-t) = \frac{1}{\sigma\sqrt{T-t}}\left(\ln(x/L) + \nu(T-t)\right)$$
$$b_2(x,T-t) = \frac{1}{\sigma\sqrt{T-t}}\left(\ln(L/x) + \nu(T-t)\right)\,.$$

Here, $\nu = r - \delta - \sigma^2/2$.

We now assume that a rebate β is paid at default time when it occurs before maturity. Assume that $\theta := \nu^2 + 2\sigma^2(r-\delta) > 0$. Then prior to the company's default (that is on the set $\{\tau > t\}$) the price of a defaultable bond equals

$$\begin{aligned}D(t,T) &= P(t,T)\big(\mathcal{N}(b_1(V_t,T-t)) - Z_t^{2\nu\sigma^{-2}}\mathcal{N}(b_2(V_t,T-t))\big)\\ &\quad + \beta V_t\big(Z_t^{\theta\sigma^{-2}+1+\zeta}\mathcal{N}(b_3(V_t,T-t)) + Z_t^{\theta\sigma^{-2}+1-\zeta}\mathcal{N}(b_4(V_t,T-t))\big),\end{aligned}$$

where $Z_t = L/V_t$, $\zeta = \sigma^{-2}\sqrt{\theta}$ and

$$b_3(V_t, T-t) = \frac{\ln(L/V_t) + \zeta\sigma^2(T-t)}{\sigma\sqrt{T-t}},$$
$$b_4(V_t, T-t) = \frac{\ln(L/V_t) - \zeta\sigma^2(T-t)}{\sigma\sqrt{T-t}}.$$

The general formulae (for L different from 1 and with an exponential barrier) can be obtained using results given in Subsection 3.9.3. See also Bielecki et al. [91].

Extensions: Zhou's Model

Zhou [877] studies the case where the dynamics of the firm's value is

$$dV_t = V_{t-}\left((\mu - \lambda\, c)dt + \sigma dW_t + dX_t\right)$$

where W is a Brownian motion, X a compound Poisson process with the jumps distributed as Y_1 where $\ln Y_1$ follows a Gaussian law with mean a and variance b^2, and $c = \exp(a + b^2/2)$. This choice of parameters implies that $V_t e^{\mu t}$ is a martingale (see $\rightarrowtail$ Subsection 8.6.3). In the first part, Zhou studies Merton's problem in that setting. In the second part, he gives an approximation for the law of the first passage if the default time is $\tau = \inf\{t : V_t \leq L\}$.

Comment 3.10.2.1 Credit risk is presented in a more detailed form in Bielecki and Rutkowski [99] and Schönbucher [765] . The reader can also refer to the survey paper of Bielecki et al. [91]. See also $\rightarrowtail$ Chapter 7.

3.11 American Options

An American option gives its owner the right to exercise at any time τ between the initial time and maturity (see Samuelson [757][1]). We refer to Elliott and Kopp [316] for a general presentation of American options and to Carr et al. [154] for a decomposition of prices. McKean [635] was the first to exhibit the relation between the evaluation problem and a free boundary problem.

[1] We reproduce the following comments, from Jarrow and Protter [480]. This is the paper that first coined the terms European and American options. According to a private communication with R.C. Merton, prior to writing the paper, P. Samuelson went to Wall Street to discuss options with industry professionals. His Wall Street contact explained that there were two types of options available, one more complex - that could be exercised any time prior to maturity, and one more simple - that could be exercised only at the maturity date, and that only the more sophisticated European mind (as opposed to the American mind) could understand the former. In response, when Samuelson wrote the paper, he used these as prefixes and reversed the ordering.

Let us consider a currency (resp. a stock) and let us assume that its dynamics under the risk-neutral probability $\mathbb{Q}$, are given by the Garman-Kohlhagen model:

$$dS_t = S_t((r - \delta)dt + \sigma dW_t)$$

where $(W_t, t \geq 0)$ is a $\mathbb{Q}$-Brownian motion, r and δ are the domestic and foreign risk-free interest rates (resp. the risk-free interest rate and the dividend rate) and σ is the currency volatility. These parameters are constant, σ is strictly positive and at least one of the positive parameters r and δ is strictly positive. We denote by $C_A(S_t, T-t)$ (resp. $P_A(S_t, T-t)$) the time-t price of an American call (resp. put) of maturity T and strike price K.

3.11.1 American Stock Options

Let us recall some well known facts on American options. The value of an **American call option** (resp. **put**) of maturity T and strike K, is

$$C_A(S_0, T) = \sup_{\tau \in \mathcal{T}(T)} \mathbb{E}_{\mathbb{Q}}(e^{-r\tau}(S_\tau - K)^+),$$

(resp. $\sup_{\tau \in \mathcal{T}(T)} \mathbb{E}_{\mathbb{Q}}(e^{-r\tau}(K - S_\tau)^+)$) where $\mathcal{T}(T)$ is the set of stopping times τ with values in $[0, T]$. Obviously, the value of an American call is greater than the value of a European call with same maturity and strike.

Lemma 3.11.1.1 *The value of an American call is equal to the value of a European call if the stock does not pay dividends before maturity ($\delta = 0$).*

PROOF: Indeed, from the convexity of $x \to (x - Ke^{-rT})^+$, the martingale property of the process $(e^{-rt}S_t, t \geq 0)$, and Jensen's inequality, the process $((e^{-rt}S_t - Ke^{-rT})^+, t \geq 0)$ is a $\mathbb{Q}$-submartingale. Hence, for any stopping time τ bounded by T,

$$\mathbb{E}_{\mathbb{Q}}((e^{-r\tau}S_\tau - Ke^{-rT})^+) \leq \mathbb{E}_{\mathbb{Q}}(e^{-rT}(S_T - K)^+).$$

The inequality

$$\mathbb{E}_{\mathbb{Q}}(e^{-r\tau}(S_\tau - K)^+) \leq \mathbb{E}_{\mathbb{Q}}((e^{-r\tau}S_\tau - Ke^{-rT})^+)$$

leads to $\sup_\tau \mathbb{E}_{\mathbb{Q}}(e^{-r\tau}(S_\tau - K)^+) \leq \mathbb{E}_{\mathbb{Q}}(e^{-rT}(S_T - K)^+)$ and the result follows (the reverse inequality is obvious). □

In the particular case of infinite maturity, an American option is called perpetual. The value of a perpetual American call $C_A(x, \infty)$ is x. Indeed, for any t,

$$x - e^{-rt}K \leq \mathbb{E}_{\mathbb{Q}}(e^{-rt}(S_t - K)^+) \leq C_A(x, \infty) \leq x$$

and the result follows when t goes to infinity. The limit of the value of a European call maturity T, when T goes to infinity is also equal to x, as can be seen from the Black-Scholes formula (see Theorem 2.3.2.1).

Exercise 3.11.1.2 The payoff of a capitalized-strike American put option is $(Ke^{rt} - S_t)^+$ if exercised at time t. Prove that the price of this option is the price of a European put, with strike $e^{rT}K$. ◁

3.11.2 American Currency Options

The exercise boundaries are defined as follows. For an American currency call (resp. put) of maturity T and for a given time $t, t \in [0, T]$,

$$\begin{cases} b_c(T-t) = \inf\{x \geq 0 \,:\, x - K = C_A(x, T-t)\}\,, \\ b_p(T-t) = \sup\{x \geq 0 \,:\, K - x = P_A(x, T-t)\}\,. \end{cases} \tag{3.11.1}$$

The **exercise boundary** for the American call (resp. put) gives for each time t before maturity the critical level at which the American option should be exercised. In the continuation region, i.e., when the underlying asset value is below (resp. above) the exercise boundary, the time value of the American call is strictly positive. In the stopping region, i.e., when the underlying asset value is above (resp. below) the exercise boundary, the time value is equal to zero and therefore it is worthwhile to exercise the option. As we recalled, for a non-dividend paying stock, it is never optimal to exercise the American call option before maturity. The exercise boundary for the call is therefore infinite before maturity. However, for currencies, it could be optimal to exercise the American call option strictly before maturity, in order to invest at the foreign interest rate instead of the domestic one. Hence, the exercise boundary given by the equation (3.11.1) is finite when $\delta > 0$.

By relying upon the proof of Proposition 2.7.1.1 for European options, the PDE that the option price satisfies in the continuation region, is obtained and is the same as in the European case:

$$\frac{\sigma^2}{2} x^2 \frac{\partial^2 C_A}{\partial x^2}(x,u) + (r-\delta) x \frac{\partial C_A}{\partial x}(x,u) - rC_A(x,u) - \frac{\partial C_A}{\partial u}(x,u) = 0\,. \tag{3.11.2}$$

Proposition 3.11.2.1 *The American currency call price satisfies the following decomposition:*

$$\begin{aligned} C_A(S_t, T-t) = C_E(S_t, T-t) &+ \delta S_t \int_t^T e^{-\delta(s-t)} \mathcal{N}(d_1(S_t, b_c(T-s), s-t))ds \\ &- rK \int_t^T e^{-r(s-t)} \mathcal{N}(d_2(S_t, b_c(T-s), s-t))ds \end{aligned} \tag{3.11.3}$$

with

$$\begin{aligned} d_1(x,y,u) &= \frac{\ln(x/y) + (r - \delta + \sigma^2/2)u}{\sigma\sqrt{u}}\,, \\ d_2(x,y,u) &= d_1(x,y,u) - \sigma\sqrt{u}\,. \end{aligned}$$

PROOF: Apply Itô's lemma to the process S and the function

$$\widetilde{C}(x,s) = e^{-r(s-t)} C_A(x, T-s)$$

on the interval $[t, T]$. Then,

$$e^{-r(T-t)} C_A(S_T, 0) = C_A(S_t, T-t) + \int_t^T \mathcal{A}\widetilde{C}(S_s, s) ds + \sigma \int_t^T S_s \frac{\partial \widetilde{C}}{\partial x}(S_s, s) dW_s \,, \tag{3.11.4}$$

where $\mathcal{A}$ is defined by:

$$\mathcal{A} = \frac{\sigma^2}{2} x^2 \frac{\partial^2}{\partial x^2} + (r-\delta) x \frac{\partial}{\partial x} + \frac{\partial}{\partial s} \,.$$

Now, in the continuation region the American call price satisfies the PDE given in equation (3.11.2) and therefore $\mathcal{A}\tilde{C}(S_s, s)$ is equal to zero. In the stopping region the American call is equal to its intrinsic value, and therefore, for $x > b_c(s)$:

$$\mathcal{A}\tilde{C}(x,s) = (r-\delta)x + r(K-x) = (rK - \delta x) \mathbb{1}_{\{x > b_c(s)\}} \,. \tag{3.11.5}$$

The last integral on the right-hand side of equation (3.11.4) is a martingale. By applying the expectation operator to this equation and by relying on the equality (3.11.5), we obtain

$$\begin{aligned} C_A(S_t, T-t) &= e^{-r(T-t)} \mathbb{E}_{\mathbb{Q}}((S_T - K)^+ | \mathcal{F}_t) \\ &\quad - \int_t^T e^{-r(s-t)} \mathbb{E}_{\mathbb{Q}}((rK - \delta S_s) \mathbb{1}_{\{S_s > b_c(T-s)\}} | \mathcal{F}_t) ds \,. \end{aligned}$$

From Subsection 2.7.1, where the Garman and Kohlhagen model was derived, the decomposition given by equation (3.11.3) is obtained. □

Along the same lines, a decomposition for the put price can be derived .

Proposition 3.11.2.2 *The American currency put price satisfies the following decomposition:*

$$\begin{aligned} P_A(S_t, T-t) &= P_E(S_t, T-t) \\ &\quad + rK \int_t^T e^{-r(s-t)} \mathcal{N}(-d_2(S_s, b_p(T-s), s-t)) ds \\ &\quad - \delta S_t \int_t^T e^{-\delta(s-t)} \mathcal{N}(-d_1(S_s, b_p(T-s), s-t)) ds \,, \end{aligned} \tag{3.11.6}$$

with d_i given in Proposition 3.11.2.1 and b_p the exercise boundary for the put defined in (3.11.1).

By relying on Barles et al. [44] for non-dividend paying stock options ($\delta = 0$), an approximation of the American put exercise boundary near expiration T can be given:

$$b_p(T-t) \approx K(1 - \sigma\sqrt{(T-t)\,|\ln(T-t)|}) \tag{3.11.7}$$

for $t < T$.

By substituting the results given by (3.11.7) into equation (3.11.6), an approximation of the American put price is obtained, for small maturities.

3.11.3 Perpetual American Currency Options

PDE Approach

When the option's maturity tends to infinity, the following ODE is obtained:

$$\frac{\sigma^2}{2} x^2 C''_A(x) + (r - \delta) x C'_A(x) - r C_A(x) = 0 \tag{3.11.8}$$

where now the following notation is used:

$$C_A(x) = C_A(x, +\infty)\,.$$

We denote by L^* the limit when T goes to infinity of the monotonic function b_c (see (3.11.1)). As seen later, L^* is finite if $\delta > 0$.

The general solution of the equation (3.11.8) is of the form $a_1 x^{\gamma_1} + a_2 x^{\gamma_2}$ where γ_1 and γ_2 are the two roots of the polynomial

$$\frac{\sigma^2}{2}\gamma^2 + \left(r - \delta - \frac{\sigma^2}{2}\right)\gamma - r \tag{3.11.9}$$

which admits a positive and a negative root. The call price being an increasing function of the exchange rate, only the positive root

$$\gamma_1 = \frac{-\nu + \sqrt{\nu^2 + 2r}}{\sigma} \tag{3.11.10}$$

will be retained, and $C_A(x) = a_1 x^{\gamma_1}$. It can be observed that $\gamma_1 > 1$. Here ν is defined (as in Section 3.3) by:

$$\nu = \frac{1}{\sigma}\left(r - \delta - \frac{\sigma^2}{2}\right). \tag{3.11.11}$$

(Note that if $\delta = 0$, then $\gamma_1 = 1$.) Now, the parameter a_1 and the boundary L^* are obtained from the boundary conditions:

$$C_A(L^*) = a_1 (L^*)^{\gamma_1} = L^* - K,\ C'_A(L^*) = a_1 \gamma_1 (L^*)^{\gamma_1 - 1} = 1$$

i.e., the option price and its derivative are continuous with respect to the underlying asset value at the exercise boundary. The continuity of the derivative at the boundary is assumed (this last property is the **smooth-fit** principle or smooth-pasting condition). It is not obvious that this property holds, see Elliott and Kopp [316] p.203. Therefore

$$a_1 = \frac{L^* - K}{(L^*)^{\gamma_1}}, \quad L^* = \frac{\gamma_1}{\gamma_1 - 1} K \geq K\,. \tag{3.11.12}$$

It follows that, in the **continuation region** (for $x < L^*$), the perpetual American call price is given by:

$$C_A(x) = (L^* - K)\left(\frac{x}{L^*}\right)^{\gamma_1}.$$

By relying on equation (3.11.12)

$$C_A(x) = \frac{K}{\gamma_1 - 1} e^{-\gamma_1 \ln\left(\frac{\gamma_1 K}{\gamma_1 - 1}\right)} x^{\gamma_1}\,. \tag{3.11.13}$$

In the stopping region, (for $x \geq L^*$): $C_A(x) = x - K$.

Martingale Approach

In order to derive the price of an American call, the martingale approach can also be used. In this framework the option's value is given by

$$C_A(S_t) = \sup_{\tau} \mathbb{E}_{\mathbb{Q}}((S_\tau - K)e^{-r(\tau - t)}|\mathcal{F}_t)\,,$$

where τ runs over all stopping times greater than t.

Let $t = 0$ and assume that the boundary is constant. By continuity of the Brownian motion if S_0 is in the continuation region (i.e., S_0 is smaller than the boundary):

$$C_A(S_0) = \sup_{L} \left[(L - K)\mathbb{E}_{\mathbb{Q}}(e^{-rT_L})\right] \tag{3.11.14}$$

where T_L is the first passage time of the underlying asset value out of the continuation region:

$$T_L = \inf\{t \geq 0 \ / \ S_t \geq L\}\,.$$

(See Elliott and Kopp, p. 196 [316] for a proof that it is possible to restrict attention to that family of stopping times.) The optimal value L^* is obtained by equating the derivative of $(L - K)\mathbb{E}_{\mathbb{Q}}(e^{-rT_L})$ with respect to L to zero, hence

$$L^* = \frac{-\mathbb{E}_{\mathbb{Q}}(e^{-rT_{L^*}})}{[\partial \mathbb{E}_{\mathbb{Q}}(e^{-rT_L})/\partial L]_{L=L^*}} + K\,. \tag{3.11.15}$$

Therefore,

$$C_A(S_0) = \frac{-\left(\mathbb{E}_{\mathbb{Q}}(e^{-rT_{L^*}})\right)^2}{[\partial \mathbb{E}_{\mathbb{Q}}(e^{-rT_L})/\partial L]_{L=L^*}}\,. \tag{3.11.16}$$

Using equation (3.3.5), the Laplace transform of the hitting time T_L is

$$\mathbb{E}_{\mathbb{Q}}(e^{-rT_L}) = e^{-(-\nu+\sqrt{\nu^2+2r})\frac{1}{\sigma}\ln(L/S_0)} = e^{-\gamma_1 \ln(L/S_0)} \tag{3.11.17}$$

where the parameter γ_1 is defined in (3.11.10). We can thus derive the value of the exercise boundary from (3.11.15) which can be written $L^* = \frac{L^*}{\gamma_1} + K$. We get $L^* = \frac{\gamma_1}{\gamma_1 - 1}K \geq K$ and by relying on equation (3.11.16), the solution given by (3.11.13) is obtained.

The same procedure allows us to derive the put price as

$$P_A(S_0) = (K - L_*)\left(\frac{S_0}{L_*}\right)^{\gamma_2}. \tag{3.11.18}$$

and the exercise boundary for the perpetual American put is constant and given by $L_* = \gamma_2 K/(\gamma_2 - 1)$, where γ_2 is the negative root of (3.11.9). Let us remark that the put-call symmetry for American options (see Detemple [251]) can also be used:

$$P_A(S_0, K, r, \delta) = C_A(K, S_0, \delta, r) \tag{3.11.19}$$

where option prices are now indexed by four arguments. This symmetry comes basically from the fact that the right to sell a foreign currency corresponds to the right to buy the domestic one, and can be proved from a change of numéraire. Let us check that formulae (3.11.18) and (3.11.19) agree. The put-call symmetry formula (3.11.19) implies that

$$P_A(S_0) = (\ell - S_0)\left(\frac{K}{\ell}\right)^{\gamma}$$

where γ is the positive root of

$$\frac{\sigma^2}{2}\gamma^2 + \left(\delta - r - \frac{\sigma^2}{2}\right)\gamma - \delta = 0$$

and $\ell = \frac{\gamma}{\gamma - 1}S_0$. Note that $\gamma > 1$ and $1 - \gamma$ satisfies

$$\frac{\sigma^2}{2}(1-\gamma)^2 + \left(r - \delta - \frac{\sigma^2}{2}\right)(1-\gamma) - r = 0$$

hence $1 - \gamma = \gamma_2$, the negative root of (3.11.9). Now,

$$P_A(S_0) = (S_0)^{1-\gamma}K^{\gamma}(\gamma - 1)^{\gamma - 1}\left(\frac{1}{\gamma}\right)^{\gamma},$$

and the relation $\gamma_2 = 1 - \gamma$ yields

$$P_A(S_0) = (S_0)^{\gamma_2}K^{1-\gamma_2}\left(\frac{1}{-\gamma_2}\right)^{\gamma_2}(1 - \gamma_2)^{\gamma_2 - 1},$$

which is (3.11.18).

By relying on the symmetrical relationship between American put and call boundaries (see Carr and Chesney [147] , Detemple [251]) the perpetual American put exercise boundary can also be obtained when T tends to infinity:

$$b_c(K, r, \delta, T - t) b_p(K, \delta, r, T - t) = K^2$$

where the exercise boundary is indexed by four arguments.

3.12 Real Options

Real options represent an important and relatively new trend in Finance and often involve the use of hitting times. Therefore, this topic will be briefly introduced in this chapter. In many circumstances, the standard NPV (Net Present Value) approach could generate wrong answers to important questions: "What are the relevant investments and when should the decision to invest be made?". This standard investment choice method consists of computing the NPV, i.e., the expected sum of the discounted difference between earnings and costs. Depending on the sign of the NPV, the criterion recommends acceptance (if it is positive) or rejection (otherwise) of the investment project. This approach is very simple and does not always model the complexity of the investment choice problem. First of all, this method presupposes that the earning and cost expectations can be estimated in a reliable way. Thus, the uncertainty inherent to many investment projects is not taken into account in an appropriate way. Secondly, this method is very sensitive to the level of the discount rate and the estimation of the this parameter is not always straightforward.

Finally, it is a static approach for a dynamical problem. Implicitly the question is: "Should the investment be undertaken now, or never?" It neglects the opportunity (one may use also the term *option*) to wait, in order to obtain more information, and to make the decision to invest or not to invest in an optimal way. In many circumstances, the timing aspects are not trivial and require specific treatment. By relying on the concept of a financial option, and more specifically on the concept of an American option (an optimal stopping theory), the investment choice problem can be tackled in a more appropriate way.

3.12.1 Optimal Entry with Stochastic Investment Costs

Mc Donald and Siegel's model [634], which corresponds to one of the seminal articles in the field of real options, is now briefly presented. As shown in their paper, some real option problems can be more complex than usual option pricing ones. They consider a firm with the following investment opportunity: at any time t, the firm can pay K_t to install the investment project which generates a sum of expected discounted future net cash-flows denoted V_t. The investment is irreversible. In their model, costs are stochastic and the maturity is infinite. It corresponds, therefore, to an extension of the perpetual American option pricing model with a stochastic strike price. See also Bellalah [68], Dixit and Pindyck [254] and Trigeorgis [820].

Let us assume that, under the historical probability $\mathbb{P}$, the dynamics of V (resp. K), the project-expected sum of discounted positive (resp. negative) instantaneous cash-flows (resp. costs) generated by the project- are given by:

$$\begin{cases} dV_t = V_t\,(\alpha_1 dt + \sigma_1 dW_t) \\ dK_t = K_t(\alpha_2 dt + \sigma_2 dB_t)\,. \end{cases}$$

The two trends α_1, α_2, the two volatilities σ_1 and σ_2, the correlation coefficient ρ of the two $\mathbb{P}$-Brownian motions W and B, and the discount rate r, are supposed to be constant. We also assume that $r > \alpha_i, i = 1, 2$.

If the investment date is t, the payoff of the real option is $(V_t - K_t)^+$. At time 0, the investment opportunity value is therefore given by

$$\begin{aligned} C_{RO}(V_0, K_0) &:= \sup_{\tau \in \mathcal{T}} \mathbb{E}_{\mathbb{P}}(e^{-r\tau}(V_\tau - K_\tau)^+) \\ &= \sup_{\tau \in \mathcal{T}} \mathbb{E}_{\mathbb{P}}\left(e^{-r\tau} K_\tau \left(\frac{V_\tau}{K_\tau} - 1\right)^+\right) \end{aligned}$$

where $\mathcal{T}$ is the set of stopping times, i.e, the set of possible investment dates.

Now, using that $K_t = K_0 e^{\alpha_2 t} e^{\sigma_2 B_t - \frac{1}{2}\sigma_2^2 t}$, the same kind of change of probability measure (change of numéraire) as in Subsection 2.7.2 leads to

$$C_{RO}(V_0, K_0) = K_0 \sup_{\tau \in \mathcal{T}} \mathbb{E}_{\mathbb{Q}}\left(e^{-(r-\alpha_2)\tau} \left(\frac{V_\tau}{K_\tau} - 1\right)^+\right).$$

Here the probability measure $\mathbb{Q}$ is defined by its Radon-Nikodým derivative with respect to $\mathbb{P}$ on the σ-algebra $\mathcal{F}_t = \sigma(W_s, B_s, s \leq t)$ by

$$\mathbb{Q}|_{\mathcal{F}_t} = \exp\left(-\frac{\sigma_2^2}{2} t + \sigma_2 B_t\right) \mathbb{P}|_{\mathcal{F}_t}\,.$$

The valuation of the investment opportunity then corresponds to that of a perpetual American option. As in Subsection 2.7.2, the dynamics of $X = V/K$ are obtained

$$dX_t/X_t = (\alpha_1 - \alpha_2)dt + \Sigma d\widehat{W}_t\,.$$

Here

$$\Sigma = \sqrt{\sigma_1^2 + \sigma_2^2 - 2\rho\sigma_1\sigma_2}$$

and $(\widehat{W}_t, t \geq 0)$ is a $\mathbb{Q}$-Brownian motion. Therefore, from the results obtained in Subsection 3.11.3 in the case of perpetual American option

$$C_{RO}(V_0, K_0) = K_0(L^* - 1)\left(\frac{V_0/K_0}{L^*}\right)^{\epsilon} \tag{3.12.1}$$

with

$$L^* = \frac{\epsilon}{\epsilon - 1}, \tag{3.12.2}$$

and

$$\epsilon = \sqrt{\left(\frac{\alpha_1 - \alpha_2}{\Sigma^2} - \frac{1}{2}\right)^2 + \frac{2(r - \alpha_2)}{\Sigma^2}} - \left(\frac{\alpha_1 - \alpha_2}{\Sigma^2} - \frac{1}{2}\right). \tag{3.12.3}$$

Let us now assume that spanning holds, that is, in this context, that there exist two assets perfectly correlated with V and K and with the same standard deviation as V and K. We can then rely on risk neutrality, and discounting at the risk-free rate.

Let us denote by α_1^* and α_2^* respectively the expected returns of assets 1 and 2 perfectly correlated respectively with V and K. Let us define δ_1 and δ_2 by

$$\delta_1 = \alpha_1^* - \alpha_1, \quad \delta_2 = \alpha_2^* - \alpha_2$$

These parameters play the rôle of the dividend yields in the exchange option context (see Section 2.7.2), and are constant in this framework (see Gibson and Schwartz [391] for stochastic convenience yields). The quantity δ_1 is an opportunity cost of delaying the investment and keeping the option to invest alive and δ_2 is an opportunity cost saved by differing installation. The trends $r - \delta_1$ (i.e., α_1 minus the risk premium associated with V which is equal to $\alpha_1^* - r$) and $r - \delta_2$ (equal to $\alpha_2 - (\alpha_2^* - r)$) should now be used instead of the trends α_1 and α_2, respectively. In this setting, r is the risk-free rate. Thus, equations (3.12.1) and (3.12.2) still give the solution, but with

$$\epsilon = \sqrt{\left(\frac{\delta_2 - \delta_1}{\Sigma^2} - \frac{1}{2}\right)^2 + \frac{2\delta_2}{\Sigma^2}} - \left(\frac{\delta_2 - \delta_1}{\Sigma^2} - \frac{1}{2}\right) \tag{3.12.4}$$

instead of equation (3.12.3). In the neo-classical framework it is optimal to invest if expected discounted earnings are higher than expected discounted costs, i.e., if X_t is higher than 1. When the risk is appropriately taken into account, the optimal time to invest is the first passage time of the process $(X_t, t \geq 0)$ for a level L^* strictly greater than 1, as shown in equation (3.12.2).

As seen above, in the real option framework usually different stochastic processes are involved (see also, for example, Loubergé et al. [604]). Results obtained by Hu and Øksendal [447] and Villeneuve [829], who consider the American option valuation with several underlyings, can therefore be very useful.

3.12.2 Optimal Entry in the Presence of Competition

If instead of a monopolistic situation, competition is introduced, by relying on Lambrecht and Perraudin [561], the value of the investment opportunity can be derived. Let us assume that the discounted sum K_t of instantaneous cost is now constant.

Two firms are involved. Only the first one behaves strategically. Both are potentially willing to invest a sum K in the same investment project. They consider only this investment project. The decision to invest is supposed to be irreversible and can be made at any time. Hence the real option is a perpetual American option. The investors are risk-neutral. Let us denote by r the constant interest rate. In this risk-neutral economy, the dynamics of S, the instantaneous cash-flows generated by the investment project, are given by

$$dS_t = S_t(\alpha dt + \sigma dW_t)\,.$$

Let us define V as the expected sum of positive instantaneous cash-flows S. The processes V and S have the same dynamics. Indeed, for $r > \alpha$:

$$\begin{aligned} V_t &= \mathbb{E}\left(\int_t^\infty e^{-r(u-t)} S_u du | \mathcal{F}_t\right) = e^{rt} \int_t^\infty e^{-(r-\alpha)u} \mathbb{E}(e^{-\alpha u} S_u | \mathcal{F}_t) du \\ &= e^{rt} \int_t^\infty e^{-(r-\alpha)u} e^{-\alpha t} S_t du = \frac{S_t}{r-\alpha}\,. \end{aligned}$$

In this model, the authors assume that firm 1 (resp. 2) completely loses the option to invest if firm 2 (resp. 1) invests first, and therefore considers the investment decision of a firm threatened by preemption.

Firm 1 behaves strategically in an incomplete information setting. This firm conjectures that firm 2 will invest when the underlying value reaches some level L_2^* and that L_2^* is an independent draw from a distribution G. The authors assume that G has a continuously differentiable density $g = G'$ with support in the interval $[L_2^D, L_2^U]$. The uncertainty in the investment level of the competitor comes from the fact that this level depends on competitor's investment costs which are not known with certainty and therefore only conjectured.

The structure of learning implied by the model is the following. Since firm 2 invests only when the underlying S hits for the first time the threshold L_2^*, firm 1 learns about firm 2 only when the underlying reaches a new supremum. Indeed, in this case, there are two possibilities. Firm 2 can either invest and firm 1 learns that the trigger level is the current S_t, but it is too late to invest for firm 1, or wait and firm 1 learns that L_2^* lies in a smaller interval than it has previously known, i.e., in $[M_t, L_2^U]$, where M_t is the supremum at time t: $M_t = \sup_{0 \le u \le t} S_u$.

In this context, firm 1 behaves strategically, in that it looks for the optimal exercise level L_1^*, i.e., the trigger value which maximizes the conditional expectation of the discounted realized payoff. Indeed, the value C_S to firm 1, the strategic firm, is therefore

$$C_S(S_t, M_t) = \sup_L \left(\frac{L}{r-\alpha} - K\right) \mathbb{E}\left(e^{-r(T_L - t)} \mathbb{1}_{\{L_2^* > L\}} | \mathcal{F}_t \vee (L_2^* > M_t)\right)$$

where the stopping time T_L is the first passage time of the process S for level L after time t:

$$T_L = \inf\{u \geqslant t,\ S_u \geqslant L\}\,.$$

The payoff is realized only if the competitor is preempted, i.e., if $L_2^* > L$. If $M_t > L_2^D$, the value to the firm depends not only on the instantaneous value S_t of the underlying, but also on M_t which represents the knowledge accumulated by firm 1 about firm 2: the fact that up until time t, firm 1 was not preempted by firm 2, i.e., $L_2^* > M_t > L_2^D$. If $M_t \leq L_2^D$, the knowledge of M_t does not represent any worthwhile information and therefore

$$C_S(S_t, M_t) = \sup_L \left(\frac{L}{r-\alpha} - K\right) \mathbb{E}(e^{-r(T_L - t)} \mathbb{1}_{\{L_2^* > L\}} | \mathcal{F}_t), \text{ if } M_t \leq L_2^D\,.$$

From now on, let us assume that $M_t > L_2^D$. Hence, by independence between the r.v. L_2^* and the stopping time $T_L = \inf\{t \geqslant 0 \,:\, S_t \geqslant L\}$

$$C_S(S_t, M_t) = \sup_L (C_{NS}(S_t, L) \mathbb{P}(L_2^* > L \mid L_2^* > M_t))\,,$$

where the value of the non strategic firm $C_{NS}(S_t, L)$ is obtained by relying on equation (3.11.17):

$$C_{NS}(S_t, L) = \left(\frac{L}{r-\alpha} - K\right)\left(\frac{S_t}{L}\right)^{\gamma}\,,$$

and from equations (3.11.10–3.11.11) $\gamma = \frac{-\nu + \sqrt{2r + \nu^2}}{\sigma} > 0$ and $\nu = \frac{\alpha - \sigma^2/2}{\sigma}$. Now, in the specific case where the lower boundary L_2^D is higher than the optimal trigger value in the monopolistic case, the solution is known:

$$C_S(S_t, M_t) = C_{NS}\left(S_t, \frac{\gamma}{\gamma - 1}(r - \alpha)K\right),\ \text{if } L_2^D \geqslant \frac{\gamma}{\gamma - 1}(r - \alpha)K\,.$$

Indeed, in this case the presence of the competition does not induce any change in the strategy of firm 1. It cannot be preempted, because the production costs of firm 2 are too high.

In the general case, when $L_2^D < \frac{\gamma}{\gamma-1}(r-\alpha)K$ and $(r-\alpha)K < L_2^U$ (otherwise the competitor will always preempt), knowing that potential candidates for L_1^* are higher than M_t:

$$C_S(S_t, M_t) = \sup_L \left(\frac{L}{r-\alpha} - K\right)\left(\frac{S_t}{L}\right)^{\gamma} \frac{\mathbb{P}(L_2^* > L)}{\mathbb{P}(L_2^* > M_t)}$$

i.e.,

$$C_S(S_t, M_t) = \sup_L \left(\left(\frac{L}{r-\alpha} - K \right) \left(\frac{S_t}{L} \right)^\gamma \frac{1-G(L)}{1-G(M_t)} \right).$$

This optimization problem implies the following result. L_1^* is the solution of the equation

$$x = \frac{\gamma + h(x)}{\gamma - 1 + h(x)}(r-\alpha)K$$

with

$$h(x) = \frac{xg(x)}{1-G(x)}.$$

The function: $y \to \frac{\gamma+y}{\gamma-1+y}$ is decreasing, hence the trigger level is smaller in presence of competition than in the monopolistic case:

$$L_1^* < \frac{\gamma}{\gamma-1}(r-\alpha)K.$$

Indeed, the threat of preemption generates incentives to invest earlier than in the monopolist case.

The value to firm 1 is

$$C_S(S_t, M_t) = \left(\frac{L_1^*}{r-\alpha} - K \right) \left(\frac{S_t}{L_1^*} \right)^\gamma \frac{1-G(L_1^*)}{1-G(M_t)}.$$

Let us now consider a specific case. If L_2^* is uniformly distributed on the interval $[L_2^D, L_2^U]$, then:

$$\begin{aligned} C_S(S_t, M_t) &= \sup_L \left[\left(\frac{L}{r-\alpha} - K \right) \left(\frac{S_t}{L} \right)^\gamma \frac{(L_2^U - L)/(L_2^U - L_2^D)}{(L_2^U - M_t)/(L_2^U - L_2^D)} \right] \\ &= \sup_L \left[\left(\frac{L}{r-\alpha} - K \right) \left(\frac{S_t}{L} \right)^\gamma \frac{L_2^U - L}{L_2^U - M_t} \right]. \end{aligned}$$

In this case

$$h(x) = \frac{x/(L_2^U - L_2^D)}{(L_2^U - x)/(L_2^U - L_2^D)} = \frac{x}{L_2^U - x}$$

and L_1^* satisfies

$$x = \frac{\gamma + \frac{x}{L_2^U - x}}{\gamma - 1 + \frac{x}{L_2^U - x}}(r-\alpha)K$$

i.e.,

$$(\gamma-2)x^2 + (1-\gamma)(L_2^U + (r-\alpha)K)x + \gamma(r-\alpha)KL_2^U = 0.$$

Hence, for $\gamma \neq 2$

$$L_1^* = \frac{(\gamma-1)(L_2^U + (r-\alpha)K) + \sqrt{\Delta}}{2(\gamma-2)}$$

with

$$\begin{aligned}\Delta &= (1-\gamma)^2(L_2^U + (r-\alpha)K)^2 - 4(\gamma-2)\gamma(r-\alpha)KL_2^U \\ &= (L_2^U - (r-\alpha)K)^2\gamma^2 - 2(L_2^U - (r-\alpha)K)^2\gamma + (L_2^U + (r-\alpha)K)^2 .\end{aligned}$$

It is straightforward to show that this discriminant is positive for any γ and therefore that L_1^* is well defined. For $\gamma = 2$, $L_1^* = \frac{2(r-\alpha)KL_2^U}{L_2^U + (r-\alpha)K}$.

3.12.3 Optimal Entry and Optimal Exit

Let us now modify the model of Lambrecht and Perraudin [561] as follows. There is no competition; the decision to invest is no longer irreversible; however, the decision to *disinvest* is irreversible and can be made at any time after the decision to invest has been taken. There are entry costs K_i and exit costs K_d.

Therefore, there are two embedded perpetual American options in such a model: First an American call that corresponds to the investment decision and a put that corresponds to the disinvestment decision.

The value to the firm VF, at initial time is therefore

$$VF(S_0) = \sup_{L_i, L_d} \left(\phi(L_i)\mathbb{E}(e^{-rT_{L_i}}) + \psi(L_d)\mathbb{E}(e^{-rT_{L_d}})\right)$$

where

$$\begin{aligned}\phi(\ell) &= \frac{\ell}{r-\alpha} - K - K_i \\ \psi(\ell) &= K - \frac{\ell}{r-\alpha} - K_d\end{aligned}$$

and where the stopping times T_{L_i} and T_{L_d} correspond respectively to the first passage time of the process S at level L_i (investment) and to the first passage time of the process S at level L_d, after T_{L_i} (disinvestment):

$$\begin{aligned}T_{L_i} &= \inf\{t \geq 0,\ S_t \geq L_i\} \\ T_{L_d} &= \inf\{t \geq T_{L_i},\ S_t \leq L_d\} .\end{aligned}$$

Indeed, the right to disinvest gives an additional value to the firm. In case of a decline of the underlying process S, for example at level L_d, by paying K_d, the firm has the right to avoid the expected discounted losses at this level: $\frac{L_d}{r-\alpha} - K$.

Hence, from Markov's property:

$$VF(S_0) = \sup_{L_i, L_d} \mathbb{E}(e^{-rT_{L_i}}) \left(\phi(L_i) + \psi(L_d)\mathbb{E}(e^{-r(T_{L_d} - T_{L_i})})\right) .$$

From Subsection 3.11.3, one gets

$$VF(S_0) = \sup_{L_i, L_d} \left(\frac{S_0}{L_i}\right)^{\gamma_1} \left[\phi(L_i) + \psi(L_d)\left(\frac{L_i}{L_d}\right)^{\gamma_2}\right]$$

with

$$\gamma_1 = \frac{-\nu + \sqrt{2r + \nu^2}}{\sigma} \geq 0, \; \gamma_2 = \frac{-\nu - \sqrt{2r + \nu^2}}{\sigma} \leq 0$$

and again

$$\nu = \frac{\alpha - \sigma^2/2}{\sigma}.$$

This optimization problem yields

$$L_d^* = \frac{\gamma_2}{\gamma_2 - 1}(r - \alpha)(K - K_d) < (r - \alpha)(K - K_d)$$

which corresponds to the standard exercise boundary of the perpetual put (see Subsection 3.11.3). It is a decreasing function of the exit cost K_d. Indeed, if this cost increases, there is less incentive to disinvest. The quantity L_i^* is a solution of

$$x = \frac{\gamma_1}{\gamma_1 - 1}(r - \alpha)(K + K_i) - \frac{\gamma_1 - \gamma_2}{\gamma_1 - 1}\left(\frac{x}{L_d^*}\right)^{\gamma_2} ((r - \alpha)(K - K_d) - L_d^*)$$

hence,

$$L_i^* \leq \frac{\gamma_1}{\gamma_1 - 1}(r - \alpha)(K + K_i)$$

i.e., the possibility to disinvest gives to the firm incentives to invest earlier than in the irreversible investment case.

The value to the firm is therefore

$$VF(S_0) = \left(\frac{S_0}{L_i^*}\right)^{\gamma_1} \left[\phi(L_i^*) + \psi(L_d^*)\left(\frac{L_i^*}{L_d^*}\right)^{\gamma_2}\right].$$

3.12.4 Optimal Exit and Optimal Entry in the Presence of Competition

Let us now assume that the firm has already invested and is in a monopolistic situation. It has the opportunity to disinvest. The decision to disinvest is not irreversible. However, even if the firm has the option to invest again after the decision to quit has been made, the monopolistic situation will be over: the firm will face competition. In this case, the firm will be threatened by preemption and the Lambrecht and Perraudin [561] setting will be used. There are exit costs K_d and entry costs K_i. Let us use the previous notation.

By relying on the last subsections the value $VF(S_t)$ to the firm is

$$V_t - K + \sup_{L_d, L_i} \left[\psi(L_d)\mathbb{E}(e^{-r(T_{L_d} - t)}|\mathcal{F}_t) + \phi(L_i)\mathbb{E}(e^{-r(T_{L_i} - t)}\mathbb{1}_{L_2^* > L_i}|\mathcal{F}_t)\right]$$

i.e., setting $t = 0$,

$$VF(S_0) = V_0 - K + \sup_{L_d, L_i} \left(\frac{S_0}{L_d}\right)^{\gamma_2} \left[\psi(L_d) + \phi(L_i)\left(\frac{L_d}{L_i}\right)^{\gamma_1} (1 - G(L_i))\right].$$

Indeed, if firm 1 cannot disinvest, its value is $V_0 - K$; however if it has the opportunity to disinvest, it adds value to the firm. Furthermore, if firm 1 decides to disinvest, as long as it is not preempted by the competition, it has the opportunity to invest again. This explains the last term on the right-hand side: the maximization of the discounted payoff generated by a perpetual American put and by a perpetual American call times the probability of avoiding preemption.

Let us remark that the value to the firm does not depend on the supremum M_t of the underlying. As long as it is active, firm 1 does not accumulate any knowledge about firm 2. The supremum M_t no longer represents the knowledge accumulated by firm 1 about firm 2. Even if $M_t > L_2^D$, it does not mean that: $L_2^* \geqslant M_t > L_2^D$. While firm 1 does not disinvest, the knowledge of M_t does not represent any worthwhile information because firm 2 cannot invest.

This optimization problem generates the following result. L_i^* is the solution of the equation:

$$x = \frac{\gamma_1 + h(x)}{\gamma_1 - 1 + h(x)}(r - \alpha)K$$

with

$$h(x) = \frac{xg(x)}{1 - G(x)},$$

and L_d^* is the solution z of the equation

$$z = \frac{\gamma_2}{\gamma_2 - 1}(r - \alpha)(K - K_d) + \frac{\gamma_1 - \gamma_2}{1 - \gamma_2}\left(\frac{z}{L_i^*}\right)^{\gamma_1} (L_i^* - (r - \alpha)(K + K_i))(1 - G(L_i^*)).$$

The value to the firm is therefore:

$$VF(S_0) = V_0 - K + \left(\frac{S_0}{L_d^*}\right)^{\gamma_2} \left[\psi(L_d^*) + \phi(L_i^*)\left(\frac{L_d^*}{L_i^*}\right)^{\gamma_1} (1 - G(L_i^*))\right].$$

A good reference concerning optimal investment and disinvestment decisions, with or without lags, is Gauthier [376].

3.12.5 Optimal Entry and Exit Decisions

Let us keep the notation of the preceding subsections and still assume risk neutrality. Furthermore, let us assume now that there is no competition. Hence, we can restrict the discussion to only one firm. If at the initial time the firm has not yet invested, it has the possibility of investing at a cost K_i at any time and of disinvesting later at a cost K_d. The number of investment and disinvestment dates is not bounded. After each investment date the option to disinvest is activated and after each disinvestment date, the option to invest is activated.

Therefore, depending on the last decision of the firm before time t (to invest or to disinvest), there are two possible states for the firm: active or inactive.

In this context, the following theorem gives the values to the firm in these states.

Theorem 3.12.5.1 *Assume that in the risk-neutral economy, the dynamics of S, the instantaneous cash-flows generated by the investment project, are given by:*

$$dS_t = S_t(\alpha dt + \sigma dW_t)\,.$$

Assume further that the discounted sum of instantaneous investment cost K is constant and that $\alpha < r$ where r is the risk-free interest rate.

If the firm is inactive, i.e., if its last decision was to disinvest, its value is

$$VF_d(S_t) = \frac{1}{\gamma_1 - \gamma_2}\left(\frac{S_t}{L_i^*}\right)^{\gamma_1}\left(\frac{L_i^*}{r-\alpha} - \gamma_2\phi(L_i^*)\right).$$

If the firm is active, i.e., if its last decision was to invest, its value is

$$VF_i(S_t) = \frac{1}{\gamma_1 - \gamma_2}\left(\frac{S_t}{L_d^*}\right)^{\gamma_2}\left(\frac{L_d^*}{r-\alpha} + \gamma_1\psi(L_d^*)\right) + \frac{S_t}{r-\alpha} - K\,.$$

Here, the optimal entry and exit thresholds, L_i^* and L_d^* are solutions of the following set of equations with unknowns (x, y)

$$\frac{1-(y/x)^{\gamma_1-\gamma_2}}{\gamma_1-\gamma_2}\left(\gamma_1 K_i - \gamma_2\left(\frac{x}{r-\alpha} - K\right) + \frac{x}{r-\alpha}\right)$$
$$= \psi(y)\left(\frac{x}{y}\right)^{\gamma_2} - K_i\left(\frac{y}{x}\right)^{\gamma_1-\gamma_2} + \frac{x}{r-\alpha} - K$$
$$\frac{1-(y/x)^{\gamma_1-\gamma_2}}{\gamma_1-\gamma_2}\left(\frac{y}{r-\alpha} + \gamma_2\psi(y)\right)$$
$$= \phi(x)\left(\frac{y}{x}\right)^{\gamma_1} + \psi(y)\left(\frac{y}{x}\right)^{\gamma_1-\gamma_2}$$

with

$$\gamma_1 = \frac{-\nu + \sqrt{2r+\nu^2}}{\sigma} \geq 0,\ \gamma_2 = \frac{-\nu - \sqrt{2r+\nu^2}}{\sigma} \leq 0$$

and

$$\nu = \frac{\alpha - \sigma^2/2}{\sigma}\,.$$

In the specific case where $K_i = K_d = 0$, the optimal thresholds are

$$L_i^* = L_d^* = (r-\alpha)K\,.$$

PROOF: In the inactive state, the value of the firm is

$$VF_d(S_t) = \sup_{L_i} \mathbb{E}\left(e^{-r(T_{L_i}-t)}(VF_i(S_{T_{L_i}}) - K_i)|\mathcal{F}_t\right)$$

where T_{L_i} is the first passage time of the process S, after time t, for the possible investment boundary L_i

$$T_{L_i} = \inf\{u \geq t,\ S_u \geq L_i\}$$

i.e., by continuity of the underlying process S:

$$VF_d(S_t) = \sup_{L_i} \mathbb{E}\left(e^{-r(T_{L_i}-t)}(VF_i(L_i) - K_i)|\mathcal{F}_t\right).$$

Along the same lines:

$$VF_i(S_t) = \sup_{L_d} \mathbb{E}\left(e^{-r(T_{L_d}-t)}\left(VF_d(L_d) + \psi(L_d)\right)|\mathcal{F}_t\right) + \frac{S_t}{r-\alpha} - K$$

where T_{L_d} is the first passage time of the process S, after time t, for the possible disinvestment boundary L_d

$$T_{L_d} = \inf\{u \geq t,\ S_u \leq L_d\}.$$

Indeed, at a given time t, without exit options, the value to the active firm would be $\frac{S_t}{r-\alpha} - K$. However, by paying K_d, it has the option to disinvest for example at level L_d. At this level, the value to the firm is $VF_d(L_d)$ plus the value of the option to quit $K - \frac{L_d}{r-\alpha}$ (the put option corresponding to the avoided losses minus the cost K_d).

Therefore

$$VF_d(S_t) = \sup_{L_i} f_d(L_i) \tag{3.12.5}$$

where the function f_d is defined by

$$f_d(x) = \left(\frac{S_t}{x}\right)^{\gamma_1} (VF_i(x) - K_i) \tag{3.12.6}$$

where

$$VF_i(S_t) = \sup_{L_d} f_i(L_d) \tag{3.12.7}$$

and

$$f_i(x) = \left(\frac{S_t}{x}\right)^{\gamma_2} (VF_d(x) + \psi(L_d)) + \frac{S_t}{r-\alpha} - K. \tag{3.12.8}$$

Let us denote by $L_i^*$and L_d^* the optimal trigger values, i.e., the values which maximize the functions f_d and f_i. An inactive (resp. active) firm will find it optimal to remain in this state as long as the underlying value S remains below

L_i^* (resp. above L_d^*) and will invest (resp. disinvest) as soon as S reaches L_i^* (resp. L_d^*).

By setting S_t equal to L_d^* in equation (3.12.5) and to L_i^* in equation (3.12.7), the following equations are obtained:

$$VF_d(L_d^*) = \left(\frac{L_d^*}{L_i^*}\right)^{\gamma_1} (VF_i(L_i^*) - K_i),$$

$$VF_i(L_i^*) = \left(\frac{L_i^*}{L_d^*}\right)^{\gamma_2} (VF_d(L_d^*) + \psi(L_d^*)) + \frac{L_i^*}{r-\alpha} - K.$$

The two unknowns $VF_d(L_d^*)$ and $VF_i(L_i^*)$ satisfy:

$$\left(1 - \left(\frac{L_d^*}{L_i^*}\right)^{\gamma_1-\gamma_2}\right) VF_d(L_d^*) = \phi(L_i^*)\left(\frac{L_d^*}{L_i^*}\right)^{\gamma_1} + \psi(L_d^*)\left(\frac{L_d^*}{L_i^*}\right)^{\gamma_1-\gamma_2} \tag{3.12.9}$$

$$\left(1 - \left(\frac{L_d^*}{L_i^*}\right)^{\gamma_1-\gamma_2}\right) VF_i(L_i^*) = \psi(L_d^*)\left(\frac{L_i^*}{L_d^*}\right)^{\gamma_2} - K_i\left(\frac{L_d^*}{L_i^*}\right)^{\gamma_1-\gamma_2} + \frac{L_i^*}{r-\alpha} - K. \tag{3.12.10}$$

Let us now derive the thresholds L_d^* and L_i^* required in order to obtain the value to the firm. From equation (3.12.8)

$$\frac{\partial f_i}{\partial x}(L_d) = \left(\frac{S_t}{L_d}\right)^{\gamma_2}\left(-\frac{\gamma_2}{L_d}(VF_d(L_d) + \psi(L_d)) + \frac{dVF_d}{dx}(L_d) - \frac{1}{r-\alpha}\right)$$

and from equation (3.12.6)

$$\frac{\partial f_d}{\partial x}(L_i) = \left(\frac{S_t}{L_i}\right)^{\gamma_1}\left(-\frac{\gamma_1}{L_i}(VF_i(L_i) - K_i) + \frac{dVF_i}{dx}(L_i)\right).$$

Therefore the equation $\frac{\partial f_i}{\partial x}(L_d) = 0$ is equivalent to

$$\frac{\gamma_2}{L_d^*}(VF_d(L_d^*) + \psi(L_d^*)) = \frac{dVF_d}{dx}(L_d^*) - \frac{1}{r-\alpha}$$

or, from equations (3.12.5) and (3.12.6):

$$\frac{\gamma_2}{L_d^*}(VF_d(L_d^*) + \psi(L_d^*)) = \frac{\gamma_1}{L_d^*}VF_d(L_d^*) - \frac{1}{r-\alpha}$$

i.e.,

$$VF_d(L_d^*) = \frac{1}{\gamma_1 - \gamma_2}\left(\frac{L_d^*}{r-\alpha} + \gamma_2\psi(L_d^*)\right). \tag{3.12.11}$$

Moreover, the equation

$$\frac{\partial f_d}{\partial x}(L_i) = 0$$

is equivalent to

$$\frac{\gamma_1}{L_i}(VF_i(L_i^*) - K_i) = \frac{dVF_i}{dx}(L_i^*)$$

i.e, by relying on equations (3.12.7) and (3.12.8)

$$\frac{\gamma_1}{L_i}(VF_i(L_i^*) - K_i) = \frac{\gamma_2}{L_i}\left(VF_i(L_i^*) - \left(\frac{L_i^*}{r-\alpha} - K\right)\right) + \frac{1}{r-\alpha}$$

i.e.,

$$VF_i(L_i^*) = \frac{1}{\gamma_1 - \gamma_2}\left(\gamma_1 K_i - \gamma_2\left(\frac{L_i^*}{r-\alpha} - K\right) + \frac{L_i^*}{r-\alpha}\right). \tag{3.12.12}$$

Therefore, by substituting $VF_d(L_d^*)$ and $VF_i(L_i^*)$, obtained in (3.12.11) and (3.12.12) respectively in equations (3.12.7) and (3.12.5), the values to the firm in the active and inactive states are derived.

Finally, by substituting in (3.12.9) the value of $VF_d(L_d^*)$ obtained in (3.12.11) and in (3.12.10) the value of $VF_i(L_i^*)$ obtained in (3.12.12), a set of two equations is derived. This set admits $L_i^*$and L_d^* as solutions.

In the specific case where $K_i = K_d = 0$, from (3.12.9) and (3.12.10) the investment and abandonment thresholds satisfy $L_i^* = L_d^*$. However we know that the investment threshold is higher than the investment cost and that the abandonment threshold is smaller $L_i^* \geq (r-\alpha)K \geq L_d^*$. Thus

$$L_i^* = L_d^* = (r-\alpha)K\,,$$

and the theorem is proved.

By relying on a differential equation approach, Dixit [253] (and also Dixit and Pyndick [254]) solve the same problem (see also Brennan and Schwartz [127] for the evaluation of mining projects). The value-matching and smooth pasting conditions at investment and abandonment thresholds generate a set of four equations, which in our notation is

$$\begin{aligned}
VF_i(L_d^*) - VF_d(L_d^*) &= -K_d \\
VF_i(L_i^*) - VF_d(L_i^*) &= K_i \\
\frac{dVF_i}{dx}(L_d^*) - \frac{dVF_d}{dx}(L_d^*) &= 0 \\
\frac{dVF_i}{dx}(L_i^*) - \frac{dVF_d}{dx}(L_i^*) &= 0\,.
\end{aligned}$$

In the probabilistic approach developed in this subsection, the first two equations correspond respectively to (3.12.7) for $S_t = L_d^*$ and to (3.12.5) for $S_t = L_i^*$.

The last two equations are obtained from the set of equations (3.12.5) to (3.12.8).

4

Complements on Brownian Motion

In the first part of this chapter, we present the definition of local time and the associated Tanaka formulae, first for Brownian motion, then for more general *continuous* semi-martingales. In the second part, we give definitions and basic properties of Brownian bridges and Brownian meander. This is motivated by the fact that, in order to study complex derivative instruments, such as passport options or Parisian options, some knowledge of local times, bridges and excursions with respect to BM in particular and more generally for diffusions, is useful. We give some applications to exotic options, in particular to Parisian options.

The main mathematical references on these topics are Chung and Williams [186], Kallenberg [505], Karatzas and Shreve [513], [RY], Rogers and Williams [742] and Yor [864, 867, 868].

4.1 Local Time

4.1.1 A Stochastic Fubini Theorem

Let X be a semi-martingale on a filtered probability space $(\Omega, \mathcal{F}, \mathbf{F}, \mathbb{P})$, μ a bounded measure on $\mathbb{R}$, and H, defined on $\mathbb{R}^+ \times \Omega \times \mathbb{R}$, a $\mathcal{P} \otimes \mathcal{B}$ bounded measurable map, where $\mathcal{P}$ is the $\mathbf{F}$-predictable σ-algebra. Then

$$\int_0^t dX_s \Big(\int \mu(da) H(s, \omega, a) \Big) = \int \mu(da) \Big(\int_0^t dX_s \, H(s, \omega, a) \Big) .$$

More precisely, both sides are well defined and are equal.

This result can be proven for $H(s, \omega, a) = h(s, \omega)\varphi(a)$, then for a general H as above by applying the MCT. We leave the details to the reader.

4.1.2 Occupation Time Formula

Theorem 4.1.2.1 (Occupation Time Formula.) *Let B be a one-dimensional Brownian motion. There exists a family of increasing processes, the*

M. Jeanblanc, M. Yor, M. Chesney, *Mathematical Methods for Financial Markets*, Springer Finance, DOI 10.1007/978-1-84628-737-4_4,

local times of B, $(L_t^x, t \geq 0; x \in \mathbb{R})$, *which may be taken jointly continuous in* (x,t), *such that, for every Borel bounded function* f

$$\int_0^t f(B_s)\,ds = \int_{-\infty}^{+\infty} L_t^x f(x)\,dx\,. \tag{4.1.1}$$

In particular, for every t *and for every Borel set* A, *the Brownian occupation time of* A *between* 0 *and* t *satisfies*

$$\nu(t,A) := \int_0^t \mathbb{1}_{\{B_s \in A\}}\,ds = \int_{-\infty}^{\infty} \mathbb{1}_A(x)\,L_t^x\,dx\,. \tag{4.1.2}$$

PROOF: To prove Theorem 4.1.2.1, we consider the left-hand side of the equality (4.1.1) as "originating" from the second order correction term in Itô's formula. Here are the details.

Let us assume that f is a continuous function with compact support. Let

$$F(x) := \int_{-\infty}^{x} dz \int_{-\infty}^{z} dy f(y) = \int_{-\infty}^{\infty} (x-y)^+ f(y)dy\,.$$

Consequently, F is C^2 and $F'(x) = \int_{-\infty}^{x} f(y)\,dy = \int_{-\infty}^{\infty} f(y)\,\mathbb{1}_{\{x>y\}}\,dy$. Itô's formula applied to F and the stochastic Fubini theorem yield

$$\int_{-\infty}^{\infty} (B_t - y)^+ f(y)dy = \int_{-\infty}^{\infty} (B_0 - y)^+ f(y)dy + \int_{-\infty}^{\infty} dy f(y) \int_0^t \mathbb{1}_{\{B_s > y\}} dB_s$$
$$+ \frac{1}{2}\int_0^t f(B_s)ds\,.$$

Therefore

$$\frac{1}{2}\int_0^t f(B_s)ds = \int_{-\infty}^{\infty} dy f(y) \left((B_t - y)^+ - (B_0 - y)^+ - \int_0^t \mathbb{1}_{\{B_s > y\}} dB_s \right) \tag{4.1.3}$$

and formula (4.1.1) is obtained by setting

$$\frac{1}{2}L_t^y = (B_t - y)^+ - (B_0 - y)^+ - \int_0^t \mathbb{1}_{\{B_s > y\}} dB_s\,. \tag{4.1.4}$$

Furthermore, it may be proven from (4.1.4), with the help of Kolmogorov's continuity criterion (see Theorem 1.1.10.6), that L_t^y may be chosen jointly continuous with respect to the two variables y and t (see [RY], Chapter VI for a detailed proof). □

Had we started from $G'(x) = -\int_x^{\infty} f(y)\,dy = -\int_{-\infty}^{\infty} f(y)\,\mathbb{1}_{\{x<y\}}\,dy$, we would have obtained the following occupation time formula

$$\int_0^t f(B_s)\,ds = \int_{-\infty}^{\infty} \widetilde{L}_t^y\, f(y)dy\,, \tag{4.1.5}$$

with

$$\frac{1}{2}\widetilde{L}_t^y = (B_t - y)^- - (B_0 - y)^- + \int_0^t \mathbb{1}_{\{B_s<y\}} dB_s\,.$$

Therefore,

$$(B_t - y)^- = (B_0 - y)^- - \int_0^t \mathbb{1}_{\{B_s<y\}} dB_s + \frac{1}{2}\widetilde{L}_t^y\,.$$

Note that $L_t^y - \widetilde{L}_t^y = B_t - B_0 - \int_0^t \mathbb{1}_{\{B_s\neq y\}} dB_s = 2\int_0^t dB_s \mathbb{1}_{\{B_s=y\}}$, hence

$$(B_t - y)^- = (B_0 - y)^- - \int_0^t \mathbb{1}_{\{B_s\leq y\}} dB_s + \frac{1}{2}L_t^y\,.$$

Furthermore, the integral $\int_0^t dB_s \mathbb{1}_{\{B_s=y\}}$ is equal to 0, because its second order moment is equal to 0; indeed:

$$\mathbb{E}\left(\int_0^t dB_s \mathbb{1}_{\{B_s=y\}}\right)^2 = \int_0^t \mathbb{P}(B_s = y)ds = 0\,.$$

Hence, $L^y = \widetilde{L}^y$.

Comments 4.1.2.2 (a) In the occupation time formula (4.1.1), the time t may be replaced by any random time τ.

(b) The concept and several constructions (different from the above) of local time in the case of Brownian motion are due to Lévy [585].

(c) Existence of local times for Markov processes whose points are regular for themselves is developed in Blumenthal and Getoor [107]. Occupation densities for general stochastic processes are discussed in Geman and Horowitz [377]. Local times for diffusions are presented in ↣ Section 5.5 and in Borodin and Salminen [109].

(d) Continuity results for Brownian local times are due to Trotter [821], and many results can be found in the collective book [37].

4.1.3 An Approximation of Local Time

The quantity L_t^x is called **the local time** of the Brownian motion at level x between 0 and t. From (4.1.1), we obtain the equality

$$L_t^x = \lim_{\epsilon\to 0}\frac{1}{2\epsilon}\int_0^t \mathbb{1}_{[x-\epsilon,x+\epsilon]}(B_s)\,ds\,,$$

where the limit holds a.s.. It can also be shown that it holds in L^2. This approximation shows in particular that $(L_t^x, t \geq 0)$, the local time at level x, is an increasing process. An important property (see [RY], Chapter VI) is that, for fixed x, the support of the random measure dL_t^x is precisely the set $\{t \geq 0 : B_t = x\}$. In other words, for $x = 0$, say, the local time (at level 0) increases only on the set of zeros of the Brownian motion B. In particular

$$\int_0^t f(B_s) dL_s^0 = f(0) L_t^0 \,.$$

Exercise 4.1.3.1 Let H be a measurable map defined on $\mathbb{R}^+ \times \Omega \times \mathbb{R}$. Prove that, for any random time τ,

$$\int_0^\tau H(s,\omega,B_s)ds = \int_{-\infty}^\infty dx \int_0^\tau H(s,\omega,x)\, d_s L_s^x \,,$$

where the notation $d_s L_s^x$ makes precise that x is fixed and the measure $d_s L_s^x$ is on $\mathbb{R}^+, \mathcal{B}(\mathbb{R}^+)$. ◁

4.1.4 Local Times for Semi-martingales

The same approach can be applied to continuous semi-martingales X (see ↣ Subsection 4.1.8). In this case, the two quantities L and $\tilde{L}$ obtained from equations (4.1.4) and (4.1.5) where B is changed to X can be different, and the continuity property does not necessarily hold. There are also different definitions of local time, the reader is referred to ↣ Section 5.5.

4.1.5 Tanaka's Formula

Tanaka's formulae are variants of Itô's formula for the absolute value and the positive and negative parts of a BM.

Proposition 4.1.5.1 (Tanaka's Formulae.) *Let B be a Brownian motion and L_t^x its local time at level x between 0 and t. For every t,*

$$(B_t - x)^+ = (B_0 - x)^+ + \int_0^t \mathbb{1}_{\{B_s > x\}}\, dB_s + \frac{1}{2} L_t^x \qquad (4.1.6)$$

$$(B_t - x)^- = (B_0 - x)^- - \int_0^t \mathbb{1}_{\{B_s \leq x\}}\, dB_s + \frac{1}{2} L_t^x \qquad (4.1.7)$$

$$|B_t - x| = |B_0 - x| + \int_0^t \text{sgn}\,(B_s - x)\, dB_s + L_t^x \tag{4.1.8}$$

where $\text{sgn}(x) = 1$ *if* $x > 0$ *and* $\text{sgn}(x) = -1$ *if* $x \leq 0$.

PROOF: The first two formulae follow directly from the definition. The last equality is obtained by summing term by term the two previous ones. □

Comment 4.1.5.2 If Itô's formula could be applied to $|B|$, without taking care of the discontinuity at 0 of the derivative of $|x|$, then arguing that BM spends Lebesgue measure zero time in a given state, one would obtain the equality of $|B_t|$ and $\int_0^t \text{sgn}\,(B_s)\, dB_s$. This is obviously absurd, since $|B_t|$ is positive and $\int_0^t \text{sgn}\,(B_s)\, dB_s$ is a centered variable. Indeed, the process $(\int_0^t \text{sgn}\,(B_s)\, dB_s,\, t \geq 0)$ is a Brownian motion, see Example 1.4.1.5. Therefore, the local time spent at level 0 by the original Brownian motion B is quite meaningful, in that Tanaka's formulae are expressions of the Doob-Meyer decomposition of the sub-martingales $(B_t - x)^+$ and $|B_t - x|$ where $\frac{1}{2}L_t^x$ and L_t^x are the corresponding increasing processes.

More generally, Tanaka's formulae may be extended to develop $f(B_t)$ as a semi-martingale when f is locally the difference of two convex functions:

$$f(B_t) = f(B_0) + \int_0^t (D_- f)(B_s)\, dB_s + \frac{1}{2}\int_{\mathbb{R}} L_t^a f''(da) \tag{4.1.9}$$

where $D_- f$ is the left derivative of f and f'' is the second derivative in the distribution sense, meaning

$$\int f''(da) g(a) = \int f(a) g''(a) da$$

for any twice differentiable function g with compact support.

Note that if f is a C^1 function, and is also C^2 on $\mathbb{R} \setminus \{a_1, \dots, a_n\}$, for a finite number of points $(a_i, i = 1, \dots, n)$,

$$f(B_t) = f(B_0) + \int_0^t f'(B_s) dB_s + \frac{1}{2}\int_0^t g(B_s) ds$$

where $g(x)dx$ is the second derivative of f in the distribution sense. In that case, there is no local time apparent in the formula.

More generally, if f is locally a difference of two convex functions, which is C^2 on $\mathbb{R} \setminus \{a_1, \dots, a_n\}$, then

$$f(B_t) = f(B_0) + \int_0^t f'(B_s) dB_s + \frac{1}{2}\int_0^t g(B_s) ds + \frac{1}{2}\sum_{i=1}^n L_t^{a_i} (f'(a_i^+) - f(a_i^-))\,.$$

Warning 4.1.5.3 Some authors (e.g., Karatzas and Shreve [513]) choose a different normalization of local times starting from the occupation formula. Hence in their version of Tanaka's formulae, a coefficient other than 1/2 appears. These different conventions should be considered with care as they may be a source of errors. On the other hand, the most common choice is the coefficient 1/2, which allows the extension of Itô's formula as in (4.1.9).

Comment 4.1.5.4 If B is a Brownian motion, a necessary and sufficient condition for $f(B)$ to be a semi-martingale is that f is locally a difference of two convex functions. In [179], Chitashvili and Mania describe the functions $f(t,x)$ such that $(f(t,B_t), t \geq 0)$ is a semi-martingale. See also Chitashvili and Mania [178], Çinlar et al. [189], Föllmer et al. [349], Kunita [546] and Wang [834] for different generalizations of Itô's formula.

Exercise 4.1.5.5 Scaling Properties of the Local Time. Prove that for any $\lambda > 0$,

$$(L^x_{\lambda^2 t};\, x,t \geq 0) \stackrel{\text{law}}{=} (\lambda L^{x/\lambda}_t;\, x,t \geq 0)\,.$$

In particular, the following equality in law holds true

$$(L^0_{\lambda^2 t}, t \geq 0) \stackrel{\text{law}}{=} (\lambda L^0_t, t \geq 0)\,. \qquad \triangleleft$$

Exercise 4.1.5.6 Let $\tau_\ell = \inf\{t > 0 : L^0_t > \ell\}$. Prove that

$$\mathbb{P}(\forall \ell \geq 0, B_{\tau_\ell} = B_{\tau_\ell -} = 0) = 1\,. \qquad \triangleleft$$

Exercise 4.1.5.7 Let $dS_t = S_t(r(t)dt + \sigma dW_t)$ where r is a deterministic function and let h be a convex function satisfying $xh'(x) - h(x) \geq 0$. Prove that $\exp(-\int_0^t r(s)ds)\, h(S_t) = R_t h(S_t)$ is a local sub-martingale.
`Hint:` Apply the Itô-Tanaka formula to obtain that

$$\begin{aligned} R(t)h(S_t) &= h(x) + \int_0^t R(u)r(u)(S_u h'(S_u) - h(S_u))du \\ &\quad + \frac{1}{2}\int h''(da) \int_0^t R(s)d_s L^a_s + \text{loc. mart.}. \end{aligned}$$

$\triangleleft$

4.1.6 The Balayage Formula

We now give some other applications of the MCT to stochastic integration, thus obtaining another kind of extension of Itô's formula.

Proposition 4.1.6.1 (Balayage Formula.) *Let Y be a continuous semi-martingale and define*

$$g_t = \sup\{s \leq t : Y_s = 0\},$$

with the convention $\sup\{\emptyset\} = 0$. *Then*

$$h_{g_t} Y_t = h_0 Y_0 + \int_0^t h_{g_s} dY_s$$

for every predictable, locally bounded process h.

PROOF: By the MCT, it is enough to show this formula for processes of the form $h_u = \mathbf{1}_{[0,\tau]}(u)$, where τ is a stopping time. In this case,

$$h_{g_t} = \mathbb{1}_{\{g_t \le \tau\}} = \mathbb{1}_{\{t \le d_\tau\}} \quad \text{where} \quad d_\tau = \inf\{s \ge \tau : Y_s = 0\}.$$

Hence,

$$h_{g_t} Y_t = \mathbb{1}_{\{t \le d_\tau\}} Y_t = Y_{t \wedge d_\tau} = Y_0 + \int_0^t \mathbb{1}_{\{s \le d_\tau\}} dY_s = h_0 Y_0 + \int_0^t h_{g_s} dY_s.$$

□

Let $Y_t = B_t$, then from the balayage formula we obtain that

$$h_{g_t} B_t = \int_0^t h_{g_s} dB_s$$

is a local martingale with increasing process $\int_0^t h_{g_s}^2 ds$.

Exercise 4.1.6.2 Let $\varphi : \mathbb{R}^+ \to \mathbb{R}$ be a locally bounded real-valued function, and L the local time of the Brownian motion at level 0. Prove that $(\varphi(L_t)B_t, t \ge 0)$ is a Brownian motion time changed by $\int_0^t \varphi^2(L_s)ds$.
`Hint:` Note that for $h_s = \varphi(L_s)$, one has $h_s = h_{g_s}$, then use the balayage formula. Note also that one could prove the result first for $\varphi \in C^1$ and then pass to the limit. ◁

4.1.7 Skorokhod's Reflection Lemma

The following real variable lemma will allow us in particular to view local times as supremum processes.

Lemma 4.1.7.1 *Let y be a continuous function. There is a unique pair of functions (z, k) such that*

(i) $k(0) = 0$, *k is an increasing continuous function*
(ii) $z(t) = -y(t) + k(t) \ge 0$
(iii) $\int_0^t \mathbb{1}_{\{z(s)>0\}} dk(s) = 0\,,$

This pair is given by

$$k^*(t) = \sup_{0<s\le t} (y(s)) \vee 0, \; z^*(t) = -y(t) + k^*(t).$$

PROOF: The pair $k^*(t) = \sup_{0<s\leq t}(y(s))\vee 0, z^*(t) = -y(t)+k^*(t)$ satisfies the required properties. Let us prove that the solution is unique. Let (z_1, k_1) and (z_2, k_2) be two pairs of solutions. Then, since $z_1 - z_2$ has bounded variation, from the integration by parts formula,

$$0 \leq (z_1 - z_2)^2(t) = 2\int_0^t (z_1(s) - z_2(s))\, d(k_1(s) - k_2(s)) .$$

From (iii), the right-hand side of the above equality is equal to

$$-2\int_0^t z_2(s)\, dk_1(s) - 2\int_0^t z_1(s)\, dk_2(s)$$

which is negative. Hence, $z_1 = z_2$. □

Note that, if y is increasing, then $z = 0$. We now give an important consequence of the Skorokhod lemma:

Theorem 4.1.7.2 (Lévy's Equivalence Theorem.) *Let B be a Brownian motion starting at 0, L its local time at level 0 and $M_t = \sup_{s\leq t} B_s$. The two-dimensional processes $(|B|, L)$ and $(M - B, M)$ have the same law, i.e.,*

$$(|B_t|, L_t\, ; t \geq 0) \stackrel{\text{law}}{=} (M_t - B_t, M_t\, ; t \geq 0) .$$

PROOF: Tanaka's formula implies that

$$|B_t| = \int_0^t \text{sgn}(B_s) dB_s + L_t^0$$

where L^0, the local time of B, is an increasing process. Therefore, $(|B|, L^0)$ is a solution of Skorokhod's lemma associated with the Brownian motion $\beta_t = -\int_0^t \text{sgn}(B_s) dB_s$. Hence, $L_t^0 = \sup_{s\leq t} \beta_s$. By denoting $M_t = \sup_{s\leq t} B_s$, we obtain the decompositions

$$\begin{aligned} |B_t| &= -\beta_t + L_t^0 \\ M_t - B_t &= (-B_t) + M_t . \end{aligned}$$

The pair $(M - B, M)$ is a solution to Skorokhod's lemma associated with the Brownian motion B, because M increases only on the set $M - B = 0$. Hence, the processes $(|B|, L)$ and $(M - B, M)$ have the same law. □

Comments 4.1.7.3 (a) We have proved, in Proposition 3.1.3.1 that, for any fixed t, $M_t \stackrel{\text{law}}{=} |B_t|$. Here, we obtain that the processes $M - B$ and $|B|$ have the same law. In particular, for fixed t, $M_t - B_t \stackrel{\text{law}}{=} |B_t|$. We also have $M_t \stackrel{\text{law}}{=} L_t$.

(b) As a consequence of Skorokhod's lemma, if $\beta_t = \int_0^t \text{sgn}(B_s)dB_s$, then it is easily shown that $\sigma(\beta_s, s \leq t) = \sigma(|B_s|, s \leq t)$. See ↣ Subsection 5.8.2 for comments. This may be contrasted with the equality obtained in 3.1.4.3:

$$\sigma(M_s - B_s, s \leq t) = \sigma(B_s, s \leq t) .$$

(c) There are various identities in law involving the BM and its maximum process. From Lévy's theorem, one obtains that

$$(|B_t| + L_t;\, t \geq 0) \stackrel{\text{law}}{=} (2M_t - B_t;\, t \geq 0) .$$

From ↣ Exercise 4.1.7.12, we obtain that, for every t, $2M_t - B_t \stackrel{\text{law}}{=} R_t$ where R is a BES^3 process (see ↣ Chapter 6 if needed). Pitman [712] has extended this result at the level of processes, proving that

$$(2M_t - B_t, M_t;\, t \geq 0) \stackrel{\text{law}}{=} (R_t, J_t;\, t \geq 0)$$

where R is a BES^3 process and $J_t = \inf_{s \geq t} R_s$ (see ↣ Section 5.7). Hence, it also holds that

$$(|B_t| + L_t, L_t;\, t \geq 0) \stackrel{\text{law}}{=} (R_t, J_t;\, t \geq 0) .$$

We now present further consequences of Lévy's theorem:

Example 4.1.7.4 Let $(\tau_\ell, \ell \geq 0)$ be the inverse of the local time $(L_t^0, t \geq 0)$ defined as $\tau_\ell = \inf\{t : L_t^0 > \ell\}$, and let T_x be the first hitting time of x. Then $(T_x, x \geq 0) \stackrel{\text{law}}{=} (\tau_x, x \geq 0)$. Indeed, from Lévy's equivalence Theorem 4.1.7.2 $(M_t, t \geq 0) \stackrel{\text{law}}{=} (L_t, t \geq 0)$. Hence the same equality holds for the inverse processes. As a consequence, we note that

$$(L_t^x,\, t \geq 0) \stackrel{\text{law}}{=} \left((L_t^0 - |x|)^+,\, t \geq 0\right) .$$

Indeed, on the one hand

$$(L_t^x,\, t \geq 0) = (L^x_{T_x + (t-T_x)^+},\, t \geq 0) \stackrel{\text{law}}{=} (L^0_{(t-T_x)^+},\, t \geq 0)$$

where L^0 and T_x are independent. On the other hand

$$\left((L^0_{\tau_\ell + (t-\tau_\ell)^+} - \ell)^+,\, t \geq 0\right) \stackrel{\text{law}}{=} (L^0_{(t-\widehat{\tau}_\ell)^+},\, t \geq 0)$$

where $\widehat{\tau}_\ell$ is independent of $(L_t^0,\, t \geq 0)$. To conclude, we use $\widehat{\tau}_\ell \stackrel{\text{law}}{=} T_\ell$, and take $\ell = |x|$.

Example 4.1.7.5 For fixed t, let θ_t be the first time at which the Brownian motion reaches its maximum over the time interval $[0,t]$:

$$\theta_t := \inf\{s \le t \,|\, B_s = M_t\} = \inf\{s \le t \,|\, B_s = \sup_{u \le t} B_u\}\,.$$

If $t = 1$, we obtain

$$\begin{aligned}(\theta_1 \le u) &= \left\{\sup_{u \le s \le 1} B_s \le \sup_{s \le u} B_s\right\}\\ &= \left\{\sup_{u \le s \le 1} (B_s - B_u) + B_u \le \sup_{s \le u} B_s\right\} = \left\{\sup_{0 \le v \le 1-u} \widehat{B}_v + B_u \le M_u\right\},\end{aligned}$$

where $\widehat{B}$ is a BM independent of $(B_s, s \le u)$. Setting $\widehat{M_u} = \sup_{s \le u} \widehat{B}_s$, we get from Lévy's Theorem 4.1.7.2 and Proposition 3.1.3.1

$$\begin{aligned}\mathbb{P}(\theta_1 \le u) &= \mathbb{P}(\widehat{M}_{1-u} \le M_u - B_u) = \mathbb{P}(|\widehat{B}_{1-u}| \le |B_u|)\\ &= \mathbb{P}(\sqrt{1-u}|\widehat{B}_1| \le \sqrt{u}|B_1|) = \mathbb{P}\left(\frac{|B_1|}{|\widehat{B}_1|} \ge \frac{\sqrt{1-u}}{\sqrt{u}}\right)\\ &= \mathbb{P}\left(C^2 \ge \frac{1-u}{u}\right) = \mathbb{P}\left(u \ge \frac{1}{1+C^2}\right)\end{aligned}$$

where C follows the standard Cauchy law (see $\rightarrowtail$ Appendix A.4.2). Hence, for $u \le 1$,

$$\mathbb{P}(\theta_1 \le u) = \frac{2}{\pi} \text{arc sin} \sqrt{u}\,.$$

Finally, by scaling, for $s \le t$,

$$\mathbb{P}(\theta_t \le s) = \frac{2}{\pi} \text{arc sin} \sqrt{\frac{s}{t}}\,,$$

therefore, θ_t is Arcsine distributed on $[0,t]$. Note the non-trivial identity in law $\theta_t \overset{\text{law}}{=} A_t^+$ where $A_t^+ = \int_0^t \mathbb{1}_{\{B_s > 0\}} ds$ (see Subsection 2.5.2). As a direct application of Lévy's equivalence theorem, we obtain

$$\theta_t \overset{\text{law}}{=} g_t = \sup\{s \le t \,:\, B_s = 0\}\,.$$

Proceeding along the same lines, we obtain the equality

$$\mathbb{P}(M_t \in dx, \theta_t \in du) = \frac{x}{\pi u \sqrt{u(t-u)}} \exp\left(-\frac{x^2}{2u}\right) \mathbb{1}_{\{0 \le x, 0 \le u \le t\}}\, du\, dx \tag{4.1.10}$$

and from the previous equalities and using the Markov property

$$\begin{aligned}\mathbb{P}(\theta_1 \le u|\mathcal{F}_u) &= \mathbb{P}(\sup_{u \le s \le 1} (B_s - B_u) + B_u \le \sup_{s \le u} B_s\,|\mathcal{F}_u)\\ &= \mathbb{P}(\widehat{M}_{1-u} \le M_u - B_u|\mathcal{F}_u) = \Psi(1-u, M_u - B_u)\,.\end{aligned}$$

Here,

$$\Psi(u,x) = \mathbb{P}(\widehat{M}_u \leq x) = \mathbb{P}(|B_u| \leq x) = \frac{2}{\sqrt{2\pi}} \int_0^{x/\sqrt{u}} \exp\left(-\frac{y^2}{2}\right) dy\,.$$

Note that, for $x > 0$, the density of M_t at x can also be obtained from the equality (4.1.10). Hence, we have the equality

$$\int_0^t du \frac{x}{\pi\sqrt{u^3(t-u)}} \exp\left(-\frac{x^2}{2u}\right) = \sqrt{\frac{2}{\pi t}} e^{-x^2/(2t)}\,. \qquad (4.1.11)$$

We also deduce from Lévy's theorem that the right-hand side of (4.1.10) is equal to $\mathbb{P}(L_t \in dx, g_t \in du)$.

Example 4.1.7.6 From Lévy's identity, it is straightforward to obtain that $\mathbb{P}(L^a_\infty = \infty) = 1$.

Example 4.1.7.7 As discussed in Pitman [713], the law of the pair (L^x_1, B_1) may be obtained from Lévy's identity: for $y > 0$,

$$\mathbb{P}(L^x_1 \in dy, B_1 \in db) = \frac{|x| + y + |b - x|}{\sqrt{2\pi}} \exp\left(-\frac{1}{2}(|x| + y + |b-x|)^2\right) dydb\,.$$

Proposition 4.1.7.8 *Let φ be a C^1 function. Then, the process*

$$\boxed{\varphi(M_t) - (M_t - B_t)\varphi'(M_t)}$$

is a local martingale.

PROOF: As a first step we assume that φ is C^2. Then, from integration by parts and using the fact that M is increasing

$$(M_t - B_t)\varphi'(M_t) = \int_0^t \varphi'(M_s)\, d(M_s - B_s) + \int_0^t (M_s - B_s)\varphi''(M_s) dM_s\,.$$

Now, we note that $\int_0^t (M_s - B_s)\varphi''(M_s)dM_s = 0$, since dM_s is carried by $\{s : M_s = B_s\}$, and that $\int_0^t \varphi'(M_s)dM_s = \varphi(M_t) - \varphi(0)$. The result follows. The general case is obtained using the MCT. □

Comment 4.1.7.9 As we mentioned in Example 1.5.4.5, any solution of Tanaka's SDE $X_t = X_0 + \int_0^t \text{sgn}(X_s)dB_s$ is a Brownian motion. We can check that there are indeed weak solutions to this equation: start with a Brownian motion X and construct the BM $B_t = \int_0^t \text{sgn}(X_s)dX_s$. This Brownian motion is equal to $|X| - L$, so B is adapted to the filtration generated by $|X|$ which is strictly smaller than the filtration generated by X. Hence, the equation $X_t = X_0 + \int_0^t \text{sgn}(X_s)dB_s$ has no strong solution. Moreover, one can find infinitely many solutions, e.g., $\epsilon_{g_t} X_t$, where ϵ is a ± 1-valued predictable process, and $g_t = \sup\{s \leq t : X_s = 0\}$.

Exercise 4.1.7.10 Prove Proposition 4.1.7.8 as a consequence of the balayage formula applied to $Y_t = M_t - B_t$. ◁

Exercise 4.1.7.11 Using the balayage formula, extend the result of Proposition 4.1.7.8 when φ' is replaced by a bounded Borel function. ◁

Exercise 4.1.7.12 Prove, using Theorem 3.1.1.2, that the joint law of the pair $(|B_t|, L_t^0)$ is

$$\mathbb{P}(|B_t| \in dx,\, L_t^0 \in d\ell) = \mathbb{1}_{\{x\geq 0\}}\mathbb{1}_{\{\ell\geq 0\}}\frac{2(x+\ell)}{\sqrt{2\pi t^3}}\exp\left(-\frac{(x+\ell)^2}{2t}\right)dx\, d\ell\,.$$

◁

Exercise 4.1.7.13 Let φ be in C_b^1. Prove that $(\varphi(L_t^0) - |B_t|\varphi'(L_t^0), t \geq 0)$ is a martingale. Let $T_a^* = \inf\{t \geq 0 : |B_t| = a\}$. Prove that $L_{T_a^*}^0$ follows the exponential law with parameter $1/a$.
`Hint:` Use Proposition 4.1.7.8 together with Lévy's Theorem. Then, compute the Laplace transform of $L_{T_a^*}^0$ by means of the optional stopping theorem. The second part may also be obtained as a particular case of ↣ Azéma's lemma 5.2.2.5. ◁

Exercise 4.1.7.14 Let y be a continuous positive function vanishing at 0: $y(0) = 0$. Prove that there exists a unique pair of functions (z, k) such that

(i) $k(0) = 0$, where k is an increasing continuous function
(ii) $z(t) + k(t) = y(t), z(t) \geq 0$
(iii) $\int_0^t \mathbb{1}_{\{z(s)>0\}}dk(s) = 0$
(iv) $\forall t, \exists d(t) \geq t,\ z(d(t)) = 0$

`Hint:` $k^*(t) = \inf_{s\geq t}(y(s))$. ◁

Exercise 4.1.7.15 Let S be a price process, assumed to be a continuous local martingale, and φ a C^1 concave, increasing function. Denote by S^* the running maximum of S. Prove that the process $X_t = \varphi(S_t^*) + \varphi'(S_t^*)(S_t - S_t^*)$ is the value of the self-financing strategy with a risky investment given by $S_t\varphi'(S_t^*)$, which satisfies the floor constraint $X_t \geq \varphi(S_t)$.
`Hint:` Using an extension of Proposition 4.1.7.8, X is a local martingale. It is easy to check that $X_t = X_0 + \int_0^t \varphi'(S_s^*)dS_s$. For an intensive study of this process in finance, see El Karoui and Meziou [305]. The equality $X_t \geq \varphi(S_t)$ follows from concavity of φ. ◁

4.1.8 Local Time of a Semi-martingale

As mentioned above, local times can also be defined in greater generality for semi-martingales. The same approach as the one used in Subsection 4.1.2 leads to the following:

Theorem 4.1.8.1 (Occupation Time Formula.) *Let X be a continuous semi-martingale. There exists a family of increasing processes (**Tanaka-Meyer local times**) $(L_t^x(X), t \geq 0 \,;\, x \in \mathbb{R})$ such that for every bounded measurable function φ*

$$\int_0^t \varphi(X_s)\, d\langle X\rangle_s = \int_{-\infty}^{+\infty} L_t^x(X)\varphi(x)\, dx\,. \tag{4.1.12}$$

There is a version of L_t^x which is jointly continuous in t and right-continuous with left limits in x. (If X is a continuous martingale, its local time may be chosen jointly continuous.) In the sequel, we always choose this version. This local time satisfies

$$L_t^x(X) = \lim_{\epsilon \to 0} \frac{1}{\epsilon} \int_0^t \mathbb{1}_{[x,x+\epsilon[}(X_s) d\langle X\rangle_s\,.$$

If Z is a continuous local martingale,

$$L_t^x(Z) = \lim_{\epsilon \to 0} \frac{1}{2\epsilon} \int_0^t \mathbb{1}_{]x-\epsilon,x+\epsilon[}(Z_s) d\langle Z\rangle_s\,.$$

The same result holds with any random time in place of t.

For a continuous semi-martingale $X = Z + A$,

$$L_t^x(X) - L_t^{x-}(X) = 2\int_0^t \mathbb{1}_{\{X_s = x\}} dX_s = 2\int_0^t \mathbb{1}_{\{X_s = x\}} dA_s\,. \tag{4.1.13}$$

In particular,

$$L_t^0(|X|) = \lim_{\epsilon \to 0} \frac{1}{\epsilon} \int_0^t \mathbb{1}_{]-\epsilon,\epsilon[}(X_s) d\langle X\rangle_s = L_t^0(X) + L_t^{0-}(X),$$

hence

$$L_t^0(|X|) = 2L_t^0(X) - 2\int_0^t \mathbb{1}_{\{X_s = 0\}} dA_s\,.$$

Example 4.1.8.2 A Non-Continuous Local Time. Let Z be a continuous martingale and X be the semi-martingale

$$X_t = aZ_t^+ - bZ_t^- = \int_0^t dZ_s(a\mathbb{1}_{\{Z_s>0\}} + b\mathbb{1}_{\{Z_s<0\}}) + \frac{a-b}{2} L_t^0(Z)\,.$$

Then, it follows from (4.1.13) that $L_t^0(X) - L_t^{0-}(X) = (a-b)L_t^0(Z)$. In particular, for the reflected BM, i.e., for X when $Z_t = B_t, a = 1, b = -1$, we get $L^0(|B|) - L^{0-}(|B|) = 2L^0(B)$. Note that $L^{0-}(|B|) = 0$, hence $L^0(|B|) = 2L^0(B)$.

Example 4.1.8.3 Let $Y_t = |B_t|$. Tanaka's formula gives:

$$|B_t| = \int_0^t \text{sgn}(B_s)dB_s + L_t$$

where $(L_t, t \geq 0)$ denotes the local time of $(B_t; t \geq 0)$ at $y = 0$. By an application of the balayage formula, we obtain

$$h_{g_t}|B_t| = \int_0^t h_{g_s}\text{sgn}(B_s)dB_s + \int_0^t h_s dL_s$$

having used the fact that $L_{g_s} = L_s$. Consequently, replacing, if necessary, h by $|h|$, we see that the process $\int_0^t |h_s| dL_s$ is the local time at 0 of $(h_{g_t} B_t, t \geq 0)$.

Tanaka-Meyer Formulae

As before we set $\text{sgn}(x) = 1$ for $x > 0$ and $\text{sgn}(x) = -1$ for $x \leq 0$. Let X be a continuous semi-martingale. For every (t, x),

$$|X_t - x| = |X_0 - x| + \int_0^t \text{sgn}\,(X_s - x)\,dX_s + L_t^x(X)\,, \tag{4.1.14}$$

$$(X_t - x)^+ = (X_0 - x)^+ + \int_0^t \mathbb{1}_{\{X_s > x\}}\,dX_s + \frac{1}{2}L_t^x(X)\,, \tag{4.1.15}$$

$$(X_t - x)^- = (X_0 - x)^- - \int_0^t \mathbb{1}_{\{X_s \leq x\}}\,dX_s + \frac{1}{2}L_t^x(X)\,. \tag{4.1.16}$$

In particular, $|X - x|, (X - x)^+$ and $(X - x)^-$ are semi-martingales.

Proposition 4.1.8.4 (Lévy's Equivalence Theorem for Drifted Brownian Motion.) *Let $B^{(\nu)}$ be a BM with drift ν, i.e., $B_t^{(\nu)} = B_t + \nu t$, and $M_t^{(\nu)} = \sup_{s \leq t} B_s^{(\nu)}$. Then*

$$(M_t^{(\nu)} - B_t^{(\nu)}, M_t^{(\nu)}; t \geq 0) \overset{\text{law}}{=} (|X_t^{(\nu)}|, L_t(X^{(\nu)}); t \geq 0) \tag{4.1.17}$$

where $X^{(\nu)}$ is the (unique) strong solution of

$$dX_t = dB_t - \nu\,\text{sgn}(X_t)\,dt, X_0 = 0\,.$$

PROOF: Let $X^{(\nu)}$ be the strong solution of

$$dX_t = dB_t - \nu \,\mathrm{sgn}(X_t)\, dt, X_0 = 0$$

(see Theorem 1.5.5.1 for the existence of X) and apply Tanaka's formula. Then,

$$|X_t^{(\nu)}| = \int_0^t \mathrm{sgn}\,(X_s^{(\nu)}) \left(dB_s - \nu \,\mathrm{sgn}\,(X_s^{(\nu)})\, ds\right) + L_t^0(X^{(\nu)})$$

where $L^0(X^{(\nu)})$ is the Tanaka-Meyer local time of $X^{(\nu)}$ at level 0. Hence, setting $\beta_t = \int_0^t \mathrm{sgn}\, X_s^{(\nu)}\, dB_s$,

$$|X_t^{(\nu)}| = (\beta_t - \nu t) + L_t^0(X^{(\nu)})$$

and the result follows from Skorokhod's lemma. □

Comments 4.1.8.5 (a) Note that the processes $|B^{(\nu)}|$ and $M^{(\nu)} - B^{(\nu)}$ do not have the same law (hence the right-hand side of (4.1.17) cannot be replaced by $(|B_t^{(\nu)}|, L_t(B^{(\nu)}), t \geq 0)$). Indeed, for $\nu > 0$, $B_t^{(\nu)}$ goes to infinity as t goes to ∞, whereas $M_t^{(\nu)} - B_t^{(\nu)}$ vanishes for some arbitrarily large values of t. Pitman and Rogers [714] extended the result of Pitman [712] and proved that

$$(|B_t^{(\nu)}| + L_t^{(\nu)},\ t \geq 0) \stackrel{\text{law}}{=} (2M_t^{(\nu)} - B_t^{(\nu)},\ t \geq 0)\,.$$

(b) The equality in law of Proposition 4.1.8.4 admits an extension to the case $dB_t^{(a)} = a_t(B_t^{(a)})dt + dB_t$ and $X^{(a)}$ the unique weak solution of

$$dX_t^{(a)} = dB_t - a_t(X_t^{(a)})\,\mathrm{sgn}(X_t^{(a)})dt, X_0^{(a)} = 0$$

where $a_t(x)$ is a bounded predictable family. The equality

$$(M^{(a)} - B^{(a)}, M^{(a)}) \stackrel{\text{law}}{=} (|X^{(a)}|, L(X^{(a)}))$$

is proved in Shiryaev and Cherny [792].

We discuss here the Itô-Tanaka formula for strict local continuous martingales, as it is given in Madan and Yor [614].

Theorem 4.1.8.6 *Let S be a positive continuous strict local martingale, τ an $\mathbf{F}^S$-stopping time, a.s. finite, and K a positive real number. Then*

$$\mathbb{E}((S_\tau - K)^+) = (S_0 - K)^+ + \frac{1}{2}\mathbb{E}(L_\tau^K) - \mathbb{E}(S_0 - S_\tau)$$

where L^K is the local time of S at level K.

PROOF: We prove that

$$M_t = \frac{1}{2}L_t^K - (S_t - K)^+ + S_t = \frac{1}{2}L_t^K + (S_t \wedge K)$$

is a uniformly integrable martingale. In a first step, from Tanaka's formula, M is a (positive) local martingale, hence a super-martingale and $\mathbb{E}(L_t^K) \le 2S_0$. Since L is an increasing process, it follows that $\mathbb{E}(L_\infty^K) \le 2S_0$ and the process M is a uniformly integrable martingale. We then apply the optimal stopping theorem at time τ. □

Comment 4.1.8.7 It is important to see that, if the discounted price process is a martingale under the e.m.m., then the put-call parity holds: indeed, taking expectation of discounted values of $(S_T - K)^+ = S_T - K + (K - S_T)^+$ leads to $C(x,T) = x - Ke^{-rT} + P(x,T)$. This is no more the case if discounted prices are strict local martingales. See Madan and Yor [614], Cox and Hobson [203], Pal and Protter [692].

4.1.9 Generalized Itô-Tanaka Formula

Theorem 4.1.9.1 *Let X be a continuous semi-martingale, f a convex function, D_-f its left derivative and $f''(dx)$ its second derivative in the distribution sense. Then,*

$$f(X_t) = f(X_0) + \int_0^t D_-f(X_s)dX_s + \frac{1}{2}\int_{\mathbb{R}} L_t^x(X)f''(dx)$$

holds.

Corollary 4.1.9.2 *Let X be a continuous semi-martingale, f a C^1 function and assume that there exists a measurable function h, integrable on any finite interval $[-a,a]$ such that $f'(y) - f'(x) = \int_x^y h(z)dz$. Then, Itô's formula*

$$f(X_t) = f(X_0) + \int_0^t f'(X_s)dX_s + \frac{1}{2}\int_0^t h(X_s)d\langle X\rangle_s$$

holds.

PROOF: In this case, f is locally the difference of two convex functions and $f''(dx) = h(x)dx$. Indeed, for every $\varphi \in C_b^\infty$,

$$\langle f'', \varphi\rangle = -\langle f', \varphi'\rangle = -\int dx f'(x)\varphi'(x) = \int dz h(z)\varphi(z) .$$

□

In particular, if f is a C^1 function, which is C^2 on $\mathbb{R}\setminus\{a_1, \dots, a_n\}$, for a finite number of points (a_i), then

$$f(X_t) = f(X_0) + \int_0^t f'(X_s)dX_s + \frac{1}{2}\int_0^t g(X_s)d\langle X^c\rangle_s .$$

Here $\mu(dx) = g(x)dx$ is the second derivative of f in the distribution sense and X^c the continuous martingale part of X (see ↣ Subsection 9.3.3).

Exercise 4.1.9.3 Let X be a semi-martingale such that $d\langle X\rangle_t = \sigma^2(t, X_t)dt$. Assuming that the law of the r.v. X_t admits a density $\varphi(t, x)$, prove that, under some regularity assumptions,

$$\mathbb{E}(d_t L_t^x) = \varphi(t,x)\sigma^2(t,x)dt . \quad \triangleleft$$

4.2 Applications

4.2.1 Dupire's Formula

In a general stochastic volatility model, with

$$dS_t = S_t\left(\alpha(t, S_t)dt + \sigma_t dB_t\right),$$

it follows that $\langle S\rangle_t = \int_0^t S_u^2\sigma_u^2 du$, therefore

$$\sigma_u^2 = \frac{d}{du}\left(\int_0^u \frac{d\langle S\rangle_s}{S_s^2}\right)$$

is $\mathbf{F}^S$-adapted. However, despite the fact that this process (the square of the volatility) is, from a mathematical point of view, adapted to the filtration of prices, it is not directly observed on the markets, due to the lack of information on prices. See ↣ Section 6.7 for some examples of stochastic volatility models. In that general setting, the volatility is a functional of prices.

Under the main assumption that the volatility is a function of time and of the current value of the underlying asset, i.e., that the underlying process follows the dynamics

$$dS_t = S_t\left(\alpha(t, S_t)dt + \sigma(t, S_t)dB_t\right) ,$$

Dupire [283, 284] and Derman and Kani [250] give a relation between the volatility and the price of European calls. The function $\sigma^2(t, x)$, called the **local volatility**, is a crucial parameter for pricing and hedging derivatives.

We recall that the **implied volatility** is the value of σ such that the price of a call is equal to the value obtained by applying the Black and Scholes formula. The interested reader can also refer to Berestycki et al. [73] where a link between local volatility and implied volatility is given. The authors also propose a calibration procedure to reconstruct a local volatility.

Proposition 4.2.1.1 (Dupire Formula.) *Assume that the European call prices $C(K,T) = \mathbb{E}(e^{-rT}(S_T - K)^+)$ for any maturity T and any strike K*

are known. If, under the risk-neutral probability, the stock price dynamics are given by

$$dS_t = S_t \left(rdt + \sigma(t, S_t)dW_t\right) \tag{4.2.1}$$

where σ is a deterministic function, then

$$\boxed{\frac{1}{2}K^2\sigma^2(T,K) = \frac{\partial_T C(K,T) + rK\partial_K C(K,T)}{\partial^2_{KK} C(K,T)}}$$

where ∂_T (resp. ∂_K) is the partial derivative operator with respect to the maturity (resp. the strike).

PROOF: (a) We note that, differentiating with respect to K the equality $e^{-rT}\mathbb{E}((S_T - K)^+) = C(K,T)$, we obtain

$$\partial_K C(K,T) = -e^{-rT}\mathbb{P}(S_T > K)$$

and that, assuming the existence of a density $\varphi(T,x)$ of S_T,

$$\varphi(T,K) = e^{rT}\partial_{KK}C(K,T)\,.$$

(b) We now follow Leblanc [572] who uses the local time technology, whereas the original proof of Dupire (see ↣ Subsection 5.4.2) does not. Tanaka's formula applied to the semi-martingale S gives

$$(S_T - K)^+ = (S_0 - K)^+ + \int_0^T \mathbb{1}_{\{S_s > K\}} dS_s + \frac{1}{2}\int_0^T dL_s^K(S)\,.$$

Therefore, using integration by parts

$$\begin{aligned} e^{-rT}(S_T - K)^+ &= (S_0 - K)^+ - r\int_0^T e^{-rs}(S_s - K)^+ ds \\ &\quad + \int_0^T e^{-rs}\mathbb{1}_{\{S_s > K\}} dS_s + \frac{1}{2}\int_0^T e^{-rs} dL_s^K(S)\,. \end{aligned}$$

Taking expectations, for every pair (K,T),

$$\begin{aligned} C(K,T) &= \mathbb{E}(e^{-rT}(S_T - K)^+) \\ &= (S_0 - K)^+ + \mathbb{E}\left(\int_0^T e^{-rs} r S_s \mathbb{1}_{\{S_s > K\}} ds\right) \\ &\quad - r\mathbb{E}\left(\int_0^T e^{-rs}(S_s - K)\mathbb{1}_{\{S_s > K\}} ds\right) + \frac{1}{2}\mathbb{E}\left(\int_0^T e^{-rs} dL_s^K(S)\right)\,. \end{aligned}$$

From the definition of the local time,

$$\mathbb{E}\left(\int_0^T e^{-rs} dL_s^K(S)\right) = \int_0^T e^{-rs} \varphi(s,K) K^2 \sigma^2(s,K) ds$$

where $\varphi(s,\cdot)$ is the density of the r.v. S_s (see Exercise 4.1.9.3). Therefore,

$$C(K,T) = (S_0 - K)^+ + rK \int_0^T e^{-rs} \mathbb{P}(S_s > K)\, ds + \frac{1}{2} \int_0^T e^{-rs} \varphi(s,K) K^2 \sigma^2(s,K) ds\,.$$

Then, by differentiating w.r.t. T, one obtains

$$\partial_T C(K,T) = rKe^{-rT} \mathbb{P}(S_T > K) + \frac{1}{2} e^{-rT} \varphi(T,K) K^2 \sigma^2(T,K)\,. \quad (4.2.2)$$

(c) We now use the result found in (a) to write (4.2.2) as

$$\partial_T C(K,T) = -rK \partial_K C(K,T) + \frac{1}{2} \sigma^2(T,K) K^2 \partial_{KK} C(K,T)$$

which is the required result. □

Comments 4.2.1.2 (a) Atlan [25] presents examples of stochastic volatility models where a local volatility can be computed.

(b) Dupire result is deeply linked with Gyöngy's theorem [414] which studies processes with given marginals. See also Brunich [133] and Hirsch and Yor [438, 439].

4.2.2 Stop-Loss Strategy

This strategy is also said to be the "all or nothing" strategy. A strategic allocation (a reference portfolio) with value V_t is given in the market. The investor would like to build a strategy, based on V, such that the value of the investment is greater than a benchmark, equal to $KP(t,T)$ where K is a constant and $P(t,T)$ is the price at time t of a zero-coupon with maturity T. We assume, w.l.g., that the initial value of V is greater than $KP(0,T)$. The stop-loss strategy relies upon the following argument: the investor takes a long position in the strategic allocation.

The first time when $V_t \leq KP(t,T)$ the investor invests his total wealth of the portfolio to buy K zero-coupon bonds. When the situation is reversed, the orders are inverted and all the wealth is invested in the strategic allocation. Hence, at maturity, the wealth is $\max(V_T, K)$. See Andreasen et al. [18], Carr and Jarrow [153] and Sondermann [804] for comments.

The well-known drawback of this method is that it cannot be applied in practice when the price of the risky asset fluctuates around the floor

$G_t = KP(t,T)$, because of transaction costs. Moreover, even in the case of constant interest rate, the strategy is not self-financing. Indeed, the value of this strategy is greater than $KP(t,T)$. If such a strategy were self-financing, and if there were a stopping time τ such that its value equalled$KP(\tau,T)$, then it would remain equal to $KP(t,T)$ after time τ, and this is obviously not the case. (See Lakner [558] for details.) It may also be noted that the discounted process $e^{-rt}\max(V_t, KP(t,T))$ is not a martingale under the risk-neutral probability measure (and the process $\max(V_t, KP(t,T))$ is not the value of a self-financing strategy). More precisely,

$$e^{-rt}\max(V_t, KP(t,T)) \stackrel{\text{mart}}{=} L_t$$

where L is the local time of $(V_t e^{-rt}, t \geq 0)$ at the level Ke^{-rT}.

Sometimes, practitioners introduce a corridor around the floor and change the strategy only when the asset price is outside this corridor. More precisely, the value of the portfolio is

$$V_t \mathbb{1}_{\{t<T_1\}} + (K-\epsilon)\mathbb{1}_{\{T_1 \leq t < T_2\}} + V_t \mathbb{1}_{\{T_2 \leq t < T_3\}} + \cdots$$

where

$$\begin{aligned} &T_1 = \inf\{t : V_t \leq K-\epsilon\},\, T_2 = \inf\{t : t > T_1, V_t \geq K+\epsilon\},\\ &T_3 = \inf\{t : t > T_2, V_t \leq K-\epsilon\} \dots \quad . \end{aligned}$$

The terminal value of the portfolio when the width of the corridor tends to 0 can be shown to converge a.s. to $\max(V_T, K) - L_T^K$, where L_T^K represents the local time of $(V_t, t \in [0,T])$ at level K.

4.2.3 Knock-out BOOST

Let (a,b) be a pair of positive real numbers with $b < a$. The **knock-out BOOST** studied in Leblanc [572] is an option which pays, at maturity, the time that the underlying asset has remained above a level b, until the first time the asset reaches the level a. We assume that the underlying follows a geometric Brownian motion, i.e., $S_t = xe^{\sigma X_t}$ where X is a BM with drift ν. In symbols, the value of this knock-out BOOST option is

$$\text{KOB}(a,b;T) = E_{\mathbb{Q}}\left(e^{-rT}\int_0^{T\wedge T_a} \mathbb{1}_{(S_s>b)}\,ds\right).$$

Let α be the level relative to X, i.e., $\alpha = \dfrac{1}{\sigma}\ln\dfrac{a}{x}$. From the occupation time formula (4.1.1) and the fact that $L^y_{T\wedge T_\alpha}(X) = 0$ for $y > \alpha$, we obtain that, for every function f

$$\mathbf{W}^{(\nu)}\left(\int_0^{T\wedge T_\alpha} f(X_s)\,ds\right) = \int_{-\infty}^{\alpha} f(y)\mathbf{W}^{(\nu)}[L^y_{T_\alpha\wedge T}]dy\,,$$

where, as in the previous chapter $\mathbf{W}^{(\nu)}$ is the law of a drifted Brownian motion (see Section 3.2). Hence, if $\beta = \frac{1}{\sigma} \ln \frac{b}{x}$,

$$\mathrm{KOB}(a, b; T) = \mathbf{W}^{(\nu)} \left(e^{-rT} \int_0^{T \wedge T_\alpha} \mathbb{1}_{\{X_s > \beta\}} \, ds \right) = e^{-rT} \int_\beta^\alpha \Psi_{\alpha,\nu}(y) \, dy$$

where $\Psi_{\alpha,\nu}(y) = \mathbf{W}^{(\nu)}(L^y_{T_\alpha \wedge T})$.

The computation of $\Psi_{\alpha,\nu}$ can be performed using Tanaka's formula. Indeed, for $y < \alpha$, using the occupation time formula,

$$\begin{aligned}
\frac{1}{2}\Psi_{\alpha,\nu}(y) &= \mathbf{W}^{(\nu)}[(X_{T_\alpha \wedge T} - y)^+] - (-y)^+ - \nu \mathbf{W}^{(\nu)} \left(\int_0^{T_\alpha \wedge T} \mathbb{1}_{\{X_s > y\}} ds \right) \\
&= \mathbf{W}^{(\nu)}[(X_{T_\alpha \wedge T} - y)^+] - (-y)^+ - \nu \int_y^\alpha \Psi_{\alpha,\nu}(z) dz \\
&= (\alpha - y)^+ \mathbf{W}^{(\nu)}(T_\alpha < T) + \mathbf{W}^{(\nu)} \left[(X_T - y)^+ \mathbb{1}_{\{T_\alpha > T\}} \right] \\
&\qquad - (-y)^+ - \nu \int_y^\alpha \Psi_{\alpha,\nu}(z) dz \,. \qquad (4.2.3)
\end{aligned}$$

Let us compute explicitly the expectation of the local time in the case $T = \infty$ and $\alpha\nu > 0$. The formula (4.2.3) reads

$$\begin{aligned}
&\frac{1}{2}\Psi_{\alpha,\nu}(y) = (\alpha - y)^+ - (-y)^+ - \nu \int_y^\alpha \Psi_{\alpha,\nu}(z) dz, \\
&\Psi_{\alpha,\nu}(\alpha) = 0 \,.
\end{aligned}$$

This gives

$$\Psi_{\alpha,\nu}(y) = \begin{cases} \frac{1}{\nu}(1 - \exp(2\nu(y - \alpha)) & \text{for } 0 \le y \le \alpha \\ \frac{1}{\nu}(1 - \exp(-2\nu\alpha)) \exp(2\nu y) & \text{for } y \le 0. \end{cases}$$

In the general case, differentiating (4.2.3) with respect to y gives for $y \le \alpha$

$$\begin{aligned}
\frac{1}{2}\Psi'_{\alpha,\nu}(y) &= -\mathbf{W}^{(\nu)}(T_\alpha < T) - \mathbf{W}^{(\nu)}(T_\alpha > T, X_T > y) + \mathbb{1}_{\{y<0\}} + \nu \Psi_{\alpha,\nu}(y) \\
&= -1 + \mathbf{W}^{(\nu)}(T_\alpha > T, X_T < y) + \mathbb{1}_{\{y<0\}} + \nu \Psi_{\alpha,\nu}(y) \\
&= -1 + \mathcal{N}(\frac{y - \nu T}{\sqrt{T}}) - e^{2\nu\alpha} \mathcal{N}(\frac{y - 2\alpha - \nu t}{\sqrt{T}}) + \mathbb{1}_{\{y<0\}} + \nu \Psi_{\alpha,\nu}(y) \,.
\end{aligned}$$

It follows that $\Psi_{\alpha,\nu}(y) =$

$$2e^{2\nu y} \int_y^\alpha e^{-2\nu x} \left(-1 + \mathcal{N}(\frac{x - \nu T}{\sqrt{T}}) - e^{2\nu\alpha} \mathcal{N}(\frac{x - 2\alpha - \nu t}{\sqrt{T}}) + \mathbb{1}_{\{x<0\}} \right) dx \,.$$

4.2.4 Passport Options

An interesting application of local time is the study of passport options. We do not present this problem here, mainly because this is related to optimization problems which are beyond the scope of this book. See Delbaen and Yor [239], Henderson [430], Henderson and Hobson [431], Shreve and Večeř [797].

4.3 Bridges, Excursions, and Meanders

Given a process $(X_t, t \geq 0)$, we shall denote by $X^{[a,b]}$, for a pair of random times $0 < a < b$, the scaled process

$$X_t^{[a,b]} = \frac{1}{\sqrt{b-a}} X_{a+t(b-a)}, \, 0 \leq t \leq 1. \tag{4.3.1}$$

In what follows, B is a BM starting from 0 with natural filtration $\mathbf{F}$.

4.3.1 Brownian Motion Zeros

Let $\mathcal{Z}(\omega)$ be the random set

$$\mathcal{Z} = \{t \geq 0 : B_t = 0\}.$$

The complementary set $\mathcal{Z}^c$ is open and is therefore a countable union of maximal open intervals. The set $\mathcal{Z}$ does not have isolated points and has zero Lebesgue measure, as a consequence of the occupation density formula (4.1.2) where $A = \{0\}$.

Exercise 4.3.1.1 Let $(\tau_\ell, \ell \geq 0)$ be the inverse of the local time at level 0, defined in Example 4.1.7.4. Prove that, if $u \in \mathcal{Z}$, then $u = \tau_s$ or $u = \tau_{s-}$ for some s.
`Hint:` if $u \in \mathcal{Z}$, either $L_{u+\epsilon} - L_u > 0$ for every ϵ, and $u = \tau_s$ for $s = L_u$, or L is constant and $u = \tau_{s-}$ for $s = L_u$. $\triangleleft$

4.3.2 Excursions

Let t be a fixed time and let $g_t = \sup\{s \leq t : B_s = 0\}$ be the **last passage time** at level 0 before time t and $d_t = \inf\{s \geq t : B_s = 0\}$ the **first passage time** at level zero after time t. The **Brownian excursion** which straddles t is the path

$$(B_{g_t+u} \, ; 0 \leq u \leq d_t - g_t).$$

The normalized excursion is taken to be the process $(B_u^{[g_t,d_t]}, 0 \leq u \leq 1)$, or sometimes it is defined as its absolute value.
It is worth noting that g_t is not an $\mathbf{F}$-stopping time, whereas d_t is an $\mathbf{F}$-stopping time.

Let us remark that, for u in the interval (g_t, d_t), the sign of B_u remains constant.

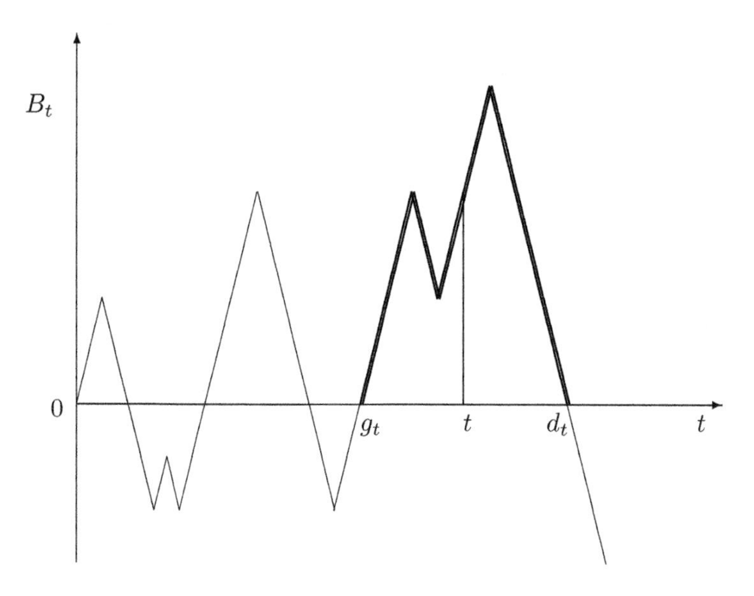

Fig. 4.1 Excursion of a Brownian motion straddling t

4.3.3 Laws of T_x, d_t and g_t

We study here the laws of the random variables T_x, d_t and g_t.

Proposition 4.3.3.1 *Let $T_x = \inf\{t : B_t = x\}$ and $M_t = \sup_{s \le t} B_s$. Then:*

$$T_x \stackrel{\text{law}}{=} x^2 T_1 \stackrel{\text{law}}{=} \left(\frac{x}{M_1}\right)^2 \stackrel{\text{law}}{=} \left(\frac{x}{B_1}\right)^2 .$$

PROOF: By scaling $T_x \stackrel{\text{law}}{=} x^2 T_1$ and $M_t \stackrel{\text{law}}{=} \sqrt{t} M_1$. Furthermore,

$$\mathbb{P}(T_1 \ge u) = \mathbb{P}(M_u \le 1) = \mathbb{P}(\sqrt{u} M_1 \le 1) = \mathbb{P}\left(\left(\frac{1}{M_1}\right)^2 \ge u\right)$$

which implies the remaining equalities, using that $B_1^2 \stackrel{\text{law}}{=} M_1^2$ (see Proposition 3.1.3.1). □

Proposition 4.3.3.2 (i) *The law of d_u is that of $u(1 + C^2)$ where C is a Cauchy random variable with density $\frac{1}{\pi}\frac{1}{1+x^2}$.*

(ii) *The variable g_t is Arcsine distributed:*

$$\mathbb{P}(g_t \in ds) = \frac{1}{\pi} \frac{1}{\sqrt{s(t-s)}} \mathbb{1}_{\{s \leq t\}} \, ds \, .$$

PROOF: By definition, $d_u = u + \inf\{v \,|\, B_{u+v} - B_u = -B_u\}$. The process $\widehat{B} = (\widehat{B}_t = B_{t+u} - B_u, t \geq 0)$ is a Brownian motion independent of B_u. Let $\widehat{T}_a$ be the first hitting time of a associated with this process $\widehat{B}$. By using results of the previous proposition and the scaling property of Brownian motion, we obtain

$$d_u \stackrel{\text{law}}{=} u + \widehat{T}_{-B_u} \stackrel{\text{law}}{=} u + B_u^2 \widehat{T}_1 \stackrel{\text{law}}{=} u + u B_1^2 \widehat{T}_1 \stackrel{\text{law}}{=} u \left(1 + \frac{B_1^2}{\widehat{B}_1^2} \right)$$

and therefore from the explicit computation of the law of $B_1^2/\widehat{B}_1^2$ (see $\rightarrowtail$ Appendix A.4.2)

$$d_u \stackrel{\text{law}}{=} u(1 + C^2) \, , \; C \text{ with density } \frac{1}{\pi} \frac{1}{1+x^2} \, .$$

From $\{g_t < u\} = \{t < d_u\}$ we deduce, for all t and u,

$$\frac{d_u}{u} \stackrel{\text{law}}{=} \frac{t}{g_t} \stackrel{\text{law}}{=} 1 + C^2 \, ;$$

consequently, g_t is Arcsine distributed. □

These results can be extended to the last time before 1 when a Brownian motion reaches level a.

Proposition 4.3.3.3 *Let* $g_1^a = \sup\{t \leq 1 \, : \, B_t = a\}$, *where* $\sup(\emptyset) = 1$. *The law of* g_1^a *is*

$$\mathbb{P}(g_1^a \in dt) = \exp\left(-\frac{a^2}{2t} \right) \frac{dt}{\pi \sqrt{t(1-t)}} \mathbb{1}_{\{0<t<1\}} \, , \qquad (4.3.2)$$

$$\mathbb{P}(g_1^a = 1) = \mathbb{P}(|G| \leq a)$$

where G is a standard Gaussian random variable. The r.v.

$$d_1^a = \inf\{u \geq 1 \, : \, B_u = a\}$$

has the same law as $1 + \frac{(a-G)^2}{\widetilde{G}^2}$ *where G and $\widetilde{G}$ are independent standard Gaussian random variables.*

PROOF: From the equality, with $t < 1$,

$$\{g_1^a \leq t\} = \{T_a \leq t\} \cap \{\widehat{g}_{1-T_a}^0 \leq t - T_a\}$$

where $\widehat{g}^0$ is relative to the Brownian motion $(\widehat{B}_u = B_{u+T_a} - B_{T_a}, u \geq 0)$, i.e., $\widehat{g}_t^0 = \sup\{s \leq t \, : \, \widehat{B}_s = 0\}$, one obtains

$$\mathbb{P}(g_1^a \le t) = \int_0^t \mathbb{P}(T_a \in du)\mathbb{P}(\widehat{g}^0_{1-u} \le t - u\} .$$

The laws of T_a and $\widehat{g}^0_{1-u}$ are known, and some easy computation leads to

$$\mathbb{P}(g_1^a \le t) = \frac{a}{\pi\sqrt{2\pi}} \int_0^t \frac{dv}{\sqrt{1-v}} \int_0^v du \frac{e^{-a^2/(2u)}}{\sqrt{u^3(v-u)}} .$$

It remains to recall that, from (4.1.11) the second integral on the right-hand side is known.

Note that the right-hand side of (4.3.2) is a sub-probability, and that the missing mass is

$$\mathbb{P}(g_1^a = 1) = \mathbb{P}(T_a \ge 1) = \mathbb{P}(|G| \le a) ,$$

where G is the standard Gaussian variable.

Let $d_t^a(B) = \inf\{u \ge t : B_u = a\}$. We obtain

$$\begin{aligned} d_t^a(B) &= t + \inf\{u \ge 0 : B_{u+t} - B_t = a - B_t\} \\ &= t + \widehat{T}_{a-B_t} \overset{\text{law}}{=} t + \frac{(a-B_t)^2}{G^2} . \end{aligned} \tag{4.3.3}$$

Here, $\widehat{T}_b = \inf\{u \ge 0 : \widehat{B}_u = b\}$, where $\widehat{B}$ is a Brownian motion independent of $\mathcal{F}_t$, and G is a standard Gaussian variable, independent of B. □

Comments 4.3.3.4 (a) Formula (4.3.2) plays an important rôle in the discussion of quantiles of Brownian motion in Yor [866] (formula (3.b) therein).

(b) We recall that we already saw the occurrence of the Arcsine law in Subsection 2.5.2 and Example 4.1.7.5.

Exercise 4.3.3.5 The aim of this exercise is to provide an explanation of the fact, obtained in Proposition 4.3.3.3, that

$$\mathbb{P}(|G| \le 1) + \int_0^1 \mathbb{P}(g_1^a \in dt) = 1 .$$

From the equality $G^2 \overset{\text{law}}{=} 2\mathbf{e}g_1$ where $\mathbf{e}$ is exponentially distributed with parameter 1 and G is a standard Gaussian variable (see ↣ Appendix A.4.2), prove that $\mathbb{P}(|G| > a) = \mathbb{E}(e^{-a^2/(2g_1)})$ and conclude. ◁

Exercise 4.3.3.6 Let

$$\begin{aligned} g_a^{(\nu)} &= \sup\{t : B_t + \nu t = a\} \\ T_a^{(\nu)} &= \inf\{t : B_t + \nu t = a\} \end{aligned}$$

Prove that

$$(T_a^{(\nu)}, g_a^{(\nu)}) \overset{\text{law}}{=} \left(\frac{1}{g_\nu^{(a)}}, \frac{1}{T_\nu^{(a)}}\right) .$$

See Bentata and Yor [72] for related results. ◁

4.3.4 Laws of (B_t, g_t, d_t)

We now study the laws of the pairs of r.v's (B_t, d_t) and (B_t, g_t) for fixed t.

Proposition 4.3.4.1 *The joint laws of the pairs* (B_t, d_t) *and* (B_t, g_t) *are given by:*

$$\mathbb{P}(B_t \in dx, d_t \in ds) = \mathbb{1}_{\{s \geq t\}} \frac{|x|}{2\pi\sqrt{t(s-t)^3}} \exp\left(-\frac{sx^2}{2t(s-t)}\right) dx\, ds\,, \quad (4.3.4)$$

$$\mathbb{P}(B_t \in dx, g_t \in ds) = \mathbb{1}_{\{s \leq t\}} \frac{|x|}{2\pi\sqrt{s(t-s)^3}} \exp\left(-\frac{x^2}{2(t-s)}\right) dx\, ds\,. \quad (4.3.5)$$

PROOF: We begin with the law of (B_t, d_t). From the Markov property we derive

$$\begin{aligned}\mathbb{P}(B_t \in dx, d_t \in ds) &= \mathbb{P}(B_t \in dx)\mathbb{P}(d_t \in ds | B_t = x)\\ &= \mathbb{P}(B_t \in dx)\mathbb{P}_x(T_0 \in ds - t)\\ &= \mathbb{P}(B_t \in dx)\mathbb{P}_0(T_x \in ds - t)\,,\end{aligned}$$

and the two expressions on the right-hand side of the latter equation are well known.

For the second law, we use time inversion for the pair (B, g). Let us define $\{\hat{B}_t = tB_{1/t}, t > 0\}$ a standard Brownian motion and let $\widehat{g}$ be related to $\widehat{B}$ via $\widehat{g}_u = \sup\{s < u : \widehat{B}_s = 0\}$. We begin with an identity in law between d_t and $g_{1/t}$:

$$\begin{aligned} d_t &= \inf\{s \geq t : B_s = 0\} = \inf\{s^{-1} \geq t : B_{1/s} = 0\}\\ &= \inf\{s^{-1} \geq t : sB_{1/s} = 0\} = \inf\{s^{-1} \geq t : \widehat{B}_s = 0\}\\ &= 1/\sup\left\{u \leq \frac{1}{t} : \widehat{B}_u = 0\right\} = \frac{1}{\widehat{g}_{1/t}}\,.\end{aligned}$$

Therefore, since $B_t = t\hat{B}_{1/t}$, we have

$$\mathbb{P}(B_t \leq x, g_t \leq s) = \mathbb{P}\left(\widehat{B}_{1/t} \leq \frac{x}{t}, \widehat{d}_{1/t} \geq \frac{1}{s}\right).$$

Denoting by $f_t(x,s)$ the density of the pair $(\widehat{B}_t, \widehat{d}_t)$, and using the first part of the proof:

$$\frac{1}{dsdx}\mathbb{P}(B_t \in dx, g_t \in ds) = \frac{\partial^2}{\partial x \partial s}\mathbb{P}\left(\widehat{B}_{1/t} \leq \frac{x}{t}, \widehat{d}_{1/t} \geq \frac{1}{s}\right) = \frac{1}{ts^2} f_{1/t}\left(\frac{x}{t}, \frac{1}{s}\right).$$

The result follows from this. □

Comment 4.3.4.2 The reader will find in Chung [183] another proof of (4.3.5) based on the following remark, which uses the law of the pair (B_t, m_t^B) established in Subsection 3.1.5:

$$\begin{aligned}\mathbb{P}(g_t \leq s, B_s \in dx, B_t \in dy) &= \mathbb{P}(B_s \in dx, B_u \neq 0, \forall u \in [s,t], B_t \in dy) \\ &= \mathbb{P}(B_s \in dx)\, \mathbb{P}_x(B_{t-s} \in dy,\, T_0 > t-s) \\ &= \mathbb{P}(B_s \in dx)\, \mathbb{P}_0(B_{t-s} + x \in dy,\, m_{t-s}^B > -x)\end{aligned}$$

$$= \frac{e^{-x^2/(2s)}}{\sqrt{2\pi s}} \frac{1}{\sqrt{2\pi(t-s)}} \left(\exp\left(-\frac{(x-y)^2}{2(t-s)}\right) - \exp\left(-\frac{(x+y)^2}{2(t-s)}\right)\right) dx\, dy\,.$$

By integrating with respect to dx, and differentiating with respect to s, the result is obtained.

Exercise 4.3.4.3 Let $t > 0$ be fixed and $\theta_t = \inf\{s \leq t \,|\, M_t = B_s\}$ where $M_t = \sup_{s \leq t} B_s$. Prove that

$$(M_t, \theta_t) \stackrel{\text{law}}{=} (|B_t|, g_t) \stackrel{\text{law}}{=} (L_t, g_t)\,.$$

Hint: Use the equalities (4.1.10) and (4.3.4) and Lévy's theorem. ◁

4.3.5 Brownian Bridge

The **Brownian bridge** $(b_t, 0 \leq t \leq 1)$ is defined as the conditioned process $(B_t, t \leq 1 | B_1 = 0)$. Note that $B_t = (B_t - tB_1) + tB_1$ where, from the Gaussian property, the process $(B_t - tB_1, t \leq 1)$ and the random variable B_1 are independent. Hence $(b_t, 0 \leq t \leq 1) \stackrel{\text{law}}{=} (B_t - tB_1,\, 0 \leq t \leq 1)$. The Brownian bridge process is a Gaussian process, with zero mean and covariance function $s(1-t), s \leq t$. Moreover, it satisfies $b_0 = b_1 = 0$.

Each of the Gaussian processes X, Y and Z where

$$X_t = (1-t) \int_0^t \frac{dB_s}{1-s}\,;\, 0 \leq t \leq 1$$

$$Z_t = tB_{(1/t)-1}\,;\, 0 \leq t \leq 1$$

$$Y_t = (1-t) B\left(\frac{t}{1-t}\right)\,;\, 0 \leq t \leq 1$$

has the same properties, and is a Brownian bridge. Note that the apparent difficulty in defining the above processes at time 0 or 1 may be resolved by extending it continuously to [0, 1].

Since $(W_{1-t} - W_1,\, t \leq 1) \stackrel{\text{law}}{=} (W_t,\, t \leq 1)$, the Brownian bridge is invariant under time reversal.

We can represent the Brownian bridge between 0 and y during the time interval $[0,1]$ as

$$(B_t - tB_1 + ty;\ t \le 1)$$

and we denote by $\mathbf{W}^{(1)}_{0\to y}$ its law on the canonical space. More generally, $\mathbf{W}^{(T)}_{x\to y}$ denotes the law of the Brownian bridge between x and y during the time interval $[0,T]$, which may be expressed as

$$\left(x + B_t - \frac{t}{T}B_T + \frac{t}{T}(y-x);\ t \le T\right),$$

where $(B_t; t \le T)$ is a standard BM starting from 0.

Theorem 4.3.5.1 *For every t, $\mathbf{W}^{(t)}_{x\to y}$ is equivalent to $\mathbf{W}_x$ on $\mathcal{F}_s$ for $s < t$.*

PROOF: Let us consider a more general case: suppose $((X_t;\ t \ge 0), (\mathcal{F}_t), \mathbb{P}_x)$ is a real valued Markov process with semigroup

$$P_t(x, dy) = p_t(x,y)dy,$$

and F_s is a non-negative $\mathcal{F}_s$-measurable functional. Then, for $s \le t$, and any function f

$$\mathbb{E}_x[F_s f(X_t)] = \mathbb{E}_x[F_s\, P_{t-s} f(X_s)]\,.$$

On the one hand

$$\begin{aligned}\mathbb{E}_x[F_s\, P_{t-s} f(X_s)] &= \mathbb{E}_x[F_s \int f(y)\, p_{t-s}(X_s, y)\, dy] \\ &= \int f(y) \mathbb{E}_x[F_s p_{t-s}(X_s,y)]\, dy\,.\end{aligned}$$

On the other hand

$$\mathbb{E}_x[F_s f(X_t)] = \mathbb{E}_x[\mathbb{E}_x[F_s|X_t] f(X_t)] = \int dy f(y) p_t(x,y) \mathbb{E}^{(t)}_{x\to y}(F_s)\,,$$

where $\mathbb{P}^{(t)}_{x\to y}$ is the probability measure associated with the bridge (for a general definition of Markov bridges, see Fitzsimmons et al. [346]) between x and y during the time interval $[0,t]$. Therefore,

$$\mathbb{E}^{(t)}_{x\to y}(F_s) = \frac{\mathbb{E}_x[F_s p_{t-s}(X_s,y)]}{p_t(x,y)}\,.$$

Thus

$$\boxed{\mathbb{P}^{(t)}_{x\to y}|_{\mathcal{F}_s} = \frac{p_{t-s}(X_s,y)}{p_t(x,y)} \mathbb{P}_x|_{\mathcal{F}_s}\,.} \tag{4.3.6}$$

□

Sometimes, we shall denote X under $\mathbb{P}^{(t)}_{x\to y}$ by $(X^{(t)}_{x\to y}(s), s \leq t)$.

If X is an n-dimensional Brownian motion and $x = y = 0$ we have, for $s < t$,

$$\boxed{\mathbf{W}^{(t)}_{0\to 0}|_{\mathcal{F}_s} = \left(\frac{t}{t-s}\right)^{n/2} \exp\left(\frac{-|X_s|^2}{2(t-s)}\right) \mathbf{W}_0|_{\mathcal{F}_s} .} \tag{4.3.7}$$

As a consequence of (4.3.7), identifying the density as the exponential martingale $\mathcal{E}(Z)$, where $Z_s = -\int_0^s \frac{X_u}{t-u}\, dX_u$, we obtain the canonical decomposition of the standard Brownian bridge (under $\mathbf{W}^{(t)}_{0\to 0}$) as:

$$X_s = B_s - \int_0^s du\, \frac{X_u}{t-u}, \quad s < t, \tag{4.3.8}$$

where $(B_s, s \leq t)$ is a Brownian motion under $\mathbf{W}^{(t)}_{0\to 0}$. (This decomposition may be related to the harness property in $\rightarrowtail$ Definition 8.5.2.1.)

Therefore, we obtain that the standard Brownian bridge b is a solution of the following stochastic equation

$$\begin{cases} db_t = -\dfrac{b_t}{1-t}\, dt + dB_t \,;\, 0 \leq t < 1 \\ \\ b_0 = 0 \,. \end{cases}$$

Proposition 4.3.5.2 *Let $X_t = \mu t + \sigma B_t$ where B is a BM, and for fixed T, $(X^{(T)}_{0\to y}(t), t \leq T)$ is the associated bridge. Then, the law of the bridge does not depend on μ, and in particular*

$$\mathbb{P}(X^{(T)}_{0\to y}(t) \in dx) = \frac{dx}{\sigma\sqrt{2\pi t}} \sqrt{\frac{T}{T-t}} \exp\left(-\frac{1}{2\sigma^2}\left(\frac{x^2}{t} + \frac{(y-x)^2}{T-t} - \frac{y^2}{T}\right)\right) \tag{4.3.9}$$

PROOF: The fact that the law does not depend on μ can be viewed as a consequence of Girsanov's theorem. The form of the density is straightforward from the computation of the joint density of (X_t, X_T), or from (4.3.6). □

Proposition 4.3.5.3 *Let $B^{(t)}_{x\to z}$ be a Brownian bridge, starting from x at time 0 and ending at z at time t, and $M^{br}_t = \sup_{0\leq s\leq t} B^{(t)}_{x\to z}(s)$. Then, for any $m > z \vee x$,*

$$\mathbb{P}^{(t)}_{x\to z}(M_t^{br} \leq m) = 1 - \exp\left(-\frac{(z+x-2m)^2}{2t} + \frac{(z-x)^2}{2t}\right).$$

In particular, let b be a standard Brownian bridge ($x = z = 0, t = 1$). Then,

$$\sup_{0\leq s\leq 1} b_s \stackrel{\text{law}}{=} \frac{1}{2}R\,,$$

where R is Rayleigh distributed with density $x \exp\left(-\frac{1}{2}x^2\right) \mathbb{1}_{\{x\geq 0\}}$. If $\ell_1^a(b)$ denotes the local time of b at level a at time 1, then for every a

$$\ell_1^a(b) \stackrel{\text{law}}{=} (R - 2|a|)^+ \,. \tag{4.3.10}$$

PROOF: Let B be a standard Brownian motion and $M_t^B = \sup_{0\leq s\leq t} B_s$. Then, for every $y > 0$ and $x \leq y$, equality (3.1.3) reads

$$\mathbb{P}(B_t \in dx\,, M_t^B \leq y) = \frac{dx}{\sqrt{2\pi t}} \exp\left(-\frac{x^2}{2t}\right) - \frac{dx}{\sqrt{2\pi t}} \exp\left(-\frac{(2y-x)^2}{2t}\right),$$

hence,

$$\begin{aligned}\mathbb{P}(M_t^B \leq y | B_t = x) &= \frac{\mathbb{P}(B_t \in dx\,, M_t^B \leq y)}{\mathbb{P}(B_t \in dx)} = 1 - \exp(-\frac{(2y-x)^2}{2t} + \frac{x^2}{2t}) \\ &= 1 - \exp\left(-\frac{2y^2 - 2xy}{t}\right).\end{aligned}$$

More generally,

$$\mathbb{P}(\sup_{s\leq t} B_s + x \leq y | B_t + x = z) = \mathbb{P}(M_t^B \leq y - x | B_t = z - x)$$

hence

$$\mathbb{P}_x(\sup_{0\leq s\leq t} B^{(t)}_{x\to z}(s) \leq y) = 1 - \exp\left(-\frac{(z+x-2y)^2}{2t} + \frac{(z-x)^2}{2t}\right).$$

The result on local time follows by conditioning w.r.t. B_1 the equality obtained in Example 4.1.7.7. □

Theorem 4.3.5.4 *Let B be a Brownian motion. For every t, the process $B^{[0,g_t]}$ defined by:*

$$B^{[0,g_t]} = \left(\frac{1}{\sqrt{g_t}} B_{ug_t}, u \leq 1\right) \tag{4.3.11}$$

is a Brownian bridge $B^{(1)}_{0\to 0}$ independent of the σ-algebra $\sigma\{g_t, B_{g_t+u}, u \geq 0\}$.

PROOF: By scaling, it suffices to prove the result for $t = 1$. Let $\widehat{B}_t = tB_{1/t}$. As in the proof of Proposition 4.3.4.1, $\widehat{d_1} = \frac{1}{g_1}$. Then,

$$\frac{1}{\sqrt{g_1}}B(ug_1) = u\sqrt{g_1}\,\widehat{B}\left(\frac{1}{ug_1}\right) = \frac{u}{\sqrt{\widehat{d}_1}}\left[\widehat{B}\left(\frac{1}{g_1}+\frac{1}{g_1}(\frac{1}{u}-1)\right) - \widehat{B}\left(\frac{1}{g_1}\right)\right]$$

$$= \frac{u}{\sqrt{\widehat{d}_1}}\left[\widehat{B}\left(\widehat{d}_1+\widehat{d}_1\left(\frac{1}{u}-1\right)\right) - \widehat{B}(\widehat{d}_1)\right].$$

Knowing that $(\widehat{B}_{\widehat{d}_1+s} - \widehat{B}_{\widehat{d}_1}; s \geq 0)$ is a Brownian motion independent of $\mathcal{F}_{\widehat{d}_1}$ and that $\widehat{B}_{\widehat{d}_1} = 0$, the process $\widetilde{B}_u = \frac{1}{\sqrt{\widehat{d}_1}}\widehat{B}_{\widehat{d}_1+\widehat{d}_1 u}$ is also a Brownian motion independent of $\mathcal{F}_{\widehat{d}_1}$. Therefore $t\widetilde{B}_{(\frac{1}{t}-1)}$ is a Brownian bridge independent of $\mathcal{F}_{\widehat{d}_1}$ and the result is proved. □

Example 4.3.5.5 Let B be a real-valued Brownian motion under $\mathbb{P}$ and

$$X_t = B_t - \int_0^t \frac{B_s}{s}ds\,.$$

This process X is an $\mathbf{F}^*$-Brownian motion where $\mathbf{F}^*$ is the filtration generated by the bridges, i.e.,

$$\mathcal{F}_t^* = \sigma\left\{B_u - \frac{u}{t}B_t,\, u \leq t\right\}.$$

Let $L_t = \exp(\lambda B_t - \frac{\lambda^2 t}{2})$ and $\mathbb{Q}_{|\mathcal{F}_t} = L_t\mathbb{P}_{|\mathcal{F}_t}$. Then $\mathbb{Q}_{|\mathcal{F}_t^*} = \mathbb{P}_{|\mathcal{F}_t^*}$.

Comments 4.3.5.6 (a) It can be proved that $|B|^{[g_1,d_1]}$ has the same law as a BES^3 bridge and is independent of

$$\sigma(B_u, u \leq g_1) \vee \sigma(B_u, u \geq d_1) \vee \sigma(\text{sgn}(B_1))\,.$$

(b) For a study of Bridges in a general Markov setting, see Fitzsimmons et al. [346].

(c) Application to fast simulation of Brownian bridge in finance can be found in Pagès [691], Metwally and Atiya [646]. We shall study Brownian bridges again when dealing with enlargements of filtrations, in ↣ Subsection 5.9.2.

Exercise 4.3.5.7 Let $T_a = \inf\{t : |X_t| = a\}$. Give the law of T_a under $\mathbf{W}_{0\to 0}^{(t)}$.

Hint: : $\mathbf{W}_{0\to 0}^{(t)}\left(f(T_a\mathbb{1}_{\{T_a<t\}})\right) = \mathbf{W}\left(f(T_a)\mathbb{1}_{\{T_a<t\}}\frac{t}{(t-T_a)^{n/2}}e^{-\frac{a^2}{2(t-T_a)}}\right)$. ◁

4.3.6 Slow Brownian Filtrations

If ζ is a random time, i.e., a random variable such that $\zeta > 0$ a.s., we define the σ-field $\mathcal{F}_\zeta^-$ of the past up to ζ as the σ-algebra generated by the variables h_ζ, where h is a generic predictable process.

Likewise, we may define $\mathcal{F}_\zeta^+$ as the σ-algebra generated by the variables h_ζ, where h is a generic $\mathbf{F}$-progressively measurable process.

In particular, we consider, as in Dellacherie et al. [241] the σ-algebras $\mathcal{F}_{g_t}^-$ and $\mathcal{F}_{g_t}^+$. The following properties are satisfied:

- Both $(\mathcal{F}_{g_t}^-, t \geq 0)$ and $(\mathcal{F}_{g_t}^+, t \geq 0)$ are increasing and are called the slow Brownian filtrations, $(\mathcal{F}_{g_t}^-, t \geq 0)$ being the strict slow Brownian filtration and $(\mathcal{F}_{g_t}^+, t \geq 0)$ the wide slow Brownian filtration.
- For fixed t, there is the double identity

$$\mathcal{F}_{g_t}^+ = \cap_{\epsilon>0} \mathcal{F}_{g_t+\epsilon}^- = \mathcal{F}_{g_t}^- \vee \sigma(\text{sgn} B_t) .$$

This shows that $\mathcal{F}_{g_t}^+$ is the σ-algebra of the immediate future after g_t and the second identity provides the independent complement $\sigma(\text{sgn}B_t)$ which needs to be added to $\mathcal{F}_{g_t}^-$ to capture $\mathcal{F}_{g_t}^+$. See Barlow et al. [50].

4.3.7 Meanders

Definition 4.3.7.1 *The Brownian meander of length 1 is the process defined by:*

$$m_u := \frac{1}{\sqrt{1-g_1}} |B_{g_1+u(1-g_1)}|; \ (u \leq 1) .$$

We begin with a very useful result:

Proposition 4.3.7.2 *The law of m_1 is the Rayleigh law whose density is*

$$x \exp(-x^2/2) \, \mathbb{1}_{\{x \geq 0\}} .$$

Consequently, $m_1 \overset{\text{law}}{=} \sqrt{2\mathbf{e}}$ holds.

PROOF: From (4.3.5),

$$\mathbb{P}(B_1 \in dx, g_1 \in ds) = \mathbb{1}_{\{s \leq 1\}} \frac{|x| \, dx \, ds}{2\pi\sqrt{s(1-s)^3}} \exp\left(-\frac{x^2}{2(1-s)}\right) .$$

We deduce, for $x > 0$,

$$\begin{aligned}
\mathbb{P}(m_1 \in dx) &= \int_{s=0}^1 \mathbb{P}(m_1 \in dx, g_1 \in ds) = \int_{s=0}^1 \mathbb{P}(\frac{|B_1|}{\sqrt{1-s}} \in dx, g_1 \in ds) \\
&= dx \, \mathbb{1}_{\{x \geq 0\}} \int_0^1 ds \frac{2x(1-s)}{2\pi\sqrt{s(1-s)^3}} \exp\left(-\frac{x^2(1-s)}{2(1-s)}\right) \\
&= 2x \, dx \, \mathbb{1}_{\{x \geq 0\}} \exp\left(-x^2/2\right) \int_0^1 ds \frac{1}{2\pi\sqrt{s(1-s)}} \\
&= x e^{-x^2/2} \mathbb{1}_{\{x \geq 0\}} dx ,
\end{aligned}$$

where we have used the fact that $\int_0^1 ds \frac{1}{\pi\sqrt{s(1-s)}} = 1$, from the property of the arcsin density. □

We continue with a more global discussion of meanders in connection with the slow Brownian filtrations. For any given t, by scaling, the law of the process

$$m_u^{(t)} = \frac{1}{\sqrt{t-g_t}} |B_{g_t+u(t-g_t)}|\,, u \leq 1$$

does not depend on t. Furthermore, this process is independent of $\mathcal{F}_{g_t}^+$ and in particular of g_t and $\mathrm{sgn}(B_t)$. All these properties extend also to the case when t is replaced by τ, any $\mathcal{F}_{g_t}^-$-stopping time.

Note that, from $|B_1| = \sqrt{1-g_1}m_1$ where m_1 and $\sqrt{1-g_1}$ are independent, we obtain from the particular case of the beta-gamma algebra (see ↣ Appendix A.4.2) $G^2 \stackrel{\text{law}}{=} 2\mathbf{e}g_1$ where $\mathbf{e}$ is exponentially distributed with parameter 1, G is a standard Gaussian variable, and g_1 and $\mathbf{e}$ are independent.

Comment 4.3.7.3 For more properties of the Brownian meander, see Biane and Yor [87] and Bertoin and Pitman [82].

4.3.8 The Azéma Martingale

We now introduce the Azéma martingale which is an $(\mathcal{F}_{g_t}^+)$-martingale and enjoys many remarkable properties.

Proposition 4.3.8.1 *Let B be a Brownian motion. The process*

$$\mu_t = (\mathrm{sgn} B_t)\sqrt{t-g_t},\ t \geq 0$$

is an $(\mathcal{F}_{g_t}^+)$-martingale. Let

$$\Psi(z) = \int_0^\infty x \exp\left(zx - \frac{x^2}{2}\right) dx = 1 + z\sqrt{2\pi}\,\mathcal{N}(z)e^{z^2/2}\,. \tag{4.3.12}$$

The process

$$\exp\left(-\frac{\lambda^2}{2}t\right)\Psi(\lambda\mu_t),\ t \geq 0$$

is an $(\mathcal{F}_{g_t}^+)$-martingale.

Proof: Following Azéma and Yor [38] closely, we project the $\mathbf{F}$-martingale B on $\mathcal{F}_{g_t}^+$. From the independence property of the meander and $\mathcal{F}_{g_t}^+$, we obtain

$$\mathbb{E}(B_t|\mathcal{F}_{g_t}^+) = \mathbb{E}(m_1^{(t)}\,\mu_t|\mathcal{F}_{g_t}^+) = \mu_t\mathbb{E}(m_1^{(t)}) = \sqrt{\frac{\pi}{2}}\,\mu_t\,. \tag{4.3.13}$$

Hence, $(\mu_t, t \geq 0)$ is an $(\mathcal{F}_{g_t}^+)$-martingale. In a second step, we project the $\mathbf{F}$-martingale $\exp(\lambda B_t - \frac{1}{2}\lambda^2 t)$ on the filtration $(\mathcal{F}_{g_t}^+)$:

$$\mathbb{E}(\exp(\lambda B_t - \frac{\lambda^2}{2}t)|\mathcal{F}_{g_t}^+) = \mathbb{E}\left(\exp(\lambda m_1^{(t)}\, \mu_t - \frac{\lambda^2}{2}t)|\mathcal{F}_{g_t}^+\right)$$

and, from the independence property of the meander and $\mathcal{F}_{g_t}^+$, we get

$$\mathbb{E}\left(\exp\left(\lambda B_t - \frac{\lambda^2}{2}t\right)|\mathcal{F}_{g_t}^+\right) = \exp\left(-\frac{\lambda^2}{2}t\right)\Psi(\lambda\mu_t)\,, \tag{4.3.14}$$

where Ψ is defined in (4.3.12) as

$$\Psi(z) = \mathbb{E}(\exp(zm_1)) = \int_0^\infty x\exp\left(zx - \frac{x^2}{2}\right)dx\,.$$

Obviously, the process in (4.3.14) is a $(\mathcal{F}_{g_t}^+)$-martingale. □

Comment 4.3.8.2 Some authors (e.g. Protter [726]) define the Azéma martingale as $\sqrt{\frac{\pi}{2}}\mu_t$, which is precisely the projection of the BM on the wide slow filtration, hence in further computations as in the next exercise, different multiplicative factors appear.

Note that the Azéma martingale is not continuous.

Exercise 4.3.8.3 Prove that the projection on the σ-algebra $\mathcal{F}_{g_t}^+$ of the **F**-martingale $(B_t^2 - t, t \geq 0)$ is $2(t - g_t) - t$, hence the process

$$\mu_t^2 - (t/2) = (t/2) - g_t$$

is an $(\mathcal{F}_{g_t}^+)$-martingale. ◁

4.3.9 Drifted Brownian Motion

We now study how our previous results are modified when working with a BM with drift. More precisely, we consider $X_t = x + \mu t + \sigma B_t$ with $\sigma > 0$. In order to simplify the proofs, we write $g^a(X)$ for $g_1^a(X) = \sup\{t \leq 1 : X_t = a\}$. The law of $g^a(X)$ may be obtained as follows

$$\begin{aligned} g^a(X) &= \sup\{t \leq 1 : \mu t + \sigma B_t = a - x\} \\ &= \sup\{t \leq 1 : \nu t + B_t = \alpha\}\,, \end{aligned}$$

where $\nu = \mu/\sigma$ and $\alpha = (a - x)/\sigma$. From Girsanov's theorem, we deduce

$$\mathbb{P}(g^a(X) \leq t) = \mathbb{E}\left(\mathbb{1}_{\{g^\alpha \leq t\}}\exp\left(\nu B_1 - \frac{\nu^2}{2}\right)\right), \tag{4.3.15}$$

where

$$g^\alpha = g_1^\alpha(B) = \sup\{t \leq 1 : B_t = \alpha\}.$$

Then, using that $|B_1| = m_1\sqrt{1-g_1}$ where m_1 is the value at time 1 of the Brownian meander,

$$\mathbb{P}(g^a(X) \le t) = \exp\Big(-\frac{\nu^2}{2}\Big)\,\mathbb{E}\Big(\mathbb{1}_{\{g^\alpha < t\}}\exp\big(\nu\epsilon m_1\sqrt{1-g^\alpha}\big)\Big) \tag{4.3.16}$$

where ϵ is a Bernoulli random variable; furthermore, the random variables g^α, ϵ, and m_1 are mutually independent. Therefore, since m_1 follows the Rayleigh law

$$\mathbb{P}(m_1 \in dy) = y\exp\left(-\frac{y^2}{2}\right)\mathbb{1}_{\{y\ge 0\}}\,dy\,,$$

we obtain

$$\begin{aligned}\mathbb{P}(g^a(X) \le t) &= \exp\Big(-\frac{\nu^2}{2}\Big)\int_0^t \frac{1}{\pi\sqrt{u(1-u)}}\exp\Big(-\frac{(a-x)^2}{2u\sigma^2}\Big)\Upsilon(\nu,u)\,du \\ &:= \Psi(x,a,t),\end{aligned} \tag{4.3.17}$$

where

$$\begin{aligned}\Upsilon(\nu,u) &= \mathbb{E}(\exp(\nu\epsilon m_1\sqrt{1-u})) \\ &= \frac{1}{2}\Big(\int_0^\infty e^{\nu y\sqrt{1-u}}ye^{-y^2/2}\,dy + \int_0^\infty e^{-\nu y\sqrt{1-u}}ye^{-y^2/2}\,dy\Big),\end{aligned}$$

that is,

$$\Upsilon(\nu,u) = \int_0^\infty \cosh(\nu y\sqrt{1-u})\,ye^{-y^2/2}\,dy.$$

Lemma 4.3.9.1 *Let $X_t = \nu t + B_t$. We have, for $t < 1$*

$$\mathbb{P}(g^a(X) > t|\mathcal{F}_t) = \mathbb{1}_{\{T_a(X)\le t\}}e^{\nu(\alpha - X_t)}H(\nu, |\alpha - X_t|, 1-t)\,,$$

where, for $y > 0$

$$H(\nu,y,s) = e^{-\nu y}\mathcal{N}\left(\frac{\nu s - y}{\sqrt{s}}\right) + e^{\nu y}\mathcal{N}\left(\frac{-\nu s - y}{\sqrt{s}}\right).$$

PROOF: From the absolute continuity relationship, we obtain, for $t < 1$

$$\mathbf{W}^{(\nu)}(g^a(X) \le t|\mathcal{F}_t) = \zeta_t^{-1}\,\mathbf{W}^{(0)}(\zeta_1\mathbb{1}_{\{g^a(X)\le t\}}|\mathcal{F}_t),$$

where

$$\zeta_t = \exp\big(\nu X_t - \frac{t\nu^2}{2}\big). \tag{4.3.18}$$

Therefore, from the equality

$$\{g^a(X) \le t\} = \{T_a(X) \le t\} \cap \{d_t^a(X) > 1\}$$

we obtain

$$\begin{aligned}&\mathbf{W}^{(0)}(\zeta_1 \mathbb{1}_{\{g^a \le t\}} | \mathcal{F}_t) \\ &= \exp\left(\nu X_t - \nu^2/2\right) \mathbb{1}_{\{T_a(X) \le t\}} \mathbf{W}^{(0)}\left(\exp[\nu(X_1 - X_t)] \mathbb{1}_{\{d_t^a(X) > 1\}} | \mathcal{F}_t\right).\end{aligned}$$

Using the independence properties of Brownian motion and equality (4.3.3), we get

$$\begin{aligned}&\mathbf{W}^{(0)}\left(\exp[\nu(X_1 - X_t)] \mathbb{1}_{\{d_t^a(X) > 1\}} | \mathcal{F}_t\right) \\ &= \mathbf{W}^{(0)}\left(\exp[\nu Z_{1-t}] \mathbb{1}_{\{T_{a-X_t}(Z) > 1-t\}} | \mathcal{F}_t\right) \\ &= \Theta(a - X_t, 1 - t)\end{aligned}$$

where $Z_t = X_1 - X_t \overset{\text{law}}{=} X_{1-t}$ is independent of $\mathcal{F}_t$ under $\mathbf{W}^{(0)}$ and

$$\Theta(x, s) := \mathbf{W}^{(0)}\left(e^{\nu X_s} \mathbb{1}_{\{T_x \ge s\}}\right) = e^{s\nu^2/2} - \mathbf{W}^{(0)}\left(e^{\nu X_s} \mathbb{1}_{\{T_x < s\}}\right).$$

By conditioning with respect to $\mathcal{F}_{T_x}$, we obtain (see Subsection 3.2.4 for the computation of H)

$$\begin{aligned}&\mathbf{W}^{(0)}\left(e^{\nu X_s} \mathbb{1}_{\{T_x < s\}}\right) \\ &= e^{\nu x} \mathbf{W}^{(0)}\left(\mathbb{1}_{\{T_x < s\}} e^{\frac{\nu^2}{2}(s - T_x)} \mathbf{W}^{(0)}\left(e^{\nu(X_s - X_{T_x}) - \frac{\nu^2}{2}(s - T_x)} | \mathcal{F}_{T_x}\right)\right) \\ &= e^{\nu x} \mathbf{W}^{(0)}\left(\mathbb{1}_{\{T_x < s\}} e^{\frac{\nu^2}{2}(s - T_x)}\right) = e^{\nu x + s\nu^2/2} H(\nu, |x|, s).\end{aligned}$$

Therefore,

$$\Theta(a - X_t, 1 - t) = e^{(1-t)\nu^2/2}(1 - e^{\nu(a - X_t)} H(\nu, |a - X_t|, 1 - t))$$

and

$$\begin{aligned}\mathbf{W}^{(\nu)}(g^a(X) \le t | \mathcal{F}_t) &= \mathbb{1}_{\{T_a(X) \le t\}} \exp\left(\frac{(t-1)\nu^2}{2}\right) \Theta(a - X_t, 1 - t) \\ &= \mathbb{1}_{\{T_a(X) \le t\}} \left(1 - e^{\nu(a - X_t)} H(\nu, |a - X_t|, 1 - t)\right).\end{aligned}$$

□

4.4 Parisian Options

In this section, our aim is to price an exotic option which we describe below, in a Black and Scholes framework: the underlying asset satisfies the stochastic differential equation

$$dS_t = S_t((r - \delta)\, dt + \sigma\, dW_t) \tag{4.4.1}$$

where W is a Brownian motion under the risk-neutral probability $\mathbb{Q}$, and w.l.g. $\sigma > 0$. In a closed form,

$$S_t = S_0 e^{\sigma X_t}$$

where $X_t = W_t + \nu t$ and $\nu = \frac{r-\delta}{\sigma} - \frac{\sigma}{2}$. The owner of an **up-and-out Parisian** option (UOPa) loses its value if the stock price reaches a level H (H is for **High**) and remains constantly above this level for a time interval longer than D (the delay). A **down-and-in** Parisian option (DIPa) is activated if the stock price falls below a **Low** level L and remains constantly below this level for a time interval longer than D. For a delay equal to zero, the Parisian option reduces to a standard barrier option. When the delay is extended beyond maturity, the UOPa option reduces to a standard European option. In the intermediate case, the option presents its "Parisian" feature and becomes a flexible financial tool which has some interesting properties: for instance, for some values of the parameters, when the underlying asset price is close to the out-barrier or when the size of the delay is small, its value is a decreasing function of the volatility. Therefore, it allows traders to bet in a simple manner on a decrease of volatility. Last but not least, as far as down-and-out barrier options are concerned, an influential agent in the market who has written such options and sees the price approaching the barrier may try to push the price further down, even momentarily and the cost of doing so may be smaller than the option payoff. In the case of Parisian options, this would be more difficult and expensive.

Parisian options, or more precisely Parisian times (the time when the option is activated or deactivated) are useful for modelling bankruptcy time; we note that following Chapter 11 of the United States Bankruptcy Code concerning reorganization of a business allows the firm to wait a certain time before being declared in bankruptcy.

For a generic continuous process Y and a given $t > 0$, we introduce $g_t^b(Y)$, the last time before t at which the process Y was at level b, i.e.,

$$g_t^b(Y) = \sup\{s \le t : Y_s = b\}.$$

For an UOPa option we need to consider the first time at which the underlying asset S is above H for a period greater than D, i.e.,

$$\begin{aligned} G_D^{+,H}(S) &= \inf\{t > 0 : (t - g_t^H(S))\mathbb{1}_{\{S_t > H\}} \ge D\} \\ &= \inf\{t > 0 : (t - g_t^h(X))\mathbb{1}_{\{X_t > h\}} \ge D\} = G_D^{+,h}(X) \end{aligned}$$

where $h = \ln(H/S_0)/\sigma$. If this stopping time occurs before the maturity then the UOPa option is worthless. The price of an UOPa call option is

$$\begin{aligned} \mathrm{UOPa}(S_0, H, D; T) &= E_{\mathbb{Q}}\left(e^{-rT}(S_T - K)^+ \mathbb{1}_{\{G_D^{+,H}(S) > T\}}\right) \\ &= E_{\mathbb{Q}}\left(e^{-rT}(S_0 e^{\sigma X_T} - K)^+ \mathbb{1}_{\{G_D^{+,h}(X) > T\}}\right) \end{aligned}$$

or, using a change of probability (see Example 1.7.5.5)

$$\mathrm{UOPa}(S_0, H, D; T) = e^{-(r+\nu^2/2)T}\mathbb{E}\left(e^{\nu W_T}(S_0 e^{\sigma W_T} - K)^+ \mathbb{1}_{\{G_D^{+,h}(W)>T\}}\right),$$

where W is a Brownian motion. The sum of the prices of an up-and-in (UIPa) and an UOPa option with the same strike and delay is obviously the price of a plain-vanilla European call.

In the same way, the value of a DIPa option with level L is defined using

$$G_D^{-,L}(S) = \inf\{t > 0 : (t - g_t^L(S))\mathbb{1}_{\{S_t<L\}} \geq D\}$$

which equals, in terms of X,

$$G_D^{-,\ell}(X) = \inf\{t > 0 : (t - g_t^\ell(X))\mathbb{1}_{\{X_t<\ell\}} \geq D\}$$

with $\ell = \frac{1}{\sigma}\ln(L/S_0)$. Then, the value of a DIPa option is equal to

$$\begin{aligned}
\mathrm{DIPa}(S_0, L, D; T) &= E_{\mathbb{Q}}\left(e^{-rT}(S_T - K)^+ \mathbb{1}_{\{G_D^{-,L}(S)<T\}}\right) \\
&= e^{-(r+\nu^2/2)T}\mathbb{E}\left(e^{\nu W_T}(S_0 e^{\sigma W_T} - K)^+ \mathbb{1}_{\{G_D^{-,\ell}(W)<T\}}\right) \\
&:= e^{-(r+\nu^2/2)T}\,{}^\star\mathrm{DIPa}(S_0, L, D; T),
\end{aligned}$$

where in this section, we define the general "star" transformation of a function f as

$${}^\star f(t) = e^{(r+\nu^2/2)t} f(t).$$

In the case $S_0 > L$, the computation of ${}^\star\mathrm{DIPa}(S_0, L, D; T)$ can be reduced to the case $L = S_0$, i.e., $\ell = 0$. Indeed, for the option to be activated, the level L has to be reached by the process S (or equivalently, the level ℓ has to be reached by the process W) before the maturity T. Therefore, introducing $T_\ell = \inf\{t : W_t = \ell\}$, we obtain

$$\begin{aligned}
&{}^\star\mathrm{DIPa}(S_0, L, D; T) = \mathbb{E}(e^{\nu W_T}(S_0 e^{\sigma W_T} - K)^+ \mathbb{1}_{\{G_D^{-,\ell}(W)<T\}}) \\
&= \mathbb{E}\left(e^{\nu(W_T - W_{T_\ell} + \ell)}(S_0 e^{\sigma(W_T - W_{T_\ell} + \ell)} - K)^+ \mathbb{1}_{\{G_D^{-,\ell}(W)<T\}}\right) \\
&= e^{\nu\ell}\mathbb{E}\left(e^{\nu Z_{T-T_\ell}}(S_0 e^{\sigma(\ell + Z_{T-T_\ell})} - K)^+ \mathbb{1}_{\{G_D^{-,0}(Z)<T-T_\ell\}}\right)
\end{aligned}$$

where $Z_t = W_{t+T_\ell} - W_{T_\ell}$ is a BM independent of T_ℓ. Let us now introduce F_ℓ, the cumulative distribution function of T_ℓ.

$$\begin{aligned}
&{}^\star\mathrm{DIPa}(S_0, L, D; T) \\
&= e^{\nu\ell}\int_0^T dF_\ell(u)\mathbb{E}(e^{\nu Z_{T-u}}(S_0 e^{\sigma(Z_{T-u}+\ell)} - K)^+ \mathbb{1}_{\{G_D^{-,0}(Z)<T-u\}}) \\
&= e^{\nu\ell}\int_0^T dF_\ell(u)\,{}^\star\mathrm{DIPa}(S_0, S_0, D; T-u).
\end{aligned}$$

We have used the fact that the computation of the law of the Parisian time below a level ℓ for a Brownian motion starting at level ℓ reduces to the law of the Parisian time below level 0 for a standard Brownian motion (starting from 0). Nevertheless, in the next subsection, we shall present a different approach.

4.4.1 The Law of $(G_D^{-,\ell}(W), W_{G_D^{-,\ell}})$

In a first step, we compute the law of the pair (Parisian time, Brownian motion at the Parisian time) for a level $\ell = 0$.

Proposition 4.4.1.1 *Let W be a Brownian motion and $G_D^- := G_D^{-,0}(W)$. The random variables G_D^- and $W_{G_D^-}$ are independent and*

$$\mathbb{P}(W_{G_D^-} \in dx) = \frac{-x}{D} \exp\left(-\frac{x^2}{2D}\right) \mathbb{1}_{\{x<0\}}\, dx, \tag{4.4.2}$$

$$\mathbb{E}\left(\exp\left(-\frac{\lambda^2}{2} G_D^-\right)\right) = \frac{1}{\Psi(\lambda\sqrt{D})} \tag{4.4.3}$$

where $\Psi(z) = \int_0^\infty x \exp\left(zx - \frac{x^2}{2}\right) dx = 1 + z\sqrt{2\pi}\mathcal{N}(z)e^{z^2/2}$.

PROOF: We have defined in Subsection 4.3.6 the wide slow Brownian filtration $(\mathcal{F}_{g_t}^+, t \geq 0)$. The r.v. G_D^- is an $(\mathcal{F}_{g_t}^+, t \geq 0)$- hence an $(\mathcal{F}_t, t \geq 0)$- stopping time. From results on meanders recalled in Subsection 4.3.7, the process

$$\left(\frac{1}{\sqrt{D}} |W_{g_{G_D^-}} + uD| \,, u \leq 1\right)$$

is a Brownian meander independent of $\mathcal{F}_{g_{G_D^-}}^+$, since $G_D^- = g_{G_D^-} + D$, the r.v. $\frac{1}{\sqrt{D}} W_{G_D^-}$ is distributed as $-m_1$, hence

$$\mathbb{P}(W_{G_D^-} \in dx) = \frac{-x}{D} \exp\left(-\frac{x^2}{2D}\right) \mathbb{1}_{\{x<0\}}\, dx,$$

and the variables G_D^- and $W_{G_D^-}$ are independent. From Proposition 4.3.8.1, the process

$$\Psi(-\lambda\mu_{t\wedge G_D^-}) \exp\left(-\frac{\lambda^2}{2}(t \wedge G_D^-)\right) \,, t \geq 0\,,$$

(where μ denotes the Azéma martingale) is a $\mathcal{F}_{g_t}^+$-local martingale. Since, for $\lambda > 0$, $0 < -\lambda\mu_{t\wedge G_D^-} < \lambda D$, this process is bounded. Hence, using the optional sampling theorem at G_D^-, we obtain

$$\mathbb{E}\left(\Psi(-\lambda\mu_{G_D^-}\right) \exp(\left(-\frac{\lambda^2}{2}\, G_D^-\right) = \Psi(0) = 1$$

and the left-hand side equals $\Psi(\lambda\sqrt{D})\,\mathbb{E}(\exp(-\frac{\lambda^2}{2}G_D^-))$. The formula

$$\mathbb{E}\left(\exp\left(-\frac{\lambda^2}{2}G_D^-\right)\right) = \frac{1}{\Psi(\lambda\sqrt{D})}$$

follows. □

From the above proposition, we can easily deduce the law of the pair $(G_D^{-,\ell}, W_{G_D^{-,\ell}})$ in the case $\ell < 0$, as we now show.

Corollary 4.4.1.2 *Let $\ell < 0$. The random variables $G_D^{-,\ell}$ and $W_{G_D^{-,\ell}}$ are independent and their laws are given by*

$$\mathbb{P}(W_{G_D^{-,\ell}} \in dx) = \frac{dx}{D}\,\mathbb{1}_{\{x<\ell\}}\,(\ell - x)\exp\left(-\frac{(x-\ell)^2}{2D}\right) \quad (4.4.4)$$

$$\mathbb{E}\left(\exp\left(-\frac{\lambda^2}{2}G_D^{-,\ell}\right)\right) = \frac{\exp(\ell\lambda)}{\Psi(\lambda\sqrt{D})}\,. \quad (4.4.5)$$

PROOF: This study may be reduced to the previous one, with the help of the stopping time $T_\ell = T_\ell(W)$. Since

$$G_D^{-,\ell} = T_\ell + \widehat{G}_D^-$$

where

$$\widehat{G}_D^- = \inf\{t \geq 0 : \mathbb{1}_{\{\widehat{W}_t \leq 0\}}(t - g_t^0(\widehat{W})) \geq D\}$$

with $\widehat{W}_t = W_{T_\ell + t} - W_{T_\ell}$, it follows, from the independence between T_ℓ and $\widehat{G}_D^-$, that

$$\mathbb{E}\left(\exp\left(-\frac{\lambda^2}{2}G_D^{-,\ell}\right)\right) = \mathbb{E}\left(\exp\left(-\frac{\lambda^2}{2}T_\ell\right)\right)\mathbb{E}\left(\exp\left(-\frac{\lambda^2}{2}\widehat{G}_D^-\right)\right).$$

The Laplace transform of the hitting time T_ℓ is known (see Proposition 3.1.6.1) and $\widehat{G}_D^- \stackrel{\text{law}}{=} G_D^-$; hence, by application of equality (4.4.3)

$$\mathbb{E}\left(\exp\left(-\frac{\lambda^2}{2}G_D^{-,\ell}\right)\right) = \frac{\exp(\ell\lambda)}{\Psi(\lambda\sqrt{D})}\,.$$

We obtain finally from (4.4.2) that

$$\begin{aligned}\mathbb{P}(W_{G_D^{-,\ell}} \in dx) &= \mathbb{P}(\widehat{W}_{\widehat{G}_D^-} - \ell \in dx - \ell) \\ &= \frac{dx}{D}\,\mathbb{1}_{\{x<\ell\}}\,(\ell - x)\exp\left(-\frac{(x-\ell)^2}{2D}\right).\end{aligned}$$

Note in particular that, since $\Psi(0) = 1$, $\mathbb{P}(G_D^{-,\ell} < \infty) = 1$. □

Proposition 4.4.1.3 *In the case* $\ell > 0$*, the random variables* $G_D^{-,\ell}$ *and* $W_{G_D^{-,\ell}}$ *are independent. Their laws are characterized by*

$$\mathbb{E}(\exp(-\lambda G_D^{-,\ell})) = e^{-\lambda D}(1 - F_\ell(D)) + \frac{1}{\Psi(\sqrt{2\lambda D})} H(\sqrt{2\lambda}, \ell, D),$$

where F_ℓ *is the cumulative distribution function of* T_ℓ *and the function* H *is defined in* (3.2.7)*, and*

$$\mathbb{P}(W_{G_D^{-,\ell}} \in dx) = \mathbb{1}_{\{x \le \ell\}} dx \left[e^{-(x-\ell)^2/(2D)} \mathbb{P}(T_\ell < D) \frac{\ell - x}{D} + \frac{1}{\sqrt{2\pi D}} \left(e^{-x^2/(2D)} - e^{-(x-2\ell)^2/(2D)} \right) \right].$$

PROOF: In the case $\ell > 0$, the first excursion below ℓ begins at $t = 0$. We now use the obvious equality

$$\mathbb{E}(\exp(-\lambda G_D^{-,\ell})) = \mathbb{E}(\mathbb{1}_{\{T_\ell < D\}} \exp(-\lambda G_D^{-,\ell})) + \mathbb{E}(\mathbb{1}_{\{T_\ell > D\}} \exp(-\lambda G_D^{-,\ell})).$$

On the set $\{T_\ell > D\}$, we have $G_D^{-,\ell} = D$. Therefore,

$$\begin{aligned}\mathbb{E}(\mathbb{1}_{\{T_\ell > D\}} \exp(-\lambda G_D^{-,\ell})) &= \exp(-\lambda D) \mathbb{P}(T_\ell > D) \\ &= \exp(-\lambda D)\,(1 - F_\ell(D)).\end{aligned}$$

Here, F_ℓ is the cumulative distribution function of T_ℓ (see formula 3.1.6). On the set $\{T_\ell < D\}$, we write, as in the proof of the previous corollary, $G_D^{-,\ell} = T_\ell + \widehat{G}_D^-$. Hence, on $(T_\ell < D)$, we have:

$$\mathbb{E}\left(\exp(-\lambda G_D^{-,\ell}) \,|\, \mathcal{F}_{T_\ell}\right) = \exp(-\lambda T_\ell)\, \mathbb{E}\left(\exp(-\lambda \widehat{G}_D^-)\right).$$

Therefore, $\mathbb{E}\big(\mathbb{1}_{\{T_\ell < D\}} \exp(-\lambda G_D^{-,\ell})\big) = \dfrac{1}{\Psi(\sqrt{2\lambda D})} \mathbb{E}(\mathbb{1}_{\{T_\ell < D\}} \exp(-\lambda T_\ell))$. The quantity $\mathbb{E}(\mathbb{1}_{\{T_\ell < D\}} \exp(-\lambda T_\ell))$ has been computed in Subsection 3.2.4, and is equal to $H(\sqrt{2\lambda}, \ell, D)$ (see formula (3.2.7)).

It follows that

$$\mathbb{E}(\exp(-\lambda G_D^{-,\ell})) = e^{-\lambda D}(1 - F_\ell(D)) + \frac{1}{\Psi(\sqrt{2\lambda D})} H(\sqrt{2\lambda}, \ell, D).$$

The law of $W_{G_D^{-,\ell}}$ can easily be deduced from the following three equalities:

$$W_{G_D^{-,\ell}} = (\ell + \widehat{W}_{\widehat{G}_D^-}) \mathbb{1}_{\{T_\ell < D\}} + W_D \mathbb{1}_{\{T_\ell > D\}}$$

$$\mathbb{P}(\ell + \widehat{W}_{\widehat{G}_D^-} \in dx, T_\ell < D) = \mathbb{P}(T_\ell < D) \mathbb{1}_{\{x \le \ell\}} (\ell - x) \exp\left(-\frac{(x-\ell)^2}{2D}\right) \frac{dx}{D}$$

$$\mathbb{P}(W_D \in dx, T_\ell > D) = \frac{dx}{\sqrt{2\pi D}} \left(\exp\left(-\frac{x^2}{2D}\right) - \exp\left(-\frac{(x-2\ell)^2}{2D}\right)\right) \mathbb{1}_{\{x \leq \ell\}} \,.$$

□

Comment 4.4.1.4 The independence property of a stopping time τ and the position of the Brownian motion at that time B_τ is a fairly rare phenomenon for Brownian stopping times; it is satisfied for $\tau = G_D^{-,\ell}$. It can be proved, for example that, if T is a bounded stopping time such that T and W_T are independent, then T is a constant. A more general study of stopping times which enjoy this independence property can be found in De Meyer et al. [246, 247]. See also the following exercise.

Exercise 4.4.1.5 Let $T_a^* = \inf\{t : |W_t| = a\}$. Prove that the r.v's T_a^* and $W_{T_a^*}$ are independent and show that $W_{T_a^*}$ is symmetric with values $\pm a$. See Section 3.5. ◁

4.4.2 Valuation of a Down-and-In Parisian Option

We have seen that the price of a down-and-in Parisian option is given by

$$\mathrm{DIPa}(S_0, L, D; T) = e^{-(r+\nu^2/2)T}\,{}^\star\mathrm{DIPa}(S_0, L, D; T)$$

where

$${}^\star\mathrm{DIPa}(S_0, L, D; T) = \mathbb{E}\left(\mathbb{1}_{\{G_D^{-,\ell} \leq T\}} \mathbb{E}\left(e^{\nu W_T}(S_0 e^{\sigma W_T} - K)^+ | \mathcal{F}_{G_D^{-,\ell}}\right)\right) .$$

From the strong Markov property

$${}^\star\mathrm{DIPa}(S_0, L, D; T) = \mathbb{E}(\mathbb{1}_{\{G_D^{-,\ell} \leq T\}} \mathcal{P}_{T - G_D^{-,\ell}}(\psi)(W_{G_D^{-,\ell}}))$$

with

$$\begin{cases} \psi(y) = e^{\nu y}(S_0 e^{\sigma y} - K)^+, \\ \mathcal{P}_t f(z) = \dfrac{1}{\sqrt{2\pi t}} \displaystyle\int_{-\infty}^{\infty} f(y) \exp\left(-\frac{(y-z)^2}{2t}\right) dy \,. \end{cases}$$

Denote by φ the density of $W_{G_D^{-,\ell}}$ and recall that $G_D^{-,\ell}$ and $W_{G_D^{-,\ell}}$ are independent. Then,

$$\begin{aligned} {}^\star\mathrm{DIPa}(S_0, L, D; T) &= \textstyle\int_{-\infty}^{\infty} \varphi(dz)\, \mathbb{E}(\mathbb{1}_{\{G_D^{-,\ell} \leq T\}} \mathcal{P}_{T - G_D^{-,\ell}}(\psi)(z)) \\ &= \textstyle\int_{-\infty}^{\infty} dy\, \psi(y)\, h_\ell(T, y) \end{aligned} \tag{4.4.6}$$

where the function h_ℓ is defined by

$$h_\ell(t, y) = \int_{-\infty}^{\infty} \varphi(dz)\, \gamma(t, y - z) \tag{4.4.7}$$

with

$$\gamma(t,x) = \mathbb{E}\left(\frac{\mathbb{1}_{\{G_D^{-,\ell} \le t\}}}{\sqrt{2\pi(t - G_D^{-,\ell})}} \exp\left(-\frac{x^2}{2(t - G_D^{-,\ell})}\right)\right).$$

Then, replacing ψ by its value, we obtain

$$^\star\mathrm{DIPa}(S_0, L; D, T) = \int_k^\infty dy\, e^{\nu y}(S_0 e^{\sigma y} - K) h_\ell(T, y) \tag{4.4.8}$$

where $k = \frac{1}{\sigma} \ln \frac{K}{S_0}$. The computation of this quantity relies on the knowledge of h_ℓ; however this function h_ℓ is only known through its time Laplace transform $\widehat{h}_\ell$ which is given in the following two theorems.

Theorem 4.4.2.1 *In the case $S_0 > L$ (i.e., $\ell < 0$) the function $t \to h_\ell(t,y)$ is characterized by its Laplace transform: for $\lambda > 0$,*

$$\widehat{h}_\ell(\lambda, y) = \frac{e^{\ell\sqrt{2\lambda}}}{D\sqrt{2\lambda}\,\Psi(\sqrt{2\lambda D})} \int_0^\infty dz\, z \exp\left(-\frac{z^2}{2D} - |y + z - \ell|\sqrt{2\lambda}\right)$$

where $\Psi(z)$ is defined in (4.3.12). *If $y > \ell$, then*

$$\widehat{h}_\ell(\lambda, y) = \frac{\Psi(-\sqrt{2\lambda D})}{\Psi(\sqrt{2\lambda D})} \frac{e^{(2\ell - y)\sqrt{2\lambda}}}{\sqrt{2\lambda}}.$$

PROOF: In the case $S_0 > L$, from (4.4.4), the density φ of $W_{G_D^{-,\ell}}$ is

$$\varphi(x) = \mathbb{P}(W_{G_D^{-,\ell}} \in dx)/dx = \frac{1}{D}(\ell - x) \exp\left(-\frac{(x-\ell)^2}{2D}\right) \mathbb{1}_{\{x \le \ell\}}.$$

The function h_ℓ is defined in terms of φ and γ. Thus, the knowledge of the Laplace transform of γ will lead to the knowledge of $\widehat{h}_\ell$.

For $\lambda > 0$, we obtain, with an obvious change of variable,

$$\begin{aligned}\int_0^\infty dt\, e^{-\lambda t}\gamma(t,x) &= \mathbb{E}\left[\int_{G_D^{-,\ell}}^\infty dt \frac{e^{-\lambda t}}{\sqrt{2\pi(t - G_D^{-,\ell})}} \exp\left(-\frac{x^2}{2(t - G_D^{-,\ell})}\right)\right] \\ &= \mathbb{E}(e^{-\lambda G_D^{-,\ell}}) \int_0^\infty dt \exp\left(-\frac{x^2}{2t}\right) \frac{e^{-\lambda t}}{\sqrt{2\pi t}}.\end{aligned} \tag{4.4.9}$$

The integral on the right of (4.4.9) is the resolvent kernel of Brownian motion and is equal to $\frac{1}{\sqrt{2\lambda}} e^{-|x|\sqrt{2\lambda}}$. By substituting this result in (4.4.9) and using the Laplace transform of $G_D^{-,\ell}$ given in (4.4.5), we obtain:

$$\int_0^\infty dt\, e^{-\lambda t}\gamma(t,x) = \frac{e^{-(|x|-\ell)\sqrt{2\lambda}}}{\sqrt{2\lambda}\,\Psi(\sqrt{2\lambda D})}. \tag{4.4.10}$$

Therefore, from the definition (4.4.7) of h_ℓ, its Laplace transform, $\widehat{h}_\ell(\lambda,y)$ is given by

$$\int_0^\infty dt\, e^{-\lambda t} h_\ell(t,y) = \int_{-\infty}^{\ell} \frac{dz}{D}(\ell-z)e^{-\frac{(\ell-z)^2}{2D}} \int_0^\infty dt\, e^{-\lambda t}\gamma(t,y-z)$$
$$= \int_0^\infty \frac{du}{D}\, u e^{-\frac{u^2}{2D}} \int_0^\infty dt\, e^{-\lambda t}\gamma(t,y+u-\ell)$$

$$= \frac{e^{\ell\sqrt{2\lambda}}}{D\sqrt{2\lambda}\,\Psi(\sqrt{2\lambda D})} \int_0^\infty du\, u \exp\left(-\frac{u^2}{2D} - |y+u-\ell|\sqrt{2\lambda}\right).$$

The corresponding integral

$$K_{\lambda,D}(a) := \frac{1}{D}\int_0^\infty du\, u \exp\left(-\frac{u^2}{2D} - |u+a|\sqrt{2\lambda}\right)$$

can be easily evaluated as follows.

▶ If $a > 0$, using the change of variables $u = z\sqrt{D}$, we obtain

$$K_{\lambda,D}(a) = \exp(-a\sqrt{2\lambda})\Psi(-\sqrt{2\lambda D})$$

and this leads to the formula for $y > \ell$.

▶ If $a < 0$, a similar method leads to

$$K_{\lambda,D}(a) = e^{a\sqrt{2\lambda}} + 2\sqrt{\pi\lambda D}e^{\lambda D}$$
$$\times\left(e^{a\sqrt{2\lambda}}\left[\mathcal{N}\left(\frac{-a}{\sqrt{D}}-\sqrt{2\lambda D}\right) - \mathcal{N}\left(-\sqrt{2\lambda D}\right)\right] - e^{-a\sqrt{2\lambda}}\mathcal{N}\left(\frac{a}{\sqrt{D}}-\sqrt{2\lambda D}\right)\right).$$

As a partial check, note that if $D = 0$, the Parisian option is a standard barrier option. The previous computation simplifies and we obtain

$$\widehat{h}_\ell(\lambda,y) = \frac{e^{\ell\sqrt{2\lambda}}}{\sqrt{2\lambda}} e^{-|\ell-y|\sqrt{2\lambda}}.$$

It is easy to invert $\widehat{h}_\ell$ and we are back to the formula (3.6.28) for the price of a DIC option obtained in Theorem 3.6.6.2 . □

Remark 4.4.2.2 The quantity $\Psi(-\sqrt{2\lambda D})$ is a Laplace transform, as well as the quantity $\frac{e^{(2\ell-y)\sqrt{2\lambda}}}{\sqrt{2\lambda}}$. Therefore, in order to invert $\widehat{h}_\ell$ in the case $y > \ell$, it suffices to invert the Laplace transform $\frac{1}{\Psi(\sqrt{2\lambda D})}$. This is not easy: see Schröder [770] for some computation.

Theorem 4.4.2.3 *In the case $S_0 < L$ (i.e., $\ell > 0$), the function $h_\ell(t,y)$ is characterized by its Laplace transform, for $\lambda > 0$,*

$$\widehat{h}_\ell(\lambda, y) = \widehat{g}(t,y) + \frac{1}{D\sqrt{2\lambda}\,\Psi(\sqrt{2\lambda D})} H(\sqrt{2\lambda}, \ell, D) \int_0^\infty dz\, z\, \exp\left(-\frac{z^2}{2D} - |y - \ell + z|\sqrt{2\lambda}\right).$$

where g is defined in the following equality (4.4.12), and H is defined in (3.2.7).

PROOF: In the case $\ell > 0$, the Laplace transform of $h_\ell(.,y)$ is more complicated. Denoting again by φ the law of $W_{G_D^{-,\ell}}$, we obtain

$$\int_0^\infty dt\, e^{-\lambda t} h_\ell(t,y) = \mathbb{E}\left(\int_{-\infty}^\infty \varphi(dz)\, e^{-\lambda G_D^{-,\ell}} \frac{1}{\sqrt{2\lambda}} \exp(-|y-z|\sqrt{2\lambda})\right).$$

Using the previous results, and the cumulative distribution function F_ℓ of T_ℓ,

$$\int_0^\infty dt\, e^{-\lambda t} h_\ell(t,y) = \tag{4.4.11}$$
$$\frac{1}{D\sqrt{2\lambda}\,\Psi(\sqrt{2\lambda D})} \int_0^\infty dz\, z\, \exp\left(-\frac{z^2}{2D} - |y-\ell+z|\sqrt{2\lambda}\right) \int_0^D F_\ell(dx)\, e^{-\lambda x}$$
$$+ \frac{e^{-\lambda D}}{2\sqrt{\lambda \pi D}} \int_{-\infty}^\ell dz \left(\exp\left(-\frac{z^2}{2D}\right) - \exp - \frac{(z-2\ell)^2}{2D}\right) e^{-|y-z|\sqrt{2\lambda}}.$$

We know from Remark 3.1.6.3 that $\frac{1}{\sqrt{2\lambda}} \exp(-|a|\sqrt{2\lambda})$ is the Laplace transform of $\frac{1}{\sqrt{2\pi t}} \exp(-\frac{a^2}{2t})$. Hence, the second term on the right-hand side of (4.4.11) is the time Laplace transform of $g(\cdot, y)$ where

$$g(t,y) = \frac{\mathbb{1}_{\{t>D\}}}{2\pi\sqrt{D(t-D)}} \int_{-\infty}^\ell e^{\frac{(y-z)^2}{2(t-D)}} \left(e^{-\frac{z^2}{2D}} - e^{-\frac{(z-2\ell)^2}{2D}}\right) dz. \tag{4.4.12}$$

We have not be able go further in the inversion of the Laplace transform.

A particular case: If $y > \ell$, the first term on the right-hand side of (4.4.11) is equal to

$$\frac{\Psi(-\sqrt{2\lambda D})}{\Psi(\sqrt{2\lambda D})} \frac{e^{-(y-\ell)\sqrt{2\lambda}}}{\sqrt{2\lambda}} \int_0^D F_\ell(dx) e^{-\lambda x}.$$

This term is the product of four Laplace transforms; however, the inverse transform of $\frac{1}{\Psi(\sqrt{2\lambda D})}$ is not identified. □

Comment 4.4.2.4 Parisian options are studied in Avellaneda and Wu [30], Chesney et al. [175], Cornwall et al. [196], Dassios [213], Gauthier [376] and Haber et al. [415]. Numerical analysis is carried out in Bernard et al. [76], Costabile [198], Labart and Lelong [556] and Schröder [770]. An approximation with an implied barrier is done in Anderluh and Van der Weide [14]. Double-sided Parisian options are presented in Anderluh and Van der Weide [15], Dassios and Wu [215, 216, 217] and Labart and Lelong [557]. The "Parisian" time models a default time in Çetin et al. [158] and in Chen and Suchanecki [162, 163]. Cumulative Parisian options are developed in Detemple [252], Hugonnier [451] and Moraux [657]. Their Parisian name is due to their birth place as well as to the meanders of the Seine River which lead many tourists to excursions around Paris.

Exercise 4.4.2.5 We have just introduced Parisian down-and-in options with a call feature, denoted here C_{DIPa}. One can also define Parisian up-and-in options P_{UIPa} with a put feature, i.e., with payoff $(K - S_T)^+ \mathbb{1}_{\{G_D^{+,L} < T\}}$. Prove the symmetry formula

$$C_{\mathrm{DIPa}}(S_0, K, L; r, \delta; D, T) = K S_0 P_{\mathrm{UIPa}}(S_0^{-1}, K^{-1}, L^{-1}, \delta, r; D, T) .$$

◁

4.4.3 PDE Approach

In Haber et al. [415] and in Wilmott [846], the following PDE approach to valuation of Parisian option is presented, in the case $\delta = 0$. The value at time t of a down-and-out Parisian option is a function of the three variables $t, S_t, t - g_t$, i.e., $\mathrm{DOPa} = \Phi(T - t, S_t, t - g_t)$ and the discounted price process $e^{-rt}\Phi(T - t, S_t, t - g_t)$ is a $\mathbb{Q}$-martingale. Using the fact that $(g_t, t \geq 0)$ is an increasing process, Itô's calculus gives

$$d[e^{-rt}\Phi(t, S_t, t - g_t)] = e^{-rt}\Big[-r\Phi dt + (\partial_t \Phi)\, dt + (\partial_x \Phi)\, dS_t + (\partial_u \Phi)\,(dt - dg_t) + \frac{1}{2}\sigma^2 S_t^2 (\partial_{xx}\Phi)\, dt\Big]$$

between two jumps of g_t. (Here, u is the third variable of the function Φ). Therefore, the dt terms must sum to 0 giving

$$\begin{cases} -r\Phi + \partial_t \Phi + x r \partial_x \Phi + \partial_u \Phi + \frac{1}{2}\sigma^2 x^2 \partial_{xx}\Phi = 0, & \text{for } u < D \\ \partial_u \Phi(t, x, 0) = 0 . & \end{cases}$$

with the boundary conditions

$$\begin{cases} \Phi(t, x, u) = \Phi(t, x, 0), & \text{for } x \geq L \\ \Phi(t, x, u) = 0, & \text{for } u \geq D,\ x < L . \end{cases}$$

4.4.4 American Parisian Options

American Parisian options are also considered. Grau [403] combined Monte Carlo simulations and PDE solvers (see also Grau and Kallsen [404]) in order to price European and American Parisian options. The PDE approaches developed by Haber et al. [415] and Wilmott [846] can also be used in order to value these options. In the same setting, where the risk-neutral dynamics of the underlying are given by (4.4.1), Chesney and Gauthier [172] developed a probabilistic approach for the pricing of American Parisian options. They derived the following result for currency options:

Proposition 4.4.4.1 *The price of an American Parisian down-and-out call* (ADOPa) *can be decomposed as follows:*

$$\begin{aligned}
&\mathrm{ADOPa}\,(S_0, L, D, T) = \mathrm{DOPa}\,(S_0, L, D, T) \\
&+ \delta S_0 \int_0^T e^{-\alpha u} \mathbb{E}\left[\mathbb{1}_{\{W_u \geq \bar{b}(u)\}} \mathbb{1}_{\{u < G_D^{-,\ell}(W)\}} \exp\left((\nu+\sigma) W_u\right)\right] du \\
&- rK \int_0^T e^{-\alpha u} \mathbb{E}\left[\mathbb{1}_{\{W_u \geq \bar{b}(u)\}} \mathbb{1}_{\{u < G_D^{-,\ell}(W)\}} \exp\left(\nu W_u\right)\right] du
\end{aligned}$$

where

$$\begin{aligned}
&\alpha = r + \frac{\nu^2}{2}, \ \nu = \frac{1}{\sigma}\left(r - \delta - \frac{\sigma^2}{2}\right), \ \bar{b}(u) = \frac{1}{\sigma} \ln\left(\frac{b_c(T-u)}{S_0}\right), \\
&\ell = \frac{1}{\sigma} \ln\left(\frac{L}{S_0}\right) \leq 0
\end{aligned}$$

and where $\{b_c(T-u), u \in [0,T]\}$ *is the exercise boundary (see Section* 3.11 *for the general definition). Here, the process* W *is a Brownian motion.*

This decomposition can also be written as follows:

$$\begin{aligned}
&\mathrm{ADOPa}\,(S_0, L, D, T) = \mathrm{DOPa}\,(S_0, L, D, T) + \delta \int_0^T \mathrm{DOPa}(S_0, b_c(T-u), u) du \\
&+ \delta \int_0^T \left(b_c(T-u) - \frac{r}{\delta} K\right) \mathrm{BinDOCPa}\,(S_0, b_c(T-u), u)\, du
\end{aligned}$$

where $\mathrm{DOPa}(S_0, b_c(T-u), u)$ is the price of the European Parisian down-and-out call option with maturity u, strike price $b_c(T-u)$, barrier L and delay D, $\mathrm{BinDOCPa}\,(S_0, b_c(T-u), u)$ is the price of a Parisian binary call (see Subsections 3.6.2 and 3.6.3 for the definitions of binary calls and binary barrier options) which generates at maturity a pay-off of one monetary unit if the underlying value is higher than the strike price and if the first instant –when the underlying price spends consecutively more than D units of time under the level L_1 – is greater than the maturity u. Otherwise, the payoff is equal to zero.

Denote by ADOPa (S_0, L, D) the price of a perpetual American Parisian option. The following proposition is obtained:

Proposition 4.4.4.2 *The price of a perpetual American Parisian down-and-out call is given by:*

$$\begin{aligned}\text{ADOPa}\,(S_0, L, D) &= \delta \int_0^{+\infty} \text{DOPa}(S_0, L_c, u)\, du \\ &\quad + \delta \int_0^{+\infty} \left(L_c - \frac{r}{\delta} K\right) \text{BinDOCPa}\,(S_0, L_c, u)\, du\end{aligned}$$

or

$$\text{ADOPa}(S_0, L, D) = \left(1 - \frac{\Psi(-\kappa\sqrt{D})}{\Psi(\kappa\sqrt{D})} e^{2\ell\kappa}\right) \frac{1}{\sigma\kappa} \left(\frac{S_0}{L_c}\right)^{\gamma} \left(\frac{\delta L_c}{\gamma - 1} - \frac{r}{\gamma} K\right)$$

with $\kappa = \sqrt{2r + \nu^2}$, $\gamma = \frac{-\nu + \sqrt{2r+\nu^2}}{\sigma}$ *and where the exercise boundary* L_c *is defined implicitly by:*

$$L_c - K = \left(1 - \frac{\Psi(-\kappa\sqrt{D})}{\Psi(\kappa\sqrt{D})} \left(\frac{L}{L_c}\right)^{2\frac{\kappa}{\sigma}}\right) \frac{1}{\sigma\kappa} \left(\frac{\delta L_c}{\gamma - 1} - \frac{r}{\gamma} K\right)$$

where the function Ψ *is defined in equation* (4.3.12).

Solutions when the excursion has already started and for the "in" barrier case are also derived. The latter case is easier to analyze. Indeed, in this setting, the option holder cannot do or decide anything before the option is activated; once the option is activated then it does not have a barrier anymore, but is just a plain vanilla American call. The exercise frontier for an American Parisian "in" barrier option is therefore the exercise frontier of the corresponding plain vanilla option, starting at the activation time.

5

Complements on Continuous Path Processes

In this chapter, we present the important notion of time change, which will be crucial when studying applications to finance in a Lévy process setting. We then introduce the operation of dual predictable projection, which will be an important tool when working with the reduced form approach in the default risk framework (of course, it has many other applications as will appear clearly in subsequent chapters). We present important facts about general homogeneous diffusions, in particular concerning their Green functions, scale functions and speed measures. These three quantities are of great interest when valuing options in a general setting. We study applications related to last passage times. A section is devoted to enlargements of filtrations, an important subject when dealing with insider trading.

The books of Borodin and Salminen [109], Itô [462], Itô and McKean [465], Karlin and Taylor [515], Karatzas and Shreve [513], Kallenberg [505], Knight [528], Øksendal [684], [RY] and Rogers and Williams [741, 742] are highly recommended. See also the review of Varadhan [826].

An excellent reference for the study of first hitting times of a fixed level for a diffusion is the book of Borodin and Salminen [109] where many results can be found. The general theory of stochastic processes is presented in Dellacherie [240], Dellacherie and Meyer [242, 244] and Dellacherie, Meyer and Maisonneuve [241]. Some results about the general theory of processes can also be found in $\rightarrowtail$ Chapter 9.

5.1 Time Changes

5.1.1 Inverse of an Increasing Process

In this paragraph, we deal with processes on a probability space but do not make any reference to a given filtration. Let us recall that by definition (see Subsection 1.1.10) an increasing process is equal to 0 at time 0; it is right

M. Jeanblanc, M. Yor, M. Chesney, *Mathematical Methods for Financial Markets*, Springer Finance, DOI 10.1007/978-1-84628-737-4_5,

continuous and of course increasing. Let A be an increasing process and let C be the **right inverse** of A, that is the increasing process defined by:

$$C_u = \inf\{t : A_t > u\} \tag{5.1.1}$$

where $\inf\{\emptyset\} = \infty$. We shall use C_u or $C(u)$ for the value at time u of the process C. The process C is right-continuous and satisfies

$$C_{u-} = \inf\{t : A_t \geq u\}$$

and $\{C_u > t\} = \{A_t < u\}$. We also have $A_{C_s} \geq s$ and $A_t = \inf\{u : C_u > t\}$. (See [RY], Chapter 0, section 4 for details.) Moreover, if A is continuous and strictly increasing, C is continuous and $C(A_t) = t$.

Proposition 5.1.1.1 *Time changing in integrals can be effected as follows: if f is a positive Borel function*

$$\int_{[0,\infty[} f(s)\, dA_s = \int_0^\infty f(C_u)\, \mathbb{1}_{\{C_u<\infty\}} du\,.$$

PROOF: For $f = \mathbb{1}_{[0,v]}$, the formula reads

$$A_v = \int_0^\infty \mathbb{1}_{\{C_u \leq v\}} du$$

and is a consequence of the definition of C. The general formula follows from the monotone class theorem. □

5.1.2 Time Changes and Stopping Times

In this section, $\mathbf{F}$ is a right-continuous filtration, and A is a right-continuous adapted increasing process with right inverse C. From the identity

$$\{C_u \leq t\} = \{A_t \geq u\}\,,$$

we see that $(C_u, u \geq 0)$ is a family of $\mathbf{F}$-stopping times. This leads us to define a **time change** C as a family $(C_u, u \geq 0)$ of stopping times such that the map $u \to C_u$ is a.s. increasing and right continuous. We denote by $\mathbf{F}_C$ the filtration $\mathbf{F}_C = (\mathcal{F}_{C_t}, t \geq 0)$. For every t the r.v. A_t is an $\mathbf{F}_C$-stopping time (indeed $\{A_t < u\} = \{C_u > t\}$).

Example 5.1.2.1 We have studied a very special case of time change while dealing with Ornstein-Uhlenbeck processes in Section 2.6. These processes are obtained from a Brownian motion by means of a deterministic time change.

Example 5.1.2.2 Let W be a Brownian motion and let

$$T_t = \inf\{s \geq 0 : W_s > t\} = \inf \left\{ s \geq 0 : \max_{u \leq s} W_u > t \right\}$$

be the right-continuous inverse of $M_t = \max_{u \leq t} W_u$. The process $(T_t, t \geq 0)$ is increasing, and right-continuous (see Subsection 3.1.2). See $\rightarrowtail$ Section 11.8 for applications.

Exercise 5.1.2.3 Let (B, W) be a two-dimensional Brownian motion and define

$$T_t = \inf\{s \geq 0 : W_s > t\}.$$

Prove that $(Y_t = B_{T_t}, t \geq 0)$ is a Cauchy process, i.e., a process with independent and stationary increments, such that Y_t has a Cauchy law with characteristic function $\exp(-t|u|)$.
Hint: $\mathbb{E}(e^{iuB_{T_t}}) = \int e^{-\frac{1}{2}u^2 T_t(\omega)} \mathbb{P}(d\omega) = \mathbb{E}(e^{-\frac{1}{2}u^2 T_t}) = e^{-t|u|}$. $\lhd$

5.1.3 Brownian Motion and Time Changes

Proposition 5.1.3.1 (Dubins-Schwarz's Theorem.) *A continuous martingale M such that*

$$\langle M \rangle_\infty = \infty$$

is a time-changed Brownian motion. In other words, there exists a Brownian motion W such that $M_t = W_{\langle M \rangle_t}$.

SKETCH OF THE PROOF: Let $A = \langle M \rangle$ and define the process W as $W_u = M_{C_u}$ where C is the inverse of A. One can then show that W is a continuous local martingale, with bracket $\langle W \rangle_u = \langle M \rangle_{C_u} = u$. Therefore, W is a Brownian motion, and replacing u by A_t in $W_u = M_{C_u}$, one obtains $M_t = W_{A_t}$. □

Comments 5.1.3.2 (a) This theorem was proved in Dubins and Schwarz [268]. It admits a partial extension due to Knight [527] to the multidimensional case: if M is a d-dimensional martingale such that $\langle M^i, M^j \rangle = 0, i \neq j$ and $\langle M^i \rangle_\infty = \infty, \forall i$, then the process $W = (M^i_{C_i(t)}, i \leq d, t \geq 0)$ is a d-dimensional Brownian motion w.r.t. its natural filtration, where the process C_i is the inverse of $\langle M^i \rangle$. See, e.g., Rogers and Williams [741]. The assumption $\langle M \rangle_\infty = \infty$ can be relaxed (See [RY], Chapter V, Theorem 1.10).

(b) Let us mention another two-dimensional extension of Dubins and Schwarz's theorem for complex valued local martingales which generalize complex Brownian motion. Getoor and Sharpe [390] introduced the notion of a continuous conformal local martingale as a process $Z = X + iY$, valued in $\mathbb{C}$, the complex plane, where X and Y are real valued continuous local martingales and Z^2 is a local martingale. A prototype is the complex-valued

Brownian motion. If Z is a continuous conformal local martingale, then, from $Z_t^2 = X_t^2 - Y_t^2 + 2iX_tY_t$, we deduce that $\langle X\rangle_t = \langle Y\rangle_t$ and $\langle X, Y\rangle_t = 0$. Hence, applying Knight's result to the two-dimensional local martingale (X, Y), there exists a complex-valued Brownian motion B such that $Z = B_{\langle X\rangle}$. In fact, in this case, B can be shown to be a Brownian motion w.r.t. $(\mathcal{F}_{\alpha_u}, u \geq 0)$, where $\alpha_u = \inf\{t : \langle X\rangle_t > u\}$. If $(Z_t, t \geq 0)$ denotes now a $\mathbb{C}$-valued Brownian motion, and $f : \mathbb{C} \to \mathbb{C}$ is holomorphic, then $(f(Z_t), t \geq 0)$ is a conformal martingale. The $\mathbb{C}$-extension of the Dubins-Schwarz-Knight theorem may then be written as:

$$f(Z_t) = \widehat{Z}_{\int_0^t |f'(Z_u)|^2 du}, t \geq 0 \tag{5.1.2}$$

where f' is the $\mathbb{C}$-derivative of f, and $\widehat{Z}$ denotes another $\mathbb{C}$-valued Brownian motion. This is an extremely powerful result due to Lévy, which expresses the conformal invariance of $\mathbb{C}$-valued Brownian motion. It is easily shown, as a consequence, using the exponential function that, if $Z_0 = a$, then $(Z_t, t \geq 0)$ shall never visit $b \neq a$ (of course, almost surely). As a consequence, (5.1.2) may be extended to any meromorphic function from $\mathbb{C}$ to itself, when $P(Z_0 \in S) = 0$ with S the set of singular points of f.

(c) See Jacod [468], Chapter 10 for a detailed study of time changes, and El Karoui and Weidenfeld [311] and Le Jan [569].

Exercise 5.1.3.3 Let f be a non-constant holomorphic function on $\mathbb{C}$ and $Z = X + iY$ a complex Brownian motion. Prove that there exists another complex Brownian motion B such that $f(Z_t) = f(Z_0) + B(\int_0^t |f'(Z_s)|^2 d\langle X\rangle_s)$ (see [RY], Chapter 5). As an example, $\exp(Z_t) = 1 + B_{\int_0^t ds\, \exp(2X_s)}$. ◁

We now come back to a study of real-valued continuous local martingales.

Lemma 5.1.3.4 *Let M be a continuous local martingale with $\langle M\rangle_\infty = \infty$, W the Brownian motion such that $M_t = W_{\langle M\rangle_t}$ and C the right-inverse of $\langle M\rangle$. If H is an adapted process such that for any t,*

$$\int_0^t H_s^2 d\langle M\rangle_s \left(= \int_0^{\langle M\rangle_t} H_{C_u}^2\, du \right) < \infty\,,$$

then

$$\int_0^t H_s dM_s = \int_0^{\langle M\rangle_t} H_{C_u} dW_u\,, \qquad \int_0^{C_t} H_s dM_s = \int_0^t H_{C_u} dW_u\,.$$

Lemma 5.1.3.5 *Let $X_t = \int_0^{C_t} H_s dW_s$, where C is a time change with respect to $\mathbf{F}$, differentiable with respect to time. Assume that $C'_t \neq 0$ for any t. Then,*

$$dX_t = H_{C_t} \sqrt{C'_t}\, dB_t\,,$$

where B is an $\mathbf{F}_C$-Brownian motion.

PROOF: From the previous lemma

$$\int_0^{C_t} H_s dW_s = \int_0^t H_{C_u} dW_{C_u},$$

hence, $dX_t = H_{C_t} dW_{C_t}$. The process $(W_{C_u}, u \geq 0)$ is a local martingale with bracket C_u. The process

$$B_t = \int_0^t \frac{1}{\sqrt{C'_u}} dW_{C_u}$$

is a Brownian motion. □

Remarks 5.1.3.6 (a) Up to an enlargement of probability space, one can generalize the previous lemma to the case where the condition $C'_t \neq 0$ does not hold, but where we keep the assumption that C is differentiable. (The proof is left to the reader.)

(b) A time-changed local martingale is not necessarily a local martingale with respect to the time-changed filtration. As seen in Example 5.1.2.2, if T_t is the first hitting time of the level t for the Brownian motion B, the process $t \to T_t$ is increasing and is a time change. However, $B_{T_t} = t$ is not a local martingale. This illustrates, although very roughly, Monroe's theorem (see Remark 5.1.3.6) which states that any semi-martingale (even discontinuous) is a time changed Brownian motion. [655, 656]

However, if X is a continuous $\mathbf{F}$-local martingale and C a continuous time change, then $(X_{C_t})_{t\geq 0}$ is a continuous $\mathbf{F}_C$-local martingale. (See [RY], Chapter V, Section 1).

Comments 5.1.3.7 (a) Changes of time are extensively used for finance purposes in the papers of Geman, Madan and Yor [379, 385, 380, 381].

(b) The "pli cacheté" of Doeblin [255] may have been one of the first papers studying time changes.

(c) Further extensions to Markov processes are found in Volkonski [831]. See also McKean's paper [637] for other aspects of this major idea and important applications to Bessel processes in ↣ Chapter 6.

Exercise 5.1.3.8 Let Y be the solution of

$$dY_t = (cY_t + kY_t^2)dt + \sqrt{Y_t} dW_t$$

Prove that $Y_t = Z(\int_0^t Y_s ds)$ where $dZ(u) = (c + kZ(u))du + d\widehat{W}_u$. ◁

Exercise 5.1.3.9 Let Z be a complex BM $Z_t = X_t + iY_t$. Consider the two martingales $|Z_t|^2 - 2t$ and $\int_0^t (X_s dY_s - Y_s dX_s)$. Prove that

$$\frac{1}{2}\left(|Z_t|^2 - 2t\right) + i\int_0^t (X_s dY_s - Y_s dX_s)$$

is a conformal martingale which can be represented as $\widehat{Z}_u = \beta_u + i\gamma_u$ time-changed by $\int_0^t |Z_s|^2 ds$ with β and γ two independent BM's. Prove that $\sigma(\beta_u, u \geq 0) = \sigma(|Z_t|, t \geq 0)$, hence γ and $|Z|$ are independent. $\triangleleft$

5.2 Dual Predictable Projections

In this section, after recalling some basic facts about optional and predictable projections, we introduce the concept of a dual predictable projection[1], which leads to the fundamental notion of predictable compensators. We recommend the survey paper of Nikeghbali [674].

Recall that a process is said to be **optional** if it is measurable with respect to the σ-algebra on $\mathbb{R}^+ \times \Omega$ generated by càdlàg **F**-adapted processes, considered as mappings on $\mathbb{R}^+ \times \Omega$, whereas a **predictable** process is measurable with respect to the σ-algebra on $\mathbb{R}^+ \times \Omega$ generated by càg **F**-adapted processes (see $\rightarrowtail$ Subsection 9.1.3 for comments).

5.2.1 Definitions

Let X be a bounded (or positive) process, and **F** a given filtration. The **optional projection** of X is the unique optional process ${}^{(o)}X$ which satisfies: for any **F**-stopping time τ

$$\mathbb{E}(X_\tau \mathbb{1}_{\{\tau<\infty\}}) = \mathbb{E}({}^{(o)}X_\tau \mathbb{1}_{\{\tau<\infty\}}) . \tag{5.2.1}$$

For any **F**-stopping time τ, let $\Gamma \in \mathcal{F}_\tau$ and apply the equality (5.2.1) to the stopping time $\tau_\Gamma = \tau \mathbb{1}_\Gamma + \infty \mathbb{1}_{\Gamma^c}$. We get the re-inforced identity:

$$\mathbb{E}(X_\tau \mathbb{1}_{\{\tau<\infty\}} | \mathcal{F}_\tau) = {}^{(o)}X_\tau \mathbb{1}_{\{\tau<\infty\}} .$$

In particular, if A is an increasing process, then, for $s \leq t$:

$$\mathbb{E}({}^{(o)}A_t - {}^{(o)}A_s | \mathcal{F}_s) = \mathbb{E}(A_t - A_s | \mathcal{F}_s) \geq 0 . \tag{5.2.2}$$

Note that, for any t, $\mathbb{E}(X_t|\mathcal{F}_t) = {}^{(o)}X_t$. However, $\mathbb{E}(X_t|\mathcal{F}_t)$ is defined almost surely for any t; thus uncountably many null sets are involved, hence, a priori, $\mathbb{E}(X_t|\mathcal{F}_t)$ is not a well-defined process whereas ${}^{(o)}X$ takes care of this difficulty.

Likewise, the **predictable projection** of X is the unique predictable process ${}^{(p)}X$ such that for any **F**-predictable stopping time τ

$$\mathbb{E}(X_\tau \mathbb{1}_{\{\tau<\infty\}}) = \mathbb{E}({}^{(p)}X_\tau \mathbb{1}_{\{\tau<\infty\}}) . \tag{5.2.3}$$

As above, this identity reinforces as

$$\mathbb{E}(X_\tau \mathbb{1}_{\{\tau<\infty\}} | \mathcal{F}_{\tau-}) = {}^{(p)}X_\tau \mathbb{1}_{\{\tau<\infty\}} ,$$

for any **F**-predictable stopping time τ (see Subsection 1.2.3 for the definition of $\mathcal{F}_{\tau-}$).

[1] See Dellacherie [240] for the notion of dual optional projection.

Example 5.2.1.1 Let τ and ϑ be two stopping times such that $\vartheta \leq \tau$ and Z a bounded r.v.. Let $X = Z\mathbb{1}_{]\!]\vartheta,\tau]\!]}$. Then, ${}^{(o)}X = U\mathbb{1}_{]\!]\vartheta,\tau]\!]}$, ${}^{(p)}X = V\mathbb{1}_{]\!]\vartheta,\tau]\!]}$ where U (resp. V) is the right-continuous (resp. left-continuous) version of the martingale $(\mathbb{E}(Z|\mathcal{F}_t), t \geq 0)$.

Let τ and ϑ be two stopping times such that $\vartheta \leq \tau$ and X a positive process. If A is an increasing optional process, then, since $\mathbb{1}_{]\!]\vartheta,\tau]\!]}(t)$ is predictable

$$\mathbb{E}\left(\int_\vartheta^\tau X_t dA_t\right) = \mathbb{E}\left(\int_\vartheta^\tau {}^{(o)}X_t dA_t\right).$$

If A is an increasing predictable process, then

$$\mathbb{E}\left(\int_\vartheta^\tau X_t dA_t\right) = \mathbb{E}\left(\int_\vartheta^\tau {}^{(p)}X_t dA_t\right).$$

The notion of interest in this section is that of **dual predictable projection**, which we define as follows:

Proposition 5.2.1.2 *Let $(A_t,\, t \geq 0)$ be an integrable increasing process (not necessarily* $\mathbf{F}$*-adapted). There exists a unique* $\mathbf{F}$*-predictable increasing process $(A_t^{(p)}, t \geq 0)$, called the dual predictable projection of A such that*

$$\mathbb{E}\left(\int_0^\infty H_s dA_s\right) = \mathbb{E}\left(\int_0^\infty H_s dA_s^{(p)}\right)$$

for any positive $\mathbf{F}$*-predictable process H.*

In the particular case where $A_t = \int_0^t a_s ds$, one has

$$A_t^{(p)} = \int_0^t {}^{(p)}a_s ds \tag{5.2.4}$$

PROOF: See Dellacherie [240], Chapter V, Dellacherie and Meyer [244], Chapter 6 paragraph (73), page 148, or Protter [727] Chapter 3, Section 5. □

This definition extends to the difference between two integrable (for simplicity) increasing processes. The terminology "dual predictable projection" refers to the fact that it is the random measure $d_t A_t(\omega)$ which is relevant when performing that operation. If X is bounded and A has integrable variation (not necessarily adapted), then

$$\mathbb{E}((X\star A^{(p)})_\infty) = \mathbb{E}(({}^{(p)}X\star A)_\infty).$$

This is equivalent to: for $s < t$,

$$\mathbb{E}(A_t - A_s|\mathcal{F}_s) = \mathbb{E}(A_t^{(p)} - A_s^{(p)}|\mathcal{F}_s). \tag{5.2.5}$$

If A is adapted (not necessarily predictable), then $(A_t - A_t^{(p)}, t \geq 0)$ is a martingale. In that case, $A_t^{(p)}$ is also called the predictable compensator of A.

More generally, from Proposition 5.2.1.2 and (5.2.5), the process ${}^{(o)}A - A^{(p)}$ is a martingale.

Proposition 5.2.1.3 *If A is increasing, the process ${}^{(o)}A$ is a sub-martingale and $A^{(p)}$ is the predictable increasing process in the Doob-Meyer decomposition of the sub-martingale ${}^{(o)}A$.*

Example 5.2.1.4 Let W be a Brownian motion, $M_t = \sup_{s \leq t} W_s$ its running maximum, and $R_t = 2M_t - W_t$. Then, from ↣ Pitman's Theorem 5.7.2.1 and its Corollary 5.7.2.2, for any positive Borel function f,

$$\mathbb{E}(f(M_t)|\mathcal{F}_t^R) = \int_0^1 dx f(R_t x) \,,$$

hence, $\mathbb{E}(2M_t|\mathcal{F}_t^R) = R_t$ and the predictable projection of $2M_t$ is R_t. On the other hand, from Pitman's theorem

$$R_t = \beta_t + \int_0^t \frac{ds}{R_s} \,,$$

where β is a Brownian motion, therefore, the dual predictable projection of $2M_t$ on $\mathcal{F}_t^R$ is $\int_0^t \frac{ds}{R_s}$. Note that the difference between these two projections is the (Brownian) martingale β.

In a general setting, the predictable projection of an increasing process A is a sub-martingale whereas the dual predictable projection is an increasing process. The predictable projection and the dual predictable projection of an increasing process A are equal if and only if $({}^{(p)}A_t, t \geq 0)$ is increasing.

It will also be convenient to introduce the following terminology:

Definition 5.2.1.5 *If ϑ is a random time, we call the* ***predictable compensator*** *associated with ϑ the dual predictable projection A^ϑ of the increasing process $\mathbb{1}_{\{\vartheta \leq t\}}$. This dual predictable projection A^ϑ satisfies*

$$\mathbb{E}(k_\vartheta) = \mathbb{E}\left(\int_0^\infty k_s dA_s^\vartheta\right) \tag{5.2.6}$$

for any positive, predictable process k.

5.2.2 Examples

In the sequel, we present examples of computation of dual predictable projections. We end up with Azéma's lemma, providing the law of the predictable compensator associated with the last passage at 0 of a BM before T, evaluated at a (stopping) time T. See also Knight [529] and ↣ Sections 5.6 and 7.4.

Example 5.2.2.1 Let $(B_s)_{s\geq 0}$ be an $\mathbf{F}-$ Brownian motion starting from 0 and $B_s^{(\nu)} = B_s + \nu s$. Let $\mathbf{G}^{(\nu)}$ be the filtration generated by the process $(|B_s^{(\nu)}|, s \geq 0)$ (which coincides with the one generated by $(B_s^{(\nu)})^2$). We now compute the decomposition of the semi-martingale $(B^{(\nu)})^2$ in the filtration $\mathbf{G}^{(\nu)}$ and the dual predictable projection (with respect to $\mathbf{G}^{(\nu)}$) of the finite variation process $\int_0^t B_s^{(\nu)} ds$.

Itô's lemma provides us with the decomposition of the process $(B^{(\nu)})^2$ in the filtration $\mathbf{F}$:

$$(B_t^{(\nu)})^2 = 2\int_0^t B_s^{(\nu)} dB_s + 2\nu \int_0^t B_s^{(\nu)} ds + t\,. \tag{5.2.7}$$

To obtain the decomposition in the filtration $\mathbf{G}^{(\nu)}$ we remark that, on the canonical space, denoting as usual by X the canonical process,

$$\mathbf{W}^{(0)}(e^{\nu X_s}|\mathcal{F}_s^{|X|}) = \cosh(\nu X_s)$$

which leads to the equality:

$$\mathbf{W}^{(\nu)}(X_s|\mathcal{F}_s^{|X|}) = \frac{\mathbf{W}^{(0)}(X_s e^{\nu X_s}|\mathcal{F}_s^{|X|})}{\mathbf{W}^{(0)}(e^{\nu X_s}|\mathcal{F}_s^{|X|})} = X_s \tanh(\nu X_s) \equiv \psi(\nu X_s)/\nu\,,$$

where $\psi(x) = x\tanh(x)$. We now come back to equality (5.2.7). Due to (5.2.4), we have just shown that:

$$\text{The dual predictable projection of } 2\nu \int_0^t B_s^{(\nu)} ds \text{ is } 2\int_0^t ds\, \psi(\nu B_s^{(\nu)})\,. \tag{5.2.8}$$

As a consequence,

$$(B_t^{(\nu)})^2 - 2\int_0^t ds\, \psi(\nu B_s^{(\nu)}) - t$$

is a $\mathbf{G}^{(\nu)}$-martingale with increasing process $4\int_0^t (B_s^{(\nu)})^2 ds$. Hence, there exists a $\mathbf{G}^{(\nu)}$-Brownian motion β such that

$$(B_t + \nu t)^2 = 2\int_0^t |B_s + \nu s| d\beta_s + 2\int_0^t ds\, \psi(\nu(B_s + \nu s)) + t\,. \tag{5.2.9}$$

Exercise 5.2.2.2 Prove that, more generally than (5.2.8), the dual predictable projection of $\int_0^t f(B_s^{(\nu)})ds$ is $\int_0^t \mathbb{E}(f(B_s^{(\nu)})|\mathcal{G}_s^{(\nu)})ds$ and that

$$\mathbb{E}(f(B_s^{(\nu)})|\mathcal{G}_s^{(\nu)}) = \frac{f(B_s^{(\nu)})e^{\nu B_s^{(\nu)}} + f(-B_s^{(\nu)})e^{-\nu B_s^{(\nu)}}}{2\cosh(\nu B_s^{(\nu)})}\,.$$

◁

Exercise 5.2.2.3 Prove that if $(\alpha_s, s \geq 0)$ is an increasing predictable process and X a positive measurable process, then

$$\left(\int_0^{\cdot} X_s d\alpha_s\right)_t^{(p)} = \int_0^t {}^{(p)}X_s d\alpha_s\,.$$

In particular

$$\left(\int_0^{\cdot} X_s ds\right)_t^{(p)} = \int_0^t {}^{(p)}X_s ds\,.$$

◁

Example 5.2.2.4 Let B be a Brownian motion and $Y_t = |B_t|$. Tanaka's formula gives

$$|B_t| = \int_0^t \mathrm{sgn}(B_s) dB_s + L_t$$

where L denotes the local time of $(B_t; t \geq 0)$ at level 0. By an application of the balayage formula (see Subsection 4.1.6), we obtain (we recall that g_t denotes the last passage time at level 0 before t)

$$h_{g_t}|B_t| = \int_0^t h_{g_s}\mathrm{sgn}(B_s) dB_s + \int_0^t h_s dL_s$$

where we have used the fact that $L_{g_s} = L_s$. Consequently, replacing, if necessary, h by $|h|$, we see that the process $\int_0^t |h_s| dL_s$ is the local time at 0 of $(h_{g_t}B_t, t \geq 0)$. Let now τ be a stopping time such that $(B_{t\wedge\tau}; t \geq 0)$ is uniformly integrable, and satisfies $\mathbb{P}(B_\tau = 0) = 0$. Then, it follows from the balayage formula that, for every predictable and bounded process h

$$\mathbb{E}\left(h_{g_\tau}|B_\tau|\right) = \mathbb{E}\left(\int_0^\tau h_s dL_s\right). \tag{5.2.10}$$

As an example, consider $\tau = T_a^* = \inf\{t : |B_t| = a\}$; we have

$$\mathbb{E}\left(h_{g_{T_a^*}}\right) = \frac{1}{a}\mathbb{E}\left(\int_0^{T_a^*} h_s dL_s\right),$$

whence we conclude that the predictable compensator $(A_t^\vartheta; t \geq 0)$ associated with $\vartheta := g_{T_a^*}$ is given by

$$A_t^\vartheta = \frac{1}{a} L_{t\wedge T_a^*}\,.$$

In the general case, applying (5.2.10) to the variable $\xi_{g_\tau} = \mathbb{E}\left(|B_\tau| | \mathcal{F}_{g_\tau}\right)$, where $(\xi_u; u \geq 0)$ is a predictable process (note that $\mathbb{P}(\xi_{g_\tau} = 0) = 0$, as a consequence of $\mathbb{P}(B_\tau = 0) = 0$) we obtain

$$\mathbb{E}(k_{g_\tau}) = \mathbb{E}\left(\int_0^\tau \frac{k_s}{\xi_s} dL_s\right) \tag{5.2.11}$$

from which we deduce that the predictable compensator associated with g_τ is

$$A_t = \int_0^{t\wedge\tau} \frac{dL_s}{\xi_s}. \tag{5.2.12}$$

In general, finding ξ may necessitate some work, but in some cases, e.g., $\tau = \inf\{t : |B_t| = \alpha_t\}$, for a continuous adapted process (α_t) such that $\alpha_t \equiv \alpha_{g_t}$, no extra computation is needed, since: $|B_\tau| = \alpha_{g_\tau}$ is $\mathcal{F}_{g_\tau}$ measurable, hence we can take: $\xi_u = \alpha_u$; finally $A_t = \int_0^{t\wedge\tau} \frac{dL_s}{\alpha_s}$.

We finish this subsection with the following interesting lemma which, in some generality, gives the law of A_τ.

Lemma 5.2.2.5 (Azéma.) *Let B be a BM and τ a stopping time such that $(B_{t\wedge\tau}; t \geq 0)$ is uniformly integrable, and satisfies $\mathbb{P}(B_\tau = 0) = 0$. Let A be the predictable compensator associated with g_τ. Then, A_τ is an exponential variable with mean 1.*

PROOF: Since, as a consequence of equality (5.2.12), $A_\tau = A_{g_\tau}$, we have for every $\lambda \geq 0$

$$\mathbb{E}\left(e^{-\lambda A_\tau}\right) = \mathbb{E}\left(\int_0^\tau e^{-\lambda A_s} dA_s\right),$$

as a consequence of (5.2.6) applied to $\vartheta = g_\tau$ and $k_t = \exp(-\lambda A_t)$. Thus, we obtain

$$\mathbb{E}\left(e^{-\lambda A_\tau}\right) = \mathbb{E}\left(\frac{1 - e^{-\lambda A_\tau}}{\lambda}\right),$$

or equivalently, $\mathbb{E}\left(e^{-\lambda A_\tau}\right) = \frac{1}{1+\lambda}$. The desired result follows immediately. □

Note that a corollary of this result provides the law of the local time of the BM at the time $T_a^* = \inf\{t : |B_t| = a\}$: $L_{T_a^*}$ is an exponential variable, with mean a.

Exercise 5.2.2.6 Let $\mathbf{F} \subset \mathbf{G}$ and let $G_t - \int_0^t \gamma_s ds$ be a $\mathbf{G}$-martingale. Recalling that ${}^{(o)}X$ is the $\mathbf{F}$-optional projection of a process X, prove that ${}^{(o)}G_t - \int_0^t {}^{(o)}\gamma_s ds$ is an $\mathbf{F}$-martingale. ◁

5.3 Diffusions

In this section, we present the main facts on linear diffusions, following closely the presentation of Chapter 2 in Borodin and Salminen [109]. We refer to

Durrett [287], Itô and McKean [465], Linetsky [595] and Rogers and Williams [742] for other studies of general diffusions.

A linear diffusion is a strong Markov process with continuous paths taking values on an interval I with left-end point $\ell \geq -\infty$ and right-end point $r \leq \infty$. We denote by ζ the life time of X (see Definition 1.1.14.1). We assume in what follows (unless otherwise stated) that all the diffusions we consider are **regular**, i.e., they satisfy $\mathbb{P}_x(T_y < \infty) > 0, \forall x, y \in I$ where $T_y = \inf\{t : X_t = y\}$.

5.3.1 (Time-homogeneous) Diffusions

In this book, we shall mainly consider diffusions which are Itô processes: let b and σ be two real-valued functions which are Lipschitz on the interval I, such that $\sigma(x) > 0$ for all x in the interval I. Then, there exists a unique solution to

$$X_t = x + \int_0^t b(X_s)ds + \int_0^t \sigma(X_s)dW_s \,, \tag{5.3.1}$$

starting at point $x \in]\ell, r[$, up to the first exit time $T_{\ell,r} = T_\ell(X) \wedge T_r(X)$. In this case, X is a time-homogeneous diffusion.

In fact, the Lipschitz assumption is not quite necessary; see Theorem 1.5.5.1 for some finer assumptions on b and σ.

Solutions of

$$X_t = x + \int_0^t b(X_s, s)ds + \int_0^t \sigma(X_s, s)dW_s \,, \tag{5.3.2}$$

with time dependent coefficients b and σ are called time-inhomogeneous diffusions; for these processes, the following results do not apply.

From now on, we shall only consider diffusions of the type (5.3.1), and we drop the term "time-homogeneous." We mention furthermore that, in general studies of diffusions (see Borodin and Salminen [109]), a rôle is also played by a killing measure; however, since we shall not use this item in our presentation, we do not introduce it.

5.3.2 Scale Function and Speed Measure

Scale Function

Definition 5.3.2.1 *Let X be a diffusion on I and $T_y = \inf\{t \geq 0 : X_t = y\}$, for $y \in I$. A scale function is an increasing function from I to $\mathbb{R}$ such that, for $x \in [a, b]$*

$$\mathbb{P}_x(T_a < T_b) = \frac{s(x) - s(b)}{s(a) - s(b)} \,. \tag{5.3.3}$$

Obviously, if s^* is a scale function, then so is $\alpha s^* + \beta$ for any (α, β), with $\alpha > 0$ and any scale function can be written as $\alpha s^* + \beta$.

Proposition 5.3.2.2 *The process $(s(X_t), 0 \leq t < T_{\ell,r})$ is a local martingale. The scale function satisfies*

$$\frac{1}{2}\sigma^2(x)s''(x) + b(x)s'(x) = 0\,.$$

PROOF: For any finite stopping time $\tau < T_{\ell,r}$, the equality

$$\mathbb{E}_x\left(\frac{s(X_\tau) - s(b)}{s(a) - s(b)}\right) = \frac{s(x) - s(b)}{s(a) - s(b)}$$

follows from the Markov property. □

In the case of diffusions of the form (5.3.1), a (differentiable) scale function is

$$s(x) = \int_c^x \exp\left(-2\int_c^u b(v)/\sigma^2(v)\,dv\right)du \tag{5.3.4}$$

for some choice of $c \in]\ell, r[$. The increasing process of $s(X)$ being

$$A_t = \int_0^t (s'\sigma)^2(X_u)du,$$

(by an application of Itô's formula), the local martingale $(s(X_t), t < T_{\ell,r})$ can be written as a time changed Brownian motion: $s(X_t) = \beta_{A_t}$.

In the case of constant coefficients with $b < 0$ (resp. $b > 0$) and $\sigma \neq 0$, the diffusion is defined on $\mathbb{R}$, $T_{\ell,r} = \infty$, and we may choose $s(x) = \exp\left(-2bx/\sigma^2\right)$, (resp. $s(x) = -\exp\left(-2bx/\sigma^2\right)$) so that s is a strictly increasing function and $s(-\infty) = 0, s(\infty) = \infty$ (resp. $s(-\infty) = -\infty, s(\infty) = 0$).

A diffusion is said to be in **natural scale** if $s(x) = x$. In this case, if $I = \mathbb{R}$, the diffusion $(X_t, t \geq 0)$ is a local martingale.

Speed Measure

The **speed measure m** is defined as the measure such that the infinitesimal generator of X can be written as

$$\mathcal{A}f(x) = \frac{d}{d\mathbf{m}}\frac{d}{ds}f(x)$$

where

$$\frac{d}{ds}f(x) = \lim_{h\to 0} \frac{f(x+h)-f(x)}{s(x+h)-s(x)},$$

and

$$\frac{d}{d\mathbf{m}}g(x) = \lim_{h\to 0} \frac{g(x+h)-g(x)}{\mathbf{m}(x,x+h)}.$$

In the case of diffusions of the form (5.3.1), the speed measure is absolutely continuous with respect to Lebesgue measure, i.e., $\mathbf{m}(dx) = m(x)dx$, hence

$$\begin{aligned}\mathcal{A}f(x) &= \frac{d}{d\mathbf{m}}\frac{d}{ds}f(x) = \frac{1}{m(x)}\frac{d}{dx}\left(\frac{1}{s'}\frac{d}{dx}f\right) \\ &= \frac{1}{m(x)\,s'(x)}f''(x) - \frac{s''(x)}{m(x)\,(s')^2(x)}f'(x) \\ &= \frac{1}{m(x)\,s'(x)}f''(x) + \frac{2\,b(x)}{m(x)\,s'(x)\,\sigma^2(x)}f'(x)\end{aligned}$$

where the last equality comes from formula (5.3.4). Since in this case the infinitesimal generator has the form

$$\mathcal{A}f(x) = \frac{1}{2}\sigma^2(x)f''(x) + b(x)f'(x),$$

the density of the speed measure is

$$\boxed{m(x) = \frac{2}{\sigma^2(x)s'(x)}.} \tag{5.3.5}$$

The density of the speed measure satisfies

$$\frac{1}{2}\left(\sigma^2(x)m(x)\right)'' + (s(x)b(x)))'(x) = 0\,.$$

It is important to consider the local martingale $s(X_t)$ only strictly before the hitting time of the boundary. The reader should keep in mind the example of the reflected Brownian motion, which is not a martingale, although $s(x) = x$ (see ↣ Proposition 6.1.2.4).

If X is a diffusion with scale function s, we have seen that $s(X_t) = \beta_{A_t}$, where β is a Brownian motion. In terms of speed measure, the increasing process A is the inverse of

$$C_u = \frac{1}{2}\int_0^u m(\beta_s)ds = \frac{1}{2}\int \mathbf{m}(dz)L_u^z(\beta)\,.$$

Remark 5.3.2.3 Beware that some authors define the speed measure with a factor 1/2, that is $\mathcal{A}f(x) = \frac{1}{2}\frac{d}{d\mathbf{m}}\frac{d}{ds}f(x)$. Our convention, without this factor 1/2 is the same as Borodin and Salminen [109].

Exercise 5.3.2.4 Prove that, if $s(X)$ is a martingale, then, equality (5.3.3) holds. ◁

5.3.3 Boundary Points

Definition 5.3.3.1 *The boundary points are classified as follows:*

- *The left-hand point ℓ is an **exit boundary** if, for any $x \in]\ell, r[$,*

$$\int_\ell^x \mathbf{m}(]y, x[)s'(y)dy < \infty$$

*and an **entrance boundary** if, for any $x \in]\ell, r[$,*

$$\int_\ell^x \mathbf{m}(]\ell, y[)s'(y)dy < \infty\,.$$

- *The right-hand point r is an **exit boundary** if, for any $x \in]\ell, r[$,*

$$\int_x^r \mathbf{m}(]x, y[)s'(y)dy < \infty$$

*and an **entrance boundary** if, for any $x \in]\ell, r[$,*

$$\int_x^r \mathbf{m}(]y, r[)s'(y)dy < \infty\,.$$

- *A boundary point which is both entrance and exit is called **non-singular**.*
- *A boundary point that is neither entrance nor exit is called **natural**.*

A diffusion reaches its non-singular boundaries with positive probability, and it is possible to start a diffusion from a non-singular boundary.

An example where 0 is an entrance boundary is given by the BES^3 process (see ↣ Chapter 6), or more generally by a BES^δ with $\delta \geq 2$. We recall that a BES^δ process with $\delta \geq 2$ does not return to 0 after it has left this point.

Definition 5.3.3.2 *Let X be a diffusion. The point ℓ is said to be **instantaneously reflecting** if $\mathbf{m}(\{\ell\}) = 0$.*

For the reflected BM $|B|$, the point 0 is instantaneously reflecting and the Lebesgue measure of the set $\{t : |B_t| = 0\}$ is zero.

Example 5.3.3.3 We present, following Borodin and Salminen [109], the computation of the scale function and speed measure for some important diffusion processes:

- **Drifted Brownian motion.**
  Suppose $X_t = B_t + \nu t$. A scale function for X is $s(x) = \exp(-2\nu x)$ for $\nu < 0$, and $s(x) = -\exp(-2\nu x)$ for $\nu > 0$. The density of the speed measure is $m(x) = 2e^{2\nu x}$. The lifetime is ∞.

- **Geometric Brownian motion.** Let $dS_t = S_t(\mu dt + \sigma dB_t)$. We have seen in Lemma 3.6.6.1 that $S_t^{1-\gamma}$ is a martingale for $\gamma = 2\mu/\sigma^2$. Hence
  - a scale function of S is $s(x) = -(x^{1-\gamma})/(1-\gamma)$ for $\gamma \neq 1$ and $\ln x$ for $\gamma = 1$,
  - the density of the speed measure is $m(x) = 2x^{\gamma-2}/\sigma^2$.

  The boundary points 0 and ∞ are natural.
  - If $\gamma > 1$, then $\lim_{t\to\infty} S_t(\omega) = \infty$, a.s.,
  - if $\gamma < 1$, then $\lim_{t\to\infty} S_t(\omega) = 0$, a.s.,
  - if $\gamma = 1$, then $\liminf_{t\to\infty} S_t(\omega) = 0, \limsup_{t\to\infty} S_t(\omega) = \infty$ a.s..

- **Reflected Brownian motion.**
  The process $X_t = |W_t|$ is a diffusion on $[0, \infty[$. The left-hand point 0 is a non-singular boundary point. The scale function is $s(x) = x$, the density of the speed measure is $m(x) = 2$.

- **Bessel processes.** A Bessel process (see ↣ Section 6.1) with dimension δ and index $\nu = \frac{\delta}{2} - 1$ is a diffusion on $]0, \infty[$, or on $[0, \infty[$ depending on the value of ν and the boundary conditions at 0.
  For all values of ν, the boundary point ∞ is natural. The boundary point 0 is
  - exit-non-entrance if $\nu \leq -1$
  - nonsingular if $-1 < \nu < 0$
  - entrance-not exit if $\nu \geq 0$.

  In the nonsingular case, the boundary condition at 0 is usually reflection or killing. A scale function for a $\text{BES}^{(\nu)}$ is $s(x) = x^{-2\nu}$ for $\nu < 0$, $s(x) = \ln x$ for $\nu = 0$ and $s(x) = -x^{-2\nu}$ for $\nu > 0$. It follows that a scale function for a $\text{BESQ}^{(\nu)}$ is
  - $s(x) = x^{-\nu}$ for $\nu < 0$,
  - $s(x) = \ln x$ for $\nu = 0$ and
  - $s(x) = -x^{-\nu}$ for $\nu > 0$.

  See ↣ Proposition 6.1.2.4 for more information.
  For $\nu > 0$, the density of the speed measure is $m(x) = \nu^{-1} x^{2\nu+1}$.

- **Affine equation.**
  Let

$$dX_t = (\alpha X_t + 1)dt + \sqrt{2}\, X_t dW_t \,, X_0 = x \,.$$

The scale function derivative is $s'(x) = x^{-\alpha}e^{1/x}$ and the speed density function is $m(x) = x^{\alpha}e^{-1/x}$.

- **OU and Vasicek processes.**
Let r be a (k,σ) Ornstein-Uhlenbeck process. A scale function derivative is $s'(x) = \exp(kx^2/\sigma^2)$. If r is a $(k,\theta;\sigma)$ Vasicek process (see Section 2.6), $s'(x) = \exp k(x-\theta)^2/\sigma^2$.

The first application of the concept of speed measure is Feller's test for non-explosion (see Definition 1.5.4.10). We shall see in the sequel that speed measures are very useful tools.

Proposition 5.3.3.4 (Feller's Test for non-explosion.) *Let b, σ belong to $C^1(\mathbb{R})$, and let X be the solution of*

$$dX_t = b(X_t)dt + \sigma(X_t)dW_t$$

with τ its explosion time. The process does not explode, i.e., $\mathbb{P}(\tau = \infty) = 1$ if and only if

$$\int_{-\infty}^{0} [s(x) - s(-\infty)]\, m(x)dx = \int_{0}^{\infty} [s(\infty) - s(x)]\, m(x)dx = \infty\,.$$

PROOF: see McKean [637], page 65. □

Comments 5.3.3.5 This proposition extends the case where the coefficients b and σ are only locally Lipschitz. Khasminskii [522] developed Feller's test for multidimensional diffusion processes (see McKean [637], page 103, Rogers and Williams [742], page 299). See Borodin and Salminen [109], Breiman [123], Freedman [357], Knight [528], Rogers and Williams [741] or [RY] for more information on speed measures.

Exercise 5.3.3.6 Let $dX_t = \theta dt + \sigma\sqrt{X_t}dW_t$, $X_0 > 0$, where $\theta > 0$ and, for $a < x < b$ let $\psi_{a,b}(x) = \mathbb{P}_x(T_b(X) < T_a(X))$. Prove that

$$\psi_{a,b}(x) = \frac{x^{1-\nu} - a^{1-\nu}}{b^{1-\nu} - a^{1-\nu}}$$

where $\nu = 2\theta/\sigma^2$. Prove also that if $\nu > 1$, then T_0 is infinite and that if $\nu < 1$, $\psi_{0,b}(x) = (x/b)^{1-\nu}$. Thus, the process $(1/X_t, t \geq 0)$ explodes in the case $\nu < 1$. ◁

5.3.4 Change of Time or Change of Space Variable

In a number of computations, it is of interest to time change a diffusion into BM by means of the scale function of the diffusion. It may also be of interest to relate diffusions of the form

$$X_t = x + \int_0^t b(X_s)ds + \int_0^t \sigma(X_s)dW_s$$

to those for which $\sigma = 1$, that is $Y_t = y + \beta_t + \int_0^t du\, \mu(Y_u)$ where β is a Brownian motion. For this purpose, one may proceed by means of a change of time or change of space variable, as we now explain.

(a) Change of Time

Let $A_t = \int_0^t \sigma^2(X_s)ds$ and assume that $|\sigma| > 0$. Let $(C_u, u \geq 0)$ be the inverse of $(A_t, t \geq 0)$. Then

$$X_{C_u} = x + \beta_u + \int_0^u dC_h\, b(X_{C_h})$$

From $h = \int_0^{C_h} \sigma^2(X_s)ds$, we deduce $dC_h = \dfrac{dh}{\sigma^2(X_{C_h})}$, hence

$$Y_u := X_{C_u} = x + \beta_u + \int_0^u dh\, \frac{b}{\sigma^2}(Y_h)$$

where β is a Brownian motion.

(b) Change of Space Variable

Assume that $\varphi(x) = \displaystyle\int_0^x \frac{dy}{\sigma(y)}$ is well defined and that φ is of class C^2. From Itô's formula

$$\begin{aligned}\varphi(X_t) &= \varphi(x) + \int_0^t \varphi'(X_s)dX_s + \frac{1}{2}\int_0^t \varphi''(X_s)\sigma^2(X_s)ds \\ &= \varphi(x) + W_t + \int_0^t ds\left(\frac{b}{\sigma}(X_s) - \frac{1}{2}\sigma'(X_s)\right).\end{aligned}$$

Hence, setting $Z_t = \varphi(X_t)$, we get

$$Z_t = z + W_t + \int_0^t \widehat{b}(Z_s)ds$$

where $\widehat{b}(z) = \frac{b}{\sigma}(\varphi^{-1}(z)) - \frac{1}{2}\sigma'(\varphi^{-1}(z))$.

Comment 5.3.4.1 See Doeblin [255] for some interesting applications.

5.3.5 Recurrence

Definition 5.3.5.1 *A diffusion X with values in I is said to be* ***recurrent*** *if*

$$\mathbb{P}_x(T_y < \infty) = 1, \ \forall x, y \in I .$$

If not, the diffusion is said to be ***transient****.*

It can be proved that the homogeneous diffusion X given by (5.3.1) on $]\ell, r[$ is recurrent if and only if $s(\ell+) = -\infty$ and $s(r-) = \infty$. (See [RY], Chapter VII, Section 3, for a proof given as an exercise.)

Example 5.3.5.2 A one-dimensional Brownian motion is a recurrent process, a Bessel process (see $\rightarrowtail$ Chapter 6) with index strictly greater than 0 is a transient process. For the (recurrent) one-dimensional Brownian motion, the times T_y are large, i.e., $\mathbb{E}_x(T_y^\alpha) < \infty$, for $x \neq y$ if and only if $\alpha < 1/2$.

5.3.6 Resolvent Kernel and Green Function

Resolvent Kernel

The resolvent of a Markov process X is the family of operators $f \rightarrow R_\lambda f$

$$R_\lambda f(x) = \mathbb{E}_x \left(\int_0^\infty e^{-\lambda t} f(X_t) dt \right) .$$

The resolvent kernel of a diffusion is the density (with respect to Lebesgue measure) of the resolvent operator, i.e., the Laplace transform in t of the transition density $p_t(x, y)$:

$$R_\lambda(x, y) = \int_0^\infty e^{-\lambda t} p_t(x, y) dt \,, \tag{5.3.6}$$

where $\lambda > 0$ for a recurrent diffusion and $\lambda \geq 0$ for a transient diffusion. It satisfies

$$\frac{1}{2}\sigma^2(x)\frac{\partial^2 R_\lambda}{\partial x^2} + b(x)\frac{\partial R_\lambda}{\partial x} - \lambda R_\lambda = 0 \quad \text{for } x \neq y$$

and $R_\lambda(x, x) = 1$. The Sturm-Liouville O.D.E.

$$\frac{1}{2}\sigma^2(x) u''(x) + b(x) u'(x) - \lambda u(x) = 0 \tag{5.3.7}$$

admits two linearly independent continuous positive solutions (the basic solutions) $\Phi_{\lambda\uparrow}(x)$ and $\Phi_{\lambda\downarrow}(x)$, with $\Phi_{\lambda\uparrow}$ increasing and $\Phi_{\lambda\downarrow}$ decreasing, which are determined up to constant factors.

A straightforward application of Itô's formula establishes that $e^{-\lambda t}\Phi_{\lambda\uparrow}(X_t)$ and $e^{-\lambda t}\Phi_{\lambda\downarrow}(X_t)$ are local martingales, for $\lambda > 0$, hence, using carefully

Doob's optional stopping theorem, we obtain the Laplace transform of the first hitting times:

$$\mathbb{E}_x\left(e^{-\lambda T_y}\right) = \begin{cases} \Phi_{\lambda\uparrow}(x)/\Phi_{\lambda\uparrow}(y) & \text{if } x < y \\ \Phi_{\lambda\downarrow}(x)/\Phi_{\lambda\downarrow}(y) & \text{if } x > y \end{cases}. \tag{5.3.8}$$

Green Function

Let $p_t^{(m)}(x, y)$ be the transition probability function relative to the speed measure $m(y)dy$:

$$\mathbb{P}_x(X_t \in dy) = p_t^{(m)}(x, y)m(y)dy\,. \tag{5.3.9}$$

It is a known and remarkable result that $p_t^{(m)}(x, y) = p_t^{(m)}(y, x)$ (see Chung [185] and page 149 in Itô and McKean [465]).

The Green function is the density with respect to the speed measure of the resolvent operator: using $p_t^{(m)}(x, y)$, the transition probability function relative to the speed measure, there is the identity

$$G_\lambda(x, y) := \int_0^\infty e^{-\lambda t} p_t^{(m)}(x, y)dt = w_\lambda^{-1}\Phi_{\lambda\uparrow}(x \wedge y)\Phi_{\lambda\downarrow}(x \vee y)\,,$$

where the Wronskian

$$w_\lambda := \frac{\Phi'_{\lambda\uparrow}(y)\Phi_{\lambda\downarrow}(y) - \Phi_{\lambda\uparrow}(y)\Phi'_{\lambda\downarrow}(y)}{s'(y)} \tag{5.3.10}$$

depends only on λ and not on y. Obviously

$$m(y)G_\lambda(x, y) = R_\lambda(x, y)\,,$$

hence

$$R_\lambda(x, y) = w_\lambda^{-1} m(y)\Phi_{\lambda\uparrow}(x \wedge y)\Phi_{\lambda\downarrow}(x \vee y)\,. \tag{5.3.11}$$

A diffusion is transient if and only if $\lim_{\lambda\to 0} G_\lambda(x, y) < \infty$ for some $x, y \in I$ and hence for all $x, y \in I$.

Comment 5.3.6.1 See Borodin and Salminen [109] and Pitman and Yor [718, 719] for an extended study. Kent [520] proposes a methodology to invert this Laplace transform in certain cases as a series expansion. See Chung [185] and Chung and Zhao [187] for an extensive study of Green functions. Many authors call Green functions our resolvent.

5.3.7 Examples

Here, we present examples of computations of functions $\Phi_{\lambda\downarrow}$ and $\Phi_{\lambda\uparrow}$ for certain diffusions.

- **Brownian motion with drift μ: $X_t = \mu t + \sigma W_t$.** In this case, the basic solutions of
$$\frac{1}{2}\sigma^2 u'' + \mu u' = \lambda u$$
are
$$\Phi_{\lambda\uparrow}(x) = \exp\left[\frac{x}{\sigma^2}\left(-\mu + \sqrt{2\lambda\sigma^2 + \mu^2}\right)\right],$$
$$\Phi_{\lambda\downarrow}(x) = \exp\left[-\frac{x}{\sigma^2}\left(\mu + \sqrt{2\lambda\sigma^2 + \mu^2}\right)\right].$$

- **Geometric Brownian motion: $dX_t = X_t(\mu dt + \sigma dW_t)$.**
The basic solutions of
$$\frac{1}{2}\sigma^2 x^2 u'' + \mu x u' = \lambda u$$
are
$$\Phi_{\lambda\uparrow}(x) = x^{\frac{1}{\sigma^2}(-\mu + \frac{\sigma^2}{2} + \sqrt{2\lambda\sigma^2 + (\mu - \sigma^2/2)^2})},$$
$$\Phi_{\lambda\downarrow}(x) = x^{-\frac{1}{\sigma^2}(\mu - \frac{\sigma^2}{2} + \sqrt{2\lambda\sigma^2 + (\mu - \sigma^2/2)^2})}.$$

- **Bessel process with index ν.** Let $dX_t = dW_t + \left(\nu + \frac{1}{2}\right)\frac{1}{X_t}dt$. For $\nu > 0$, the basic solutions of
$$\frac{1}{2}u'' + \left(\nu + \frac{1}{2}\right)\frac{1}{x}u' = \lambda u$$
are
$$\Phi_{\lambda\uparrow}(x) = x^{-\nu} I_\nu(x\sqrt{2\lambda}),\ \Phi_{\lambda\downarrow}(x) = x^{-\nu} K_\nu(x\sqrt{2\lambda}),$$
where I_ν and K_ν are the classical Bessel functions with index ν (see $\rightarrowtail$ Appendix A.5.2).

- **Affine Equation.**
Let
$$dX_t = (\alpha X_t + \beta)dt + \sqrt{2} X_t dW_t,$$
with $\beta \neq 0$. The basic solutions of
$$x^2 u'' + (\alpha x + \beta)u' = \lambda u$$
are

$$\Phi_{\lambda\uparrow}(x) = \left(\frac{\beta}{x}\right)^{(\nu+\mu)/2} M\left(\frac{\nu+\mu}{2}, 1+\mu, \frac{\beta}{x}\right),$$
$$\Phi_{\lambda\downarrow}(x) = \left(\frac{\beta}{x}\right)^{(\nu+\mu)/2} U\left(\frac{\nu+\mu}{2}, 1+\mu, \frac{\beta}{x}\right)$$

where M and U denote the Kummer functions (see ↣ A.5.4 in the Appendix) and $\mu = \sqrt{\nu^2 + 4\lambda}$, $1+\nu = \alpha$.

- **Ornstein-Uhlenbeck and Vasicek Processes.** Let $k > 0$ and

$$dX_t = k(\theta - X_t)dt + \sigma dW_t\,, \tag{5.3.12}$$

a Vasicek process. The basic solutions of

$$\frac{1}{2}\sigma^2 u'' + k(\theta - x)u' = \lambda u$$

are

$$\Phi_{\lambda\uparrow}(x) = \exp\left(\frac{k\,(x-\theta)^2}{2\sigma^2}\right) D_{-\lambda/k}\left(-\frac{x-\theta}{\sigma}\sqrt{2k}\right),$$
$$\Phi_{\lambda\downarrow}(x) = \exp\left(\frac{k\,(x-\theta)^2}{2\sigma^2}\right) D_{-\lambda/k}\left(\frac{x-\theta}{\sigma}\sqrt{2k}\right).$$

Here, D_ν is the parabolic cylinder function with index ν (see ↣ Appendix A.5.4).

Comment 5.3.7.1 For OU processes, i.e., in the case $\theta = 0$ in equation (5.3.12), Ricciardi and Sato [732] obtained, for $x > a$, that the density of the hitting time of a is

$$-ke^{k(x^2-a^2)/2} \sum_{n=1}^{\infty} \frac{D_{\nu_{n,a}}(x\sqrt{2k})}{D'_{\nu_{n,a}}(a\sqrt{2k})} e^{-k\nu_{n,a}t}$$

where $0 < \nu_{1,a} < \cdots < \nu_{n,a} < \cdots$ are the zeros of $\nu \to D_\nu(-a)$. The expression $D'_{\nu_{n,a}}$ denotes the derivative of $D_\nu(a)$ with respect to ν, evaluated at the point $\nu = \nu_{n,a}$. Note that the formula in Leblanc et al. [573] for the law of the hitting time of a is only valid for $a = 0, \theta = 0$. See also the discussion in Subsection 3.4.1.

Extended discussions on this topic are found in Alili et al. [10], Göing-Jaeschke and Yor [398, 397], Novikov [678], Patie [697] or Borodin and Salminen [109].

- **CEV Process.**
The constant elasticity of variance process (See ↣ Section 6.4) follows

$$dS_t = S_t(\mu dt + S_t^\beta dW_t)\,.$$

In the case $\beta < 0$, the basic solutions of

$$\frac{1}{2}x^{2\beta+2}u''(x) + \mu x u'(x) = \lambda u(x)$$

are

$$\Phi_{\lambda\uparrow}(x) = x^{\beta+1/2}e^{\epsilon x/2}M_{k,m}(x),\ \Phi_{\lambda\downarrow}(x) = x^{\beta+1/2}e^{\epsilon x/2}W_{k,m}(x)$$

where M and W are the Whittaker functions (see ↣ Subsection A.5.7) and

$$\epsilon = \text{sgn}(\mu\beta),\ m = -\frac{1}{4\beta},\ k = \epsilon\left(\frac{1}{2} + \frac{1}{4\beta}\right) - \frac{\lambda}{2|\mu\beta|}\,.$$

See Davydov and Linetsky [225].

Exercise 5.3.7.2 Prove that the process

$$X_t = \exp(aB_t + bt)\left(x + \int_0^t ds \exp(-aB_s - bs)\right)$$

satisfies

$$X_t = x + a\int_0^t X_u dB_u + \int_0^t \left(\left(\frac{a^2}{2} + b\right)X_u + 1\right)du\,.$$

(See Donati-Martin et al. [258] for further properties of this process, and application to Asian options.) More generally, consider the process

$$dY_t = (aY_t + b)dt + (cY_t + d)dW_t\,,$$

where $c \neq 0$. Prove that, if $X_t = cY_t + d$, then

$$dX_t = (\alpha X_t + \beta)dt + X_t dW_t$$

with $\alpha = a/c, \beta = b - da/c$. From $T_\alpha(Y^y) = T_{c\alpha+d}(X^{cx+d})$, deduce the Laplace transform of first hitting times for the process Y. ◁

5.4 Non-homogeneous Diffusions

5.4.1 Kolmogorov's Equations

Let

$$Lf(s,x) = b(s,x)\partial_x f(s,x) + \frac{1}{2}\sigma^2(s,x)\partial^2_{xx} f(s,x)\,.$$

A **fundamental solution** of

$$\partial_s f(s,x) + Lf(s,x) = 0 \tag{5.4.1}$$

is a positive function $p(x,s;y,t)$ defined for $0 \le s < t$, $x,y \in \mathbb{R}$, such that for any function $\varphi \in C_0(\mathbb{R})$ and any $t > 0$ the function

$$f(s,x) = \int_{\mathbb{R}} \varphi(y) p(s,x;t,y) dy$$

is bounded, is of class $C^{1,2}$, satisfies (5.4.1) and obeys $\lim_{s\uparrow t} f(s,x) = \varphi(x)$.

If b and σ are real valued bounded and continuous functions $\mathbb{R}^+ \times \mathbb{R}$ such that

(i) $\sigma^2(t,x) \ge c > 0$,
(ii) there exists $\alpha \in]0,1]$ such that for all (x,y), for all $s,t \ge 0$,

$$|b(t,x) - b(s,y)| + |\sigma^2(t,x) - \sigma^2(s,y)| \le K(|t-s|^\alpha + |x-y|^\alpha),$$

then the equation

$$\partial_s f(s,x) + Lf(s,x) = 0$$

admits a strictly positive fundamental solution p. For fixed (y,t) the function $u(s,x) = p(s,x;t,y)$ is of class $C^{1,2}$ and satisfies the backward Kolmogorov equation that we present below. If in addition, the functions $\partial_x b(t,x)$, $\partial_x \sigma(t,x)$, $\partial_{xx}\sigma(t,x)$ are bounded and Hölder continuous, then for fixed (x,s) the function $v(t,y) = p(s,x;t,y)$ is of class $C^{1,2}$ and satisfies the forward Kolmogorov equation that we present below.

Note that a time-inhomogeneous diffusion process can be treated as a homogeneous process. Instead of X, consider the space-time diffusion process (t, X_t) on the enlarged state space $\mathbb{R}^+ \times \mathbb{R}^d$.

We give Kolmogorov's equations for the general case of inhomogeneous diffusions

$$dX_t = b(t,X_t)dt + \sigma(t,X_t)dW_t.$$

Proposition 5.4.1.1 *The transition probability density $p(s,x;t,y)$ defined for $s < t$ as $\mathbb{P}_{x,s}(X_t \in dy) = p(s,x;t,y)dy$ satisfies the two partial differential equations (recall δ_x is the Dirac measure at x):*

- *The **backward Kolmogorov equation:***

$$\begin{cases} \dfrac{\partial}{\partial s} p(s,x;t,y) + \dfrac{1}{2}\sigma^2(s,x)\dfrac{\partial^2}{\partial x^2} p(s,x;t,y) + b(s,x)\dfrac{\partial}{\partial x} p(s,x;t,y) = 0, \\ \lim_{s\to t} p(s,x;t,y)dy = \delta_x(dy). \end{cases}$$

- *The **forward Kolmogorov equation***

$$\begin{cases} \dfrac{\partial}{\partial t} p(s,x;t,y) - \dfrac{1}{2}\dfrac{\partial^2}{\partial y^2}\left(p(s,x;t,y)\sigma^2(t,y)\right) + \dfrac{\partial}{\partial y}\left(p(s,x;t,y)b(t,y)\right) = 0, \\ \lim_{t\to s} p(s,x;t,y)dy = \delta_x(dy). \end{cases}$$

SKETCH OF THE PROOF: The backward equation is really straightforward to derive. Let φ be a C^2 function with compact support. For any fixed t, the martingale $\mathbb{E}(\varphi(X_t)|\mathcal{F}_s)$ is equal to $f(s, X_s) = \int_{\mathbb{R}} \varphi(y)p(s, X_s; t, y)dy$ since X is a Markov process. An application of Itô's formula to $f(s, X_s)$ leads to its decomposition as a semi-martingale. Since it is in fact a true martingale its bounded variation term must be equal to zero. This result being true for every φ, it provides the backward equation.

The forward equation is in a certain sense the dual of the backward one. Recall that if φ is a C^2 function with compact support, then

$$\mathbb{E}_{s,x}(\varphi(X_t)) = \int_{\mathbb{R}} \varphi(y)p(s, x; t, y)dy\,.$$

From Itô's formula, for $t > s$

$$\varphi(X_t) = \varphi(X_s) + \int_s^t \varphi'(X_u)dX_u + \frac{1}{2}\int_s^t \varphi''(X_u)\sigma^2(u, X_u)du\,.$$

Hence, taking (conditional) expectations

$$\begin{aligned}\mathbb{E}_{s,x}(\varphi(X_t)) &= \varphi(x) + \int_s^t \mathbb{E}_{s,x}\left(\varphi'(X_u)b(u, X_u) + \frac{1}{2}\sigma^2(u, X_u)\varphi''(X_u)\right)du \\ &= \varphi(x) + \int_s^t du \int_{\mathbb{R}} \left(\varphi'(y)b(u, y) + \frac{1}{2}\sigma^2(u, y)\varphi''(y)\right)p(s, x; u, y)dy\,.\end{aligned}$$

From the integration by parts formula (in the sense of distributions if the coefficients are not smooth enough) and since φ and φ' vanish at ∞:

$$\begin{aligned}\int_{\mathbb{R}} \varphi(y)p(s, x; t, y)dy &= \varphi(x) - \int_s^t du \int_{\mathbb{R}} \varphi(y)\frac{\partial}{\partial y}\left(b(u, y)p(s, x; u, y)\right)dy \\ &\quad + \frac{1}{2}\int_s^t du \int_{\mathbb{R}} \varphi(y)\frac{\partial^2}{\partial y^2}\left(\sigma^2(u, y)p(s, x; u, y)\right)dy\,.\end{aligned}$$

Differentiating with respect to t, we obtain that

$$\frac{\partial}{\partial t}p(s, x; t, y) = -\frac{\partial}{\partial y}(b(t, y)p(s, x; t, y)) + \frac{1}{2}\frac{\partial^2}{\partial y^2}\left(\sigma^2(t, y)p(s, x; t, y)\right)\,.$$

□

Note that for homogeneous diffusions, the density

$$p(x; t, y) = \mathbb{P}_x(X_t \in dy)/dy$$

satisfies the backward Kolmogorov equation

$$\frac{1}{2}\sigma^2(x)\frac{\partial^2 p}{\partial x^2}(x; t, y) + b(x)\frac{\partial p}{\partial x}(x; t, y) = \frac{\partial}{\partial t}p(x; t, y)\,.$$

Comments 5.4.1.2 (a) The Kolmogorov equations are the topic studied by Doeblin [255] in his now celebrated " pli cacheté n⁰ 11668".

(b) We refer to Friedman [361] p.141 and 148, Karatzas and Shreve [513] p.328, Stroock and Varadhan [812] and Nagasawa [663] for the multidimensional case and for regularity assumptions for uniqueness of the solution to the backward Kolmogorov equation. See also Itô and McKean [465], p.149 and Stroock [810].

5.4.2 Application: Dupire's Formula

Under the assumption that the underlying asset follows

$$dS_t = S_t(rdt + \sigma(t, S_t)dW_t)$$

under the risk-neutral probability, Dupire [284, 283] established a formula relating the local volatility $\sigma(t,x)$ and the value $C(T,K)$ of a European Call where K is the strike and T the maturity, i.e.,

$$\frac{1}{2}K^2\sigma^2(T,K) = \frac{\partial_T C(T,K) + rK\partial_K C(T,K)}{\partial^2_{KK} C(T,K)}\,.$$

We have established this formula using a local-time methodology in Subsection 4.2.1; here we present the original proof of Dupire as an application of the Kolmogorov backward equation. Let $f(T,x)$ be the density of the random variable S_T, i.e.,

$$f(T,x)dx = \mathbb{P}(S_T \in dx)\,.$$

Then,

$$C(T,K) = e^{-rT}\int_0^\infty (x-K)^+ f(T,x)dx = e^{-rT}\int_K^\infty (x-K)f(T,x)dx$$

$$= e^{-rT}\int_K^\infty dx f(T,x)\int_K^x dy = e^{-rT}\int_K^\infty dy\int_y^\infty f(T,x)dx\,. \quad (5.4.2)$$

By differentiation with respect to K,

$$\frac{\partial C}{\partial K}(T,K) = -e^{-rT}\int_0^\infty \mathbb{1}_{\{x>K\}} f(T,x)dx = -e^{-rT}\int_K^\infty f(T,x)dx\,,$$

hence, differentiating again

$$\frac{\partial^2 C}{\partial K^2}(T,K) = e^{-rT} f(T,K) \quad (5.4.3)$$

which allows us to obtain the law of the underlying asset from the prices of the European options. For notational convenience, we shall now write $C(t,x)$

instead of $C(T,K)$. From (5.4.3), $f(t,x) = e^{rt}\frac{\partial^2 C}{\partial x^2}(t,x)$, hence differentiating both sides of this equality w.r.t. t gives

$$\frac{\partial}{\partial t}f = re^{rt}\frac{\partial^2 C}{\partial x^2} + e^{rt}\frac{\partial^2}{\partial x^2}\frac{\partial}{\partial t}C\,.$$

The density f satisfies the forward Kolmogorov equation

$$\frac{\partial f}{\partial t}(t,x) - \frac{1}{2}\frac{\partial^2}{\partial x^2}\left(x^2\sigma^2(t,x)f(t,x)\right) + \frac{\partial}{\partial x}(rxf(t,x)) = 0\,,$$

or

$$\frac{\partial f}{\partial t} = \frac{1}{2}\frac{\partial^2}{\partial x^2}\left(x^2\sigma^2 f\right) - rf - rx\frac{\partial}{\partial x}f\,. \tag{5.4.4}$$

Replacing f and $\frac{\partial f}{\partial t}$ by their expressions in terms of C in (5.4.4), we obtain

$$re^{rt}\frac{\partial^2 C}{\partial x^2} + e^{rt}\frac{\partial^2}{\partial x^2}\frac{\partial}{\partial t}C = e^{rt}\frac{1}{2}\frac{\partial^2}{\partial x^2}\left(x^2\sigma^2\frac{\partial^2 C}{\partial x^2}\right) - re^{rt}\frac{\partial^2 C}{\partial x^2} - rxe^{rt}\frac{\partial}{\partial x}\frac{\partial^2 C}{\partial x^2}$$

and this equation can be simplified as follows

$$\begin{aligned}\frac{\partial^2}{\partial x^2}\frac{\partial}{\partial t}C &= \frac{1}{2}\frac{\partial^2}{\partial x^2}\left(x^2\sigma^2\frac{\partial^2 C}{\partial x^2}\right) - 2r\frac{\partial^2 C}{\partial x^2} - rx\frac{\partial}{\partial x}\frac{\partial^2 C}{\partial x^2}\\ &= \frac{1}{2}\frac{\partial^2}{\partial x^2}\left(x^2\sigma^2\frac{\partial^2 C}{\partial x^2}\right) - r\left(2\frac{\partial^2 C}{\partial x^2} + x\frac{\partial}{\partial x}\frac{\partial^2 C}{\partial x^2}\right)\\ &= \frac{1}{2}\frac{\partial^2}{\partial x^2}\left(x^2\sigma^2\frac{\partial^2 C}{\partial x^2}\right) - r\frac{\partial^2}{\partial x^2}\left(x\frac{\partial C}{\partial x}\right),\end{aligned}$$

hence,

$$\frac{\partial^2}{\partial x^2}\frac{\partial C}{\partial t} = \frac{\partial^2}{\partial x^2}\left(\frac{1}{2}x^2\sigma^2\frac{\partial^2 C}{\partial x^2} - rx\frac{\partial C}{\partial x}\right).$$

Integrating twice with respect to x shows that there exist two functions α and β, depending only on t, such that

$$\frac{1}{2}x^2\sigma^2(t,x)\frac{\partial^2 C}{\partial x^2}(t,x) = rx\frac{\partial C}{\partial x}(t,x) + \frac{\partial C}{\partial t}(t,x) + \alpha(t)x + \beta(t)\,.$$

Assuming that the quantities

$$\begin{cases} x^2\sigma^2(t,x)\dfrac{\partial^2 C}{\partial x^2}(t,x) = e^{-rt}x^2\sigma^2(t,x)f(t,x)\\ x\dfrac{\partial C}{\partial x}(t,x) = -e^{-rt}x\displaystyle\int_x^\infty f(t,y)dy\\ \dfrac{\partial C}{\partial t}(t,x)\end{cases}$$

go to 0 as x goes to infinity, we obtain $\lim_{x\to\infty}\alpha(t)x + \beta(t) = 0, \forall t$, hence $\alpha(t) = \beta(t) = 0$ and

$$\frac{1}{2}x^2\sigma^2(t,x)\frac{\partial^2 C}{\partial x^2}(t,x) = rx\frac{\partial C}{\partial x}(t,x) + \frac{\partial C}{\partial t}(t,x).$$

The value of $\sigma(t,x)$ in terms of the call prices follows. □

5.4.3 Fokker-Planck Equation

Proposition 5.4.3.1 *Let $dX_t = b(t,X_t)dt + \sigma(t,X_t)dB_t$, and assume that h is a deterministic function such that $X_0 > h(0)$, $\tau = \inf\{t \geq 0 : X_t \leq h(t)\}$ and*

$$g(t,x)dx = \mathbb{P}(X_t \in dx, \tau > t).$$

The function $g(t,x)$ satisfies the Fokker-Planck equation

$$\frac{\partial}{\partial t}g(t,x) = -\frac{\partial}{\partial x}\left(b(t,x)g(t,x)\right) + \frac{1}{2}\frac{\partial^2}{\partial x^2}\left(\sigma^2(t,x)g(t,x)\right); \; x > h(t)$$

and the boundary conditions

$$\lim_{t\to 0} g(t,x)dx = \delta(x - X_0)$$
$$g(t,x)|_{x=h(t)} = 0.$$

PROOF: The proof follows that of the backward Kolmogorov equation.

▶ We first note that

$$\begin{aligned}\mathbb{E}(\varphi(X_{t\wedge\tau})) &= \mathbb{E}(\varphi(X_t)\mathbb{1}_{\{t\leq\tau\}}) + \mathbb{E}(\varphi(X_\tau)\mathbb{1}_{\{\tau<t\}}) \\ &= \int_{\mathbb{R}} \varphi(x)g(t,x)dx + \mathbb{E}(\varphi(h(\tau))\mathbb{1}_{\{\tau<t\}}) \\ &= \int_{\mathbb{R}} \varphi(x)g(t,x)dx + \int_0^t \varphi(h(u))\mu(du)\end{aligned}$$

where μ is the law of τ.

▶ If φ is a C^2 function with compact support,

$$\varphi(X_{t\wedge\tau}) = \varphi(X_{s\wedge\tau}) + \int_{s\wedge\tau}^{t\wedge\tau} \varphi'(X_u)dX_u + \frac{1}{2}\int_{s\wedge\tau}^{t\wedge\tau} \varphi''(X_u)\sigma^2(u,X_u)du,$$

hence,

$$\begin{aligned}\mathbb{E}(\varphi(X_{t\wedge\tau})) &= \mathbb{E}(\varphi(X_{s\wedge\tau})) + \mathbb{E}\left(\int_s^t \mathbb{1}_{\{u<\tau\}}\varphi'(X_u)b(u,X_u)du\right) \\ &\quad + \frac{1}{2}\mathbb{E}\left(\int_s^t \mathbb{1}_{\{u<\tau\}}\varphi''(X_u)\sigma^2(u,X_u)du\right) \\ &= \int \varphi(x)g(s,x)dx + \int_0^s \varphi(h(v))\mu(dv) \\ &\quad + \int_s^t du \int_{\mathbb{R}} dx \left(\varphi'(x)b(u,x) + \frac{1}{2}\varphi''(x)\sigma^2(u,x)\right) g(x,u).\end{aligned}$$

This identity holds for any function φ of class C^2, therefore, using integration by parts for the last integral, and differentiation with respect to t, we get the result. The law of τ is obtained by integration w.r.t. x. □

Using the Fokker-Planck equation, Iyengar [466], He et al. [426] and Zhou [876] established the following result.

Proposition 5.4.3.2 *Let $X_t^i = \alpha_i t + \sigma_i W_t^i$ where W^1, W^2 are two correlated Brownian motions, with correlation ρ, and let m_t^i be the running minimum of X^i. The probability density*

$$\mathbb{P}(X_t^1 \in dx_1, X_t^2 \in dx_2, m_t^1 \in dm_1, m_t^2 \in dm_2) = p(x_1, x_2, t; m_1, m_2) dx_1 dx_2 dm_1 dm_2$$

is given by

$$p(x_1, x_2, t;\, m_1, m_2) = \frac{e^{a_1 x_1 + a_2 x_2 + bt}}{\sigma_1 \sigma_2 \sqrt{1-\rho^2}}\, h(x_1, x_2, t; m_1, m_2) \tag{5.4.5}$$

with

$$h(x_1, x_2, t; m_1, m_2) = \frac{2}{\beta t} e^{-(r^2+r_0^2)/(2t)} \sum_{n=1}^{\infty} \sin\left(\frac{n\pi\theta_0}{\beta}\right) \sin\left(\frac{n\pi\theta}{\beta}\right) I_{(n\pi)/\beta}\left(\frac{r r_0}{t}\right)$$

where I_ν is the modified Bessel function of index ν and

$$a_1 = \frac{\alpha_1\sigma_2 - \rho\alpha_2\sigma_1}{(1-\rho^2)\sigma_1^2\sigma_2}, \quad a_2 = \frac{\alpha_2\sigma_1 - \rho\alpha_1\sigma_2}{(1-\rho^2)\sigma_1\sigma_2^2}$$

$$b = -\alpha_1 a_1 - \alpha_2 a_2 + \frac{1}{2}\left(\sigma_1^2 a_1^2 + \sigma_2^2 a_2^2\right) + \rho\sigma_1\sigma_2 a_1 a_2$$

$$\beta = \tan^{-1}\left(-\frac{\sqrt{1-\rho^2}}{\rho}\right), \quad for\, \rho < 0$$

$$= \pi - \tan^{-1}\left(\frac{\sqrt{1-\rho^2}}{\rho}\right), \quad for\, \rho > 0$$

$$z_1 = \frac{1}{\sqrt{1-\rho^2}}\left[\left(\frac{x_1 - m_1}{\sigma_1}\right) - \rho\left(\frac{x_2 - m_2}{\sigma_2}\right)\right], \quad z_2 = \frac{x_2 - m_2}{\sigma_2}$$

$$z_{10} = \frac{1}{\sqrt{1-\rho^2}}\left[-\frac{m_1}{\sigma_1} + \rho\frac{m_2}{\sigma_2}\right], \quad z_{20} = -\frac{m_2}{\sigma_2}$$

$$r = \sqrt{z_1^2 + z_2^2}, \quad \tan\theta = \frac{z_2}{z_1}, \quad \theta \in [0, \beta]$$

$$r_0 = \sqrt{z_{10}^2 + z_{20}^2}, \quad \tan\theta_0 = \frac{z_{20}}{z_{10}}, \quad \theta_0 \in [0, \beta].$$

The joint law with the maximum M_i is

$$\mathbb{P}(X_t^1 \in dx_1, X_t^2 \in dx_2, m_t^1 \geq m_1, M_t^2 \leq M_2)$$
$$= p(x_1, -x_2, t; m_1, -M_2, \alpha_1, -\alpha_2, \sigma_1, \sigma_2, -\rho)dx_1 dx_2$$

where $p(x_1, x_2, t; m_1, m_2; \alpha_1, \alpha_2, \sigma_1, \sigma_2, \rho)$ is the density given in (5.4.5).

Comments 5.4.3.3 (a) The knowledge of the multidimensional laws of such variables is important in the structural approach of credit risk. However, the complexity of the above formula makes it difficult to implement. Let us mention that the wrong formula given in Bielecki and Rutkowski [99] in the first edition has been corrected in the second printing. See also the recent paper of Patras [698] where a proof using probabilistic and geometric tools is given and Blanchet-Scalliet and Patras [106] for application to counterparty risk.

(b) Recently, Rogers and Shepp [739] have studied the correlation $c(\rho)$ of the maxima of correlated BMs. Denoting by $M_t^i = \sup_{s \leq t} W_s^i$ the running supremum of the BM W^i, they established that

$$c(\rho) = (\cos \alpha) \int_0^\infty du \frac{\cosh(\alpha u)}{\sinh(u\pi/2)} \tanh(u\gamma)$$

where α is given in terms of the correlation coefficient ρ between the BMs as $\alpha = \arcsin(\rho) \in [\pi/2, \pi/2]$ and $2\gamma = \alpha + \pi/2$.

The proof relies on three steps: the first one is to compute the joint law of (M_Θ^1, M_Θ^2) for Θ an exponential random variable with parameter λ, independent of (W^1, W^2). If

$$F(x_1, x_2) = \mathbb{P}(x_1 \leq M_\Theta^1, x_2 \leq M_\Theta^2),$$

then it is easy to check that

$$c(\rho) = \lambda \int_0^\infty \int_0^\infty f(x_1, x_2) dx_1 dx_2$$

In a second step, the authors note that, since $\mathbb{P}(M_\Theta^1 > x_i) = e^{-\sqrt{2\lambda} x_i}$, then

$$F(x_1, x_2) = e^{-\sqrt{2\lambda} x_1} + e^{-\sqrt{2\lambda} x_2} - \mathbb{P}(M_\Theta^1 < x_1, M_\Theta^2 < x_2).$$

They introduce $X_t^i = M_t^i - W_t^i$ and obtain

$$\mathbb{P}(M_\Theta^1 < x_1, M_\Theta^2 < x_2) = \mathbb{P}(\tau \leq \Theta | X_0^1 = x_1, X_0^2 = x_2)$$

where $\tau = \inf\{t : X_t^1 X_t^2 = 0\}$. The last step consists of the computation of

$$\widehat{F}(x_1, x_2) = \mathbb{P}(\tau \leq \Theta | X_0^1 = x_1, X_0^2 = x_2) = \mathbb{E}(e^{-\lambda \tau} | X_0^1 = x_1, X_0^2 = x_2)$$

which satisfies

$$2\lambda \widehat{f}(x_1, x_2) = (\partial^2_{x_1 x_1} + 2\rho \partial_{x_1} \partial_{x_2} + \partial^2_{x_1 x_1}) \widehat{f}(x_1, x_2)$$

with the boundary condition $\widehat{f} = 1$ on the axes.

5.4.4 Valuation of Contingent Claims

Suppose $V_f(x,T)$ is the value of a contingent claim with payoff $f(S_T)$, i.e., $V_f(x,T) = \mathbb{E}_{\mathbb{Q}}(e^{-rT}f(S_T))$ where

$$dS_t = S_t((r-\kappa)dt + \sigma(S_t)dW_t), \; S_0 = x$$

under the risk-adjusted probability $\mathbb{Q}$. In terms of the transition probability of S relative to the Lebesgue measure, that is $\mathbb{Q}(S_T \in dy) = p_T(x,y)dy$ the value of the claim is:

$$V_f(x,T) = \mathbb{E}_{\mathbb{Q}}(e^{-rT}f(S_T)) = e^{-rT}\int_0^\infty f(y)p_T(x,y)dy\,.$$

Therefore, the quantity $e^{-rT}p_T(x,y)$ can be interpreted as the price of a security with the Dirac measure payoff δ_y. It is called the price of an Arrow-Debreu security or the pricing kernel. The Laplace transform of V_f with respect to the maturity is

$$\widehat{V}_f(x,\lambda) = \int_0^\infty e^{-\lambda T}V_f(x,T)dT\,.$$

This can be written as

$$\widehat{V}_f(x,\lambda) = \int_0^\infty dTe^{-\lambda T}e^{-rT}\int_0^\infty dyf(y)p_T(x,y) = \frac{1}{\lambda}\mathbb{E}_{\mathbb{Q}}(e^{-r\mathbf{e}}f(S_{\mathbf{e}}))$$

where $\mathbf{e}$ is an exponential random variable with parameter λ which is independent of $(S_t, t \geq 0)$; this is the so-called exponential weighing, or Canadization, an expression due to Carr [146], who uses this tool to price options. In terms of an Arrow-Debreu security, we obtain that

$$\widehat{V}_f(x,\lambda) = \int_0^\infty f(y)\widehat{A}(y,\lambda)dy\,.$$

Here, $\widehat{A}$ is the Laplace transform of the price of an Arrow-Debreu security,

$$\widehat{A}(y,\lambda) = \int_0^\infty e^{-\lambda t}e^{-rt}p_t(x,y)dt = R_{\lambda+r}(x,y)\,.$$

We have seen in (5.3.11) that the resolvent is given in terms of the fundamental solutions of the ODE (5.3.7), hence

$$\widehat{V}_f(x,\lambda) =$$
$$w_\nu^{-1}\left(\Phi_{\nu\downarrow}(x)\int_0^x m(y)f(y)\Phi_{\nu\uparrow}(y)dy + \Phi_{\nu\uparrow}(x)\int_x^\infty m(y)f(y)\Phi_{\nu\downarrow}(y)dy\right)$$

where $\nu = r+\lambda$.

5.5 Local Times for a Diffusion

5.5.1 Various Definitions of Local Times

We assume that $(X_t, t \geq 0)$ is a regular diffusion on $\mathbb{R}$, with a C^1 scale function s and speed measure $\mathbf{m}$. As discussed in Itô and McKean [465], Borodin and Salminen [109] and [RY], there exists a jointly continuous family of local times $\ell_t^x(X)$, sometimes called **Itô-McKean local times** or **diffusion local times**, defined by the following occupation density formula

$$\int_0^t du\, f(X_u) = \int_{\mathbb{R}} \mathbf{m}(dx)\, f(x) \ell_t^x(X) \tag{5.5.1}$$

for all positive Borel functions f.

The process $(Y_t = s(X_t), t \geq 0)$ is a local martingale, and, as such (see formula (4.1.16)), it admits a **Tanaka-Meyer local time** $(L_t^y(Y), t \geq 0)$ at level y, which is characterized by the property that

$$\left((Y_t - y)^+ - \frac{1}{2} L_t^y(Y)\,,\, t \geq 0\right)$$

is a local martingale.

Assuming that $\mathbf{m}(dx) = m(x)dx$, there exists an **occupation local time** λ_t^x which is defined via the occupation time formula

$$\int_0^t f(X_u) du = \int_{\mathbb{R}} dx f(x) \lambda_t^x(X)\,.$$

Lemma 5.5.1.1 *Let X be a diffusion, s a scale function and $Y = s(X)$. For all x and $t \geq 0$, one has*

$$L_t^x(X) = \frac{1}{s'(x)} L_t^{s(x)}(Y),\ L_t^{s(x)}(Y) = 2\ell_t^x(X)\,.$$

Hence, $\ell_t^x(X)$ is the Tanaka-Meyer diffusion local time of $s(X)$ at level $s(x)$. Assuming that the density m exists,

$$\lambda_t^x = m(x) \ell_t^x\,.$$

PROOF: Let $L^y(Y)$ be the Tanaka-Meyer local time of $Y = s(X)$.

$$\begin{aligned}\int_{\mathbb{R}} f(y) L_t^y(Y) dy &= \int_0^t f(Y_u) d\langle Y \rangle_u = \int_0^t f(s(X_u))(s'(X_u))^2 d\langle X \rangle_u \\ &= \int_{\mathbb{R}} f(s(x))\,(s'(x))^2 L_t^x(X) dx \\ &= \int_{\mathbb{R}} f(y)\, s'(s^{-1}(y)) L_t^{s^{-1}(y)}(X) dy\,.\end{aligned}$$

Hence

$$L_t^y(Y) = s'(s^{-1}(y)) L_t^{s^{-1}(y)}(X)$$

so that

$$L_t^{s(x)}(Y) = s'(x) L_t^x(X) . \tag{5.5.2}$$

From the definition of $L_t^x(X)$, and recalling that $m(x)\sigma^2(x) = \frac{2}{s'(x)}$ (see equality 5.3.5), one obtains, on the one hand

$$\int_0^t d\langle X\rangle_u f(X_u) = \int_{\mathbb{R}} f(x) L_t^x(X) dx .$$

On the other hand,

$$\begin{aligned} \int_0^t d\langle X\rangle_u f(X_u) &= \int_0^t \sigma^2(X_u) f(X_u) du \\ &= \int_{\mathbb{R}} m(x)\sigma^2(x) f(x) \ell_t^x(X) dx = \int_{\mathbb{R}} \frac{2}{s'(x)} f(x) \ell_t^x dx \end{aligned}$$

and it follows that (see formula (5.3.2))

$$L_t^x(X) = \frac{2}{s'(x)} \ell_t^x(X) ,$$

hence, from (5.5.2), $L_t^{s(x)}(Y) = 2\ell_t^x(X)$. □

We recall that (see equality (5.3.9), there exists a density $p^{(m)}$ such that

$$\mathbb{E}_{x_0}(f(X_u)) = \int m(dx) p_u^{(m)}(x_0, x) f(x) .$$

Consequently

$$\mathbb{E}_{x_0}(\ell_t^x(X)) = \int_0^t du \, p_u^{(m)}(x_0, x) .$$

5.5.2 Some Diffusions Involving Local Time

Example 5.5.2.1 Skew Brownian Motion. The skew BM with parameter α is a process Y satisfying $Y_t = W_t + \alpha L_t^0(Y)$ where W is a Brownian motion, $L^0(Y)$ is the Tanaka-Meyer local time of the process Y at level 0, and $\alpha \leq 1/2$. Note that this process, which turns out to be a continuous strong Markov process, is not an Itô process. In order to prove the existence of the skew Brownian motion, we look for a function φ of the form $\beta y^+ - \gamma y^-$ for two constants β and γ such that $\varphi(Y_t)$ is a martingale, which solves an SDE. Using Tanaka's formula, we obtain

$$\begin{aligned}\varphi(Y_t) &= \beta\left(\int_0^t \mathbb{1}_{\{Y_s>0\}}dY_s + \frac{1}{2}L_t^0(Y)\right) - \gamma\left(-\int_0^t \mathbb{1}_{\{Y_s\leq 0\}}dY_s + \frac{1}{2}L_t^0(Y)\right)\\ &= \beta\left(\int_0^t \mathbb{1}_{\{Y_s>0\}}dW_s + \frac{1}{2}L_t^0(Y)\right)\\ &\qquad - \gamma\left(-\int_0^t \mathbb{1}_{\{Y_s\leq 0\}}dW_s - \alpha L_t^0(Y) + \frac{1}{2}L_t^0(Y)\right)\\ &= \int_0^t \left(\beta\mathbb{1}_{\{Y_s>0\}} + \gamma\mathbb{1}_{\{Y_s\leq 0\}}\right)dW_s + \frac{1}{2}\left(\beta-\gamma+2\alpha\gamma\right)L_t^0(Y).\end{aligned}$$

Hence, for $\beta - \gamma + 2\alpha\gamma = 0$, $\beta > 0$ and $\gamma > 0$, the process $X_t = \varphi(Y_t)$ is a martingale solution of the stochastic differential equation

$$dX_t = (\beta\mathbb{1}_{X_t>0} + \gamma\mathbb{1}_{X_t\leq 0})dW_t\,. \tag{5.5.3}$$

This SDE has no strong solution for β and γ strictly positive but has a unique strictly weak solution (see Barlow [47]).

The process Y is such that $|Y|$ is a reflecting Brownian motion. Indeed,

$$dY_t^2 = 2Y_t(dW_t + \alpha dL_t^0(Y)) + dt = 2Y_t dW_t + dt\,.$$

Walsh [833] proved that, conversely, the only continuous diffusions whose absolute values are reflected BM's are the skew BM's. It can be shown that for fixed $t > 0$, $Y_t \overset{\text{law}}{=} \epsilon|W_t|$ where W is a BM independent of the Bernoulli r.v. ϵ, $\mathbb{P}(\epsilon = 1) = p, \mathbb{P}(\epsilon = -1) = 1 - p$ where $p = \frac{1}{2(1-\alpha)}$.

The relation (4.1.13) between $L_t^0(Y)$ and $L_t^{0-}(Y)$ reads

$$L_t^0(Y) - L_t^{0-}(Y) = 2\int_0^t \mathbb{1}_{\{Y_s=0\}}dY_s\,.$$

The integral $\int_0^t \mathbb{1}_{\{Y_s=0\}}dW_s$ is null and $\int_0^t \mathbb{1}_{\{Y_s=0\}}dL_s^0(Y) = L_t^0(Y)$, hence

$$L_t^0(Y) - L_t^{0-}(Y) = 2\alpha L_t^0(Y)$$

that is $L_t^{0-}(Y) = L_t^0(Y)(1-2\alpha)$, which proves the nonexistence of a skew BM for $\alpha > 1/2$.

Comment 5.5.2.2 For several studies of skew Brownian motion, and more generally of processes Y satisfying

$$Y_t = \int_0^t \sigma(Y_s)dB_s + \int \nu(dy)L_t^y(Y)$$

we refer to Barlow [47], Harrison and Shepp [424], Ouknine [687], Le Gall [567], Lejay [575], Stroock and Yor [813] and Weinryb [838].

Example 5.5.2.3 Sticky Brownian Motion. Let $x > 0$. The solution of

$$X_t = x + \int_0^t \mathbb{1}_{\{X_s>0\}} dW_s + \theta \int_0^t \mathbb{1}_{\{X_s=0\}} ds \tag{5.5.4}$$

with $\theta > 0$ is called sticky Brownian motion with parameter θ. From Tanaka's formula,

$$X_t^- = -\theta \int_0^t \mathbb{1}_{\{X_s=0\}} ds + \frac{1}{2} L_t(X) .$$

The process $\theta \int_0^t \mathbb{1}_{\{X_s=0\}} ds$ is increasing, hence, from Skorokhod's lemma, $L_t(X) = 2\theta \int_0^t \mathbb{1}_{\{X_s=0\}} ds$ and $X_t^- = 0$. Hence, we may write the equation (5.5.4) as

$$X_t = x + \int_0^t \mathbb{1}_{\{X_s>0\}} dW_s + \frac{1}{2} L_t(X)$$

which enables us to write

$$X_t = \beta\left(\int_0^t \mathbb{1}_{\{X_s>0\}} ds\right)$$

where $(\beta(u), u \geq 0)$ is a reflecting BM starting from x. See Warren [835] for a thorough study of sticky Brownian motion.

Exercise 5.5.2.4 Let $\theta > 0$ and X be the sticky Brownian motion with $X_0 = 0$.

(1) Prove that $L_t^x(X) = 0$, for every $x < 0$; then, prove that $X_t \geq 0$, a.s.

(2) Let $A_t^+ = \int_0^t ds\, \mathbb{1}_{\{X_s>0\}}$, $A_t^0 = \int_0^t ds\, \mathbb{1}_{\{X_s=0\}}$, and define their inverses $\alpha_u^+ = \inf\{t : A_t^+ > u\}$ and $\alpha_u^0 = \inf\{t : A_t^0 > u\}$. Identify the law of $(X_{\alpha_u^+}, u \geq 0)$.

(3) Let G be a Gaussian variable, with unit variance and 0 expectation. Prove that, for any $u > 0$ and $t > 0$

$$\alpha_u^+ \stackrel{\text{law}}{=} u + \frac{1}{\theta}\sqrt{u}\,|G|\,;\; A_t^+ \stackrel{\text{law}}{=} \left(\sqrt{t + \frac{G^2}{4\theta^2}} - \frac{|G|}{2\theta}\right)^2$$

deduce that

$$A_t^0 \stackrel{\text{law}}{=} \frac{|G|}{\theta}\sqrt{t + \frac{G^2}{4\theta^2}} - \frac{G^2}{2\theta^2}$$

and compute $\mathbb{E}(A_t^0)$.

Hint: The process $X_{\alpha_u^+} = W_u^+ + \theta A^0_{\alpha_u^+}$ where W_u^+ is a BM and $A^0_{\alpha_u^+}$ is an increasing process, constant on $\{u : X_{\alpha_u^+} > 0\}$, solves Skorokhod equation. Therefore it is a reflected BM. The obvious equality $t = A_t^+ + A_t^0$ leads to $\alpha_u^+ = u + A^0_{\alpha_u^+}$, and $aA^0_{\alpha_u^+} \stackrel{\text{law}}{=} L_u^0$. ◁

5.6 Last Passage Times

We now present the study of the law (and the conditional law) of some last passage times for diffusion processes. In this section, W is a standard Brownian motion and its natural filtration is $\mathbf{F}$. These random times have been studied in Jeanblanc and Rutkowski [486] as theoretical examples of default times, in Imkeller [457] as examples of insider private information and, in a pure mathematical point of view, in Pitman and Yor [715] and Salminen [754].

5.6.1 Notation and Basic Results

If τ is a random time, then, it is easy to check that the process $\mathbb{P}(\tau > t|\mathcal{F}_t)$ is a super-martingale. Therefore, it admits a Doob-Meyer decomposition.

Lemma 5.6.1.1 *Let τ be a positive random time and*

$$\mathbb{P}(\tau > t|\mathcal{F}_t) = M_t - A_t$$

the Doob-Meyer decomposition of the super-martingale $Z_t = \mathbb{P}(\tau > t|\mathcal{F}_t)$. Then, for any predictable positive process H,

$$\mathbb{E}(H_\tau) = \mathbb{E}\left(\int_0^\infty dA_u H_u\right).$$

PROOF: For any process H of the form $H = \Lambda_s \mathbb{1}_{]s,t]}$ with $\Lambda_s \in b\mathcal{F}_s$, one has

$$\mathbb{E}(H_\tau) = \mathbb{E}(\Lambda_s \mathbb{1}_{]s,t]}(\tau)) = \mathbb{E}(\Lambda_s(A_t - A_s)).$$

The result follows from MCT. □

Comment 5.6.1.2 The reader will find in Nikeghbali and Yor [676] a multiplicative decomposition of the super-martingale Z as $Z_t = n_t D_t$ where D is a decreasing process and n a local martingale, and applications to enlargement of filtration.

We now show that, in a diffusion setup, A_t and M_t may be computed explicitly for some random times τ.

5.6.2 Last Passage Time of a Transient Diffusion

Proposition 5.6.2.1 *Let X be a transient homogeneous diffusion such that $X_t \to +\infty$ when $t \to \infty$, and s a scale function such that $s(+\infty) = 0$ (hence, $s(x) < 0$ for $x \in \mathbb{R}$) and $\Lambda_y = \sup\{t : X_t = y\}$ the last time that X hits y. Then,*

$$\mathbb{P}_x(\Lambda_y > t|\mathcal{F}_t) = \frac{s(X_t)}{s(y)} \wedge 1.$$

PROOF: We follow Pitman and Yor [715] and Yor [868], p.48, and use that under the hypotheses of the proposition, one can choose a scale function such that $s(x) < 0$ and $s(+\infty) = 0$ (see Sharpe [784]).

Observe that

$$\begin{aligned}
\mathbb{P}_x\big(\Lambda_y > t|\mathcal{F}_t\big) &= \mathbb{P}_x\Big(\inf_{u\geq t} X_u < y \,\Big|\, \mathcal{F}_t\Big) = \mathbb{P}_x\Big(\sup_{u\geq t}(-s(X_u)) > -s(y) \,\Big|\, \mathcal{F}_t\Big) \\
&= \mathbb{P}_{X_t}\Big(\sup_{u\geq 0}(-s(X_u)) > -s(y)\Big) = \frac{s(X_t)}{s(y)} \wedge 1,
\end{aligned}$$

where we have used the Markov property of X, and the fact that if M is a continuous local martingale with $M_0 = 1$, $M_t \geq 0$, and $\lim_{t\to\infty} M_t = 0$, then

$$\sup_{t\geq 0} M_t \stackrel{\text{law}}{=} \frac{1}{U},$$

where U has a uniform law on $[0, 1]$ (see Exercise 1.2.3.10). □

Lemma 5.6.2.2 *The* $\mathbf{F}^X$*-predictable compensator* A *associated with the random time* Λ_y *is the process* A *defined as* $A_t = -\frac{1}{2s(y)} L_t^{s(y)}(Y)$, *where* $L(Y)$ *is the local time process of the continuous martingale* $Y = s(X)$.

PROOF: From $x \wedge y = x - (x-y)^+$, Proposition 5.6.2.1 and Tanaka's formula, it follows that

$$\frac{s(X_t)}{s(y)} \wedge 1 = M_t + \frac{1}{2s(y)} L_t^{s(y)}(Y) = M_t + \frac{1}{s(y)} \ell_t^y(X)$$

where M is a martingale. The required result is then easily obtained. □

We deduce the law of the last passage time:

$$\begin{aligned}
\mathbb{P}_x(\lambda_y > t) &= \left(\frac{s(x)}{s(y)} \wedge 1\right) + \frac{1}{s(y)} \mathbb{E}_x(\ell_t^y(X)) \\
&= \left(\frac{s(x)}{s(y)} \wedge 1\right) + \frac{1}{s(y)} \int_0^t du\, p_u^{(m)}(x, y)\,.
\end{aligned}$$

Hence, for $x < y$

$$\begin{aligned}
\mathbb{P}_x(\Lambda_y \in dt) &= -\frac{dt}{s(y)} p_t^{(m)}(x, y) = -\frac{dt}{s(y)m(y)} p_t(x, y) \\
&= -\frac{\sigma^2(y)s'(y)}{2s(y)} p_t(x, y) dt\,. \qquad (5.6.1)
\end{aligned}$$

For $x > y$, we have to add a mass at point 0 equal to

$$1 - \left(\frac{s(x)}{s(y)} \wedge 1\right) = 1 - \frac{s(x)}{s(y)} = \mathbb{P}_x(T_y < \infty)\,.$$

Example 5.6.2.3 Last Passage Time for a Transient Bessel Process: For a Bessel process of dimension $\delta > 2$ and index ν (see $\rightarrowtail$ Chapter 6), starting from 0,

$$\mathbb{P}_0^\delta(\Lambda_a < t) = \mathbb{P}_0^\delta(\inf_{u \geq t} R_u > a) = \mathbb{P}_0^\delta(\sup_{u \geq t} R_u^{-2\nu} < a^{-2\nu})$$
$$= \mathbb{P}_0^\delta\left(\frac{R_t^{-2\nu}}{U} < a^{-2\nu}\right) = \mathbb{P}_0^\delta(a^{2\nu} < UR_t^{2\nu}) = \mathbb{P}_0^\delta\left(\frac{a^2}{R_1^2 U^{1/\nu}} < t\right).$$

Thus, the r.v. $\Lambda_a = \frac{a^2}{R_1^2 U^{1/\nu}}$ is distributed as $\frac{a^2}{2\gamma(\nu+1)\beta_{\nu,1}} \stackrel{\text{law}}{=} \frac{a^2}{2\gamma(\nu)}$ where $\gamma(\nu)$ is a gamma variable with parameter ν:

$$\mathbb{P}(\gamma(\nu) \in dt) = \mathbb{1}_{\{t \geq 0\}} \frac{t^{\nu-1}e^{-t}}{\Gamma(\nu)} dt.$$

Hence,

$$\mathbb{P}_0^\delta(\Lambda_a \in dt) = \mathbb{1}_{\{t \geq 0\}} \frac{1}{t\Gamma(\nu)} \left(\frac{a^2}{2t}\right)^\nu e^{-a^2/(2t)} dt. \tag{5.6.2}$$

We might also find this result directly from the general formula (5.6.1) and apply formula (6.2.3) for the expression of the density.

Proposition 5.6.2.4 *For H a positive predictable process*

$$\mathbb{E}_x(H_{\Lambda_y} | \Lambda_y = t) = \mathbb{E}_x(H_t | X_t = y)$$

and, for $y > x$,

$$\mathbb{E}_x(H_{\Lambda_y}) = \int_0^\infty \mathbb{E}_x(\Lambda_y \in dt)\, \mathbb{E}_x(H_t | X_t = y).$$

In the case $x > y$,

$$\mathbb{E}_x(H_{\Lambda_y}) = H_0 \left(1 - \frac{s(x)}{s(y)}\right) + \int_0^\infty \mathbb{E}_x(\Lambda_y \in dt)\, \mathbb{E}_x(H_t | X_t = y).$$

PROOF: We have shown in the previous Proposition 5.6.2.1 that

$$\mathbb{P}_x(\Lambda_y > t | \mathcal{F}_t) = \frac{s(X_t)}{s(y)} \wedge 1.$$

From Itô-Tanaka's formula

$$\frac{s(X_t)}{s(y)} \wedge 1 = \frac{s(x)}{s(y)} \wedge 1 + \int_0^t \mathbb{1}_{\{X_u > y\}} d\frac{s(X_u)}{s(y)} - \frac{1}{2} L_t^{s(y)}(s(X)).$$

It follows, using Lemma 5.6.1.1 that

$$\begin{aligned}\mathbb{E}_x(H_{\Lambda_x}) &= \frac{1}{2}\mathbb{E}_x\left(\int_0^\infty H_u\, d_u L_u^{s(y)}(s(X))\right)\\ &= \frac{1}{2}\mathbb{E}_x\left(\int_0^\infty \mathbb{E}_x(H_u|X_u = y)\, d_u L_u^{s(y)}(s(X))\right).\end{aligned}$$

Therefore, replacing H_u by $H_u g(u)$, we get

$$\mathbb{E}_x\left(H_{\Lambda_x} g(\Lambda_x)\right) = \frac{1}{2}\mathbb{E}_x\left(\int_0^\infty g(u)\,\mathbb{E}_x\left(H_u|X_u = y\right)\, d_u L_u^{s(y)}(s(X))\right). \quad (5.6.3)$$

Consequently, from (5.6.3), we obtain

$$\begin{aligned}\mathbb{P}_x\left(\Lambda_y \in du\right) &= \frac{1}{2} d_u \mathbb{E}_x\left(L_u^{s(y)}(s(X))\right)\\ \mathbb{E}_x\left(H_{\Lambda_y}|\Lambda_y = t\right) &= \mathbb{E}_x(H_t|X_t = y).\end{aligned}$$

□

Remark 5.6.2.5 In the literature, some studies of last passage times employ time inversion. See an example in the next Exercise 5.6.2.6.

Exercise 5.6.2.6 Let X be a drifted Brownian motion with positive drift ν and Λ_y^ν its last passage time at level y. Prove that

$$\mathbb{P}_x(\Lambda_y^{(\nu)} \in dt) = \frac{\nu}{\sqrt{2\pi t}} \exp\left(-\frac{1}{2t}(x - y + \nu t)^2\right) dt,$$

and

$$\mathbb{P}_x(\Lambda_y^{(\nu)} = 0) = \begin{cases} 1 - e^{-2\nu(x-y)}, & \text{for } x > y \\ 0 & \text{for } x < y. \end{cases}$$

Prove, using time inversion that, for $x = 0$,

$$\Lambda_y^{(\nu)} \overset{\text{law}}{=} \frac{1}{T_\nu^{(y)}}$$

where

$$T_a^{(b)} = \inf\{t : B_t + bt = a\}$$

See Madan et al. [611]. ◁

5.6.3 Last Passage Time Before Hitting a Level

Let $X_t = x + \sigma W_t$ where the initial value x is positive and σ is a positive constant. We consider, for $0 < a < x$ the last passage time at the level a before hitting the level 0, given as $g_{T_0}^a(X) = \sup\{t \leq T_0 : X_t = a\}$, where

$$T_0 = T_0(X) = \inf\{t \geq 0 : X_t = 0\}.$$

(In a financial setting, T_0 can be interpreted as the time of bankruptcy.) Then, setting $\alpha = (a-x)/\sigma$, $T_{-x/\sigma}(W) = \inf\{t : W_t = -x/\sigma\}$ and $d_t^\alpha(W) = \inf\{s \geq t : W_s = \alpha\}$

$$\mathbb{P}_x\big(g^a_{T_0}(X) \leq t|\mathcal{F}_t\big) = \mathbb{P}\big(d_t^\alpha(W) > T_{-x/\sigma}(W)|\mathcal{F}_t\big)$$

on the set $\{t < T_{-x/\sigma}(W)\}$. It is easy to prove that

$$\mathbb{P}\big(d_t^\alpha(W) < T_{-x/\sigma}(W)|\mathcal{F}_t\big) = \Psi(W_{t\wedge T_{-x/\sigma}(W)}, \alpha, -x/\sigma),$$

where the function $\Psi(\cdot, a, b) : \mathbb{R} \to \mathbb{R}$ equals, for $a > b$,

$$\Psi(y,a,b) = \mathbb{P}_y(T_a(W) > T_b(W)) = \begin{cases} (a-y)/(a-b) & \text{for } b < y < a, \\ 1 & \text{for } a < y, \\ 0 & \text{for } y < b. \end{cases}$$

(See Proposition 3.5.1.1 for the computation of Ψ.) Consequently, on the set $\{T_0(X) > t\}$ we have

$$\mathbb{P}_x\big(g^a_{T_0}(X) \leq t|\mathcal{F}_t\big) = \frac{(\alpha - W_{t\wedge T_0})^+}{a/\sigma} = \frac{(\alpha - W_t)^+}{a/\sigma} = \frac{(a - X_t)^+}{a}. \quad (5.6.4)$$

As a consequence, applying Tanaka's formula, we obtain the following result.

Lemma 5.6.3.1 *Let $X_t = x + \sigma W_t$, where $\sigma > 0$. The* **F***-predictable compensator associated with the random time $g^a_{T_0(X)}$ is the process A defined as $A_t = \frac{1}{2\alpha} L^\alpha_{t\wedge T_{-x/\sigma}(W)}(W)$, where $L^\alpha(W)$ is the local time of the Brownian Motion W at level $\alpha = (a-x)/\sigma$.*

5.6.4 Last Passage Time Before Maturity

In this subsection, we study the last passage time at level a of a diffusion process X before the fixed horizon (maturity) T. We start with the case where $X = W$ is a Brownian motion starting from 0 and where the level a is null:

$$g_T = \sup\{t \leq T : W_t = 0\}.$$

Lemma 5.6.4.1 *The* **F***-predictable compensator associated with the random time g_T equals*

$$A_t = \sqrt{\frac{2}{\pi}} \int_0^{t\wedge T} \frac{dL_s}{\sqrt{T-s}},$$

where L is the local time at level 0 of the Brownian motion W.

PROOF: It suffices to give the proof for $T = 1$, and we work with $t < 1$. Let G be a standard Gaussian variable. Then

$$\mathbb{P}\Big(\frac{a^2}{G^2} > 1 - t\Big) = \Phi\Big(\frac{|a|}{\sqrt{1-t}}\Big),$$

where $\Phi(x) = \sqrt{\frac{2}{\pi}} \int_0^x \exp(-\frac{u^2}{2}) du$. For $t < 1$, the set $\{g_1 \leq t\}$ is equal to $\{d_t > 1\}$. It follows from (4.3.3) that

$$\mathbb{P}(g_1 \leq t | \mathcal{F}_t) = \Phi\left(\frac{|W_t|}{\sqrt{1-t}}\right).$$

Then, the Itô-Tanaka formula combined with the identity

$$x\Phi'(x) + \Phi''(x) = 0$$

leads to

$$\begin{aligned}
\mathbb{P}(g_1 \leq t | \mathcal{F}_t) &= \int_0^t \Phi'\left(\frac{|W_s|}{\sqrt{1-s}}\right) d\left(\frac{|W_s|}{\sqrt{1-s}}\right) + \frac{1}{2}\int_0^t \frac{ds}{1-s} \Phi''\left(\frac{|W_s|}{\sqrt{1-s}}\right) \\
&= \int_0^t \Phi'\left(\frac{|W_s|}{\sqrt{1-s}}\right) \frac{\mathrm{sgn}(W_s)}{\sqrt{1-s}} dW_s + \int_0^t \frac{dL_s}{\sqrt{1-s}} \Phi'\left(\frac{|W_s|}{\sqrt{1-s}}\right) \\
&= \int_0^t \Phi'\left(\frac{|W_s|}{\sqrt{1-s}}\right) \frac{\mathrm{sgn}(W_s)}{\sqrt{1-s}} dW_s + \sqrt{\frac{2}{\pi}} \int_0^t \frac{dL_s}{\sqrt{1-s}}.
\end{aligned}$$

It follows that the **F**-predictable compensator associated with g_1 is

$$A_t = \sqrt{\frac{2}{\pi}} \int_0^t \frac{dL_s}{\sqrt{1-s}}, \ (t < 1).$$

□

These results can be extended to the last time before T when the Brownian motion reaches the level α, i.e., $g_T^\alpha = \sup\{t \leq T : W_t = \alpha\}$, where we set $\sup(\emptyset) = T$. The predictable compensator associated with g_T^α is

$$A_t = \sqrt{\frac{2}{\pi}} \int_0^{t \wedge T} \frac{dL_s^\alpha}{\sqrt{T-s}},$$

where L^α is the local time of W at level α.

We now study the case where $X_t = x + \mu\, t + \sigma\, W_t$, with constant coefficients μ and $\sigma > 0$. Let

$$\begin{aligned}
g_1^a(X) &= \sup\{t \leq 1 : X_t = a\} \\
&= \sup\{t \leq 1 : \nu t + W_t = \alpha\}
\end{aligned}$$

where $\nu = \mu/\sigma$ and $\alpha = (a - x)/\sigma$. From Lemma 4.3.9.1, setting

$$V_t = \alpha - \nu t - W_t = (a - X_t)/\sigma,$$

we obtain

$$\mathbb{P}(g_1^a(X) \le t|\mathcal{F}_t) = (1 - e^{\nu V_t} H(\nu, |V_t|, 1-t))\mathbb{1}_{\{T_0(V)\le t\}},$$

where

$$H(\nu, y, s) = e^{-\nu y}\mathcal{N}\left(\frac{\nu s - y}{\sqrt{s}}\right) + e^{\nu y}\mathcal{N}\left(\frac{-\nu s - y}{\sqrt{s}}\right).$$

Using Itô's lemma, we obtain the decomposition of $1 - e^{\nu V_t} H(\nu, |V_t|, 1-t)$ as a semi-martingale $M_t + C_t$.

We note that C increases only on the set $\{t : X_t = a\}$. Indeed, setting $g_1^a(X) = g$, for any predictable process H, one has

$$\mathbb{E}(H_g) = \mathbb{E}\left(\int_0^\infty dC_s H_s\right)$$

hence, since $X_g = a$,

$$0 = \mathbb{E}(\mathbb{1}_{X_g \neq a}) = \mathbb{E}\left(\int_0^\infty dC_s \mathbb{1}_{X_s \neq a}\right).$$

Therefore, $dC_t = \kappa_t dL_t^a(X)$ and, since L increases only at points such that $X_t = a$ (i.e., $V_t = 0$), one has

$$\kappa_t = H_x'(\nu, 0, 1-t).$$

The martingale part is given by $dM_t = m_t dW_t$ where

$$m_t = e^{\nu V_t}\left(\nu H(\nu, |V_t|, 1-t) - \mathrm{sgn}(V_t)\, H_x'(\nu, |V_t|, 1-t)\right).$$

Therefore, the predictable compensator associated with $g_1^a(X)$ is

$$\int_0^t \frac{H_x'(\nu, 0, 1-s)}{e^{\nu V_s} H(\nu, 0, 1-s)} dL_s^a.$$

Exercise 5.6.4.2 The aim of this exercise is to compute, for $t < T < 1$, the quantity $\mathbb{E}(h(W_T)\mathbb{1}_{\{T<g_1\}}|\mathcal{G}_t)$, which is the price of the claim $h(S_T)$ with barrier condition $\mathbb{1}_{\{T<g_1\}}$.

Prove that

$$\mathbb{E}(h(W_T)\mathbb{1}_{\{T<g_1\}}|\mathcal{F}_t) = \mathbb{E}(h(W_T)|\mathcal{F}_t) - \mathbb{E}\Big(h(W_T)\Phi\Big(\frac{|W_T|}{\sqrt{1-T}}\Big)\,\Big|\,\mathcal{F}_t\Big),$$

where

$$\Phi(x) = \sqrt{\frac{2}{\pi}}\int_0^x \exp\left(-\frac{u^2}{2}\right) du.$$

Define $k(w) = h(w)\Phi(|w|/\sqrt{1-T})$. Prove that $\mathbb{E}\big(k(W_T)\,\big|\,\mathcal{F}_t\big) = \widetilde{k}(t, W_t)$, where

$$\begin{aligned}\widetilde{k}(t,a) &= \mathbb{E}\Big(k(W_{T-t}+a)\Big)\\ &= \frac{1}{\sqrt{2\pi(T-t)}}\int_{\mathbb{R}} h(u)\Phi\Big(\frac{|u|}{\sqrt{1-T}}\Big)\exp\Big(-\frac{(u-a)^2}{2(T-t)}\Big)du.\end{aligned}$$

◁

5.6.5 Absolutely Continuous Compensator

From the preceding computations, the reader might think that the **F**-predictable compensator is always singular w.r.t. the Lebesgue measure. This is not the case, as we show now. We are indebted to Michel Émery for this example.

Let W be a Brownian motion and let $\tau = \sup\{t \leq 1 : W_1 - 2W_t = 0\}$, that is the last time before 1 when the Brownian motion is equal to half of its terminal value at time 1. Then,

$$\{\tau \leq t\} = \left\{\inf_{t \leq s \leq 1} 2W_s \geq W_1 \geq 0\right\} \cup \left\{\sup_{t \leq s \leq 1} 2W_s \leq W_1 \leq 0\right\}.$$

▶ The quantity

$$\mathbb{P}(\tau \leq t, W_1 \geq 0|\mathcal{F}_t) = \mathbb{P}\left(\inf_{t \leq s \leq 1} 2W_s \geq W_1 \geq 0|\mathcal{F}_t\right)$$

can be evaluated using the equalities

$$\begin{aligned}\left\{\inf_{t \leq s \leq 1} W_s \geq \frac{W_1}{2} \geq 0\right\} &= \left\{\inf_{t \leq s \leq 1}(W_s - W_t) \geq \frac{W_1}{2} - W_t \geq -W_t\right\} \\ &= \left\{\inf_{0 \leq u \leq 1-t}(\widetilde{W}_u) \geq \frac{\widetilde{W}_{1-t}}{2} - \frac{W_t}{2} \geq -W_t\right\},\end{aligned}$$

where $(\widetilde{W}_u = W_{t+u} - W_t, u \geq 0)$ is a Brownian motion independent of $\mathcal{F}_t$. It follows that

$$\mathbb{P}\left(\inf_{t \leq s \leq 1} W_s \geq \frac{W_1}{2} \geq 0|\mathcal{F}_t\right) = \Psi(1-t, W_t),$$

where

$$\begin{aligned}\Psi(s,x) &= \mathbb{P}\left(\inf_{0 \leq u \leq s} \widetilde{W}_u \geq \frac{\widetilde{W}_s}{2} - \frac{x}{2} \geq -x\right) = \mathbb{P}\left(2M_s - W_s \leq \frac{x}{2}, W_s \leq \frac{x}{2}\right) \\ &= \mathbb{P}\left(2M_1 - W_1 \leq \frac{x}{2\sqrt{s}}, W_1 \leq \frac{x}{2\sqrt{s}}\right).\end{aligned}$$

▶ The same kind of computation leads to

$$\mathbb{P}\left(\sup_{t \leq s \leq 1} 2W_s \leq W_1 \leq 0|\mathcal{F}_t\right) = \Psi(1-t, -W_t).$$

▶ The quantity $\Psi(s,x)$ can now be computed from the joint law of the maximum and of the process at time 1; however, we prefer to use Pitman's theorem (see ↣ Section 5.7): let $\widetilde{U}$ be a r.v. uniformly distributed on $[-1,+1]$ independent of $R_1 := 2M_1 - W_1$, then

$$\mathbb{P}(2M_1 - W_1 \leq y, W_1 \leq y) = \mathbb{P}(R_1 \leq y, \widetilde{U}R_1 \leq y)$$
$$= \frac{1}{2}\int_{-1}^{1} \mathbb{P}(R_1 \leq y, uR_1 \leq y)du\,.$$

For $y > 0$,

$$\frac{1}{2}\int_{-1}^{1} \mathbb{P}(R_1 \leq y, uR_1 \leq y)du = \frac{1}{2}\int_{-1}^{1} \mathbb{P}(R_1 \leq y)du = \mathbb{P}(R_1 \leq y)\,.$$

For $y < 0$

$$\int_{-1}^{1} \mathbb{P}(R_1 \leq y, uR_1 \leq y)du = 0\,.$$

Therefore

$$\mathbb{P}(\tau \leq t|\mathcal{F}_t) = \Psi(1-t, W_t) + \Psi(1-t, -W_t) = \rho\left(\frac{|W_t|}{\sqrt{1-t}}\right)$$

where

$$\rho(y) = \mathbb{P}(R_1 \leq y) = \sqrt{\frac{2}{\pi}}\int_0^y x^2 e^{-x^2/2}dx\,.$$

Then $Z_t = \mathbb{P}(\tau > t|\mathcal{F}_t) = 1-\rho(\frac{|W_t|}{\sqrt{1-t}})$. We can now apply Tanaka's formula to the function ρ. Noting that $\rho'(0) = 0$, the contribution to the Doob-Meyer decomposition of Z of the local time of W at level 0 is 0. Furthermore, the increasing process A of the Doob-Meyer decomposition of Z is given by

$$dA_t = \left(\frac{1}{2}\rho''\left(\frac{|W_t|}{\sqrt{1-t}}\right)\frac{1}{1-t} + \frac{1}{2}\rho'\left(\frac{|W_t|}{\sqrt{1-t}}\right)\frac{|W_t|}{\sqrt{(1-t)^3}}\right)dt$$
$$= \frac{1}{1-t}\frac{|W_t|}{\sqrt{1-t}}e^{-W_t^2/2(1-t)}dt\,.$$

We note that A may be obtained as the dual predictable projection on the Brownian filtration of the process $A_s^{(W_1)}, s \leq 1$, where $(A_s^{(x)}, s \leq 1)$ is the compensator of τ under the law of the Brownian bridge $\mathbb{P}^{(1)}_{0\to x}$.

5.6.6 Time When the Supremum is Reached

Let W be a Brownian motion, $M_t = \sup_{s\leq t} W_s$ and let τ be the time when the supremum on the interval $[0, 1]$ is reached, i.e.,

$$\tau = \inf\{t \leq 1 : W_t = M_1\} = \sup\{t \leq 1 : M_t - W_t = 0\}\,.$$

Let us denote by ζ the positive continuous semimartingale

$$\zeta_t = \frac{M_t - W_t}{\sqrt{1-t}},\ t < 1.$$

Let $F_t = \mathbb{P}(\tau \le t|\mathcal{F}_t)$. Since $F_t = \Phi(\zeta_t)$, (where $\Phi(x) = \sqrt{\frac{2}{\pi}}\int_0^x \exp(-\frac{u^2}{2})du$, see Example 4.1.7.5) using Itô's formula, we obtain the canonical decomposition of F as follows:

$$\begin{aligned} F_t &= \int_0^t \Phi'(\zeta_u)\, d\zeta_u + \frac{1}{2}\int_0^t \Phi''(\zeta_u)\,\frac{du}{1-u} \\ &\overset{(i)}{=} -\int_0^t \Phi'(\zeta_u)\frac{dW_u}{\sqrt{1-u}} + \sqrt{\frac{2}{\pi}}\int_0^t \frac{dM_u}{\sqrt{1-u}} \overset{(ii)}{=} U_t + \tilde{F}_t, \end{aligned}$$

where $U_t = -\int_0^t \Phi'(\zeta_u)\dfrac{dW_u}{\sqrt{1-u}}$ is a martingale and $\widetilde{F}$ a predictable increasing process. To obtain (i), we have used that $x\Phi' + \Phi'' = 0$; to obtain (ii), we have used that $\Phi'(0) = \sqrt{2/\pi}$ and also that the process M increases only on the set

$$\{u \in [0,t] : M_u = W_u\} = \{u \in [0,t] : \zeta_u = 0\}.$$

5.6.7 Last Passage Times for Particular Martingales

Proposition 5.6.7.1 *Let X be a continuous positive local martingale such that $X_0 = x$, and $\lim_{t\to\infty} X_t = 0$. Let $\Sigma_t = \sup_{s\le t} X_s$ the (continuous) supremum process. We consider the last passage time of the process X at the level Σ_∞:*

$$\begin{aligned} g &= \sup\{t \ge 0 : \quad X_t = \Sigma_\infty\} \\ &= \sup\{t \ge 0 : \quad \Sigma_t - X_t = 0\}. \end{aligned} \tag{5.6.5}$$

Consider the supermartingale

$$Z_t = \mathbb{P}(g > t \mid \mathcal{F}_t).$$

Then:

(i) *the multiplicative decomposition of the supermartingale Z reads*

$$Z_t = \frac{X_t}{\Sigma_t},$$

(ii) *The Doob-Meyer (additive decomposition) of Z is:*

$$Z_t = m_t - \log(\Sigma_t), \tag{5.6.6}$$

where m is the $\mathbf{F}$-martingale

$$m_t = \mathbb{E}\left[\log \Sigma_\infty|\mathcal{F}_t\right].$$

PROOF: We recall the Doob's maximal identity 1.2.3.10. Applying (1.2.2) to the martingale $(Y_t := X_{T+t}, t \ge 0)$ for the filtration $\mathbf{F}^T := (\mathcal{F}_{t+T}, t \ge 0)$, where T is a $\mathbf{F}$-stopping time, we obtain that

$$\mathbb{P}\left(\Sigma^T > a|\mathcal{F}_T\right) = \left(\frac{X_T}{a}\right) \wedge 1, \tag{5.6.7}$$

where

$$\Sigma^T := \sup_{u \geq T} X_u.$$

Hence $\frac{X_T}{\Sigma^T}$ is a uniform random variable on $(0,1)$, independent of $\mathcal{F}_T$. The multiplicative decomposition of Z follows from

$$\mathbb{P}(g > t \mid \mathcal{F}_t) = \mathbb{P}\left(\sup_{u \geq t} X_u \geq \Sigma_t \mid \mathcal{F}_t\right) = \left(\frac{X_t}{\Sigma_t}\right) \wedge 1 = \frac{X_t}{\Sigma_t}$$

From the integration by parts formula applied to $\frac{X_t}{\Sigma_t}$, and using the fact that X, hence Σ are continuous, we obtain

$$dZ_t = \frac{dX_t}{\Sigma_t} - X_t \frac{d\Sigma_t}{(\Sigma_t)^2}$$

Since $d\Sigma_t$ charges only the set $\{t : X_t = \Sigma_t\}$, one has

$$dZ_t = \frac{dX_t}{\Sigma_t} - \frac{d\Sigma_t}{\Sigma_t} = \frac{dX_t}{\Sigma_t} - d(\ln \Sigma_t)$$

From the uniqueness of the Doob-Meyer decomposition, we obtain that the predictable increasing part of the submartingale Z is $\ln \Sigma_t$, hence

$$Z_t = m_t - \ln \Sigma_t$$

where m is a martingale. The process Z is of class (D), hence m is a uniformly integrable martingale. From $Z_\infty = 0$, one obtains that $m_t = \mathbb{E}(\ln \Sigma_\infty | \mathcal{F}_t)$. □

Remark 5.6.7.2 From the Doob-Meyer (additive) decomposition of Z, we have $1 - Z_t = (1 - m_t) + \ln \Sigma_t$. From Skorokhod's reflection lemma presented in Subsection 4.1.7 we deduce that

$$\ln \Sigma_t = \sup_{s \leq t} m_s - 1$$

We now study the Azéma supermartingale associated with the random time L, a last passage time or the end of a predictable set Γ, i.e.,

$$L(\omega) = \sup\{t : (t, \omega) \in \Gamma\}$$

(See ↣ Section 5.9.4 for properties of these times in an enlargement of filtration setting).

Proposition 5.6.7.3 *Let L be the end of a predictable set. Assume that all the $\mathbf{F}$-martingales are continuous and that L avoids the $\mathbf{F}$-stopping times. Then, there exists a continuous and nonnegative local martingale N, with $N_0 = 1$ and $\lim_{t\to\infty} N_t = 0$, such that:*

$$Z_t = \mathbb{P}(L > t \mid \mathcal{F}_t) = \frac{N_t}{\Sigma_t}$$

where $\Sigma_t = \sup_{s\le t} N_s$. The Doob-Meyer decomposition of Z is

$$Z_t = m_t - A_t$$

and the following relations hold

$$\begin{aligned} N_t &= \exp\left(\int_0^t \frac{dm_s}{Z_s} - \frac{1}{2}\int_0^t \frac{d\langle m\rangle_s}{Z_s^2}\right) \\ \Sigma_t &= \exp(A_t) \\ m_t &= 1 + \int_0^t \frac{dN_s}{\Sigma_s} = \mathbb{E}(\ln S_\infty | \mathcal{F}_t) \end{aligned}$$

PROOF: As recalled previously, the Doob-Meyer decomposition of Z reads $Z_t = m_t - A_t$ with m and A continuous, and dA_t is carried by $\{t : Z_t = 1\}$. Then, for $t < T_0 := \inf\{t : Z_t = 0\}$

$$-\ln Z_t = -\left(\int_0^t \frac{dm_s}{Z_s} - \frac{1}{2}\int_0^t \frac{d\langle m\rangle_s}{Z_s^2}\right) + A_t$$

From Skorokhod's reflection lemma (see Subsection 4.1.7) we deduce that

$$A_t = \sup_{u\le t}\left(\int_0^u \frac{dm_s}{Z_s} - \frac{1}{2}\int_0^u \frac{d\langle m\rangle_s}{Z_s^2}\right)$$

Introducing the local martingale N defined by

$$N_t = \exp\left(\int_0^t \frac{dm_s}{Z_s} - \frac{1}{2}\int_0^t \frac{d\langle m\rangle_s}{Z_s^2}\right),$$

it follows that

$$Z_t = \frac{N_t}{\Sigma_t}$$

and

$$\Sigma_t = \sup_{u\le t} N_u = \exp\left(\sup_{u\le t}\left(\int_0^u \frac{dm_s}{Z_s} - \frac{1}{2}\int_0^u \frac{d\langle m\rangle_s}{Z_s^2}\right)\right) = e^{A_t}$$

□

The three following exercises are from the work of Bentata and Yor [72].

Exercise 5.6.7.4 Let M be a positive martingale, such that $M_0 = 1$ and $\lim_{t\to\infty} M_t = 0$. Let $a \in [0,1[$ and define $G_a = \sup\{t : M_t = a\}$. Prove that

$$\mathbb{P}(G_a \leq t|\mathcal{F}_t) = \left(1 - \frac{M_t}{a}\right)^+$$

Assume that, for every $t > 0$, the law of the r.v. M_t admits a density $(m_t(x), x \geq 0)$, and $(t, x) \to m_t(x)$ may be chosen continuous on $(0,\infty)^2$ and that $d\langle M\rangle_t = \sigma_t^2 dt$, and there exists a jointly continuous function $(t, x) \to \theta_t(x) = \mathbb{E}(\sigma_t^2 | M_t = x)$ on $(0,\infty)^2$. Prove that

$$\mathbb{P}(G_a \in dt) = \left(1 - \frac{M_0}{a}\right)\delta_0(dt) + \mathbb{1}_{\{t>0\}}\frac{1}{2a}\theta_t(a)m_t(a)dt$$

Hint: Use Tanaka's formula to prove that the result is equivalent to $d_t\mathbb{E}(L_t^a(M)) = dt\theta_t(a)m_t(a)$ where L is the Tanaka-Meyer local time (see Subsection 5.5.1). ◁

Exercise 5.6.7.5 Let B be a Brownian motion and

$$T_a^{(\nu)} = \inf\{t : B_t + \nu t = a\}$$
$$G_a^{(\nu)} = \sup\{t : B_t + \nu t = a\}$$

Prove that

$$(T_a^{(\nu)}, G_a^{(\nu)}) \stackrel{\text{law}}{=} \left(\frac{1}{G_\nu^{(a)}}, \frac{1}{T_\nu^{(a)}}\right)$$

Give the law of the pair $(T_a^{(\nu)}, G_a^{(\nu)})$. ◁

Exercise 5.6.7.6 Let X be a transient diffusion, such that

$$\mathbb{P}_x(T_0 < \infty) = 0, x > 0$$
$$\mathbb{P}_x(\lim_{t\to\infty} X_t = \infty) = 1, x > 0$$

and note s the scale function satisfying $s(0^+) = -\infty, s(\infty) = 0$. Prove that for all $x, t > 0$,

$$\mathbb{P}_x(G_y \in dt) = \frac{-1}{2s(y)}p_t^{(m)}(x,y)dt$$

where $p^{(m)}$ is the density transition w.r.t. the speed measure m. ◁

5.7 Pitman's Theorem about $(2M_t - W_t)$

5.7.1 Time Reversal of Brownian Motion

In our proof of Pitman's theorem, we shall need two results about time reversal of Brownian motion which are of interest by themselves:

Lemma 5.7.1.1 *Let W be a Brownian motion, L its local time at level 0 and $\tau_\ell = \inf\{t : L_t \geq \ell\}$. Then*

$$(W_u, u \leq \tau_\ell | \tau_\ell = t) \stackrel{\text{law}}{=} (W_u, u \leq t | L_t = \ell, W_t = 0)$$

As a consequence,

$$(W_{\tau_\ell - u}, u \leq \tau_\ell) \stackrel{\text{law}}{=} (W_u, u \leq \tau_\ell)$$

PROOF: Assuming the first property, we show how the second one is deduced. The scaling property allows us to restrict attention to the case $\ell = 1$. Since the law of the Brownian bridge is invariant under time reversal (see Section 4.3.5), we get that

$$(W_u, u \leq t | W_t = 0) \stackrel{\text{law}}{=} (W_{t-u}, u \leq t | W_t = 0).$$

This identity implies

$$((W_u, u \leq t), L_t | W_t = 0) \stackrel{\text{law}}{=} ((W_{t-u}, u \leq t), L_t | W_t = 0).$$

Therefore

$$(W_u, u \leq \tau_1 | \tau_1 = t) \stackrel{\text{law}}{=} (W_u, u \leq t | L_t = 1, W_t = 0)$$

$$\stackrel{\text{law}}{=} (W_{t-u}, u \leq t | L_t = 1, W_t = 0) \stackrel{\text{law}}{=} (W_{\tau_1 - u}, u \leq \tau_1 | \tau_1 = t).$$

We conclude that

$$(W_{\tau_1 - u}; u \leq \tau_1)(W_u; u \leq \tau_1).$$

□

The second result about time reversal is a particular case of a general result for Markov processes due to Nagasawa. We need some references to the Bessel process of dimension 3 (see ↣ Chapter 6).

Theorem 5.7.1.2 (Williams' Time Reversal Result.) *Let W be a BM, T_a the first hitting time of a by W and R a Bessel process of dimension 3 starting from 0, and Λ_a its last passage time at level a. Then*

$$(a - W_{T_a - t}, t \leq T_a) \stackrel{\text{law}}{=} (R_t, t \leq \Lambda_a).$$

PROOF: We refer to [RY], Chapter VII. □

5.7.2 Pitman's Theorem

Here again, the Bessel process of dimension 3 (denoted as BES^3) plays an essential rôle (see ↣ Chapter 6).

Theorem 5.7.2.1 (Pitman's Theorem.) *Let W be a Brownian motion and $M_t = \sup_{s \le t} W_s$. The following identity in law holds*

$$(2M_t - W_t, M_t; t \ge 0) \stackrel{\text{law}}{=} (R_t, J_t; t \ge 0)$$

where $(R_t; t \ge 0)$ is a BES^3 *process starting from 0 and $J_t = \inf_{s \ge t} R_s$.*

PROOF: We note that it suffices to prove the identity in law between the first two components, i.e.,

$$(2M_t - W_t; t \ge 0) \stackrel{\text{law}}{=} (R_t; t \ge 0)\,. \tag{5.7.1}$$

Indeed, the equality (5.7.1) implies

$$(2M_t - W_t, \inf_{s \ge t}(2M_s - W_s); t \ge 0) \stackrel{\text{law}}{=} \left(R_t, \inf_{s \ge t} R_s; t \ge 0\right)\,.$$

We prove below that $M_t = \inf_{s \ge t}(2M_s - W_s)$. Hence, the equality

$$(2M_t - W_t, M_t; t \ge 0) \stackrel{\text{law}}{=} (R_t, J_t; t \ge 0)\,.$$

holds.

▶ We prove $M_t = \inf_{s \ge t}(2M_s - W_s)$ in two steps. First, note that for $s \ge t$, $2M_s - W_s \ge M_s \ge M_t$ hence $M_t \le \inf_{s \ge t}(2M_s - W_s)$.

In a second step, we introduce $\theta_t = \inf\{s \ge t : M_s = W_s\}$. Since the increasing process M increases only when $M = W$, it is obvious that $M_t = M_{\theta_t}$. From $M_{\theta_t} = 2M_{\theta_t} - W_{\theta_t} \ge \inf_{s \ge \theta_t}(2M_s - W_s)$ we deduce that $M_t = \inf_{s \ge \theta_t}(2M_s - W_s) \ge \inf_{s \ge t}(2M_s - W_s)$. Therefore, the equality $M_t = \inf_{s \ge t}(2M_s - W_s)$ holds.

▶ We now prove the desired result (5.7.1) with the help of Lévy's identity: the two statements

$$(2M_t - W_t; t \ge 0) \stackrel{\text{law}}{=} (R_t; t \ge 0)$$

and

$$(|W_t| + L_t; t \ge 0) \stackrel{\text{law}}{=} (R_t; t \ge 0)\,,$$

are equivalent (we recall that L denotes the local time at 0 of W). Hence, we only need to prove that, for every ℓ,

$$(|W_t| + L_t; t \le \tau_\ell) \stackrel{\text{law}}{=} (R_t; t \le \Lambda_\ell) \tag{5.7.2}$$

where

$$\tau_\ell = \inf\{t : L_t \ge \ell\} \quad \text{and} \quad \Lambda_\ell = \sup\{t : R_t = \ell\}.$$

Accordingly, using Lemma 5.7.1.1, the equality (5.7.2) is equivalent to:

$$(|W_{\tau_\ell - t}| + (\ell - L_{\tau_\ell - t}); t \leq \tau_\ell) \stackrel{law}{=} (R_t; t \leq \Lambda_\ell).$$

By Lévy's identity, this is equivalent to:

$$(\ell - W_{T_\ell - t}; t \leq T_\ell) \stackrel{\text{law}}{=} (R_t; t \leq \Lambda_\ell)$$

which is precisely Williams' time reversal theorem.

□

Corollary 5.7.2.2 *Let $\widetilde{R}_t = 2M_t - W_t$, $\mathcal{R}_t = \sigma\{\widetilde{R}_s; s \leq t\}$, and let T be an $(\mathcal{R}_t)$ stopping time. Then, conditionally on $\mathcal{R}_T$, the r.v. M_T (and, consequently, the r.v. $M_T - W_T$) is uniformly distributed on $[0, \widetilde{R}_T]$. Hence, $\frac{M_T - W_T}{\widetilde{R}_T}$ is uniform on $[0,1]$ and independent of $\mathcal{R}_T$.*

PROOF: Using Pitman's theorem, the statement of the corollary is equivalent to: if $(R^a_s; s \geq 0)$ is a BES^3_a process, $\inf_{s \geq 0} R^a_s$ is uniform on $[0, a]$, which follows from the useful lemma of Exercise 1.2.3.10.

Consequently for $x < y$

$$\mathbb{P}(M_u \leq x | \widetilde{R}_u = y) = \mathbb{P}(Uy \leq x) = x/y.$$

□

The property featured in the corollary entails an intertwining property between the semigroups of BM and BES^3 which is detailed in the following exercise.

Exercise 5.7.2.3 Denote by (P_t) and (Q_t) respectively the semigroups of the Brownian motion and of the BES^3. Prove that $Q_t \Lambda = \Lambda P_t$ where

$$\Lambda : f \to \Lambda f(r) = \frac{1}{2r} \int_{-r}^{+r} dx f(x).$$

◁

Exercise 5.7.2.4 With the help of Corollary 5.7.2.2 and the Cameron-Martin formula, prove that the process $2M^{(\mu)}_t - W^{(\mu)}_t$, where $W^{(\mu)}_t = W_t + \mu t$, is a diffusion whose generator is $\frac{1}{2}\frac{d^2}{dx^2} + \mu \coth \mu x \frac{d}{dx}$. ◁

5.8 Filtrations

In the Black-Scholes model with constant coefficients, i.e.,

$$dS_t = S_t(\mu dt + \sigma dW_t),\ S_0 = x \tag{5.8.1}$$

where μ, σ and x are constants, the filtration $\mathbf{F}^S$ generated by the asset prices

$$\mathcal{F}_t^S := \sigma(S_s, s \leq t)$$

is equal to the filtration $\mathbf{F}^W$ generated by W. Indeed, the solution of (5.8.1) is

$$S_t = x \exp\left(\left(\mu - \frac{\sigma^2}{2}\right)t + \sigma W_t\right) \tag{5.8.2}$$

which leads to

$$W_t = \frac{1}{\sigma}\left(\ln \frac{S_t}{S_0} - \left(\mu - \frac{\sigma^2}{2}\right)t\right). \tag{5.8.3}$$

From (5.8.2), any function of S_t is a function of W_t, and $\mathcal{F}_t^S \subset \mathcal{F}_t^W$. From (5.8.3) the reverse inclusion holds.

This result remains valid for μ and σ deterministic functions, as long as $\sigma(t) > 0, \forall t$.

However, in general, the source of randomness is not so easy to identify; likewise models which are chosen to calibrate the data may involve more complicated filtrations. We present here a discussion of such set-ups. Our present aim is not to give a general framework but to study some particular cases.

5.8.1 Strong and Weak Brownian Filtrations

Amongst continuous-time processes, Brownian motion is undoubtedly the most studied process, and many characterizations of its law are known. It may thus seem a little strange that, deciding whether or not a filtration $\mathbf{F}$, on a given probability space $(\Omega, \mathcal{F}, \mathbb{P})$, is the natural filtration $\mathbf{F}^B$ of a Brownian motion $(B_t, t \geq 0)$ is a very difficult question and that, to date, no necessary and sufficient criterion has been found.

However, the following necessary condition can already discard a number of unsuitable "candidates," in a reasonably efficient manner: in order that $\mathbf{F}$ be a Brownian filtration, it is necessary that there exists an $\mathbf{F}$-Brownian motion β such that all $\mathbf{F}$-martingales may be written as $M_t = c + \int_0^t m_s d\beta_s$ for some $c \in \mathbb{R}$ and some predictable process m which satisfies $\int_0^t ds\, m_s^2 < \infty$. If needed, the reader may refer to $\rightarrowtail$ Section 9.5 for the general definition of the predictable representation property (PRP). This leads us to the following definition.

Definition 5.8.1.1 *A filtration* $\mathbf{F}$ *on* $(\Omega, \mathcal{F}, \mathbb{P})$ *such that* $\mathcal{F}_0$ *is* $\mathbb{P}$ *a.s. trivial is said to be **weakly Brownian** if there exists an* $\mathbf{F}$*-Brownian motion* β *such that* β *has the predictable representation property with respect to* $\mathbf{F}$.

A filtration $\mathbf{F}$ *on* $(\Omega, \mathcal{F}, \mathbb{P})$ *such that* $\mathcal{F}_0$ *is* $\mathbb{P}$ *a.s. trivial is said to be **strongly Brownian** if there exists an* $\mathbf{F}$*-BM* β *such that* $\mathcal{F}_t = \mathcal{F}_t^\beta$.

Implicitly, in the above definition, we assume that β is one-dimensional, but of course, a general discussion with d-dimensional Brownian motion can be developed.

Note that a strongly Brownian filtration is weakly Brownian since the Brownian motion enjoys the PRP. Since the mid-nineties, the study of weak Brownian filtrations has made quite some progress, starting with the proof by Tsirel'son [823] that the filtration of Walsh's Brownian motion as defined in Walsh [833] (see also Barlow and Yor [50]) taking values in $N \geq 3$ rays is weakly Brownian, but not strongly Brownian. See, in particular, the review paper of Émery [327] and notes and comments in Chapter V of [RY].

▶ We first show that weakly Brownian filtrations are left globally invariant under locally equivalent changes of probability. We start with a weakly Brownian filtration $\mathbf{F}$ on a probability space $(\Omega, \mathcal{F}, \mathbb{P})$ and we consider another probability $\mathbb{Q}$ on $(\Omega, \mathcal{F})$ such that $\mathbb{Q}|_{\mathcal{F}_t} = L_t \mathbb{P}|_{\mathcal{F}_t}$.

Proposition 5.8.1.2 *If* $\mathbf{F}$ *is weakly Brownian under* $\mathbb{P}$ *and* $\mathbb{Q}$ *is locally equivalent to* $\mathbb{P}$*, then* $\mathbf{F}$ *is also weakly Brownian under* $\mathbb{Q}$*.*

PROOF: Let M be an $(\mathbf{F}, \mathbb{Q})$-local martingale, then ML is an $(\mathbf{F}, \mathbb{P})$-local martingale, hence $N_t := M_t L_t = c + \int_0^t n_s d\beta_s$ for some Brownian motion β defined on $(\Omega, \mathcal{F}, \mathbf{F}, \mathbb{P})$, independently from M. Similarly, $dL_s = \ell_s d\beta_s$. Therefore, we have

$$\begin{aligned} M_t = \frac{N_t}{L_t} &= N_0 + \int_0^t \frac{dN_s}{L_s} - \int_0^t \frac{N_s dL_s}{L_s^2} + \int_0^t \frac{N_s d\langle L\rangle_s}{L_s^3} - \int_0^t \frac{d\langle N, L\rangle_s}{L_s^2} \\ &= c + \int_0^t \frac{n_s}{L_s} d\beta_s - \int_0^t \frac{N_s \ell_s}{L_s^2} d\beta_s + \int_0^t \frac{N_s \ell_s^2}{L_s^3} ds - \int_0^t \frac{n_s \ell_s}{L_s^2} ds \\ &= c + \int_0^t \left(\frac{n_s}{L_s} - \frac{N_s \ell_s}{L_s^2} \right) \left(d\beta_s - \frac{d\langle \beta, L\rangle_s}{L_s} \right). \end{aligned}$$

Thus, $(\widetilde{\beta}_t := \beta_t - \int_0^t \frac{d\langle \beta, L\rangle_s}{L_s};\ t \geq 0)$, the Girsanov transform of the original Brownian motion β, allows the representation of all $(\mathbf{F}, \mathbb{Q})$-martingales. □

▶ We now show that weakly Brownian filtrations are left globally invariant by "nice" time changes. Again, we consider a weakly Brownian filtration $\mathbf{F}$ on a probability space $(\Omega, \mathcal{F}, \mathbb{P})$. Let $A_t = \int_0^t a_s ds$ where $a_s > 0$, $d\mathbb{P} \otimes ds$ a.s., be a strictly increasing, $\mathbf{F}$ adapted process, such that $A_\infty = \infty$, $\mathbb{P}$ a.s..

Proposition 5.8.1.3 *If* $\mathbf{F}$ *is weakly Brownian under* $\mathbb{P}$ *and* τ_u *is the right-inverse of the strictly increasing process* $A_t = \int_0^t a_s ds$*, then* $(\mathcal{F}_{\tau_u}, u \geq 0)$ *is also weakly Brownian under* $\mathbb{P}$*.*

PROOF: It suffices to be able to represent any $(\mathcal{F}_{\tau_u}, u \geq 0)$-square integrable martingale in terms of a given $(\mathcal{F}_{\tau_u}, u \geq 0)$-Brownian motion $\widetilde{\beta}$. Consider $\widetilde{M}$ a square integrable $(\mathcal{F}_{\tau_u}, u \geq 0)$-martingale. From our hypothesis, we know that $\widetilde{M}_\infty = c + \int_0^\infty m_s d\beta_s$, where β is an **F**-Brownian motion and m is an **F**-predictable process such that $\mathbb{E}\left(\int_0^\infty ds\, m_s^2\right) < \infty$. Thus, we may write

$$\widetilde{M}_\infty = c + \int_0^\infty \frac{m_s}{\sqrt{a_s}} \sqrt{a_s}\, d\beta_s\,. \tag{5.8.4}$$

It remains to define $\widetilde{\beta}$ the $(\mathcal{F}_{\tau_u}, u \geq 0)$-Brownian motion which satisfies $\int_0^t \sqrt{a_s} d\beta_s := \widetilde{\beta}_{A_t}$. Going back to (5.8.4), we obtain

$$\widetilde{M}_\infty = c + \int_0^\infty \frac{m_{\tau_u}}{\sqrt{a_{\tau_u}}}\, d\widetilde{\beta}_u\,.$$

□

These two properties do not extend to strongly Brownian filtrations. In particular, **F** may be strongly Brownian under $\mathbb{P}$ and only weakly Brownian under $\mathbb{Q}$ (see Dubins et al. [267], Barlow et al. [48]).

5.8.2 Some Examples

In what follows, we shall sometimes write Brownian filtration for strongly Brownian filtration.

Let **F** be a Brownian filtration, M an **F**-martingale and $\mathbf{F}^M = (\mathcal{F}_t^M)$ the natural filtration of M.

(a) Reflected Brownian Motion. Let B be a Brownian motion and $\widetilde{B}_t = \int_0^t \text{sgn}(B_s) dB_s$. The process $\widetilde{B}$ is a Brownian motion in the filtration $\mathbf{F}^{|B|}$. From $L_t = \sup_{s \leq t}(-\widetilde{B}_s)$, it follows that $\mathcal{F}_t^{\widetilde{B}} = \mathcal{F}_t^{|B|}$, hence, $\mathbf{F}^{|B|}$ is strongly Brownian and different from **F** since the r.v. $\text{sgn}(B_t)$ is independent of $(|B_s|, s \leq t)$.

(b) Discontinuous Martingales Originating from a Brownian Setup. We give an example where there exists $\mathbf{F}^M$-discontinuous martingales. Let $M_t := \int_0^t \mathbb{1}_{\{B_s<0\}} dB_s$. Tanaka's formula leads to

$$B_t^- = -\int_0^t \mathbb{1}_{\{B_s<0\}} dB_s + \frac{1}{2} L_t\,.$$

The natural filtration of M, i.e., $\mathbf{F}^M$ is equal to the natural filtration of the process $(B_t^-, t \geq 0)$. The $\mathbf{F}^M$-martingale

$$\mathbb{E}\left(B_t^+ - \frac{1}{2} L_t | \mathcal{F}_t^M\right) = -\frac{1}{2} L_t + \mathbb{1}_{\{B_t>0\}} \sqrt{t - g_t}\, \mathbb{E}(m_1)\,,$$

(where we use here the notation of Section 4.3) is discontinuous, thus $\mathbf{F}^M$ is not even weakly Brownian. We refer to Williams [841] for a discussion.

(c) A Note about the PRP. Let $\mathbf{F}$ be a filtration and suppose that for a given $\mathbf{F}$-martingale M, any $\mathbf{F}$-martingale $(N_t, t \geq 0)$ vanishing at 0 can be written as $N_t = \int_0^t n_s dM_s$. This does not imply that $\sigma(M_s, s \leq t)$ equals $\mathcal{F}_t$ (in fact this is at the heart of the distinction between strongly and weakly Brownian filtrations). For example, let $\widetilde{B}_t = \int_0^t \text{sgn}\,(B_s)\, dB_s$. As we have seen in the first example above, $\mathcal{F}_t^{\widetilde{B}} = \sigma(|B_s|, s \leq t)$ and is strictly smaller than $\mathbf{F}$. Nevertheless, any $\mathbf{F}$-martingale $(N_t, t \geq 0)$ with $N_0 = 0$ can be represented as

$$N_t = \int_0^t \nu_s dB_s = \int_0^t \nu_s \text{sgn}\,(B_s)\, \text{sgn}\,(B_s)\, dB_s = \int_0^t n_s d\widetilde{B}_s\,,$$

where $n_s = \nu_s\, \text{sgn}\,(B_s)$.

(d) Another Example. Let $Y_t = \int_0^t B_s dW_s$ where W and B are independent Brownian motions. From

$$Y_t = \int_0^t |B_s| \text{sgn}(B_s) dW_s = \int_0^t |B_s| d\widehat{W}_s$$

where $\widehat{W}_t = \int_0^t \text{sgn}(B_s) dW_s$, it follows that

$$\mathcal{F}_t^Y = \sigma\{|B_s|, \widehat{W}_s, s \leq t\} = \sigma\{\widehat{B}_s, \widehat{W}_s, s \leq t\}\,,$$

where $\widehat{B}_t = \int_0^t \text{sgn}(B_s) dB_s$ is a BM independent of $\widehat{W}$. Any $\mathbf{F}^Y$-martingale can be written as

$$y + \int_0^t \varphi_s d\widehat{B}_s + \int_0^t \psi_s d\widehat{W}_s\,,$$

for two $\mathbf{F}^Y$-predictable processes ψ and φ.

(e) Filtration Generated by a Stochastic Integral with Non-vanishing Integrator. Let $X_t = \int_0^t H_s dW_s$ where W is a $\mathbf{G}$-Brownian motion for some filtration $\mathbf{G}$, and H is a strictly positive continuous $\mathbf{G}$-adapted process (we do not require that $\mathbf{G}$ is the natural filtration of W). Then $\mathcal{F}_t^X = \sigma(H_s, W_s;\, s \leq t)$.

The case where the integrator may vanish is not so easy. Here are other examples.

(f) Tsirel'son's drift. Let

$$\mathbf{W}^{(T)}|_{\mathcal{F}_t} = \exp\left(\int_0^t T(s, X_.) dX_s - \frac{1}{2}\int_0^t T^2(s, X_.) ds\right) \mathbf{W}|_{\mathcal{F}_t}$$

where T is Tsirel'son drift (see Example 1.5.5.6). The process

$$X_t^{(T)} = X_t - \int_0^t T(s, X_{\bullet})ds$$

is a $\mathbf{W}^{(T)}$-Brownian motion whose filtration is strictly smaller than $\mathbf{F}$; however, $\mathbf{F}$ is the natural filtration of a $\mathbf{W}^{(T)}$-Brownian motion.

More generally, if

$$\mathbf{W}^b|_{\mathcal{F}_t} = \exp\left(\int_0^t b(s, X_{\bullet})dX_s - \frac{1}{2}\int_0^t b^2(s, X_{\bullet})ds\right)\mathbf{W}|_{\mathcal{F}_t},$$

the process

$$X_t^b = X_t - \int_0^t b(s, X.)ds$$

is a $\mathbf{W}^b, \mathbf{F}$-Brownian motion. Dubins et al. [267] established that there exist infinitely many b's such that $\mathbf{F}$ is not the natural filtration of a $\mathbf{W}^b$ Brownian motion, i.e., $\mathbf{F}$ is not strongly Brownian under $\mathbf{W}^b$. See also Emery and Schachermayer [329].

(g) Let W and B be two independent Brownian motions, and let $Z = BW$. From $B_tW_t = \int_0^t (B_s dW_s + W_s dB_s)$ one obtains that $B_t^2 + W_t^2$ is measurable w.r.t $\mathcal{F}_t^Z$. Hence, the random variables $\frac{1}{\sqrt{2}}|B_t + W_t|$ and $\frac{1}{\sqrt{2}}|B_t - W_t|$ are $\mathcal{F}_t^Z$-measurable. The processes $\beta_t^{(\pm)} = \frac{1}{\sqrt{2}}(B_t \pm W_t)$ are independent Brownian motions. The filtration $\mathbf{F}^Z$ is generated by two independent reflected BMs, hence from **a)** above, it is generated by two independent Brownian motions.

Exercise 5.8.2.1 Let B and W be two independent Brownian motions and $Y_t = aB_t + bW_t$. Prove that $\sigma(Y_s, s \leq t) \subset \sigma(B_s, W_s, s \leq t)$ and that the inclusion is strict.

Let N_1 and N_2 be two independent Poisson processes and $Y_t = aN_{1,t} + bN_{2,t}$, where $a \neq b$. Prove that $\sigma(Y_s, s \leq t) = \sigma(N_{1,s}, N_{2,s}, s \leq t)$. ◁

Exercise 5.8.2.2 Let B and W be two independent Brownian motions, a and b two strictly positive numbers with $a \neq b$ and $Y_t = aB_t^2 + bW_t^2$. Prove that $\sigma(Y_s, s \leq t) = \sigma(B_s^2, W_s^2, s \leq t)$.

Generalize this result to the case $Y_t = \sum_{i=1}^n a_i (B_t^i)^2$ where $a_i > 0$ and $a_i \neq a_j$ for $i \neq j$. Prove that the filtration of Y is that of an n-dimensional Brownian motion.

`Hint:` Compute the bracket of Y and iterate this procedure. ◁

Example 5.8.2.3 Example of a martingale with respect to two different probabilities:

Let $B = (B_1, B_2)$ be a two-dimensional BM, and $R_t^2 = B_1^2(t) + B_2^2(t)$. The process

$$L_t = \exp\left(\int_0^t (B_1(s)dB_1(s) + B_2(s)dB_2(s)) - \frac{1}{2}\int_0^t R_s^2 ds\right)$$

is a martingale. Let $\mathbb{Q}|_{\mathcal{F}_t} = L_t \mathbb{P}|_{\mathcal{F}_t}$. The process

$$X_t = \int_0^t (B_2(s)dB_1(s) - B_1(s)dB_2(s))$$

is a $\mathbb{P}$ (and a $\mathbb{Q}$) martingale. The process R^2 is a BESQ under $\mathbb{P}$ and a CIR under $\mathbb{Q}$ (see $\rightarrowtail$ Chapter 6). See also Example 1.7.3.10.

Comment 5.8.2.4 In [328], Emery and Schachermayer show that there exists an absolutely continuous strictly increasing time-change such that the time-changed filtration is no longer Brownian.

5.9 Enlargements of Filtrations

In general, if $\mathbf{G}$ is a filtration larger than $\mathbf{F}$, it is not true that an $\mathbf{F}$-martingale remains a martingale in the filtration $\mathbf{G}$ (an interesting example is Azéma's martingale μ (see Subsection 4.3.8): this discontinuous $\mathbf{F}^\mu$-martingale is not an $\mathcal{F}^B$-martingale, it is not even a $\mathcal{F}^B$-semi-martingale; see $\rightarrowtail$ Example 9.4.2.3).

In the seminal paper [461], Itô studies the definition of the integral of a non-adapted process of the form $f(B_1, B_s)$ for some function f, with respect to a Brownian motion B. From the end of the seventies, Barlow, Jeulin and Yor started a systematic study of the problem of enlargement of filtrations: namely which $\mathbf{F}$-martingales M remain $\mathbf{G}$-semi-martingales and if it is the case, what is the semi-martingale decomposition of M in $\mathbf{G}$?

Up to now, four lecture notes volumes have been dedicated to this question: Jeulin [493], Jeulin and Yor [497], Yor [868] and Mansuy and Yor [622]. See also related chapters in the books of Protter [727] and Dellacherie, Maisonneuve and Meyer [241]. Some important papers are Brémaud and Yor [126], Barlow [45], Jacod [469, 468] and Jeulin and Yor [495].

These results are extensively used in finance to study two specific problems occurring in insider trading: existence of arbitrage using strategies adapted w.r.t. the large filtration, and change of prices dynamics, when an $\mathbf{F}$-martingale is no longer a $\mathbf{G}$-martingale.

We now study mathematically the two situations.

5.9.1 Immersion of Filtrations

Let $\mathbf{F}$ and $\mathbf{G}$ be two filtrations such that $\mathbf{F} \subset \mathbf{G}$. Our aim is to study some conditions which ensure that $\mathbf{F}$-martingales are $\mathbf{G}$-semi-martingales, and one can ask in a first step whether all **F**-martingales are **G**-martingales. This last property is equivalent to $\mathbb{E}(D|\mathcal{F}_t) = \mathbb{E}(D|\mathcal{G}_t)$, for any t and $D \in L^1(\mathcal{F}_\infty)$.

Let us study a simple example where $\mathbf{G} = \mathbf{F} \vee \sigma(D)$ where $D \in L^1(\mathcal{F}_\infty)$ and D is not $\mathcal{F}_0$-measurable. Obviously $\mathbb{E}(D|\mathcal{G}_t) = D$ is a **G**-martingale and $\mathbb{E}(D|\mathcal{F}_t)$ is a **F**-martingale. However $\mathbb{E}(D|\mathcal{G}_0) \neq \mathbb{E}(D|\mathcal{F}_0)$, and some **F**-martingales are not **G**-martingales.

The filtration **F** is said to be **immersed** in **G** if any square integrable **F**-martingale is a **G**-martingale (Tsirel'son [824], Émery [327]). This is also referred to as the $(\mathcal{H})$ hypothesis by Brémaud and Yor [126] which was defined as:

$(\mathcal{H})$ Every **F**-square integrable martingale is a **G**-square integrable martingale.

Proposition 5.9.1.1 *Hypothesis $(\mathcal{H})$ is equivalent to any of the following properties:*

$(\mathcal{H}1)$ $\forall\, t \geq 0$, *the σ-fields $\mathcal{F}_\infty$ and $\mathcal{G}_t$ are conditionally independent given $\mathcal{F}_t$.*
$(\mathcal{H}2)$ $\forall\, t \geq 0$, $\forall\, G_t \in L^1(\mathcal{G}_t)$, $\mathbb{E}(G_t|\mathcal{F}_\infty) = \mathbb{E}(G_t|\mathcal{F}_t)$.
$(\mathcal{H}3)$ $\forall\, t \geq 0$, $\forall\, F \in L^1(\mathcal{F}_\infty)$, $\mathbb{E}(F|\mathcal{G}_t) = \mathbb{E}(F|\mathcal{F}_t)$.

In particular, $(\mathcal{H})$ holds if and only if every **F***-local martingale is a* **G***-local martingale.*

PROOF:
▶ $(\mathcal{H}) \Rightarrow (\mathcal{H}1)$. Let $F \in L^2(\mathcal{F}_\infty)$ and assume that hypothesis $(\mathcal{H})$ is satisfied. This implies that the martingale $F_t = \mathbb{E}(F|\mathcal{F}_t)$ is a **G**-martingale such that $F_\infty = F$, hence $F_t = \mathbb{E}(F|\mathcal{G}_t)$. It follows that for any t and any $G_t \in L^2(\mathcal{G}_t)$:

$$\mathbb{E}(FG_t|\mathcal{F}_t) = \mathbb{E}(G_t\mathbb{E}(F|\mathcal{G}_t)|\mathcal{F}_t) = \mathbb{E}(G_t\mathbb{E}(F|\mathcal{F}_t)|\mathcal{F}_t) = \mathbb{E}(G_t|\mathcal{F}_t)\mathbb{E}(F|\mathcal{F}_t)$$

which is equivalent to $(\mathcal{H}1)$.
▶ $(\mathcal{H}1) \Rightarrow (\mathcal{H})$. Let $F \in L^2(\mathcal{F}_\infty)$ and $G_t \in L^2(\mathcal{G}_t)$. Under $(\mathcal{H}1)$,

$$\mathbb{E}(F\mathbb{E}(G_t|\mathcal{F}_t)) = \mathbb{E}(\mathbb{E}(F|\mathcal{F}_t)\mathbb{E}(G_t|\mathcal{F}_t)) \overset{\mathcal{H}1}{=} \mathbb{E}(\mathbb{E}(FG_t|\mathcal{F}_t)) = \mathbb{E}(FG_t)$$

which is $(\mathcal{H})$.
▶ $(\mathcal{H}2) \Rightarrow (\mathcal{H}3)$. Let $F \in L^2(\mathcal{F}_\infty)$ and $G_t \in L^2(\mathcal{G}_t)$. If $(\mathcal{H}2)$ holds, then it is easy to prove that, for $F \in L^2(\mathcal{F}_\infty)$,

$$\mathbb{E}(G_t\mathbb{E}(F|\mathcal{F}_t)) = \mathbb{E}(F\mathbb{E}(G_t|\mathcal{F}_t)) \overset{\mathcal{H}2}{=} \mathbb{E}(FG_t) = \mathbb{E}(G_t\mathbb{E}(F|\mathcal{G}_t)),$$

which implies $(\mathcal{H}3)$. The general case follows by approximation.
▶ Obviously $(\mathcal{H}3)$ implies $(\mathcal{H})$. □

In particular, under $(\mathcal{H})$, if W is an **F**-Brownian motion, then it is a **G**-martingale with bracket t, since such a bracket does not depend on the filtration. Hence, it is a **G**-Brownian motion.

A trivial (but useful) example for which $(\mathcal{H})$ is satisfied is $\mathbf{G} = \mathbf{F} \vee \mathbf{F}^1$ where $\mathbf{F}$ and $\mathbf{F}^1$ are two filtrations such that $\mathcal{F}_\infty$ is independent of $\mathcal{F}^1_\infty$.

We now present two propositions, in which setup the immersion property is preserved under change of probability.

Proposition 5.9.1.2 *Assume that the filtration* $\mathbf{F}$ *is immersed in* $\mathbf{G}$ *under* $\mathbb{P}$*, and let* $\mathbb{Q}|_{\mathcal{G}_t} = L_t\mathbb{P}|_{\mathcal{G}_t}$ *where* L *is assumed to be* $\mathbf{F}$*-adapted. Then,* $\mathbf{F}$ *is immersed in* $\mathbf{G}$ *under* $\mathbb{Q}$*.*

PROOF: Let N be a $(\mathbf{F},\mathbb{Q})$-martingale, then $(N_tL_t, t \geq 0)$ is a $(\mathbf{F},\mathbb{P})$-martingale, and since $\mathbf{F}$ is immersed in $\mathbf{G}$ under $\mathbb{P}$, $(N_tL_t, t \geq 0)$ is a $(\mathbf{G},\mathbb{P})$-martingale which implies that N is a $(\mathbf{G},\mathbb{Q})$-martingale. □

In the next proposition, we do not assume that the Radon-Nikodým density is $\mathbf{F}$-adapted.

Proposition 5.9.1.3 *Assume that* $\mathbf{F}$ *is immersed in* $\mathbf{G}$ *under* $\mathbb{P}$*, and define* $\mathbb{Q}|_{\mathcal{G}_t} = L_t\mathbb{P}|_{\mathcal{G}_t}$ *and* $\Lambda_t = \mathbb{E}(L_t|\mathcal{F}_t)$*. Assume that all* $\mathbf{F}$*-martingales are continuous and that the* $\mathbf{G}$*-martingale* L *is continuous. Then,* $\mathbf{F}$ *is immersed in* $\mathbf{G}$ *under* $\mathbb{Q}$ *if and only if the* $(\mathbf{G},\mathbb{P})$*-local martingale*

$$\int_0^t \frac{dL_s}{L_s} - \int_0^t \frac{d\Lambda_s}{\Lambda_s} := \mathcal{L}(L)_t - \mathcal{L}(\Lambda)_t$$

is orthogonal to the set of all $(\mathbf{F},\mathbb{P})$*-local martingales.*

PROOF: We prove that any $(\mathbf{F},\mathbb{Q})$-martingale is a $(\mathbf{G},\mathbb{Q})$-martingale. Every $(\mathbf{F},\mathbb{Q})$-martingale M^Q may be written as

$$M^Q_t = M^P_t - \int_0^t \frac{d\langle M^P, \Lambda\rangle_s}{\Lambda_s}$$

where M^P is an $(\mathbf{F},\mathbb{P})$-martingale. By hypothesis, M^P is a $(\mathbf{G},\mathbb{P})$-martingale and, from Girsanov's theorem, $M^P_t = N^Q_t + \int_0^t \frac{d\langle M^P, L\rangle_s}{L_s}$ where N^Q is an $(\mathbf{F},\mathbb{Q})$-martingale. It follows that

$$\begin{aligned} M^Q_t &= N^Q_t + \int_0^t \frac{d\langle M^P, L\rangle_s}{L_s} - \int_0^t \frac{d\langle M^P, \Lambda\rangle_s}{\Lambda_s} \\ &= N^Q_t + \int_0^t d\langle M^P, \mathcal{L}(L) - \mathcal{L}(\Lambda)\rangle_s \,. \end{aligned}$$

Thus M^Q is an $(\mathbf{G},\mathbb{Q})$ martingale if and only if $\langle M^P, \mathcal{L}(L) - \mathcal{L}(\Lambda)\rangle_s = 0$. □

Exercise 5.9.1.4 Assume that hypothesis $(\mathcal{H})$ holds under $\mathbb{P}$. Let

$$\mathbb{Q}|_{\mathcal{G}_t} = L_t\mathbb{P}|_{\mathcal{G}_t}; \quad \mathbb{Q}|_{\mathcal{F}_t} = \widehat{L}_t\mathbb{P}|_{\mathcal{F}_t} \,.$$

Prove that hypothesis $(\mathcal{H})$ holds under $\mathbb{Q}$ if and only if:

$$\forall X \geq 0, \ X \in \mathcal{F}_\infty, \quad \frac{\mathbb{E}(XL_\infty|\mathcal{G}_t)}{L_t} = \frac{\mathbb{E}(X\widehat{L}_\infty|\mathcal{F}_t)}{\widehat{L}_t}$$

See Nikeghbali [674]. ◁

5.9.2 The Brownian Bridge as an Example of Initial Enlargement

Rather than studying ab initio the general problem of initial enlargement, we discuss an interesting example. Let us start with a BM $(B_t, t \geq 0)$ and its natural filtration $\mathbf{F}^B$. Define a new filtration as $\mathcal{G}_t = \mathcal{F}_t^B \vee \sigma(B_1)$. In this filtration, the process $(B_t, t \geq 0)$ is no longer a martingale. It is easy to be convinced of this by looking at the process $(\mathbb{E}(B_1|\mathcal{G}_t), t \leq 1)$: this process is identically equal to B_1, not to B_t, hence $(B_t, t \geq 0)$ is not a $\mathbf{G}$-martingale. However, $(B_t, t \geq 0)$ is a $\mathbf{G}$-semi-martingale, as follows from the next proposition

Proposition 5.9.2.1 *The decomposition of B in the filtration $\mathbf{G}$ is*

$$B_t = \beta_t + \int_0^{t\wedge 1} \frac{B_1 - B_s}{1-s} ds$$

where β is a $\mathbf{G}$-Brownian motion.

PROOF: We have seen, in (4.3.8), that the canonical decomposition of Brownian bridge under $\mathbf{W}^{(1)}_{0\to 0}$ is

$$X_t = \beta_t - \int_0^t ds \frac{X_s}{1-s}, \quad t \leq 1 .$$

The same proof implies that the decomposition of B in the filtration $\mathbf{G}$ is

$$B_t = \beta_t + \int_0^{t\wedge 1} \frac{B_1 - B_s}{1-s} ds .$$

□

It follows that if M is an $\mathbf{F}$-local martingale such that $\int_0^1 \frac{1}{\sqrt{1-s}} d|\langle M, B\rangle|_s$ is finite, then

$$M_t = \widehat{M}_t + \int_0^{t\wedge 1} \frac{B_1 - B_s}{1-s} d\langle M, B\rangle_s$$

where $\widehat{M}$ is a $\mathbf{G}$-local martingale.

Comments 5.9.2.2 (a) As we shall see in ↣ Subsection 11.2.7, Proposition 5.9.2.1 can be extended to integrable Lévy processes: if X is a Lévy process which satisfies $\mathbb{E}(|X_t|) < \infty$ and $\mathbf{G} = \mathbf{F}^X \vee \sigma(X_1)$, the process

$$X_t - \int_0^{t\wedge 1} \frac{X_1 - X_s}{1-s} ds,$$

is a **G**-martingale.

(b) The singularity of $\frac{B_1-B_t}{1-t}$ at $t=1$, i.e., the fact that $\frac{B_1-B_t}{1-t}$ is not square integrable between 0 and 1 prevents a Girsanov measure change transforming the $(\mathbb{P}, \mathbf{G})$ semi-martingale B into a $(\mathbb{Q}, \mathbf{G})$ martingale. Let

$$dS_t = S_t(\mu dt + \sigma dB_t)$$

and enlarge the filtration with S_1 (or equivalently, with B_1). In the enlarged filtration, setting $\zeta_t = \frac{B_1-B_t}{1-t}$, the dynamics of S are

$$dS_t = S_t((\mu + \sigma\zeta_t)dt + \sigma d\beta_t),$$

and there does not exist an e.m.m. such that the discounted price process $(e^{-rt}S_t, t \leq 1)$ is a **G**-martingale. However, for any $\epsilon \in]0,1]$, there exists a uniformly integrable **G**-martingale L defined as

$$dL_t = \frac{\mu - r + \sigma\zeta_t}{\sigma} L_t d\beta_t,\ t \leq 1-\epsilon, \quad L_0 = 1,$$

such that, setting $d\mathbb{Q}|_{\mathcal{G}_t} = L_t d\mathbb{P}|_{\mathcal{G}_t}$, the process $(e^{-rt}S_t, t \leq 1-\epsilon)$ is a $(\mathbb{Q}, \mathbf{G})$-martingale.

This is the main point in the theory of insider trading where the knowledge of the terminal value of the underlying asset creates an arbitrage opportunity, which is effective at time 1.

5.9.3 Initial Enlargement: General Results

Let **F** be a Brownian filtration generated by B. We consider $\mathcal{F}_t^{(L)} = \mathcal{F}_t \vee \sigma(L)$ where L is a real-valued random variable. More precisely, in order to satisfy the usual hypotheses, redefine

$$\mathcal{F}_t^{(L)} = \cap_{\epsilon>0} \{\mathcal{F}_{t+\epsilon} \vee \sigma(L)\}.$$

We recall that there exists a family of regular conditional distributions $\lambda_t(\omega, dx)$ such that $\lambda_t(\cdot, A)$ is a version of $\mathbb{E}(\mathbb{1}_{\{L\in A\}}|\mathcal{F}_t)$ and for any ω, $\lambda_t(\omega, \cdot)$ is a probability on $\mathbb{R}$.

Proposition 5.9.3.1 (Jacod's Criterion.) *Suppose that, for each $t < T$, $\lambda_t(\omega, dx) \ll \nu(dx)$ where ν is the law of L. Then, every* **F***-semi-martingale $(X_t, t < T)$ is also an $\mathcal{F}_t^{(L)}$-semi-martingale.*

Moreover, if $\lambda_t(\omega, dx) = p_t(\omega, x)\nu(dx)$ and if X is an **F***-martingale, its decomposition in the filtration $\mathcal{F}_t^{(L)}$ is*

$$X_t = \widetilde{X}_t + \int_0^t \frac{d\langle p.(L), X\rangle_s}{p_s(L)}.$$

In a more general setting (see Yor [868]), for a bounded Borel function f, let $(\lambda_t(f), t \geq 0)$ be the continuous version of the martingale $(\mathbb{E}(f(L)|\mathcal{F}_t), t \geq 0)$. There exists a predictable kernel $\lambda_t(dx)$ such that

$$\lambda_t(f) = \int \lambda_t(dx) f(x) .$$

From the predictable representation property applied to the martingale $\mathbb{E}(f(L)|\mathcal{F}_t)$, there exists a predictable process $\widehat{\lambda}(f)$ such that

$$\lambda_t(f) = \mathbb{E}(f(L)) + \int_0^t \widehat{\lambda}_s(f) dB_s .$$

Proposition 5.9.3.2 *We assume that there exists a predictable kernel* $\widehat{\lambda}_t(dx)$ *such that*

$$dt \text{ a.s.,} \quad \widehat{\lambda}_t(f) = \int \widehat{\lambda}_t(dx) f(x) .$$

Assume furthermore that $dt \times d\mathbb{P}$ *a.s. the measure* $\widehat{\lambda}_t(dx)$ *is absolutely continuous with respect to* $\lambda_t(dx)$*:*

$$\widehat{\lambda}_t(dx) = \rho(t,x) \lambda_t(dx) .$$

Then, if X *is an* $\mathbf{F}$*-martingale, there exists a* $\mathbf{F}^{(L)}$*-martingale* $\widehat{X}$ *such that*

$$X_t = \widehat{X}_t + \int_0^t \rho(s,L) d\langle X, B\rangle_s .$$

Sketch of the proof: Let X be an $\mathbf{F}$-martingale, f a given bounded Borel function and $F_t = \mathbb{E}(f(L)|\mathcal{F}_t)$. From the hypothesis

$$\begin{aligned} F_t &= \mathbb{E}(f(L)) + \int_0^t \widehat{\lambda}_s(f) dB_s \\ &= \mathbb{E}(f(L)) + \int_0^t \left(\int \rho(s,x) \lambda_s(dx) f(x) \right) dB_s . \end{aligned}$$

Then, for $A_s \in \mathcal{F}_s$, $s < t$:

$$\begin{aligned} \mathbb{E}(\mathbb{1}_{A_s} f(L)(X_t - X_s)) &= \mathbb{E}(\mathbb{1}_{A_s}(F_t X_t - F_s X_s)) = \mathbb{E}(\mathbb{1}_{A_s}(\langle F, X\rangle_t - \langle F, X\rangle_s)) \\ &= \mathbb{E}\left(\mathbb{1}_{A_s} \int_s^t d\langle X, B\rangle_u \, \widehat{\lambda}_u(f) \right) \\ &= \mathbb{E}\left(\mathbb{1}_{A_s} \int_s^t d\langle X, B\rangle_u \int \lambda_u(dx) f(x) \rho(u,x) \right) . \end{aligned}$$

Therefore, $V_t = \int_0^t \rho(u,L) \, d\langle X, B\rangle_u$ satisfies

$$\mathbb{E}(\mathbb{1}_{A_s} f(L)(X_t - X_s)) = \mathbb{E}(\mathbb{1}_{A_s} f(L)(V_t - V_s)) .$$

It follows that, for any $G_s \in \mathcal{F}_s^{(L)}$,

$$\mathbb{E}(\mathbb{1}_{G_s}(X_t - X_s)) = \mathbb{E}(\mathbb{1}_{G_s}(V_t - V_s)),$$

hence, $(X_t - V_t, t \geq 0)$ is an $\mathbf{F}^{(L)}$-martingale. □

Let us write the result of Proposition 5.9.3.2 in terms of Jacod's criterion. If $\lambda_t(dx) = p_t(x)\nu(dx)$, then

$$\lambda_t(f) = \int p_t(x)f(x)\nu(dx).$$

Hence,

$$d\langle \lambda.(f), B\rangle_t = \widehat{\lambda}_t(f)dt = \int dx f(x)\, d_t\langle p.(x), B\rangle_t$$

and

$$\widehat{\lambda}_t(dx) = d_t\langle p.(x), B\rangle_t = \frac{d_t\langle p.(x), B\rangle_t}{p_t(x)} p_t(x)dx$$

therefore,

$$\widehat{\lambda}_t(dx)dt = \frac{d_t\langle p.(x), B\rangle_t}{p_t(x)} \lambda_t(dx).$$

In the case where $\lambda_t(dx) = \Phi(t,x)dx$, with $\Phi > 0$, it is possible to find ψ such that

$$\Phi(t,x) = \Phi(0,x)\exp\left(\int_0^t \psi(s,x)dB_s - \frac{1}{2}\int_0^t \psi^2(s,x)ds\right)$$

and it follows that $\widehat{\lambda}_t(dx) = \psi(t,x)\lambda_t(dx)$. Then, if X is an $\mathbf{F}$-martingale of the form $X_t = x + \int_0^t x_s dB_s$, the process $(X_t - \int_0^t ds\, x_s\, \psi(s,L), t \geq 0)$ is an $\mathbf{F}^{(L)}$-martingale.

Example 5.9.3.3 We now give some examples taken from Mansuy and Yor [622] in a Brownian set-up for which we use the preceding. Here, B is a standard Brownian motion.

▶ **Enlargement with B_1.** We compare the results obtained in Subsection 5.9.2 and the method presented in Subsection 5.9.3. Let $L = B_1$. From the Markov property

$$\mathbb{E}(g(B_1)|\mathcal{F}_t) = \mathbb{E}(g(B_1 - B_t + B_t)|\mathcal{F}_t) = F_g(B_t, 1-t)$$

where $F_g(y, 1-t) = \int g(x)p_{1-t}(y,x)dx$ and $p_s(y,x) = \frac{1}{\sqrt{2\pi s}}\exp\left(-\frac{(x-y)^2}{2s}\right)$. It follows that $\lambda_t(dx) = \frac{1}{\sqrt{2\pi(1-t)}}\exp\left(-\frac{(x-B_t)^2}{2(1-t)}\right)dx$. Then

$$\lambda_t(dx) = p_t^x \mathbb{P}(B_1 \in dx)$$

with

$$p_t^x = \frac{1}{\sqrt{(1-t)}} \exp\left(-\frac{(x-B_t)^2}{2(1-t)} + \frac{x^2}{2}\right).$$

From Itô's formula,

$$dp_t^x = p_t^x \frac{x - B_t}{1-t} dB_t .$$

It follows that $d\langle p^x, B\rangle_t = p_t^x \frac{x-B_t}{1-t} dt$, hence

$$B_t = \widetilde{B}_t + \int_0^t \frac{x - B_s}{1-s} ds .$$

Note that, in the notation of Proposition 5.9.3.2, one has

$$\widehat{\lambda}_t(dx) = \frac{x-B_t}{1-t} \frac{1}{\sqrt{2\pi(1-t)}} \exp\left(-\frac{(x-B_t)^2}{2(1-t)}\right) dx .$$

► **Enlargement with $M^B = \sup_{s \leq 1} B_s$.** From Exercise 3.1.6.7,

$$\mathbb{E}(f(M^B)|\mathcal{F}_t) = F(1-t, B_t, M_t^B)$$

where $M_t^B = \sup_{s \leq t} B_s$ with

$$F(s,a,b) = \sqrt{\frac{2}{\pi s}} \left(f(b) \int_0^{b-a} e^{-u^2/(2s)} du + \int_b^\infty f(u) e^{-(u-a)^2/(2s)} du \right)$$

and

$$\lambda_t(dy) = \sqrt{\frac{2}{\pi(1-t)}} \left\{ \delta_y(M_t^B) \int_0^{M_t^B - B_t} \exp\left(-\frac{u^2}{2(1-t)}\right) du \right.$$
$$\left. + \mathbb{1}_{\{y > M_t^B\}} \exp\left(-\frac{(y-B_t)^2}{2(1-t)}\right) dy \right\} .$$

Hence, by differentiation w.r.t. $x (= B_t)$, i.e., more precisely, by applying Itô's formula

$$\widehat{\lambda}_t(dy) = \sqrt{\frac{2}{\pi(1-t)}} \left\{ \delta_y(M_t^B) \exp\left(-\frac{(M_t^B - B_t)^2}{2(1-t)}\right) \right.$$
$$\left. + \mathbb{1}_{\{y > M_t^B\}} \frac{y - B_t}{1-t} \exp\left(-\frac{(y-B_t)^2}{2(1-t)}\right) \right\} .$$

It follows that

$$\rho(t,x) = \mathbb{1}_{\{x>M_t^B\}} \frac{x-B_t}{1-t} + \mathbb{1}_{\{M_t^B=x\}} \frac{1}{\sqrt{1-t}} \frac{\Phi'}{\Phi}\left(\frac{x-B_t}{\sqrt{1-t}}\right)$$

with $\Phi(x) = \sqrt{\frac{2}{\pi}} \int_0^x e^{-\frac{u^2}{2}} du$.

More examples can be found in Jeulin [493] and Mansuy and Yor [622]. Matsumoto and Yor [629] consider the case where $L = \int_0^\infty ds \exp(2(B_s - \nu s))$. See also Baudoin [61].

Exercise 5.9.3.4 Assume that the hypotheses of Proposition 5.9.3.1 hold and that $1/p_\infty(\cdot, L)$ is integrable with expectation $1/c$. Prove that under the probability R defined as

$$dR|_{\mathcal{F}_\infty} = c/p_\infty(\cdot, L) d\mathbb{P}|_{\mathcal{F}_\infty}$$

the r.v. L is independent of $\mathcal{F}_\infty$. This fact plays an important rôle in Grorud and Pontier [411]. ◁

5.9.4 Progressive Enlargement

We now consider a different case of enlargement, more precisely the case where τ is a finite random time, i.e., a finite non-negative random variable, and we denote

$$\mathcal{F}_t^\tau = \cap_{\epsilon>0} \{\mathcal{F}_{t+\epsilon} \vee \sigma(\tau \wedge (t+\epsilon))\} .$$

Proposition 5.9.4.1 *For any $\mathbf{F}^\tau$-predictable process H, there exists an $\mathbf{F}$-predictable process h such that $H_t \mathbb{1}_{\{t \leq \tau\}} = h_t \mathbb{1}_{\{t \leq \tau\}}$. Under the condition $\forall t, \mathbb{P}(\tau \leq t | \mathcal{F}_t) < 1$, the process $(h_t, t \geq 0)$ is unique.*

PROOF: We refer to Dellacherie [245] and Dellacherie et al. [241], page 186. The process h may be recovered as the ratio of the $\mathbf{F}$-predictable projections of $H_t \mathbb{1}_{\{t<\tau\}}$ and $\mathbb{1}_{\{t<\tau\}}$:

$$h_t = \frac{\mathbb{E}(H_t \mathbb{1}_{\{t<\tau\}} | \mathcal{F}_t)}{\mathbb{P}(t < \tau | \mathcal{F}_t)} . \qquad \square$$

Immersion Setting

Let us first investigate the case where the $(\mathcal{H})$ hypothesis holds.

Lemma 5.9.4.2 *In the progressive enlargement setting, $(\mathcal{H})$ holds between $\mathbf{F}$ and $\mathbf{F}^\tau$ if and only if one of the following equivalent conditions holds:*

$$\begin{array}{ll} \text{(i) } \forall (t,s),\, s \leq t, & \mathbb{P}(\tau \leq s | \mathcal{F}_\infty) = \mathbb{P}(\tau \leq s | \mathcal{F}_t), \\ \text{(ii) } \forall t, & \mathbb{P}(\tau \leq t | \mathcal{F}_\infty) = \mathbb{P}(\tau \leq t | \mathcal{F}_t). \end{array} \qquad (5.9.1)$$

PROOF: If (ii) holds, then (i) holds too. If (i) holds, $\mathcal{F}_\infty$ and $\sigma(t \wedge \tau)$ are conditionally independent given $\mathcal{F}_t$. The property follows. This result can also be found in Dellacherie and Meyer [243]. □

Note that, if $(\mathcal{H})$ holds, then (ii) implies that the process $\mathbb{P}(\tau \leq t|\mathcal{F}_t)$ is decreasing.

Example: assume that $\mathbf{F} \subset \mathbf{G}$ where $(\mathcal{H})$ holds for $\mathbf{F}$ and $\mathbf{G}$. Let τ be a $\mathbf{G}$-stopping time. Then, $(\mathcal{H})$ holds for $\mathbf{F}$ and $\mathbf{F}^\tau$.

General Setting

We denote by Z^τ the $\mathbf{F}$-super-martingale $\mathbb{P}(\tau > t|\mathcal{F}_t)$, also called the Azéma supermartingale (introduced in [35]). We assume in what follows

(A) The random time τ avoids the $\mathbf{F}$-stopping times, i.e., $\mathbb{P}(\tau = \vartheta) = 0$ for any $\mathbf{F}$-stopping time ϑ.

Under **(A)**, the $\mathbf{F}$-dual predictable projection of the process $D_t := \mathbb{1}_{\tau \leq t}$, denoted A^τ, is continuous. Indeed, if ϑ is a jump time of A^τ, it is predictable, and

$$\mathbb{E}(A^\tau_\vartheta - A^\tau_{\vartheta-}) = \mathbb{E}(\mathbb{1}_{\tau=\vartheta}) = 0\,;$$

the continuity of A^τ follows.

Proposition 5.9.4.3 *The canonical decomposition of the semi-martingale Z^τ is*

$$Z^\tau_t = \mathbb{E}(A^\tau_\infty|\mathcal{F}_t) - A^\tau_t = \mu^\tau_t - A^\tau_t$$

where $\mu^\tau_t := \mathbb{E}(A^\tau_\infty|\mathcal{F}_t)$.

PROOF: From the definition of the dual predictable projection, for any predictable process H, one has

$$\mathbb{E}(H_\tau) = \mathbb{E}\left(\int_0^\infty H_u dA^\tau_u\right).$$

Let t be fixed and $F_t \in \mathcal{F}_t$. Then, the process $H_u = F_t \mathbb{1}_{\{t<u\}}, u \geq 0$ is $\mathbf{F}$-predictable. Then

$$\mathbb{E}(F_t \mathbb{1}_{\{t<\tau\}}) = \mathbb{E}(F_t(A^\tau_\infty - A^\tau_t))\,.$$

It follows that $\mathbb{E}(A^\tau_\infty|\mathcal{F}_t) = Z^\tau_t + A^\tau_t$. □

Comment 5.9.4.4 It can be proved that the martingale

$$\mu^\tau_t := \mathbb{E}(A^\tau_\infty|\mathcal{F}_t) = A^\tau_t + Z^\tau_t$$

is BMO (see Definition 1.2.3.9).

It is proved in Yor [860] that if X is an $\mathbf{F}$-martingale then the processes $X_{t\wedge\tau}$ and $X_t(1-D_t)$ are $\mathbf{F}^\tau$ semi-martingales. Furthermore, the decompositions of the $\mathbf{F}$-martingales in the filtration $\mathbf{F}^\tau$ are known up to time τ (Jeulin and Yor [495]).

Proposition 5.9.4.5 *Every $\mathbf{F}$-martingale M stopped at time τ is an $\mathbf{F}^\tau$-semi-martingale with canonical decomposition*

$$M_{t\wedge\tau} = \widetilde{M}_t + \int_0^{t\wedge\tau} \frac{d\langle M, \mu^\tau\rangle_s}{Z^\tau_{s-}},$$

where $\widetilde{M}$ is an $\mathbf{F}^\tau$-local martingale. The process

$$\mathbb{1}_{\{\tau\le t\}} - \int_0^{t\wedge\tau} \frac{1}{Z^\tau_{s-}} dA^\tau_s$$

is an $\mathbf{F}^\tau$-martingale.

PROOF: Let H be an $\mathbf{F}^\tau$-predictable process. There exists an $\mathbf{F}$-predictable process h such that $H_t\mathbb{1}_{\{t\le\tau\}} = h_t\mathbb{1}_{\{t\le\tau\}}$, hence, if M is an $\mathbf{F}$-martingale, for $s<t$,

$$\begin{aligned}
&\mathbb{E}(H_s(M_{t\wedge\tau} - M_{s\wedge\tau})) = \mathbb{E}(H_s\mathbb{1}_{\{s<\tau\}}(M_{t\wedge\tau} - M_{s\wedge\tau}))\\
&= \mathbb{E}(h_s\mathbb{1}_{\{s<\tau\}}(M_{t\wedge\tau} - M_{s\wedge\tau}))\\
&= \mathbb{E}\left(h_s(\mathbb{1}_{\{s<\tau\le t\}}(M_\tau - M_s) + \mathbb{1}_{\{t<\tau\}}(M_t - M_s))\right)
\end{aligned}$$

From the definition of Z,

$$\mathbb{E}\left(h_s\mathbb{1}_{\{s<\tau\le t\}}M_\tau\right) = -\mathbb{E}\left(h_s\int_s^t M_u dZ_u\right)$$

and, noting that

$$\int_s^t M_u dZ_u - M_s Z_s + Z_t M_t = \int_s^t Z_u dM_u + \langle M, Z\rangle_t - \langle M, Z\rangle_s$$

we get, from the martingale property of M

$$\begin{aligned}
&\mathbb{E}(H_s(M_{t\wedge\tau} - M_{s\wedge\tau})) = \mathbb{E}(h_s(\langle M, \mu^\tau\rangle_t - \langle M, \mu^\tau\rangle_s))\\
&= \mathbb{E}\left(h_s\int_s^t \frac{d\langle M, \mu^\tau\rangle_u}{Z^\tau_{u-}} Z^\tau_{u-}\right) = \mathbb{E}\left(h_s\int_s^t \frac{d\langle M, \mu^\tau\rangle_u}{Z^\tau_{u-}}\mathbb{E}(\mathbb{1}_{\{u<\tau\}}|\mathcal{F}_u)\right)\\
&= \mathbb{E}\left(h_s\int_s^t \frac{d\langle M, \mu^\tau\rangle_u}{Z^\tau_{u-}}\mathbb{1}_{\{u<\tau\}}\right) = \mathbb{E}\left(h_s\int_{s\wedge\tau}^{t\wedge\tau} \frac{d\langle M, \mu^\tau\rangle_u}{Z^\tau_{u-}}\right).
\end{aligned}$$

The result follows. □

Pseudo-stopping Times

As we have mentioned, if $(\mathcal{H})$ holds, the process $(Z_t^\tau, t \geq 0)$ is a decreasing process. The converse is not true. The decreasing property of Z^τ is closely related with the definition of pseudo-stopping times, a notion developed from D. Williams example (see Example 5.9.4.8 below).

Definition 5.9.4.6 *A random time τ is a pseudo-stopping time if, for any bounded **F**-martingale M, $\mathbb{E}(M_\tau) = M_0$.*

Proposition 5.9.4.7 *The random time τ is a pseudo-stopping time if and only if one of the following equivalent properties holds:*

- *For any local **F**-martingale m, the process $(m_{t\wedge\tau}, t \geq 0)$ is a local $\mathbf{F}^\tau$-martingale,*
- $A_\infty^\tau = 1$,
- $\mu_t^\tau = 1, \forall t \geq 0$,
- *The process Z^τ is a decreasing **F**-predictable process.*

PROOF: We refer to Nikeghbali and Yor [675]. □

Example 5.9.4.8 The first example of a pseudo-stopping time was given by Williams [844]. Let B be a Brownian motion and define the stopping time $T_1 = \inf\{t : B_t = 1\}$ and the random time $\theta = \sup\{t < T_1 : B_t = 0\}$. Set

$$\tau = \sup\{s < \theta : B_s = M_s^B\}$$

where M_s^B is the running maximum of the Brownian motion. Then, τ is a pseudo-stopping time. Note that $\mathbb{E}(B_\tau)$ is not equal to 0; this illustrates the fact we cannot take any martingale in Definition 5.9.4.6. The martingale $(B_{t\wedge T_1}, t \geq 0)$ is neither bounded, nor uniformly integrable. In fact, since the maximum M_θ^B $(=B_\tau)$ is uniformly distributed on $[0,1]$, one has $\mathbb{E}(B_\tau) = 1/2$.

Honest Times

For a general random time τ, it is not true that **F**-martingales are $\mathbf{F}^\tau$-semi-martingales. Here is an example: due to the separability of the Brownian filtration, there exists a bounded random variable τ such that $\mathcal{F}_\infty = \sigma(\tau)$. Hence, $\mathcal{F}_{\tau+t}^\tau = \mathcal{F}_\infty, \forall t$ so that the $\mathbf{F}^\tau$-martingales are constant after τ. Consequently, **F**-martingales are not $\mathbf{F}^\tau$-semi-martingales.

On the other hand, there exists an interesting class of random times τ such that **F**-martingales are $\mathbf{F}^\tau$-semi-martingales.

Definition 5.9.4.9 *A random time is honest if it is the end of a predictable set, i.e., $\tau(\omega) = \sup\{t : (t,\omega) \in \Gamma\}$, where Γ is an **F**-predictable set.*

In particular, an honest time is $\mathcal{F}_\infty$-measurable. If X is a transient diffusion, the last passage time Λ_a (see Proposition 5.6.2.1) is honest. Jeulin [493] established that an $\mathcal{F}_\infty$-measurable random time is honest if and only if it is equal, on $\{\tau < t\}$, to an $\mathcal{F}_t$-measurable random variable.

A key point in the proof of the next Proposition 5.9.4.10 is the following description of $\mathbf{F}^\tau$-predictable processes: if τ, an $\mathcal{F}_\infty$-measurable random time, is honest, and if H is an $\mathbf{F}^\tau$-predictable process, then there exist two $\mathbf{F}$-predictable processes h and $\widetilde{h}$ such that

$$H_t = h_t \mathbb{1}_{\{\tau > t\}} + \widetilde{h}_t \mathbb{1}_{\{\tau \le t\}} .$$

(See Jeulin [493] for a proof.)

Proposition 5.9.4.10 *Let τ be honest. Then, if X is an $\mathbf{F}$-local martingale, there exists an $\mathbf{F}^\tau$-local martingale $\widetilde{X}$ such that*

$$X_t = \widetilde{X}_t + \int_0^{t\wedge\tau} \frac{d\langle X, \mu^\tau\rangle_s}{Z^\tau_{s-}} - \int_\tau^{\tau\vee t} \frac{d\langle X, \mu^\tau\rangle_s}{1 - Z^\tau_{s-}} .$$

PROOF: Let M be an $\mathbf{F}$-martingale which belongs to $\mathbf{H}^1$ and $G_s \in \mathcal{F}^\tau_s$. We define a $\mathbf{G}^\tau$ predictable process H as $H_u = \mathbb{1}_{G_s}\mathbb{1}_{]s,t]}(u)$. For $s < t$, one has, using the decomposition of $\mathbf{G}^\tau$ predictable processes:

$$\begin{aligned}\mathbb{E}(\mathbb{1}_{G_s}(M_t - M_s)) &= \mathbb{E}\left(\int_0^\infty H_u dM_u\right)\\ &= \mathbb{E}\left(\int_0^\tau h_u dM_u\right) + \mathbb{E}\left(\int_\tau^\infty \widetilde{h}_u dM_u\right) .\end{aligned}$$

Noting that $\int_0^t \widetilde{h}_u dM_u$ is a martingale yields $\mathbb{E}\left(\int_0^\infty \widetilde{h}_u dM_u\right) = 0$,

$$\begin{aligned}\mathbb{E}(\mathbb{1}_{G_s}(M_t - M_s)) &= \mathbb{E}\left(\int_0^\tau (h_u - \widetilde{h}_u) dM_u\right)\\ &= \mathbb{E}\left(\int_0^\infty dA^\tau_v \int_0^v (h_u - \widetilde{h}_u) dM_u\right) .\end{aligned}$$

By integration by parts, with $N_t = \int_0^t (h_u - \widetilde{h}_u) dM_u$, we get

$$\mathbb{E}(\mathbb{1}_{G_s}(M_t - M_s)) = \mathbb{E}(N_\infty A^\tau_\infty) = \mathbb{E}\left(\int_0^\infty (h_u - \widetilde{h}_u) d\langle M, \mu^\tau\rangle_u\right) .$$

Now, it remains to note that

$$\mathbb{E}\left(\int_0^\infty H_u\left(\frac{d\langle M,\mu^\tau\rangle_u}{Z_{u-}}\mathbb{1}_{\{u\le\tau\}}-\frac{d\langle M,\mu^\tau\rangle_u}{1-Z_{u-}}\mathbb{1}_{\{u>\tau\}}\right)\right)$$
$$=\mathbb{E}\left(\int_0^\infty\left(h_u\frac{d\langle M,\mu^\tau\rangle_u}{Z_{u-}}\mathbb{1}_{\{u\le\tau\}}-\widetilde{h}_u\frac{d\langle M,\mu^\tau\rangle_u}{1-Z_{u-}}\mathbb{1}_{\{u>\tau\}}\right)\right)$$
$$=\mathbb{E}\left(\int_0^\infty\left(h_u d\langle M,\mu^\tau\rangle_u-\widetilde{h}_u d\langle M,\mu^\tau\rangle_u\right)\right)$$
$$=\mathbb{E}\left(\int_0^\infty(h_u-\widetilde{h}_u)d\langle M,\mu^\tau\rangle_u\right)$$

to conclude the result in the case $M\in\mathbf{H}^1$. The general result follows by localization. □

Example 5.9.4.11 Let W be a Brownian motion, and $\tau=g_1$, the last time when the BM reaches 0 before time 1, i.e., $\tau=\sup\{t\le 1: W_t=0\}$. Using the computation of Z^{g_1} in Subsection 5.6.4 and Proposition 5.9.4.10, we obtain the decomposition of the Brownian motion in the enlarged filtration

$$W_t=\widetilde{W}_t-\int_0^t\mathbb{1}_{[0,\tau]}(s)\frac{\Phi'}{1-\Phi}\left(\frac{|W_s|}{\sqrt{1-s}}\right)\frac{\operatorname{sgn}(W_s)}{\sqrt{1-s}}ds$$
$$+\mathbb{1}_{\{\tau\le t\}}\operatorname{sgn}(W_1)\int_\tau^t\frac{\Phi'}{\Phi}\left(\frac{|W_s|}{\sqrt{1-s}}\right)ds$$

where $\Phi(x)=\sqrt{\frac{2}{\pi}}\int_0^x\exp(-u^2/2)du$.

Comments 5.9.4.12 (a) The $(\mathcal{H})$ hypothesis was studied by Brémaud and Yor [126] and Mazziotto and Szpirglas [632], and in a financial setting by Kusuoka [552], Elliott et al. [315] and Jeanblanc and Rutkowski [486, 487].

(b) An incomplete list of authors concerned with enlargement of filtration in finance for insider trading is: Amendinger [12], Amendinger et al. [13], Baudoin [61], Corcuera et al. [194], Eyraud-Loisel [338], Florens and Fougère [347], Gasbarra et al. [374], Grorud and Pontier [410], Hillairet [436], Imkeller [457], Imkeller et al. [458], Karatzas and Pikovsky [512], Kohatsu-Higa [532, 533] and Kohatsu-Higa and Øksendal [534].

(c) Enlargement theory is also used to study asymmetric information, see e. g. Föllmer et al. [353] and progressive enlargement is an important tool for the study of default in the reduced form approach by Bielecki et al. [91, 92, 93], Elliott et al.[315] and Kusuoka [552] (see ↣ Chapter 7).

(d) See also the papers of Ankirchner et al. [19] and Yoeurp [858].

(e) Note that the random time τ presented in Subsection 5.6.5 is not the end of a predictable set, hence, is not honest. However, $\mathbf{F}$-martingales are semi-martingales in the progressive enlarged filtration: it suffices to note that $\mathbf{F}$-martingales are semi-martingales in the filtration initially enlarged with W_1.

5.10 Filtering the Information

A priori, one might think somewhat naïvely that the drift term in the historical dynamics of the asset plays no rôle in contingent claims valuation. Nevertheless, working in the filtration generated by the asset shows the importance of this parameter. We present here some results, linked with filtering theory. However, we do not present the theory in detail, and the reader can refer to Lipster and Shiryaev [598] and Brémaud [124] for processes with jumps.

5.10.1 Independent Drift

Suppose that $dB_t^{(Y)} = Y dt + dB_t$, $B_0^{(Y)} = 0$ where Y is some r.v. independent of B and with law ν. The following proposition describes the distribution of $B^{(Y)}$.

Proposition 5.10.1.1 *The law of $B^{(Y)}$ is $\mathbf{W}^{h_\nu}$ defined as*

$$\mathbf{W}^{h_\nu}|_{\mathcal{F}_t} = h_\nu(X_t, t)\, \mathbf{W}|_{\mathcal{F}_t} .$$

Here, $h_\nu(x,t) = \int \nu(dy) \exp(yx - \frac{y^2}{2}t)$.

PROOF: Let F be a functional on $C([0,t], \mathbb{R})$. Using the independence between Y and B, and the Cameron-Martin theorem, we get

$$\begin{aligned}\mathbb{E}[F(B_s^{(Y)}, s \le t)] = \mathbb{E}[F(sY + B_s, s \le t)] &= \int \nu(dy)\mathbb{E}[F(sy + B_s, s \le t)] \\ &= \int \nu(dy)\mathbb{E}\left[F(B_s, s \le t) \exp\left(yB_t - \frac{y^2}{2}t\right)\right] \\ &= \mathbb{E}[F(B_s; s \le t) h_\nu(B_t, t)] .\end{aligned}$$

□

We now give the canonical decomposition of $B^{(Y)}$ in its own filtration. Let $\mathbf{W}^{h_\nu}|_{\mathcal{F}_t} = h_\nu(X_t, t)\, \mathbf{W}|_{\mathcal{F}_t} = L_t\, \mathbf{W}|_{\mathcal{F}_t}$. Therefore, the bracket $\langle X, L\rangle_t$ is equal to $\int_0^t \partial_x h_\nu(X_s, s)\, ds$, and, from Girsanov's theorem,

$$\beta_t = X_t - \int_0^t ds \frac{\partial_x h_\nu}{h_\nu}(X_s, s)$$

is a $\mathbf{W}^{h_\nu}$-martingale, more precisely a $\mathbf{W}^{h_\nu}$-Brownian motion and

$$X_t = \beta_t + \int_0^t ds \frac{\partial_x h_\nu}{h_\nu}(X_s, s) .$$

The canonical decomposition of $B^{(Y)}$ is

$$B_t^{(Y)} = \gamma_t + \int_0^t ds \, \frac{\partial_x h_\nu}{h_\nu}(B_s^{(Y)}, s) .$$

where γ is a BM with respect to the natural filtration of $B^{(Y)}$.

The next proposition describes the conditional law of Y, given $B^{(Y)}$.

Proposition 5.10.1.2 *If $f : \mathbb{R} \to \mathbb{R}^+$ is a Borel function, then*

$$\pi_t(f) := \mathbb{E}(f(Y)|B_s^{(Y)}, s \le t) = \frac{h_{(f \cdot \nu)}(B_t^{(Y)}, t)}{h_\nu(B_t^{(Y)}, t)}$$

where $h_{(f\cdot\nu)}(x,t) = \int \nu(dy) f(y) \exp(yx - \frac{y^2}{2}t)$ and

$$\pi_t(f) = 1 + \int_0^t \left(\partial_x \frac{h_{(f \cdot \nu)}}{h_\nu} \right) (B_s^{(Y)}, s) d\gamma_s .$$

PROOF: On the one hand

$$\mathbb{E}(f(Y)F(B_s^{(Y)}, s \le t) = \mathbb{E}(F(B_s, s \le t)\, h_{(f\cdot\nu)}(B_t, t)) . \tag{5.10.1}$$

On the other hand, if

$$\Phi(B_s^{(Y)}, s \le t) = \mathbb{E}(f(Y)|B_s^{(Y)}, s \le t) ,$$

the left-hand side of (5.10.1) is equal to

$$\mathbb{E}\left(\Phi(B_s^{(Y)}, s \le t) F(B_s^{(Y)}, s \le t)\right) = \mathbb{E}\left(\Phi(B_s, s \le t) F(B_s, s \le t) h_\nu(B_t, t)\right) . \tag{5.10.2}$$

It follows that

$$\pi_t(f) = \Phi(B_s^{(Y)}, s \le t) = \frac{h_{(f \cdot \nu)}(B_t^{(Y)}, t)}{h_\nu(B_t^{(Y)}, t)} .$$

The expression of $\pi_t(f)$ as a stochastic integral follows directly from this expression of $\pi_t(f)$ (and the martingale property of $\pi_t(f)$). □

5.10.2 Other Examples of Canonical Decomposition

The above result can be generalized to the case where

$$dX_t = dW_t + (f(t)\tilde{W}_t + h(t)X_t)dt$$

where $\tilde{W}$ is independent of W. In that case, studied by Föllmer et al. [353], the canonical decomposition of X is

$$X_t = \beta_t + \int_0^t (f(u)k_u(X_v; v \leq u) + h(u)X_u)\, du$$

where

$$k_u(X_s; s \leq u) = \frac{1}{\Psi'(u)} \int_0^u \Psi(v)\,(f(v)dX_v - f(v)h(v)X_v dv)$$

with Ψ the fundamental solution of the Sturm-Liouville equation

$$\Psi''(t) = f^2(t)\Psi(t)$$

with boundary conditions $\Psi(0) = 0, \Psi'(0) = 1$.

5.10.3 Innovation Process

The following formula plays an important rôle in filtering theory and will be illustrated below.

Proposition 5.10.3.1 *Let $dX_t = Y_t dt + dW_t$, where W is an $\mathbf{F}$-Brownian motion and Y an $\mathbf{F}$-adapted process. Define $\widehat{Y}_t = \mathbb{E}(Y_t|\mathcal{F}_t^X)$, the optional projection of Y on $\mathbf{F}^X$. Then, the process*

$$Z_t := X_t - \int_0^t \widehat{Y}_s ds$$

is an $\mathbf{F}^X$-Brownian motion, called the innovation process.

PROOF: Note that, for $t > s$,

$$\begin{aligned}
\mathbb{E}(Z_t|\mathcal{F}_s^X) &= \mathbb{E}(X_t|\mathcal{F}_s^X) - \mathbb{E}\left(\int_0^t \widehat{Y}_u du|\mathcal{F}_s^X\right) \\
&= \mathbb{E}(W_t|\mathcal{F}_s^X) + \mathbb{E}\left(\int_0^t Y_u du|\mathcal{F}_s^X\right) - \int_0^s \widehat{Y}_u du - \mathbb{E}\left(\int_s^t \widehat{Y}_u du|\mathcal{F}_s^X\right).
\end{aligned}$$

From the inclusion $\mathcal{F}_t^X \subset \mathcal{F}_t$ and the fact that W is an $\mathbf{F}$-martingale, we obtain $\mathbb{E}(W_t|\mathcal{F}_s^X) = \mathbb{E}(W_s|\mathcal{F}_s^X)$. Therefore, by using

$$\int_s^t \mathbb{E}(Y_u|\mathcal{F}_s^X)du = \int_s^t \mathbb{E}(\widehat{Y}_u|\mathcal{F}_s^X)du$$

we obtain

$$\begin{aligned}
\mathbb{E}(Z_t|\mathcal{F}_s^X) &= \mathbb{E}(W_s|\mathcal{F}_s^X) + \mathbb{E}\left(\int_0^t Y_u du|\mathcal{F}_s^X\right) - \int_0^s \widehat{Y}_u du - \mathbb{E}\left(\int_s^t \widehat{Y}_u du|\mathcal{F}_s^X\right) \\
&= \mathbb{E}(X_s|\mathcal{F}_s^X) + \int_s^t \mathbb{E}(Y_u|\mathcal{F}_s^X)du - \int_0^s \widehat{Y}_u du - \mathbb{E}\left(\int_s^t \widehat{Y}_u du|\mathcal{F}_s^X\right) \\
&= X_s + \int_s^t \mathbb{E}(\widehat{Y}_u|\mathcal{F}_s^X)du - \int_0^s \widehat{Y}_u du - \mathbb{E}\left(\int_s^t \widehat{Y}_u du|\mathcal{F}_s^X\right) \\
&= X_s - \int_0^s \widehat{Y}_u du\,.
\end{aligned}$$

□

Proposition 5.10.3.1 is in fact a particular case of the more general result that follows, which is of interest if Z is not $\mathbf{F}$-adapted.

Proposition 5.10.3.2 *Let Z be a measurable process such that $\mathbb{E}(\int_0^t |Z_u|du)$ is finite for every t. Then, $\mathbb{E}(\int_0^t Z_u du|\mathcal{F}_t)$ is an $\mathbf{F}$-semi-martingale which decomposes as $M_t + \int_0^t du\, \mathbb{E}(Z_u|\mathcal{F}_u)$, where M is a martingale.*

PROOF: We leave the proof to the reader. □

Example 5.10.3.3 As an example, take $Z_u = B_1, \forall u$, with B a Brownian motion. Then

$$\mathbb{E}\left(\int_0^t duB_1|\mathcal{F}_t\right) = tB_t = M_t + \int_0^t duB_u \,.$$

Comment 5.10.3.4 The paper of Pham and Quenez [711] and the paper of Lefebvre et al. [574] study the problem of optimal consumption under partial observation, by means of filtering theory. See also Nakagawa [665] for an application to default risk.

6

A Special Family of Diffusions: Bessel Processes

Bessel processes are intensively used in finance, to model the dynamics of asset prices, of the spot rate and of the stochastic volatility, or as a computational tool. In particular, computations for the celebrated Cox-Ingersoll-Ross (CIR) and Constant Elasticity Variance (CEV) models can be carried out using Bessel processes.

We present here some main facts about Bessel, CIR, and CEV processes in the spirit of the survey in Göing-Jaeschke and Yor [398], and we apply these facts to Asian options; more generally, applications to finance can be found in, among others, Aquilina and Rogers [22], Chen and Scott [164], Dassios and Nagaradjasarma [214], Deelstra and Parker [229], Delbaen [230], Davydov and Linetsky [225], Delbaen and Shirakawa [238], Dufresne [279], Duffie and Singleton [275], Grasselli [402], Heath and Platen [428], Leblanc [571, 572], Linetsky [594], Shirakawa [790] and in Szatzschneider [816, 817].

6.1 Definitions and First Properties

6.1.1 The Euclidean Norm of the n-Dimensional Brownian Motion

Let $\beta = (\beta_1, \beta_2, \dots, \beta_n)$ be an n-dimensional BM where $n \in \mathbb{N}$ and define the process R as $R_t = ||\beta_t||$, i.e., $R_t^2 = \sum_{i=1}^n (\beta_i)^2(t)$. Itô's formula leads to $dR_t^2 = \sum_{i=1}^n 2\beta_i(t)d\beta_i(t) + n\,dt$.

Note that for any $t > 0$, $\mathbb{P}(R_t = 0) = 0$, hence the process W defined as

$$dW_t = \frac{1}{R_t} \sum_{i=1}^n \beta_i(t) d\beta_i(t)$$

is a real-valued Brownian motion (see Exercise 1.5.3.5) and the process R satisfies

$$d(R_t^2) = 2R_t dW_t + n\,dt\,.$$

M. Jeanblanc, M. Yor, M. Chesney, *Mathematical Methods for Financial Markets*, Springer Finance, DOI 10.1007/978-1-84628-737-4_6,

Hence, setting $\rho_t = R_t^2$

$$d\rho_t = 2\sqrt{\rho_t} dW_t + n\, dt\,,$$

and, using Itô's formula, which may be easily justified for $n > 1$, one obtains

$$dR_t = dW_t + \frac{n-1}{2}\frac{dt}{R_t}\,. \tag{6.1.1}$$

The case $n = 1$ requires more care: if B is a one dimensional Brownian motion, then Tanaka's formula (4.1.8) asserts that $|B_t| = \int_0^t \text{sgn}(B_s)dB_s + L_t^0$, i.e., $dR_t = d\beta_t + dL_t^0$, which is the analog of equation (6.1.1).

We shall say that R is a Bessel process (BES) of dimension n, and ρ is a squared Bessel process (BESQ) of dimension n. The reason for the Bessel terminology is that many quantities involving Bessel processes may be expressed in terms of Bessel functions.

Exercise 6.1.1.1 Develop $(\epsilon + B_t^2)^{1/2}$ using Itô's formula. Letting ϵ go to 0, prove that $|B_t| = W_t + L_t^0$, where L^0 is the local time at 0 of B. Prove (6.1.1) using the same method for $n > 1$.
Hint:

$$(\epsilon + B_t^2)^{1/2} = \epsilon^{1/2} + \frac{1}{2}\int_0^t 2B_s \frac{dB_s}{\sqrt{\epsilon + B_s^2}} + \frac{\epsilon}{2}\int_0^t \frac{ds}{(\epsilon + B_s^2)^{3/2}}\,.$$

◁

6.1.2 General Definitions

Let W be a real-valued Brownian motion. Using the elementary inequality $|\sqrt{x} - \sqrt{y}| \leq \sqrt{|x-y|}$, for $x, y \geq 0$, the existence Theorem 1.5.5.1 proves that for every $\delta \geq 0$, not necessarily an integer, and $x \geq 0$, the equation

$$d\rho_t = \delta\, dt + 2\sqrt{|\rho_t|}\, dW_t, \quad \rho_0 = x \tag{6.1.2}$$

admits a unique strong solution. The solution is called the squared Bessel process of dimension δ, in short BESQ$^\delta$. In the particular case when $x = 0$ and $\delta = 0$, the obvious solution $\rho \equiv 0$ is the unique solution. From the comparison Theorem 1.5.5.9, if $0 \leq \delta \leq \delta'$ and if ρ and ρ' are squared Bessel processes with dimensions δ and δ' starting at the same point, then $0 \leq \rho_t \leq \rho'_t$ a.s.. Therefore, ρ satisfies $\rho_t \geq 0$ for all t and a posteriori the absolute value under the square root in (6.1.2) is not needed.

Definition 6.1.2.1 (BESQ$^\delta$) *For every $\delta \geq 0$ and $x \geq 0$, the unique strong solution to the equation*

$$\rho_t = x + \delta t + 2\int_0^t \sqrt{\rho_s}\, dW_s, \quad \rho_t \geq 0$$

*is called a **squared Bessel process with dimension** δ, starting at x and is denoted by* BESQ$_x^\delta$.

In particular, this process is a diffusion, and so is any positive power of this process.

Definition 6.1.2.2 (BES$^\delta$) *Let ρ be a* BESQ$^\delta_x$. *The process $R = \sqrt{\rho}$ is called a* ***Bessel process of dimension*** *δ, starting at $r = \sqrt{x}$ and is denoted* BES$^\delta_r$. *We also parametrize the family of Bessel processes with the* ***index*** *ν given by $\nu = (\delta/2) - 1$ instead of the dimension δ. We shall write* BES$^{(\nu)}$ *instead of* BES$^\delta$.

It follows from the note after the previous definition that BES$^\delta$ is a diffusion.

- For $\delta > 1$, the BES$^\delta_r$ is the solution of

$$R_t = r + W_t + \frac{\delta - 1}{2} \int_0^t \frac{1}{R_s} ds . \tag{6.1.3}$$

  In terms of the index, the BES$^{(\nu)}$process is the solution of

$$R_t = r + W_t + \left(\nu + \frac{1}{2}\right) \int_0^t \frac{1}{R_s} ds .$$

- For $\delta < 1$, the integral $\int_0^t \frac{ds}{R_s}$ does not converge and

$$R_t = r + W_t + \frac{\delta - 1}{2} \text{p.v.} \int_0^t \frac{1}{R_s} ds ,$$

  where the principal value is defined as

$$\text{p.v.} \int_0^t \frac{1}{R_s} ds = \int_0^\infty x^{\delta - 2} (\ell_t^x - \ell_t^0) dx$$

  and the family of diffusion local times is defined via the occupation time formula and the speed measure (see Section 5.5):

$$\int_0^t \phi(R_s) ds = \int_0^\infty \phi(x) \ell_t^x x^{\delta - 1} dx .$$

  For a study of principal values, see Chapter 10 in Yor [868].

- For $\delta = 1$

$$R_t = r + W_t + \frac{1}{2} L_t^0(R) ,$$

  where $L^0(R)$ is the local time of R at level 0.

We use the superscript $^{(\nu)}$ for the index ν, whereas there is no bracket in the superscript when we refer to the dimension δ. Important particular cases are $\nu = 1/2$ which corresponds to $\delta = 3$, $\nu = 0$ which corresponds to $\delta = 2$ and $\nu = -\frac{1}{2}$ which corresponds to $\delta = 1$ (reflected BM).

We now present scale functions of the Bessel processes as diffusions.

Definition 6.1.2.3 *Let B be a one-dimensional Brownian motion. A process with the same law as $|B|$ is called a reflected Brownian motion.*

Proposition 6.1.2.4 *Let ρ be a squared Bessel process with dimension δ. A scale function is*

$$-x^{1-(\delta/2)} \text{ for } \delta > 2;\ \ln x \text{ for } \delta = 2;\ x^{1-(\delta/2)} \text{ for } \delta < 2\,.$$

Let R be a Bessel process with dimension δ. A scale function is

$$-x^{2-\delta} \text{ for } \delta > 2;\ \ln x \text{ for } \delta = 2;\ x^{2-\delta} \text{ for } \delta < 2\,.$$

PROOF: The result can be checked from an application of Itô's lemma. □

Warning 6.1.2.5 A blind application of the result "the scale function of a Bessel process with dimension 1 is $s(x) = x$" would lead to the false conclusion that $|W|$ is a local martingale.

Example 6.1.2.6 Strict Local Martingale. Here, we give an example of a local martingale which is not a martingale, i.e., a **strict local martingale.** Let M be a continuous martingale such that $M_0 = 1$ and define $T_0 = \inf\{t : M_t = 0\}$. We assume that $\mathbb{P}(T_0 < \infty) = 1$. We introduce the probability measure $\mathbb{Q}$ as $\mathbb{Q}|_{\mathcal{F}_t} = M_{t\wedge T_0}\mathbb{P}|_{\mathcal{F}_t}$. It follows that

$$\mathbb{Q}(T_0 < t) = \mathbb{E}_{\mathbb{P}}(\mathbb{1}_{T_0<t} M_{t\wedge T_0}) = 0\,, \tag{6.1.4}$$

i.e., $\mathbb{Q}(T_0 = \infty) = 1$. The process X defined by $(X_t = M_t^{-1}, t \geq 0)$ is a $\mathbb{Q}$-local martingale and is positive. It is not a martingale: indeed its expectation is not constant

$$\mathbb{E}_{\mathbb{Q}}(X_t) = \mathbb{E}_{\mathbb{P}}\left(\frac{M_{t\wedge T_0}}{M_t}\right) = \mathbb{P}(t < T_0) \neq 1 = X_0\,.$$

From Girsanov's theorem, the process $\widetilde{M}_t = M_t - \int_0^t \frac{d\langle M\rangle_s}{M_s}$ is a $\mathbb{Q}$-local martingale. In the case $M_t = B_t$, we get

$$B_t = \beta_t + \int_0^t \frac{ds}{B_s}$$

where β is a $\mathbb{Q}$-Brownian motion. Hence, the process B is a $\mathbb{Q}$-Bessel process of dimension 3. See Wong and Heyde [848], Elworthy et al. [319], Kotani [539], Kotani and Sin [799] and Watanabe [837] for comments on strict local martingales.

6.1.3 Path Properties

Proposition 6.1.3.1 *Let ρ be a δ-dimensional squared Bessel process. For $\delta = 0$, the point 0 is absorbing (the process remains at* 0 *as soon as it reaches it). For* $0 < \delta < 2$, *the* BESQ$^\delta$ *is reflected instantaneously.*

PROOF: In the case $\delta = 0$, the point 0 is reached a.s.. It is obvious that the point is absorbing. In the case $0 < \delta < 2$, the process ρ is a semi-martingale. The occupation time formula leads to

$$t \geq \int_0^t \mathbb{1}_{\{\rho_s > 0\}} ds = \int_0^t \mathbb{1}_{\{\rho_s > 0\}} (4\rho_s)^{-1} d\langle \rho \rangle_s$$
$$= \int_0^\infty (4a)^{-1} L_t^a(\rho) da \,.$$

Hence, the local time at 0 is identically equal to 0 (otherwise, the integral on the right-hand side is not convergent). From the study of the local time and the fact that $L_t^{0-}(\rho) = 0$, we obtain

$$L_t^0(\rho) = 2\delta \int_0^t \mathbb{1}_{\{\rho_s = 0\}} ds \,.$$

Therefore, the time spent by ρ in 0 has zero Lebesgue measure. □

- **Bessel process with dimension $\delta > 2$:** It follows from the properties of the scale function that: for $\delta > 2$, the BESQ$^\delta_x$ will never reach 0 and is a transient process (ρ_t goes to infinity as t goes to infinity),
  $\mathbb{P}_x(R_t > 0, \, \forall t > 0) = 1$,
  $\mathbb{P}_x(R_t \to \infty, \, t \to \infty) = 1$.
- **Bessel process with dimension $\delta = 2$:** The BES^2_x will never reach 0:
  $\mathbb{P}_x(R_t > 0, \, \forall t > 0) = 1$,
  $\mathbb{P}_x(\sup_t R_t = \infty, \inf_t R_t = 0) = 1$.
- **Bessel process with dimension $0 < \delta < 2$:** It follows from the properties of the scale function that for $0 \leq \delta < 2$ the process R reaches 0 in finite time and that the point 0 is an entrance boundary (see Definition 5.3.3.1). One has, for $a > 0$, $\mathbb{P}(R_t > 0, \, \forall t > a) = 0$.

6.1.4 Infinitesimal Generator

Bessel Processes

A Bessel process R with index $\nu \geq 0$ (i.e., with dimension $\delta = 2(\nu+1) \geq 2$) is a diffusion process which takes values in $\mathbb{R}^+$ and has infinitesimal generator

$$\mathcal{A} = \frac{1}{2}\frac{d^2}{dx^2} + \frac{2\nu+1}{2x}\frac{d}{dx} = \frac{1}{2}\frac{d^2}{dx^2} + \frac{\delta-1}{2x}\frac{d}{dx} \,,$$

i.e., for any $f \in C^2(]0,\infty[)$, and $R_0 = r > 0$ the process

$$f(R_t) - \int_0^t \mathcal{A}f(R_s)ds,\ t \geq 0$$

is a local martingale. In particular, if R is a $\mathrm{BES}^{(\nu)}$, the process $1/(R_t)^{2\nu}$ is a local martingale. Hence the scale function is $s(x) = -x^{-2\nu}$ for $\nu \geq 0$. For $\delta > 1$, a BES^δ_r satisfies $\mathbb{E}_r\left(\int_0^t ds (R_s)^{-1}\right) < \infty$, for every $r \geq 0$.

The BES^1 is a reflected Brownian motion $R_t = |\beta_t| = W_t + L_t$ where W and β are Brownian motions and L is the local time at 0 of Brownian motion β.

Squared Bessel Processes

The infinitesimal generator of the squared Bessel process ρ is

$$\mathcal{A} = 2x\frac{d^2}{dx^2} + \delta\frac{d}{dx}$$

hence, for any $f \in C^2(]0,\infty[)$, the process

$$f(\rho_t) - \int_0^t \mathcal{A}f(\rho_s)ds$$

is a local martingale.

Proposition 6.1.4.1 (Scaling Properties.) *If* $(\rho_t, t \geq 0)$ *is a* BESQ^δ_x*, then* $(\frac{1}{c}\rho_{ct}, t \geq 0)$ *is a* $\mathrm{BESQ}^\delta_{x/c}$.

PROOF: From

$$\rho_t = x + 2\int_0^t \sqrt{\rho_s}\, dW_s + \delta\, t\,,$$

we deduce that

$$\frac{1}{c}\rho_{ct} = \frac{x}{c} + \frac{2}{c}\int_0^{ct} \sqrt{\rho_s}\, dW_s + \frac{\delta}{c}ct = \frac{x}{c} + 2\int_0^{ct} \left(\frac{\rho_s}{c}\right)^{1/2} \frac{1}{\sqrt{c}}\, dW_s + \delta t\,.$$

Setting $u_t = \dfrac{1}{c}\rho_{ct}$, we obtain using a simple change of variable

$$u_t = \frac{x}{c} + 2\int_0^t \sqrt{u_s}\, d\widetilde{W}_s + \delta\, t$$

where $(\widetilde{W}_t = \frac{1}{\sqrt{c}} W_{tc}, t \geq 0)$ is a Brownian motion. □

$\delta = 2(1+\nu)$			
$\delta = 2$	0 is polar	$\ln R$ is a strict local-martingale	R is a semi-martingale
$\delta > 2$	0 is polar	$R^{-2\nu}$ is a strict local-martingale	R is a semi-martingale
$2 > \delta > 1$	R reflects at 0	$R^{-2\nu}$ is a sub-martingale	R is a semi-martingale
$\delta = 1$	R reflects at 0	R is a sub-martingale	R is a semi-martingale
$1 > \delta > 0$	R reflects at 0	$R^{-2\nu}$ is a sub-martingale	R is not a semi-martingale
$\delta = 0$	0 is absorbing	R^2 is a martingale	R is a semi-martingale

Fig. 6.1 Bessel processes

Comment 6.1.4.2 Delbaen and Schachermayer [237] allow general admissible integrands as trading strategies, and prove that the three-dimensional Bessel process admits arbitrage possibilities. Pal and Protter [692], Yen and Yor [856] establish pathological behavior of asset price processes modelled by continuous strict local martingales, in particular the reciprocal of a three-dimensional Bessel process under a risk-neutral measure.

6.1.5 Absolute Continuity

On the canonical space $\Omega = C(\mathbb{R}^+, \mathbb{R}^+)$, we denote by R the canonical map $R_t(\omega) = \omega(t)$, by $\mathcal{R}_t = \sigma(R_s, s \leq t)$ the canonical filtration and by $\mathbb{P}_r^{(\nu)}$ (resp. $\mathbb{P}_r^{\delta}$) the law of the Bessel process of index ν (resp. of dimension δ), starting at r, i.e., such that $\mathbb{P}_r^{(\nu)}(R_0 = r) = 1$. The law of BESQ^{δ} starting at x on the canonical space $C(\mathbb{R}^+, \mathbb{R}^+)$ is denoted by $\mathbb{Q}_x^{\delta}$.

Proposition 6.1.5.1 *The following absolute continuity relation between the laws of a* $\text{BES}^{(\nu)}$ *(with* $\nu \geq 0$*) and a* $\text{BES}^{(0)}$ *holds*

$$\mathbb{P}_r^{(\nu)}|_{\mathcal{R}_t} = \left(\frac{R_t}{r}\right)^{\nu} \exp\left(-\frac{\nu^2}{2}\int_0^t \frac{ds}{R_s^2}\right) \mathbb{P}_r^{(0)}|_{\mathcal{R}_t} . \tag{6.1.5}$$

PROOF: Under $\mathbb{P}^{(0)}$, the canonical process R which is a Bessel process with dimension 2, satisfies

$$dR_t = dW_t + \frac{1}{2R_t}dt .$$

Itô's formula applied to the process

$$D_t = \left(\frac{R_t}{r}\right)^{\nu} \exp\left(-\frac{\nu^2}{2}\int_0^t \frac{ds}{R_s^2}\right)$$

leads to

$$dD_t = \nu D_t (R_t)^{-1} dW_t ,$$

therefore, the process D is a local martingale. We prove now that it is a martingale.

Obviously, $\sup_{t\leq T} D_t \leq \sup_{t\leq T}(R_t/r)^{\nu}$. The process R^2 is a squared Bessel process of dimension 2, and is equal in law to $B^2 + \tilde{B}^2$ where B and $\tilde{B}$ are independent BMs. It follows that R_t^k is integrable for $k \geq 2$. The process R is a submartingale as a sum of a martingale and an increasing process, and Doob's inequality (1.2.1) implies that

$$\mathbb{E}\left[\left(\sup_{t\leq T} R_t\right)^k\right] \leq C_k \mathbb{E}[R_T^k].$$

Hence, the process D is a martingale. From Girsanov's theorem, it follows that the process Z defined by

$$dZ_t = dW_t - \frac{\nu}{R_t}dt = dR_t - \frac{1}{R_t}\left(\nu + \frac{1}{2}\right)dt$$

is a Brownian motion under $\mathbb{P}_r^{(\nu)}$ where $\mathbb{P}_r^{(\nu)}|_{\mathcal{R}_t} = D_t \mathbb{P}_r^{(0)}|_{\mathcal{R}_t}$. □

If the index $\nu = -\mu$ is negative (i.e., $\mu > 0$), then the absolute continuity relation holds before T_0, the first hitting time of 0:

$$\mathbb{P}_r^{(\nu)}|_{\mathcal{R}_t \cap \{t<T_0\}} = \left(\frac{R_t}{r}\right)^{\nu} \exp\left(-\frac{\nu^2}{2}\int_0^t \frac{ds}{R_s^2}\right) \bullet \mathbb{P}_r^{(0)}|_{\mathcal{R}_t} . \tag{6.1.6}$$

Comparison with equality (6.1.5) shows that,

$$\text{for } \nu < 0, \quad \mathbb{P}_r^{(-\nu)}|_{\mathcal{R}_t} = \frac{R_{t\wedge T_0}^{-2\nu}}{r^{-2\nu}} \bullet \mathbb{P}_r^{(\nu)}|_{\mathcal{R}_t} . \tag{6.1.7}$$

In particular, this shows that the BES^3 process ($\mu = 1/2$) is an h-transform of Brownian motion killed at 0, which simply means

$$\boxed{\mathbb{P}_r^{(1/2)}|_{\mathcal{R}_t} = \frac{R_{t\wedge T_0}}{r} \cdot \mathbb{P}_r^{(-1/2)}|_{\mathcal{R}_t}}$$

or, if $\mathbf{W}_a$ denotes the law of the BM starting from $a > 0$ and T_0 the hitting time of 0,

$$\mathbb{P}_a^3\big|_{\mathcal{F}_t} = \frac{X_{t\wedge T_0}}{a} \cdot \mathbf{W}_a\big|_{\mathcal{F}_t}. \tag{6.1.8}$$

(See Dellacherie et al. [241].)

Comment 6.1.5.2 The absolute continuity relationship (6.1.5) has been of some use in a number of problems, see, e.g., Kendall [518] for the computation of the shape distribution for triangles, Geman and Yor [383] for the pricing of Asian options, Hirsch and Song [437] in connection with the flows of Bessel processes and Werner [839] for the computation of Brownian intersection exponents.

Exercise 6.1.5.3 With the help of the explicit expression for the semi-group of BM killed at time T_0 (3.1.9), deduce the semi-group of BES^3 from formula (6.1.8). ◁

Exercise 6.1.5.4 Let S be the solution of

$$dS_t = S_t^2 dW_t$$

where W is a Brownian motion. Prove that $X = 1/S$ is a Bessel process of dimension 3.

This kind of SDE will be extended to different choices of volatilities in ↣ Section 6.4. ◁

Exercise 6.1.5.5 Let R be a BES^3 process starting from 1. Compute $\mathbb{E}(R_t^{-1})$.
Hint: From the absolute continuity relationship

$$\mathbb{E}(R_t^{-1}) = \mathbf{W}_1(T_0 > t) = \mathbb{P}(|G| < 1/\sqrt{t})$$

where G is a standard Gaussian r.v.. ◁

Exercise 6.1.5.6 Let R and $\widetilde{R}$ be two independent BES^3 processes. The process $Y_t = R_t^{-1} - (\widetilde{R}_t)^{-1}$ is a local martingale with null expectation. Prove that Y is a strict local martingale.
Hint: Let T_n be a localizing sequence of stopping times for $1/R$. If Y were a martingale, $1/\widetilde{R}_{t\wedge T_n}$ would also be a martingale. The expectation of $1/\widetilde{R}_{t\wedge T_n}$ can be computed and depends on t. ◁

Exercise 6.1.5.7 Let R and $\widetilde{R}$ be two independent BES^3 processes. Prove that the filtration generated by the process $Y_t = R_t^{-1} - \widetilde{R}_t^{-1}$ is the filtration generated by the processes R and $\widetilde{R}$.
`Hint`: Indeed, the bracket of Y, i.e., $\int_0^t (\frac{1}{R_s^4} + \frac{1}{\widetilde{R}_s^4})ds$ is adapted w.r.t. the filtration $(\mathcal{Y}_t, t \geq 0)$ generated by Y. Hence the process $(\frac{1}{R_t^4} + \frac{1}{\widetilde{R}_t^4})$ is $\mathcal{Y}$-adapted. Now, if a and b are given, there exists a unique pair (x, y) of positive numbers such that $x - y = a$, $x^4 + y^4 = b$ (this pair can even be given explicitly, noting that $x^4 + y^4 - (x-y)^4 = 2xy(xy - 2(x-y)^2)$). This completes the proof. ◁

6.2 Properties

6.2.1 Additivity of BESQ's

An important property, due to Shiga and Watanabe [788], is the additivity of the BESQ family. Let us denote by $\mathbb{P} * \mathbb{Q}$ the convolution of $\mathbb{P}$ and $\mathbb{Q}$, two probabilities on $C(\mathbb{R}^+, \mathbb{R}^+)$.

Proposition 6.2.1.1 *The sum of two independent squared Bessel processes with respective dimension δ and δ', starting respectively from x and x' is a squared Bessel process with dimension $\delta + \delta'$, starting from $x + x'$:*

$$\mathbb{Q}_x^\delta * \mathbb{Q}_y^{\delta'} = \mathbb{Q}_{x+y}^{\delta+\delta'} .$$

PROOF: Let X and Y be two independent BESQ processes starting at x (resp. at y) and with dimension δ (resp. δ') and $Z = X + Y$. We want to show that Z is distributed as $\mathbb{Q}_{x+y}^{\delta+\delta'}$. Note that the result is obvious from the definition when the dimensions are integers (this is what D. Williams calls the "Pythagoras" property). In the general case

$$Z_t = x + y + (\delta + \delta')t + 2\int_0^t \left(\sqrt{X_s}\, dB_s + \sqrt{Y_s}\, dB'_s\right) ,$$

where (B, B') is a two-dimensional Brownian motion. This process satisfies $\int_0^t \mathbb{1}_{\{Z_s=0\}} ds = 0$. Let $\widehat{B}$ be a third Brownian motion independent of (B, B'). The process W defined as

$$W_t = \int_0^t \mathbb{1}_{\{Z_s>0\}} \left(\frac{\sqrt{X_s}\, dB_s + \sqrt{Y_s} dB'_s}{\sqrt{Z_s}}\right)$$

is a Brownian motion (it is a martingale with increasing process equal to t). The process Z satisfies

$$Z_t = x + y + (\delta + \delta')t + 2\int_0^t \sqrt{Z_s}\, dW_s ,$$

and this equation admits a unique solution in law. □

6.2.2 Transition Densities

Bessel and squared Bessel processes are Markov processes and their transition densities are known. Expectation under $\mathbb{Q}_x^\delta$ will be denoted by $\mathbb{Q}_x^\delta[\cdot]$. We also denote by ρ the canonical process (a squared Bessel process) under the $\mathbb{Q}^\delta$-law. From Proposition 6.2.1.1, the Laplace transform of ρ_t satisfies

$$\mathbb{Q}_x^\delta[\exp(-\lambda\rho_t)] = \mathbb{Q}_x^1[\exp(-\lambda\rho_t)]\ \left[\mathbb{Q}_0^1[\exp(-\lambda\rho_t)]\right]^{\delta-1}$$

and since, under $\mathbb{Q}_x^1$, the r.v. ρ_t is the square of a Gaussian variable, one gets, using Exercise 1.1.12.3,

$$\mathbb{Q}_x^1[\exp(-\lambda\rho_t)] = \frac{1}{\sqrt{1+2\lambda t}}\exp\left(-\frac{\lambda x}{1+2\lambda t}\right).$$

Therefore

$$\boxed{\mathbb{Q}_x^\delta[\exp(-\lambda\rho_t)] = \frac{1}{(1+2\lambda t)^{\delta/2}}\exp\left(-\frac{\lambda x}{1+2\lambda t}\right).} \tag{6.2.1}$$

Inverting the Laplace transform yields the transition density $q_t^{(\nu)}$ of a BESQ$^{(\nu)}$ for $\nu > -1$ as

$$\boxed{q_t^{(\nu)}(x,y) = \frac{1}{2t}\left(\frac{y}{x}\right)^{\nu/2}\exp\left(-\frac{x+y}{2t}\right) I_\nu\left(\frac{\sqrt{xy}}{t}\right),} \tag{6.2.2}$$

and the Bessel process of index ν has a transition density $p_t^{(\nu)}$ defined by

$$\boxed{p_t^{(\nu)}(x,y) = \frac{y}{t}\left(\frac{y}{x}\right)^{\nu}\exp\left(-\frac{x^2+y^2}{2t}\right) I_\nu\left(\frac{xy}{t}\right)} \tag{6.2.3}$$

where I_ν is the usual modified Bessel function with index ν. (See $\rightarrowtail$ Appendix A.5.2 for the definition of modified Bessel functions.) For $x = 0$, the transition probability of the BESQ$^{(\nu)}$ (resp. of the BES$^{(\nu)}$) is

$$q_t^{(\nu)}(0,y) = (2t)^{-(\nu+1)}[\Gamma(\nu+1)]^{-1}y^\nu \exp\left(-\frac{y}{2t}\right),$$

$$p_t^{(\nu)}(0,y) = 2^{-\nu}t^{-(\nu+1)}[\Gamma(\nu+1)]^{-1}y^{2\nu+1}\exp\left(-\frac{y^2}{2t}\right). \tag{6.2.4}$$

In the case $\delta = 0$ (i.e., $\nu = -1$), the semi-group of BESQ^0 is

$$Q_t^0(x,\cdot) = \exp\left(-\frac{x}{2t}\right)\epsilon_0 + \widehat{Q}_t(x,\cdot)$$

where ϵ_0 is the Dirac measure at 0 and $\widehat{Q}_t(x,dy)$ has density

$$q_t^0(x,y) = \frac{1}{2t}\left(\frac{y}{x}\right)^{-1/2}\exp\left(-\frac{x+y}{2t}\right) I_1\left(\frac{\sqrt{xy}}{t}\right),$$

while the semi-group for BES^0 is

$$P_t^0(x,\cdot) = \exp\left(-\frac{x^2}{2t}\right)\epsilon_0 + \widehat{P}_t(x,\cdot)$$

where $\widehat{P}_t(x,dy)$ has density

$$p_t^0(x,y) = \frac{x}{t}\exp\left(-\frac{x^2+y^2}{2t}\right) I_1\left(\frac{xy}{t}\right).$$

Remark 6.2.2.1 From the equality (6.2.3), we can check that, if R is a BES^δ starting from x, then $R_t^2 \stackrel{\text{law}}{=} tZ$ where Z has a $\chi^2(\delta, \frac{x}{t})$ law. (See Exercise 1.1.12.5 for the definition of χ^2.)

Comment 6.2.2.2 Carmona [140] presents an extension of squared Bessel processes with time varying dimension $\delta(t)$, as the solution of

$$dX_t = \delta(t)dt + 2\sqrt{X_t}dW_t.$$

Here, δ is a function with positive values. The Laplace transform of X_t is

$$\mathbb{E}_x(\exp(-\lambda X_t)) = \exp\left(-\lambda\frac{x}{1+2\lambda t} - \int_0^t \frac{\lambda\delta(u)}{1+2\lambda(t-u)}du\right).$$

See Shirakawa [790] for applications to interest rate models.

Comment 6.2.2.3 The negative moments of a squared Bessel process have been computed in Yor [863], Aquilina and Rogers [22] and Dufresne [279]

$$\mathbb{Q}_x^{(\nu)}(\rho_t^{-a}) = \frac{\Gamma(\nu+1-a)}{\Gamma(\nu+1)}\exp\left(-\frac{x}{2t}\right)(2t)^{-a} M\left(\nu+1-a, \nu+1, \frac{x}{2t}\right)$$

where M is the Kummer function given in $\rightarrowtail$ Appendix A.5.6.

Exercise 6.2.2.4 Let ρ be a 0-dimensional squared Bessel process starting at x, and T_0 its first hitting time of 0. Prove that $1/T_0$ follows the exponential law with parameter $x/2$.
Hint: Deduce the probability that $T_0 \leq t$ from knowledge of $\mathbb{Q}_x^0(e^{-\lambda\rho_t})$. $\triangleleft$

Exercise 6.2.2.5 (from Azéma and Yor [38].) Let X be a BES^3 starting from 0. Prove that $1/X$ is a local martingale, but not a martingale. Establish that, for $u < 1$,

$$\mathbb{E}\left(\frac{1}{X_1}|\mathcal{R}_u\right) = \frac{1}{X_u}\sqrt{\frac{2}{\pi}}\,\Phi(\frac{X_u}{1-u}),$$

where $\Phi(a) = \int_0^a dy\, e^{-y^2/2}$. Such a formula " measures" the non-martingale property of the local martingale $(1/X_t, t \leq 1)$. In general, the quantity $\mathbb{E}(Y_t|\mathcal{F}_s)/Y_s$ for $s < t$, or even its mean $\mathbb{E}(Y_t/Y_s)$, could be considered as a measure of the non-martingale property of Y. ◁

6.2.3 Hitting Times for Bessel Processes

Expectation under $\mathbb{P}_a^{(\nu)}$ will be denoted by $\mathbb{P}_a^{(\nu)}(\cdot)$. We assume here that $\nu > 0$, i.e., $\delta > 2$.

Proposition 6.2.3.1 *Let a, b be positive numbers and $\lambda > 0$.*

$$\mathbb{P}_a^{(\nu)}(e^{-\lambda T_b}) = \left(\frac{b}{a}\right)^\nu \frac{K_\nu(a\sqrt{2\lambda})}{K_\nu(b\sqrt{2\lambda})}, \quad \text{for } b \leq a\,, \tag{6.2.5}$$

$$\mathbb{P}_a^{(\nu)}(e^{-\lambda T_b}) = \left(\frac{b}{a}\right)^\nu \frac{I_\nu(a\sqrt{2\lambda})}{I_\nu(b\sqrt{2\lambda})}, \quad \text{for } a \leq b\,, \tag{6.2.6}$$

where K_ν and I_ν are modified Bessel functions, defined in $\rightarrowtail$ *Appendix A.5.2.*

PROOF: The proof is an application of (5.3.8) (see Kent [519]). Indeed, for a Bessel process the solutions of the Sturm-Liouville equation

$$\frac{1}{2}xu''(x) + \left(\nu + \frac{1}{2}\right)u'(x) - \lambda x u(x) = 0$$

are

$$\Phi_{\lambda\uparrow}(r) = c_1 I_\nu(r\sqrt{2\lambda})r^{-\nu},\ \Phi_{\lambda\downarrow}(r) = c_2 K_\nu(r\sqrt{2\lambda})r^{-\nu}$$

where c_1, c_2 are two constants. □

Note that, for $a > b$, using the asymptotic of $K_\nu(x)$, when $x \to \infty$, we may deduce from (6.2.5) that $\mathbb{P}_a^{(\nu)}(T_b < \infty) = (b/a)^{2\nu}$. Another proof may be given using the fact that the process $M_t = (1/R_t)^{\delta-2}$ is a local martingale, which converges to 0, and the result follows from Lemma 1.2.3.10.

Here is another consequence of Proposition 6.2.3.1, in particular of formula (6.2.5): for a three-dimensional Bessel process ($\nu = 1/2$) starting from 0, from equality (A.5.3) in Appendix which gives the value of the Bessel function of index 1/2,

$$\mathbb{P}_0^3\left(\exp\left(-\frac{\lambda^2}{2}T_b\right)\right) = \frac{\lambda b}{\sinh \lambda b}\,.$$

For a three-dimensional Bessel process starting from a

$$\mathbb{P}_a^3\left(\exp\left(-\frac{\lambda^2}{2}T_b\right)\right) = \frac{b}{a}\frac{\sinh\lambda a}{\sinh\lambda b}, \qquad \text{for } b \geq a, \tag{6.2.7}$$

$$\mathbb{P}_a^3\left(\exp\left(-\frac{\lambda^2}{2}T_b\right)\right) = \frac{b}{a}\exp\left(-(a-b)\lambda\right), \quad \text{for } b < a.$$

Inverting the Laplace transform, we obtain the density of T_b, the hitting time of b for a three-dimensional Bessel process starting from 0:

$$\mathbb{P}_0^3(T_b \in dt) = \sum_{n\geq 1}(-1)^{n+1}\frac{\pi^2 n^2}{b^2}e^{-n^2\pi^2 t/(2b^2)}\,dt\,.$$

For $b < a$, it is simple to find the density of the hitting time T_b for a BES_a^3. The absolute continuity relationship (6.1.8) yields the equality

$$\mathbb{E}_a^3(\phi(T_b)) = \frac{1}{a}\mathbf{W}_a(\phi(T_b)X_{T_b\wedge T_0})$$

which holds for $b < a$. Consequently

$$\mathbb{P}_a^3(T_b > t) = \mathbb{P}_a^3(\infty > T_b > t) + \mathbb{P}_a^3(T_b = \infty) = \frac{b}{a}\mathbf{W}_0(T_{a-b} > t) + 1 - \frac{b}{a}$$
$$= \frac{b}{a}\mathbb{P}(a - b > \sqrt{t}|G|) + 1 - \frac{b}{a}\,.$$

where G stands for a standard Gaussian r.v. under $\mathbb{P}$. Hence

$$\mathbb{P}_a^3(T_b > t) = \frac{b}{a}\sqrt{\frac{2}{\pi}}\int_0^{(a-b)/\sqrt{t}} e^{-y^2/2}\,dy + 1 - \frac{b}{a}\,.$$

Note that $\mathbb{P}_a^3(T_b < \infty) = \frac{b}{a}$. The density of T_b is

$$\mathbb{P}_a^3(T_b \in dt)/dt = (a-b)\frac{1}{\sqrt{2\pi t^3}}\frac{b}{a}\exp\left(-\frac{(a-b)^2}{2t}\right)\,.$$

Thanks to results on time reversal (see Williams [840], Pitman and Yor [715]) we have, for R a transient Bessel process starting at 0, with dimension $\delta > 2$ and index $\nu > 0$, denoting by Λ_1 the last passage time at 1,

$$(R_t, t < \Lambda_1) \stackrel{\text{law}}{=} (\widehat{R}_{T_0 - u}, u \leq T_0(\widehat{R})) \tag{6.2.8}$$

where $\widehat{R}$ is a Bessel process, starting from 1, with dimension $\widehat{\delta} = 2(1-\nu) < 2$. Using results on last passage times (see Example 5.6.2.3), it follows that

$$T_0(\widehat{R}) \stackrel{\text{law}}{=} \frac{1}{2\gamma(\nu)} \tag{6.2.9}$$

where $\gamma(\nu)$ has a gamma law with parameter ν.

Comment 6.2.3.2 See Pitman and Yor [716] and Biane et al. [86] for more comments on the laws of Bessel hitting times. In the case $a < b$, the density of T_b under $\mathbb{P}_a^3$ is given as a series expansion in Ismail and Kelker [460] and Borodin and Salminen [109]. This may be obtained from (6.2.7).

6.2.4 Lamperti's Theorem

We present a particular example of the relationship between exponentials of Lévy processes and semi-stable processes studied by Lamperti [562] who proved that (powers of) Bessel processes are the only semi-stable one-dimensional diffusions. See also Yor [863, 865] and DeBlassie [228].

Theorem 6.2.4.1 *The exponential of Brownian motion with drift* $\nu \in \mathbb{R}^+$ *can be represented as a time-changed* $\mathrm{BES}^{(\nu)}$. *More precisely,*

$$\exp(W_t + \nu t) = R^{(\nu)}\left(\int_0^t \exp[2(W_s + \nu s)]\, ds\right)$$

where $(R^{(\nu)}(t), t \geq 0)$ *is a* $\mathrm{BES}^{(\nu)}$.

Remark that, thanks to the scaling property of the Brownian motion, this result can be extended to $\exp(\sigma W_t + \nu t)$. In that case

$$\exp(\sigma W_t + \nu t) = R^{(\nu/\sigma^2)}\left(\sigma^2 \int_0^t \exp[2(\sigma W_s + \nu s)]\, ds\right).$$

PROOF: Introduce the increasing process $A_t = \int_0^t \exp[2(W_s + \nu s)]\, ds$ and C its inverse $C_u = \inf\{t \geq 0 : A_t \geq u\}$. From

$$\exp[2(W_s + \nu s)] = \exp[2s(\nu + W_s/s)],$$

it can be checked that $A_\infty = \infty$ a.s., hence $C_u < \infty, \forall u < \infty$ and $C_\infty = \infty, a.s.$. By definition of C, we get $A_{C_t} = t = \int_0^{C_t} \exp[2(W_s + \nu s)]\, ds$. By differentiating this equality, we obtain $dt = \exp[2(W_{C_t} + \nu C_t)] dC_t$. The continuous process $\widetilde{W}$ defined by

$$\widetilde{W}_u := \int_0^{C_u} \exp(W_s + \nu s)\, dW_s$$

is a martingale with increasing process $\int_0^{C_u} \exp[2(W_s + \nu s)]\, ds = u$. Therefore, $\widetilde{W}$ is a Brownian motion. From the definition of C,

$$\widetilde{W}_{A_t} = \int_0^t \exp(W_s + \nu s)\, dW_s.$$

This identity may be written in a differential form

$$d\widetilde{W}_{A_t} = \exp(W_t + \nu t)\, dW_t.$$

We now prove that $R_u := \exp(W_{C_u} + \nu C_u)$ is a Bessel process. Itô's formula gives

$$d[\exp(W_t + \nu t)] = \exp(W_t + \nu t)\,(dW_t + \nu dt) + \frac{1}{2}\exp(W_t + \nu t)dt$$
$$= d\widetilde{W}_{A_t} + \left(\nu + \frac{1}{2}\right)\exp(W_t + \nu t)dt\,.$$

This equality can be written in an integral form

$$\exp(W_t + \nu t) = 1 + \widetilde{W}_{A_t} + \int_0^t \left(\nu + \frac{1}{2}\right)\exp(W_s + \nu s)ds$$
$$= 1 + \widetilde{W}_{A_t} + \int_0^t \left(\nu + \frac{1}{2}\right)\frac{\exp 2(W_s + \nu s)}{\exp(W_s + \nu s)}ds$$
$$= 1 + \widetilde{W}_{A_t} + \int_0^t \left(\nu + \frac{1}{2}\right)\frac{dA_s}{\exp(W_s + \nu s)}\,.$$

Therefore

$$\exp(W_{C_u} + \nu C_u) = 1 + \widetilde{W}_u + \left(\nu + \frac{1}{2}\right)\int_0^u \frac{ds}{\exp(W_{C_s} + \nu C_s)}\,.$$

Hence,

$$d\exp(W_{C_u} + \nu C_u) = d\widetilde{W}_u + \left(\nu + \frac{1}{2}\right)\frac{du}{\exp(W_{C_u} + \nu C_u)}$$

that is,

$$dR_u = d\widetilde{W}_u + \left(\nu + \frac{1}{2}\right)\frac{du}{R_u}\,.$$

The result follows from the uniqueness of the solution to the SDE associated with the $\mathrm{BES}^{(\nu)}$(see Definition 6.1.2.2), and from $R_{A_t} = \exp(W_t + \nu t)$. □

Remark 6.2.4.2 From the obvious equality

$$\exp(\sigma B_t + \nu t) = \left(\exp\left[\frac{\sigma}{2}B_t + \frac{\nu}{2}t\right]\right)^2,$$

it follows that the exponential of a Brownian motion with drift is also a time-changed Bessel squared process. This remark is closely related to the following exercise.

Exercise 6.2.4.3 Prove that the power of a Bessel process is another Bessel process time-changed:

$$q[R_t^{(\nu)}]^{1/q} = R^{(\nu q)}\left(\int_0^t \frac{ds}{[R_s^{(\nu)}]^{2/p}}\right)$$

where $\frac{1}{p} + \frac{1}{q} = 1, \nu > -\frac{1}{q}$. ◁

6.2.5 Laplace Transforms

In this section, we give explicit formulae for some Laplace transforms related to Bessel processes.

Proposition 6.2.5.1 *The joint Laplace transform of the pair* $\left(R_t^2, \int_0^t \frac{ds}{R_s^2}\right)$ *satisfies*

$$\mathbb{P}_r^{(\nu)}\left\{\exp\left(-aR_t^2 - \frac{\mu^2}{2}\int_0^t \frac{ds}{R_s^2}\right)\right\} = \mathbb{P}_r^{(\gamma)}\left\{\left(\frac{R_t}{r}\right)^{\nu-\gamma} \exp(-aR_t^2)\right\} \quad (6.2.10)$$

$$= \frac{r^{\gamma-\nu}}{\Gamma(\alpha)}\int_0^\infty dv\, v^{\alpha-1}(1+2(v+a)t)^{-(1+\gamma)} \exp\left(-\frac{r^2(v+a)}{1+2(v+a)t}\right)$$

where $\gamma = \sqrt{\mu^2+\nu^2}$ *and* $\alpha = \frac{1}{2}(\gamma-\nu) = \frac{1}{2}(\sqrt{\mu^2+\nu^2}-\nu)$.

PROOF: From the absolute continuity relationship (6.1.5)

$$\begin{aligned}
&\mathbb{P}_r^{(\nu)}\left\{\exp\left(-aR_t^2 - \frac{\mu^2}{2}\int_0^t \frac{ds}{R_s^2}\right)\right\} \\
&= \mathbb{P}_r^{(0)}\left[\left(\frac{R_t}{r}\right)^{\nu} \exp\left(-aR_t^2 - \frac{\mu^2+\nu^2}{2}\int_0^t \frac{ds}{R_s^2}\right)\right] \\
&= \mathbb{P}_r^{(\gamma)}\left\{\left(\frac{R_t}{r}\right)^{\nu-\gamma} \exp(-aR_t^2)\right\}.
\end{aligned}$$

The quantity $\mathbb{P}_r^{(\gamma)}\left[\left(\frac{R_t}{r}\right)^{\nu-\gamma} \exp(-aR_t^2)\right]$ can be computed as follows. From

$$\frac{1}{x^\alpha} = \frac{1}{\Gamma(\alpha)}\int_0^\infty dv \exp(-vx)v^{\alpha-1} \quad (6.2.11)$$

(see ↣ formula (A.5.8) in the appendix), it follows that

$$\mathbb{P}_r^{(\gamma)}\left(\frac{1}{(R_t)^{2\alpha}}\right) = \frac{1}{\Gamma(\alpha)}\int_0^\infty dv\, v^{\alpha-1}\mathbb{P}_r^{(\gamma)}[\exp(-vR_t^2)].$$

Therefore, for any $\alpha \geq 0$, the equality

$$\mathbb{P}_r^{(\gamma)}\left(\frac{1}{(R_t)^{2\alpha}}\right) = \frac{1}{\Gamma(\alpha)}\int_0^{1/2t} dv\, v^{\alpha-1}(1-2tv)^{\gamma-\alpha}\exp(-r^2 v)$$

follows from the identity (6.2.1) written in terms of $\text{BES}^{(\gamma)}$,

$$\mathbb{P}_r^{(\gamma)}[\exp(-vR_t^2)] = \frac{1}{(1+2vt)^{1+\gamma}}\exp\left(-\frac{r^2 v}{1+2vt}\right), \quad (6.2.12)$$

and a change of variable.

Using equality (6.2.11) again,

$$\mathbb{P}_r^{(\gamma)}\left[\left(\frac{1}{R_t}\right)^{2\alpha}\exp(-aR_t^2)\right] = \frac{1}{\Gamma(\alpha)}\int_0^\infty dv\, v^{\alpha-1}\mathbb{P}_r^{(\gamma)}[\exp(-(v+a)R_t^2)] := I\,.$$

The identity (6.2.12) shows that

$$I = \frac{1}{\Gamma(\alpha)}\int_0^\infty dv\, v^{\alpha-1}(1+2(v+a)t)^{-(1+\gamma)}\exp\left(-\frac{r^2(v+a)}{1+2(v+a)t}\right).$$

Therefore

$$\begin{aligned}&\mathbb{P}_r^{(\nu)}\left[\exp\left(-aR_t^2 - \frac{\mu^2}{2}\int_0^t \frac{ds}{R_s^2}\right)\right]\\ &= \frac{r^{\gamma-\nu}}{\Gamma(\alpha)}\int_0^\infty dv\, v^{\alpha-1}(1+2(v+a)t)^{-(1+\gamma)}\exp\left(-\frac{r^2(v+a)}{1+2(v+a)t}\right)\end{aligned}$$

where $\alpha = \frac{1}{2}(\gamma-\nu) = \frac{1}{2}(\sqrt{\mu^2+\nu^2}-\nu)$. □

We state the following translation of Proposition 6.2.5.1 in terms of BESQ processes:

Corollary 6.2.5.2 *The quantity*

$$\mathbb{Q}_x^{(\nu)}\left[\exp\left(-a\rho_t - \frac{\mu^2}{2}\int_0^t \frac{ds}{\rho_s}\right)\right] = \mathbb{Q}_x^{(\gamma)}\left[\left(\frac{\rho_t}{x}\right)^{(\nu-\gamma)/2}\exp(-a\rho_t)\right] \qquad (6.2.13)$$

where $\gamma = \sqrt{\mu^2+\nu^2}$ *is given by* (6.2.10).

Another useful result is that of the Laplace transform of the pair $(\rho_t, \int_0^t \rho_s ds)$ under $\mathbb{Q}_x^\delta$.

Proposition 6.2.5.3 *For a* BESQ^δ, *we have for every* $\lambda > 0, b \neq 0$

$$\begin{aligned}&\mathbb{Q}_x^\delta\left[\exp(-\lambda\rho_t - \frac{b^2}{2}\int_0^t \rho_s ds)\right] \qquad (6.2.14)\\ &= \left(\cosh(bt) + 2\lambda b^{-1}\sinh(bt)\right)^{-\delta/2}\exp\left(-\frac{1}{2}xb\frac{1+2\lambda b^{-1}\coth(bt)}{\coth(bt)+2\lambda b^{-1}}\right).\end{aligned}$$

PROOF: Let ρ be a BESQ^δ process starting from x:

$$d\rho_t = 2\sqrt{\rho_t}dW_t + \delta\, dt\,.$$

Let $F : \mathbb{R}^+ \to \mathbb{R}$ be a locally bounded function. The process Z defined by

$$Z_u := \exp\left[\int_0^u F(s)\sqrt{\rho_s}dW_s - \frac{1}{2}\int_0^u F^2(s)\rho_s ds\right]$$

is a local martingale. Furthermore,

$$Z_u = \exp\left[\frac{1}{2}\int_0^u F(s)d(\rho_s - \delta s) - \frac{1}{2}\int_0^u F^2(s)\rho_s ds\right].$$

If F has bounded variation, an integration by parts leads to

$$\int_0^u F(s)d\rho_s = F(u)\rho_u - F(0)\rho_0 - \int_0^u \rho_s dF(s)$$

and

$$Z_u = \exp\left[\frac{1}{2}\left(F(u)\rho_u - F(0)x - \int_0^u (\delta F(s) + F^2(s)\rho_s)ds + \rho_s dF(s)\right)\right].$$

Let t be fixed. We now consider only processes indexed by u with $u \leq t$.

Let b be given and choose $F = \dfrac{\Phi'}{\Phi}$ where Φ is the decreasing solution of

$$\Phi'' = b^2\Phi, \text{ on } [0,t]; \quad \Phi(0) = 1; \quad \Phi'(t) = -2\lambda\Phi(t),$$

where $\Phi'(t)$ is the left derivative of Φ at t. Then,

$$Z_u = \exp\left[\frac{1}{2}(F(u)\rho_u - F(0)x - \delta \ln \Phi(u)) - \frac{b^2}{2}\int_0^u \rho_s ds\right]$$

is a bounded local martingale, hence a martingale. Moreover,

$$Z_t = \exp\left[-\lambda\rho_t - \frac{1}{2}(\Phi'(0)x + \delta \ln \Phi(t)) - \frac{b^2}{2}\int_0^t \rho_s ds\right]$$

and $1 = \mathbb{E}(Z_t)$, hence the left-hand side of equality (6.2.14) is equal to

$$(\Phi(t))^{\delta/2} \exp\left(\frac{x}{2}\Phi'(0)\right).$$

The general solution of $\Phi'' = b^2\Phi$ is $\Phi(s) = c_1 \sinh(bs) + c_2 \cosh(bs)$, and the constants $c_i, i = 1,2$ are determined from the boundary conditions. The boundary condition $\Phi(0) = 1$ implies $c_2 = 1$ and the condition $\Phi'(t) = -2\lambda\Phi(t)$ implies

$$c_1 = -\frac{b\sinh(bt) + 2\lambda\cosh(bt)}{b\cosh(bt) + 2\lambda\sinh(bt)}.$$

□

Remark 6.2.5.4 The transformation $F = \frac{\Phi'}{\Phi}$ made in the proof allows us to link the Sturm-Liouville equation satisfied by Φ to a Ricatti equation satisfied by F. This remark is also valid for the general computation made in the following Exercise 6.2.5.8 .

Corollary 6.2.5.5 *As a particular case of the equality* (6.2.14), *one has*

$$\mathbb{Q}_0^\delta \left[\exp(-\frac{b^2}{2}\int_0^t \rho_s ds)\right] = (\cosh(bt))^{-\delta/2}.$$

Consequently, the density of $\int_0^1 \rho_s ds$ *is*

$$f(u) = 2^{\delta/2}\sum_{n=0}^{\infty}\alpha_n(\delta/2)\frac{2n+\delta/2}{\sqrt{2\pi u^3}}e^{-\frac{1}{2u}(2n+\delta/2)^2},$$

where $\alpha_n(x)$ *are the coefficients of the series expansion*

$$(1+a)^{-x} = \sum_{n=0}^{\infty}\alpha_n(x)a^n.$$

PROOF: From the equality (6.2.14),

$$\mathbb{E}\left(\exp\left(-\frac{b^2}{2}\int_0^1 \rho_s ds\right)\right) = (\cosh(b))^{-\delta/2} = e^{-b\delta/2}\frac{2^{\delta/2}}{(1+e^{-2b})^{\delta/2}}$$
$$= 2^{\delta/2}\sum_{n=0}^{\infty}\alpha_n(\delta/2)e^{-2bn}e^{-b\delta/2}.$$

Using

$$e^{-ba} = \int_0^\infty dt \frac{a}{\sqrt{2\pi t^3}}e^{-a^2/(2t)-b^2t/2}dt$$

we obtain

$$\mathbb{E}\left(\exp\left(-\frac{b^2}{2}\int_0^1 \rho_s ds\right)\right) = 2^{\delta/2}\sum_n \alpha_n\left(\frac{\delta}{2}\right)\int_0^\infty \frac{2n+\delta/2}{\sqrt{2\pi t^3}}e^{-(2n+\delta/2)^2/(2t)-b^2t/2}dt$$

and the result follows. □

Comment 6.2.5.6 See Pitman and Yor [720] for more results of this kind.

Exercise 6.2.5.7 Prove, using the same method as in Proposition 6.2.5.1 that

$$\mathbb{P}_r^{(\nu)}\left[\frac{1}{R_t^\alpha}\exp\left(-\frac{\mu^2}{2}\int_0^t\frac{ds}{R_s^2}\right)\right] = \mathbb{P}_r^{(0)}\left[\frac{R_t^\nu}{r^\nu R_t^\alpha}\exp\left(-\frac{\mu^2+\nu^2}{2}\int_0^t\frac{ds}{R_s^2}\right)\right]$$
$$= \mathbb{P}_r^{(\gamma)}\left[\frac{R_t^{\nu-\gamma-\alpha}}{r^{\nu-\gamma}}\right],$$

and compute the last quantity. ◁

Exercise 6.2.5.8 We shall now extend the result of Proposition 6.2.5.3 by computing the Laplace transform

$$\mathbb{Q}_x^\delta \left[\exp\left(-\lambda \int_0^t du \phi(u) \rho_u \right) \right].$$

In fact, let μ be a positive, diffuse Radon measure on $\mathbb{R}^+$. The Sturm-Liouville equation $\Phi'' = \mu\Phi$ has a unique solution Φ_μ, which is positive, decreasing on $[0, \infty[$ and such that $\Phi_\mu(0) = 1$. Let $\Psi_\mu(t) = \Phi_\mu(t) \int_0^t \frac{ds}{\Phi_\mu^2(s)}$.

1. Prove that the function Ψ_μ is a solution of the Sturm-Liouville equation, such that $\Psi_\mu(0) = 0, \Psi'_\mu(0) = 1$, and the pair (Φ_μ, Ψ_μ) satisfies the Wronskian relation
$$W(\Phi_\mu, \Psi_\mu) = \Phi_\mu \Psi'_\mu - \Phi'_\mu \Psi_\mu = 1.$$
2. Prove that, for every $t \geq 0$:
$$\mathbb{Q}_x^\delta \left(\exp\left(-\frac{1}{2} \left(\int_0^t \rho_s d\mu(s) + \lambda \rho_t \right) \right) \right)$$
$$= \frac{1}{(\Psi'_\mu(t) + \lambda \Psi_\mu(t))^{\delta/2}} \exp\left(\frac{x}{2} \left(\Phi'_\mu(0) - \frac{\Phi'_\mu(t) + \lambda \Phi_\mu(t)}{\Psi'_\mu(t) + \lambda \Psi_\mu(t)} \right) \right),$$
and
$$\mathbb{Q}_x^\delta \left(\exp\left(-\frac{1}{2} \int_0^\infty \rho_s d\mu(s) \right) \right) = (\Phi_\mu(\infty))^{\delta/2} \exp\left(\frac{x}{2} \Phi'_\mu(0) \right).$$
3. Compute the solution of the Sturm Liouville equation for
$$\mu(ds) = \frac{\lambda}{(a+s)^2} ds, \quad (\lambda, a > 0)$$
(one can find a solution of the form $(a+s)^\alpha$ where α is to be determined).

See [RY] or Pitman and Yor [717] for details of the proof and Carmona [140] and Shirakawa [790] for extension to the case of Bessel processes with time-dependent dimension. ◁

6.2.6 BESQ Processes with Negative Dimensions

As an application of the absolute continuity relationship (6.1.6), one obtains

Lemma 6.2.6.1 *Let $\delta \in]-\infty, 2[$ and Φ a positive function. Then, for any $x > 0$*
$$\mathbb{Q}_x^\delta \left(\Phi(\rho_t) \mathbb{1}_{\{T_0 > t\}} \right) = x^{1-\frac{\delta}{2}} \mathbb{Q}_x^{4-\delta} \left(\Phi(\rho_t) (\rho_t)^{\frac{\delta}{2}-1} \right).$$

Definition 6.2.6.2 *The solution to the equation*

$$dX_t = \delta\, dt + 2\sqrt{|X_t|}dW_t\,,\ X_0 = x$$

where $\delta \in \mathbb{R}$, $x \in \mathbb{R}$ is called the square of a δ-dimensional Bessel process starting from x.

This equation has a unique strong solution (see [RY], Chapter IX, Section 3). Let us assume that $X_0 = x > 0$ and $\delta < 0$. The comparison theorem establishes that this process is smaller than the process with $\delta = 0$, hence, the point 0 is reached in finite time. Let T_0 be the first time when the process X hits the level 0. We have

$$\widetilde{X}_t := X_{T_0+t} = \delta t + 2\int_{T_0}^{T_0+t} \sqrt{|X_s|}dW_s,\ t \geq 0\,.$$

Setting $\gamma = -\delta$ and $\widetilde{W}_s = -(W_{s+T_0} - W_{T_0})$, we obtain

$$-\widetilde{X}_t = \gamma t + 2\int_0^t \sqrt{|\widetilde{X}_s|}d\widetilde{W}_s,\ t \geq 0\,,$$

hence, if $Y_t = -\widetilde{X}_t$ we get

$$Y_t = \gamma t + 2\int_0^t \sqrt{|Y_s|}d\widetilde{W}_s,\ t \geq 0\,.$$

This is the SDE satisfied by a BESQ_0^γ, hence $-X_{T_0+t}$ is a BESQ_0^γ. A BESQ_x^δ process with $x < 0$ and $\delta < 0$ behaves as minus a $\text{BESQ}_{-x}^{-\delta}$ and never becomes strictly positive.

One should note that the additivity property for BESQ with arbitrary (non-positive) dimensions does not hold. Indeed, let $\delta > 0$ and consider

$$X_t = 2\int_0^t \sqrt{|X_s|}d\beta_s + \delta t$$
$$Y_t = 2\int_0^t \sqrt{|Y_s|}d\gamma_s - \delta t$$

where β and γ are independent BM's, then, if additivity held, $(X_t + Y_t,\ t \geq 0)$ would be a BESQ_0^0, hence it would be equal to 0, so that $X = -Y$, which is absurd.

Proposition 6.2.6.3 *The probability transition of a $BESQ_x^{-\gamma}$, $\gamma > 0$ and $x \geq 0$ is $\mathbb{Q}_x^{-\gamma}(X_t \in dy) = q_t^{-\gamma}(x,y)dy$ where*

$$\begin{aligned} q_t^{-\gamma}(x,y) &= q_t^{4+\gamma}(y,x)\,, \quad \textit{for } y > 0 \\ &= k(x,y,\gamma,t)e^{-a-b}\int_0^\infty \frac{(z+1)^m}{z^m} e^{-bz-\frac{a}{z}}dz\,, \quad \textit{for } y < 0 \end{aligned}$$

where

$$k(x,y,\gamma,t) = \left(\Gamma\left(\frac{\gamma}{2}\right)\right)^{-2} \frac{2^{-\gamma}}{\gamma} x^{1+m}|y|^{m-1}t^{-\gamma-1}$$

and

$$m = \frac{\gamma}{2}\,,\ a = \frac{|y|}{2t}\,,\ b = \frac{x}{2t}\,.$$

PROOF: We decompose the process X before and after its hitting time T_0 as follows:

$$\mathbb{Q}_x^{-\gamma}[f(X_t)] = \mathbb{Q}_x^{-\gamma}[f(X_t)\mathbb{1}_{\{t<T_0\}}] + \mathbb{Q}_x^{-\gamma}[f(X_t)\mathbb{1}_{\{t>T_0\}}]\,.$$

From the time reversal, using Lemma 6.2.6.1 and noting that

$$\left(\frac{y}{x}\right)^{-1-\frac{\gamma}{2}} q_t^{4-\gamma}(x,y) = q_t^{4+\gamma}(y,x)$$

we obtain

$$(X_t, t \le T_0) \stackrel{\text{law}}{=} (\widetilde{X}_{\gamma_x - t}, t \le \gamma_x)$$

where $\widetilde{X}$ is a $\mathrm{BESQ}_0^{4+\gamma}$ process and $\gamma_x = \sup\{t : \widetilde{X}_t = x\}$. It follows that

$$\begin{aligned}\mathbb{Q}_x^{-\gamma}[f(X_t)\mathbb{1}_{\{t<T_0\}}] &= \mathbb{Q}_0^{4+\gamma}[f(\widetilde{X}_{\gamma_x - t})\mathbb{1}_{\{t<\gamma_x\}}] \\ &= \int_t^\infty q_x(s)\mathbb{Q}_0^{4+\gamma}[f(X_{s-t})|X_s = x]ds\end{aligned}$$

where $q_x(s)ds = \mathbb{Q}_0^{4+\gamma}(\gamma_x \in ds)$. Then, some standard computation leads to

$$\mathbb{Q}_0^{4+\gamma}[f(X_{s-t})|X_s = x] = \int f(y)\frac{q_{s-t}^{4+\gamma}(0,y)\,q_t^{4+\gamma}(y,x)}{q_s^{4+\gamma}(0,x)}dy,\ s > t\,.$$

From the study of the last passage times (see equality (5.6.2)) one obtains

$$q_x(s) = \frac{1}{s\Gamma(\nu)}\left(\frac{x}{2s}\right)^\nu e^{-x/(2s)} = (2+\gamma)q_s^{4+\gamma}(0,x)\,. \tag{6.2.15}$$

Hence

$$\mathbb{Q}_x^{-\gamma}[f(X_t)\mathbb{1}_{\{t<T_0\}}] = \int_t^\infty ds(2+\gamma)\left(\int_0^\infty f(y)q_{s-t}^{4+\gamma}(0,y)q_t^{4+\gamma}(y,x)dy\right)$$

and using Fubini's theorem

$$\mathbb{Q}_x^{-\gamma}[f(X_t)\mathbb{1}_{\{t<T_0\}}] = \int_0^\infty f(y)q_t^{4+\gamma}(y,x)dy\,.$$

The computation of $\mathbb{Q}_x^{-\gamma}[f(X_t)\mathbb{1}_{\{t>T_0\}}]$ is carried out using (6.2.15) and $\mathbb{Q}_x^{-\gamma}[T_0 \in ds] = q_x(s)ds$. One obtains

$$\mathbb{Q}_x^{-\gamma}[f(X_t)\mathbb{1}_{\{t>T_0\}}] = \int_0^t (\gamma+2)q_s^{4+\gamma}(0,x)q_{t-s}^{\gamma}(0,-y)ds\,.$$

□

Exercise 6.2.6.4 Prove that, for $\delta > 2$, the default of martingality of $R^{2-\delta}$ (where R is a Bessel process of dimension δ starting from x) is given by

$$\mathbb{E}(R_0^{2-\delta} - R_t^{2-\delta}) = x^{2-\delta}\mathbb{P}^{4-\delta}(T_0 \leq t)\,.$$

`Hint:` Prove that

$$\mathbb{E}(R_t^{2-\delta}) = x^{2-\delta}\mathbb{P}^{4-\delta}(t < T_0)\,.$$

◁

6.2.7 Squared Radial Ornstein-Uhlenbeck

The above attempt to deal with negative dimension has shown a number of drawbacks. From now on, we shall maintain positive dimensions.

Definition 6.2.7.1 *The solution to the SDE*

$$dX_t = (a - bX_t)dt + 2\sqrt{|X_t|}\, dW_t$$

where $a \in \mathbb{R}^+, b \in \mathbb{R}$ is called a squared radial Ornstein-Uhlenbeck process with dimension a. We shall denote by ${}^b\mathbb{Q}_x^a$ its law, and $\mathbb{Q}_x^a = {}^0\mathbb{Q}_x^a$.

Proposition 6.2.7.2 *The following absolute continuity relationship between a squared radial Ornstein-Uhlenbeck process and a squared Bessel process holds:*

$$ {}^b\mathbb{Q}_x^a|_{\mathcal{F}_t} = \exp\left(-\frac{b}{4}(X_t - x - at) - \frac{b^2}{8}\int_0^t X_s ds\right)\mathbb{Q}_x^a|_{\mathcal{F}_t}\,.$$

PROOF: This is a straightforward application of Girsanov's theorem. We have $\int_0^t X_s dW_s = \frac{1}{2}(X_t - x - at)$. □

Exercise 6.2.7.3 Let X be a Bessel process with dimension $\delta < 2$, starting at $x > 0$ and $T_0 = \inf\{t : X_t = 0\}$. Using time reversal theorem (see (6.2.8)), prove that the density of T_0 is

$$\frac{1}{t\Gamma(\alpha)}\left(\frac{x^2}{2t}\right)^{\alpha} e^{-x^2/(2t)}$$

where $\alpha = (4 - \delta)/2 - 1$, i.e., T_0 is a multiple of the reciprocal of a Gamma variable. ◁

6.3 Cox-Ingersoll-Ross Processes

In the finance literature, the CIR processes have been considered as term structure models. As we shall show, they are closely connected to squared Bessel processes, in fact to squared radial OU processes.

6.3.1 CIR Processes and BESQ

From Theorem 1.5.5.1 on the solutions to SDE, the equation

$$dr_t = k(\theta - r_t)\,dt + \sigma\sqrt{|r_t|}dW_t\,, r_0 = x \tag{6.3.1}$$

admits a unique solution which is strong. For $\theta = 0$ and $x = 0$, the solution is $r_t = 0$, and from the comparison Theorem 1.5.5.9, we deduce that, in the case $k\theta > 0$, $r_t \geq 0$ for $x \geq 0$. In that case, we omit the absolute value and consider the positive solution of

$$dr_t = k(\theta - r_t)\,dt + \sigma\sqrt{r_t}dW_t\,, r_0 = x. \tag{6.3.2}$$

This solution is called a **Cox-Ingersoll-Ross (CIR)** process or a square-root process (See Feller [342]). For $\sigma = 2$, this process is the square of the norm of a δ-dimensional OU process, with dimension $\delta = k\theta$ (see Subsection 2.6.5 and the previous Subsection 6.2.6), but this equation also makes sense even if δ is not an integer.

We shall denote by ${}^k\mathbb{Q}^{k\theta,\sigma}$ the law of the CIR process solution of the equation (6.3.1). In the case $\sigma = 2$, we simply write ${}^k\mathbb{Q}^{k\theta,2} = {}^k\mathbb{Q}^{k\theta}$. Now, the elementary change of time $A(t) = 4t/\sigma^2$ reduces the study of the solution of (6.3.2) to the case $\sigma = 2$: indeed, if $Z_t = r(4t/\sigma^2)$, then

$$dZ_t = k'(\theta - Z_t)\,dt + 2\sqrt{Z_t}dB_t$$

with $k' = 4k/\sigma^2$ and B a Brownian motion.

Proposition 6.3.1.1 *The CIR process (6.3.2) is a BESQ process transformed by the following space-time changes:*

$$\boxed{r_t = e^{-kt}\rho\left(\frac{\sigma^2}{4k}(e^{kt} - 1)\right)}$$

where $(\rho(s), s \geq 0)$ is a $BESQ^\delta$ process, with dimension $\delta = \dfrac{4k\theta}{\sigma^2}$.

PROOF: The proof is left as an exercise for the reader. A more general case will be presented in the following Theorem 6.3.5.1. □

It follows that for $2k\theta \geq \sigma^2$, a CIR process starting from a positive initial point stays strictly positive. For $0 \leq 2k\theta < \sigma^2$, a CIR process starting from a positive initial point hits 0 with probability $p \in]0,1[$ in the case $k < 0$ ($\mathbb{P}(T_0^x < \infty) = p$) and almost surely if $k \geq 0$ ($\mathbb{P}(T_0^x < \infty) = 1$). In the case $0 < 2k\theta$, the boundary 0 is instantaneously reflecting, whereas in the case $2k\theta < 0$, the process r starting from a positive initial point reaches 0 almost surely. Let $T_0 = \inf\{t : r_t = 0\}$ and set $Z_t = -r_{T_0+t}$. Then,

$$dZ_t = (-\delta + \lambda Z_t)dt + \sigma\sqrt{|Z_t|}dB_t$$

where B is a BM. We know that $Z_t \geq 0$, thus r_{T_0+t} takes values in $\mathbb{R}^-$.

Absolute Continuity Relationship

A routine application of Girsanov's theorem (See Example 1.7.3.5 or Proposition 6.2.7.2) leads to (for $k\theta > 0$)

$$^{k}\mathbb{Q}_x^{k\theta}|_{\mathcal{F}_t} = \exp\left(\frac{k}{4}\left[x + k\theta t - \rho_t\right] - \frac{k^2}{8}\int_0^t \rho_s\, ds\right) \mathbb{Q}_x^{k\theta}|_{\mathcal{F}_t}\,. \tag{6.3.3}$$

Comments 6.3.1.2 (a) From an elementary point of view, if the process r reaches 0 at time t, the formal equality between dr_t and $k\theta dt$ explains that the increment of r_t is positive if $k\theta > 0$. Again formally, for $k > 0$, if at time t, the inequality $r_t > \theta$ holds (resp. $r_t < \theta$), then the drift $k(\theta - r_t)$ is negative (resp. positive) and, at least in mean, r is decreasing (resp. increasing). Note also that $\mathbb{E}(r_t) \to \theta$ when t goes to infinity. This is the mean-reverting property.

(b) Here, we have used the notation r for the CIR process. As shown above, this process is close to a BESQ ρ (and not to a BES R).

(c) Dufresne [281] has obtained explicit formulae for the moments of the r.v. r_t and for the process $(I_t = \int_0^t r_s ds, t \geq 0)$. Dassios and Nagaradjasarma [214] present an explicit computation of the joint moments of r_t and I_t, and, in the case $\theta = 0$, the joint density of the pair (r_t, I_t).

6.3.2 Transition Probabilities for a CIR Process

From the expression of a CIR process as a time-changed squared Bessel process given in Proposition 6.3.1.1, using the transition density of the squared Bessel process given in (6.2.2), we obtain the transition density of the CIR process.

Proposition 6.3.2.1 *Let r be a CIR process following* (6.3.2). *The transition density $^{k}\mathbb{Q}^{k\theta,\sigma}(r_{t+s} \in dy | r_s = x) = f_t(x,y)dy$ is given by*

$$f_t(x,y) = \frac{e^{kt}}{2c(t)}\left(\frac{ye^{kt}}{x}\right)^{\nu/2} \exp\left(-\frac{x + ye^{kt}}{2c(t)}\right) I_\nu\left(\frac{1}{c(t)}\sqrt{xye^{kt}}\right)\mathbb{1}_{\{y\geq 0\}}\,,$$

where $c(t) = \frac{\sigma^2}{4k}(e^{kt} - 1)$ *and* $\nu = \frac{2k\theta}{\sigma^2} - 1$.

PROOF: From the relation $r_t = e^{-kt}\rho_{c(t)}$, where ρ is a $\mathrm{BESQ}^{(\nu)}$, we obtain

$$^{k}\mathbb{Q}^{k\theta,\sigma}(r_{t+s} \in dy | r_s = x) = e^{kt} q_{c(t)}^{(\nu)}(x, ye^{kt})dy\,.$$

Denoting by $(r_t(x); t \geq 0)$ the CIR process with initial value $r_0 = x$, the random variable $Y_t = r_t(x)e^{kt}[c(t)]^{-1}$ has density

$$\begin{aligned}\mathbb{P}(Y_t \in dy)/dy &= c(t)e^{-kt} f_t(x, yc(t)e^{-kt})\mathbb{1}_{\{y>0\}} \\ &= \frac{e^{-\alpha/2}}{2\alpha^{\nu/2}} e^{-y/2} y^{\nu/2} I_\nu(\sqrt{y\alpha})\mathbb{1}_{\{y\geq 0\}}\end{aligned}$$

where $\alpha = x/c(t)$. □

Remark 6.3.2.2 This density is that of a noncentral chi-square $\chi^2(\delta, \alpha)$ with $\delta = 2(\nu + 1)$ degrees of freedom, and non-centrality parameter α. Using the notation of Exercise 1.1.12.5, we obtain

$$^k\mathbb{Q}_x^{k\theta,\sigma}(r_t < y) = \chi^2\left(\frac{4k\theta}{\sigma^2}, \frac{x}{c(t)}; \frac{ye^{kt}}{c(t)}\right),$$

where the function $\chi^2(\delta, \alpha; \cdot)$, defined in Exercise 1.1.12.5, is the cumulative distribution function associated with the density

$$\begin{aligned} f(x;\delta,\alpha) &= 2^{-\delta/2} \exp\left(-\frac{1}{2}(\alpha + x)\right) x^{\frac{\delta}{2}-1} \sum_{n=0}^{\infty} \left(\frac{\alpha}{4}\right)^n \frac{x^n}{n!\Gamma(n + \delta/2)} \mathbb{1}_{\{x>0\}}, \\ &= \frac{e^{-\alpha/2}}{2\alpha^{\nu/2}} e^{-x/2} x^{\nu/2} I_\nu(\sqrt{x\alpha}) \mathbb{1}_{\{x>0\}}. \end{aligned}$$

6.3.3 CIR Processes as Spot Rate Models

The Cox-Ingersoll-Ross model for the short interest rate has been the object of many studies since the seminal paper of Cox et al. [206] where the authors assume that the riskless rate r follows a square root process under the historical probability given by

$$dr_t = \tilde{k}(\tilde{\theta} - r_t)\, dt + \sigma\sqrt{r_t} d\tilde{W}_t .$$

Here, $\tilde{k}(\tilde{\theta} - r)$ defines a mean reverting drift pulling the interest rate towards its long-term value $\tilde{\theta}$ with a speed of adjustment equal to $\tilde{k}$. In the risk-adjusted economy, the dynamics are supposed to be given by:

$$dr_t = (\tilde{k}(\tilde{\theta} - r_t) - \lambda r_t)dt + \sigma\sqrt{r_t} dW_t$$

where $(W_t = \widetilde{W}_t + \frac{\lambda}{\sigma}\int_0^t \sqrt{r_s}ds, t \geq 0)$ is a Brownian motion under the risk-adjusted probability $\mathbb{Q}$ where λ denotes the market price of risk.

Setting $k = \tilde{k} + \lambda$, $\theta = \tilde{k}(\tilde{\theta}/k)$, the $\mathbb{Q}$ -dynamics of r are

$$dr_t = k(\theta - r_t)dt + \sigma\sqrt{r_t} dW_t .$$

Therefore, we shall establish formulae under general dynamics of this form, already given in (6.3.2).

Even though no closed-form expression as a functional of W can be written for r_t, it is remarkable that the Laplace transform of the process, i.e.,

$$^k\mathbb{Q}_x^{k\theta,\sigma}\left[\exp\left(-\int_0^t du\, \phi(u) r_u\right)\right]$$

is known (see Exercise 6.2.5.8).

Theorem 6.3.3.1 *Let r be a CIR process, the solution of*

$$\boxed{dr_t = k(\theta - r_t)dt + \sigma\sqrt{r_t}dW_t\,.}$$

The conditional expectation and the conditional variance of the r.v. r_t are given by, for $s < t$,

$$^{k}\mathbb{Q}^{k\theta,\sigma}_{x}(r_t\,|\mathcal{F}_s) = r_s e^{-k(t-s)} + \theta(1 - e^{-k(t-s)}),$$

$$\mathrm{Var}(r_t\,|\mathcal{F}_s) = r_s\,\frac{\sigma^2(e^{-k(t-s)} - e^{-2k(t-s)})}{k} + \frac{\theta\sigma^2(1 - e^{-k(t-s)})^2}{2k}.$$

PROOF: From the definition, for $s \leq t$, one has

$$r_t = r_s + k\int_s^t (\theta - r_u)du + \sigma\int_s^t \sqrt{r_u}\,dW_u\,.$$

Itô's formula leads to

$$\begin{aligned} r_t^2 &= r_s^2 + 2k\int_s^t (\theta - r_u)r_u du + 2\sigma\int_s^t (r_u)^{3/2}dW_u + \sigma^2\int_s^t r_u du \\ &= r_s^2 + (2k\theta + \sigma^2)\int_s^t r_u du - 2k\int_s^t r_u^2 du + 2\sigma\int_s^t (r_u)^{3/2}dW_u. \end{aligned}$$

It can be checked that the stochastic integrals involved in both formulae are martingales: indeed, from Proposition 6.3.1.1, r admits moments of any order. Therefore, the expectation of r_t is given by

$$\mathbb{E}(r_t) = {}^{k}\mathbb{Q}^{k\theta,\sigma}_{x}(r_t) = r_0 + k\left(\theta t - \int_0^t \mathbb{E}(r_u)du\right).$$

We now introduce $\Phi(t) = \mathbb{E}(r_t)$. The integral equation

$$\Phi(t) = r_0 + k(\theta t - \int_0^t \Phi(u)du)$$

can be written in differential form $\Phi'(t) = k(\theta - \Phi(t))$ where Φ satisfies the initial condition $\Phi(0) = r_0$. Hence

$$\mathbb{E}[r_t] = \theta + (r_0 - \theta)e^{-kt}.$$

In the same way, from

$$\mathbb{E}(r_t^2) = r_0^2 + (2k\theta + \sigma^2)\int_0^t \mathbb{E}(r_u)du - 2k\int_0^t \mathbb{E}(r_u^2)du\,,$$

setting $\Psi(t) = \mathbb{E}(r_t^2)$ leads to $\Psi'(t) = (2k\theta + \sigma^2)\Phi(t) - 2k\Psi(t)$, hence

$$\text{Var}\,[r_t] \;=\; \frac{\sigma^2}{k}\,(1-e^{-kt})[r_0 e^{-kt} + \frac{\theta}{2}(1-e^{-kt})]\,.$$

Thanks to the Markovian character of r, the conditional expectation can also be computed:

$$\mathbb{E}(r_t\,|\mathcal{F}_s) = \theta + (r_s - \theta)e^{-k(t-s)} = r_s\, e^{-k(t-s)} + \theta(1-e^{-k(t-s)}),$$
$$\text{Var}(r_t\,|\mathcal{F}_s) = r_s\,\frac{\sigma^2(e^{-k(t-s)} - e^{-2k(t-s)})}{k} + \frac{\theta\sigma^2(1-e^{-k(t-s)})^2}{2k}.$$

□

Note that, if $k > 0$, $\mathbb{E}(r_t) \to \theta$ as t goes to infinity, this is the reason why the process is said to be mean reverting.

Comment 6.3.3.2 Using an induction procedure, or with computations done for squared Bessel processes, all the moments of r_t can be computed. See Dufresne [279].

Exercise 6.3.3.3 If r is a CIR process and $Z = r^\alpha$, prove that

$$dZ_t = \left(\alpha Z_t^{1-1/\alpha}(k\theta + (\alpha-1)\sigma^2/2) - Z_t\alpha k\right)dt + \alpha Z_t^{1-1/(2\alpha)}\sigma dW_t\,.$$

In particular, for $\alpha = -1$, $dZ_t = Z_t(k - Z_t(k\theta - \sigma^2))dt - Z_t^{3/2}\sigma dW_t$ is the so-called 3/2 model (see Section 6.4 on CEV processes and the book of Lewis [587]). ◁

6.3.4 Zero-coupon Bond

We now address the problem of the valuation of a zero-coupon bond, i.e., we assume that the dynamics of the interest rate are given by a CIR process under the risk-neutral probability and we compute $\mathbb{E}\left(\exp\left(-\int_t^T r_u\,du\right)|\mathcal{F}_t\right)$.

Proposition 6.3.4.1 *Let r be a CIR process defined as in* (6.3.2) *by*

$$dr_t = k(\theta - r_t)\,dt + \sigma\sqrt{r_t}dW_t\,,$$

and let ${}^k\mathbb{Q}^{k\theta,\sigma}$ be its law. Then, for any pair (λ,μ) of positive numbers

$${}^k\mathbb{Q}_x^{k\theta,\sigma}\left(\exp\left(-\lambda r_T - \mu\int_0^T r_u\,du\right)\right) = \exp[-A_{\lambda,\mu}(T) - xG_{\lambda,\mu}(T)]$$

with

$$G_{\lambda,\mu}(s) = \frac{\lambda(\gamma + k + e^{\gamma s}(\gamma - k)) + 2\mu(e^{\gamma s} - 1)}{\sigma^2\lambda(e^{\gamma s}-1) + \gamma(e^{\gamma s}+1) + k(e^{\gamma s}-1)}$$
$$A_{\lambda,\mu}(s) = -\frac{2k\theta}{\sigma^2}\ln\left(\frac{2\gamma e^{(\gamma+k)s/2}}{\sigma^2\lambda(e^{\gamma s}-1) + \gamma(e^{\gamma s}+1) + k(e^{\gamma s}-1)}\right)$$

where $\gamma = \sqrt{k^2 + 2\sigma^2\mu}$.

PROOF: We seek $\varphi : \mathbb{R} \times [|0, T] \to \mathbb{R}^+$ such that the process

$$\varphi(r_t, t) \exp\left(-\mu \int_0^t r_s ds\right)$$

is a martingale. Using Itô's formula, and assuming that φ is regular, this necessitates that φ satisfies the equation

$$-\frac{\partial \varphi}{\partial t} = -x\mu\varphi + k(\theta - x)\frac{\partial \varphi}{\partial x} + \frac{1}{2}\sigma^2 x \frac{\partial^2 \varphi}{\partial x^2}. \tag{6.3.4}$$

Furthermore, if φ satisfies the boundary condition $\varphi(x, T) = e^{-\lambda x}$, we obtain

$${}^k\mathbb{Q}_x^{k\theta,\sigma}\left(\exp\left(-\lambda r_T - \mu \int_0^T r_u\, du\right)\right) = \varphi(x, 0)$$

It remains to prove that there exist two functions A and G such that $\varphi(x, t) = \exp(-A(T-t) - xG(T-t))$ is a solution of the PDE (6.3.4), where A and G satisfy $A(0) = 0, G(0) = \lambda$. Some involved calculation leads to the proposition. □

Corollary 6.3.4.2 *In particular, taking $\lambda = 0$,*

$$\begin{aligned} &{}^k\mathbb{Q}_x^{k\theta,\sigma}\left(\exp(-\mu \int_0^t r_s ds)\right) \\ &= e^{k^2\theta t/\sigma^2}\left(\cosh\frac{\gamma t}{2} + \frac{k}{\gamma}\sinh\frac{\gamma t}{2}\right)^{-2k\theta/\sigma^2} \exp\left(\frac{-2\mu x}{k + \gamma \coth\frac{\gamma t}{2}}\right) \end{aligned}$$

where $\gamma^2 = k^2 + 2\mu\sigma^2$.

These formulae may be considered as extensions of Lévy's area formula for planar Brownian motion. See Pitman and Yor [716].

Corollary 6.3.4.3 *Let r be a CIR process defined as in* (6.3.2) *under the risk-neutral probability. Then, the price at time t of a zero-coupon bond maturing at T is*

$${}^k\mathbb{Q}_x^{k\theta,\sigma}\left(\exp\left(-\int_t^T r_u\, du\right)\Big|\mathcal{F}_t\right) = \exp[-A(T-t) - r_t G(T-t)] = B(r_t, T-t)$$

with

$$B(r, s) = \exp(-A(s) - rG(s))$$

and

$$G(s) = \frac{2(e^{\gamma s}-1)}{(\gamma+k)(e^{\gamma s}-1)+2\gamma} = \frac{2}{k+\gamma\coth(\gamma s/2)}$$

$$A(s) = -\frac{2k\theta}{\sigma^2}\ln\left(\frac{2\gamma e^{(\gamma+k)s/2}}{(\gamma+k)(e^{\gamma s}-1)+2\gamma}\right)$$

$$= -\frac{2k\theta}{\sigma^2}\left[\frac{ks}{2}+\ln\left(\cosh\frac{\gamma s}{2}+\frac{k}{\gamma}\sinh\frac{\gamma s}{2}\right)^{-1}\right],$$

where $\gamma = \sqrt{k^2+2\sigma^2}$.

The dynamics of the zero-coupon bond $P(t,T) = B(r_t, T-t)$ *are, under the risk-neutral probability*

$$d_t P(t,T) = P(t,T)\left(r_t dt + \sigma(T-t, r_t)dW_t\right)$$

with $\sigma(s,r) = -\sigma G(s)\sqrt{r}$.

PROOF: The expression of the price of a zero-coupon bond follows from the Markov property and the previous proposition with $A = A_{0,1}, G = G_{0,1}$. Use Itô's formula and recall that the drift term in the dynamics of the zero-coupon bond price is of the form $P(t,T)r_t$. □

Corollary 6.3.4.4 *The Laplace transform of the r.v.* r_T *is*

$${}^k\mathbb{Q}_x^{k\theta,\sigma}(e^{-\lambda r_T}) = \left(\frac{1}{2\lambda\tilde{c}+1}\right)^{2k\theta/\sigma^2}\exp\left(-\frac{\lambda\tilde{c}\tilde{x}}{2\lambda\tilde{c}+1}\right)$$

with $\tilde{c} = c(T)e^{-kT}$ *and* $\tilde{x} = x/c(T)$, $c(T) = \frac{\sigma^2}{4k}(e^{kT}-1)$.

PROOF: The corollary follows from Proposition 6.3.4.1 with $\mu = 0$. It can also be obtained using the expression of r_T as a time-changed BESQ (see Proposition 6.3.1.1). □

One can also use that the Laplace transform of a $\chi^2(\delta,\alpha)$ distributed random variable is

$$\left(\frac{1}{2\lambda+1}\right)^{\delta/2}\exp\left(-\frac{\lambda\alpha}{2\lambda+1}\right),$$

and that, setting $c(t) = \sigma^2(e^{kt}-1)/(4k)$, the random variable $r_t e^{kt}/c(t)$ is $\chi^2(\delta,\alpha)$ distributed, where $\alpha = x/c(t)$.

Comment 6.3.4.5 One may note the "affine structure" of the model: the Laplace transform of the value of the process at time T is the exponential of an affine function of its initial value $\mathbb{E}_x(e^{-\lambda r_T}) = e^{-A(T)-xG(T)}$. For a complete characterization and description of affine term structure models, see Duffie et al. [272].

Exercise 6.3.4.6 Prove that if

$$dr_t^i = (\delta_i - kr_t^i)dt + \sigma\sqrt{r_t^i}\,dW_t^i,\ i = 1,2$$

where W^i are independent BMs, then the sum $r^1 + r^2$ is a CIR process. ◁

6.3.5 Inhomogeneous CIR Process

Theorem 6.3.5.1 *If r is the solution of*

$$dr_t = (a - \lambda(t)r_t)dt + \sigma\sqrt{r_t}dW_t\,, r_0 = x \tag{6.3.5}$$

where λ is a continuous function and $a > 0$, then

$$(r_t, t \geq 0) \stackrel{\text{law}}{=} \left(\frac{1}{\ell(t)}\rho\left(\frac{\sigma^2}{4}\int_0^t \ell(s)ds\right), t \geq 0\right)$$

where $\ell(t) = \exp\left(\int_0^t \lambda(s)ds\right)$ and ρ is a squared Bessel process with dimension $4a/\sigma^2$.

PROOF: Let us introduce $Z_t = r_t \exp(\int_0^t \lambda(s)ds) = r_t\ell(t)$. From the integration by parts formula,

$$Z_t = x + a\int_0^t \ell(s)ds + \sigma\int_0^t \sqrt{\ell(s)}\,\sqrt{Z_s}\,dW_s\,.$$

Define the increasing function $C(u) = \frac{\sigma^2}{4}\int_0^u \ell(s)\,ds$ and its inverse $A(t) = \inf\{u : C(u) = t\}$. Apply a change of time so that

$$Z_{A(t)} = x + a\int_0^{A(t)} \ell(s)ds + \sigma\int_0^{A(t)} \sqrt{\ell(s)}\,\sqrt{Z_s}dW_s\,.$$

The process $\sigma\int_0^{A(t)} \sqrt{\ell(s)}\,\sqrt{Z_s}dW_s$ is a local martingale with increasing process $\sigma^2\int_0^{A(t)} \ell(s)\,Z_s ds = 4\int_0^t Z_{A(u)}du$, hence,

$$\rho_t := Z_{A(t)} = x + \frac{4a}{\sigma^2}t + 2\int_0^t \sqrt{Z_{A(u)}}\,dB_u = \rho_0 + \frac{4a}{\sigma^2}t + 2\int_0^t \sqrt{\rho_u}\,dB_u \tag{6.3.6}$$

where B is a Brownian motion. □

Proposition 6.3.2.1 admits an immediate extension.

Proposition 6.3.5.2 *The transition density of the inhomogeneous process (6.3.5) is*

$$\mathbb{P}(r_t \in dy | r_s = x) \tag{6.3.7}$$

$$= \frac{\ell(s,t)}{2\ell^*(s,t)} \exp\left(-\frac{x + y\ell(s,t)}{2\ell^*(s,t)}\right) \left(\frac{y\ell(s,t)}{x}\right)^{\nu/2} I_\nu \left(\frac{\sqrt{xy\ell(s,t)}}{\ell^*(s,t)}\right) dy$$

where $\nu = (2a)/\sigma^2 - 1$, $\ell(s,t) = \exp\left(\int_s^t \lambda(u)du\right)$, $\ell^*(s,t) = \frac{\sigma^2}{4} \int_s^t \ell(s,u)du$.

Comment 6.3.5.3 Maghsoodi [615], Rogers [736] and Shirakawa [790] study the more general model

$$dr_t = (a(t) - b(t)r_t)dt + \sigma(t)\sqrt{r_t}dW_t$$

under the "constant dimension condition"

$$\frac{a(t)}{\sigma^2(t)} = \text{constant}.$$

As an example (see e.g. Szatzschneider [817]), let

$$dX_t = (\delta + \beta(t)X_t)dt + 2\sqrt{X_t}dW_t$$

and choose $r_t = \varphi(t)X_t$ where φ is a given positive C^1 function. Then

$$dr_t = \left(\delta\varphi(t) + \left[\beta(t) + \frac{\varphi'(t)}{\varphi(t)}\right] r_t\right) dt + 2\sqrt{\varphi(t)r_t}dW_t$$

satisfies this constant dimension condition.

6.4 Constant Elasticity of Variance Process

The **Constant Elasticity of Variance (CEV) process** has dynamics

$$dZ_t = Z_t(\mu dt + \sigma Z_t^\beta dW_t). \tag{6.4.1}$$

The CEV model reduces to the geometric Brownian motion for $\beta = 0$ and to a particular case of the square-root model for $\beta = -1/2$ (See equation (6.3.2)). In what follows, we do not consider the case $\beta = 0$.

Cox [205] studied the case $\beta < 0$, Emanuel and MacBeth [320] the case $\beta > 0$ and Delbaen and Shirakawa [238] the case $-1 < \beta < 0$. Note that the choice of parametrization is

$$dS_t = S_t(\mu dt + \sigma S_t^{\theta/2} dW_t)$$

in Cox, and

$$dS_t = S_t \mu dt + \sigma S_t^\rho dW_t$$

in Delbaen and Shirakawa.

In [428], Heath and Platen study a model where the numéraire portfolio follows a CEV process.

The CEV process is intensively studied by Davydov and Linetsky [225, 227] and Linetsky [592]; the main part of the following study is taken from their work. See also Beckers [64], Forde [354] and Lo et al. [599, 601]. Occupation time densities for CEV processes are presented in Leung and Kwok [582] in the case $\beta < 0$. Atlan and Leblanc [27] and Campi et al. [138, 139] present a model where the default time is related with the first time when a CEV process with $\beta < 0$ reaches 0.

One of the interesting properties of the CEV model is that (for $\beta < 0$) a stock price increase implies that the variance of the stock's return decreases (this is known as the leverage effect). The SABR model introduced in Hagan et al. [417, 418] to fit the volatility surface, corresponds to the case

$$dX_t = \alpha_t X_t^\beta dW_t, \; d\alpha_t = \nu \alpha_t dB_t$$

where W and B are correlated Brownian motions with correlation ρ. This model was named the **s**tochastic **a**lpha-**b**eta-**r**ho model, hence the acronym SABR.

6.4.1 Particular Case $\mu = 0$

Let S follow the dynamics

$$dS_t = \sigma S_t^{1+\beta} dW_t$$

which is the particular case $\mu = 0$ in (6.4.1). Let $T_0(S) = \inf\{t : S_t = 0\}$.

▶ **Case** $\beta > 0$. We define $X_t = \frac{1}{\sigma\beta} S_t^{-\beta}$ for $t < T_0(S)$ and $X_t = \partial$, where ∂ is a cemetery point for $t \geq T_0(S)$. The process X satisfies

$$\begin{aligned} dX_t &= \frac{1}{2}\frac{\beta+1}{\beta}\frac{1}{X_t}dt - dW_t \\ &= \left(\nu + \frac{1}{2}\right)\frac{1}{X_t}dt - dW_t \,. \end{aligned}$$

It is a Bessel process of index $\nu = 1/(2\beta)$ (and dimension $\delta = 2 + \frac{1}{\beta} > 2$). The process X does not reach 0 and does not explode, hence the process S enjoys the same properties. From $S_t = (\sigma\beta X_t)^{-1/\beta} = kX_t^{-2\nu}$, we deduce that the process S is a strict local martingale (see Table 6.1, page 338). The density of the r.v. S_t is obtained from the density of a Bessel process (6.2.3):

$$\mathbb{P}_x(S_t \in dy) = \frac{1}{\sigma^2 \beta t} x^{1/2} y^{-2\beta-3/2} \exp\left(-\frac{x^{-2\beta} + y^{-2\beta}}{2\sigma^2\beta^2 t}\right) I_\nu\left(\frac{x^{-\beta}y^{-\beta}}{\sigma^2\beta^2 t}\right) dy \,.$$

The functions $\Phi_{\lambda\uparrow}$ and $\Phi_{\lambda\downarrow}$ are

$$\Phi_{\lambda\uparrow}(x) = \sqrt{x}K_\nu\left(\frac{x^{-\beta}}{\sigma\beta}\sqrt{2\lambda}\right), \; \Phi_{\lambda\downarrow}(x) = \sqrt{x}I_\nu\left(\frac{x^{-\beta}}{\sigma\beta}\sqrt{2\lambda}\right)$$

with $\nu = 1/(2\beta)$.

▶ **Case** $\beta < 0$. One defines $X_t = -\frac{1}{\sigma\beta}S_t^{-\beta}$ and one checks that, on the set $t < T_0(S)$,

$$dX_t = dW_t + \frac{1}{2}\frac{\beta+1}{\beta}\frac{1}{X_t}dt\,.$$

Therefore, X is a Bessel process of negative index $1/(2\beta)$ which reaches 0, hence S reaches 0 too.

The formula for the density of the r.v. X_t is still valid as long as the dimension δ of the Bessel process X is positive, i.e., for $\delta = 2 + \frac{1}{\beta} > 0$ (or $\beta < -\frac{1}{2}$) and one obtains

$$\mathbb{P}_x(S_t \in dy) = \frac{1}{\sigma^2(-\beta)t}x^{1/2}y^{-2\beta-3/2}\exp\left(-\frac{x^{-2\beta}+y^{-2\beta}}{2\sigma^2\beta^2 t}\right)I_\nu\left(\frac{x^{-\beta}y^{-\beta}}{\sigma^2\beta^2 t}\right)dy\,.$$

For $-\frac{1}{2} < \beta < 0$, the process X with negative dimension ($\delta < 0$), reaches 0 (see Subsection 6.2.6). Here, we stop the process after it first hits 0, i.e., we set

$$\begin{aligned} S_t &= (\sigma\beta X_t)^{1/(-\beta)}\,, \quad \text{for} \quad t \le T_0(X)\,, \\ S_t &= 0, \quad \text{for } t > T_0(X) = T_0(S)\,. \end{aligned}$$

The density of S_t is now given from the one of a Bessel process of positive dimension $4-(2+\frac{1}{\beta}) = 2-\frac{1}{\beta}$, (see Subsection 6.2.6), i.e., with positive index $-\frac{1}{2\beta}$. Therefore, for any $\beta \in [-1/2, 0[$ and $y > 0$,

$$\mathbb{P}_x(S_t \in dy) = \frac{x^{1/2}y^{-2\beta-3/2}}{\sigma^2(-\beta)t}\exp\left(-\frac{x^{-2\beta}+y^{-2\beta}}{2\sigma^2\beta^2 t}\right)I_{|\nu|}\left(\frac{x^{-\beta}y^{-\beta}}{\sigma^2\beta^2 t}\right)dy\,.$$

It is possible to prove that

$$\Phi_{\lambda\uparrow}(x) = \sqrt{x}I_{|\nu|}\left(\frac{x^{-\beta}}{\sigma|\beta|}\sqrt{2\lambda}\right), \; \Phi_{\lambda\downarrow}(x) = \sqrt{x}K_{|\nu|}\left(\frac{x^{-\beta}}{\sigma|\beta|}\sqrt{2\lambda}\right).$$

In the particular case $\beta = 1$, we obtain that the solution of $dS_t = \sigma S_t^2 dW_t$ is $S_t = 1/(\sigma R_t)$, where R is a BES^3 process.

6.4.2 CEV Processes and CIR Processes

Let S follow the dynamics

$$dS_t = S_t(\mu dt + \sigma S_t^\beta dW_t)\,, \tag{6.4.2}$$

where $\mu \neq 0$. For $\beta \neq 0$, setting $Y_t = \frac{1}{4\beta^2} S_t^{-2\beta}$, we obtain

$$dY_t = k(\theta - Y_t)dt + \widehat{\sigma}\sqrt{Y_t}dW_t$$

with $k = 2\mu\beta, \theta = \sigma^2 \frac{2\beta+1}{4k\beta}$, $\widehat{\sigma} = -\text{sgn}(\beta)\sigma$ and $k\theta = \sigma^2 \frac{2\beta+1}{4\beta}$, i.e., Y follows a CIR dynamics; hence S is the power of a (time-changed) CIR process.

▶ Let us study the particular case $k > 0, \theta > 0$, which is obtained either for $\mu > 0, \beta > 0$ or for $\mu < 0, \beta < -1/2$.

In the case $\mu > 0, \beta > 0$, one has $k\theta \geq \sigma^2/2$ and the point 0 is not reached by the process Y. In the case $\beta > 0$, from $S_t^{2\beta} = \frac{1}{4\beta^2 Y_t}$, we obtain that S does not explode and does not reach 0.

In the case $\mu < 0, \beta < -1/2$, one has $0 < k\theta < \sigma^2/2$, hence, the point 0 is reached and is a reflecting boundary for the process Y; from $S_t^{-2\beta} = 4\beta^2 Y_t$, we obtain that S reaches 0 and is reflected.

▶ The other cases can be studied following the same lines (see Lemma 6.4.4.1 for related results).

Note that the change of variable $Z_t = \frac{1}{\sigma|\beta|} S_t^{-\beta}$ reduces the CEV process to a "Bessel process with linear drift":

$$dZ_t = \left(\frac{\beta+1}{2\beta Z_t} - \mu\beta Z_t\right) dt + d\widehat{W}_t$$

where $\widehat{W}_t = -(\text{sgn}\beta)W_t$. Thus, such a process is the square root of a CIR process (as proved before!).

6.4.3 CEV Processes and BESQ Processes

We also extend the result obtained in Subsection 6.4.1 for $\mu = 0$ to the general case.

Lemma 6.4.3.1 *For $\beta > 0$, or $\beta < -\frac{1}{2}$, a CEV process is a deterministic time-change of a power of a BESQ process:*

$$\left(S_t = e^{\mu t}\left(\rho_{c(t)}\right)^{-1/(2\beta)}, t \geq 0\right) \tag{6.4.3}$$

where ρ is a BESQ *with dimension $\delta = 2 + \frac{1}{\beta}$ and $c(t) = \frac{\beta\sigma^2}{2\mu}(e^{2\mu\beta t} - 1)$.*

If $0 > \beta > -\frac{1}{2}$

$$\left(S_t = e^{\mu t}\left(\rho_{c(t)}\right)^{-1/(2\beta)}, t \leq T_0\right)$$

where T_0 *is the first hitting time of 0 for the* BESQ ρ.

For any β *and* $y > 0$, *one has*

$$\mathbb{P}_x(S_t \in dy) = \frac{|\beta|}{c(t)} e^{\mu(2\beta+1/2)t} \exp\left(-\frac{1}{2c(t)}\left(x^{-2\beta} + y^{-2\beta}e^{2\mu\beta t}\right)\right)$$
$$\times\, x^{1/2} y^{-2\beta-3/2}\, I_{1/(2\beta)}\left(\frac{1}{\gamma(t)} x^{-\beta} y^{-\beta} e^{\mu\beta t}\right) dy\,.$$

PROOF: This follows either by a direct computation or by using the fact that the process $Y = \frac{1}{4\beta^2} S^{-2\beta}$ is a CIR(k, θ) process which satisfies

$$Y_t = e^{-kt}\rho(\sigma^2 \frac{e^{kt} - 1}{4k}),\ t \geq 0\,,$$

where $\rho(\cdot)$ is a BESQ process with index $\frac{2k\theta}{\sigma^2} - 1 = \frac{1}{2\beta}$. This implies that

$$S_t = e^{\mu t}\left(4\beta^2 \rho\left(\sigma^2 \frac{e^{2\mu\beta t} - 1}{8\beta\mu}\right)\right)^{-1/(2\beta)},\ t \geq 0\,.$$

It remains to transform ρ by scaling to obtain formula (6.4.3). The density of the r.v. S_t follows from the knowledge of densities of Bessel processes with negative dimensions (see Proposition 6.2.6.3). □

Proposition 6.4.3.2 *Let* S *be a CEV process starting from* x *and introduce* $\delta = 2 + \frac{1}{\beta}$. *Let* $\chi^2(\delta, \alpha; \cdot)$ *be the cumulative distribution function of a* χ^2 *law with* δ *degrees of freedom, and non-centrality parameter* α *(see Exercise 1.1.12.5). We set* $c(t) = \frac{\beta\sigma^2}{2\mu}(e^{2\mu\beta t} - 1)$ *and* $\widehat{y} = \frac{1}{c(t)} y^{-2\beta} e^{2\mu\beta t}$.

For $\beta > 0$, *the cumulative distribution function of* S_t *is*

$$\mathbb{P}_x(S_t \leq y) = 1 - \chi^2\left(\delta, \frac{x^{-2\beta}}{c(t)}; \widehat{y}\right)$$
$$= 1 - \sum_{n=1}^{\infty} g\left(n, \frac{x^{-2\beta}}{2c(t)}\right) G\left(n + \frac{1}{2\beta}, \frac{1}{2c(t)} y^{-2\beta} e^{2\mu\beta t}\right)$$

where

$$g(\alpha, u) = \frac{u^{\alpha-1}}{\Gamma(\alpha)} e^{-u}, \quad G(\alpha, u) = \int_{v \geq u} g(\alpha, v) \mathbb{1}_{v \geq 0} dv\,.$$

For $-\frac{1}{2} > \beta$ *(i.e.,* $\delta > 0, \beta < 0$*) the cumulative distribution function of* S_t *is*

$$\mathbb{P}_x(S_t \leq y) = \chi^2\left(\delta, \frac{x^{-2\beta}}{c(t)}; \widehat{y}\right).$$

For $0 > \beta > -\frac{1}{2}$ (i.e., $\delta < 0$), the cumulative distribution function of S_t is

$$\mathbb{P}_x(S_t \leq y) = 1 - \sum_{n=1}^{\infty} g(n - \frac{1}{2\beta}, \frac{1}{2c(t)} x^{-2\beta})\, G\left(n, \frac{1}{2c(t)} y^{-2\beta} e^{2\mu\beta t}\right).$$

PROOF: Let $\delta = 2 + \frac{1}{\beta}$, $x_0 = x^{-2\beta}$ and $c(t) = \frac{\beta\sigma^2}{2\mu}(e^{2\mu\beta t} - 1)$.

▶ If ρ is a BESQ^δ_x with $\delta \geq 0$, then $\rho_t \overset{\text{law}}{=} tY$, where $Y \overset{\text{law}}{=} \chi^2(\delta, \frac{x}{t})$ (see Remark 6.2.2.1). Hence, $\rho_{c(t)} \overset{\text{law}}{=} c(t)Z$, where $Z \overset{\text{law}}{=} \chi^2\left(2 + \frac{1}{\beta}, \frac{x^{-2\beta}}{2c(t)}\right)$. The formula given in the Proposition follows from a standard computation.

▶ For $\delta < 0$ (i.e., $0 > \beta > -1/2$), from Lemma 6.2.6.1

$$\begin{aligned}\mathbb{P}_x(S_t \geq y, T_0(S) \geq t) &= \mathbb{Q}^\delta_{x_0}(\rho_{c(t)} \geq (e^{-\mu t} y)^{-2\beta} \mathbb{1}_{\{c(t) < T_0(\rho)\}}) \\ &= x\mathbb{Q}^{4-\delta}_{x_0}\left((\rho_{c(t)})^{1/(2\beta)} \mathbb{1}_{\{\rho(c(t)) \geq (e^{-\mu t} y)^{-2\beta}\}}\right).\end{aligned}$$

Setting $w = \frac{1}{2c(t)} y^{-2\beta} e^{2\mu\beta t}$ and $z = \frac{1}{2c(t)} x^{-2\beta}$

$$\begin{aligned}\mathbb{P}_x(S_t \geq y) &= x(c(t))^{1/(2\beta)} \mathbb{E}\left(Z^{1/(2\beta)} \mathbb{1}_{\{Z \geq 2w\}}\right) \\ &= x e^{-z} (2c(t))^{1/(2\beta)} \sum_{n=0}^{\infty} \frac{z^n}{n! \Gamma(n + 1 - 1/(2\beta))} \int_{v \geq w} v^n e^{-v} dv \\ &= e^{-z} \sum_{n=0}^{\infty} \frac{z^{n - \frac{1}{2\beta}}}{n! \Gamma(n + 1 - 1/(2\beta))} \int_{v \geq w} v^n e^{-v} dv \\ &= \sum_{n=0}^{\infty} g\left(1 + n - \frac{1}{2\beta}, z\right) G(n+1, w) = \sum_{n=1}^{\infty} g\left(n - \frac{1}{2\beta}, z\right) G(n, w).\end{aligned}$$

□

6.4.4 Properties

From the results on Bessel processes, we obtain (recall that in the case of negative dimension we stop the processes at level 0).

Lemma 6.4.4.1 *For $\beta < 0$, the boundary $\{0\}$ is reached a.s..*
For $\beta < -1/2$, $\{0\}$ is instantaneously reflecting.
For $-1/2 < \beta < 0$, $\{0\}$ is an absorbing point.
For $0 < \beta$, $\{0\}$ is an unreachable boundary.

From the knowledge of the density of S_T, we can check that, for $x > 0$

$$\mathbb{P}_x(S_T > 0) = \begin{cases} \gamma(-\nu, \zeta(T))/\Gamma(-\nu), & \beta < 0 \\ 1, & \beta > 0\,, \end{cases}$$

$$\mathbb{E}_x(S_T) = \begin{cases} xe^{\mu T}, & \beta < 0 \\ xe^{\mu T}\gamma(\nu, \zeta(T))/\Gamma(\nu) & \beta > 0 \end{cases}$$

where $\nu = 1/(2\beta)$, $\gamma(\nu, \zeta) = \int_0^\zeta t^{\nu-1}e^{-t}dt$ is the incomplete gamma function, and

$$\zeta(T) = \begin{cases} \dfrac{\mu}{\beta\sigma^2(e^{2\mu\beta T} - 1)} x^{-2\beta}, & \mu \neq 0 \\ \dfrac{1}{2\beta^2\sigma^2 T} x^{-2\beta}, & \mu = 0\,. \end{cases}$$

Note that, in the case $\beta > 0$, the expectation of $e^{-\mu T}S_T$ is not S_0, hence, the process $(e^{-\mu t}S_t, t \geq 0)$ is a strict local martingale. We have already noticed this fact in Subsection 6.4.1 devoted to the study of the case $\mu = 0$, using the results on Bessel processes.

Using (6.2.9), we deduce:

Proposition 6.4.4.2 *Let X be a CEV process, with $\beta > -1$. Then*

$$\mathbb{P}_x(T_0 < t) = G\left(\frac{-1}{2\beta}, \zeta(t)\right).$$

6.4.5 Scale Functions for CEV Processes

The derivative of the scale function and the density of the speed measure are

$$s'(x) = \begin{cases} \exp\left(\frac{\mu}{\sigma^2\beta} x^{-2\beta}\right), & \beta \neq 0 \\ x^{-2\mu/\sigma^2} & \beta = 0 \end{cases}$$

and

$$m(x) = \begin{cases} 2\sigma^{-2}x^{-2-2\beta}\exp\left(-\frac{\mu}{\sigma^2\beta} x^{-2\beta}\right), & \beta \neq 0 \\ 2\sigma^{-1}x^{-2+2\mu/\sigma^2} & \beta = 0\,. \end{cases}$$

The functions $\Phi_{\lambda\downarrow}$ and $\Phi_{\lambda\uparrow}$ are solutions of

$$\frac{1}{2}\sigma^2 x^{2+2\beta}u'' + \mu x u' - \lambda u = 0$$

and are given by:

► for $\beta > 0$,

$$\Phi_{\lambda\uparrow}(x) = x^{\beta+1/2}\exp\left(\frac{\epsilon}{2}cx^{-2\beta}\right) W_{k,n}(cx^{-2\beta}),$$

$$\Phi_{\lambda\downarrow}(x) = x^{\beta+1/2}\exp\left(\frac{\epsilon}{2}cx^{-2\beta}\right) M_{k,n}(cx^{-2\beta}),$$

▶ for $\beta < 0$,

$$\Phi_{\lambda\uparrow}(x) = x^{\beta+1/2} \exp\left(\frac{\epsilon}{2} c x^{-2\beta}\right) M_{k,n}(cx^{-2\beta}),$$
$$\Phi_{\lambda\downarrow}(x) = x^{\beta+1/2} \exp\left(\frac{\epsilon}{2} c x^{-2\beta}\right) W_{k,n}(cx^{-2\beta}),$$

where

$$c = \frac{|\mu|}{|\beta|\sigma^2}, \quad \epsilon = \mathrm{sign}(\mu\beta),$$
$$n = \frac{1}{4|\beta|}, \quad k = \epsilon\left(\frac{1}{2} + \frac{1}{4\beta}\right) - \frac{\lambda}{2|\mu\beta|},$$

and W and M are the classical Whittaker functions (see $\rightarrowtail$ Subsection A.5.7).

6.4.6 Option Pricing in a CEV Model

European Options

We give the value of a European call, first derived by Cox [205], in the case where $\beta < 0$. The interest rate is supposed to be a constant r. The previous computation leads to

$$\mathbb{E}_{\mathbb{Q}}(e^{-rT}(S_T - K)^+) = S_0 \sum_{n=1}^{\infty} g(n,z) G\left(n - \frac{1}{2\beta}, w\right) - K e^{-rT} \sum_{n=1}^{\infty} g\left(n - \frac{1}{2\beta}, z\right) G(n,w)$$

where g, G are defined in Proposition 6.4.3.2 (with $\mu = r$), $z = \frac{1}{2c(T)} S_0^{-2\beta}$ and $w = \frac{1}{2c(T)} K^{-2\beta} e^{2r\beta T}$.

In the case $\beta > 0$,

$$\mathbb{E}_{\mathbb{Q}}(e^{-rT}(S_T - K)^+) = S_0 \sum_{n=1}^{\infty} g\left(n + \frac{1}{2\beta}, z\right) G(n,w) - K e^{-rT} \sum_{n=1}^{\infty} g(n,z) G\left(n + \frac{1}{2\beta}, w\right).$$

We recall that, in that case $(S_t e^{-rt}, t \geq 0)$ is not a martingale, but a strict local martingale.

Barrier and Lookback Options

Boyle and Tian [121] and Davydov and Linetsky [225] study barrier and lookback options.

See Lo et al. [601], Schroder [772], Davydov and Linetsky [227] and Linetsky [592, 593] for option pricing when the underlying asset follows a CEV model.

Perpetual American Options

The price of a perpetual American option can be obtained by solving the associated PDE. The pair (b, V) (exercise boundary, value function) of an American perpetual put option satisfies

$$\begin{aligned}\frac{\sigma^2}{2} x^{2(\beta+1)} V''(x) + rxV'(x) &= rV(x), \ x > b \\ V(x) &\geq (K-x)^+ \\ V(x) &= K - x, \text{ for } x \leq b \\ V'(b) &= -1 .\end{aligned}$$

The solution is

$$V(x) = Kx \int_x^\infty \frac{1}{y^2} \exp\left(\frac{r}{\sigma^2 \beta}(y^{-2\beta} - b^{-2\beta})\right) dy, \ x > b$$

where b is the unique solution of the equation $V(b) = K - b$. (See Ekström [295] for details and properties of the price.)

6.5 Some Computations on Bessel Bridges

In this section, we present some computations for Bessel bridges, which are useful in finance. Let $t > 0$ and $\mathbb{P}^\delta$ be the law of a δ-dimensional Bessel process on the canonical space $C([0,t], \mathbb{R}^+)$ with the canonical process now denoted by $(R_t, t \geq 0)$. There exists a regular conditional distribution for $\mathbb{P}^\delta_x(\,\boldsymbol{.}\,|R_t)$, namely a family $\mathbb{P}^{\delta,t}_{x\to y}$ of probability measures on $C([0,t], \mathbb{R}^+)$ such that for any Borel set A

$$\mathbb{P}^\delta_x(A) = \int \mathbb{P}^{\delta,t}_{x\to y}(A) \mu_t(dy)$$

where μ_t is the law of R_t under $\mathbb{P}^\delta_x$. A continuous process with law $\mathbb{P}^{\delta,t}_{x\to y}$ is called a Bessel bridge from x to y over $[0,t]$.

6.5.1 Bessel Bridges

Proposition 6.5.1.1 *For a Bessel process R, for each pair $(\nu, \mu) \in \mathbb{R}^+ \times \mathbb{R}^+$, and for each $t > 0$, one has*

$$\mathbb{P}^{(\nu)}_x \left[\exp\left(-\frac{\mu^2}{2} \int_0^t \frac{ds}{R_s^2}\right) | R_t = y\right] = \frac{I_\gamma(xy/t)}{I_\nu(xy/t)}, \tag{6.5.1}$$

where $\gamma = \sqrt{\nu^2 + \mu^2}$.

Equivalently, for a squared Bessel process, for each pair $(\nu, \mu) \in \mathbb{R}^+ \times \mathbb{R}^+$, and for each $t > 0$, one has

$$\mathbb{Q}_x^{(\nu)}\left[\exp\left(-\frac{\mu^2}{2}\int_0^t \frac{ds}{\rho_s}\right)|\rho_t = y\right] = \frac{I_\gamma(\sqrt{xy}/t)}{I_\nu(\sqrt{xy}/t)}, \tag{6.5.2}$$

and, for any $b \in \mathbb{R}$,

$$\mathbb{Q}_x^{(\nu)}\left[\exp\left(-\frac{b^2}{2}\int_0^t \rho_s ds\right)|\rho_t = y\right] \tag{6.5.3}$$

$$= \frac{bt}{\sinh(bt)}\exp\left\{\frac{x+y}{2t}(1 - bt\coth bt)\right\}\frac{I_\nu[b\sqrt{xy}/\sinh bt]}{I_\nu[\sqrt{xy}/t]}.$$

PROOF: On the one hand, from (6.2.3) and (6.2.10), for any $a > 0$ and any bounded Borel function f,

$$\mathbb{P}_x^{(\nu)}\left[f(R_t)\exp\left(-\frac{\mu^2}{2}\int_0^t \frac{ds}{R_s^2}\right)\right] = \mathbb{P}_x^{(\gamma)}\left[\left(\frac{R_t}{x}\right)^{\nu-\gamma} f(R_t)\right]$$

$$= \int_0^\infty dy \left(\frac{y}{x}\right)^{\nu-\gamma} f(y)\frac{y}{t}\left(\frac{y}{x}\right)^\gamma \exp\left(-\frac{x^2+y^2}{2t}\right) I_\gamma(xy/t).$$

On the other hand, from (6.2.3), this expression equals

$$\int_0^\infty dy f(y)\mathbb{P}_x^{(\nu)}\left[\exp\left(-\frac{\mu^2}{2}\int_0^t \frac{ds}{R_s^2}\right)|R_t = y\right]\frac{y}{t}\left(\frac{y}{x}\right)^\nu \exp\left(-\frac{x^2+y^2}{2t}\right) I_\nu(xy/t)$$

and the result follows by identification. The squared Bessel bridge case follows from (6.2.14) and the knowledge of transition probabilities. □

Example 6.5.1.2 The normalized Brownian excursion $|B|^{[g_1,d_1]}$ is a BES^3 bridge from $x = 0$ to $y = 0$ with $t = 1$ (see equation (4.3.1) for the notation).

6.5.2 Bessel Bridges and Ornstein-Uhlenbeck Processes

We shall apply the previous computation in order to obtain the law of hitting times for an OU process. We have studied OU processes in Subsection 2.6.3. Here, we are concerned with hitting time densities for OU processes with parameter k:

$$dX_t = dW_t - kX_t dt.$$

Proposition 6.5.2.1 *Let X be an OU process starting from x and $T_0(X)$ its first hitting time of 0: $T_0(X) := \inf\{t : X_t = 0\}$.*

Then $\mathbb{P}_x(T_0(X) \in dt) = f(x,t)dt$, where

$$f(x,t) = \frac{|x|}{\sqrt{2\pi}}\exp\left(\frac{kx^2}{2}\right)\exp\left(\frac{k}{2}(t - x^2\coth(kt))\right)\left(\frac{k}{\sinh(kt)}\right)^{3/2}.$$

PROOF: The absolute continuity relationship established in Subsection 2.6.3 reads

$$\mathbb{P}_x^{k,0}|_{\mathcal{F}_t} = \exp\left[-\frac{k}{2}(W_t^2 - t - x^2) - \frac{k^2}{2}\int_0^t W_s^2 ds\right] \mathbf{W}_x|_{\mathcal{F}_t}$$

and holds with the fixed time t replaced by the stopping time T_a, restricted to the set $T_a < \infty$, leading to

$$\mathbb{P}_x^{k,0}(T_a \in dt) = \exp\left[-\frac{k}{2}(a^2 - t - x^2)\right] \mathbf{W}_x\left(\mathbb{1}_{\{T_a \in dt\}} \exp\left(-\frac{k^2}{2}\int_0^t W_s^2 ds\right)\right).$$

Hence,

$$\mathbb{P}_x^{k,0}(T_a \in dt) = \exp\left[-\frac{k}{2}(a^2 - t - x^2)\right]$$
$$\times \mathbf{W}_{a-x}\left(\mathbb{1}_{\{T_0 \in dt\}} \exp\left(-\frac{k^2}{2}\int_0^t (a - W_s)^2 ds\right)\right).$$

Let us set $y = |a - x|$. Under $\mathbf{W}_y$, the process $(W_s, s \leq T_0)$ conditioned by $T_0 = t$ is a BES^3 bridge of length t, starting at y and ending at 0, therefore

$$\mathbf{W}_{a-x}\left(\mathbb{1}_{\{T_0 \in dt\}} \exp\left(-\frac{k^2}{2}\int_0^t (a - W_s)^2 ds\right)\right)$$
$$= \mathbb{P}_y^3\left[\exp\left(-\frac{k^2}{2}\int_0^t (a - \epsilon R_s)^2 ds\right) \Big| R_t = 0\right] \mathbf{W}_y(T_0 \in dt)$$

where $\epsilon = \text{sgn}(a - x)$. For $a = 0$, the computation reduces to that of

$$\mathbb{P}_y^3\left[\exp\left(-\frac{k^2}{2}\int_0^t R_s^2 ds\right) \Big| R_t = 0\right]$$

which, from (6.5.3) and (A.5.3) is equal to

$$\left(\frac{kt}{\sinh(kt)}\right)^{3/2} \exp\left(-\frac{y^2}{2t}(kt\coth(kt) - 1)\right).$$

□

Comment 6.5.2.2 The law of the hitting time for a general level a requires the knowledge of the joint law of $\left(\int_0^t R_s ds, \int_0^t R_s^2 ds\right)$ under the BES^3-bridge law. See Alili et al. [10], Göing-Jaeschke and Yor [397] and Patie [697].

Exercise 6.5.2.3 As an application of the absolute continuity relationship (6.3.3) between a BESQ and a CIR, prove that

$$ {}^k\mathbb{Q}_{x \to y}^{\delta,t} = \frac{\exp\left(-\frac{k^2}{2}\int_0^t \rho_s ds\right)}{\mathbb{Q}_{x \to y}^{\delta,t}\left[\exp\left(-\frac{k^2}{2}\int_0^t \rho_s ds\right)\right]} \mathbb{Q}_{x \to y}^{\delta,t} \tag{6.5.4}$$

where ${}^k\mathbb{Q}^{\delta,(t)}_{x\to y}$ denotes the bridge for $(\rho_u, 0 \le u \le t)$ obtained by conditioning ${}^k\mathbb{Q}^\delta_x$ by $(\rho_t = y)$. For more details, see Pitman and Yor [716]. ◁

6.5.3 European Bond Option

Let $P(t,T^*) = B(r_t, T^* - t)$ be the price of a zero-coupon bond maturing at time T^* when the interest rate $(r_t, t \ge 0)$ follows a risk-neutral CIR dynamics, where $B(r,s) = \exp[-A(s) - rG(s)]$. The functions A and G are defined in Corollary 6.3.4.3. The price C_{EB} of a T-maturity European call option on a zero-coupon bond maturing at time T^*, where $T^* > T$, with a strike price equal to K is

$$C_{EB}(r_t, T-t) = \mathbb{E}_Q\left[\exp\left(-\int_t^T r_s ds\right)(P(T,T^*) - K)^+ |\mathcal{F}_t\right].$$

We present the computation of this quantity.

Proposition 6.5.3.1 *Let us assume that the risk-neutral dynamics of* r *are*

$$dr_t = k(\theta - r_t)dt + 2\sqrt{r_t}dW_t\,, r_0 = x\,.$$

Then,

$$C_{EB}(r, T-t) = B(r, T^* - t)\,\Psi_1 - KB(r, T-t)\,\Psi_2 \qquad (6.5.5)$$

where

$$\begin{aligned}
\Psi_1 &= \chi^2\left(\frac{4k\theta}{\sigma^2}, \frac{2\phi^2 r\exp(\gamma(T^*-t))}{\phi+\psi+G(T^*-T)}; 2r^*(\phi+\psi+G(T^*-T))\right),\\
\Psi_2 &= \chi^2\left(\frac{4k\theta}{\sigma^2}, \frac{2\phi^2 r\exp(\gamma(T^*-t))}{\phi+\psi}; 2r^*(\phi+\psi)\right),\\
\phi &= \frac{2\gamma}{\sigma^2(\exp(\gamma(T^*-t))-1)}, \quad \psi = \frac{k+\gamma}{\sigma^2},\\
\gamma &= \sqrt{k^2+2\sigma^2}, \quad r^* = -\frac{A(T^*-T)+\ln(K)}{G(T^*-T)}.
\end{aligned}$$

Here, χ^2 *is the non-central chi-squared distribution function defined in Exercise* 1.1.12.5. *The real number* r^* *is the critical interest rate defined by* $K = B(r^*, T^*-T)$ *for* $T^* > T$ *below which the European call will be exercised at maturity* T.

Finally notice that K is constrained to be strictly less than $A(T^* - T)$, the maximum value of the discount bond at time T, otherwise exercising would of course never be done.

PROOF: From $P(T,T^*) = B(r_T, T^* - T)$, we obtain

$$C_{EB}(x,T) = \mathbb{E}_Q \left[\exp\left(-\int_0^T r_s ds\right) P(T,T^*) \mathbb{1}_{\{r_T \le r^*\}} \right]$$
$$- K\mathbb{E}_Q \left[\exp\left(-\int_0^T r_s ds\right) \mathbb{1}_{\{r_T \le r^*\}} \right],$$

where $B(r^*, T^* - T) = K$, that is

$$r^* = -\frac{A(T^* - T) + \ln K}{G(T^* - T)}.$$

We observe now that the processes

$$L_1(t) = (P(0,T))^{-1} \exp\left(-\int_0^t r_s ds\right) P(t,T)$$
$$L_2(t) = (P(0,T^*))^{-1} \exp\left(-\int_0^t r_s ds\right) P(t,T^*)$$

are positive $\mathbb{Q}$-martingales with expectations equal to 1, from the definition of $P(t,\cdot)$. Hence, using change of numéraire techniques, we can define two probabilities $\mathbb{Q}_1$ and $\mathbb{Q}_2$ by

$$\mathbb{Q}_i|_{\mathcal{F}_t} = L_i(t)\, \mathbb{Q}|_{\mathcal{F}_t}, \; i = 1, 2.$$

Therefore,

$$C_{EB}(x,T) = P(0,T^*)\mathbb{Q}_2(r_T \le r^*) - KP(0,T)\mathbb{Q}_1\left(r_T \le r^*\right).$$

We characterize the law of the r.v. r_T under $\mathbb{Q}_1$ from its Laplace transform

$$\mathbb{E}_{\mathbb{Q}_1}(e^{-\lambda r_T}) = \mathbb{E}_{\mathbb{Q}}\left(e^{-\lambda r_T} \exp\left(-\int_0^T r_s ds\right)\right) (P(0,T))^{-1}.$$

From Corollary 6.3.4.3,

$$\mathbb{E}_{\mathbb{Q}_1}(e^{-\lambda r_T}) = \exp\left(-A_{\lambda,1} + A_{0,1} - x(G_{\lambda,1} - G_{0,1})\right).$$

Let $\tilde{c}_1 = \dfrac{\sigma^2}{2}\dfrac{e^{\gamma T} - 1}{D}$ with $D = \gamma(e^{\gamma T} + 1) + k(e^{\gamma T} - 1)$. Then,

$$\exp\left(-A_{\lambda,1} + A_{0,1}\right) = \left(\frac{1}{1 + 2\lambda \tilde{c}_1}\right)^{2k\theta/\sigma^2}.$$

Some computations and the equality $\gamma^2 = k^2 + 2\sigma^2$ yield

$$\begin{aligned} G_{\lambda,1} - G_{0,1} &= \frac{\lambda}{D^2(1 + 2\lambda\tilde{c}_1)}\left[D(\gamma + k + e^{\gamma T}(\gamma - k)) - 2\sigma^2(e^{\gamma T} - 1)^2\right] \\ &= \frac{\lambda}{D^2(1 + 2\lambda\tilde{c}_1)}\left[\gamma^2(e^{\gamma T} + 1)^2 - k^2(e^{\gamma T} - 1)^2 - 2\sigma^2(e^{\gamma T} - 1)^2\right] \\ &= \lambda\frac{4\gamma^2 e^{\gamma T}}{D(1 + 2\lambda\tilde{c}_1)} = \lambda\frac{\tilde{c}_1\tilde{x}_1}{(1 + 2\lambda\tilde{c}_1)} \end{aligned}$$

where

$$\tilde{x}_1 = \frac{8x\gamma^2 e^{\gamma T}}{\sigma^2(e^{\gamma T}-1)[\gamma(e^{\gamma T}+1)+k(e^{\gamma T}-1)]}.$$

Hence,

$$\mathbb{E}_{\mathbb{Q}_1}(e^{-\lambda r_T}) = \left(\frac{1}{2\lambda\tilde{c}_1+1}\right)^{2k\theta/\sigma^2} \exp\left(-\frac{\lambda\tilde{c}_1\tilde{x}_1}{2\lambda\tilde{c}_1+1}\right)$$

and, from Corollary 6.3.4.4, the r.v. $r_T/\tilde{c}_1$ follows, under $\mathbb{Q}_1$, a χ^2 law with $2k\theta/\sigma^2$ degrees of freedom and non-centrality $\tilde{x}_1$.

The same kind of computation establishes that the r.v. $r_T/\tilde{c}_2$ follows, under $\mathbb{Q}_2$, a χ^2 law with parameters $\tilde{c}_2, \tilde{x}_2$ given by

$$\tilde{c}_2 = \frac{\sigma^2}{2}\frac{e^{\gamma T}-1}{\gamma(e^{\gamma T}+1)+(k+\sigma^2 G(T^*-T))(e^{\gamma T}-1)},$$

$$\tilde{x}_2 = \frac{8x\gamma^2 e^{\gamma T}}{\sigma^2(e^{\gamma T}-1)\left(\gamma(e^{\gamma T}+1)+(k+\sigma^2 G(T^*-T))(e^{\gamma T}-1)\right)}.$$

□

Comment 6.5.3.2 Maghsoodi [615] presents a solution of bond option valuation in the extended CIR term structure.

6.5.4 American Bond Options and the CIR Model

Consider the problem of the valuation of an American bond put option in the context of the CIR model.

Proposition 6.5.4.1 *Let us assume that*

$$dr_t = k(\theta - r_t)dt + \sigma\sqrt{r_t}dW_t\,,\; r_0 = x\,. \tag{6.5.6}$$

The American put price decomposition reduces to

$$P_A(r, T-t) = P_E(r, T-t) + K\mathbb{E}_Q\left[\int_t^T \exp\left(-\int_t^u r_s ds\right) r_u \mathbb{1}_{\{r_u \geq b(u)\}} du \,\middle|\, r_t = r\right] \tag{6.5.7}$$

where $\mathbb{Q}$ is the risk-neutral probability and $b(\cdot)$ is the put exercise boundary.

PROOF: We give a decomposition of the American put price into two components: the European put price given by Cox-Ingersoll-Ross, and the American premium. We shall proceed along the lines used for the American stock options. Let t be fixed and denote

$$R_u^t = \exp\left(-\int_t^u r_v\, dv\right).$$

We further denote by $P_A(r, T-u)$ the value at time u of an American put with maturity T on a zero-coupon bond paying one monetary unit at time T^* with $T \le T^*$. In a similar way, to the function F defined in the context of American stock options, P_A is differentiable and only piecewise twice differentiable in the variable r. Itô's formula leads to

$$\begin{aligned} R_T^t P_A(r_T, 0) = P_A(r_t, T-t) &+ \int_t^T R_u^t\, \mathcal{L}(P_A)(r_u, T-u)\, du \\ &- \int_t^T R_u^t(r_u P_A(r_u, T-u) - \partial_u P_A(r_u, T-u)\,)du \\ &+ \int_t^T R_u^t\, \partial_r P_A(r_u, T-u)\sigma\sqrt{r_u}dW_u \end{aligned} \tag{6.5.8}$$

where the infinitesimal generator $\mathcal{L}$ of the process solution of equation (6.5.6) is defined by

$$\mathcal{L} = \frac{1}{2}\sigma^2 r \frac{\partial^2}{\partial r^2} + k(\theta - r)\frac{\partial}{\partial r}\,.$$

Taking expectations and noting that the stochastic integral on the right-hand side of (6.5.8) has 0 expectation, we obtain

$$\begin{aligned} \mathbb{E}_{\mathbb{Q}}(R_T^t P_A(r_T, 0)|\mathcal{F}_t) &= P_A(r_t, T-t) + \mathbb{E}_{\mathbb{Q}}\left(\int_t^T R_u^t \mathcal{L}(P_A)(r_u, T-u)\, du|\mathcal{F}_t\right) \\ &+ \mathbb{E}_{\mathbb{Q}}\left(\int_t^T R_u^t(r_u P_A(r_u, T-u) - \partial_u P_A(r_u, T-u))du|\mathcal{F}_t\right)\,. \end{aligned} \tag{6.5.9}$$

In the continuation region, P_A satisfies the same partial differential equation as the European put

$$\begin{aligned} \mathcal{A}(r, T-u) := \mathcal{L}(P_A)(r, T-u) + rP_A(r, T-u) - \partial_u P_A(r, T-u)) = 0, \\ \forall u \in [t, T[, \forall r \in]0, b(u)] \end{aligned}$$

where $b(u)$ is the level of the exercise boundary at time u:

$$b(u) := \inf\{\alpha \in \mathbb{R}^+ \mid P_A(\alpha, T-u) = K - B(\alpha, T-u)\}$$

Here, $B(r,s) = \exp(-A(s) - rG(s))$ determines the price of the zero-coupon bond with time to maturity s and current value of the spot interest rate r (see Corollary 6.3.4.3). Therefore, the quantity $\mathcal{A}(r, T-u)$ is different from zero only in the stopping region, i.e., when $r_u \ge b(u)$, or, since B is a decreasing function of r (the function G is positive), when $B(r_u, T-u) \le B(b(u), T-u)$. Equation (6.5.9) can be rewritten

$$\begin{aligned} &\mathbb{E}_{\mathbb{Q}}\left(R_T^t(K - B(r_T, T^* - T))^+|\mathcal{F}_t\right) = P_A(r_t, T-t) \\ &+ \mathbb{E}_{\mathbb{Q}}\left(\int_t^T R_u^t\left(-\mathcal{L}(B(r_u, T-u)) + r_u(K - B(r_u, T-u))\right.\right. \\ &+ \left.\left. \partial_u B(r_u, T-u)\right)\mathbb{1}_{\{r_u \ge b(u)\}}du|r_t = r\right)\,. \end{aligned} \tag{6.5.10}$$

Another way to derive the latter equation from equation (6.5.9) makes use of the martingale property of the process

$$R_u^t P_A(r_u, T-u), u \in [t, T]$$

under the risk-adjusted probability $\mathbb{Q}$, in the continuation region. Notice that the bond value satisfies the same PDE as the bond option. Therefore

$$-\mathcal{L}(B(r_u, T-u)) + r_u(K - B(r_u, T-u)) + \partial_u B(r_u, T-u) = 0\,.$$

Finally, equation (6.5.10) can be rewritten as follows

$$P_A(r, T-t) = P_E(r, T-t) + K \int_t^T \mathbb{E}_{\mathbb{Q}}\left(R_u^t r_u \mathbb{1}_{\{r_u \geq b(u)\}} | \mathcal{F}_t\right) du\,.$$

□

Jamshidian [474] computed the early exercise premium given by the latter equation, using the forward risk-adjusted probability measure. Under this equivalent measure, the expected future spot rate is equal to the forward rate, and the expected future bond price is equal to the future price. This enables the discount factor to be pulled out of the expectation and the expression for the early exercise premium to be simplified.

Here, we follow another direction (see Chesney et al. [171]) and show how analytic expressions for the early exercise premium and the American put price can be derived by relying on known properties of Bessel bridges [716]. Let us rewrite the American premium as follows

$$\mathbb{E}_{\mathbb{Q}}\left(\int_t^T R_u^t r_u \mathbb{1}_{\{r_u \geq b(u)\}} du \,\Big|\, r_t = r\right)$$
$$= \mathbb{E}_{\mathbb{Q}}\left(\int_t^T du\, \mathbb{E}_{\mathbb{Q}}\left(R_u^t | r_u\right) r_u \mathbb{1}_{\{r_u \geq b(u)\}} \,\Big|\, r_t = r\right).$$

The probability density of the interest rate r_u conditional on its value at time t is known. Therefore, the problem of the valuation of the American premium rests on the computation of the inner expectation. More generally let us consider the following Laplace transform (as in Scott [777])

$$\mathbb{E}_{\mathbb{Q}}\left(R_u^t \big| r_t, r_u\right)\,.$$

Defining the process Z by a simple change of time

$$Z_s = r_{4s/\sigma^2}\,, \tag{6.5.11}$$

we obtain

$$\mathbb{E}_{\mathbb{Q}}\left(R_u^t | (r_t, r_u)\right)$$
$$= \mathbb{E}_{\mathbb{Q}}\left(\exp\left(-\frac{4}{\sigma^2}\int_{t\sigma^2/4}^{u\sigma^2/4} Z_v dv\right) | Z\left(\frac{t\sigma^2}{4}\right), Z\left(\frac{u\sigma^2}{4}\right)\right).$$

In the risk-adjusted economy, the spot rate is given by the equation (6.3.2) and from the time change (6.5.11), the process Z is a CIR process whose volatility is equal to 2. Setting $\delta = \frac{4k\theta}{\sigma^2}$, $\kappa = -\frac{4k}{\sigma^2}$, we obtain

$$dZ_t = (\kappa Z_t + \delta)\,dt + 2\sqrt{Z_t}dW_t\,,$$

where $(W_t, t \geq 0)$ is a $\mathbb{Q}$-Brownian motion. We now use the absolute continuity relationship (6.5.4) between CIR bridges and Bessel bridges, where the notation ${}^{\kappa}\mathbb{Q}^{\delta,T}_{x\to y}$ is defined

$$\begin{aligned}
&\mathbb{E}\left[R^t_u \middle| r_t = x, r_u = y\right] \\
&= {}^{\kappa}\mathbb{Q}^{\delta,\sigma^2(u-t)/4}_{x\to y}\left[\exp\left(-\frac{4}{\sigma^2}\int_{t\sigma^2/4}^{u\sigma^2/4} Z_s ds\right)\right] \\
&= \frac{\mathbb{Q}^{\delta,\sigma^2(u-t)/4}_{x\to y}\left[\exp\left(-\left(\frac{4}{\sigma^2}+\frac{\kappa^2}{2}\right)\int_{t\sigma^2/4}^{u\sigma^2/4} Z_s ds\right)\right]}{\mathbb{Q}^{\delta,\sigma^2(u-t)/4}_{x\to y}\left[\exp\left(-\frac{\kappa^2}{2}\int_{t\sigma^2/4}^{u\sigma^2/4} Z_s ds\right)\right]}\,.
\end{aligned}$$

From the results (6.5.3) on Bessel bridges and some obvious simplifications

$$\begin{aligned}
&\mathbb{E}\left[\exp\left(-\int_t^u r_s ds\right)\middle| r_t = x, r_u = y\right] \\
&= \frac{c\sinh(\kappa\gamma)\exp\left(\frac{x+y}{2\gamma}(1 - c\gamma\coth(c\gamma))\right) I_\nu\left(\frac{c\sqrt{xy}}{\sinh c\gamma}\right)}{\kappa\sinh(c\gamma)\exp\left(\frac{x+y}{2\gamma}(1 - \kappa\gamma\coth(\kappa\gamma))\right) I_\nu\left(\frac{\kappa\sqrt{xy}}{\sinh \kappa\gamma}\right)} \\
&:= m(x,t,y,u)
\end{aligned}$$

where $\gamma = \frac{\sigma^2(u-t)}{4}$, $\nu = \frac{\delta}{2} - 1$, $c = \sqrt{\frac{8}{\sigma^2} + \kappa^2}$ and I_ν is the modified Bessel function. We obtain the American put price

$$P_A(r, T-t) = P_E(r, T-t) + K\int_t^T du \int_{b(u)}^{\infty} m(r,t,y,u)\, f_{u-t}(r,y)dy\,, \quad (6.5.12)$$

where f is the density defined in Proposition 6.3.2.1.

Note that, as with formula (6.5.7), formula (6.5.12) necessitates, in order to be implemented, knowledge of the boundary $b(u)$.

6.6 Asian Options

Asian options on the underlying S have a payoff, paid at maturity T, equal to $(\frac{1}{T}\int_0^T S_u\, du - K)^+$. The expectation of this quantity is usually difficult to evaluate. The fact that the payoff is based on an average price is an attractive feature, especially for commodities where price manipulations are possible. Furthermore, Asian options are often cheaper than the vanilla ones. Here, we work in the Black and Scholes framework where the underlying asset follows

$$dS_t = S_t(rdt + \sigma dW_t),$$

or

$$S_t = S_0 \exp[\sigma(W_t + \nu t)],$$

where W is a BM under the e.m.m. The price of an Asian option is

$$C^{\text{Asian}}(S_0, K) = \mathbb{E}\left(e^{-rT}\left(\frac{S_0}{T}\int_0^T e^{\sigma(W_s+\nu s)}\,ds - K\right)^+\right).$$

For any real ν, we denote $A_T^{(\nu)} = \int_0^T \exp[2(W_s + \nu s)]ds$ and $A_T = A_T^{(0)}$.

The scaling property of BM leads to $S_s = S_0 \exp(\sigma\nu s)\exp(2\widetilde{W}_{\sigma^2 s/4})$, where $(\widetilde{W}_u := \frac{\sigma}{2}W_{4u/\sigma^2}, u \geq 0)$ is a Brownian motion. Using $\widetilde{W}$, we see that

$$\int_0^T S_s\,ds = \frac{4}{\sigma^2}S_0\int_0^{\sigma^2 T/4}\exp[2(\widetilde{W}_u + \mu u)]\,du \overset{\text{law}}{=} \frac{4}{\sigma^2}S_0 A^{(\mu)}_{\sigma^2 T/4}$$

with $\mu = \dfrac{2\nu}{\sigma}$.

6.6.1 Parity and Symmetry Formulae

Assume here that

$$dS_t = S_t((r-\delta)dt + \sigma dW_t).$$

We denote $A_T^S = \frac{1}{T}\int_0^T S_u du$ and $C_{fi}^{\text{Asian}} = C_{fi}^{\text{Asian}}(S_0, K; r, \delta)$ the price of a call Asian option with a fixed-strike, whose payoff is $(\frac{1}{T}A_T^S - K)^+$ and $C_{f\ell}^{\text{Asian}} = C_{f\ell}^{\text{Asian}}(S_0, \lambda; r, \delta)$ the price of a call Asian option with floating strike, with payoff $(\lambda S_T - \frac{1}{T}A_T^S)^+$.

The payoff of a put Asian option with a fixed-strike is $(K - \frac{1}{T}A_T^S)^+$, the price of this option is denoted by $P_{fi}^{\text{Asian}} = P_{fi}^{\text{Asian}}(S_0, K; r, \delta)$. The price of a put Asian option with floating strike, with payoff $(\frac{1}{T}A_T^S - \lambda S_T)^+$ is $P_{f\ell}^{\text{Asian}} = P_{f\ell}^{\text{Asian}}(S_0, \lambda; r, \delta)$.

Proposition 6.6.1.1 *The following parity relations hold*

(i) $P_{f\ell}^{\text{Asian}} = C_{f\ell}^{\text{Asian}} + \dfrac{1}{(r-\delta)T}(e^{-\delta T} - e^{-rT})S_0 - \lambda S_0 e^{-\delta T}$,

(ii) $P_{fi}^{\text{Asian}} = C_{fi}^{\text{Asian}} + \dfrac{1}{(r-\delta)T}(e^{-\delta T} - e^{-rT})S_0 - Ke^{-rT}$.

PROOF: Obvious. □

We present a symmetry result.

Proposition 6.6.1.2 *The following symmetry relations hold*

$$C_{f\ell}^{\text{Asian}}(S_0, \lambda; r, \delta) = P_{fi}^{\text{Asian}}(S_0, \lambda S_0; \delta, r),$$
$$C_{fi}^{\text{Asian}}(S_0, K; r, \delta) = P_{f\ell}^{\text{Asian}}(S_0, K/S_0; , \delta, r).$$

PROOF: This is a standard application of change of numéraire techniques.

$$\begin{aligned} C_{f\ell}^{\text{Asian}}(S_0, \lambda; r, \delta) &= \mathbb{E}(e^{-rT}(\lambda S_T - \frac{1}{T} A_T^S)^+) \\ &= \mathbb{E}\left(e^{-rT} \frac{S_T}{S_0} S_0 \left(\lambda - \frac{1}{T}\int_0^T \frac{S_u}{S_T} du\right)^+\right) \\ &= \widehat{\mathbb{E}}\left(e^{-\delta T} S_0 \left(\lambda - \frac{1}{T}\int_0^T \frac{S_u}{S_T} du\right)^+\right) \end{aligned}$$

where $\widehat{\mathbb{P}}|_{\mathcal{F}_T} = e^{-(r-\delta)T} \frac{S_T}{S_0} \mathbb{P}|_{\mathcal{F}_T}$. From Cameron-Martin's Theorem, the process Z defined as $Z_t = W_t - \sigma t$ is a $\widehat{\mathbb{P}}$-Brownian motion. From

$$\frac{1}{T}\frac{A_T^S}{S_T} = \frac{1}{T}\int_0^T \frac{S_u}{S_T} du = \frac{1}{T}\int_0^T e^{(r-\delta+\frac{1}{2}\sigma^2)(u-T)+\sigma(Z_u - Z_T)} du,$$

the law of $\frac{A_T^S}{S_T}$ under $\widehat{\mathbb{P}}$ is equal to the law of

$$\frac{1}{T}\int_0^T e^{(r-\delta+\frac{1}{2}\sigma^2)(u-T)-\sigma Z_{T-u}} du \overset{\text{law}}{=} \frac{1}{T}\int_0^T e^{(\delta-r-\frac{1}{2}\sigma^2)s+\sigma Z_s} ds.$$

The second formula is obtained using the call-put parity. □

Comment 6.6.1.3 The second symmetry formula of Proposition 6.6.1.2 is extended to the exponential Lévy framework by Fajardo and Mordecki [339] and by Eberlein and Papapantoleon [293]. The relation between floating and fixed strike, due to Henderson and Wojakowski [432] is extended to the exponential Lévy framework in Eberlein and Papapantoleon [292].

6.6.2 Laws of $A_\Theta^{(\nu)}$ and $A_t^{(\nu)}$

Here, we follow Leblanc [571] and Yor [863].

Let W be a Brownian motion, and $A_t = \int_0^t e^{2W_s} ds$, $A_t^\dagger = \int_0^t e^{W_s} ds$. Let f, g, h be Borel functions and

$$k(t) = \mathbb{E}\left(f(A_t)\, g(A_t^\dagger)\, h(e^{W_t})\right).$$

Lemma 6.6.2.1 *The Laplace transform of k is given by*

$$\int_0^\infty k(t)e^{-\theta^2 t/2}dt = \int_0^\infty du\, f(u)\mathbb{P}_1^{(\theta)}\left(\frac{h(R_u)g(K_u)}{R_u^{2+\theta}}\right)$$

where R is a Bessel process with index θ, starting from 1 and $K_t = \int_0^t \frac{du}{R_u}$.

PROOF: Let C be the inverse of the increasing process A. We have seen, in the proof of Lamperti's theorem (applied here in the particular case $\nu = 0$) that $dC_u = \frac{1}{R_u^2}du$ where $R_u = \exp W_{C_u}$ is a Bessel process of index 0 (of dimension 2) starting from 1. The change of variable $t = C_u$ leads to

$$A_{C_u} = t,\ A^\dagger_{C_u} = K_u = \int_0^u \frac{ds}{R_s},\ \exp W_{C_u} = R_u$$

and

$$\begin{aligned}\int_0^\infty k(t)e^{-\theta^2 t/2}dt &= \mathbb{E}\left(\int_0^\infty dt\, e^{-\theta^2 t/2}\, f(A_t)\, g(A_t^\dagger)\, h(e^{W_t})\right)\\ &= \mathbb{E}\left(\int_0^\infty \frac{du}{R_u^2} e^{-\theta^2 C_u/2}\, f(u)\, g(K_u)\, h(R_u)\right).\end{aligned}$$

The absolute continuity relationship (6.1.5) leads to

$$\int_0^\infty k(t)e^{-\theta^2 t/2}dt = \int_0^\infty du\, f(u)\, \mathbb{P}_1^{(\theta)}\left(\frac{h(R_u)g(K_u)}{R_u^{2+\theta}}\right).$$

□

As a first corollary, we give the joint law of $(\exp W_\Theta, A_\Theta)$, where Θ is an exponential random variable with parameter $\theta^2/2$, independent of W.

Corollary 6.6.2.2

$$\mathbb{P}(\exp W_\Theta \in d\rho, A_\Theta \in du) = \frac{\theta^2}{2\rho^{2+\theta}}\, p_u^{(\theta)}(1,\rho)\mathbb{1}_{\{u>0\}}\mathbb{1}_{\{\rho>0\}}d\rho du$$

where $p^{(\theta)}$ *is the transition density of a* $\text{BES}^{(\theta)}$.

PROOF: It suffices to apply the result of Lemma 6.6.2.1 with $g = 1$. □

We give a closed form expression for $\mathbb{P}(A_t^{(\nu)} \in du | B_t + \nu t = x)$. A straightforward application of Cameron-Martin's theorem proves that this expression does not depend on ν and we shall denote it by $a(t; x, u)du$.

Proposition 6.6.2.3 *If* $a(t;x,u)du = \mathbb{P}(A_t \in du | B_t = x)$, *then*

$$\frac{1}{\sqrt{2\pi t}}\exp\left(-\frac{x^2}{2t}\right)a(t;x,u) = \frac{1}{u}\exp\left(-\frac{1}{2u}(1+e^{2x})\right)\Psi_{e^x/u}(t) \qquad (6.6.1)$$

where

$$\Psi_r(t) = \frac{r}{(2\pi^3 t)^{1/2}} \exp\left(\frac{\pi^2}{2t}\right) \Upsilon_r(t), \tag{6.6.2}$$

$$\Upsilon_r(t) = \int_0^\infty dy \exp(-y^2/2t) \exp[-r(\cosh y)] \sinh(y) \sin\left(\frac{\pi y}{t}\right). \tag{6.6.3}$$

PROOF: Consider two positive Borel functions f and g. On the one hand,

$$\begin{aligned}
&\mathbb{E}\left[\int_0^\infty dt \exp\left(-\frac{\mu^2 t}{2}\right) f(\exp(W_t))\, g(A_t)\right] \\
&= \int_0^\infty dt \exp\left(-\frac{\mu^2 t}{2}\right) \int_{-\infty}^\infty \frac{dx}{\sqrt{2\pi t}} f(e^x) \exp\left(-\frac{x^2}{2t}\right) \int_0^\infty du\, g(u) a(t; x, u).
\end{aligned}$$

On the other hand, from Proposition 6.6.2.2, this quantity equals

$$\begin{aligned}
&\int_0^\infty du\, g(u) \int_0^\infty \frac{d\rho}{\rho^{\mu+2}} f(\rho)\, p_u^{(\mu)}(1, \rho) \\
&= \int_{-\infty}^\infty dx \exp[-(\mu+1)x]\, f(e^x) \int_0^\infty du\, g(u) p_u^{(\mu)}(1, e^x).
\end{aligned}$$

We obtain

$$\frac{1}{\sqrt{2\pi t}} \int_0^\infty dt \exp\left(-\frac{1}{2}\left(\mu^2 t + \frac{x^2}{t}\right)\right) a(t; x, u) = \exp\left(-(\mu+1)x\right)\, p_u^{(\mu)}(1, e^x). \tag{6.6.4}$$

Using the equality

$$I_\nu(r) = \int_0^\infty e^{-\frac{\nu^2 u}{2}} \Psi_r(u) du,$$

the explicit form of the density $p_u^{(\mu)}$ and the definition of Ψ, we write the right-hand side of (6.6.4) as

$$\frac{1}{u} \exp\left(-\frac{1}{2u}\left(1 + e^{2x}\right)\right) \int_0^\infty dt\, \Psi_{e^x/u}(t) \exp\left(-\frac{\mu^2 t}{2}\right).$$

□

Corollary 6.6.2.4 *The law of* $A_t^{(\nu)}$ *is* $\mathbb{P}(A_t^{(\nu)} \in du) = \varphi(t, u)du$, *where*

$$\varphi(t, u) = u^{\nu-1} \frac{1}{(2\pi^3 t)^{1/2}} \exp\left(\frac{\pi^2}{2t} - \frac{1}{2u} - \frac{\nu^2 t}{2}\right) \int_0^\infty dy\, y^\nu \exp(-\frac{1}{2} u y^2) \Upsilon_y(t) \tag{6.6.5}$$

where Υ *is defined in* (6.6.3).

PROOF: From the previous proposition

$$\mathbb{P}(A_t^{(\nu)} \in du) = \int_{\mathbb{R}} f(t,x)a(t,x,u)dx\,du = \varphi(t,u)du$$

where

$$f(t,x)dx = \mathbb{P}(B_t + \nu t \in dx) = \frac{1}{\sqrt{2\pi t}} \exp\left(-\frac{1}{2t}(x-\nu t)^2\right) dx\,.$$

Using the expression of $a(t,x,u)$, some obvious simplifications and a change of variable, we get

$$\varphi(t,u) = \frac{1}{u\sqrt{2\pi t}} \int_{-\infty}^{\infty} dx\, e^{-(x-\nu t)^2/2t} \exp\left(-\frac{1+e^{2x}}{2u}\right) \Psi_{e^x/u}(t)\sqrt{2\pi t}\, e^{x^2/(2t)}$$

$$= \frac{1}{u(2\pi^3 t)^{1/2}} \exp\left(\frac{\pi^2}{2t} - \frac{1}{2u} - \frac{\nu^2 t}{2}\right) \int_{-\infty}^{\infty} dx \frac{e^{x(\nu+1)}}{u} \exp\left(-\frac{e^{2x}}{2u}\right) \Upsilon_{e^x/u}(t)$$

$$= \frac{1}{u(2\pi^3 t)^{1/2}} \exp\left(\frac{\pi^2}{2t} - \frac{1}{2u} - \frac{\nu^2 t}{2}\right) \int_0^{\infty} dy\, (yu)^{\nu} \exp\left(-\frac{1}{2}uy^2\right) \Upsilon_y(t)\,.$$

□

One can invert the Laplace transform of the pair (e^{W_t}, A_t) given in Corollary 6.6.2.2 and we obtain:

Corollary 6.6.2.5 *Let W be a Brownian motion and $A_t = \int_0^t e^{2W_s} ds$. Then, for any positive Borel function f,*

$$\mathbb{E}(f(e^{W_t}, A_t)) = \frac{1}{(2\pi^3 t)^{1/2}} \int_0^{\infty} dy \int_0^{\infty} dv\, f(y, \frac{1}{v})\, e^{-v(1+y^2)/2} \Upsilon_{yv}(t) \quad (6.6.6)$$

where the function Υ was defined in (6.6.3).

Exercise 6.6.2.6 Prove that the density of the pair $(\exp(W_\Theta + \nu\Theta), A_\Theta^{(\nu)})$, where $A_t^{(\nu)} = \int_0^t e^{2(W_s+\nu s)} ds$, is

$$\frac{\theta^2}{2x^{2+\lambda}} x^{\nu}\, p_a^{(\lambda)}(1,x) \mathbb{1}_{\{x>0\}} \mathbb{1}_{\{a>0\}} dx da$$

with $\lambda^2 = \theta^2 + \nu^2$. ◁

Proposition 6.6.2.7 *Let W be a BM, $A_t = \int_0^t e^{2W_s} ds$ and $A_t^\dagger = \int_0^t e^{W_s} ds$. The Laplace transform of $(W_\Theta, A_\Theta, A_\Theta^\dagger)$, where Θ is an exponential random variable with parameter $\theta^2/2$, independent of W, is*

$$H_{a,b,c}(\theta) = \mathbb{E}\left(\exp(-aW_\Theta - \frac{b^2}{2} A_\Theta - cA_\Theta^\dagger)\right)$$

$$= \frac{\theta^2}{2} \frac{4^{1+\theta}}{\Gamma(1+\theta)} \int_0^{\infty} dt e^{-ct} \int_0^{\infty} dy\, J_4^{y+a/4,\, b/2,\, 2\theta}(t)\, y^{\theta}$$

where J_x (here, $x = 4$) was computed in Proposition 6.2.5.3 *as the Laplace transform of the pair* $(\rho_t, \int_0^t \rho_s ds)$ *for a squared Bessel process with index* ν, *starting from* x *as*

$$J_x^{a,b,\nu}(t) = \left(\cosh(bt) + 2ab^{-1}\sinh(bt)\right)^{-\nu-1} \exp\left(-\frac{1}{2}xb\frac{1+2ab^{-1}\coth(bt)}{\coth(bt)+2ab^{-1}}\right).$$

PROOF: We start with the formula established in Lemma 6.6.2.1:

$$\int_0^\infty k(t)e^{-\theta^2 t/2}dt = \mathbb{P}^{(\theta)}\left(\int_0^\infty du\, f(u)\frac{h(R_u)g(K_u)}{R_u^{2+\theta}}\right).$$

If R is a Bessel process with index θ, then, from Exercice 6.2.4.3 with $q = 2$, $R_t = \frac{1}{4}\widehat{R}^2(\int_0^t \frac{ds}{R_s})$, where $\widehat{R}$ is a BES with index 2θ. Using the notation of Lemma 6.6.2.1 and introducing the inverse H of the increasing process K as $H_t = \inf\{u : K_u = t\}$,

$$K_{H_t} = t = \int_0^{H_t} \frac{ds}{R_s}.$$

By differentiation, we obtain $dH_t = R_{H_t}dt$ and $H_t = \frac{1}{4}\int_0^t \widehat{R}_s^2 ds$. It follows that

$$\int_0^\infty k(t)e^{-\theta^2 t/2}dt = 4^{1+\theta}\int_0^\infty dt\, g(t)\,\mathbb{P}^{(2\theta)}\left(\frac{h(4^{-1}\widehat{R}_t^2)f(4^{-1}\int_0^t \widehat{R}_s^2 ds)}{\widehat{R}_t^{2(1+\theta)}}\right).$$

In particular, for $f(x) = e^{-4bx}, g(x) = e^{-cx}$ and $h(x) = e^{-4ax}$, we obtain

$$\begin{aligned}&\mathbb{E}(\exp(-aW_\Theta - bA_\Theta - cA_\Theta^\dagger))\\ &= 4^{1+\theta}\int_0^\infty dt\, e^{-ct}\,\mathbb{P}^{2\theta}\left(\frac{\exp(-a\widehat{R}_t^2 - b\int_0^t \widehat{R}_s^2 ds)}{\widehat{R}_t^{2(1+\theta)}}\right).\end{aligned}$$

Using the identity $r^{-\gamma} = \frac{1}{\Gamma(\gamma)}\int_0^\infty dy\, e^{-ry}y^{\gamma-1}$ we transform the quantity $\frac{1}{\widehat{R}_t^{2(1+\theta)}}$. The initial condition $R_0 = 1$ implies $\widehat{R}_0 = 2$ and, denoting $\rho_t = \widehat{R}_t^2$,

$$\begin{aligned}&\mathbb{P}^{(2\theta)}\left(\frac{\exp(-a\widehat{R}_t^2 - b\int_0^t \widehat{R}_s^2 ds)}{\widehat{R}_t^{2(1+\theta)}}\right)\\ &= \frac{1}{\Gamma(1+\theta)}\mathbb{Q}_4^{(2\theta)}\left(\int dy\, y^\theta \exp(-(a+y)\rho_t - b\int_0^t \rho_s ds)\right).\end{aligned}$$

From 6.2.14 we know the Laplace transform of the pair $(\rho_t, \int_0^t \rho_s ds)$, where ρ is a BESQ of index 2θ, starting from $\rho_0 = 4$. □

Exercise 6.6.2.8 Check that the distribution of $A_\Theta^{(\nu)}$ is that of $B/(2\Gamma)$ where B has a Beta$(1,\alpha)$ law and Γ a Gamma$(\beta, 1)$ law with

$$\alpha = \frac{\nu+\gamma}{2},\ \beta = \frac{\gamma-\nu}{2},\ \gamma = \sqrt{2\lambda+\nu^2}.$$

See Dufresne [282]. ◁

6.6.3 The Moments of A_t

The moments of $A_t^{(\nu)}$ exist because

$$te^{-2t\nu^- + 2m_t} \leq A_t^{(\nu)} \leq te^{2t\nu^+ + 2M_t}$$

where $m_t = \inf_{s \leq t} W_s$, $M_t = \sup_{s \leq t} W_s$.

Elementary arguments allow us to compute the moments of $A_t = A_t^{(0)}$.

Proposition 6.6.3.1 *The moments of the random variable A_t are given by $\mathbb{E}(A_t^n) = \frac{1}{4^n}\mathbb{E}(P_n(e^{2W_t}))$, where P_n is the polynomial*

$$P_n(z) = 2^n(-1)^n \left(\frac{1}{n!} + 2\sum_{j=1}^{n} \frac{n!(-z)^j}{(n-j)!\,(n+j)!} \right).$$

PROOF: Let $\mu \geq 0$ and $\lambda > \varphi(\mu + n)$, where $\varphi(x) = \frac{1}{2}x^2$. Then,

$$\begin{aligned}
\Phi_n(t,\mu) &:= \mathbb{E}\left(\left(\int_0^t ds\, e^{W_s} \right)^n e^{\mu W_t} \right) \\
&= n! \int_0^t ds_1 \int_0^{s_1} ds_2 \cdots \int_0^{s_{n-1}} ds_n \mathbb{E}\left[\exp(W_{s_1} + \cdots + W_{s_n} + \mu W_t)\right].
\end{aligned}$$

The expectation under the integral sign is easily computed, using the independent increments property of the BM as well as the Laplace transform

$$\mathbb{E}\left[\exp(W_{s_1} + \cdots + W_{s_n} + \mu W_t)\right] =$$

$$\exp\left[\varphi(\mu)(t - s_1) + \varphi(\mu+1)(s_1 - s_2) + \cdots + \varphi(\mu+n)s_n\right].$$

It follows, by integrating successively the exponential functions that

$$\int_0^\infty dt e^{-\lambda t}\mathbb{E}\left[\left(\int_0^t ds\, e^{W_s} \right)^n e^{\mu W_t} \right] = \frac{n!}{\prod_{j=0}^{n} (\lambda - \varphi(\mu + j))}.$$

Setting, for fixed j, $c_j^{(\mu)} = \prod_{0 \leq k \neq j \leq n} (\varphi(\mu + j) - \varphi(\mu + k))^{-1}$, the use of the formula

$$\frac{1}{\prod_{j=0}^{n}(\lambda - \varphi(\mu + j))} = \sum_{j=0}^{n} c_j(\mu) \frac{1}{\lambda - \varphi(\mu + j)}$$

and the invertibility of the Laplace transform lead to

$$\mathbb{E}\left[\left(\int_0^t ds\, e^{W_s} \right)^n e^{\mu W_t} \right] = \mathbb{E}(e^{\mu W_t} P_n^{(\mu)}(e^{W_t}))$$

where $P_n^{(\mu)}$ is the sequence of polynomials

$$P_n^{(\mu)}(z) = n! \sum_{j=0}^{n} c_j^{(\mu)} z^j .$$

In particular, we obtain

$$\mathbb{E}\left[\left(\int_0^t dse^{W_s}\right)^n\right] = \mathbb{E}(P_n^{(0)}(e^{W_t}))$$

and, from the scaling property of the BM,

$$\alpha^{2n}\mathbb{E}\left[\left(\int_0^t dse^{\alpha W_s}\right)^n\right] = \mathbb{E}(P_n^{(0)}(e^{\alpha W_t})) .$$

□

Therefore, we have obtained the moments of $A_t^{(0)}$. The general case follows using Girsanov's theorem.

$$\mathbb{E}([A_t^{(\nu)}]^n) = \frac{n!}{2^{2n}}\left(\sum_{j=0}^{n} c_j^{(\nu/2)} \exp\left[(2j^2 + 2j\nu)t\right]\right) .$$

Nevertheless, knowledge of the moments is not enough to characterize the law of A_t. Recall the following result:

Proposition 6.6.3.2 (Carleman's Criterion.) *If a random variable X satisfies $\sum(m_{2n})^{-1/2n} = \infty$ where $m_{2n} = \mathbb{E}(X^{2n})$, then its distribution is determined by its moments.*

However, this criterion does not apply to the moments of $A_t^{(\nu)}$ (see Geman and Yor [383]). The moments of A_t do not characterize its law (see Hörfelt [446] and Nikeghbali [673]). On the other hand, Dufresne [279] proved that the law of $1/A_t^{(\nu)}$ is determined by its moments. We recall that the log-normal law is not determined by its moments.

Comment 6.6.3.3 A computation of positive and negative moments can be found in Donati-Martin et al. [259]. See also Dufresne [279, 282], Ramakrishnan [728] and Schröder [771].

6.6.4 Laplace Transform Approach

We now return to the computation of the price of an Asian option, i.e., to the computation of

$$\Psi(T, K) = \mathbb{E}\left[\left(\int_0^T \exp[2(W_s + \mu s)]\, ds - K\right)^+\right] . \qquad (6.6.7)$$

Indeed, as seen in the beginning of Section 6.6, one can restrict attention to the case $\sigma = 2$ since

$$\begin{aligned} C^{\text{Asian}}(S_0, K) &= e^{-rT}\frac{S_0}{T}\mathbb{E}\left[\left(\int_0^T e^{\sigma(W_s+\nu s)}\,ds - \frac{KT}{S_0}\right)^+\right] \\ &= e^{-rT}\frac{4S_0}{\sigma^2 T}\Psi\left(\frac{\sigma^2 T}{4}, \frac{KT\sigma^2}{4S_0}\right). \end{aligned}$$

The Geman and Yor method consists in computing the Laplace transform (with respect to the maturity) of Ψ, i.e.,

$$\begin{aligned} \Phi(\lambda) &= \int_0^\infty dt\, e^{-\lambda t}\Psi(t, K) = \mathbb{E}\left[\int_0^\infty dt\, e^{-\lambda t}\left(\int_0^t e^{2(W_s+\mu s)}\,ds - K\right)^+\right] \\ &= \mathbb{E}\left(\int_0^\infty dt\, e^{-\lambda t}(A_t^{(\mu)} - K)^+\right). \end{aligned}$$

Lamperti's result (Theorem 6.2.4.1) and the change of time $A_t^{(\mu)} = u$ yield to

$$\Phi(\lambda) = \mathbb{P}_1^{(\mu)}\left[\int_0^\infty du\, e^{-\lambda C_u}\frac{1}{(R_u)^2}(u-K)^+\right]$$

where C is the inverse of A. From the absolute continuity of Bessel laws (6.1.5)

$$\mathbb{P}_1^{(\mu)}|_{\mathcal{F}_t} = (R_t)^{\mu-\gamma}\exp\left(-\frac{\mu^2-\gamma^2}{2}C_t\right)\mathbb{P}_1^{(\gamma)}|_{\mathcal{F}_t}$$

with γ given by $\lambda = \frac{1}{2}(\gamma^2 - \mu^2)$ and

$$\Phi(\lambda) = \mathbb{P}_1^{(\gamma)}\left(\int_0^\infty du\, \frac{1}{R_u^{2+\gamma-\mu}}(u-K)^+\right).$$

The transition density of a Bessel process, given in (6.2.3), now leads to

$$\Phi(\lambda) = \int_K^\infty du \frac{u-K}{u}\int_0^\infty \frac{d\rho}{\rho^{1-\mu}}\exp\left(-\frac{1+\rho^2}{2u}\right)I_\mu\left(\frac{\rho}{u}\right).$$

It remains to invert the Laplace transform. □

Comment 6.6.4.1 Among the papers devoted to the study of the law of the integral of a geometric BM and Asian options we refer to Buecker and Kelly-Lyth [135], Carr and Schröder [156], Donati-Martin et al. [258], Dufresne [279, 280, 282], Geman and Yor [382, 383], Linetsky [594], Lyasoff [606], Yor [863], Schröder [767, 769], and Večeř and Xu [827]. Nielsen and Sandmann [672] study the pricing of Asian options under stochastic interest rates. In this book, we shall not consider Asian options on interest rates. A reference is Poncet and Quittard-Pinon [722].

Exercise 6.6.4.2 Compute the price of an Asian option in a Bachelier framework, i.e., compute

$$\mathbb{E}\left(\left(\int_0^T (\nu s + \sigma W_s) ds - K\right)^+\right).$$

◁

Exercise 6.6.4.3 Prove that, for fixed t,

$$A_t^{(\nu)} \stackrel{\text{law}}{=} \int_0^t e^{2(\nu(t-s)+W_t-W_s}ds := Y_t^{(\nu)}$$

and that, as a process

$$dY_t^{(\nu)} = (2(\nu+1)Y_t^{(\nu)} + 1)dt + 2Y_t^{(\nu)}dW_t\,.$$

See Carmona et al. [141], Donati-Martin et al. [258], Dufresne [277] and Proposition 11.2.1.7. ◁

6.6.5 PDE Approach

A second approach to the evaluation problem, studied in Stanton [805], Rogers and Shi [740] and Alziary et al. [11] among others, is based on PDE methods and the important fact that the pair (S_t, Y_t) is Markovian where

$$Y_t := \frac{1}{S_t}\left(\frac{1}{T}\int_0^t S_u\, du - K\right).$$

The value C_t^{Asian} of an Asian option is a function of the three variables: t, S_t and Y_t, i.e., $C_t^{\text{Asian}} = S_t \mathcal{A}(t, Y_t)$ and, from the martingale property of $e^{-rt}C_t^{\text{Asian}}$, we obtain that $\mathcal{A}$ is the solution of

$$\frac{\partial \mathcal{A}}{\partial t} + \left(\frac{1}{T} - ry\right)\frac{\partial \mathcal{A}}{\partial y} + \frac{1}{2}\sigma^2 y^2 \frac{\partial^2 \mathcal{A}}{\partial y^2} = 0 \tag{6.6.8}$$

with the boundary condition $\mathcal{A}(T, y) = y^+$.

Furthermore, the hedging portfolio is $\mathcal{A}(t, Y_t) - Y_t\mathcal{A}'_y(t, Y_t)$. Indeed

$$\begin{aligned} d(e^{-rt}C_t^{\text{Asian}}) &= \sigma S_t e^{-rt}\left(\mathcal{A}(t, Y_t) - Y_t\mathcal{A}'_y(t, Y_t)\right) dW_t \\ &= \left(\mathcal{A}(t, Y_t) - Y_t\mathcal{A}'_y(t, Y_t)\right) d\tilde{S}_t\,. \end{aligned}$$

6.7 Stochastic Volatility

6.7.1 Black and Scholes Implied Volatility

In a Black and Scholes model, the prices of call options with different strikes and different maturities are computed with the same value of the volatility. However, given the observed prices of European calls $C_{\text{obs}}(S_0, K, T)$ on an underlying with value S_0, with maturity T and strike K, the Black and Scholes **implied volatility** is defined as the value σ_{imp} of the volatility which, substituted in the Black and Scholes formula, gives equality between the Black and Scholes price and the observed price, i.e.,

$$\mathcal{BS}(S_0, \sigma_{\text{imp}}, K, T) = C_{\text{obs}}(S_0, K, T)\,.$$

Now, this parameter σ_{imp} depends on S_0, T and K as we just wrote. If the Black and Scholes assumption were satisfied, this parameter would be constant for all maturities and all strikes, and, for fixed S_0, the volatility surface $\sigma_{\text{imp}}(T, K)$ would be flat. This is not what is observed. For currency options, the profile is often symmetric in moneyness $m = K/S_0$. This is the well-known smile effect (see Hagan et al. [417]). We refer to the work of Crépey [207] for more information on smiles and implied volatilities. A way to produce smiles is to introduce stochastic volatility. Stochastic volatility models are studied in details in the books of Lewis [587], Fouque et al. [356].

Here, we present some attempts to solve the problem of option pricing for models with stochastic volatility.

6.7.2 A General Stochastic Volatility Model

This section is devoted to some examples of models with stochastic volatility. Let us mention that a model

$$dS_t = S_t(\mu_t dt + \sigma_t d\widetilde{W}_t)$$

where[1] $\widetilde{W}$ is a BM and μ, σ are $\mathbf{F}^{\widetilde{W}}$-adapted processes is not called a stochastic volatility model, this name being generally reserved for the case where the volatility induces a new source of noise. The main models of stochastic volatility are of the form

$$dS_t = S_t(\mu_t dt + \sigma(t, Y_t) d\widetilde{W}_t)$$

where μ is $\mathbf{F}^{\widetilde{W}}$-adapted and

$$dY_t = a(t, Y_t)dt + b(t, Y_t)d\widetilde{B}_t$$

[1] Throughout our discussion, we shall use tildes and hats for intermediary BMs, whereas W and $W^{(1)}$ will denote our final pair of independent BMs.

(or, equivalently,

$$df(Y_t) = \alpha(t, Y_t)dt + \beta(t, Y_t)d\widetilde{B}_t$$

for a smooth function f) where $\widetilde{B}$ is a Brownian motion, correlated with $\widetilde{W}$ (with correlation ρ). The independent case ($\rho = 0$) is an interesting model, but more realistic ones involve a correlation $\rho \neq 0$.Some authors add a jump component to the dynamics of Y (see e.g., the model of Barndorff-Nielsen and Shephard [55]).

In the Hull and White model [453], $\sigma(t, y) = y$ and Y follows a geometric Brownian motion. We shall study this model in the next section. In the Scott model [775],

$$dS_t = S_t(\mu_t dt + Y_t d\widetilde{W}_t)$$

where $Z = \ln(Y)$ follows a Vasicek process:

$$dZ_t = \beta(a - Z_t)dt + \lambda d\widetilde{B}_t$$

where a, β and λ are constant. Heston [433] relies on a square root process for the square of the volatility.

Obviously, if the volatility is not a traded asset, the model is incomplete, and there exist infinitely many e.m.m's, which may be derived as follows. In a first step, one introduces a BM $\widehat{W}$, independent of $\widetilde{W}$ such that $\widetilde{B}_t = \rho\widetilde{W}_t + \sqrt{1-\rho^2}\,\widehat{W}_t$. Then, any $\sigma(\widetilde{B}_s, \widetilde{W}_s, s \leq t) = \sigma(\widetilde{W}_s, \widehat{W}_s, s \leq t)$-martingale can be written as a stochastic integral with respect to the pair $(\widetilde{W}, \widehat{W})$. Therefore, any Radon-Nikodým density satisfies

$$dL_t = L_t(\phi_t d\widetilde{W}_t + \gamma_t d\widehat{W}_t)\,,$$

for some pair of predictable processes ϕ, γ. We then look for conditions on the pair (ϕ, γ) such that the discounted price SR is a martingale under $\mathbb{Q} = L\mathbb{P}$, that is if the process LSR is a $\mathbb{P}$-local martingale. This is the case if and only if $-r + \mu_t - \sigma(t, Y_t)\phi_t = 0$. This involves no restriction on the coefficient γ other than $\int_0^t \gamma_s^2 ds < \infty$ and the local martingale property of the process LSR.

6.7.3 Option Pricing in Presence of Non-normality of Returns: The Martingale Approach

We give here a second motivation for introducing stochastic volatility models. The property of non-normality of stock or currency returns which has been observed and studied in many articles is usually taken into account by relying either on a stochastic volatility or on a mixed jump diffusion process for the price dynamics (see ↣ Chapter 10).

Strong underlying assumptions concerning the information arrival dynamics determine the choice of the model. Indeed, a stochastic volatility model will be adapted to a continuous information flow and mixed jump-diffusion processes will correspond to possible discontinuities in this flow of information. In this context, option valuation is difficult. Not only is standard risk-neutral valuation usually no longer possible, but these models rest on the valuation of more parameters. In spite of their complexity, some semi-closed-form solutions have been obtained.

Let us consider the following dynamics for the underlying (a currency):

$$dS_t = S_t \left(\mu dt + \sigma_t d\widetilde{B}_t\right) , \tag{6.7.1}$$

and for its volatility:

$$d\sigma_t = \sigma_t \left(f(\sigma_t)dt + \gamma d\widetilde{W}_t\right) . \tag{6.7.2}$$

Here, $(\widetilde{B}_t, t \geq 0)$ and $(\widetilde{W}_t, t \geq 0)$ are two correlated Brownian motions under the historical probability, and the parameter γ is a constant.

In the Hull and White model [453] the underlying is a stock and the function f is constant, hence σ follows a geometric Brownian motion. In the Scott model [775], this function has the form: $f(\sigma) = \beta(a - ln(\sigma)) + \gamma^2/2$, where the parameters a, β and γ are constant.

The standard Black and Scholes approach of riskless arbitrage is not enough to produce a unique option pricing function C_E. Indeed, the volatility is not a traded asset and there is no asset perfectly correlated with it. The three assets required in order to eliminate the two sources of uncertainty and to create a riskless portfolio will be the foreign bond and, for example, two options of different maturities. Therefore, it will be impossible to determine the price of an option without knowing the price of another option on the same underlying (See Scott [775]).

Under any risk-adjusted probability $\mathbb{Q}$, the dynamics of the underlying spot price and of the volatility are given by

$$\begin{cases} dS_t = S_t \left((r-\delta)dt + \sigma_t d\widehat{B}_t\right) , \\ d\sigma_t = \sigma_t(f(\sigma_t) - \Phi_t^\sigma)dt + \gamma\sigma_t dW_t \end{cases} \tag{6.7.3}$$

where $(\widehat{B}_t, t \geq 0)$ and $(W_t, t \geq 0)$ are two correlated $\mathbb{Q}$-Brownian motions and where Φ_t^σ, the risk premium associated with the volatility, is unknown since the volatility is not traded.

The expression of the underlying price at time T

$$S_T = S_0 \exp\left((r-\delta)T - \frac{1}{2}\int_0^T \sigma_u^2 du + \int_0^T \sigma_u d\widehat{B}_u\right) , \tag{6.7.4}$$

which follows from (6.7.3), will be quite useful in pricing European options as is now detailed.

We first start with the case of 0 correlation between $\widehat{B}$ and W.

Proposition 6.7.3.1 *Assume that the dynamics of the underlying spot price (for instance a currency) and of the volatility are given, under the risk-adjusted probability, by equations (6.7.3), with a zero correlation coefficient, where the risk premium Φ^σ associated with the volatility is assumed constant. The European call price can be written as follows:*

$$C_E(S_0, \sigma_0, T) = \int_0^{+\infty} \mathcal{BS}(S_0 e^{-\delta T}, a, T) dF(a) \tag{6.7.5}$$

where $\mathcal{BS}$ is the Black and Scholes price:

$$\mathcal{BS}(x, a, T) = x\mathcal{N}(d_1(x, a)) - Ke^{-rT}\mathcal{N}(d_2(x, a)) \, .$$

Here,

$$d_1(x, a) = \frac{\ln(x/K) + rT + a/2}{\sqrt{a}}, \quad d_2(x, a) = d_1(x, a) - \sqrt{a} \, ,$$

δ is the foreign interest rate and F is the distribution function (under the risk-adjusted probability) of the cumulative squared volatility Σ_T defined as

$$\Sigma_T = \int_0^T \sigma_u^2 \, du \, . \tag{6.7.6}$$

PROOF: The stochastic integral which appears on the right-hand side of (6.7.4) is a stochastic time-changed Brownian motion:

$$\int_0^t \sigma_u d\widehat{B}_u = B^*_{\Sigma_t}$$

where $(B^*_s, s \geq 0)$ is a $\mathbb{Q}$-Brownian motion. Therefore,

$$S_T = S_0 \exp\left((r - \delta)T + B^*_{\Sigma_T} - \frac{\Sigma_T}{2} \right)$$

and the conditional law of $\ln(S_T/S_0)$ given Σ_T is

$$\mathcal{N}\left((r - \delta)T - \frac{\Sigma_T}{2}, \Sigma_T \right) .$$

By conditioning with respect to Σ_T, returns are normally distributed, and the Black and Scholes formula can be used for the computation of the conditional expectation

$$\widehat{C}_T := \mathbb{E}_{\mathbb{Q}}(e^{-rT}(S_T - K)^+ | \Sigma_T) \, .$$

Formula (6.7.5) is therefore obtained from

$$\mathbb{E}_{\mathbb{Q}}(e^{-rT}(S_T - K)^+) = \mathbb{E}_{\mathbb{Q}}(\widehat{C}_T) = \mathbb{E}_{\mathbb{Q}}(\mathcal{BS}(S_0, \Sigma_T, K)).$$

In order to use this result, the risk-adjusted distribution function F of Σ_T is needed. One method of approximating the option price is the Monte Carlo method. By simulating the instantaneous volatility process σ over discrete intervals from 0 to T, the random variable Σ_T can be simulated. Each value of Σ_T is substituted into the Black and Scholes formula with Σ_T in place of $\sigma^2 T$. The sample mean converges in probability to the option price as the number of simulations increases to infinity. The estimates for the parameters of the volatility process could be computed by methods described in Scott [775] and in Chesney and Scott [177] for currencies: the method of moments and the use of ARMA processes. The risk premium associated with the volatility, i.e., Φ^σ, which is assumed to be a constant, should also be estimated.

6.7.4 Hull and White Model

Hull and White [453] consider a stock option ($\delta = 0$); they assume that the volatility follows a geometric Brownian motion and that it is uncorrelated with the stock price and has zero systematic risk. Hence, the drift $f(\sigma_t)$ of the volatility is a constant k and Φ_σ is taken to be null:

$$d\sigma_t = \sigma_t(kdt + \gamma dW_t)\,. \tag{6.7.7}$$

They also introduce the following random variable: $V_T = \Sigma_T/T$. In this framework, they obtain another version of equation (6.7.5):

$$C_E(S_0, \sigma_0, T) = \int_0^{+\infty} \mathcal{BS}(S_0, vT, T) dG(v) \tag{6.7.8}$$

where G is the distribution function of V_T.

Approximation

By relying on a Taylor expansion, the left-hand side of (6.7.8) may be approximated as follows: introduce $\mathcal{C}(y) = BS(S_0, yT, T)$, then

$$C_E(S_0, \sigma_0, T) \approx \mathcal{C}(\nabla_T) + \frac{1}{2}\frac{d^2\mathcal{C}}{dy^2}(\nabla_T)\mathrm{Var}(V_T) + \cdots$$

where

$$\nabla_T := \frac{\mathbb{E}(\Sigma_T)}{T} = \mathbb{E}(V_T)\,.$$

We recall the following results:

$$\begin{cases} \mathbb{E}(V_T) = \dfrac{e^{\kappa T} - 1}{\kappa T}\sigma_0^2 \\ \mathbb{E}(V_T^2) = \left(\dfrac{2e^{(2\kappa + \vartheta^2)T}}{(\kappa + \vartheta^2)(2\kappa + \vartheta^2)T^2} + \dfrac{2}{\kappa T^2}\left(\dfrac{1}{2\kappa + \vartheta^2} - \dfrac{e^{\kappa T}}{\kappa + \vartheta^2}\right)\right)\sigma_0^4 \end{cases} \tag{6.7.9}$$

where κ and ϑ are respectively the drift and the volatility of the squared volatility:

$$\kappa = 2k + \gamma^2, \quad \vartheta = 2\gamma\,.$$

When κ is zero, i.e., when the squared volatility is a martingale, the following approximation is obtained:

$$C_E(S_0, \sigma_0, T) \approx \mathcal{C}(\nabla_T) + \frac{1}{2}\frac{d^2\mathcal{C}}{dy^2}(\nabla_T)\mathrm{Var}(V_T) + \frac{1}{6}\frac{d^3\mathcal{C}}{dy^3}(\nabla_T)\mathrm{Skew}(V_T) + \cdots$$

where $\mathrm{Skew}(V_T)$ is the third central moment of V_T. The first three moments are obtained by relying on the following formulae (note that the first two moments are the limits obtained from (6.7.9) as κ goes to 0) :

$$\begin{aligned}
\mathbb{E}(V_T) &= \sigma_0^2\,,\\
\mathbb{E}(V_T^2) &= \frac{2(e^{\vartheta^2 T} - \vartheta^2 T - 1)}{\vartheta^4 T^2}\sigma_0^4\,,\\
\mathbb{E}(V_T^3) &= \frac{e^{3\vartheta^2 T} - 9e^{\vartheta^2 T} + 6\vartheta^2 T + 8}{3\vartheta^6 T^3}\sigma_0^6\,.
\end{aligned}$$

Closed-form Solutions in the Case of Uncorrelated Processes

As we have explained above in Proposition 6.7.3.1, in the case of uncorrelated Brownian motions, the knowledge of the law of Σ_T yields the price of a European option, at least in a theoretical way. In fact, this law is quite complicated, and a closed-form result is given as a double integral.

Proposition 6.7.4.1 *Let $(\sigma_t, t \geq 0)$ be the GBM solution of (6.7.7) and $\Sigma_T = \int_0^T \sigma_s^2 ds$. The density of Σ_T is $\mathbb{Q}(\Sigma_T \in dx)/dx = g(x)$ where*

$$\begin{aligned}
g(x) &= \frac{1}{x}\left(\frac{x\gamma^2}{\sigma_0^2}\right)^{\nu} \frac{1}{(2\pi^3\gamma^2 T)^{1/2}} \exp\left(\frac{\pi^2}{2\gamma^2 T} - \frac{\sigma_0^2}{2\gamma^2 x} - \frac{\nu^2\gamma^2 T}{2}\right)\\
&\quad \times \int_0^\infty dy\, y^\nu \exp\left(-\frac{1}{2\sigma_0^2}\gamma^2 x y^2\right)\Upsilon_y(\gamma^2 T)
\end{aligned}$$

where the function Υ_y is defined in (6.6.3) and where $\nu = \frac{k}{\gamma^2} - \frac{1}{2}$.

PROOF: From the definition of σ_t, we have

$$\sigma_t^2 = \sigma_0^2 e^{2(\gamma W_t + (k - \frac{\gamma^2}{2})t)}\,,$$

hence $\mathbb{Q}(\Sigma_T \in dx) = \frac{1}{\sigma_0^2} h(T, \frac{x}{\sigma_0^2})\, dx$, where

$$h(T, x)dx = \mathbb{Q}\left(\int_0^T dt\, e^{2(\gamma W_t + (k - \frac{\gamma^2}{2})t)} \in dx\right).$$

From the scaling property of the Brownian motion, setting $\nu = (k-\gamma^2/2)\gamma^{-2}$,

$$\int_0^T e^{2(\gamma W_t+(k-\frac{\gamma^2}{2})t)}dt \overset{\text{law}}{=} \int_0^T e^{2(W^*_{t\gamma^2}+t\gamma^2\nu)}dt = \frac{1}{\gamma^2}\int_0^{\gamma^2 T} e^{2(W^*_s+s\nu)}ds\,,$$

where W^* is a $\mathbb{Q}$-Brownian motion. Therefore,

$$h(T,x) = \gamma^2\varphi(\gamma^2 T, x\gamma^2)$$

where

$$\varphi(t,x)dx = \mathbb{Q}\left(\int_0^t ds\, e^{2(W^*_s+s\nu)} \in dx\right)\,.$$

Now, using (6.6.5),

$$\varphi(t,x) = x^{\nu-1}\frac{1}{(2\pi^3 t)^{1/2}}\exp\left(\frac{\pi^2}{2t}-\frac{1}{2x}-\frac{\nu^2 t}{2}\right)\int_0^\infty dy\, y^\nu \exp(-\frac{1}{2}xy^2)\Upsilon_y(t)\,.$$

□

6.7.5 Closed-form Solutions in Some Correlated Cases

We now present the formula for a European call in a closed form (up to the computation of some integrals). We consider the general case where the correlation between the two Brownian motions $\widetilde{W},\widetilde{B}$ given in (6.7.1, 6.7.2) under the historical probability (hence between the risk-neutral Brownian motions $W,\widehat{B}$ defined in (6.7.3)) is equal to ρ. Thus, we write the risk-neutral dynamics of the stock price ($\delta = 0$) and the volatility process as

$$\begin{cases} dS_t = S_t\left(rdt + \sigma_t(\sqrt{1-\rho^2}\,dW^{(1)}_t + \rho\,dW_t)\right), \\ d\sigma_t = \sigma_t(kdt+\gamma\,dW_t), \end{cases} \tag{6.7.10}$$

where the two Brownian motions $(W^{(1)},W)$ are independent. In an explicit form

$$\ln(S_T/S_0) = rt - \frac{1}{2}\int_0^t \sigma_s^2 ds + \rho\int_0^t \sigma_s dW_s + \sqrt{1-\rho^2}\int_0^t \sigma_s dW^{(1)}_s \tag{6.7.11}$$

We first present the case where the two Brownian motions $W,\widehat{B}$ are independent, i.e., when $\rho = 0$.

Proposition 6.7.5.1 ***Case*** $\rho = 0$: *Assume that, under the risk-neutral probability*

$$\begin{cases} dS_t = S_t\left(r\,dt + \sigma_t\,dW^{(1)}_t\right), \\ d\sigma_t = \sigma_t(k\,dt+\gamma\,dW_t), \end{cases} \tag{6.7.12}$$

where the two Brownian motions $(W^{(1)}, W)$ *are independent. The price of a European call, with strike* K *is*

$$C = S_0 f_1(T) - Ke^{-rT} f_2(T)$$

where the functions f_j *are defined as*

$$f_1(T) := \mathbb{E}(\mathcal{N}(d_1(\Sigma_T))) = \int_0^\infty \mathcal{N}(d_1(x))g(x)dx$$

and

$$f_2(T) := \mathbb{E}(\mathcal{N}(d_2(\Sigma_T))) = \int_0^\infty \mathcal{N}(d_2(x))g(x)dx$$

where the function g *is defined in Proposition 6.7.4.1.*

PROOF: The solution of the system (6.7.12) is:

$$\begin{cases} S_t = S_0 e^{rt} \exp\left(\int_0^t \sigma_s dW_s^{(1)} - \frac{1}{2}\int_0^t \sigma_s^2 ds\right), \\ \sigma_t = \sigma_0 \exp(\gamma W_t + (k - \gamma^2/2)t) = \sigma_0 \exp(\gamma W_t + \gamma^2 \nu t), \end{cases} \tag{6.7.13}$$

where $\nu = k/\gamma^2 - 1/2$. Conditionally on $\mathbf{F}^W$, the process $\ln S$ is Gaussian, and $\ln(S_T/S_0)$ is a Gaussian variable with mean $rT - \Sigma_T/2$ and variance Σ_T where $\Sigma_T = \int_0^T \sigma_s^2 ds$. It follows that

$$C = S_0 \mathbb{E}\left(\mathcal{N}(d_1(\Sigma_T))\right) - Ke^{-rt}\mathbb{E}\left(\mathcal{N}(d_2(\Sigma_T))\right)$$

By relying on Proposition 6.7.4.1, the result is obtained. □

We now consider the case $\rho \neq 0$, but $k = 0$, i.e., the volatility is a martingale.

Proposition 6.7.5.2 ***(Case*** $k = 0$***)*** *Assume that*

$$\begin{cases} dS_t = S_t\left(rdt + \sigma_t(\sqrt{1-\rho^2}\, dW_t^{(1)} + \rho\, dW_t)\right), \\ d\sigma_t = \gamma \sigma_t\, dW_t\,. \end{cases}$$

with $\rho \neq 1$. *The price of a European call with strike* K *is given by*

$$C = S_0 f_1^*(\gamma^2 T) - Ke^{-rT} f_2^*(\gamma^2 T)$$

where the functions f_j^*, *defined as* $f_j^*(t) := \mathbb{E}(f_j(e^{W_t}, A_t))$, $j = 1, 2$, *where* $A_t = \int_0^t e^{2W_s} ds$, *are obtained as in Equation (6.6.6).Here, the functions* $f_j(x, y)$ *are:*

$$f_1(x, y) = e^{-\gamma^2 T/8} \frac{1}{\sqrt{x}} e^{\rho\sigma_0(x-1)/\gamma} e^{-\sigma_0^2 \rho^2 y/(2\gamma^2)} \mathcal{N}(d_1^*(x, y)),$$

$$f_2(x, y) = e^{-\gamma^2 T/8} \frac{1}{\sqrt{x}} \mathcal{N}(d_2^*(x, y)),$$

where

$$d_1^*(x,y) = \frac{\gamma}{\sigma_0\sqrt{(1-\rho^2)y}}\left(\ln\frac{S_0}{K} + rT + \frac{\rho\sigma_0(x-1)}{\gamma} + \frac{\sigma_0^2 y(1-\rho^2)}{2\gamma^2}\right),$$
$$d_2^*(x,y) = d_1^*(x,y) - \frac{\sigma_0}{\gamma}\sqrt{(1-\rho^2)y}\,.$$

PROOF: In that case, one notices that $\int_0^t \sigma_s dW_s = \frac{1}{\gamma}(\sigma_t - \sigma_0)$ where

$$\sigma_t = \sigma_0 \exp(\gamma W_t - \gamma^2 t/2)\,.$$

Hence, from equation (6.7.4),

$$S_t = S_0 \exp\left(rt - \frac{1}{2}\int_0^t \sigma_s^2 ds + \frac{\rho}{\gamma}(\sigma_t - \sigma_0) + \sqrt{1-\rho^2}\int_0^t \sigma_s dW_s^{(1)}\right),$$

and conditionally on $\mathbf{F}^W$ the law of $\ln(S_t/S_0)$ is Gaussian with mean

$$rt - \frac{1}{2}\int_0^t \sigma_s^2 ds + \frac{\rho}{\gamma}(\sigma_t - \sigma_0)$$

and variance

$$(1-\rho^2)\int_0^t \sigma_s^2 ds = (1-\rho^2)\Sigma_t = (1-\rho^2)\sigma_0^2\int_0^t e^{2(\gamma W_s - \gamma^2 s/2)}ds\,.$$

The price of a call option is

$$\mathbb{E}_{\mathbb{Q}}(e^{-rT}(S_T-K)^+) = \mathbb{E}_{\mathbb{Q}}(e^{-rT}S_T \mathbb{1}_{\{S_T \geq K\}}) - Ke^{-rT}\mathbb{Q}(S_T \geq K)\,.$$

We now recall that, if Z is a Gaussian random variable with mean m and variance β, then

$$\mathbb{P}(e^Z \geq k) = \mathcal{N}\left(\frac{1}{\sqrt{\beta}}(m - \ln k)\right)$$
$$\mathbb{E}(e^Z \mathbb{1}_{\{Z\geq k\}}) = e^{m+\beta/2}\mathcal{N}\left(\frac{\beta + m - \ln k}{\sqrt{\beta}}\right).$$

It follows that $\mathbb{Q}(S_T \geq K) = \mathbb{E}_{\mathbb{Q}}\left(\mathcal{N}(d_2(\sigma_T, \Sigma_T))\right)$ where

$$d_2(u,v) = \frac{1}{\sqrt{(1-\rho^2)v}}\left(\ln\frac{S_0}{K} + rT - \frac{1}{2}v + \frac{\rho}{\gamma}(u - \sigma_0)\right) = d_2^*\left(\frac{u}{\sigma_0}, \frac{\gamma^2}{\sigma_0^2}v\right).$$

Using the scaling property and by relying on Girsanov's theorem, setting $\tau = T\gamma^2$ and $A_t^* = \int_0^t e^{2W_s^*}ds$, we obtain

$$\mathbb{Q}(S_T \geq K) = \mathbb{E}_{\mathbb{Q}^*}\left(\mathcal{N}\left(d_2\left(\sigma_0 e^{W_\tau^*}, \frac{\sigma_0^2}{\gamma^2}A_\tau^*\right)\right)e^{-\frac{1}{2}W_\tau^* - \tau/8}\right)$$
$$= \mathbb{E}_{\mathbb{Q}^*}(f_2(e^{W_\tau^*}, A_\tau^*))\,,$$

where

$$\frac{d\mathbb{Q}^*}{d\mathbb{Q}}|_{\mathcal{F}_t} = e^{-\frac{\nu^2 t}{2} - \nu W_t}$$

and $W_t^* = W_t + \nu t$ is a $\mathbb{Q}^*$ Brownian motion. Indeed the scaling property of BM implies that

$$\Sigma_T \stackrel{\text{law}}{=} \frac{\sigma_0^2}{\gamma^2} \int_0^{\gamma^2 T} \exp\left(2(W_s + \nu s)\right) ds = \frac{\sigma_0^2}{\gamma^2} \int_0^{\gamma^2 T} \exp\left(2W_s^*\right) ds = \frac{\sigma_0^2}{\gamma^2} A_{\mathcal{T}}^* .$$

where $\nu = -1/2$, because $k = 0$.
For the term $\mathbb{E}(e^{-rT} S_T \mathbb{1}_{\{S_T \geq K\}})$, we obtain easily

$$\begin{aligned}
&\mathbb{E}(e^{-rT} S_T \mathbb{1}_{\{S_T \geq K\}}) \\
&= S_0 \mathbb{E}_{\mathbb{Q}^*} \left(\mathcal{N} \left(d_1(\sigma_0 e^{W_{\mathcal{T}}^*}, \frac{\sigma_0^2}{\gamma^2} A_{\mathcal{T}}^*) \right) \right. \\
&\qquad\qquad \left. \times \exp\left(-\frac{W_{\mathcal{T}}^*}{2} - \frac{\mathcal{T}}{8} + \frac{\rho\sigma_0}{\gamma}(e^{W_{\mathcal{T}}^*} - 1) - \frac{\rho^2\sigma_0^2}{2\gamma^2} A_{\mathcal{T}}^* \right) \right) \\
&= S_0 \mathbb{E}_{\mathbb{Q}^*}(f_1(e^{W_{\mathcal{T}}^*}, A_{\mathcal{T}}^*)) .
\end{aligned}$$

The result is obtained. □

6.7.6 PDE Approach

We come back to the simple Hull and White model 6.7.12, where

$$\begin{cases} dS_t = S_t \left(r\,dt + \sigma_t\,dW_t^{(1)} \right), \\ d\sigma_t = \sigma_t(k\,dt + \gamma\,dW_t), \end{cases} \tag{6.7.14}$$

for two independent Brownian motions. Itô's lemma allows us to obtain the equation:

$$dC_E = \frac{\partial C_E}{\partial x} dS_t + \frac{\partial C_E}{\partial \sigma} d\sigma_t + \left[\frac{\partial C_E}{\partial t} + \frac{1}{2}\sigma_t^2 S_t^2 \frac{\partial^2 C_E}{\partial x^2} + \frac{1}{2}\gamma^2\sigma_t^2 \frac{\partial^2 C_E}{\partial \sigma^2} \right] dt .$$

Then, the martingale property of the discounted price leads to the following equation:

$$\frac{1}{2}\sigma^2 x^2 \frac{\partial^2 C_E}{\partial x^2} + \frac{1}{2}\gamma^2\sigma^2 \frac{\partial^2 C_E}{\partial \sigma^2} + rx\frac{\partial C_E}{\partial x} + \sigma k \frac{\partial C_E}{\partial \sigma} + \frac{\partial C_E}{\partial t} - rC_E = 0 .$$

6.7.7 Heston's Model

In Heston's model [433], the underlying process follows a geometric Brownian motion under the risk-neutral probability $\mathbb{Q}$ (with $\delta = 0$):

$$dS_t = S_t \left(rdt + \sigma_t d\widehat{B}_t \right) ,$$

and the squared volatility follows a square-root process. The model allows arbitrary correlation between volatility and spot asset returns. The dynamics of the volatility are given by:

$$d\sigma_t^2 = \kappa(\theta - \sigma_t^2)dt + \gamma\sigma_t dW_t .$$

The parameters κ, θ and γ are constant, and $\kappa\theta > 0$, so that the square-root process remains positive. The Brownian motions $(W_t, t \geq 0)$ and $(\widehat{B}_t, t \geq 0)$ have correlation coefficient equal to ρ. Setting $X_t = \ln S_t$ and $Y_t = \sigma_t^2$ these dynamics can be written under the risk-neutral probability $\mathbb{Q}$ as

$$\begin{cases} dX_t = \left(r - \dfrac{\sigma_t^2}{2} \right) dt + \sigma_t d\widehat{B}_t , \\ dY_t = \kappa(\theta - Y_t)dt + \gamma\sqrt{Y_t}dW_t . \end{cases} \tag{6.7.15}$$

As usual, the computation of the value of a call reduces to the computation of

$$\mathbb{E}_{\mathbb{Q}}(S_T \mathbb{1}_{S_T \geq K}) - Ke^{-rT}\mathbb{Q}(S_T \geq K) = S_0\widehat{\mathbb{Q}}(S_T \geq K) - Ke^{-rT}\mathbb{Q}(S_T \geq K)$$

where, under $\widehat{\mathbb{Q}}$, X follows $dX_t = (r + \sigma_t^2/2)dt + \sigma_t dB_t$, where B is a $\widehat{\mathbb{Q}}$-Brownian motion. In this setting, the price of the European call option is:

$$C_E(S_0, \sigma_0, T) = S_0\widehat{\Gamma} - Ke^{-rT}\Gamma ,$$

with

$$\widehat{\Gamma} = \widehat{\mathbb{Q}}(S_T \geq K),\, \Gamma = \mathbb{Q}(S_T \geq K) .$$

We present the computation for Γ, then the computation for $\widehat{\Gamma}$ follows from a simple change of parameters. The law of X is not easy to compute; however, the characteristic function of X_T, i.e.,

$$f(x, \sigma, u) = \mathbb{E}_{\mathbb{Q}}(e^{iuX_T} | X_0 = x, \sigma_0 = \sigma)$$

can be computed using the results on affine models. From Fourier transform inversion, Γ is given by

$$\Gamma = \frac{1}{2} + \frac{1}{\pi} \int_0^{+\infty} \mathrm{Re}\left(\frac{e^{-iu\ln(K)} f(x, \sigma, u)}{iu} \right) du .$$

As in Heston [433], one can check that

$$\mathbb{E}_{\mathbb{Q}}(e^{iuX_T} | X_t = x, \sigma_t = \sigma) = \exp(C(T - t, u) + D(T - t, u)\sigma + iux) .$$

The coefficients C and D are given by

$$C(s,u) = i\,(rus) + \frac{\kappa\theta}{\gamma^2}\left((\kappa - i(\rho\gamma u) + d)s - 2\ln\left(\frac{1-ge^{ds}}{1-g}\right)\right)$$

$$D(s,u) = \frac{\kappa - i(\rho\gamma u) + d}{\gamma^2}\,\frac{1-e^{ds}}{1-ge^{ds}}$$

where

$$g = \frac{\kappa - \rho\gamma u + d}{\kappa - i(\rho\gamma u) - d}, \quad d = \sqrt{(\kappa - i(\rho\gamma u))^2 + \gamma^2(iu + u^2)}\,.$$

6.7.8 Mellin Transform

Instead of using a time-change methodology, one can use some transform of the option price. The Mellin transform of a function f is $\int_0^\infty dk\, k^\alpha f(k)$; for example, for a call option price it is $e^{-rT}\int_0^\infty dk\, k^\alpha \mathbb{E}_\mathbb{Q}(S_T - k)^+$. This Mellin transform can be given in terms of the moments of S_T:

$$\int_0^\infty dk\, k^\alpha \mathbb{E}_\mathbb{Q}(S_T - k)^+ = \mathbb{E}_\mathbb{Q}\left(\int_0^{S_T} dk\, k^\alpha (S_T - k)\right)$$

$$= \mathbb{E}_\mathbb{Q}\left(S_T\frac{S_T^{\alpha+1}}{\alpha+1} - \frac{S_T^{\alpha+2}}{\alpha+2}\right) = \frac{1}{(\alpha+1)(\alpha+2)}\mathbb{E}_\mathbb{Q}(S_T^{\alpha+2})\,.$$

The value of S_T is given in (6.7.4), hence, if $\widehat{B}$ is independent of σ

$$\mathbb{E}_\mathbb{Q}(S_T^\beta) = (xe^{(r-\delta)T})^\beta$$

$$\mathbb{E}_\mathbb{Q}\left(\exp\left(\beta\int_0^T \sigma_s d\widehat{B}_s - \frac{\beta^2}{2}\int_0^T \sigma_s^2 ds + \frac{\beta}{2}(\beta-1)\int_0^T \sigma_s^2 ds\right)\right)$$

$$= (xe^{(r-\delta)T})^\beta \mathbb{E}_\mathbb{Q}\left(\exp\left(\frac{1}{2}\beta(\beta-1)\int_0^T \sigma_s^2 ds\right)\right).$$

Assume now that the square of the volatility follows a CIR process. It remains to apply Proposition 6.3.4.1 and to invert the Mellin transform (see Patterson [699] for inversion of Mellin transforms and Panini and Srivastav [693] for application of Mellin transforms to option pricing).

Part II

Jump Processes

7

Default Risk: An Enlargement of Filtration Approach

In this chapter, our goal is to present results that cover the *reduced form* methodology of credit risk modelling (the structural approach was presented in Section 3.10). In the first part, we provide a detailed analysis of the relatively simple case where the flow of information available to an agent reduces to observations of the random time which models the default event. The focus is on the evaluation of conditional expectations with respect to the filtration generated by a default time by means of the hazard function. In the second part, we study the case where an additional information flow – formally represented by some filtration $\mathbf{F}$ – is present; we then use the conditional survival probability, also called the hazard process. We present the intensity approach and discuss the link between both approaches. After a short introduction to CDS's, we end the chapter with a study of hedging defaultable claims.

For a complete study of credit risk, the reader can refer to Bielecki and Rutkowski [99], Bielecki et al. [91, 89], Cossin and Pirotte [197], Duffie [271], Duffie and Singleton [276], Lando [563, 564], Schönbucher [765] and to the collective book [408]. The book by Frey et al. [359] contains interesting chapters devoted to credit risk. The first part of this chapter is mainly based on the notes of Bielecki et al. for the Cimpa school in Marrakech [95].

7.1 A Toy Model

We begin with the simple case where a riskless asset, with deterministic interest rate $(r(s); s \geq 0)$, is the only asset available in the market. We denote as usual by $R(t) = \exp\left(-\int_0^t r(s)ds\right)$ the discount factor. The time-t price of a zero-coupon bond with maturity T is

$$P(t,T) = \exp\left(-\int_t^T r(s)ds\right).$$

M. Jeanblanc, M. Yor, M. Chesney, *Mathematical Methods for Financial Markets*, Springer Finance, DOI 10.1007/978-1-84628-737-4_7,

Default occurs at time τ (where τ is assumed to be a positive random variable, constructed on a probability space $(\Omega, \mathcal{G}, \mathbb{P})$). We denote by F the right-continuous cumulative distribution function of the r.v. τ defined as $F(t) = \mathbb{P}(\tau \leq t)$ and we assume that $F(t) < 1$ for any $t \leq T$, where T is a finite horizon (the maturity date); otherwise there would exist $t_0 \leq T$ such that $F(t_0) = 1$, and default would occur a.s. before t_0, which is an uninteresting case to study.

We emphasize that the risk associated with the default is not hedgeable in this model. Indeed, a random payoff of the form $\mathbb{1}_{\{T<\tau\}}$ cannot be perfectly hedged with deterministic zero-coupon bonds which are the only tradeable assets. To hedge the risk, we shall assume later on that some defaultable asset is traded, e.g., a defaultable zero-coupon bond or a Credit Default Swap (CDS).

7.1.1 Defaultable Zero-coupon with Payment at Maturity

A **defaultable zero-coupon** bond (DZC in short) - or a corporate bond - with maturity T and constant rebate δ paid at maturity, consists of:

- The payment of one monetary unit at time T if (and only if) default has not occurred before time T, i.e., if $\tau > T$.
- A payment of δ monetary units, made at maturity, if (and only if) $\tau < T$. We assume $0 < \delta < 1$. In case of default, the loss is $1 - \delta$.

Value of a Defaultable Zero-coupon Bond

The time-t value of the defaultable zero-coupon bond is defined as the expectation of the discounted payoff, given the information that the default has occurred in the past or not.

▶ If the default has occurred before time t, the payment of δ will be made at time T and the price of the DZC is $\delta P(t,T)$: in that case, the payoff is hedgeable with δ default-free zero-coupon bonds.

▶ If the default has not yet occurred at time t, the holder does not know when it will occur. Then, the value $D(t,T)$ of the DZC is the conditional expectation of the discounted payoff $P(t,T)\,[\mathbb{1}_{\{T<\tau\}} + \delta \mathbb{1}_{\{\tau \leq T\}}]$ given the information that the default has not occurred. We denote by $\widetilde{D}(t,T)$ this predefault value (i.e., the value of the defaultable zero-coupon bond on the set $\{t < \tau\}$) given by

$$\begin{aligned}\widetilde{D}(t,T) &= P(t,T)\mathbb{E}\big((\mathbb{1}_{\{T<\tau\}} + \delta\mathbb{1}_{\{\tau\le T\}})\,\big|t<\tau\big)\\ &= P(t,T)\big(1-(1-\delta)P(\tau\le T\big|t<\tau)\big)\\ &= P(t,T)\left(1-(1-\delta)\frac{\mathbb{P}(t<\tau\le T)}{\mathbb{P}(t<\tau)}\right)\\ &= P(t,T)\left(1-(1-\delta)\frac{F(T)-F(t)}{1-F(t)}\right). \end{aligned} \tag{7.1.1}$$

Note that $\widetilde{D}(t,T)$ is a deterministic function. In fact, this quantity is a net present value and is equal to the price of the ZC, minus the discounted expected loss, computed under the historical probability.

We summarize these results, writing

$$D(t,T) = \mathbb{1}_{\{\tau\le t\}}P(t,T)\delta + \mathbb{1}_{\{t<\tau\}}\widetilde{D}(t,T)$$

Note that the value of the DZC is discontinuous at time τ, unless $F(T)=1$ (or $\delta = 1$). In the case $F(T)=1$, the default appears with probability one before maturity and the DZC is equivalent to a payment of δ at maturity. If $\delta = 1$, the DZC is in fact a default-free zero coupon bond.

Formula (7.1.1) can be read as

$$\widetilde{D}(t,T) = P(t,T) - \text{DLGD} \times \text{DP}$$

where the **Discounted Loss Given Default** (DLGD) is $P(t,T)(1-\delta)$ and the conditional **Default Probability** (DP) is

$$DP = \frac{\mathbb{P}(t<\tau\le T)}{\mathbb{P}(t<\tau)} = \mathbb{P}(\tau\le T|t<\tau)\,.$$

We now consider the general case when the payment is a function of the default time, say $\delta(\tau)$; then, the time-t value of this defaultable zero-coupon is

$$D(t,T) = \mathbb{1}_{\{t<\tau\}}\widetilde{D}(t,T) + \mathbb{1}_{\{\tau\le t\}}P(t,T)\delta(\tau)$$

where the predefault time-t value $\widetilde{D}(t,T)$ satisfies

$$\begin{aligned}\widetilde{D}(t,T) &= P(t,T)\mathbb{E}(\,\mathbb{1}_{\{T<\tau\}} + \delta(\tau)\mathbb{1}_{\{\tau\le T\}}\big|t<\tau)\\ &= P(t,T)\left[\frac{\mathbb{P}(T<\tau)}{\mathbb{P}(t<\tau)} + \frac{1}{\mathbb{P}(t<\tau)}\int_t^T \delta(s)dF(s)\right].\end{aligned}$$

Hazard Function

We introduce the **survival distribution function**

$$G(t) = \mathbb{P}(\tau>t) = 1-F(t)$$

and the **hazard function** Γ defined by $\Gamma(t) = -\ln(G(t))$.

In the case where F is continuous, $\Gamma(t) = \int_0^t \frac{dF(s)}{G(s)}$, and if F admits a derivative f, the derivative of Γ is

$$\gamma(t) = \frac{f(t)}{G(t)} = \lim_{h\to 0} \frac{1}{h\,\mathbb{P}(\tau > t)}\mathbb{P}(t < \tau \leq t+h) = \mathbb{P}(\tau \in dt | \tau > t)/dt\,.$$

In this case,

$$G(t) = e^{-\Gamma(t)} = \exp\left(-\int_0^t \gamma(s)ds\right).$$

It follows that

$$d_t\widetilde{D}(t,T) = (r(t)+\gamma(t))\widetilde{D}(t,T)dt - P(t,T)\gamma(t)\delta(t)dt\,,$$

where the notation $d_t\widetilde{D}(t,T)$ denotes the differential of $\widetilde{D}$ with respect to t. In the case where $\delta = 0$

$$\widetilde{D}(t,T) = \exp\left(-\int_t^T (r+\gamma)(s)ds\right).$$

Hence, the spot rate has to be adjusted by means of a spread (equal to γ) in order to evaluate DZCs.

If γ and δ are constant, the credit spread $s(t,T)$ is, on the set $\{t < \tau\}$,

$$s(t,T) = \frac{1}{T-t}\ln\frac{P(t,T)}{\widetilde{D}(t,T)} = \gamma - \frac{1}{T-t}\ln\left(1+\delta(e^{\gamma(T-t)}-1)\right)$$

and goes to $\gamma(1-\delta)$ when t goes to T.

Remark 7.1.1.1 Note that, if τ is a random time with differentiable cumulative distribution function F, setting $\Lambda(t) = -\ln(1-F(t))$ allows us to interpret τ as the first jump time of an inhomogeneous Poisson process with intensity the derivative of Λ. However, at this point, we insist that no hidden Poisson process occurs in our framework, which only involves the time τ and quantities related with it.

Exercise 7.1.1.2 Compute the dynamics of D. $\triangleleft$

7.1.2 Defaultable Zero-coupon with Payment at Hit

Here, a defaultable zero-coupon bond with maturity T consists of:

- The payment of one monetary unit at time T if (and only if) default has not yet occurred.
- A payment of $\delta(\tau)$ monetary units, where δ is a deterministic function, made at time τ if (and only if) $\tau < T$.

Value of a Defaultable Zero-coupon

Obviously, if the default has occurred before time t, the value of the DZC is null (this was not the case for payment of the rebate at maturity). Therefore, $D(t,T) = \mathbb{1}_{\{t<\tau\}}\widetilde{D}(t,T)$ where $\widetilde{D}(t,T)$ is a deterministic function, called the predefault price. The predefault time-t value $\widetilde{D}(t,T)$ satisfies

$$\begin{aligned}\widetilde{D}(t,T) &= \mathbb{E}(P(t,T)\,\mathbb{1}_{\{T<\tau\}} + P(t,\tau)\delta(\tau)\mathbb{1}_{\{\tau\le T\}}|t<\tau)\\ &= \frac{\mathbb{P}(T<\tau)}{\mathbb{P}(t<\tau)}P(t,T) + \frac{1}{\mathbb{P}(t<\tau)}\int_t^T P(t,s)\delta(s)dF(s)\,.\end{aligned}$$

Hence,

$$G(t)\widetilde{D}(t,T) = G(T)P(t,T) - \int_t^T P(t,s)\delta(s)dG(s)\,.$$

The process $t \to D(t,T)$ admits a discontinuity at time τ and the size of the jump is $-\widetilde{D}(\tau,T)$.

A Particular Case

If F is differentiable,

$$\widetilde{D}(t,T) = P^d(t,T) + \int_t^T P^d(t,s)\delta(s)\gamma(s)ds$$

with $P^d(t,s) = \exp\left(-\int_t^s[r(u)+\gamma(u)]du\right)$. The defaultable interest rate is $r+\gamma$ and is, as expected, greater than r (the value of a DZC with $\delta = 0$ is smaller than the value of a default-free zero-coupon). The dynamics of $\widetilde{D}(t,T)$ are

$$d_t\widetilde{D}(t,T) = \Big((r(t)+\gamma(t))\widetilde{D}(t,T) - \delta(t)\gamma(t)\Big)dt\,.$$

It is interesting to recall (see Subsection 2.3.5) that, if X is the price of an asset which pays a deterministic dividend rate $\kappa(t)$ in a financial market with a deterministic interest rate ρ, then

$$dX(t) = \rho(t)X(t)dt - \kappa(t)dt\,.$$

The dynamics of the predefault price can be interpreted as the price in a default-free market of an asset paying dividend rate $\gamma(t)\delta(t)$, with interest rate $\rho = r+\gamma$. The quantity $\gamma(t)$ in the dividend rate represents the probability that the dividend $\delta(t)$ is paid in the time interval $[t, t+dt]$.

7.2 Toy Model and Martingales

We now present the results of the previous section in a different form, following closely Dellacherie ([240], page 122). We denote by $(D_t, t \geq 0)$ the right-continuous increasing process $D_t = \mathbb{1}_{\{t \geq \tau\}}$ and by $\mathbf{D}$ its natural filtration. The filtration $\mathbf{D}$ is the smallest filtration which makes τ a stopping time (we work with the usual completed filtration). The σ-algebra $\mathcal{D}_t$ is generated by the sets $\{\tau \leq s\}$ for $s \leq t$ (or by the r.v. $\tau \wedge t$). As a consequence, and this is a key point, every $\mathcal{D}_t$-measurable random variable H is of the form $H = h(\tau \wedge t) = h(\tau)\mathbb{1}_{\{\tau \leq t\}} + h(t)\mathbb{1}_{\{t < \tau\}}$ where h is a Borel function.

We assume in this section that the cumulative distribution function F is continuous.

7.2.1 Key Lemma

We now give some elementary tools to compute the conditional expectation w.r.t. $\mathcal{D}_t$, as presented in Brémaud [124], Dellacherie [240], and Elliott [313].

Lemma 7.2.1.1 *If X is any integrable, $\mathcal{G}$-measurable r.v.*

$$\mathbb{E}(X|\mathcal{D}_s)\mathbb{1}_{\{s<\tau\}} = \mathbb{1}_{\{s<\tau\}} \frac{\mathbb{E}(X\mathbb{1}_{\{s<\tau\}})}{\mathbb{P}(s < \tau)} . \tag{7.2.1}$$

PROOF: The r.v. $\mathbb{E}(X|\mathcal{D}_s)$ is $\mathcal{D}_s$-measurable. Therefore, it can be written in the form $\mathbb{E}(X|\mathcal{D}_s) = h(\tau \wedge s) = h(\tau)\mathbb{1}_{\{s \geq \tau\}} + h(s)\mathbb{1}_{\{s<\tau\}}$ for some function h. By multiplying both sides by $\mathbb{1}_{\{s<\tau\}}$, and taking the expectation, we obtain

$$\begin{aligned} \mathbb{E}[\mathbb{1}_{\{s<\tau\}}\mathbb{E}(X|\mathcal{D}_s)] &= \mathbb{E}[\mathbb{E}(\mathbb{1}_{\{s<\tau\}}X|\mathcal{D}_s)] = \mathbb{E}[\mathbb{1}_{\{s<\tau\}}X] \\ &= \mathbb{E}(h(s)\mathbb{1}_{\{s<\tau\}}) = h(s)\mathbb{P}(s < \tau) . \end{aligned}$$

Hence, $h(s) = \dfrac{\mathbb{E}(X\mathbb{1}_{\{s<\tau\}})}{\mathbb{P}(s < \tau)}$, which is the desired result. □

Remark 7.2.1.2 If the cumulative distribution function F is continuous, then τ is a $\mathbf{D}$-totally inaccessible stopping time. (See Dellacherie and Meyer [244] IV, 107.)

7.2.2 The Fundamental Martingale

Proposition 7.2.2.1 *The process $(M_t, t \geq 0)$ defined as*

$$M_t = D_t - \int_0^{\tau \wedge t} \frac{dF(s)}{G(s)} = D_t - \int_0^t (1 - D_s)\frac{dF(s)}{G(s)} = D_t - \Gamma(t \wedge \tau)$$

is a $\mathbf{D}$-martingale.

PROOF: Let $s < t$. Then, an application of (7.2.1) with $X = \mathbb{1}_{\{\tau \le t\}}$ yields

$$\mathbb{E}(D_t - D_s|\mathcal{D}_s) = \mathbb{1}_{\{s<\tau\}}\mathbb{E}(\mathbb{1}_{\{s<\tau\le t\}}|\mathcal{D}_s) = \mathbb{1}_{\{s<\tau\}}\frac{F(t)-F(s)}{G(s)}, \qquad (7.2.2)$$

On the other hand, the quantity

$$C := \mathbb{E}\left[\int_s^t (1-D_u)\frac{dF(u)}{G(u)} \,|\, \mathcal{D}_s\right],$$

is given by

$$\begin{aligned} C &= \int_s^t \frac{dF(u)}{G(u)}\mathbb{E}\left[\mathbb{1}_{\{\tau>u\}}|\mathcal{D}_s\right] \\ &= \mathbb{1}_{\{\tau>s\}}\int_s^t \frac{dF(u)}{G(u)}\left(1-\frac{F(u)-F(s)}{G(s)}\right) = \mathbb{1}_{\{\tau>s\}}\left(\frac{F(t)-F(s)}{G(s)}\right) \end{aligned}$$

which, from (7.2.2) proves the result. □

7.2.3 Hazard Function

From Proposition 7.2.2.1, the Doob-Meyer decomposition of the submartingale D is $M_t + \Gamma(t \wedge \tau)$. The predictable increasing process $A_t = \Gamma(t \wedge \tau)$ is called the **compensator** of D (it is the dual predictable projection of the increasing process D, see Section 5.2).

We now assume that F is differentiable with derivative f, therefore the process

$$M_t = D_t - \int_0^{\tau\wedge t}\gamma(s)ds = D_t - \int_0^t \gamma(s)(1-D_s)ds$$

is a martingale; the deterministic positive function $\gamma(s) = \dfrac{f(s)}{1-F(s)}$ is called **the intensity of** τ. We can now write the dynamics of the value of a defaultable zero-coupon bond with recovery δ paid at hit.

Proposition 7.2.3.1 *The dynamics of a DZC with recovery δ paid at hit is*

$$d_tD(t,T) = (r(t)D(t,T) - \delta(t)\gamma(t)(1-D_t))\,dt - \widetilde{D}(t,T)dM_t\,. \qquad (7.2.3)$$

PROOF: From $D(t,T) = \mathbb{1}_{\{t<\tau\}}\widetilde{D}(t,T) = (1-D_t)\widetilde{D}(t,T)$ and the dynamics of $\widetilde{D}(t,T)$, we obtain, from the integration by parts formula,

$$\begin{aligned} d_tD(t,T) &= (1-D_t)d\widetilde{D}(t,T) - \widetilde{D}(t,T)dD_t \\ &= (1-D_t)\Big((r(t)+\gamma(t))\widetilde{D}(t,T) - \delta(t)\gamma(t)\Big)\,dt - \widetilde{D}(t,T)dD_t \\ &= (r(t)D(t,T) - \delta(t)\gamma(t)(1-D_t))\,dt - \widetilde{D}(t,T)dM_t\,. \end{aligned}$$

□

We can also write (7.2.3) as

$$d_t D(t,T) = (r(t)D(t,T) - \delta(t)\gamma(t)(1-D_t))\,dt - D(t^-,T)dM_t\,.$$

We have just detailed the additive decomposition of $(D_t, t \geq 0)$, or of $1-D$; here now is the multiplicative decomposition of $(1-D_t, t \geq 0)$.

Proposition 7.2.3.2 *The process $L_t := \mathbb{1}_{\{\tau>t\}} \exp(\Gamma(t))$ is a* **D***-martingale. In particular, for $t < T$,*

$$\mathbb{E}(\mathbb{1}_{\{\tau>T\}}|\mathcal{D}_t) = \mathbb{1}_{\{\tau>t\}} \exp\left(-\int_t^T \gamma(s)ds\right).$$

The multiplicative decomposition of the supermartingale $1-D$ is

$$1 - D_t = L_t \exp\left(-\Gamma(t)\right)$$

PROOF: We shall give three different arguments, each of which constitutes a proof.

a) Since the function γ is deterministic, for $t > s$

$$\mathbb{E}(L_t|\mathcal{D}_s) = \exp(\Gamma(t))\,\mathbb{E}(\mathbb{1}_{\{t<\tau\}}|\mathcal{D}_s)\,.$$

From the equality (7.2.1)

$$\mathbb{E}(\mathbb{1}_{\{t<\tau\}}|\mathcal{D}_s) = \mathbb{1}_{\{\tau>s\}}\frac{G(t)}{G(s)} = \mathbb{1}_{\{\tau>s\}} \exp\left(-\Gamma(t) + \Gamma(s)\right).$$

Hence,

$$\mathbb{E}(L_t|\mathcal{D}_s) = \mathbb{1}_{\{\tau>s\}} \exp\left(\Gamma(s)\right) = L_s.$$

b) Another method is to apply the integration by parts formula for bounded variation processes (see ↣ Subsection 8.3.4 if needed) to the process

$$L_t = (1-D_t)\exp\left(\Gamma(t)\right).$$

Then,

$$\begin{aligned} dL_t &= -e^{\Gamma(t)}\,dD_t + \gamma(t)e^{\Gamma(t)}(1-D_t)dt \\ &= -e^{\Gamma(t)}\,dM_t\,. \end{aligned}$$

c) A third (more sophisticated) method is to note that L is the exponential martingale of M (see ↣ Subsection 8.4.4), i.e., the solution of the SDE

$$dL_t = -L_{t-}dM_t\,,\; L_0 = 1.$$

□

Exercise 7.2.3.3 In this exercise, F is only assumed continuous on the right, and $G(t^-)$ is the left-limit of G at point t. Prove that the process $(M_t, t \geq 0)$ defined as

$$M_t = D_t - \int_0^{\tau \wedge t} \frac{dF(s)}{G(s-)} = D_t - \int_0^t (1 - D_{s-}) \frac{dF(s)}{G(s-)}$$

is a **D**-martingale. ◁

7.2.4 Incompleteness of the Toy Model, non Arbitrage Prices

In order to study the completeness of the financial market, we first need to define the tradeable assets.

If the market consists only of the risk-free zero-coupon bond, the market is incomplete (one cannot hedge defaultable claims, i.e., the payoff of which belongs to $\mathcal{D}_T$), hence, there exists infinitely many e.m.m's. The discounted asset prices are constant, hence the set $\mathcal{Q}$ of equivalent martingale measures is the set of probabilities equivalent to the historical one. For every $\mathbb{Q} \in \mathcal{Q}$, we denote by $F_\mathbb{Q}$ the cumulative distribution function of τ under $\mathbb{Q}$, i.e.,

$$F_\mathbb{Q}(t) = \mathbb{Q}(\tau \leq t)\,.$$

Assuming that $F_\mathbb{P} = \mathbb{P}(\tau \leq t)$ is continuous, then, since $\mathbb{Q} \sim \mathbb{P}$, $F_\mathbb{Q}$ is also continuous. The range of prices is defined as the set of prices which do not induce arbitrage opportunities. In the case of constant interest rate, the range of prices for a contingent claim H to be delivered at maturity is equal to the interval

$$\left]\inf_{\mathbb{Q}\in\mathcal{Q}} e^{-rT}\mathbb{E}_\mathbb{Q}(H),\ \sup_{\mathbb{Q}\in\mathcal{Q}} e^{-rT}\mathbb{E}_\mathbb{Q}(H)\right[$$

For a DZC with a constant rebate δ paid at maturity, the range of prices is equal to the set

$$\{\mathbb{E}_\mathbb{Q}(e^{-rT}(\mathbb{1}_{\{T<\tau\}} + \delta \mathbb{1}_{\{\tau \leq T\}})), \mathbb{Q} \in \mathcal{Q}\}\,.$$

This set is exactly the interval $]\delta e^{-rT}, e^{-rT}[$. Indeed, it is obvious that the range of prices is included in the interval $]\delta e^{-rT}, e^{-rT}[$. Now, in the set $\mathcal{Q}$, one can select a sequence of probabilities $\mathbb{Q}_n$ which converges weakly to the Dirac measure at point 0 (resp. at point T) (the bounds are obtained as limit cases: the default appears at time 0^+, or never).

7.2.5 Predictable Representation Theorem

Every square integrable **D**-martingale terminates at $h(\tau)$, hence is of the form $\mathbb{E}(h(\tau)|\mathcal{D}_t)$. We now prove that this martingale can be written as a stochastic integral with respect to the fundamental martingale M.

Proposition 7.2.5.1 *The martingale $H_t = \mathbb{E}(h(\tau)|\mathcal{D}_t)$ admits the representation*

$$\mathbb{E}(h(\tau)|\mathcal{D}_t) = \mathbb{E}(h(\tau)) - \int_0^{t\wedge\tau} (H_{s-} - h(s))\, dM_s\,.$$

Consequently, every square integrable **D***-martingale $(X_t \geq 0)$ can be written as $X_t = X_0 + \int_0^t x_s dM_s$ where $(x_s, s \geq 0)$ is a* **D***-predictable process.*

PROOF: From Lemma 7.2.1.1

$$\begin{aligned} H_t &= h(\tau)\mathbb{1}_{\{\tau\leq t\}} + \mathbb{1}_{\{t<\tau\}} \frac{\mathbb{E}(h(\tau)\mathbb{1}_{\{t<\tau\}})}{\mathbb{P}(t<\tau)} \\ &= h(\tau)\mathbb{1}_{\{\tau\leq t\}} + \mathbb{1}_{\{t<\tau\}}(G((t))^{-1}\mathbb{E}(h(\tau)\mathbb{1}_{\{t<\tau\}}) \\ &= \int_0^t h(s)dD_s + \mathbb{1}_{\{t<\tau\}}(G(t))^{-1}\int_t^\infty h(s)f(s)ds\,. \end{aligned}$$

It remains to apply integration by parts (see ↣ equation 9.1.1) to obtain

$$dH_t = (h(t) - H_{t-})(dD_t - (1-D_t)\frac{f(t)}{G(t)}dt) = (h(t) - H_{t-})dM_t$$

□

7.2.6 Risk-neutral Probability Measures

If DZCs with rebate paid at maturity are traded, their prices are *given by the market*, and the equivalent martingale measure $\mathbb{Q}$, *chosen by the market*, is such that, on the set $\{t<\tau\}$,

$$D(t,T) = P(t,T)\mathbb{E}_{\mathbb{Q}}\big([\mathbb{1}_{T<\tau} + \delta\mathbb{1}_{t<\tau\leq T}]\,|t<\tau\big)\,.$$

Therefore, we can characterize the cumulative distribution function of τ under $\mathbb{Q}$ from the market prices of the DZC as follows.

Zero Recovery

If a DZC of maturity T with zero recovery is traded at a price $D(t,T)$ which belongs to the interval $]0, P(t,T)[$, then, under any risk-neutral probability $\mathbb{Q}$, the discounted value of $D(t,T)$ is a martingale (for the moment, we do not know that the market is complete, so we cannot claim that the e.m.m. is unique), and the following equality holds

$$D(t,T) = \mathbb{E}_{\mathbb{Q}}(P(t,T)\mathbb{1}_{\{T<\tau\}}|\mathcal{D}_t) = P(t,T)\mathbb{1}_{\{t<\tau\}}\exp\left(-\int_t^T \lambda^{\mathbb{Q}}(s)ds\right)$$

where $\lambda^{\mathbb{Q}}(s) = \frac{dF_{\mathbb{Q}}(s)/ds}{1 - F_{\mathbb{Q}}(s)}$. It is obvious that if $D(t,T)$ belongs to the range of viable prices $]0, P(t,T)[$, then the function $\lambda^{\mathbb{Q}}$ is strictly positive (and the converse holds true). The process $D_t - \int_0^{t\wedge\tau} \lambda^{\mathbb{Q}}(t)dt$ is a $\mathbb{Q}$-martingale and $\lambda^{\mathbb{Q}}$ is the $\mathbb{Q}$-intensity of τ. Therefore, the value of $\int_t^T \lambda^{\mathbb{Q}}(s)ds$ is known for every T as long as there are DZC bonds for each maturity, and the unique risk-neutral intensity can be obtained from the prices of DZCs as

$$r(t) + \lambda^{\mathbb{Q}}(t) = -\frac{\partial}{\partial T} \ln D(t,T)|_{T=t} \,.$$

Remark 7.2.6.1 It is important to note that there is no relation between the risk-neutral intensity and the historical one. The risk-neutral intensity can be greater (resp. smaller) than the historical one. The historical intensity can be deduced from observations of default time, the risk-neutral one is obtained from the prices of traded defaultable claims.

Fixed Payment at Maturity

If the prices of DZCs with different maturities and the same δ are known, then from (7.1.1)

$$\frac{P(0,T) - D(0,T)}{P(0,T)(1-\delta)} = F_{\mathbb{Q}}(T)$$

where $F_{\mathbb{Q}}(t) = \mathbb{Q}(\tau \leq t)$, so that the law of τ is known under the e.m.m.. However, as noticed in Hull and White [454], *extracting default probabilities from bond prices is in practice, usually complicated. First, the recovery rate is usually non-zero. Second, most corporate bonds are not zero-coupon bonds.*

Payment at Hit

In this case the cumulative function can be obtained using the derivative of the defaultable zero-coupon price with respect to the maturity. Indeed, writing for short $\partial_T D$ for $\frac{\partial}{\partial T} D$, the partial derivative of the value of the DZC at time 0 with respect to the maturity, and assuming that $G = 1 - F$ is differentiable, we obtain

$$\partial_T D(0,T) = g(T)P(0,T) - G(T)P(0,T)r(T) - \delta(T)g(T)P(0,T)\,,$$

where $g(t) = G'(t)$. Therefore, solving this equation leads to

$$\mathbb{Q}(\tau > t) = G(t) = \Delta(t)\left[1 + \int_0^t \partial_T D(0,s) \frac{1}{P(0,s)(1-\delta(s))}(\Delta(s))^{-1}ds\right]\,,$$

where $\Delta(t) = \exp\left(\int_0^t \frac{r(u)}{1-\delta(u)}du\right)$.

7.2.7 Partial Information: Duffie and Lando's Model

Duffie and Lando [273] study the case where $\tau = \inf\{t : V_t \leq m\}$ for V a diffusion process which satisfies

$$dV_t = \mu(t, V_t)dt + \sigma(t, V_t)dW_t \, .$$

Here the process W is a Brownian motion. If the information at hand is the Brownian filtration, and if V is adapted w.r.t. W (i.e., is a strong solution of the SDE), the time τ is a stopping time w.r.t. this Brownian filtration $\mathbf{F}$, and hence is predictable and does not admit an intensity, i.e., there is no process λ such that $D_t - \int_0^t (1 - D_s)\lambda_s ds$ is an $\mathbf{F}$-martingale. We assume now that the agents do not observe the process V, but they have only the minimal information $\mathcal{D}_t$, i.e., they know when the default appears. We assume for simplicity a null interest rate. In the case where the cumulative distribution of the hitting time τ admits a derivative f, we have established above that the value of a defaultable zero-coupon of maturity T is, when the default has not yet occurred, $\exp\left(-\int_t^T \lambda(s)ds\right)$ where $\lambda(s) = \frac{f(s)}{1-F(s)}$. A more general case is presented in Subsection 7.6.1.

Remark 7.2.7.1 Duffie and Lando showed that the value of the DZC is $\exp(-\int_t^T \widehat{\lambda}(s)ds)$, where

$$\widehat{\lambda}(t) = \frac{1}{2}\sigma^2(t,0)\frac{\partial \varphi}{\partial x}(t,0)$$

and $\varphi(t,x)$ is the conditional density of V_t when $T_0 > t$, i.e., the differential w.r.t. x of $\dfrac{\mathbb{P}(V_t \leq x, T_0 > t)}{\mathbb{P}(T_0 > t)}$, where $T_0 = \inf\{t; \, V_t = 0\}$. Even in the case where V is a homogeneous diffusion, i.e., $dV_t = \mu(V_t)dt + \sigma(V_t)dW_t$, the equality between Duffie and Lando's result and ours is not obvious. See Elliott et al. [315] for a proof based on time reversal properties.

7.3 Default Times with a Given Stochastic Intensity

We now present a case where some additional information is available in the market, i.e., a stochastic process plays the rôle of the hazard function. We construct a default time from this process on an enlarged probability space.

7.3.1 Construction of Default Time with a Given Stochastic Intensity

Let $(\Omega, \mathcal{G}, \mathbf{F}, \mathbb{P})$ be a filtered probability space. *A positive* $\mathbf{F}$*-adapted process* λ *is given.* We assume that there exists a r.v. Θ, constructed on Ω, independent of $\mathcal{F}_\infty$, with the exponential law of parameter 1: $\mathbb{P}(\Theta \geq t) = e^{-t}$. We define the random time τ as the first time when the process $(\Lambda_t := \int_0^t \lambda_s \, ds, t \geq 0)$ is above the random level Θ, i.e.,

$$\tau = \inf \{t \geq 0 \, : \, \Lambda_t \geq \Theta\}.$$

In particular, $\{\tau \geq s\} = \{\Lambda_s \leq \Theta\}$. We assume here that $\Lambda_t < \infty, \forall t$, and that $\Lambda_\infty = \infty$.

We shall refer to this construction as **the Cox process model.**

Comment 7.3.1.1 The choice of an exponential law for the r.v. Θ has no real importance. In Wong [849], the time of default is given as

$$\tau = \inf\{t \, : \, \Lambda_t \geq \Sigma\}$$

where Σ a non-negative r.v. independent of $\mathcal{F}_\infty$. Under some regularity assumptions, this model reduces to the previous one as we discuss now. We recall that if X is a r.v. with continuous distribution function F, the r.v. $F(X)$ is uniformly distributed on $[0,1]$. Let $\Psi(x) = 1 - e^{-x}$ be the cumulative distribution function of the exponential law. If Φ, the cumulative distribution function of Σ, is strictly increasing, the r.v. $\Phi(\Sigma)$ is uniformly distributed and $\Theta = \Psi^{-1}(\Phi(\Sigma))$ is an exponential r.v. Then,

$$\tau = \inf\{t \, : \, \Phi(\Lambda_t) \geq \Phi(\Sigma)\} = \inf\{t \, : \, \Psi^{-1}[\Phi(\Lambda_t)] \geq \Theta\}$$

and

$$F_t = \mathbb{P}(\tau \leq t | \mathcal{F}_t) = \mathbb{P}(\Lambda_t \geq \Sigma | \mathcal{F}_t) = 1 - \exp\left(-\Psi^{-1}(\Phi(\Lambda_t))\right) .$$

It remains to write $\Psi^{-1}(\Phi(\Lambda_t))$ as $\int_0^t \gamma_s ds$ to recover the previous case. The case where Φ is not strictly increasing has to be solved carefully, see for example Bélanger et al. [67].

7.3.2 Conditional Expectation with Respect to $\mathcal{F}_t$

Lemma 7.3.2.1 *The conditional distribution function of τ given the σ-algebra $\mathcal{F}_t$ is, for $t \geq s$*

$$\mathbb{P}(\tau > s | \mathcal{F}_t) = \exp(-\Lambda_s) .$$

PROOF: The proof follows from the equality $\{\tau > s\} = \{\Lambda_s < \Theta\}$. From the independence assumption and the $\mathcal{F}_t$-measurability of Λ_s for $s \leq t$, we obtain

$$\mathbb{P}(\tau > s | \mathcal{F}_t) = \mathbb{P}(\Lambda_s < \Theta \, | \, \mathcal{F}_t) = \exp(-\Lambda_s).$$

□

In particular, for $t \geq s$, $\mathbb{P}(\tau > s | \mathcal{F}_t) = \mathbb{P}(\tau > s | \mathcal{F}_s)$, and, letting $t \to \infty$, we obtain

$$G_s := \mathbb{P}(\tau > s | \mathcal{F}_s) = \mathbb{P}(\tau > s | \mathcal{F}_\infty) = e^{-\Lambda_s} . \tag{7.3.1}$$

Here, the Azéma supermartingale $\mathbb{P}(\tau > t | \mathcal{F}_t)$ is a decreasing process.

Remarks 7.3.2.2 (a) For $t < s$, we obtain $\mathbb{P}(\tau > s|\mathcal{F}_t) = \mathbb{E}(\exp(-\Lambda_s)|\mathcal{F}_t)$.

(b) If the process λ is not positive, the process Λ is not increasing and we obtain, for $s < t$,

$$\mathbb{P}(\tau > s|\mathcal{F}_t) = \mathbb{P}\left(\sup_{u \le s} \Lambda_u < \Theta\right) = \exp\left(-\sup_{u \le s} \Lambda_u\right).$$

7.3.3 Enlargements of Filtrations

Write as before $D_t = \mathbb{1}_{\{\tau \le t\}}$ and $\mathcal{D}_t = \sigma(D_s\,;\, s \le t)$. Introduce the filtration $\mathcal{G}_t = \mathcal{F}_t \vee \mathcal{D}_t$, that is, the enlarged filtration generated by the underlying filtration $\mathbf{F}$ and the process D. (We denote by $\mathbf{F}$ the original **F**iltration and by $\mathbf{G}$ the enlar**G**ed one.) We shall frequently write $\mathbf{G} = \mathbf{F} \vee \mathbf{D}$. The filtration $\mathbf{G}$ is the smallest one which contains $\mathbf{F}$ and such that τ is a stopping time.

It is easy to describe the events which belong to the σ-algebra $\mathcal{G}_t$ on the set $\{\tau > t\}$. Indeed, if $G_t \in \mathcal{G}_t$, then $G_t \cap \{\tau > t\} = \widehat{G}_t \cap \{\tau > t\}$ for some event $\widehat{G}_t \in \mathcal{F}_t$.

Therefore, from the monotone class theorem, any $\mathcal{G}_t$-measurable random variable Y_t satisfies $\mathbb{1}_{\{\tau > t\}} Y_t = \mathbb{1}_{\{\tau > t\}} y_t$, where y_t is an $\mathcal{F}_t$-measurable random variable.

7.3.4 Conditional Expectations with Respect to $\mathcal{G}_t$

Lemma 7.3.4.1 (Key Lemma) *Let Y be an integrable r.v.. Then,*

$$\mathbb{1}_{\{\tau > t\}} \mathbb{E}(Y|\mathcal{G}_t) = \mathbb{1}_{\{\tau > t\}} \frac{\mathbb{E}(Y \mathbb{1}_{\{\tau > t\}}|\mathcal{F}_t)}{\mathbb{E}(\mathbb{1}_{\{\tau > t\}}|\mathcal{F}_t)} = \mathbb{1}_{\{\tau > t\}} e^{\Lambda_t} \mathbb{E}(Y \mathbb{1}_{\{\tau > t\}}|\mathcal{F}_t).$$

PROOF: From the remarks on $\mathcal{G}_t$-measurability, if $Y_t = \mathbb{E}(Y|\mathcal{G}_t)$, there exists an $\mathcal{F}_t$-measurable r.v. y_t such that

$$\mathbb{1}_{\{\tau > t\}} \mathbb{E}(Y|\mathcal{G}_t) = \mathbb{1}_{\{\tau > t\}} y_t\,.$$

Taking conditional expectation w.r.t. $\mathcal{F}_t$ of both members of the above equality, we obtain $y_t = \dfrac{\mathbb{E}(Y \mathbb{1}_{\{\tau > t\}}|\mathcal{F}_t)}{\mathbb{E}(\mathbb{1}_{\{\tau > t\}}|\mathcal{F}_t)}$. The last equality in the proposition follows from $\mathbb{E}(\mathbb{1}_{\{\tau > t\}}|\mathcal{F}_t) = e^{-\Lambda_t}$. □

Corollary 7.3.4.2 *If X is an integrable $\mathcal{F}_T$-measurable random variable, then, for $t < T$*

$$\mathbb{E}(X \mathbb{1}_{\{T < \tau\}}|\mathcal{G}_t) = \mathbb{1}_{\{\tau > t\}} e^{\Lambda_t} \mathbb{E}(X e^{-\Lambda_T}|\mathcal{F}_t)\,. \tag{7.3.2}$$

PROOF: The conditional expectation $\mathbb{E}(X\mathbb{1}_{\{\tau>T\}}|\mathcal{G}_t)$ is equal to 0 on the $\mathcal{G}_t$-measurable set $\{\tau < t\}$, whereas for any $X \in L^1(\mathcal{F}_T)$,

$$\mathbb{E}(X\mathbb{1}_{\{\tau>T\}}|\mathcal{F}_t) = \mathbb{E}(X\mathbb{1}_{\{\tau>T\}}|\mathcal{F}_T|\mathcal{F}_t) = \mathbb{E}(Xe^{-\Lambda_T}|\mathcal{F}_t).$$

The result follows from Lemma 7.3.4.1. □

We now compute the expectation of an **F**-predictable process evaluated at time τ, and we give the **F**-Doob-Meyer decomposition of the increasing process D. We denote by $F_t := \mathbb{P}(\tau \le t|\mathcal{F}_t)$ the conditional distribution function.

Lemma 7.3.4.3 (i) *If h is an **F**-predictable (bounded) process, then*

$$\mathbb{E}(h_\tau|\mathcal{F}_t) = \mathbb{E}\left(\int_0^\infty h_u\lambda_u e^{-\Lambda_u}\,du \,\Big|\, \mathcal{F}_t\right) = \mathbb{E}\left(\int_0^\infty h_u dF_u \,\Big|\, \mathcal{F}_t\right)$$

and

$$\mathbb{E}(h_\tau|\mathcal{G}_t) = e^{\Lambda_t}\mathbb{E}\left(\int_t^\infty h_u\lambda_u dF_u \,\Big|\, \mathcal{F}_t\right)\mathbb{1}_{\{\tau>t\}} + h_\tau\mathbb{1}_{\{\tau\le t\}}. \tag{7.3.3}$$

In particular

$$\mathbb{E}(h_\tau) = \mathbb{E}\left(\int_0^\infty h_u\lambda_u\, e^{-\Lambda_u}\,du\right) = \mathbb{E}\left(\int_0^\infty h_u dF_u\right).$$

(ii) *The process $(D_t - \int_0^{t\wedge\tau}\lambda_s ds, t\ge 0)$ is a **G**-martingale.*

PROOF: Let $B_v \in \mathcal{F}_v$ and h the elementary **F**-predictable process defined as $h_t = \mathbb{1}_{\{t>v\}}B_v$. Then,

$$\begin{aligned}\mathbb{E}(h_\tau|\mathcal{F}_t) &= \mathbb{E}\Big(\mathbb{1}_{\{\tau>v\}}B_v|\mathcal{F}_t\Big) = \mathbb{E}\Big(\mathbb{E}(\mathbb{1}_{\{\tau>v\}}B_v|\mathcal{F}_\infty)\,\Big|\,\mathcal{F}_t\Big)\\ &= \mathbb{E}\Big(B_v\mathbb{P}(v<\tau|\mathcal{F}_\infty)\,\Big|\,\mathcal{F}_t\Big) = \mathbb{E}\Big(B_v e^{-\Lambda_v}|\mathcal{F}_t\Big).\end{aligned}$$

It follows that

$$\mathbb{E}(h_\tau|\mathcal{F}_t) = \mathbb{E}\left(B_v\int_v^\infty \lambda_u e^{-\Lambda_u}\,du \,\Big|\, \mathcal{F}_t\right) = \mathbb{E}\left(\int_0^\infty h_u\lambda_u e^{-\Lambda_u}du \,\Big|\, \mathcal{F}_t\right)$$

and (i) is derived from the monotone class theorem. Equality 7.3.3 follows from the key Lemma 7.3.4.1.

The martingale property (ii) follows from the integration by parts formula. Indeed, let $s<t$. Then, on the one hand from the key Lemma

$$\begin{aligned}\mathbb{E}(D_t - D_s|\mathcal{G}_s) &= \mathbb{P}(s<\tau\le t|\mathcal{G}_s) = \mathbb{1}_{\{s<\tau\}}\frac{\mathbb{P}(s<\tau\le t|\mathcal{F}_s)}{\mathbb{P}(s<\tau|\mathcal{F}_s)}\\ &= \mathbb{1}_{\{s<\tau\}}\left(1 - e^{\Lambda_s}\mathbb{E}\left(e^{-\Lambda_t}|\mathcal{F}_s\right)\right).\end{aligned}$$

On the other hand, from part (i), for $s < t$,

$$\mathbb{E}\left(\int_{s\wedge\tau}^{t\wedge\tau} \lambda_u du \,\Big|\, \mathcal{G}_s\right) = \mathbb{E}\left(\Lambda_{t\wedge\tau} - \Lambda_{s\wedge\tau}|\mathcal{G}_s\right) = \mathbb{E}(\psi_\tau|\mathcal{G}_t)$$
$$= \mathbb{1}_{\{s<\tau\}} e^{\Lambda_s} \mathbb{E}\left(\int_s^\infty \psi_u \lambda_u e^{-\Lambda_u} du \,\Big|\, \mathcal{F}_s\right)$$

where $\psi_u = \Lambda_{t\wedge u} - \Lambda_{s\wedge u} = \mathbb{1}_{\{s<u\}}(\Lambda_{t\wedge u} - \Lambda_s)$. Consequently,

$$\begin{aligned}\int_s^\infty \psi_u \lambda_u e^{-\Lambda_u} du &= \int_s^t (\Lambda_u - \Lambda_s)\lambda_u e^{-\Lambda_u} du + (\Lambda_t - \Lambda_s)\int_t^\infty \lambda_u e^{-\Lambda_u} du \\ &= \int_s^t \Lambda_u \lambda_u e^{-\Lambda_u} du - \Lambda_s \int_s^\infty \lambda_u e^{-\Lambda_u} du + \Lambda_t e^{-\Lambda_t} \\ &= \int_s^t \Lambda_u \lambda_u e^{-\Lambda_u} du - \Lambda_s e^{-\Lambda_s} + \Lambda_t e^{-\Lambda_t} \\ &= e^{-\Lambda_s} - e^{-\Lambda_t}.\end{aligned}$$

It follows that

$$\mathbb{E}(D_t - D_s|\mathcal{G}_s) = \mathbb{E}\left(\int_{s\wedge\tau}^{t\wedge\tau} \lambda_u du \,\Big|\, \mathcal{G}_s\right),$$

hence the martingale property of the process $D_t - \int_0^{t\wedge\tau} \lambda_u du$. □

Comment 7.3.4.4 Property (ii) shows that $\lambda_t \mathbb{1}_{\{t<\tau\}}$ is the **G**-intensity of τ (see ↣ Section 7.7).

Remark 7.3.4.5 As we said before in Subsection 7.2.4, if no defaultable assets are traded, the defaultable market is incomplete. A change of probability can affect the law of the r.v. Θ, hence, can affect the value of the intensity. We shall study this problem in a more general hazard process framework in the following section.

7.3.5 Conditional Expectations of $\mathcal{F}_\infty$-Measurable Random Variables

Lemma 7.3.5.1 *Let X be an integrable $\mathcal{F}_\infty$-measurable r.v.. Then*

$$\mathbb{E}(X|\mathcal{G}_t) = \mathbb{E}(X|\mathcal{F}_t). \tag{7.3.4}$$

PROOF: Let X be an integrable $\mathcal{F}_\infty$-measurable r.v.. The σ-algebra $\mathcal{G}_t$ is generated by r.v's of the form $B_t h(\tau \wedge t)$ where $B_t \in \mathcal{F}_t$ and $h = \mathbb{1}_{[0,a]}$. Therefore, to prove that $\mathbb{E}(X|\mathcal{G}_t) = \mathbb{E}(X|\mathcal{F}_t)$, it suffices to check that

$$\mathbb{E}(B_t h(\tau\wedge t)X) = \mathbb{E}(B_t h(\tau\wedge t)\mathbb{E}(X|\mathcal{F}_t))$$

for any $B_t \in \mathcal{F}_t$ and any $h = \mathbb{1}_{[0,a]}$. For $t \leq a$, the equality is obvious. For $t > a$, we have from the equality (7.3.1)

$$\begin{aligned}\mathbb{E}\big(B_t \mathbb{1}_{\{\tau\leq a\}}\mathbb{E}(X|\mathcal{F}_t)\big) &= \mathbb{E}\big(\mathbb{1}_{\{\tau\leq a\}}\mathbb{E}(B_t X|\mathcal{F}_t)\big) = \mathbb{E}\big(B_t X \mathbb{E}(\mathbb{1}_{\{\tau\leq a\}}|\mathcal{F}_t)\big)\\ &= \mathbb{E}\big(B_t X \mathbb{E}(\mathbb{1}_{\{\tau\leq a\}}|\mathcal{F}_\infty)\big) = \mathbb{E}\big(B_t X \mathbb{1}_{\{\tau\leq a\}}\big).\end{aligned}$$

The result follows. □

Remark 7.3.5.2 The equality (7.3.4) implies that every **F**-square integrable martingale is a **G**-martingale. However, equality (7.3.4) does not apply to any $\mathcal{G}_\infty$-measurable r.v.; in particular, since τ is a **G**-stopping time and not an **F**-stopping time, $\mathbb{P}(\tau \leq t|\mathcal{G}_t) = \mathbb{1}_{\{\tau\leq t\}}$ is not equal to $\mathbb{P}(\tau \leq t|\mathcal{F}_t)$.

7.3.6 Correlated Defaults: Copula Approach

An approach to modelling dependent credit risks is the use of copulas.

If $X_i, i = 1, \ldots, n$ are random variables with cumulative distribution function F_i, and if the F_i's are strictly increasing, then, setting $U_i = F_i(X_i)$

$$\mathbb{P}(X_i \leq x_i, \forall i) = \mathbb{P}(F_i(X_i) \leq F_i(x_i), \forall i) = \mathbb{P}(U_i \leq F_i(x_i), \forall i)$$

hence, the joint law of the X_i can be characterized in terms of the joint law of the vector U. Note that the r.v. U_i has a uniform law, and that, a priori the U_i's are not independent.

Basic Definitions

Definition 7.3.6.1 *A mapping C defined on $[0,1]^n$ is a copula if it satisfies:*
(i) *$C(u_1, \ldots, u_n)$ is increasing with respect to each component u_i,*
(ii) *$C(1, \ldots, u_i, \ldots, 1) = u_i$, for every i, for every $u_i \in [0,1]$,*
(iii) *for every $a, b \in [0,1]^n$ with $a \leq b$ (i.e., $a_i \leq b_i$, $\forall i$)*

$$\sum_{i_1=1}^{2} \cdots \sum_{i_n=1}^{2} (-1)^{i_1+\cdots+i_n} C(u_{1,i_1}, \ldots, u_{n,i_n}) \geq 0,$$

where $u_{j,1} = a_j$, $u_{j,2} = b_j$.

A copula is the cumulative distribution function of an n-uple $(U_1, \cdots, U_n)$ where the U_i are r.v's with uniform law on the interval $[0,1]$. The **survival copula** is $\widehat{C}(u_1, \ldots, u_n) = \mathbb{P}(U_1 > u_1, \ldots, U_n > u_n)$.

Property (iii) indicates that the probability that the n-tuple belongs to the rectangle $\prod_{i=1}^{n}]a_i, b_i]$ is non-negative.

Theorem 7.3.6.2 (Sklar's Theorem.) *Let F be an n-dimensional cumulative distribution with margins F_i. Then there exists a copula C such that*

$$F(x) = C(F_1(x_1), \ldots, F_n(x_n)).$$

The copula of the n-dimensional random variable $(X_1, \dots, X_n)$ gives information on the dependence of the marginal one-dimensional random variables X_i. From the definition, if the marginal distributions are strictly increasing,

$$\begin{aligned} C(u_1, \dots, u_n) &= F(F_1^{-1}(u_1), \dots, F_n^{-1}(u_n)) \\ &= \mathbb{P}(X_1 \leq F_1^{-1}(u_1), \dots, X_n \leq F_n^{-1}(u_n)) \\ &= \mathbb{P}(F_1(X_1) \leq u_1, \dots, F_n(X_n) \leq u_n) . \end{aligned}$$

A particular copula is the independence copula $C(u_1, \dots, u_n) = \prod_{i=1}^n u_i$, where the different random variables U_i are independent and uniformly distributed.

One of the copulas used by practioners is the Gaussian copula, given by

$$C(v_1, \dots, v_n) = N_\Sigma^n \left(N^{-1}(v_1), \dots, N^{-1}(v_n)\right),$$

where N_Σ^n is the c.d.f for the n-variate central normal distribution with the linear correlation matrix Σ, and N^{-1} is the inverse of the c.d.f. for the univariate standard normal distribution.

Copula and Threshold

As in Subsection 7.3.1, we define $\tau_i = \inf\{t : \Lambda_i(t) \geq \Theta_i\}$ where Λ_i are $\mathbf{F}$-adapted, increasing processes and Θ_i are random variables, independent of $\mathcal{F}_\infty$, with a survival copula

$$\widehat{C}(u_1, \dots, u_n) = \mathbb{P}(\Phi_i(\Theta_i) > u_i, \forall i \leq n)$$

where Φ_i is the cumulative distribution function of Θ_i, assumed to be strictly increasing. Then, the joint law of default times may be characterized by

$$\begin{aligned} \mathbb{P}(\tau_i > t_i, \forall i \leq n) &= \mathbb{P}(\Lambda_i(t_i) < \Theta_i, \forall i \leq n) \\ &= \mathbb{P}(\Phi_i(\Lambda_i(t_i)) < \Phi_i(\Theta_i), \forall i \leq n) \\ &= \mathbb{E}\left(\widehat{C}(\Psi_1(t_1), \dots, \Psi_n(t_n))\right) \end{aligned}$$

where $\Psi_i(t) = \Phi_i(\Lambda_i(t))$. We have also, for $t_i > t, \forall i$,

$$\begin{aligned} \mathbb{P}(\tau_i > t_i, \forall i \leq n | \mathcal{F}_t) &= \mathbb{P}(\Lambda_i(t_i) < \Theta_i, \forall i \leq n | \mathcal{F}_t) \\ &= \mathbb{E}\left(\widehat{C}(\Psi_1(t_1), \dots, \Psi_n(t_n)) | \mathcal{F}_t\right) . \end{aligned}$$

In particular, if $\tau = \inf_i(\tau_i)$ and $Y \in \mathcal{F}_T$

$$\begin{aligned} \mathbb{E}(Y \mathbb{1}_{\{\tau_i > T\}} | \mathcal{G}_t) &= \mathbb{1}_{\{\tau > t\}} \frac{\mathbb{E}(Y \mathbb{1}_{\{\tau_i > T\}} \mathbb{1}_{\{\tau > t\}} | \mathcal{F}_t)}{\mathbb{P}(\tau > t | \mathcal{F}_t)} \\ &= \mathbb{1}_{\{\tau > t\}} \mathbb{E}\left(Y \frac{\widehat{C}(\Psi_1(t), \dots, \Psi_i(T), \dots \Psi_n(t))}{\widehat{C}(\Psi_1(t), \dots, \Psi_i(t), \dots \Psi_n(t))} | \mathcal{F}_t\right) . \end{aligned}$$

Comment 7.3.6.3 The reader is refered to various works, e.g., Bouyé et al [115], Coutant et al. [200] and Nelsen [668] for definitions and properties of copulas and to Frey and McNeil [358], Embrechts et al. [323] and Li [588] for financial applications.

7.3.7 Correlated Defaults: Jarrow and Yu's Model

Let us define $\tau_i = \inf\{t : \Lambda_i(t) \geq \Theta_i\}, i = 1, 2$ where $\Lambda_i(t) = \int_0^t \lambda_i(s)ds$ and Θ_i are independent random variables with exponential law of parameter 1. As in Jarrow and Yu [481], we consider the case where λ_1 is a constant and

$$\lambda_2(t) = \lambda_2 + (\alpha_2 - \lambda_2)\mathbb{1}_{\{\tau_1 \leq t\}} = \lambda_2 \mathbb{1}_{\{t<\tau_1\}} + \alpha_2 \mathbb{1}_{\{\tau_1 \leq t\}} .$$

Assume for simplicity that $r = 0$. Our aim is to compute the value of a defaultable zero-coupon with default time τ_i, with a rebate δ_i paid at maturity:

$$D_i(t,T) = \mathbb{E}(\mathbb{1}_{\{\tau_i>T\}} + \delta_i \mathbb{1}_{\{\tau_i<T\}}|\mathcal{G}_t), \text{ for } \mathcal{G}_t = \mathcal{D}_t^1 \vee \mathcal{D}_t^2 .$$

We compute the joint law of the pair (τ_1, τ_2) given by the survival probability $G(s,t) = \mathbb{P}(\tau_1 > s, \tau_2 > t)$.

▶ **Case** $t \leq s$: For $t \leq s < \tau_1$, one has $\Lambda_2(t) = \lambda_2 t$. Hence, the following equality

$$\begin{aligned}\{\tau_1 > s\} \cap \{\tau_2 > t\} &= \{\tau_1 > s\} \cap \{\Lambda_2(t) < \Theta_2\} = \{\tau_1 > s\} \cap \{\lambda_2 t < \Theta_2\} \\ &= \{\lambda_1 s < \Theta_1\} \cap \{\lambda_2 t < \Theta_2\}\end{aligned}$$

leads to

$$\text{for } t \leq s,\ \mathbb{P}(\tau_1 > s, \tau_2 > t) = e^{-\lambda_1 s} e^{-\lambda_2 t} . \tag{7.3.5}$$

▶ **Case** $t > s$

$$\begin{aligned}\{\tau_1 > s\} \cap \{\tau_2 > t\} &= \{\{t > \tau_1 > s\} \cap \{\tau_2 > t\}\} \cup \{\{\tau_1 > t\} \cap \{\tau_2 > t\}\} \\ \{t > \tau_1 > s\} \cap \{\tau_2 > t\} &= \{t > \tau_1 > s\} \cap \{\Lambda_2(t) < \Theta_2\} \\ &= \{t > \tau_1 > s\} \cap \{\lambda_2 \tau_1 + \alpha_2 (t - \tau_1) < \Theta_2\} .\end{aligned}$$

The independence between Θ_1 and Θ_2 implies that the r.v. $\tau_1 = \Theta_1/\lambda_1$ is independent of Θ_2, hence

$$\begin{aligned}\mathbb{P}(t > \tau_1 > s, \tau_2 > t) &= \mathbb{E}\left(\mathbb{1}_{\{t>\tau_1>s\}} e^{-(\lambda_2\tau_1 + \alpha_2(t-\tau_1))}\right) \\ &= \int du\, \mathbb{1}_{\{t>u>s\}} e^{-(\lambda_2 u + \alpha_2(t-u))} \lambda_1 e^{-\lambda_1 u}\end{aligned}$$

$$= \frac{\lambda_1 e^{-\alpha_2 t}}{\lambda_1 + \lambda_2 - \alpha_2} \lambda_1 e^{-\alpha_2 t} \left(e^{-s(\lambda_1+\lambda_2-\alpha_2)} - e^{-t(\lambda_1+\lambda_2-\alpha_2)}\right) ;$$

Setting $\Delta = \lambda_1 + \lambda_2 - \alpha_2$, it follows that

$$\mathbb{P}(\tau_1 > s, \tau_2 > t) = \frac{1}{\Delta}\lambda_1 e^{-\alpha_2 t}\left(e^{-s\Delta} - e^{-t\Delta}\right) + e^{-\lambda_1 t}e^{-\lambda_2 t}. \tag{7.3.6}$$

In particular, for $s = 0$,

$$\mathbb{P}(\tau_2 > t) = \frac{1}{\Delta}\left(\lambda_1\left(e^{-\alpha_2 t} - e^{-(\lambda_1+\lambda_2)t}\right) + \Delta e^{-\lambda_1 t}\right).$$

▶ The computation of D_1 reduces to that of

$$\mathbb{P}(\tau_1 > T|\mathcal{G}_t) = \mathbb{P}(\tau_1 > T|\mathcal{F}_t \vee \mathcal{D}_t^1)$$

where $\mathcal{F}_t = \mathcal{D}_t^2$. From the key Lemma 7.3.4.1,

$$\mathbb{P}(\tau_1 > T|\mathcal{D}_t^2 \vee \mathcal{D}_t^1) = \mathbb{1}_{\{t<\tau_1\}}\frac{\mathbb{P}(\tau_1 > T|\mathcal{D}_t^2)}{\mathbb{P}(\tau_1 > t|\mathcal{D}_t^2)}.$$

Therefore, using equalities (7.3.5) and (7.3.6)

$$D_1(t,T) = \delta_1 + \mathbb{1}_{\{\tau_1>t\}}(1-\delta_1)e^{-\lambda_1(T-t)}.$$

▶ The computation of D_2 follows from

$$\mathbb{P}(\tau_2 > T|\mathcal{D}_t^1 \vee \mathcal{D}_t^2) = \mathbb{1}_{\{t<\tau_2\}}\frac{\mathbb{P}(\tau_2 > T|\mathcal{D}_t^1)}{\mathbb{P}(\tau_2 > t|\mathcal{D}_t^1)}$$

and

$$\mathbb{P}(\tau_2 > T|\mathcal{D}_t^1) = \mathbb{1}_{\{\tau_1>t\}}\frac{\mathbb{P}(\tau_1 > t, \tau_2 > T)}{\mathbb{P}(\tau_1 > t)} + \mathbb{1}\{\tau_1 \le t\}\mathbb{P}(\tau_2 > T|\tau_1).$$

Some easy computations lead to

$$\begin{aligned} D_2(t,T) = \delta_2 + (1-\delta_2)\mathbb{1}_{\{\tau_2>t\}}\Big(&\mathbb{1}_{\{\tau_1\le t\}}e^{-\alpha_2(T-t)} \\ &+\mathbb{1}_{\{\tau_1>t\}}\frac{1}{\Delta}(\lambda_1 e^{-\alpha_2(T-t)} + (\lambda_2-\alpha_2)e^{-(\lambda_1+\lambda_2)(T-t)})\Big). \end{aligned}$$

7.4 Conditional Survival Probability Approach

We present now a more general model. We deal with two kinds of information: the information from the asset's prices, denoted by $(\mathcal{F}_t, t \ge 0)$, and the information from the default time τ, i.e., the knowledge of the time when the default occurred in the past, if it occurred. More precisely, this latter information is modelled by the filtration $\mathbf{D} = (\mathcal{D}_t, t \ge 0)$ generated by the default process $D_t = \mathbb{1}_{\{\tau\le t\}}$.

At the intuitive level, $\mathbf{F}$ is generated by the prices of some assets, or by other economic factors (e.g., long and short interest rates). This filtration can also be a subfiltration of that of the prices. The case where $\mathbf{F}$ is the trivial filtration is exactly what we have studied in the toy model. Though in typical examples $\mathbf{F}$ is chosen to be the Brownian filtration, most theoretical results do not rely on such a specification. We denote $\mathcal{G}_t = \mathcal{F}_t \vee \mathcal{D}_t$.

Special attention is paid here to the hypothesis $(\mathcal{H})$, which postulates the immersion property of $\mathbf{F}$ in $\mathbf{G}$. We establish a representation theorem, in order to understand the meaning of a complete market in a defaultable world, and we deduce the hedging strategies for some defaultable claims. The main part of this section can be found in the surveys of Jeanblanc and Rutkowski [486, 487].

7.4.1 Conditional Expectations

The conditional law of τ with respect to the information $\mathcal{F}_t$ is characterized by $\mathbb{P}(\tau \leq u|\mathcal{F}_t)$. Here, we restrict our attention to the survival conditional distribution (the Azéma's supermartingale)

$$G_t := \mathbb{P}(\tau > t|\mathcal{F}_t).$$

The super-martingale $(G_t, t \geq 0)$ admits a decomposition as $Z - A$ where Z is an $\mathbf{F}$-martingale and A an $\mathbf{F}$-predictable increasing process. We assume in this section that $G_t > 0$ for any t and that G is continuous.

It is straightforward to establish that every $\mathcal{G}_t$-random variable is equal, on the set $\{\tau > t\}$, to an $\mathcal{F}_t$-measurable random variable.

Lemma 7.4.1.1 (Key Lemma.) *Let X be an $\mathcal{F}_T$-measurable integrable r.v. Then, for $t < T$*

$$\boxed{\mathbb{E}(X\mathbb{1}_{\{T<\tau\}}|\mathcal{G}_t) = \mathbb{1}_{\{\tau>t\}}\frac{\mathbb{E}(X\mathbb{1}_{\{\tau>T\}}|\mathcal{F}_t)}{\mathbb{E}(\mathbb{1}_{\{\tau>t\}}|\mathcal{F}_t)} = \mathbb{1}_{\{\tau>t\}}(G_t)^{-1}\mathbb{E}(XG_T|\mathcal{F}_t).} \tag{7.4.1}$$

PROOF: The proof is exactly the same as that of Corollary 7.3.4.2. Indeed,

$$\mathbb{1}_{\{\tau>t\}}\mathbb{E}(X\mathbb{1}_{\{T<\tau\}}|\mathcal{G}_t) = \mathbb{1}_{\{\tau>t\}}x_t$$

where x_t is $\mathcal{F}_t$-measurable. Taking conditional expectation w.r.t. $\mathcal{F}_t$ of both sides, we deduce

$$x_t = \frac{\mathbb{E}(X\mathbb{1}_{\{\tau>T\}}|\mathcal{F}_t)}{\mathbb{E}(\mathbb{1}_{\{\tau>t\}}|\mathcal{F}_t)} = \mathbb{1}_{\{\tau>t\}}(G_t)^{-1}\mathbb{E}(XG_T|\mathcal{F}_t).$$

□

Lemma 7.4.1.2 *Let h be an* **F***-predictable process. Then,*

$$\mathbb{E}(h_\tau \mathbb{1}_{\{\tau \le T\}} | \mathcal{G}_t) = h_\tau \mathbb{1}_{\{\tau \le t\}} - \mathbb{1}_{\{\tau > t\}} (G_t)^{-1} \mathbb{E}\left(\int_t^T h_u dG_u | \mathcal{F}_t \right) . \tag{7.4.2}$$

In terms of the increasing process A of the Doob-Meyer decomposition of G,

$$\mathbb{E}(h_\tau \mathbb{1}_{\{\tau \le T\}} | \mathcal{G}_t) = h_\tau \mathbb{1}_{\{\tau \le t\}} + \mathbb{1}_{\{\tau > t\}} (G_t)^{-1} \mathbb{E}\left(\int_t^T h_u dA_u | \mathcal{F}_t \right) .$$

PROOF: The proof follows the same lines as that of Lemma 7.3.4.3. □

As we shall see, this elementary result will allow us to compute the value of defaultable claims and of credit derivatives as CDS's. We are not interested with the case where h is a **G**-predictable process, mainly because every **G**-predictable process is equal, on the set $\{t \le \tau\}$, to an **F**-predictable process.

Lemma 7.4.1.3 *The process* $M_t := D_t - \int_0^{t\wedge\tau} \frac{dA_s}{G_s}$ *is a* **G***-martingale.*

PROOF: The proof is based on the key lemma and Fubini's theorem. We leave the details to the reader. □

In other words, the process $\int_0^{t\wedge\tau} \frac{dA_s}{G_s}$ is the **G**-predictable compensator of D.

Comment 7.4.1.4 The hazard process $\Gamma_t := -\ln G_t$ is often introduced. In the case of the Cox Process model, the hazard process is increasing and is equal to Λ.

7.5 Conditional Survival Probability Approach and Immersion

We discuss now the hypothesis on the modelling of default time that we require, under suitable conditions, to avoid arbitrages in the defaultable market. We recall that the immersion property, also called $(\mathcal{H})$-hypothesis (see Subsection 5.9.1) states that any square integrable **F**-martingale is a **G**-martingale. In the first part, we justify that, under some financial conditions, the $(\mathcal{H})$-hypothesis holds. Then, we present some consequences of this condition and we establish a representation theorem and give an important application to hedging.

7.5.1 $(\mathcal{H})$-Hypothesis and Arbitrages

If r is the interest rate, we denote, as usual $R_t = \exp(-\int_0^t r_s ds)$.

Proposition 7.5.1.1 *Let S be the dynamics of a default-free price process, represented as a semi-martingale on $(\Omega, \mathcal{G}, \mathbb{P})$ and $\mathcal{F}_t^S = \sigma(S_s, s \leq t)$ its natural filtration. Assume that the interest rate r is $\mathbf{F}^S$-adapted and that there exists a unique probability $\mathbb{Q}$, equivalent to $\mathbb{P}$ on $\mathcal{F}_T^S$, such that the discounted process $(S_t R_t, 0 \leq t \leq T)$ is an $\mathbf{F}^S$-martingale under the probability $\mathbb{Q}$. Assume also that there exists a probability $\widetilde{\mathbb{Q}}$, equivalent to $\mathbb{P}$ on $\mathcal{G}_T = \mathcal{F}_T^S \vee \mathcal{D}_T$, such that $(S_t R_t, 0 \leq t \leq T)$ is a $\mathbf{G}$-martingale under the probability $\widetilde{\mathbb{Q}}$. Then, $(\mathcal{H})$ holds under $\widetilde{\mathbb{Q}}$ and the restriction of $\widetilde{\mathbb{Q}}$ to $\mathcal{F}_T^S$ is equal to $\mathbb{Q}$.*

PROOF: We give a "financial proof." Under our hypothesis, any $\mathbb{Q}$-square integrable $\mathcal{F}_T^S$-measurable r.v. X can be thought of as the value of a contingent claim. Since the same claim exists in the larger market, which is assumed to be arbitrage free, the discounted value of that claim is a $(\mathbf{G}, \tilde{\mathbb{Q}})$-martingale. From the uniqueness of the price for a hedgeable claim, for any contingent claim $X \in \mathcal{F}_T^S$ and any $\mathbf{G}$-e.m.m. $\tilde{\mathbb{Q}}$,

$$\mathbb{E}_{\mathbb{Q}}(X R_T | \mathcal{F}_t^S) = \mathbb{E}_{\tilde{\mathbb{Q}}}(X R_T | \mathcal{G}_t) .$$

In particular, $\mathbb{E}_{\mathbb{Q}}(Z) = \mathbb{E}_{\tilde{\mathbb{Q}}}(Z)$ for any $Z \in \mathcal{F}_T^S$ (take $X = Z R_T^{-1}$ and $t = 0$), hence the restriction of any e.m.m. $\tilde{\mathbb{Q}}$ to the σ-algebra $\mathcal{F}_T^S$ equals $\mathbb{Q}$. Moreover, since every square integrable $(\mathbf{F}^S, \mathbb{Q})$-martingale can be written as $\mathbb{E}_{\mathbb{Q}}(Z | \mathcal{F}_t^S) = \mathbb{E}_{\tilde{\mathbb{Q}}}(Z | \mathcal{G}_t)$, we obtain that every square integrable $(\mathbf{F}^S, \tilde{\mathbb{Q}})$-martingale is a $(\mathbf{G}, \widetilde{\mathbb{Q}})$-martingale. □

Some Consequences

We recall that the hypothesis $(\mathcal{H})$, studied in Subsection 5.9.4, reads in this particular setting:

$$\forall t, \quad \mathbb{P}(\tau \leq t | \mathcal{F}_\infty) = \mathbb{P}(\tau \leq t | \mathcal{F}_t) := F_t. \tag{7.5.1}$$

In particular, if $(\mathcal{H})$ holds, then F is increasing. Furthermore, if F is continuous, then the predictable increasing process of the Doob-Meyer decomposition of $G = 1 - F$ is equal to F and

$$D_t - \int_0^{t \wedge \tau} \frac{dF_s}{G_s}$$

is a $\mathbf{G}$-martingale.

Remarks 7.5.1.2 (a) If τ is $\mathcal{F}_\infty$-measurable, then equality (7.5.1) is equivalent to: τ is an **F**-stopping time. Moreover, if **F** is the Brownian filtration, then τ is predictable and the Doob-Meyer decomposition of G is $G_t = 1 - F_t$, where F is the predictable increasing process.

(b) Though the hypothesis $(\mathcal{H})$ does not necessarily hold true, in general, it is satisfied when τ is constructed through a Cox process approach (see Section 7.3).

(c) This hypothesis is quite natural under the historical probability, and is stable under particular changes of measure. However, Kusuoka provides an example where $(\mathcal{H})$ holds under the historical probability and does not hold after a particular change of probability. This counterexample is linked with dependency between different defaults (see $\rightarrowtail$ Subsection 7.5.3).

(d) Hypothesis $(\mathcal{H})$ holds in particular if τ is independent of $\mathcal{F}_\infty$. See Greenfeld's thesis [406] for a study of derivative claims in that simple setting.

Comment 7.5.1.3 Elliott et al. [315] pay attention to the case when F is increasing. Obviously, if $(\mathcal{H})$ holds, then F is increasing, however, the reverse does not hold. The increasing property of F is equivalent to the fact that every **F**-martingale, stopped at time τ, is a **G**-martingale. Nikeghbali and Yor [675] proved that this is equivalent to $\mathbb{E}(m_\tau) = m_0$ for any bounded **F**-martingale m (see Proposition 5.9.4.7). It is worthwhile noting that in $\rightarrowtail$ Subsection 7.6.1, the process F is not increasing.

7.5.2 Pricing Contingent Claims

Assume that the default-free market consists of **F**-adapted prices and that the default-free interest rate is **F**-adapted. Whether the defaultable market is complete or not will be studied in the following section. Let $\mathbb{Q}^*$ be the e.m.m. chosen by the market and G^* the $\mathbb{Q}^*$-survival probability, assumed to be continuous. From (7.4.1) the discounted price of the defaultable contingent claim $X \in \mathcal{F}_T$ is

$$R_t \mathbb{E}_{\mathbb{Q}^*}(X \mathbb{1}_{\{\tau > T\}} R_T | \mathcal{G}_t) = \mathbb{1}_{\{t<\tau\}} (G_t^*)^{-1} \mathbb{E}_{\mathbb{Q}^*}(X G_T^* R_T | \mathcal{F}_t) .$$

If $(\mathcal{H})$ holds under $\mathbb{Q}^*$ and if G^* is differentiable, then $G_t^* = \exp(-\int_0^t \lambda_s^* ds)$ and

$$\mathbb{E}_{\mathbb{Q}^*}(X R_T \mathbb{1}_{\{\tau > T\}} | \mathcal{G}_t) = \mathbb{1}_{\{t<\tau\}} \mathbb{E}_{\mathbb{Q}^*} \left(X \exp\left(- \int_t^T (r_s + \lambda_s^*) ds \right) | \mathcal{F}_t \right) . \tag{7.5.2}$$

Particular Case: Defaultable Zero-coupon Bond

The price at time t of a default-free bond paying 1 at maturity t satisfies

$$P(t,T) = \mathbb{E}_{\mathbb{Q}^*}\left(\exp\left(-\int_t^T r_s\,ds\right)\Big|\mathcal{F}_t\right),$$

for any e.m.m. $\mathbb{Q}^*$ (recall that the market is incomplete as long as no defaultable asset is traded). Now, if a defaultable zero-coupon bond is traded in the market with a price - given by the market - equal to $D(t,T)$, we have, for a particular $\mathbb{Q}^*$,

$$\begin{aligned} D(t,T) &= \mathbb{E}_{\mathbb{Q}^*}\left(\mathbb{1}_{\{T<\tau\}}\exp\left(-\int_t^T r_s\,ds\right)\Big|\mathcal{G}_t\right) \\ &= \mathbb{1}_{\{\tau>t\}}\mathbb{E}_{\mathbb{Q}^*}\left(\exp\left(-\int_t^T [r_s+\lambda_s^*]\,ds\right)\Big|\mathcal{F}_t\right). \end{aligned}$$

The value of λ^* is obtained from the price of the defaultable zero coupons, and is related to the conditional law of τ given $\mathcal{F}_t$ as we have explained in Subsection 7.3.1.

7.5.3 Correlated Defaults: Kusuoka's Example

Kusuoka [552] assumes that the default times $\tau_i, i = 1,2$ are independent exponential random variables, i.e., they have joint density

$$f(x,y) = \lambda_1\lambda_2 e^{-(\lambda_1 x+\lambda_2 y)}\mathbb{1}_{\{x\geq 0, y\geq 0\}}$$

under the historical probability $\mathbb{P}$. The reference filtration $\mathbf{F}$ is trivial and the observation filtration is $\mathbf{G} = \mathbf{D}^1 \vee \mathbf{D}^2$ where $\mathbf{D}^i$ is the natural filtration of the process $D_t^i = \mathbb{1}_{\{\tau_i\leq t\}}$. Define a measure $\mathbb{Q}$ as

$$\mathbb{Q}|_{\mathcal{G}_T} = \eta_T\,\mathbb{P}|_{\mathcal{G}_T}$$

where

$$d\eta_t = \eta_{t-}(\kappa_t^1 dM_t^1 + \kappa_t^2 dM_t^2),\ \eta_0 = 1,$$

the $(\mathbb{P},\mathbf{G})$-martingales $(M^i, i=1,2)$ are defined by $M_t^i = D_t^i - \lambda_i(t\wedge\tau_i)$ and

$$\kappa_t^i = \mathbb{1}_{\{\tau_j\leq t\}}\left(\frac{\alpha_i}{\lambda_i}-1\right),\quad \text{for } i\neq j, i,j=1,2.$$

Using Girsanov's Theorem (see ↣ Section 9.4), the processes

$$\widetilde{M}_t^i = D_t^i - \int_0^{t\wedge\tau_i}\lambda_s^i ds$$

where

$$\lambda_t^1 = \lambda_1\mathbb{1}_{\{\tau_2>t\}} + \alpha_1\mathbb{1}_{\{\tau_2\leq t\}}\quad \lambda_t^2 = \lambda_2\mathbb{1}_{\{\tau_1>t\}} + \alpha_2\mathbb{1}_{\{\tau_1\leq t\}}$$

are $(\mathbb{Q}, \mathbf{G})$-martingales. The two default times are no longer independent under $\mathbb{Q}$. Furthermore, the $(\mathcal{H})$ hypothesis does not hold under $\mathbb{Q}$ between $\mathbf{D}^2$ and $\mathbf{D}^1 \vee \mathbf{D}^2$, in particular

$$\mathbb{Q}(\tau_1 > T | \mathcal{D}_t^1 \vee \mathcal{D}_t^2) \neq \mathbb{1}_{\{\tau_1 > t\}} \mathbb{E}_{\mathbb{Q}} \left(\exp \left(-\int_t^T \lambda_u^1 du \right) | \mathcal{D}_t^2 \right).$$

7.5.4 Stochastic Barrier

We assume that the $(\mathcal{H})$-hypothesis holds under $\mathbb{P}$ and that F is strictly increasing and continuous. Then, there exists a continuous strictly increasing $\mathbf{F}$-adapted process Γ such that

$$\mathbb{P}(\tau > t | \mathcal{F}_\infty) = e^{-\Gamma_t}.$$

Our goal is to show that there exists a random variable Θ, independent of $\mathcal{F}_\infty$, with exponential law of parameter 1, such that $\tau = \inf\{t \geq 0 : \Gamma_t \geq \Theta\}$. Let us set $\Theta := \Gamma_\tau$. Then

$$\{t < \Theta\} = \{t < \Gamma_\tau\} = \{C_t < \tau\},$$

where C is the right inverse of Γ, so that $\Gamma_{C_t} = t$. Therefore

$$\mathbb{P}(\Theta > u | \mathcal{F}_\infty) = e^{-\Gamma_{C_u}} = e^{-u}.$$

We have thus established the required properties, namely, that the exponential probability law of Θ and its independence of the σ-algebra $\mathcal{F}_\infty$. Furthermore, $\tau = \inf\{t : \Gamma_t > \Gamma_\tau\} = \inf\{t : \Gamma_t > \Theta\}$.

Comment 7.5.4.1 This result is extended to a multi-default setting for a trivial filtration in Norros [677] and Shaked and Shanthikumar [782].

7.5.5 Predictable Representation Theorems

We still assume that the $(\mathcal{H})$ hypothesis holds and that F is absolutely continuous w.r.t. Lebesgue measure with density f. We recall that

$$M_t := D_t - \int_0^{t \wedge \tau} \lambda_s ds$$

is a $\mathbf{G}$-martingale where $\lambda_s = f_s / G_s$. Kusuoka [552] establishes the following representation theorem:

Theorem 7.5.5.1 *Suppose that* $\mathbf{F}$ *is a Brownian filtration generated by the Brownian motion* W*. Then, under the hypothesis* $(\mathcal{H})$*, every* $\mathbf{G}$*-square integrable martingale* $(H_t, t \geq 0)$ *admits a representation as the sum of a stochastic integral with respect to the Brownian motion* W *and a stochastic integral with respect to the discontinuous martingale* M*:*

$$H_t = H_0 + \int_0^t \phi_s dW_s + \int_0^t \psi_s dM_s$$

where, for any t, $\mathbb{E}\left(\int_0^t \phi_s^2 ds\right) < \infty$ and $\mathbb{E}\left(\int_0^t \psi_s^2 \lambda_s ds\right) < \infty$.

In the case of a $\mathbf{G}$-martingale of the form $\mathbb{E}(X\mathbb{1}_{\{T<\tau\}}|\mathcal{G}_t)$ where X is $\mathcal{F}_T$-measurable, or for $\mathbb{E}(h_\tau|\mathcal{G}_t)$ where h is predictable, one can be more precise.

Proposition 7.5.5.2 *Suppose that hypothesis $(\mathcal{H})$ holds, that G is continuous and that every $\mathbf{F}$-martingale is continuous.*

Then, the martingale $H_t = \mathbb{E}(h_\tau|\,\mathcal{G}_t)$, where h is an $\mathbf{F}$-predictable process such that $\mathbb{E}(|h_\tau|) < \infty$, admits the following decomposition as the sum of a $\mathbf{G}$-continuous martingale and a $\mathbf{G}$-purely discontinuous martingale:

$$H_t = m_0^h + \int_0^{t\wedge\tau} G_u^{-1} dm_u^h + \int_{]0,t\wedge\tau]} (h_u - J_u)\, dM_u\,. \tag{7.5.3}$$

Here m^h is the continuous $\mathbf{F}$-martingale

$$m_t^h = \mathbb{E}\left(\int_0^\infty h_u dF_u \,|\, \mathcal{F}_t\right),$$

and $J_t = G_t^{-1}(m_t^h - \int_0^t h_u dF_u)$. Moreover, $J_u = H_u$ on the set $\{u < \tau\}$.

PROOF: From (7.3.3) we know that

$$\begin{aligned} H_t &= \mathbb{E}(h_\tau|\,\mathcal{G}_t) = \mathbb{1}_{\{\tau\leq t\}} h_\tau + \mathbb{1}_{\{\tau>t\}} G_t^{-1} \mathbb{E}\left(\int_t^\infty h_u dF_u \,|\, \mathcal{F}_t\right) \\ &= \mathbb{1}_{\{\tau\leq t\}} h_\tau + \mathbb{1}_{\{\tau>t\}} J_t\,. \end{aligned} \tag{7.5.4}$$

From the fact that G is a decreasing continuous process and m^h a continuous martingale, and using the integration by parts formula, we deduce, after some easy computations, that

$$dJ_t = G_t^{-1} dm_t^h + J_t G_t\, d(G_t^{-1}) - h_t G_t^{-1} dF_t = G_t^{-1} dm_t^h + J_t G_t^{-1} dF_t - h_t G_t^{-1} dF_t\,.$$

Therefore,

$$dJ_t = G_t^{-1} dm_t^h + (J_t - h_t)\frac{dF_t}{G_t}$$

or, in an integrated form,

$$J_t = m_0^h + \int_0^t G_u^{-1} dm_u^h + \int_0^t (J_u - h_u)\frac{dF_u}{G_u}.$$

Note that, from (7.5.4), $J_u = H_u$ for $u < \tau$. Therefore, on $\{t < \tau\}$,

$$H_t = m_0^h + \int_0^{t\wedge\tau} G_u^{-1} dm_u^h + \int_0^{t\wedge\tau} (J_u - h_u)\frac{dF_u}{G_u}\,.$$

From (7.5.4), the jump of H at time τ is $h_\tau - J_\tau = h_\tau - H_{\tau-}$. Therefore, (7.5.3) follows. Since hypothesis $(\mathcal{H})$ holds, the processes $(m_t^h, t \geq 0)$ and $(\int_0^t G_u^{-1} dm_u^h, t \geq 0)$ are also **G**-martingales. Hence, the stopped process $(\int_0^{t\wedge\tau} G_u^{-1} dm_u^h, t \geq 0)$ is a **G**-martingale. □

7.5.6 Hedging Contingent Claims with DZC

We assume that $(\mathcal{H})$ holds under the risk-neutral probability $\mathbb{Q}$ chosen by the market, and that the process $G_t = \mathbb{Q}(\tau > t|\mathcal{F}_t)$ is continuous.

We suppose moreover that:

- The *default-free market* including the default-free zero-coupon with constant interest rate and the risky asset S, is complete and arbitrage free, and
$$dS_t = S_t(rdt + \sigma_t dW_t)\,.$$
Here W is a $\mathbb{Q}$-BM.
- A defaultable zero-coupon with maturity T and price $D(t,T)$ is traded on the market.
- The market which consists of the default-free zero-coupon $P(t,T)$, the defaultable zero-coupon $D(t,T)$ and the risky asset S is arbitrage free (in particular, $D(t,T)$ belongs to the range of prices $]0, P(t,T)[$).

We now make precise the hedging of a defaultable claim and check that any $\mathcal{G}_T$-measurable square integrable contingent claim is hedgeable.

The market price of the DZC and the e.m.m. $\mathbb{Q}$ are related by
$$\begin{aligned} D(t,T)e^{-rt} &= \mathbb{E}_\mathbb{Q}(e^{-rT}\mathbb{1}_{\{T<\tau\}}|\mathcal{G}_t) \\ &= \mathbb{1}_{\{t<\tau\}} G_t^{-1} \mathbb{E}_\mathbb{Q}(e^{-rT} G_T|\mathcal{F}_t) = L_t m_t \end{aligned} \tag{7.5.5}$$
where $m_t = \mathbb{E}_\mathbb{Q}(e^{-rT} G_T|\mathcal{F}_t)$ is an **F**-martingale and $L_t = \mathbb{1}_{\{t<\tau\}} G_t^{-1}$.

We recall that a triple (a, b, c) of **G**-predictable processes is a hedging strategy for the contingent claim $Z \in \mathcal{G}_T$ if, denoting by
$$Z_t = a_t e^{rt} + b_t S_t + c_t D(t,T)$$
the time-t value of this strategy, the self-financing relation
$$dZ_t = a_t r e^{rt} dt + b_t dS_t + c_t d_t D(t,T)$$
holds and
$$Z_T = a_T e^{rT} + b_T S_T + c_T D(T,T) = Z\,.$$
From the no-arbitrage hypothesis, we obtain
$$Z_t e^{-rt} = \mathbb{E}_\mathbb{Q}(Z_T e^{-rT}|\mathcal{G}_t)\,.$$

Terminal Payoff

In a first step we study the case of a terminal payoff of the particular form $Z = X\mathbb{1}_{\{T<\tau\}}$ where $X \in L^2(\mathcal{F}_T)$. We compute $\mathbb{E}_{\mathbb{Q}}(X\mathbb{1}_{\{T<\tau\}}e^{-rT}|\mathcal{G}_t)$, and we give the hedging strategy for $X\mathbb{1}_{\{T<\tau\}}$ based on the riskless asset, the risky asset and the defaultable zero-coupon bond.

Theorem 7.5.6.1 *The hedging strategy (a, b, c) for the defaultable contingent claim $X\mathbb{1}_{\{T<\tau\}}$, based on the riskless bond, the asset and the defaultable zero-coupon satisfies*

$$c_t D(t, T) = e^{rt}\mathbb{E}_{\mathbb{Q}}(Xe^{-rT}\mathbb{1}_{\{T<\tau\}}|\mathcal{G}_t) .$$

Hence, $a_t e^{rt} + b_t S_t = 0$.

More precisely, let $(V_t^X - v_t^X S_t, v_t^X)$ be the hedging strategy for the default-free contingent claim XG_T, and $(V_t - v_t S_t, v_t)$ the hedging strategy for the default-free contingent claim G_T, i.e.,

$$\begin{aligned} e^{-rt}V_t^X &= \mathbb{E}_{\mathbb{Q}}(XG_T e^{-rT}|\mathcal{F}_t) = x + \int_0^t v_s^X d(e^{-rs}S_s) \\ e^{-rt}V_t &= \mathbb{E}_{\mathbb{Q}}(G_T e^{-rT}|\mathcal{F}_t) = x + \int_0^t v_s d(e^{-rs}S_s) . \end{aligned} \tag{7.5.6}$$

Then, on $\{t < \tau\}$

(i) $c_t = \dfrac{V_t^X}{V_t}$,

(ii) $b_t = G_t^{-1}\left(v_t^X - \dfrac{V_t^X}{V_t}v_t\right)$,

(iii) $a_t = -G_t^{-1}\left(v_t^X - \dfrac{V_t^X}{V_t}v_t\right)e^{-rt}S_t$.

Obviously, $a_t = b_t = c_t = 0$ on $\{\tau \le t\}$.

PROOF: Let us reduce attention to the simple case $r = 0$. The time-t price of the defaultable claim $X\mathbb{1}_{\{T<\tau\}}$ is Z_t which is defined by

$$Z_t = \mathbb{E}_{\mathbb{Q}}(X\mathbb{1}_{\{T<\tau\}}|\mathcal{G}_t) = L_t V_t^X$$

where $L_t = \mathbb{1}_{\{t<\tau\}}G_t^{-1}$ and $V_t^X = \mathbb{E}_{\mathbb{Q}}(XG_T|\mathcal{F}_t)$. The price of the DZC satisfies $D(t, T) = L_t V_t$. Hence $Z_t = \mathbb{E}_{\mathbb{Q}}(X\mathbb{1}_{\{T<\tau\}}|\mathcal{G}_t) = \dfrac{V_t^X}{V_t}D(t, T)$.

We now check that there exists a triple (a, b, c) determining a self-financing portfolio such that $a_t e^{rt} + b_t S_t = 0$ (in particular, $c_t D(t, T) = Z_t$, hence the process c satisfies (i)). The self-financing condition reads

$$b_t S_t \sigma dW_t + c_t(L_{t-}dV_t + V_t dL_t) = dZ_t = L_{t-}dV_t^X + V_t^X dL_t ,$$

where we have used the fact that the martingales L and V (resp. L and V^X) are orthogonal. From the choice of c, this equality reduces to

$$b_t S_t \sigma dW_t + c_t L_{t-} dV_t = L_{t-} dV_t^X ,$$

i.e.,

$$b_t + c_t L_{t-} v_t = L_{t-} v_t^X .$$

Hence the form of b given in the theorem. □

Comments 7.5.6.2 (a) It is worthwhile comparing this result with the results of Subsection 2.4.5.

(b) See Bielecki et al.[92] for a more complete study of hedging strategies, using a martingale approach, and trading strategies satisfying $a_t e^{rt} + b_t S_t = 0$.

Rebate Part

The representation theorem also provides a hedging strategy for a rebate h paid at hit. We assume that F is differentiable, with derivative f. We denote by C_t^h the price of the default-free contingent claim which consists of a dividend hf paid between time t and T, i.e.,

$$C_t^h e^{-rt} = \mathbb{E}_{\mathbb{Q}}\left(\int_t^T e^{-ru} f_u h_u du | \mathcal{F}_t\right) ,$$

and by μ^h the associated hedging strategy:

$$C_t^h e^{-rt} + \int_0^t e^{-ru} f_u h_u du = C_0^h + \int_0^t \mu_s^h d(e^{-rs} S_s) .$$

Proposition 7.5.6.3 *Let (a, b, c) be the hedging strategy for the rebate part, i.e., the self-financing strategy such that*

$$e^{-rt}(a_t e^{rt} + b_t S_t + c_t D(t,T)) = \mathbb{E}_{\mathbb{Q}}(h_\tau \mathbb{1}_{\{\tau \le T\}} e^{-r\tau} | \mathcal{G}_t) .$$

Then, on the set $\{t < \tau\}$, one has $c_t D(t,T) = e^{rt} \mathbb{E}_{\mathbb{Q}}(h_\tau \mathbb{1}_{\{\tau \le T\}} e^{-r\tau} | \mathcal{G}_t) - h_t$. More precisely, the hedging strategy before the default time of the rebate part, paid at hit, consists of:

(i) $c_t = \frac{1}{V_t}(C_t^h - G_t^{-1} h_t)$,

(ii) $b_t = G_t^{-1}\left(\mu_t^h - \frac{1}{V_t} v_t C_t^h\right) + \frac{1}{V_t} v_t h_t$,

(iii) $a_t e^{rt} = S_t \left(\frac{1}{V_t} v_t h_t - G_t^{-1}\left(\mu_t^h + \frac{1}{V_t} v_t C_t^h\right)\right) + h_t$

where V and v are defined in (7.5.6).

PROOF: We give the proof for the case $r = 0$.

We compute the quantity $\mathbb{E}_{\mathbb{Q}}(h_\tau \mathbb{1}_{\{\tau \le T\}} | \mathcal{G}_t)$, which corresponds to the price of the rebate, when the compensation is paid at hit. The representation theorem 7.5.5.2 states that

$$\mathbb{E}_{\mathbb{Q}}(h_\tau \mathbb{1}_{\{\tau \le T\}}|\mathcal{G}_t) = C_0^h + \int_0^{t\wedge\tau} G_u^{-1}\mu_u^h dS_u + \int_{[0,t\wedge\tau[} (h_u - J_{u-})dM_u\,,$$

where, on the set $\{t < \tau\}$,

$$J_t = \mathbb{E}_{\mathbb{Q}}(h_\tau \mathbb{1}_{\{\tau \le T\}}|\mathcal{G}_t) = G_t^{-1}\mathbb{E}_{\mathbb{Q}}\left(\int_t^T h_u dF_u|\mathcal{F}_t\right) = G_t^{-1}C_t^h\,.$$

The value of a DZC given in (7.5.5) can be written $D(t,T) = \mathbb{1}_{\{t<\tau\}}G_t^{-1}V_t$. In particular,

$$d_t D(t,T) = -D(t-,T)dM_t + L_t dV_t = -D(t-,T)dM_t + L_t v_t dS_t$$

It follows that

$$\begin{aligned}\mathbb{E}_{\mathbb{Q}}\left(h_\tau \mathbb{1}_{\{\tau \le T\}}|\mathcal{G}_t\right) = C_0^h &+ \int_0^{t\wedge\tau} G_u^{-1}\mu_u^h dS_u \\ &- \int_{[0,t\wedge\tau[} (h_u - G_u^{-1}C_u^h)\frac{1}{D(u,T)}[d_u D(u,T) - v_u L_u dS_u]\end{aligned}$$

which leads to, using that $D(t,T) = G_t^{-1}V_t$ on the set $\{t < \tau\}$,

$$\begin{aligned}\mathbb{E}_{\mathbb{Q}}\left(h_\tau \mathbb{1}_{\{\tau \le T\}}|\mathcal{G}_t\right) = V_0^h &+ \int_0^{t\wedge\tau} \left[G_u^{-1}\left(v_u^h - C_u^h\frac{v_u^h}{V_u}\right) + \frac{v_u h_u}{V_u}\right] dS_u \\ &- \int_{[0,t\wedge\tau[} (h_u - G_u^{-1}C_u^h)\frac{1}{G_u V_u}d_u D(u,T)\,.\end{aligned}$$

□

Comments 7.5.6.4 (a) Under the $(\mathcal{H})$ hypothesis, the Kusuoka representation theorem and the form of the dynamics of a DZC imply that if the default-free market is complete, the defaultable market is complete as soon as a defaultable zero-coupon is traded.

(b) See also Bélanger et al. [67], Jeanblanc and Rutkowski [488] and Bielecki et al. [89] for an extensive study of hedging strategies and Bielecki et al. [90] and ↣ Section 7.9 for a PDE approach, based on Itô's calculus for processes with jumps.

7.6 General Case: Without the $(\mathcal{H})$-Hypothesis

7.6.1 An Example of Partial Observation

As pointed out by Jamshidian [476], *"one may wish to apply the general theory perhaps as an intermediate step, to a subfiltration that is not equal to the default-free filtration. In that case,* **F** *rarely satisfies hypothesis* $(\mathcal{H})$*".* We present here a simple case of such a situation. Assume that

$$dV_t = V_t(\mu dt + \sigma dW_t),\ V_0 = v$$

i.e., $V_t = ve^{\sigma(W_t+\nu t)} = ve^{\sigma X_t}$, with $\nu = (\mu - \sigma^2/2)/\sigma$ and $X_t = W_t + \nu t$. We denote by $\mathcal{F}_t^W = \sigma(W_s, s \leq t)$ the natural filtration of the Brownian motion (this is also the natural filtration of X).

The default time is assumed to be the first hitting time of α with $\alpha < v$, i.e.,

$$\tau = \inf\{t : V_t \leq \alpha\} = \inf\{t : X_t \leq a\}$$

where $a = \sigma^{-1}\ln(\alpha/v)$. Here, the reference filtration $\mathbf{F}$ is the filtration of the observations of V at discrete times $t_1, \cdots t_n$ where $t_n \leq t < t_{n+1}$, i.e.,

$$\mathcal{F}_t = \sigma(V_{t_1}, \ldots, V_{t_n}, t_i \leq t)$$

and we compute $F_t = \mathbb{P}(\tau \leq t|\mathcal{F}_t)$. Let us recall that (See Subsection 3.2.2)

$$\mathbb{P}(\inf_{s\leq t} X_s > z) = \Phi(\nu, t, z)\,, \tag{7.6.1}$$

where

$$\begin{cases} \Phi(\nu,t,z) = \begin{cases} \mathcal{N}\left(\dfrac{\nu t - z}{\sqrt{t}}\right) - e^{2\nu z}\mathcal{N}\left(\dfrac{z+\nu t}{\sqrt{t}}\right), & \text{for } z < 0,\ t > 0, \\ 0, & \text{for } z \geq 0,\ t \geq 0, \end{cases} \\ \Phi(\nu, 0, z) = 1, \quad \text{for } z < 0. \end{cases}$$

We now divide the study into three steps:

▶ **On** $t < t_1$**.** In that case, for $a < 0$, we obtain

$$\begin{aligned} F_t &= \mathbb{P}(\tau \leq t) = \mathbb{P}\left(\inf_{s\leq t} X_s \leq a\right) \\ &= 1 - \Phi(\nu, t, a) = \mathcal{N}\left(\frac{a-\nu t}{\sqrt{t}}\right) + e^{2\nu a}\mathcal{N}\left(\frac{a+\nu t}{\sqrt{t}}\right). \end{aligned}$$

▶ **On** $t_1 < t < t_2$**.**

$$\begin{aligned} F_t &= \mathbb{P}(\tau \leq t|X_{t_1}) = 1 - \mathbb{P}(\tau > t|X_{t_1}) \\ &= 1 - \mathbb{E}\left(\mathbb{1}_{\{\inf_{s<t_1} X_s > a\}}\, \mathbb{P}\left(\inf_{t_1\leq s<t} X_s > a|\mathcal{F}_{t_1}^W\right)|X_{t_1}\right) \end{aligned}$$

The independence and stationarity of the increments of X yield

$$\mathbb{P}\left(\inf_{t_1\leq s<t} X_s > a|\mathcal{F}_{t_1}^W\right) = \Phi(\nu, t-t_1, a - X_{t_1})\,.$$

Hence

$$F_t = 1 - \Phi(\nu, t - t_1, a - X_{t_1})\mathbb{P}(\inf_{s<t_1} X_s > a | X_{t_1}) .$$

From Exercise 3.2.2.1, for $X_{t_1} > a$, we obtain (we omit the parameter ν in the definition of Φ)

$$F_t = 1 - \Phi(t - t_1, a - X_{t_1}) \left[1 - \exp\left(-\frac{2\,a}{t_1}\,(a - X_{t_1})\right)\right] . \tag{7.6.2}$$

The case $X_{t_1} \leq a$ corresponds to default and, therefore, for $X_{t_1} \leq a$, $F_t = 1$.

The process F is continuous and increasing in $[t_1, t_2[$.

When t approaches t_1 from above, one has $\lim_{t\to t_1^+} \Phi(t - t_1, a - X_{t_1}) = 1$ for $X_{t_1} > a$, hence $F_{t_1^+} = \exp\left[-\frac{2a}{t_1}\,(a - X_{t_1})\right]$. For $X_{t_1} > a$, the jump of F at t_1 is

$$\Delta F_{t_1} = \exp\left[-\frac{2a}{t_1}\,(a - X_{t_1})\right] - 1 + \Phi(t_1, a).$$

For $X_{t_1} \leq a$, $\Phi(t - t_1, a - X_{t_1}) = 0$ by the definition of $\Phi(\cdot)$ and

$$\Delta F_{t_1} = \Phi(t_1, a).$$

► **General Observation Times $t_i < t < t_{i+1} < T$, $i \geq 2$.**
For $t_i < t < t_{i+1}$,

$$\begin{aligned}\mathbb{P}(\tau > t | X_{t_1}, \ldots, X_{t_i}) &= \mathbb{P}\left(\inf_{s\leq t_i} X_s > a\, \mathbb{P}\left(\inf_{t_i\leq s<t} X_s > a | \mathcal{F}_{t_i}\right) | X_{t_1}, \ldots, X_{t_i}\right)\\ &= \Phi(t - t_i, a - X_{t_i})\mathbb{P}\left(\inf_{s\leq t_i} X_s > a | X_{t_1}, \ldots, X_{t_i}\right).\end{aligned}$$

We write K_i for the second term on the right-hand side

$$\begin{aligned}K_i &= \mathbb{P}\left(\inf_{s\leq t_i} X_s > a | X_{t_1}, \ldots, X_{t_i}\right)\\ &= \mathbb{P}\left(\inf_{s\leq t_{i-1}} X_s > a\, \mathbb{P}\left(\inf_{t_{i-1}\leq s<t_i} X_s > a | \mathcal{F}_{t_{i-1}} \vee X_{t_i}\right) | X_{t_1}, \ldots, X_{t_i}\right).\end{aligned}$$

Obviously,

$$\begin{aligned}\mathbb{P}\left(\inf_{t_{i-1}\leq s<t_i} X_s > a | \mathcal{F}_{t_{i-1}} \vee X_{t_i}\right) &= \mathbb{P}\left(\inf_{t_{i-1}\leq s<t_i} X_s > a | X_{t_{i-1}}, X_{t_i}\right)\\ &= \exp\left(-\frac{2}{t_i - t_{i-1}}(a - X_{t_{i-1}})(a - X_{t_i})\right).\end{aligned}$$

Therefore,

$$K_i = K_{i-1} \exp\left(-\frac{2}{t_i - t_{i-1}}(a - X_{t_{i-1}})(a - X_{t_i})\right). \tag{7.6.3}$$

Hence,

$$\begin{aligned}\mathbb{P}(\tau \le t|\mathcal{F}_t) &= 1 \qquad \text{if } X_{t_j} < a \text{ for at least one } t_j,\, t_j < t\,,\\ &= 1 - \Phi(t - t_i, a - X_{t_i})K_i, \quad \text{otherwise}\end{aligned}$$

where

$$K_i = k(t_1, X_{t_1}, 0)k(t_2 - t_1, X_{t_1}, X_{t_2}) \cdots k(t_i - t_{i-1}, X_{t_{i-1}}, X_{t_i})$$

and $k(s, x, y) = 1 - \exp\left(-\frac{2}{s}(a - x)(a - y)\right)$.

Comment 7.6.1.1 It is also possible, as in Duffie and Lando [273], to assume that the observation at time $[t]$ is only $V_{[t]} + \epsilon$ where ϵ is a noise, modelled as a random variable independent of V. Another example, related to Parisian stopping times is presented in Çetin et al. [158].

Exercise 7.6.1.2 Prove that the process ζ defined by $\zeta_t = \sum_{i, t_i \le t} \Delta F_{t_i}$ is an **F**-martingale.
`Hint:` This is a trivial check. For details, see Jeanblanc and Valchev [489]. ◁

7.6.2 Two Defaults, Trivial Reference Filtration

We present some results on the case of two default times in the particular case where the reference filtration **F** is the trivial filtration. We denote by **G** the filtration $\mathbf{H}^1 \vee \mathbf{H}^2$ and by $G(u, v) = \mathbb{P}(\tau_1 > u, \tau_2 > v)$ the survival probability of the pair (τ_1, τ_2) and by $F_i(s) = \mathbb{P}(\tau_i \le s) = \int_0^s f_i(u)du$ the marginal cumulative distribution functions. We assume that G is twice continuously differentiable. Our aim is to study the $(\mathcal{H})$ hypothesis between $\mathbf{H}^1$ and **G**.

▶ **Filtration $\mathbf{H}^i$** From Proposition 7.2.2.1, for any $i = 1, 2$, the process

$$M_t^i = H_t^i - \int_0^{t \wedge \tau_i} \frac{f_i(s)}{1 - F_i(s)} ds \tag{7.6.4}$$

is an $\mathbf{H}^i$-martingale.

▶ **Filtration G**

Lemma 7.6.2.1 *The $\mathbf{H}^2$ Doob-Meyer decomposition of $F_t^{1|2} := \mathbb{P}(\tau_1 \le t|\mathcal{H}_t^2)$ is*

$$dF_t^{1|2} = \left(\frac{G(t,t)}{G(0,t)} - \frac{\partial_2 G(t,t)}{\partial_2 G(0,t)}\right) dM_t^2 + \left(H_t^2 \partial_1 h(t, \tau_2) - (1 - H_t^2)\frac{\partial_1 G(t,t)}{G(0,t)}\right) dt$$

where

$$h(t, v) = 1 - \frac{\partial_2 G(t, v)}{\partial_2 G(0, v)}\,.$$

PROOF: Some easy computation enables us to write

$$\begin{aligned} F_t^{1|2} &= H_t^2 \mathbb{P}(\tau_1 \le t|\tau_2) + (1 - H_t^2)\frac{\mathbb{P}(\tau_1 \le t < \tau_2)}{\mathbb{P}(\tau_2 > t)} \\ &= H_t^2 h(t, \tau_2) + (1 - H_t^2)\frac{G(0,t) - G(t,t)}{G(0,t)}, \end{aligned} \tag{7.6.5}$$

Introducing the deterministic function $\psi(t) = 1 - G(t,t)/G(0,t)$, the sub-martingale $F_t^{1|2}$ has the form

$$F_t^{1|2} = H_t^2 h(t, \tau_2) + (1 - H_t^2)\psi(t) \tag{7.6.6}$$

The function $t \to \psi(t)$ and the process $t \to h(t, \tau_2)$ are continuous and of finite variation, hence the integration by parts formula leads to

$$\begin{aligned} dF_t^{1|2} &= h(t,\tau_2)dH_t^2 + H_t^2 \partial_1 h(t,\tau_2)dt + (1 - H_t^2)\psi'(t)dt - \psi(t)dH_t^2 \\ &= (h(t,\tau_2) - \psi(t))\, dH_t^2 + \left(H_t^2 \partial_1 h(t,\tau_2) + (1 - H_t^2)\psi'(t)\right) dt \\ &= \left(\frac{G(t,t)}{G(0,t)} - \frac{\partial_2 G(t,\tau_2)}{\partial_2 G(0,\tau_2)}\right) dH_t^2 + \left(H_t^2 \partial_1 h(t,\tau_2) + (1 - H_t^2)\psi'(t)\right) dt\,. \end{aligned}$$

Now, we note that

$$\begin{aligned} \int_0^T \left(\frac{G(t,t)}{G(0,t)} - \frac{\partial_2 G(t,\tau_2)}{\partial_2 G(0,\tau_2)}\right) dH_t^2 &= \left(\frac{G(\tau_2,\tau_2)}{G(0,\tau_2)} - \frac{\partial_2 G(\tau_2,\tau_2)}{\partial_2 G(0,\tau_2)}\right) 1_{\{\tau_2 \le t\}} \\ &= \int_0^T \left(\frac{G(t,t)}{G(0,t)} - \frac{\partial_2 G(t,t)}{\partial_2 G(0,t)}\right) dH_t^2 \end{aligned}$$

and substitute it into the expression of $dF^{1|2}$:

$$dF_t^{1|2} = \left(\frac{G(t,t)}{G(0,t)} - \frac{\partial_2 G(t,t)}{\partial_2 G(0,t)}\right) dH_t^2 + \left(H_t^2 \partial_1 h(t,\tau_2) + (1 - H_t^2)\psi'(t)\right) dt\,.$$

From

$$dH_t^2 = dM_t^2 - \left(1 - H_t^2\right) \frac{\partial_2 G(0,t)}{G(0,t)} dt\,,$$

where M^2 is an $\mathbb{H}^2$-martingale, we get the Doob-Meyer decomposition of $F^{1|2}$:

$$\begin{aligned} dF_t^{1|2} = &\left(\frac{G(t,t)}{G(0,t)} - \frac{\partial_2 G(t,t)}{\partial_2 G(0,t)}\right) dM_t^2 \\ &- \left(1 - H_t^2\right) \left(\frac{G(t,t)}{G(0,t)} - \frac{\partial_2 G(t,t)}{\partial_2 G(0,t)}\right) \frac{\partial_2 G(0,t)}{G(0,t)} dt \\ &+ \left(H_t^2 \partial_1 h(t,\tau_2) + (1 - H_t^2)\psi'(t)\right) dt \end{aligned}$$

and, after the computation of $\psi'(t)$, one obtains

$$dF_t^{1|2} = \left(\frac{G(t,t)}{G(0,t)} - \frac{\partial_2 G(t,t)}{\partial_2 G(0,t)}\right) dM_t^2 + \left(H_t^2 \partial_1 h(t,\tau_2) - (1-H_t^2)\frac{\partial_1 G(t,t)}{G(0,t)}\right) dt\,.$$

□

As a consequence:

Lemma 7.6.2.2 *The $(\mathcal{H})$ hypothesis is satisfied for $\mathbf{H}^1$ and $\mathbf{G}$ if and only if*

$$\frac{G(t,t)}{G(0,t)} = \frac{\partial_2 G(t,t)}{\partial_2 G(0,t)}\,.$$

Proposition 7.6.2.3 *The process*

$$H_t^1 - \int_0^{t\wedge\tau_1} \frac{a(s)}{1-F^{1|2}(s)} ds\,,$$

where $a(t) = H_t^2 \partial_1 h(t,\tau_2) - (1-H_t^2)\frac{\partial_1 G(t,t)}{G(0,t)}$ and $h(t,s) = 1 - \frac{\partial_2 G(t,s)}{\partial_2 G(0,s)}$, is a $\mathbf{G}$-martingale.

PROOF: The result follows from Lemma 7.4.1.3 and the form of the Doob-Meyer decomposition of $F^{1|2}$. □

7.6.3 Initial Times

In order that the prices of the default-free assets do not induce arbitrage opportunities, one needs to prove that $\mathbf{F}$-martingales remain $\mathbf{G}$-semi-martingales. We have seen in Proposition 5.9.4.10 that this is the case when the random time τ is honest. However, in the credit risk setting, default times are not honest (see for example the Cox model). Hence, we have to give another condition. We shall assume that the conditions of Proposition 5.9.3.1 are satisfied. This will imply that $\mathbf{F}$-martingales are $\mathbf{F} \vee \sigma(\tau)$- semi-martingales, and of course $\mathbf{G}$-semi-martingales.

For any positive random time τ, and for every t, we write $q_t(\omega, dT)$ the regular conditional distribution of τ, and

$$G_t^T(\omega) = \mathbb{Q}(\tau > T|\mathcal{F}_t)(\omega) = q_t(\omega,]T,\infty[).$$

For simplicity, we introduce the following (non-standard) definition:

Definition 7.6.3.1 (Initial Times) *The positive random time τ is called an initial time if there exists a probability measure η on $\mathcal{B}(\mathbb{R}^+)$ such that*

$$q_t(\omega, dT) \ll \eta(dT).$$

Then, there exists a family of positive **F**-adapted processes $(\alpha_t^u, t \geq 0)$ such that

$$G_t^T = \int_T^\infty \alpha_t^u \eta(du) .$$

From the martingale property of $(G_t^T, t \geq 0)$, i.e., for every T, for every $s \leq t$, $G_s^T = \mathbb{E}(G_t^T | \mathcal{F}_s)$, it is immediate to check that for any $u \geq 0$, $(\alpha_t^u, t \geq 0)$ is a positive **F**-martingale. Note that $\mathbb{Q}(\tau \in du) = \alpha_0^u \eta(du)$, hence $\alpha_0^u = 1$.

Remark that in this framework, we can write the conditional survival process $G_t := G_t^t$ as

$$G_t = \mathbb{Q}(\tau > t | \mathcal{F}_t) = \int_t^\infty \alpha_t^u \eta(du) = \int_0^\infty \alpha_{u \wedge t}^u \eta(du) - \int_0^t \alpha_u^u \eta(du) = M_t - \widetilde{A}_t$$

where M is an **F**-martingale (indeed, $(\alpha_{u\wedge t}^u)_t$ is a stopped **F**-martingale) and $\widetilde{A}$ an **F**-predictable increasing process. The process $(G_t^T, t \geq 0)$ being a martingale, it admits a representation as

$$G_t^T = G_0^T + \int_0^t g_s^T dW_s$$

where, for any T, the process $(g_s^T, s \geq 0)$ is **F**-predictable. In the case where $\eta(du) = \varphi(u)du$, using the Itô-Kunita-Ventzel formula (see Theorem 1.5.3.2), we obtain:

Lemma 7.6.3.2 *The Doob-Meyer decomposition of the conditional survival process* $(G_t, t \geq 0)$ *is*

$$G_t^t = 1 + \int_0^t g_s^s \, dW_s - \int_0^t \alpha_s^s \, \varphi(s) \, ds . \tag{7.6.7}$$

Lemma 7.6.3.3 *The process*

$$M_t := H_t - \int_0^t (1 - H_s) G_s^{-1} \alpha_s^s \, \varphi(s) \, ds$$

is a **G**-*martingale.*

PROOF: This follows directly from the Doob-Meyer decomposition of G given in Lemma 7.6.3.2. □

Using a method similar to Proposition 5.9.4.10, assuming that G is continuous, it is possible to prove (see Jeanblanc and Le Cam [482] for details) thatif X is a square integrable **F**-martingale

$$Y_t = X_t - \int_0^{t\wedge\tau} \frac{d\,\langle X, G\rangle_u}{G_u} - \int_{t\wedge\tau}^t \left. \frac{d\left\langle X, \alpha^\theta\right\rangle_u}{\alpha_{u-}^\theta} \right|_{\theta=\tau} \tag{7.6.8}$$

is a **G**-martingale. Note that, the first integral, which describes the bounded variation part before τ, is the same as in progressive enlargement of filtration – even without the honesty hypothesis – (see Proposition 5.9.4.10), and that the second integral, which describes the bounded variation part after τ, is the same as in Proposition 5.9.3.1.

Exercise 7.6.3.4 Prove that, if τ is an initial time with $\mathbb{E}_{\mathbb{Q}}(1/\alpha^{\tau}_{\infty}) < \infty$, there exists a probability $\widehat{\mathbb{Q}}$ equivalent to $\mathbb{Q}$ under which τ and $\mathcal{F}_\infty$ are independent.
`Hint:` Use Exercise 5.9.3.4. ◁

Exercise 7.6.3.5 Let τ be an initial time which avoids **F**-stopping times. Prove that the $(\mathcal{H})$-hypothesis holds if and only if $\alpha^u_t = \alpha^u_{t\wedge u}$. ◁

Exercise 7.6.3.6 Let $(K^u_t, t \geq 0)$ be a family of **F**-predictable processes indexed by $u \geq 0$ (i.e., for any $u \geq 0$, $t \to K^u_t$ is **F**-predictable).
Prove that $\mathbb{E}\left(K^{\tau}_t | \mathcal{F}_t\right) = \int_0^\infty K^u_t \alpha^u_t \eta(du)$. ◁

7.6.4 Explosive Defaults

Let

$$dX_t = (\theta - k(t)X_t)dt + \sigma\sqrt{X_t}dW_t$$

where the parameters are chosen so that $\mathbb{P}(T_0 < \infty) = 1$ where T_0 is the first hitting time of 0 for the process X. Andreasen [17] defines the default time as in the Cox process modelling presented in Subsection 7.3.1, setting the process $(\lambda_t, t \geq 0)$ equal to $1/X_t$ before T_0 and equal to $+\infty$ after time T_0. Note that the default time is not a totally inaccessible stopping time (obviously, $\mathbb{P}(\tau = T_0)$ is not null).

The survival probability is, for $\mathcal{F}_t = \mathcal{F}^X_t$,

$$\mathbb{P}(\tau > T | \mathcal{G}_t) = \mathbb{1}_{\{t<\tau\}} \mathbb{E}\left(\exp\left(-\int_t^T \lambda_s ds\right) | \mathcal{F}_t\right) := \mathbb{1}_{\{t<\tau\}} L(t, T, X_t)\,.$$

The process

$$L(t, T, X_t) \exp\left(-\int_0^t \lambda_s ds\right) = L(t, T, X_t) \exp\left(-\int_0^t X_s^{-1} ds\right)$$

is a local-martingale, hence

$$\partial_t L + (\theta - k(t)x)\partial_x L + \frac{1}{2}\sigma^2 x \partial_{xx} L - \frac{1}{x} L = 0\,,$$

and, taking into account the boundary conditions

$$L(t, T, 0) = 0, \quad L(T, T, x) = 1, \quad L(t, T, \infty) = 1$$

one gets

$$L(t,T,x) = \frac{\Gamma(\beta-\gamma)}{\Gamma(\beta)} M\left(\gamma, \beta, -\frac{x}{2K(t,T)}\right)\left(\frac{x}{2K(t,T)}\right)^{\gamma}$$

where M is the Kummer function (see $\rightarrowtail$ Appendix A.5.6) and

$$\gamma = \frac{-(\theta-\sigma^2/2) + \sqrt{(\theta-\sigma^2/2)^2 + 2\sigma^2}}{\sigma^2},$$
$$\beta = 2(\gamma + \theta/\sigma^2)$$

and

$$K(t,T) = \frac{\sigma^2}{4}\int_t^T \exp\left(\int_t^u k(s)ds\right) du\,.$$

Comment 7.6.4.1 Campi et al. [138] work with a model where the default is the first time when the process X hits the barrier 0, with

$$dX_t = X_{t-}(\mu dt + X_t^{\beta} dW_t - dN_t)\,.$$

Here N is a Poisson process. In other words $X_t = Y_t \mathbb{1}_{\{t<\tau\}}$ where Y is a CEV process and τ is an exponentially distributed random variable, which is independent of Y.

7.7 Intensity Approach

7.7.1 Definition

In the **intensity approach**, the default time τ is a $\mathbf{G}$-stopping time for a given filtration $\mathbf{G}$. From the Doob-Meyer Theorem, there exists a unique $\mathbf{G}$-predictable increasing process $\Lambda^{\mathbf{G}}$ such that the process $M_t = D_t - \Lambda_t^{\mathbf{G}}$ is a $\mathbf{G}$-martingale. This process $\Lambda^{\mathbf{G}}$ satisfies $\Lambda_t^{\mathbf{G}} = \Lambda_{t\wedge\tau}^{\mathbf{G}}$. The continuity of $\Lambda^{\mathbf{G}}$ is equivalent to the fact that τ is a $\mathbf{G}$-totally inaccessible stopping time.

In what follows, we assume that $\Lambda^{\mathbf{G}}$ is absolutely continuous w.r.t. Lebesgue measure, i.e., $\Lambda_t^{\mathbf{G}} = \int_0^t \lambda_s^{\mathbf{G}} ds$. Then, the process $\lambda^{\mathbf{G}}$, called the $\mathbf{G}$-intensity of τ, vanishes after τ.

Comment 7.7.1.1 Note that, in the Cox Process approach with $\mathbf{F}$-adapted intensity λ, or in the conditional survival probability approach (see Section 7.4) where λ denotes the $\mathbf{F}$-adapted process given by $\lambda_s ds = dF_s/G_s$, we have $\lambda_t^{\mathbf{G}} = \mathbb{1}_{\{\tau<t\}}\lambda_t$.

Lemma 7.7.1.2 *The process* $L_t = \mathbb{1}_{\{t<\tau\}} \exp\left(\Lambda_t^{\mathbf{G}}\right)$ *is a local martingale.*

PROOF: From Itô calculus (See $\rightarrowtail$ Subsection 8.3.4)

$$dL_t = \exp\left(\Lambda_t^{\mathbf{G}}\right)(-dD_t + (1-D_{t-})\lambda_t^{\mathbf{G}} dt) = -\exp\left(\Lambda_t^{\mathbf{G}}\right) dM_t\,. \qquad \square$$

7.7.2 Valuation Formula

Proposition 7.7.2.1 *For every integrable r.v.* $X \in \mathcal{G}_T$:

$$\mathbb{E}(X\mathbb{1}_{\{T<\tau\}}|\mathcal{G}_t) = \mathbb{1}_{\{\tau>t\}}\left(V_t - \mathbb{E}(\Delta V_\tau \mathbb{1}_{\{\tau\le T\}}|\mathcal{G}_t)\right)$$

where $V_t = e^{\Lambda_t^{\mathbf{G}}}\mathbb{E}\left(Xe^{-\Lambda_T^{\mathbf{G}}}|\mathcal{G}_t\right)$.

PROOF: By application of the integration by parts formula to the product $U_t = V_t(1-D_t)$, we obtain

$$dU_t = -V_t dD_t + (1-D_t)dV_t - \Delta V_t\,\Delta D_t$$

Writing $V_t = e^{\Lambda_t^{\mathbf{G}}} m_t$, where $m_t = \mathbb{E}\left(Xe^{-\Lambda_T^{\mathbf{G}}}|\mathcal{G}_t\right)$ is a **G**-martingale, one gets

$$dV_t = e^{\Lambda_t^{\mathbf{G}}} dm_t + m_t\lambda_t^{\mathbf{G}} e^{\Lambda_t^{\mathbf{G}}} dt = e^{\Lambda_t^{\mathbf{G}}} dm_t + V_t\lambda_t^{\mathbf{G}} dt\,,$$

hence,

$$dU_t = -V_t(dD_t - \lambda_t^{\mathbf{G}} dt) + (1-D_t)e^{\Lambda_t^{\mathbf{G}}} dm_t - \Delta V_t\,\Delta D_t\,.$$

By integration, we obtain $U_t = \mathbb{E}(\Delta V_\tau \mathbb{1}_{\{t<\tau\le T\}} + U_T|\mathcal{G}_t)$. It remains to note that $U_T = \mathbb{1}_{\{T<\tau\}}X$, and the result follows. □

Exercise 7.7.2.2 Assume that τ is an exponential r.v. with parameter λ and that **G** is the filtration generated by $\mathbb{1}_{\{\tau\le t\}}$. Check that $\lambda_t^{\mathbf{G}} = \mathbb{1}_{\{t<\tau\}}\lambda$. Let $Y_t = \mathbb{E}\left(\exp -(T\wedge\tau)|\mathcal{G}_t\right)$. Compute the jump of Y at time τ and deduce $\mathbb{E}(\mathbb{1}_{\{t<\tau\le T\}}|\mathcal{G}_t)$ using Proposition 7.7.2.1. Of course, here, the conditional survival methodology is more powerful. ◁

7.8 Credit Default Swaps

We present briefly Credit Default Swaps (CDS). The reader can consult Bielecki et al. [94, 97], Brigo and Alphonsi [128] and Jamshidian [477] for more details. We assume here that the interest rate r is constant.

A *credit default swap* (CDS) with a constant rate κ and *recovery at default* δ is a contract between the buyer and the seller. The buyer (i.e., the buyer of protection against a reference entity which defaults at time τ), pays a premium κ to the seller till $\tau\wedge T$ where T is the maturity of the CDS: the amount κdt is paid during the time dt. If the default occurs before T, the seller pays, at the default time, $\delta(\tau)$ to the buyer. Note that we assume here that the premium is paid in continuous time, for simplicity; usually, the premium is paid at some predetermined dates T_i.

The function $\delta : [0,T] \to \mathbb{R}$ represents the *default protection*, and κ is the *CDS rate* (also termed the *spread, premium* or *annuity* of the CDS).

The price of the CDS is the expectation of the difference of the discounted payoffs and is given by the formula

$$S_t(\kappa, \delta, T; r) = e^{rt}\, \mathbb{E}_{\mathbb{Q}}\Big(e^{-r\tau}\delta(\tau)\mathbb{1}_{\{\tau \le T\}} - \int_{t\wedge\tau}^{T\wedge\tau} e^{-ru}\,\kappa du \Big| \mathcal{G}_t\Big).$$

Here, $\mathbf{G}$ is the information available for the protection buyer. We shall note in short $S_t(\kappa)$ this price. Of course, it vanishes after time τ. Note that this price does not remain positive: at date t, the buyer is not allowed to cancel the contract; if $S_t < 0$, the buyer has to pay $-S_t(\kappa)$ to the seller (or to the new owner of this CDS) to do so.

At time 0, the price of a CDS is null: the spread κ is chosen so that

$$\kappa = \frac{\mathbb{E}_{\mathbb{Q}}\left(e^{-r\tau}\delta(\tau)\mathbb{1}_{\{\tau \le T\}}\right)}{\mathbb{E}_{Q}\left(\int_0^{T\wedge\tau} e^{-ru}\, du\right)}.$$

7.8.1 Dynamics of the CDS's Price in a single name setting

In this subsection, $\mathbf{G} = \mathbf{F} \vee \mathbf{H}$, and $G_t = \mathbb{Q}(\tau > t|\mathcal{F}_t)$.

Proposition 7.8.1.1 *The price of a CDS equals, for any $t \in [0, T]$,*

$$S_t(\kappa) = \mathbb{1}_{\{t<\tau\}}\, \frac{1}{e^{-rt}G_t}\, \mathbb{E}_{\mathbb{Q}}\left(\int_t^T e^{-ru}G_u(\delta_u\lambda_u - \kappa)\, du \,\Big|\, \mathcal{F}_t\right) \qquad (7.8.1)$$

PROOF: This relies on a direct and simple application of the key Lemma 7.4.1.1. □

The dynamics of the CDS's price can now be obtained:

Proposition 7.8.1.2 *If the immersion property holds, then*

$$dS_t(\kappa) = -S_{t^-}(\kappa)\, dM_t + (1 - H_t)\big(rS_t(\kappa) + \kappa - \lambda_t\delta_t\big)\, dt + (1 - H_t)e^{rt}G_t^{-1}\, dn_t \qquad (7.8.2)$$

where $n_t = \mathbb{E}_{\mathbb{Q}}\Big(\int_0^T e^{-ru}G_u(\delta_u\lambda_u - \kappa)\, du \,\Big|\, \mathcal{F}_t\Big)$.

PROOF: Apply Itô's formula. □

Comment 7.8.1.3 Note that the risk-neutral dynamics of the CDS's discounted price do not constitute a martingale. This price is indeed an ex-dividend price. The cum-dividend price $S^{\text{cum}}(\kappa)$ satisfies, for every $t \in [0, T]$,

$$d_tS_t^{\text{cum}}(\kappa) = rS_t^{\text{cum}}(\kappa)\, dt + \big(\delta_t - S_{t-}(\kappa)\big)\, dM_t + (1 - H_t)G_t^{-1}e^{rt}\, dn_t\,.$$

The case where immersion is not satisfied is presented in [97].

7.8.2 Dynamics of the CDS's Price in a multi-name setting

An important feature of price's dynamics is that the price of a CDS depends strongly of the choice of the observation filtration. We present here the case where a second firm can default at time τ_2 and where the two default times are correlated, in the simple case of a trivial filtration $\mathbf{F}$ and a null interest rate. Let us denote by $G(t,s) = \mathbb{P}(\tau_1 > t, \tau_2 > s)$ the survival joined law of the two defaults, assumed to be regular. Let

$$S(\kappa)_t = \mathbb{1}_{\{t<\tau_1\}}\mathbb{E}(\delta(\tau_1) - \kappa(\tau \wedge T - t)^+ | \mathcal{G}_t)$$

be the price of a CDS written on τ_1, with spread κ and recovery the function δ, where $\mathbf{G} = \mathbf{H}^1 \vee \mathbf{H}^2$. Then, $\mathbb{1}_{\{t<\tau_1\}}S_t(\kappa) = \mathbb{1}_{\{t<\tau_1\}}\widetilde{S}_t(\kappa)$, where an easy computation leads to

$$\widetilde{S}_t(\kappa) = \frac{1}{G(t,t)}\left(-\int_t^T \delta(u)\partial_1 G(u,t)\,du - \kappa\int_t^T G(u,t)\,du\right). \qquad (7.8.3)$$

Hence the dynamics of the *pre-default ex-dividend price* $\widetilde{S}_t$ are

$$d\widetilde{S}_t(\kappa) = \left((\widetilde{\lambda}_1(t) + \widetilde{\lambda}_2(t))\widetilde{S}_t(\kappa) + \kappa_1 - \widetilde{\lambda}_1(t)\delta(t) - \widetilde{\lambda}_2(t)S_{t|2}(\kappa_1)\right)\,dt,$$

where for $i = 1,2$ the function $\widetilde{\lambda}_i(t) = -\frac{\partial_i G(t,t)}{G(t,t)}$ is the (deterministic) *pre-default intensity* of τ_i and $S_{t|2}(\kappa)$ is given by the expression

$$S_{t|2}(\kappa) = \frac{-1}{\partial_2 G(t,t)}\left(\int_t^T \delta(u)f(u,t)\,du + \kappa\int_t^T \partial_2 G(u,t)\,du\right).$$

In the financial interpretation, $S_{1|2}(t)$ is the ex-dividend price at time t of a CDS on the first credit name, under the assumption that the default τ_2 occurs at time t and the first name has not yet defaulted (recall that simultaneous defaults are excluded).

Let us now consider the event $\{\tau_2 \le t < \tau_1\}$. It is not difficult to show that in that case the ex-dividend price of a CDS equals

$$\widehat{S}_t(\kappa) = \frac{1}{\partial_2 G(t,\tau_2)}\left(-\int_t^T \delta(u)f(u,\tau_2)\,du - \kappa\int_t^T \partial_2 G(u,\tau_2)\,du\right). \qquad (7.8.4)$$

Consequently, on the event $\{\tau_2 \le t < \tau_1\}$ we obtain

$$d\widehat{S}_t(\kappa) = \left(\lambda^{1|2}(t,\tau_2)\left(\widehat{S}_t(\kappa) - \delta(t)\right) + \kappa\right)\,dt,$$

where $\lambda^{1|2}(t,s) = -\frac{f(t,s)}{\partial_2 G(t,s)}$, evaluated at $s = \tau_2$, represents the value of the default intensity process of τ_1 with respect to the filtration $\mathbf{H}^2$ on the event $\{\tau_2 < t\}$.

7.9 PDE Approach for Hedging Defaultable Claims

We briefly present a PDE approach for defaultable claims. We assume that Y^1 is the price of the savings account, with deterministic interest rate. A default free asset is supposed to be traded in the market with price dynamics

$$dY_t^2 = Y_t^2\big(\mu_{2,t}dt + \sigma_{2,t}\,dW_t\big)\,. \tag{7.9.1}$$

A defaultable asset with price dynamics

$$dY_t^3 = Y_{t-}^3\left((\mu_{3,t} - \kappa_{3,t}\lambda_t\mathbb{1}_{\{t\leq\tau\}})\,dt + \sigma_{3,t}\,dW_t + \kappa_{3,t}\,dD_t\right), \tag{7.9.2}$$

is also traded in the market. All the processes are assumed to be $\mathbf{G}$-adapted, where $\mathbf{G}$ is the filtration generated by W and D. Here, W is a Brownian motion, $D_t = \mathbb{1}_{\{\tau\leq t\}}$ is the default process (see Section 7.4) and the process $M_t = D_t - \int_0^t(1-D_s)\lambda_s ds$ is assumed to be a $\mathbf{G}$-martingale. The Brownian motion W is assumed to be a $\mathbf{G}$-Brownian motion, hence, the $(\mathcal{H})$ hypothesis holds between $\mathbf{F}^W$ and $\mathbf{G}$.

Our aim is to replicate a contingent claim of the form

$$Y = \mathbb{1}_{\{T<\tau\}}g_0(Y_T^2, Y_T^3) + \mathbb{1}_{\{T\geq\tau\}}g_1(Y_T^2, Y_T^3) = G(Y_T^2, Y_T^3, D_T),$$

which settles at time T.

7.9.1 Defaultable Asset with Total Default

We first assume that the third asset is subject to total default, i.e., $\kappa_3 = -1$,

$$dY_t^3 = Y_{t-}^3\left(\mu_{3,t}\,dt + \sigma_{3,t}\,dW_t - dM_t\right).$$

A first step is to find some condition on the coefficients such that the market is arbitrage free.

Equivalent Martingale Measure

Let $Y^{i,1} = Y^i/Y^1$ be the relative price of the ith-asset in terms of the first one. Our goal is now to find a martingale measure $\mathbb{Q}^1$ (if it exists) for the relative prices $Y^{2,1}$ and $Y^{3,1}$. The dynamics of $Y^{3,1}$ under $\mathbb{P}$ are

$$dY_t^{3,1} = Y_{t-}^{3,1}\Big\{\big(\mu_{3,t} - r(t)\big)dt + \sigma_{3,t}\,dW_t - dM_t\Big\}.$$

Let $\mathbb{Q}^1$ be any probability measure equivalent to $\mathbb{P}$ on $(\Omega, \mathcal{G}_T)$. The associated Radon-Nikodým density L satisfies (we use Kusuoka's representation Theorem 7.5.5.1)

$$dL_t = L_{t-}(\theta_t\,dW_t + \zeta_t\,dM_t)$$

for some $\mathbf{G}$-predictable processes θ and ζ.

From Girsanov's theorem (see ↣ Section 9.4), the processes $\widehat{W}$ and $\widehat{M}$, given by

$$\widehat{W}_t = W_t - \int_0^t \theta_u \, du, \quad \widehat{M}_t = M_t - \int_0^t \mathbb{1}_{\{u<\tau\}} \lambda_u \zeta_u \, du, \tag{7.9.3}$$

are $\mathbf{G}$-martingales under $\mathbb{Q}^1$. To ensure that $Y^{2,1}$ is a $\mathbb{Q}^1$-martingale, we have to choose

$$\theta_t = \frac{r_t - \mu_{2,t}}{\sigma_{2,t}} .$$

For the process $Y^{3,1}$ to be a $\mathbb{Q}^1$-martingale, it is necessary and sufficient that ζ satisfies

$$\lambda_t \zeta_t = \mu_{3,t} - r_t + \sigma_{3,t} \frac{r_t - \mu_{2,t}}{\sigma_{2,t}} .$$

To ensure that $\mathbb{Q}^1$ is a probability measure equivalent to $\mathbb{P}$, we require that $\zeta_t > -1$. The unique martingale measure $\mathbb{Q}^1$ is then given by

$$L_t = \mathcal{E}_t \left(\int_0^{\cdot} \theta_u \, dW_u \right) \mathcal{E}_t \left(\int_0^{\cdot} \zeta_u \, dM_u \right) .$$

Proposition 7.9.1.1 *Assume that the process $\theta_t = \frac{r_t - \mu_{2,t}}{\sigma_{2,t}}$ is bounded, and*

$$\zeta_t = \frac{1}{\lambda_t} \left(\mu_{3,t} - r_t + \sigma_{3,t} \frac{r_t - \mu_{2,t}}{\sigma_{2,t}} \right) > -1. \tag{7.9.4}$$

Then the market is arbitrage free and complete. The dynamics of the relative prices under the unique martingale measure $\mathbb{Q}^1$ are

$$\begin{aligned} dY_t^{2,1} &= Y_t^{2,1} \sigma_{2,t} \, d\widehat{W}_t, \\ dY_t^{3,1} &= Y_{t-}^{3,1} \big(\sigma_{3,t} \, d\widehat{W}_t - d\widehat{M}_t \big). \end{aligned}$$

This means that any $\mathbb{Q}^1$-integrable contingent claim $Y = G(Y_T^2, Y_T^3; D_T)$ is attainable, and its arbitrage price equals

$$\pi_t(Y) = R_t^{-1} \, \mathbb{E}(Y R_T \,|\, \mathcal{G}_t), \quad \forall t \in [0,T] \tag{7.9.5}$$

where $R_t = \exp(-\int_0^t r_s ds)$.

7.9.2 PDE for Valuation

Since our goal is to develop the PDE approach, it will be essential to postulate the Markovian property of our model. We assume that the coefficients μ_i, σ_i, λ are regular functions of (t, Y_t^2, Y_{t-}^3).

Lemma 7.9.2.1 *The process (Y^1, Y^2, Y^3, D) is a $\mathbf{G}$-Markov process under the martingale measure $\mathbb{Q}^1$. For any attainable claim $Y = G(Y_T^1, Y_T^2, Y_T^3; D_T)$ there exists a function $C : [0,T] \times \mathbb{R}^3 \times \{0,1\} \to \mathbb{R}$ such that*

$$\pi_t(Y) = C(t, Y_t^1, Y_t^2, Y_t^3; D_t) .$$

Note that, since Y^1_T is deterministic, up to a change of notation one can restrict attention to claims of the form $G(Y^2_T, Y^3_T; D_T)$, the price of which is $C(t, Y^2_t, Y^3_t; D_t)$. We find it convenient to introduce the *pre-default* pricing function $C(\cdot\,;0) = C(t, y_2, y_3; 0)$ and the *post-default* pricing function $C(\cdot\,;1) = C(t, y_2, y_3; 1)$. In fact, since $Y^3_t = 0$ if $D_t = 1$, it suffices to study the post-default function $C(t, y_2; 1) = C(t, y_2, 0; 1)$. Also, we write

$$b = (r - \mu_3)\sigma_3 - (\mu_3 - r)\sigma_2.$$

Let $\lambda > 0$ be the default intensity under $\mathbb{P}$, and let $\zeta > -1$ be given by formula (7.9.4) where we do not indicate the dependence on t. (In fact, b, σ, λ and ζ are Markovian coefficients.) We denote by $\partial_i, i = 1, 2$ the partial derivative with respect to y_i.

Proposition 7.9.2.2 *Assume that the functions $C(\cdot\,;0)$ and $C(\cdot\,;1)$ belong to the class $\mathrm{C}^{1,2}([0,T] \times \mathbb{R}^+ \times \mathbb{R}^+, \mathbb{R})$. Then $C(t, y_2, y_3; 0)$ satisfies the PDE*

$$\partial_t C(\cdot\,;0) + r y_2 \partial_2 C(\cdot\,;0) + (r + \zeta) y_3 \partial_3 C(\cdot\,;0) + \frac{1}{2} \sum_{i,j=2}^{3} \sigma_i \sigma_j y_i y_j \partial_{ij} C(\cdot\,;0)$$
$$- r C(\cdot\,;0) + \left(\lambda + \frac{b}{\sigma_2}\right) \big[C(t, y_2; 1) - C(t, y_2, y_3; 0)\big] = 0$$

subject to the terminal condition $C(T, y_2, y_3; 0) = G(y_2, y_3; 0)$, and $C(t, y_2; 1)$ satisfies the PDE

$$\partial_t C(\cdot\,;1) + r y_2 \partial_2 C(\cdot\,;1) + \frac{1}{2} \sigma_2^2 y_2^2 \partial_{22} C(\cdot\,;1) - r C(\cdot\,;1) = 0$$

subject to the terminal condition $C(T, y_2; 1) = G(y_2, 0; 1)$.

PROOF: For simplicity, we write $C_t = \pi_t(Y)$. Let us define

$$\Delta C(t, y_2, y_3) = C(t, y_2; 1) - C(t, y_2, y_3; 0).$$

Then the jump $\Delta C_t = C_t - C_{t-}$ can be represented as follows:

$$\Delta C_t = \mathbb{1}_{\{\tau = t\}} \left(C(t, Y^2_t; 1) - C(t, Y^2_t, Y^3_{t-}; 0)\right) = \mathbb{1}_{\{\tau = t\}}\, \Delta C(t, Y^2_t, Y^3_{t-}).$$

We typically omit the variables $(t, Y^2_{t-}, Y^3_{t-}, D_{t-})$ in expressions $\partial_t C$, $\partial_i C$, ΔC, etc. We shall also make use of the fact that for any Borel measurable function g we have

$$\int_0^t g(u, Y^2_u, Y^3_{u-})\, du = \int_0^t g(u, Y^2_u, Y^3_u)\, du$$

since Y^3_u and Y^3_{u-} differ for at most one value of u (for each ω).

Let $\xi_t = \mathbb{1}_{\{t < \tau\}} \lambda_t$. An application of Itô's formula yields

$$
\begin{aligned}
dC_t &= \partial_t C\,dt + \sum_{i=2}^{3} \partial_i C\,dY_t^i + \frac{1}{2}\sum_{i,j=2}^{3} \sigma_i\sigma_j Y_t^i Y_t^j \partial_{ij} C\,dt \\
&\quad + \Big(\Delta C + Y_{t-}^3\,\partial_3 C\Big)\,dD_t \\
&= \partial_t C\,dt + \sum_{i=2}^{3} \partial_i C\,dY_t^i + \frac{1}{2}\sum_{i,j=2}^{3} \sigma_i\sigma_j Y_t^i Y_t^j \partial_{ij} C\,dt \\
&\quad + \Big(\Delta C + Y_{t-}^3\,\partial_3 C\Big)\big(dM_t + \xi_t\,dt\big),
\end{aligned}
$$

and this in turn implies that

$$
\begin{aligned}
dC_t &= \partial_t C\,dt + \sum_{i=2}^{3} Y_t^i \partial_i C\big(\mu_i\,dt + \sigma_i\,dW_t\big) + \frac{1}{2}\sum_{i,j=2}^{3} \sigma_i\sigma_j Y_t^i Y_t^j \partial_{ij} C\,dt \\
&\quad + \Delta C\,dM_t + \Big(\Delta C + Y_{t-}^3\,\partial_3 C\Big)\xi_t\,dt \\
&= \left(\sum_{i=2}^{3} \mu_i Y_t^i \partial_i C + \frac{1}{2}\sum_{i,j=2}^{3} \sigma_i\sigma_j Y_t^i Y_t^j \partial_{ij} C + \Big(\Delta C + Y_t^3 \partial_3 C\Big)\xi_t \right\} dt \\
&\quad + \partial_t C + \left(\sum_{i=2}^{3} \sigma_i Y_t^i \partial_i C\right) dW_t + \Delta C\,dM_t .
\end{aligned}
$$

We now use the integration by parts formula to derive dynamics of the relative price $\widehat{C}_t = C_t(Y_t^1)^{-1}$. We find that

$$
\begin{aligned}
Y_t^1 d\widehat{C}_t &= \sum_{i=2}^{3} \sigma_i Y_t^i \partial_i C\,dW_t + \Delta C\,dM_t + (\partial_t C - rC_t)\,dt \\
&\quad + \left\{\sum_{i=2}^{3} \mu_i Y_t^i \partial_i C + \frac{1}{2}\sum_{i,j=2}^{3} \sigma_i\sigma_j Y_t^i Y_t^j \partial_{ij} C + \Big(\Delta C + Y_t^3 \partial_3 C\Big)\xi_t \right\} dt .
\end{aligned}
$$

Hence, using (7.9.3), we obtain

$$
\begin{aligned}
Y_t^1 d\widehat{C}_t &= \sum_{i=2}^{3} \sigma_i Y_t^i \partial_i C\,d\widehat{W}_t + \Delta C\,d\widehat{M}_t - rC_t\,dt + \partial_t C \\
&\quad + \left\{\sum_{i=2}^{3} \mu_i Y_t^i \partial_i C + \frac{1}{2}\sum_{i,j=2}^{3} \sigma_i\sigma_j Y_t^i Y_t^j \partial_{ij} C + \Big(\Delta C + Y_t^3 \partial_3 C\Big)\xi_t \right\} dt \\
&\quad + \left(\sum_{i=2}^{3} \sigma_i Y_t^i \theta \partial_i C + \zeta\xi_t \Delta C - \sigma_1 \sum_{i=2}^{3} \sigma^i Y_t^i \partial_i C\right) dt.
\end{aligned}
$$

This means that the process $\widehat{C}$ admits the following decomposition under $\mathbb{Q}^1$

$$
\begin{aligned}
Y_t^1 d\widehat{C}_t &= -rC_t\,dt + \left(\sum_{i=2}^{3}\sigma_i Y_t^i\theta\,\partial_i C + \zeta\xi_t \Delta C + \partial_t C\right) dt \\
&+ \left(\sum_{i=2}^{3}\mu_i Y_t^i \partial_i C + \frac{1}{2}\sum_{i,j=2}^{3}\sigma_i\sigma_j Y_t^i Y_t^j \partial_{ij} C + \left(\Delta C + Y_t^3 \partial_3 C\right)\xi_t\right) dt \\
&+ d(\text{a } \mathbb{Q}^1\text{-martingale})\,.
\end{aligned}
$$

From (7.9.5), it follows that the process $\widehat{C}$ is a martingale under $\mathbb{Q}^1$. Therefore, the continuous finite variation part in the above decomposition necessarily vanishes, and thus we get

$$
\begin{aligned}
0 = -rC_t + \partial_t C &+ \sum_{i=2}^{3}\mu_i Y_t^i \partial_i C + \frac{1}{2}\sum_{i,j=2}^{3}\sigma_i\sigma_j Y_t^i Y_t^j \partial_{ij} C \\
&+ \left(\Delta C + Y_t^3 \partial_3 C\right)\xi_t + \sum_{i=2}^{3}\sigma_i Y_t^i \theta \partial_i C + \zeta\xi_t \Delta C\,.
\end{aligned}
$$

Finally, we conclude that

$$
\begin{aligned}
&\partial_t C + rY_t^2 \partial_2 C + (r+\xi_t)\, Y_t^3 \partial_3 C + \frac{1}{2}\sum_{i,j=2}^{3}\sigma_i\sigma_j Y_t^i Y_t^j \partial_{ij} C \\
&- rC_t + (1+\zeta)\xi_t \Delta C = 0.
\end{aligned}
$$

Recall that $\xi_t = \mathbb{1}_{\{t<\tau\}}\lambda$. It is thus clear that the pricing functions $C(\cdot,0)$ and $C(\cdot;1)$ satisfy the PDEs given in the statement of the proposition. □

Hedging Strategy

The next result provides a replicating strategy for Y.

Proposition 7.9.2.3 *The replicating strategy ϕ for the claim Y is given by the formulae*

$$
\begin{aligned}
\phi_t^3 Y_{t-}^3 &= -\Delta C(t, Y_t^2, Y_{t-}^3) = C(t, Y_t^2, Y_{t-}^3; 0) - C(t, Y_t^2; 1), \\
\phi_t^2 \sigma_2 Y_t^2 &= \sigma_3 \Delta C + \sum_{i=2}^{3} Y_t^i \sigma_i \partial_i C, \\
\phi_t^1 Y_t^1 &= C - \phi_t^2 Y_t^2 - \phi_t^3 Y_t^3.
\end{aligned}
$$

PROOF: As a by-product of our previous computations, we obtain

$$
d\widehat{C}_t = (Y_t^1)^{-1}\sum_{i=2}^{3}\sigma_i Y_{t-}^i \partial_i C\, d\widehat{W}_t + (Y_{t-}^1)^{-1}\Delta C\, d\widehat{M}_t.
$$

The self-financing strategy that replicates Y is determined by two components ϕ^2, ϕ^3 and the following relationship:

$$d\widehat{C}_t = \phi_t^2\, dY_t^{2,1} + \phi_t^3\, dY_t^{3,1} = \phi_t^2 Y_t^{2,1} \sigma_2\, d\widehat{W}_t + \phi_t^3 Y_{t-}^{3,1} \left(\sigma_3\, d\widehat{W}_t - d\widehat{M}_t \right).$$

By identification, we obtain $\phi_t^3 Y_{t-}^{3,1} = (Y_t^1)^{-1} \Delta C$ and

$$\phi_t^2 \sigma_2 Y_t^2 - \sigma_3 \Delta C = \sum_{i=2}^{3} Y_{t-}^i \sigma_i \partial_i C.$$

This yields the claimed formulae. □

Corollary 7.9.2.4 *In the case of a total default claim, the hedging strategy satisfies the condition* $\phi_t^1 Y_t^1 + \phi_t^2 Y_t^2 = 0$.

PROOF: A total default corresponds to the assumption that $G(y_2, y_3, 1) = 0$. We then have $C(t, y_2; 1) = 0$, and thus $\phi_t^3 Y_{t-}^3 = C(t, Y_t^1, Y_t^2, Y_{t-}^3; 0)$ for every $t \in [0, T]$. Hence, the equality $\phi_t^1 Y_t^1 + \phi_t^2 Y_t^2 = 0$ holds for every $t \in [0, T]$ and ensures that the wealth of a replicating portfolio jumps to zero at default time. □

7.9.3 General Case

Proposition 7.9.3.1 *Let* $\sigma_2 \neq 0$ *and let* Y^1, Y^2, Y^3 *satisfy*

$$\begin{aligned} dY_t^1 &= rY_t^1\, dt, \\ dY_t^2 &= Y_t^2 (\mu_2\, dt + \sigma_2\, dW_t), \\ dY_t^3 &= Y_{t-}^3 (\mu_3\, dt + \sigma_3\, dW_t + \kappa_3\, dM_t). \end{aligned}$$

Assume that $\kappa_3 \neq 0$, $\kappa_3 > -1$. *The market is complete and arbitrage free if and only if* $\sigma_2(r - \mu_3) = \sigma_3(r - \mu_2)$.

PROOF: We leave this to the reader. □

Corollary 7.9.3.2 *In case of constant coefficients, the risk-neutral intensity is equal to the historical intensity.*

PROOF: This follows from the determination of the unique risk-neutral probability, which transforms the Brownian motion W into $\widehat{W}$ where

$$d\widehat{W}_t = W_t - \frac{\mu_2 - r}{\sigma_2} dt = W_t - \frac{\mu_3 - r}{\sigma_3} dt,$$

but does not change the martingale M. □

Proposition 7.9.3.3 *The price of a contingent claim* $Y = G(Y_T^2, Y_T^3, D_T)$ *can be represented as* $\pi_t(Y) = C(t, Y_t^2, Y_t^3, D_t)$, *where the pricing functions* $C(\cdot\,;0)$ *and* $C(\cdot\,;1)$ *satisfy the following PDEs:*

$$\begin{aligned}\partial_t C(t,y_2,y_3;0) &+ ry_2\partial_2 C(t,y_2,y_3;0) + y_3\,(r-\kappa_3\lambda)\,\partial_3 C(t,y_2,y_3;0)\\ &+\frac{1}{2}\sum_{i,j=2}^{3}\sigma_i\sigma_j y_i y_j \partial_{ij}C(t,y_2,y_3;0)\\ &+\lambda\big(C(t,y_2,y_3(1+\kappa_3);1)-C(t,y_2,y_3;0)\big) = rC(t,y_2,y_3;0)\end{aligned}$$

and

$$\begin{aligned}&\partial_t C(t,y_2,y_3;1) + ry_2\partial_2 C(t,y_2,y_3;1) + ry_3\partial_3 C(t,y_2,y_3;1) - rC(t,y_2,y_3;1)\\ &+\frac{1}{2}\sum_{i,j=2}^{3}\sigma_i\sigma_j y_i y_j\partial_{ij}C(t,y_2,y_3;1) = 0\end{aligned}$$

subject to the terminal conditions

$$C(T,y_2,y_3;0) = G(y_2,y_3;0), \quad C(T,y_2,y_3;1) = G(y_2,y_3;1).$$

The replicating strategy ϕ *comprises*

$$\begin{aligned}\phi_t^2 &= \frac{1}{\sigma_2 Y_t^2}\sum_{i=2}^{3}\sigma_i y_i \partial_i C(t, Y_t^2, Y_{t-}^3, D_{t-})\\ &\quad - \frac{\sigma_3}{\sigma_2\kappa_3 Y_t^2}\big(C(t,Y_t^2,Y_{t-}^3(1+\kappa_3);1) - C(t,Y_t^2,Y_{t-}^3;0)\big),\\ \phi_t^3 &= \frac{1}{\kappa_3 Y_{t-}^3}\big(C(t,Y_t^2,Y_{t-}^3(1+\kappa_3);1) - C(t,Y_t^2,Y_{t-}^3;0)\big),\end{aligned}$$

with ϕ_t^1 *given by* $\phi_t^1 Y_t^1 + \phi_t^2 Y_t^2 + \phi_t^3 Y_t^3 = C_t$.

PROOF: This is obtained by lengthy computations, as in Proposition 7.9.2.2. We leave the details to the reader. □

Hedging of a Survival Claim

We shall illustrate Proposition 7.9.3.3 by means of an example. Consider a survival claim of the form

$$Y = G(Y_T^2, Y_T^3, D_T) = \mathbb{1}_{\{T<\tau\}}\, g(Y_T^3).$$

Then the post-default pricing function $C^g(\cdot\,;1)$ vanishes identically, and the pre-default pricing function $C^g(\cdot\,;0)$ solves the PDE

$$\partial_t C^g(\cdot\,;0) + r y_2 \partial_2 C^g(\cdot\,;0) + y_3\left(r - \kappa_3\lambda\right)\partial_3 C^g(\cdot\,;0)$$
$$+ \frac{1}{2}\sum_{i,j=2}^{3} \sigma_i\sigma_j y_i y_j \partial_{ij} C^g(\cdot\,;0) - (r+\lambda) C^g(\cdot\,;0) = 0$$

with the terminal condition $C^g(T, y_2, y_3; 0) = g(y_3)$. Denote $\alpha = r - \kappa_3\lambda$ and $\beta = \lambda(1 + \kappa_3)$.

It is not difficult to check that $C^g(t, y_2, y_3; 0) = e^{\beta(T-t)} C^{\alpha,g,3}(t, y_3)$ is a solution of the above equation, where the function $w(t, y) = C^{\alpha,g,3}(t, y)$ is the solution of the standard Black and Scholes PDE

$$\partial_t w + y\alpha\partial_y w + \frac{1}{2}\sigma_3^2 y^2 \partial_{yy} w - \alpha w = 0$$

with the terminal condition $w(T, y) = g(y)$, that is, the price of the contingent claim $g(Y_T)$ in the Black and Scholes framework with interest rate α and volatility parameter equal to σ_3.

Let C_t be the current value of the contingent claim Y, so that

$$C_t = \mathbb{1}_{\{t<\tau\}} e^{\beta(T-t)} C^{\alpha,g,3}(t, Y_t^3).$$

The hedging strategy of the survival claim is, on the event $\{t < \tau\}$,

$$\phi_t^3 Y_t^3 = -\frac{1}{\kappa_3} e^{-\beta(T-t)} C^{\alpha,g,3}(t, Y_t^3) = -\frac{1}{\kappa_3} C_t,$$
$$\phi_t^2 Y_t^2 = \frac{\sigma_3}{\sigma_2}\left(Y_t^3 e^{-\beta(T-t)} \partial_y C^{\alpha,g,3}(t, Y_t^3) - \phi_t^3 Y_t^3\right).$$

Obviously, there is no need for the strategy on the set $\{\tau < t\}$.

8

Poisson Processes and Ruin Theory

We give in this chapter the main results on Poisson processes, which are basic examples of jump processes. Despite their elementary properties they are building blocks of jump process theory. We present various generalizations such as inhomogeneous Poisson processes and compound Poisson processes. These processes are not used to model financial prices, due to the simple character of their jumps and are in practice mixed with Brownian motion, as we shall present in ↣ Chapter 10. However, they represent the main model in insurance theory. We end this chapter with two sections about point processes and marked point processes.

The reader can refer to Çinlar [188], Cocozza-Thivent [190], Karlin and Taylor [515] and the last chapter in Shreve [795] for the study of standard Poisson processes, to Brémaud [124] for general Poisson processes, and to Jacod and Shiryaev [471], Kallenberg [504], Kingman [523], Last and Brandt [565], Neveu [669], Prigent [725] and Protter [727] for point processes, and to Mikosch [651, 652] for applications.

8.1 Counting Processes and Stochastic Integrals

A **counting process** is a process which increases in unit steps at isolated times and is constant between these times. It can be constructed as follows. Let $(T_n, n \geq 0)$ be a sequence of random variables defined on the same probability space $(\Omega, \mathcal{F}, \mathbb{P})$ such that

$$T_0 = 0, \quad T_n < T_{n+1} \text{ for } T_n < \infty .$$

This sequence models the times when jumps occur. We define the family of random variables, for $t \geq 0$,

$$N_t = \begin{cases} n \text{ if } t \in [T_n, T_{n+1}[\\ +\infty \text{ otherwise,} \end{cases}$$

M. Jeanblanc, M. Yor, M. Chesney, *Mathematical Methods for Financial Markets*, Springer Finance, DOI 10.1007/978-1-84628-737-4_8,

or, equivalently,

$$N_t = \sum_{n \geq 1} \mathbb{1}_{\{T_n \leq t\}} = \sum_{n \geq 0} n \mathbb{1}_{\{T_n \leq t < T_{n+1}\}}, \; N_0 = 0 .$$

This counting process $(N_t, t \geq 0)$, associated with the sequence $(T_n, n \geq 0)$, is increasing and right-continuous. We denote by N_{t-} the left-limit of N_s when $s \to t, s < t$ and by $\Delta N_s = N_s - N_{s-}$ the jump process of N. The **explosion time** is the r.v. $T = \sup_n T_n$. In what follows, we reduce our attention to the case $T = \infty$.

Let $\mathbf{F}$ be a given filtration. A counting process is $\mathbf{F}$-adapted if and only if the random variables $(T_n, n \geq 1)$ are $\mathbf{F}$-stopping times. In that case, for any n, the set $\{N_t \leq n\} = \{T_{n+1} > t\}$ belongs to $\mathcal{F}_t$.

The natural filtration of N denoted by $\mathbf{F}^N$ where $\mathcal{F}_t^N = \sigma(N_s, s \leq t)$ is the smallest filtration $\mathbf{F}^N$ which satisfies the usual hypotheses and such that N is $\mathbf{F}^N$-adapted.

The **stochastic integral** $\int_0^t C_s dN_s$ is defined pathwise as a Stieltjes integral for every bounded measurable process (not necessarily $\mathbf{F}^N$-adapted) $(C_t, t \geq 0)$ by

$$(C \star N)_t := \int_0^t C_s dN_s = \int_{]0,t]} C_s dN_s := \sum_{n=1}^{\infty} C_{T_n} \mathbb{1}_{\{T_n \leq t\}} .$$

We emphasize that the integral $\int_0^t C_s dN_s$ is here an integral over the time interval $]0, t]$, where the upper limit t is included and the lower limit 0 excluded. This integral is finite since there is a finite number of jumps during the time interval $]0, t]$. We shall also write

$$\int_0^t C_s dN_s = \sum_{s \leq t} C_s \Delta N_s$$

where the right-hand side contains only a finite number of non-zero terms. The integral $\int_0^\infty C_s dN_s$ is defined as $\int_0^\infty C_s dN_s = \sum_{n=1}^\infty C_{T_n}$, when the right-hand side converges.

We shall also use the differential notation $d(C \star N)_t := C_t dN_t$.

We can associate a **random measure** to any counting process as follows. For any Borel set $\Lambda \subset \mathbb{R}^+$, for any ω, set

$$\mu(\omega, \Lambda) = \#\{n \geq 1 : T_n(\omega) \in \Lambda\} .$$

For any ω, the map $\Lambda \to \mu(\omega, \Lambda)$ defines a positive measure on $\mathbb{R}^+$. One can note that $\mu(\omega, dt) = \sum_n \delta_{T_n(\omega)}(dt)$.

The random variable N_t can be written as

$$N_t(\omega) = \mu(\omega,]0,t]) = \int_{]0,t]} \mu(\omega, ds)$$

and the Stieltjes (or stochastic) integral as $\int_0^t C_s dN_s = \int_0^t C_s \mu(ds)$.

8.2 Standard Poisson Process

8.2.1 Definition and First Properties

The **standard Poisson process** is a counting process such that the random variables $(T_{n+1} - T_n,\ n \geq 0)$ are independent and identically distributed with exponential law of parameter λ with $\lambda > 0$. Hence, the explosion time is infinite and

$$\mathbb{P}(N_t = n) = e^{-\lambda t} \frac{(\lambda t)^n}{n!}.$$

The standard Poisson process can be redefined as follows (see e.g., Çinlar [188]): it is a counting process without explosion (i.e., $T = \infty$) such that
- for every $s, t \geq 0$ the r.v. $N_{t+s} - N_t$ is independent of $\mathcal{F}_t^N$,
- for every s, t, the r.v. $N_{t+s} - N_t$ has the same law as N_s.

or, in an equivalent way, a counting process without explosion whose increments are independent and stationary.

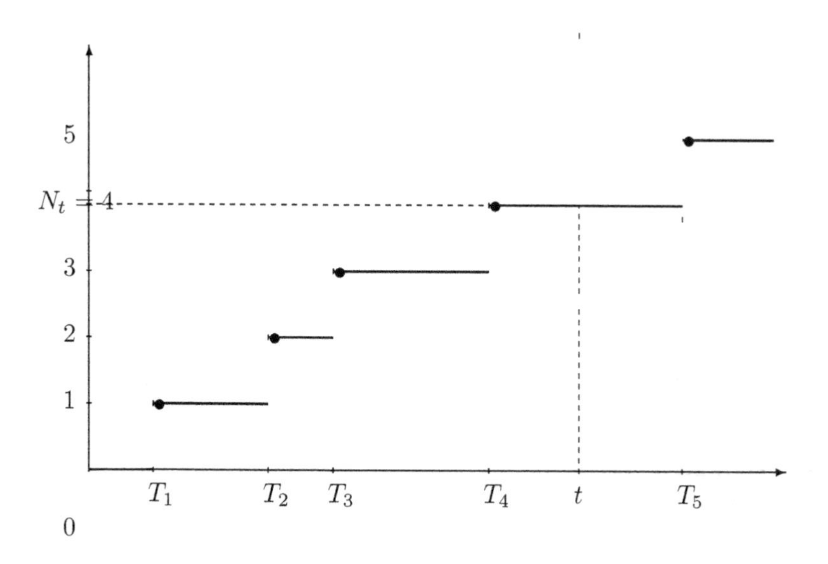

Fig. 8.1 Poisson process

Definition 8.2.1.1 *Let* $\mathbf{F}$ *be a given filtration and* λ *a positive constant. The process* N *is an* $\mathbf{F}$*-Poisson process with intensity* λ *if* N *is an* $\mathbf{F}$*-adapted process, such that for all positive numbers* (t,s)*, the r.v.* $N_{t+s} - N_t$ *is independent of* $\mathcal{F}_t$ *and follows the Poisson law with parameter* λs*.*

The random measure μ associated with a Poisson process is such that $\mu(\Lambda)$ is almost surely finite for any bounded set Λ (the number of jumps in any finite interval of time is almost surely finite), and $\mathbb{E}(\mu(\Lambda)) = \lambda|\Lambda|$ where $|\Lambda|$ is the Lebesgue measure of the set Λ.

We now recall some properties of Poisson processes.

• The time T_n when the nth-jump of N occurs is the sum of n independent exponential r.v's, hence it has a Gamma law with parameters (n,λ):

$$\mathbb{P}(T_n \in dt) = \frac{(\lambda t)^{n-1}}{(n-1)!}\lambda e^{-\lambda t}\,\mathbb{1}_{\{t>0\}}dt,$$

and its Laplace transform, for $\mu > -\lambda$, is given by

$$\mathbb{E}(e^{-\mu T_n}) = \left(\frac{\lambda}{\lambda+\mu}\right)^n.$$

• From the properties of the Poisson distribution, it follows that for every $t > 0$,

$$\mathbb{E}(N_t) = \lambda t, \quad \mathrm{Var}\,(N_t) = \lambda t$$

and for every $x > 0$, $t \geq 0, u, \alpha \in \mathbb{R}$

$$\mathbb{E}(x^{N_t}) = e^{\lambda t(x-1)}\,;\ \mathbb{E}(e^{iuN_t}) = e^{\lambda t(e^{iu}-1)}\,;\ \mathbb{E}(e^{\alpha N_t}) = e^{\lambda t(e^{\alpha}-1)}\,. \tag{8.2.1}$$

• Conditionally on $(N_t = n)$, the law of $(T_1, T_2, \ldots, T_n)$ is a multinomial distribution on $[0,t]$.

• Let, for t fixed and $i \geq 1$, $T_i^{(t)} := T_{N_t+i} - t$ where T_{N_t+i} is the time of the i-th jump which occurs after t. The sequence of times $(T_i^{(t)}, i \geq 1)$ has the same law as $(T_i, i \geq 1)$. This property is called the lack of memory of the Poisson process.

Exercise 8.2.1.2 Let N be a Poisson process. Prove that $N_t t^{-1} \to \lambda$ a.s. when t goes to infinity. ◁

Exercise 8.2.1.3 Let N be a Poisson process and T_n its n-th jump time. Prove that

$$\mathbb{P}(T_n \geq s|\mathcal{F}_t) = \mathbb{1}_{s\leq T_n \leq t} + \mathbb{1}_{t<T_n}\int_{s-t}^{\infty} \frac{\lambda(\lambda u)^{n-1-N_t}}{(n-1-N_t)!}e^{-\lambda u}\,\mathbb{1}_{\{u\geq 0\}}\,du\,.$$

◁

8.2.2 Martingale Properties

From the independence of the increments of the Poisson process, we derive the following martingale properties:

Proposition 8.2.2.1 *Let N be an* $\mathbf{F}$*-Poisson process. For each $\alpha \in \mathbb{R}$, for each bounded Borel function h, the following processes are* $\mathbf{F}$*-martingales:*

$$\begin{array}{ll}
\text{(i)} & M_t := N_t - \lambda t, \\
\text{(ii)} & M_t^2 - \lambda t = (N_t - \lambda t)^2 - \lambda t, \qquad (8.2.2)\\
\text{(iii)} & \exp(\alpha N_t - \lambda t(e^\alpha - 1)), \qquad (8.2.3)\\
\text{(iv)} & \exp\left(\int_0^t h(s)dN_s - \lambda \int_0^t (e^{h(s)} - 1)ds\right), \\
\text{(v)} & \int_0^t h(s)dM_s \,, \\
\text{(vi)} & \left(\int_0^t h(s)dM_s\right)^2 - \lambda \int_0^t h^2(s)ds \,.
\end{array}$$

PROOF: Let $s < t$. From the independence of the increments of the Poisson process, we obtain:

(i) $\mathbb{E}(M_t - M_s|\mathcal{F}_s) = \mathbb{E}(N_t - N_s) - \lambda(t-s) = 0$, hence M is a martingale.

(ii) The martingale property of M and the independence of the increments of the Poisson process imply

$$\begin{aligned}
\mathbb{E}(M_t^2 - M_s^2|\mathcal{F}_s) &= \mathbb{E}[(M_t - M_s)^2|\mathcal{F}_s] = \mathbb{E}[(N_t - N_s - \lambda(t-s))^2|\mathcal{F}_s] \\
&= \mathbb{E}[(N_t - N_s)^2] - \lambda^2(t-s)^2 \\
&= \mathbb{E}[N_{t-s}^2] - \lambda^2(t-s)^2 = \mathrm{Var} N_{t-s} \,,
\end{aligned}$$

hence,

$$\mathbb{E}(M_t^2 - M_s^2|\mathcal{F}_s) = \lambda(t-s) \,,$$

and the process $(M_t^2 - \lambda t, t \geq 0)$ is a martingale.

(iii) From the form of the Laplace transform of N_t given in (8.2.1) and the independence of the increments, $\mathbb{E}[\exp[\alpha(N_t - N_s) - \lambda(t-s)(e^\alpha - 1)]\,|\mathcal{F}_s] = 1$, hence the martingale property of the process in (iii).

Assertions (iv-v-vi) can be proved first for elementary functions h of the form $h = \sum_i a_i \mathbb{1}_{]t_i, t_{i+1}]}$ and by then passing to the limit for general bounded Borel functions h. □

Exercise 8.2.2.2 Prove that, for any $\beta > -1$, any bounded Borel function h, and any bounded Borel function φ valued in $]-1, \infty[$, the processes

$$\exp[\ln(1+\beta)N_t - \lambda\beta t] = (1+\beta)^{N_t} e^{-\lambda\beta t},$$

$$\exp\left(\int_0^t h(s)dN_s + \lambda \int_0^t (1-e^{h(s)})ds\right)$$
$$= \exp\left(\int_0^t h(s)dM_s + \lambda \int_0^t (1+h(s)-e^{h(s)})ds\right),$$

$$\exp\left(\int_0^t \ln(1+\varphi(s))dN_s - \lambda \int_0^t \varphi(s)ds\right)$$
$$= \exp\left(\int_0^t \ln(1+\varphi(s))dM_s + \lambda \int_0^t (\ln(1+\varphi(s)) - \varphi(s))ds\right),$$

are martingales.
Hint: These formulae are "avatars" of those of Proposition 8.2.2.1. ◁

Exercise 8.2.2.3 Prove (without using the following Proposition!) that the process $(\int_0^t N_{s-}dM_s, t \geq 0)$ is a martingale, and that the process $\int_0^t N_s dM_s$ is not a martingale. ◁

Definition 8.2.2.4 *The martingale* $(M_t = N_t - \lambda t, t \geq 0)$ *is called the* ***compensated process*** *of* N*, and* λ *the* ***intensity*** *of the process* N.

Remarks 8.2.2.5 (a) Note that the process M is a discontinuous martingale with bounded variation.

(b) We give an example of a martingale which is not square integrable. Let $X_t = \int_0^t \frac{1}{\sqrt{s}} dM_s$. The process X is a martingale, however, it is not square integrable.

The previous Proposition 8.2.2.1 can be generalized to predictable integrands:

Proposition 8.2.2.6 *Let* N *be an* $\mathbf{F}$*-Poisson process and let* H *be an* $\mathbf{F}$*-predictable bounded process. Then the following processes are martingales:*

$$\left.\begin{array}{l}
\text{(i)}\ \ (H\star M)_t = \int_0^t H_s dM_s = \int_0^t H_s dN_s - \lambda \int_0^t H_s ds \\
\text{(ii)}\ \ (H\star M)_t^2 - \lambda \int_0^t H_s^2 ds \\
\text{(iii)}\ \exp\left(\int_0^t H_s dN_s + \lambda \int_0^t (1-e^{H_s})ds\right) =: \mathcal{E}(H\star M)_t \\
\qquad = 1 + \int_0^t \mathcal{E}(H\star M)_{s-} H_s dM_s
\end{array}\right\} \quad (8.2.4)$$

PROOF: One establishes (8.2.4) for predictable processes $(H_t, t \geq 0)$ of the form $H_t = K_S \mathbb{1}_{]S,T]}(t)$ where S and T are two stopping times and K_S is $\mathcal{F}_S$-measurable. In that case,

$$\int_0^t H_s dM_s = K_S(M_{T\wedge t} - M_{S\wedge t})$$

and the martingale property follows. Then, one passes to the limit. The same procedure can be applied to prove that the two processes (ii) and (iii) of (8.2.4) are martingales. □

We have used in (iii) the notation $\mathcal{E}(H \star M)_t$ for the Doléans-Dade exponential of the martingale $\int H_s dM_s$.

Comments 8.2.2.7 (a) If H satisfies $\mathbb{E}(\int_0^t |H_s|ds) < \infty$, the process in (i) is still a martingale.

(b) The results of Exercise 8.2.2.3 are now quite clear: in general, the martingale property (8.2.4) does not extend from predictable to adapted processes H. Indeed, from the definition of the stochastic integral w.r.t. N, and the fact that for every fixed s, $N_s - N_{s-} = 0, \mathbb{P}\, a.s.$,

$$\int_0^t (N_s - N_{s-})dM_s = \int_0^t (N_s - N_{s-})dN_s - \lambda \int_0^t (N_s - N_{s-})ds$$
$$= N_t - \lambda \int_0^t (N_s - N_{s-})ds = N_t\,.$$

Hence, the left-hand side, where one integrates the adapted (unpredictable) process $N_s - N_{s-}$ with respect to the martingale M, is not a martingale. Equivalently, the process

$$\int_0^t N_s dM_s = \int_0^t N_{s-} dM_s + N_t\,,$$

is not a martingale.

(c) Property (i) of Proposition 8.2.2.6 enables us to prove that the jump times $(T_i, i \geq 1)$ are not predictable. Indeed, if T_1 were a predictable stopping time, then the process $(\mathbb{1}_{\{t<T_1\}}, t \geq 0)$ would be predictable, however $\int_0^t \mathbb{1}_{\{s<T_1\}} dM_s = -\lambda(t \wedge T_1)$ is not a martingale. More generally, assume that T_i is predictable. Then, $(\int_0^t \mathbb{1}_{[T_i]}(s)dM_s, t \geq 0)$ would be a martingale and

$$\mathbb{E}\left(\int_0^t \mathbb{1}_{[T_i]}(s)dN_s\right) = \mathbb{E}\left(\mathbb{1}_{T_i \leq t}\left(N_{T_i} - N_{T_i-}\right)\right) = \mathbb{P}(T_i \leq t)$$

would be equal to $\mathbb{E}\left(\int_0^t \mathbb{1}_{[T_i]}(s)\lambda ds\right) = 0$, which is absurd.

Remark 8.2.2.8 Note that (i) and (ii) of Proposition 8.2.2.1 imply that the process $(M_t^2 - N_t; t \geq 0)$ is a martingale. Hence, there exist (at least) two increasing processes A such that $(M_t^2 - A_t, t \geq 0)$ is a martingale. The increasing process $(\lambda t, t \geq 0)$ is the predictable quadratic variation of M (denoted $\langle M \rangle$), whereas the increasing process $(N_t, t \geq 0)$ is the optional quadratic variation of M (denoted $[M]$). For any $\mu \in [0,1]$, the process $(\mu N_t + (1-\mu)\lambda t; t \geq 0)$ is increasing and the process

$$M_t^2 - (\mu N_t + (1-\mu)\lambda t) = M_t^2 - \lambda t - \mu(N_t - \lambda t)$$

is a martingale. (See ↣ Section 9.2 for the definition of quadratic variation if needed.)

8.2.3 Infinitesimal Generator

Proposition 8.2.3.1 *The Poisson process is a process with independent and stationary increments, and hence is a Markov process; its infinitesimal generator $\mathcal{L}$ is given by*

$$\mathcal{L}(f)(x) = \lambda[f(x+1) - f(x)],$$

where f is a bounded Borel function.

PROOF: The Markov property follows from

$$\mathbb{E}(f(N_t)|\mathcal{F}_s^N) = \mathbb{E}(f(N_t - N_s + N_s)|\mathcal{F}_s^N) = F(t-s, N_s)$$

where $F(u,x) = \mathbb{E}(f(x+N_u))$ and $t \geq s$. We recall the definition of the infinitesimal generator:

$$\mathcal{L}(f)(x) = \lim_{t\to 0} \frac{1}{t}(\mathbb{E}(f(x+N_t)) - f(x)).$$

Hence, from $\mathbb{E}(f(x+N_t)) = \sum_{n=0}^{\infty} f(x+n)\mathbb{P}(N_t = n)$, we obtain

$$\frac{1}{t}(\mathbb{E}(f(x+N_t)) - f(x)) = e^{-\lambda t} \sum_{n=0}^{\infty} \frac{f(n+x)-f(x)}{t} \frac{(\lambda t)^n}{n!}.$$

From

$$e^{-\lambda t} \sum_{n\geq 2} \frac{(\lambda t)^n}{n!} \leq \frac{\lambda^2 t^2}{2}$$

the limit of $\frac{1}{t}(\mathbb{E}(f(x+N_t)) - f(x))$ when t goes to 0 is equal to the limit of $e^{-\lambda t}\frac{\lambda t}{t}(f(x+1) - f(x))$, that is to $\lambda(f(x+1) - f(x))$. □

Therefore, for any bounded Borel function f, the process

$$C_t^f = f(N_t) - f(0) - \int_0^t \mathcal{L}(f)(N_s)ds$$

is a martingale (see Proposition 1.1.14.2). Using that

$$f(N_t) - f(0) = \int_0^t (f(N_{s-}+1) - f(N_{s-}))\, dN_s\,, \tag{8.2.5}$$

the martingale $(C_t^f, t \geq 0)$ can be written as a stochastic integral with respect to the compensated martingale $(M_t = N_t - \lambda t, t \geq 0)$ as

$$C_t^f = \int_0^t [f(N_{s-}+1) - f(N_{s-})] dM_s\,.$$

Comment 8.2.3.2 Processes with independent and stationary increments are called Lévy processes, the reader may refer to $\rightarrowtail$ Chapter 11 for a more extended study.

Exercise 8.2.3.3 Extend formula (8.2.5) to functions f defined on $\mathbb{R}^+ \times \mathbb{N}$ that are C^1 with respect to the first variable, and prove that if β is a constant with $\beta > -1$ and $L_t = \exp(\log(1+\beta)N_t - \lambda\beta t)$, then $dL_t = L_{t^-}\beta dM_t$.

More generally, let $L_t = (1+a)^{N_t}e^{-\lambda a t}$ for $a \in \mathbb{R}$. Prove that L satisfies $dL_t = L_{t^-}adM_t$, i.e.,

$$L_t = 1 + \int_0^t L_{s^-}\, adM_s = 1 + a\int_0^t L_{s^-}\, dN_s - \lambda a \int_0^t L_{s^-}\, ds\,.$$

Note that, for $a < -1$, L_t takes values in $\mathbb{R}$. The process L is the Doléans-Dade exponential of the martingale aM. ◁

Exercise 8.2.3.4 Let $T > 0$ be fixed and let $\varphi : [0,T] \to \mathbb{R}$ be a bounded Borel function and N a Poisson process. Prove that there exist a predictable process h and a constant c such that

$$\exp\left(\int_0^T \varphi(s)dN_s\right) = c + \int_0^T h_s dN_s\,.$$

`Hint:` Set $Z_t = \int_0^t \varphi(s)dN_s$. Then,

$$de^{Z_t} = \left(e^{Z_{t^-}+\varphi(t)} - e^{Z_{t^-}}\right)dN_t\,.$$

The reader may be interested to compare this simple result with the predictable representation theorem in Subsection 8.3.5. ◁

8.2.4 Change of Probability Measure: An Example

If N is a Poisson process with constant intensity λ, then, from Exercises 8.2.2.2 and 8.2.3.3, for $\beta > -1$, the process L defined by

$$L_t = (1+\beta)^{N_t}e^{-\lambda\beta t}$$

is a strictly positive martingale with expectation equal to 1. Let $\mathbb{Q}$ be the probability defined via $\mathbb{Q}|_{\mathcal{F}_t} = L_t\mathbb{P}|_{\mathcal{F}_t}$. From

$$\mathbb{E}_{\mathbb{Q}}(x^{N_t}) = \mathbb{E}_{\mathbb{P}}(L_t x^{N_t}) = e^{-\lambda\beta t}\mathbb{E}_{\mathbb{P}}([(1+\beta)x]^{N_t}) = \exp((1+\beta)\lambda t(x-1))$$

we deduce that the r.v. N_t follows the Poisson law with parameter $(1+\beta)\lambda t$ under $\mathbb{Q}$. Let $t_1 < \cdots < t_i < t_{i+1} < \cdots < t_n$ and let $(x_i, i \le n)$ be a sequence of positive real numbers. The equalities

$$\mathbb{E}_{\mathbb{Q}}\left(\prod_{i=1}^{n} x_i^{N_{t_{i+1}}-N_{t_i}}\right) = \mathbb{E}_{\mathbb{P}}\left(e^{-\lambda\beta t}\prod_{i=1}^{n}((1+\beta)x_i)^{N_{t_{i+1}}-N_{t_i}}\right)$$
$$= e^{-\lambda\beta t}\prod_{i=1}^{n} e^{-\lambda(t_{i+1}-t_i)}\, e^{\lambda(t_{i+1}-t_i)(1+\beta)x_i}$$
$$= \prod_{i=1}^{n} e^{(1+\beta)\lambda(t_{i+1}-t_i)(x_i-1)}$$

establish that, under $\mathbb{Q}$, $N_{t_{i+1}} - N_{t_i} \stackrel{\text{law}}{=} N_{t_{i+1}-t_i}$ is a Poisson r.v. with parameter $(1+\beta)\lambda(t_{i+1}-t_i)$ and that N has independent increments. Therefore, the process N is a $\mathbb{Q}$-Poisson process with intensity equal to $(1+\beta)\lambda$. Let us state this result as a proposition:

Proposition 8.2.4.1 *Let Π^λ be the probability on the canonical space which makes the coordinate process a Poisson process with intensity λ. Then, the following absolute continuity relationship holds:*

$$\boxed{\Pi^{(1+\beta)\lambda}|_{\mathcal{F}_t} = \left((1+\beta)^{N_t} e^{-\lambda\beta t}\right)\, \Pi^{\lambda}|_{\mathcal{F}_t}\,.}$$

Comment 8.2.4.2 One should note the analogy between the change of intensity of Poisson processes and the change of drift of a BM under a change of probability. However, let us point out a major difference. If $\mathbb{Q}$ is equivalent to $\mathbb{P}$, we know that if B is a $\mathbb{P}$-BM and $\widehat{B}$ is the martingale part of B under $\mathbb{Q}$, then $B_t^2 - t$ is a $\mathbb{P}$-martingale and $\widehat{B}_t^2 - t$ is a $\mathbb{Q}$-martingale (in other words the brackets are the same, i.e., $\langle B\rangle = \langle \widehat{B}\rangle$). If $\mathbb{Q}$ is equivalent to $\mathbb{P}$, and $M_t = N_t - \lambda t$ the compensated martingale associated with a Poisson process, the process $M_t^2 - \lambda t$ is a $\mathbb{P}$-martingale and the $\mathbb{P}$-(predictable) bracket of M is λt. We have proved above that the $\mathbb{Q}$-(predictable) bracket of $\widehat{M}_t = N_t - (1+\beta)\lambda t$ is $(1+\beta)\lambda t$. Hence, the predictable bracket is no longer the same under a change of probability. See ↣ Section 9.4 for a general Girsanov theorem and ↣ Subsection 11.3.1 for the case of Lévy processes.

8.2.5 Hitting Times

Let $x > 0$ and $T_x = \inf\{t, N_t \geq x\}$. Then, for $n - 1 < x \leq n$, the hitting time $T_x = \inf\{t, N_t \geq n\} = \inf\{t, N_t = n\}$ is equal to the time of the nth-jump of N, and hence has a Gamma (n, λ) law.

Exercise 8.2.5.1 Let $X_t = N_t + ct$. Compute $\mathbb{P}(\inf_{s \leq t} X_s \leq a)$. One should distinguish the cases $c > 0$ and $c < 0$. ◁

8.3 Inhomogeneous Poisson Processes

8.3.1 Definition

Instead of considering a constant intensity λ as before, now $(\lambda(t), t \geq 0)$ is an $\mathbb{R}^+$-valued Borel function satisfying $\int_0^t \lambda(u)du < \infty, \forall t$ and $\int_0^\infty \lambda(u)du = \infty$. An **inhomogeneous Poisson process** N with intensity λ is a counting process with independent increments which satisfies, for $t > s$,

$$\mathbb{P}(N_t - N_s = n) = e^{-\Lambda(s,t)} \frac{(\Lambda(s,t))^n}{n!} \tag{8.3.1}$$

where $\Lambda(s,t) = \Lambda(t) - \Lambda(s) = \int_s^t \lambda(u)du$, and $\Lambda(t) = \int_0^t \lambda(u)du$.

If $(T_n, n \geq 1)$ is the sequence of successive jump times associated with N, the law of T_n is:

$$\mathbb{P}(T_n \leq t) = \frac{1}{(n-1)!} \int_0^t \exp(-\Lambda(s)) \, (\Lambda(s))^{n-1} \, d\Lambda(s) \, .$$

It can easily be shown that an inhomogeneous Poisson process with deterministic intensity is an inhomogeneous Markov process. Moreover, since N_t has a Poisson law with parameter $\Lambda(t)$, one has $\mathbb{E}(N_t) = \Lambda(t), \text{Var}(N_t) = \Lambda(t)$. For any real numbers u and α, for any $t \geq 0$,

$$\begin{aligned} \mathbb{E}(e^{iuN_t}) &= \exp((e^{iu} - 1)\Lambda(t)), \\ \mathbb{E}(e^{\alpha N_t}) &= \exp((e^{\alpha} - 1)\Lambda(t)) \, . \end{aligned}$$

An inhomogeneous Poisson process can be constructed as a deterministic time changed Poisson process, i.e., if $\widehat{N}$ is a Poisson process with constant intensity equal to 1, then $N_t = \widehat{N}_{\Lambda(t)}$ is an inhomogeneous Poisson process with intensity Λ.

We emphasize that we shall use the term Poisson process only when dealing with the standard Poisson process, i.e., when $\Lambda(t) = \lambda t$.

8.3.2 Martingale Properties

The martingale properties of a standard Poisson process can be extended to an inhomogeneous Poisson process:

Proposition 8.3.2.1 *Let N be an inhomogeneous Poisson process with deterministic intensity λ and $\mathbf{F}^N$ its natural filtration. The process*

$$M_t = N_t - \int_0^t \lambda(s)ds, \, t \geq 0$$

is an $\mathbf{F}^N$-martingale. The increasing function $\Lambda(t) := \int_0^t \lambda(s)ds$ is called the (deterministic) **compensator** *of N.*

Let ϕ be an $\mathbf{F}^N$-predictable process such that $\mathbb{E}(\int_0^t |\phi_s|\lambda(s)ds) < \infty$ for every t. Then, the process $(\int_0^t \phi_s dM_s, t \geq 0)$ is an $\mathbf{F}^N$-martingale. In particular,

$$\mathbb{E}\left(\int_0^t \phi_s \, dN_s\right) = \mathbb{E}\left(\int_0^t \phi_s \lambda(s) ds\right). \tag{8.3.2}$$

As in the constant intensity case, for any bounded $\mathbf{F}^N$-predictable process H, the following processes are martingales:

$$\begin{aligned}
&(i) && (H\star M)_t = \int_0^t H_s dM_s = \int_0^t H_s dN_s - \int_0^t \lambda(s) H_s ds\,, \\
&(ii) && (H\star M)_t^2 - \int_0^t \lambda(s) H_s^2 ds\,, \\
&(iii) && \exp\left(\int_0^t H_s dN_s - \int_0^t \lambda(s)(e^{H_s} - 1)ds\right).
\end{aligned}$$

8.3.3 Watanabe's Characterization of Inhomogeneous Poisson Processes

The study of inhomogeneous Poisson processes can be generalized to the case where the intensity is not absolutely continuous with respect to the Lebesgue measure. In this case, Λ is an increasing, right-continuous, deterministic function with value zero at time zero, and it satisfies $\Lambda(\infty) = \infty$. If N is a counting process with independent increments and if (8.3.1) holds, the process $(N_t - \Lambda(t), t \geq 0)$ is a martingale and for any bounded predictable process ϕ, the equality $\mathbb{E}(\int_0^t \phi_s \, dN_s) = \mathbb{E}(\int_0^t \phi_s d\Lambda(s))$ is satisfied for any t. This result admits a converse.

Proposition 8.3.3.1 (Watanabe's Characterization.) *Let N be a counting process and Λ an increasing, continuous function with value zero at time zero. Let us assume that the process $(M_t := N_t - \Lambda(t),\ t \geq 0)$ is a martingale. Then N is an inhomogeneous Poisson process with compensator Λ. It is a Poisson process if $\Lambda(t) = \lambda t$.*

PROOF: Let $s < t$ and $\theta > 0$.

$$\begin{aligned}
e^{\theta N_t} - e^{\theta N_s} &= \sum_{s<u\leq t} e^{\theta N_u} - e^{\theta N_{u-}} \\
&= \sum_{s<u\leq t} e^{\theta N_{u-}}(e^\theta - 1)\Delta N_u = (e^\theta - 1)\int_{]s,t]} e^{\theta N_{u-}} dN_u \\
&= (e^\theta - 1)\left(\int_{]s,t]} e^{\theta N_{u-}} dM_u + \int_{]s,t]} e^{\theta N_u} d\Lambda(u)\right).
\end{aligned}$$

By relying on the fact that the first integral is a martingale,

$$\begin{aligned}\mathbb{E}(e^{\theta N_t} - e^{\theta N_s}|\mathcal{F}_s) &= (e^\theta - 1)\mathbb{E}\left(\int_{]s,t]} e^{\theta N_u} d\Lambda(u)|\mathcal{F}_s\right) \\ &= (e^\theta - 1)\int_{]s,t]} \mathbb{E}\,(e^{\theta N_u}|\mathcal{F}_s)d\Lambda(u)\,.\end{aligned}$$

Let s be fixed and define $\phi(t) = \mathbb{E}(e^{\theta N_t}|\mathcal{F}_s)$. Then, for $t > s$,

$$\phi(t) = \phi(s) + (e^\theta - 1)\int_s^t \phi(u)d\Lambda(u)\,.$$

Solving this equation leads to

$$\phi(t) = e^{\theta N_s}\exp\left[(e^\theta - 1)\int_s^t d\Lambda(u)\right]\,.$$

This shows that the process N has independent increments and that, for $s < t$, the r.v. $N_t - N_s$ has a Poisson law with parameter $\Lambda(t) - \Lambda(s)$. □

8.3.4 Stochastic Calculus

In this section, M is the compensated martingale of an inhomogeneous Poisson process N with deterministic intensity $(\lambda(s), s \geq 0)$. From now on, we restrict our attention to integrals of predictable processes, even if the stochastic integral is defined in a more general setting.

Integration by Parts Formula

Let x and y be two predictable processes and define two processes X and Y as

$$X_t = x + \int_0^t x_s dN_s, \quad Y_t = y + \int_0^t y_s dN_s\,.$$

The jumps of X (resp. of Y) occur at the same times as the jumps of N and $\Delta X_s = x_s \Delta N_s, \Delta Y_s = y_s \Delta N_s$. The processes X and Y are of finite variation and are constant between two jumps. Then, it is easy to check that

$$X_t Y_t = xy + \sum_{s\leq t} \Delta(XY)_s = xy + \sum_{s\leq t} X_{s-}\Delta Y_s + \sum_{s\leq t} Y_{s-}\Delta X_s + \sum_{s\leq t} \Delta X_s\,\Delta Y_s$$

We shall write this equality as

$$\boxed{X_t Y_t = xy + \int_0^t Y_{s-}dX_s + \int_0^t X_{s-}dY_s + [X,Y]_t}$$

where (note that $(\Delta N_t)^2 = \Delta N_t$)

$$[X,Y]_t := \sum_{s\leq t} \Delta X_s \, \Delta Y_s = \sum_{s\leq t} x_s y_s \Delta N_s = \int_0^t x_s y_s dN_s \,.$$

More generally (a general discussion is proposed in ↣ Chapter 9 and 10), if

$$\begin{aligned} dX_t &= h_t dt + x_t dN_t, \; X_0 = x \\ dY_t &= \widetilde{h}_t dt + y_t dN_t, \; Y_0 = y \,, \end{aligned}$$

one still gets

$$X_t Y_t = xy + \int_0^t Y_{s-} dX_s + \int_0^t X_{s-} dY_s + [X,Y]_t$$

where

$$[X,Y]_t = \int_0^t x_s y_s dN_s \,.$$

In particular, if $dX_t = x_t dM_t$ and $dY_t = y_t dM_t$, the process $X_t Y_t - [X,Y]_t$ is a local martingale.

Itô's Formula

For Poisson processes, Itô's formula is obvious as we now explain. We shall give an extension of this formula for more general processes in the following Chapter 9.

Let N be a Poisson process and f a bounded Borel function. The trivial equality

$$f(N_t) = f(N_0) + \sum_{0<s\leq t} f(N_s) - f(N_{s-}) \tag{8.3.3}$$

is the main step in obtaining Itô's formula for a Poisson process.

We can write the right-hand side of (8.3.3) as a stochastic integral:

$$\begin{aligned} \sum_{0<s\leq t} f(N_s) - f(N_{s-}) &= \sum_{0<s\leq t} \left[f(N_{s-}+1) - f(N_{s-}) \right] \Delta N_s \\ &= \int_0^t \left[f(N_{s-}+1) - f(N_{s-}) \right] dN_s \,, \end{aligned}$$

hence, the canonical decomposition of the semi-martingale $f(N)$ as the sum of a martingale and an absolutely continuous adapted process is

$$f(N_t) = f(N_0) + \int_0^t [f(N_{s-}+1) - f(N_{s-})] dM_s + \int_0^t [f(N_{s-}+1) - f(N_{s-})] \lambda ds \,.$$

It is straightforward to generalize this result. Let

$$X_t = x + \int_0^t x_s dN_s = x + \sum_{T_n \leq t} x_{T_n},$$

with x a predictable process. The process $(X_t, t \geq 0)$ has at time T_n, a jump of size $(\Delta X)_{T_n} = x_{T_n}$, and is constant between two consecutive jumps. The obvious identity

$$F(X_t) = F(X_0) + \sum_{s \leq t} F(X_s) - F(X_{s-}),$$

holds for any bounded function F. The number of jumps before t is a.s. finite, and the sum is well defined. This formula can be written in an equivalent form:

$$\begin{aligned}F(X_t) - F(X_0) &= \sum_{s \leq t} (F(X_s) - F(X_{s-}))\, \Delta N_s \\ &= \int_0^t (F(X_s) - F(X_{s-}))\, dN_s = \int_0^t (F(X_{s-} + x_s) - F(X_{s-}))\, dN_s\end{aligned}$$

where the integral on the right-hand side is a Stieltjes integral. More generally again, we have the following result

Proposition 8.3.4.1 *Let h be an adapted process, x a predictable process and*

$$dX_t = h_t dt + x_t dM_t = (h_t - x_t \lambda(t))dt + x_t dN_t$$

where N is an inhomogeneous Poisson process. Let $F \in C^{1,1}(\mathbb{R}^+ \times \mathbb{R})$. Then

$$\begin{aligned}F(t, X_t) &= \int_0^t [F(s, X_{s-} + x_s) - F(s, X_{s-})] dM_s \qquad (8.3.4)\\ &+ \int_0^t (\partial_t F(s, X_s) + \partial_x F(s, X_s) h_s)\, ds \\ &+ \int_0^t (F(s, X_{s-} + x_s) - F(s, X_{s-}) - \partial_x F(s, X_{s-}) x_s)\, \lambda(s) ds\,.\end{aligned}$$

PROOF: Indeed, between two jumps of the process N, $dX_t = (h_t - \lambda(t)x_t)dt$, and for $T_n < s < t < T_{n+1}$,

$$F(t, X_t) = F(s, X_s) + \int_s^t \partial_t F(u, X_u) du + \int_s^t \partial_x F(u, X_u)(h_u - x_u \lambda(u)) du\,.$$

At jump times T_n, one has $F(T_n, X_{T_n}) = F(T_n, X_{T_n-}) + \Delta F(\cdot, X)_{T_n}$. Hence,

$$\begin{aligned}F(t, X_t) = F(0, X_0) &+ \int_0^t \partial_t F(s, X_s) ds + \int_0^t \partial_x F(s, X_s)(h_s - x_s \lambda(s))\, ds \\ &+ \sum_{s \leq t} (F(s, X_s) - F(s, X_{s-}))\,.\end{aligned}$$

This formula can be written as

$$
\begin{aligned}
F(t,X_t) - F(0,X_0) &= \int_0^t \partial_t F(s,X_s)ds + \int_0^t \partial_x F(s,X_s)(h_s - x_s\lambda(s))ds \\
&\quad + \int_0^t [F(s,X_s) - F(s,X_{s-})]dN_s \\
&= \int_0^t \partial_t F(s,X_s)ds + \int_0^t \partial_x F(s,X_{s-})dX_s \\
&\quad + \int_0^t [F(s,X_s) - F(s,X_{s-}) - \partial_x F(s,X_{s-})x_s]dN_s \\
&= \int_0^t \partial_t F(s,X_s)ds + \int_0^t \partial_x F(s,X_{s-})dX_s \\
&\quad + \int_0^t [F(s,X_{s-}+x_s) - F(s,X_{s-}) - \partial_x F(s,X_{s-})x_s]dN_s \,.
\end{aligned}
$$

One can also write

$$
\begin{aligned}
F(t,X_t) = F(0,X_0) &+ \int_0^t \partial_t F(s,X_s)ds + \int_0^t \partial_x F(s,X_{s-})dX_s \\
&+ \sum_{s\leq t} [F(s,X_s) - F(s,X_{s-}) - \partial_x F(s,X_{s-})x_s \Delta N_s] \,.
\end{aligned}
$$

which is easy to memorize. The first three terms on the right-hand side are obtained from "ordinary" calculus, the fourth term takes into account the jumps of the left-hand side and of the stochastic integral on the right-hand side.

Remarks 8.3.4.2 (a) In the "ds" integrals, we can write X_{s-} or X_s, since, for any bounded Borel function f,

$$\int_0^t f(X_{s-})ds = \int_0^t f(X_s)ds \,.$$

Note that since dN_s a.s. $N_s = N_{s-} + 1$, one has

$$\int_0^t f(N_{s-})dN_s = \int_0^t f(N_s - 1)dN_s \,.$$

However, we systematically use the form $\int_0^t f(N_{s-})dN_s$, even though the integral $\int_0^t f(N_s - 1)dN_s$ has a meaning. The reason is that

$$\int_0^t f(N_{s-})dM_s = \int_0^t f(N_{s-})dN_s - \lambda \int_0^t f(N_s)ds$$

is a martingale, whereas $\int_0^t f(N_s - 1)dM_s$ is not.

(b) We have named Itô's formula a formula allowing us to write the process $F(t, X_t)$ as a sum of stochastic integrals, as in equation (8.3.5). In fact, the aim of Itô's formula is to give, under some suitable conditions on F, the canonical decomposition of the semi-martingale $F(t, X_t)$.

Exercise 8.3.4.3 Let N be a Poisson process with intensity λ. Prove that, if $S_t = S_0 e^{\mu t + \sigma N_t}$, then

$$dS_t = S_{t-}(\mu dt + (e^\sigma - 1)dN_t)$$

and that S is a martingale iff $\mu = -\lambda(e^\sigma - 1)$. Prove that, for $a + 1 > 0$, the process $(L_t = \exp(N_t \ln(1 + a) - \lambda a t), t \geq 0)$ is a martingale and that, if $\mathbb{Q}|_{\mathcal{F}_t} = L_t \mathbb{P}|_{\mathcal{F}_t}$, the process N is a $\mathbb{Q}$-Poisson process with intensity $\lambda(1 + a)$. Note the progression made from Exercise 8.2.3.3. ◁

Exercise 8.3.4.4 The aim of this exercise is to prove that the linear equation $dZ_t = Z_{t-}\mu dM_t$, $Z_0 = 1$ with $\mu > -1$ has a unique solution. Assume that Z^1 and Z^2 are two solutions. W.l.g., we can assume that Z^2 is strictly positive. Prove that Z^1/Z^2 satisfies an ordinary differential equation with unique solution equal to 1. ◁

8.3.5 Predictable Representation Property

Proposition 8.3.5.1 *Let $\mathbf{F}^N$ be the natural filtration of the standard Poisson process N and let $H \in L^2(\mathcal{F}_\infty^N)$ be a square integrable random variable. Then, there exists a unique $\mathbf{F}^N$-predictable process $(h_t, t \geq 0)$ such that*

$$H = \mathbb{E}(H) + \int_0^\infty h_s dM_s$$

and $\mathbb{E}(\int_0^\infty h_s^2 ds) < \infty$.

PROOF: The family of exponential random variables

$$Y = \exp\left(\int_0^\infty \varphi(s)dN_s - \lambda \int_0^\infty (e^{\varphi(s)} - 1)ds\right),$$

where φ is a bounded deterministic function with compact support, is total in $L^2(\mathcal{F}_\infty^N)$. Any Y in this family can be written as a stochastic integral with respect to dM. Indeed, from Exercise 8.2.2.2 the process

$$Y_t = \exp\left(\int_0^t \varphi(s)dN_s - \lambda \int_0^t (e^{\varphi(s)} - 1)ds\right) = \mathbb{E}(Y|\mathcal{F}_t^N)$$

is a martingale, and is the solution of

$$dY_t = Y_{t-}(e^{\varphi(t)} - 1)dM_t\,,$$

so that,

$$Y = 1 + \int_0^\infty Y_{s-}(e^{\varphi(s)} - 1)dM_s\,.$$

Hence, with the notation of the statement, $h_s = Y_{s-}(e^{\varphi(s)} - 1)$. For more general random variables, the result follows by passing to the limit, owing to the isometry formula

$$\mathbb{E}\left(\int_0^\infty h_s dM_s\right)^2 = \lambda \mathbb{E}\left(\int_0^\infty h_s^2 ds\right)\,.$$

□

Comment 8.3.5.2 This result goes back to Brémaud and Jacod [125], Chou and Meyer [180], Davis [219].

8.3.6 Multidimensional Poisson Processes

Definition 8.3.6.1 *A process $(N^1, \ldots, N^d)$ is a d-dimensional* **F***-Poisson process if each component N^j is a right-continuous adapted process such that $N_0^j = 0$ and if there exist positive constants λ_j such that for every $t \geq s \geq 0$ and every integer n_j*

$$\mathbb{P}\left[\bigcap_{j=1}^d (N_t^j - N_s^j = n_j)|\mathcal{F}_s\right] = \prod_{j=1}^d e^{-\lambda_j(t-s)}\frac{(\lambda_j(t-s))^{n_j}}{n_j!}\,.$$

Note that the processes $(N^j, j = 1, \ldots, d)$ are independent; more generally, for any s, the processes $\left((N_{s+t}^j - N_s^j, j = 1, \ldots, d), t \geq 0\right)$ are independent and also independent of $\mathcal{F}_s$.

Proposition 8.3.6.2 *An* **F***-adapted process N is a d-dimensional* **F***-Poisson process if and only if:*

(i) *each N^j is an* **F***-Poisson process,*
(ii) *no two N^j's jump simultaneously $\mathbb{P}$ a.s..*

PROOF: We give the proof for $d = 2$.

(a) We assume (i) and (ii). For any pair (f, g) of bounded Borel functions, the process

$$X_t = \exp\left(\int_0^t f(s)dN_s^1 + \int_0^t g(s)dN_s^2\right)$$

satisfies

$$X_t = 1 + \sum_{0<s\leq t} \Delta X_s = 1 + \sum_{0<s\leq t} X_{s-}\left[\exp(f(s)\Delta N_s^1 + g(s)\Delta N_s^2) - 1\right].$$

From condition (ii)

$$X_t = 1 + \sum_{0<s\le t} X_{s-}\left[(e^{f(s)}-1)\Delta N^1_s + (e^{g(s)}-1)\Delta N^2_s\right],$$

hence, from the martingale property of the compensated process $N^i_t - \lambda_i t$:

$$\begin{aligned}\mathbb{E}(X_t) &= 1 + \mathbb{E}\left[\int_0^t X_{s-}\left((e^{f(s)}-1)\lambda_1 + (e^{g(s)}-1)\lambda_2\right)ds\right] \\ &= 1 + \int_0^t \mathbb{E}[X_s]\left((e^{f(s)}-1)\lambda_1 + (e^{g(s)}-1)\lambda_2\right)ds\,.\end{aligned}$$

Therefore, solving this equation, we find

$$\begin{aligned}\mathbb{E}(X_t) &= \exp\left(\int_0^t (e^{f(s)}-1)\lambda_1 ds\right)\exp\left(\int_0^t (e^{g(s)}-1)\lambda_2 ds\right) \\ &= \mathbb{E}\left[\exp\left(\int_0^t f(s)dN^1_s\right)\right]\mathbb{E}\left[\exp\left(\int_0^t g(s)dN^2_s\right)\right].\end{aligned}$$

The result follows.

(b) Conversely, if N is a d-dimensional Poisson process, then (i) and (ii) hold. □

Comment 8.3.6.3 Another proof follows from the predictable representation theorem valid for M^1 and M^2 individually. Let $H^i \in L^2(\mathcal{F}^i_\infty)$ for $i = 1, 2$. From $H^i = \mathbb{E}(H^i) + \int_0^\infty h^i_s dM^i_s$ and the integration by parts formula, we deduce that $\mathbb{E}(H^1H^2) = \mathbb{E}(H^1)\mathbb{E}(H^2)$ if and only if $[M^i, M^j] = 0$.

In order to construct correlated Poisson processes, one can proceed as follows. Let $(N^i, i = 1, 2, 3)$ be independent Poisson processes. Then the processes $\hat{N} = N^1 + N^2$ and $\tilde{N} = N^1 + N^3$ are correlated Poisson processes.

Exercise 8.3.6.4 Let $(N^i, i = 1, 2)$ be two independent Poisson processes. Prove that $N = N^1 + N^2$ is a Poisson process. Compute the compensator of N. Let $\tau^i = \inf\{t : N^i_t = 1\}$ and $\tau = \inf\{t : N_t = 1\}$. Compute $\mathbb{P}(\tau = \tau^1)$. ◁

8.4 Stochastic Intensity Processes

8.4.1 Doubly Stochastic Poisson Processes

Let $\mathbf{F}$ be a given filtration, where $\mathcal{F}_0$ is not the trivial σ-algebra; let N be a counting process which is $\mathbf{F}$-adapted and let λ be a positive process such that for any t, λ_t is $\mathcal{F}_0$-measurable and $\int_0^t \lambda_s ds < \infty, \mathbb{P}$ a.s.. Let $\Lambda(s,t) = \int_s^t \lambda_u du$. If

$$\mathbb{E}(e^{i\alpha(N_t-N_s)}|\mathcal{F}_s) = \exp\left((e^{i\alpha}-1)\Lambda(s,t)\right)$$

for any $t > s$ and any α, then N is called a **doubly stochastic Poisson process**. In that case,

$$\mathbb{P}(N_t - N_s = k|\mathcal{F}_s) = \exp(-\Lambda(s,t))\frac{(\Lambda(s,t))^n}{n!}$$

and the process $(N_t - \int_0^t \lambda_u du, t \geq 0)$ is an $\mathbf{F}$-martingale.

Comment 8.4.1.1 Doubly stochastic intensity processes are used in finance to model the intensity of default process (see Schönbucher [765]).

8.4.2 Inhomogeneous Poisson Processes with Stochastic Intensity

Definition 8.4.2.1 *Let* $\mathbf{F}$ *be a given filtration,* N *an* $\mathbf{F}$*-adapted counting process, and* $(\lambda_t, t \geq 0)$ *a positive* $\mathbf{F}$*-progressively measurable process such that for every* t, $\Lambda_t := \int_0^t \lambda_s ds < \infty$, $\mathbb{P}$ *a.s..*

The process N *is an* ***inhomogeneous Poisson process with stochastic intensity*** λ *if for every positive* $\mathbf{F}$*-predictable process* $(\phi_t, t \geq 0)$ *the following equality is satisfied:*

$$\mathbb{E}\left(\int_0^\infty \phi_s \, dN_s\right) = \mathbb{E}\left(\int_0^\infty \phi_s \lambda_s ds\right).$$

Therefore $(M_t = N_t - \Lambda_t, \, t \geq 0)$ is an $\mathbf{F}$-local martingale and an $\mathbf{F}$-martingale if for every t, $\mathbb{E}(\Lambda_t) < \infty$.

Proposition 8.4.2.2 *Let* N *be an inhomogeneous Poisson process with stochastic intensity* λ*. Then, for any* $\mathbf{F}$*-predictable process* ϕ *such that* $\forall t$, $\mathbb{E}(\int_0^t |\phi_s|\lambda_s ds) < \infty$*, the process* $(\int_0^t \phi_s dM_s, \, t \geq 0)$ *is an* $\mathbf{F}$*-martingale.*

The intensity depends in an important manner of the reference filtration. For example, the $\mathbf{F}^N$-intensity of N is $\mathbb{E}(\lambda_s|\mathcal{F}_s^N)$, i.e.,

$$N_t - \int_0^t \mathbb{E}(\lambda_s|\mathcal{F}_s^N)ds$$

is an $\mathbf{F}^N$-martingale. This is a particular case of the general filtering formula given in Proposition 5.10.3.1.

An inhomogeneous Poisson process N with stochastic intensity λ_t can be viewed as a time change of a standard Poisson process $\tilde{N}$, i.e., $N_t = \tilde{N}_{\Lambda_t}$.

8.4.3 Itô's Formula

The formula obtained in Subsection 8.3.4 can be generalized to inhomogeneous Poisson processes with stochastic intensities.

8.4.4 Exponential Martingales

We now extend Exercise 8.2.3.3 to more general Doléans-Dade exponentials:

Proposition 8.4.4.1 *Let N be an inhomogeneous Poisson process with stochastic intensity $(\lambda_t,\, t \geq 0)$, and $(\mu_t,\, t \geq 0)$ a predictable process such that, for any t, $\int_0^t |\mu_s|\lambda_s\, ds < \infty$. Let $(T_n, n \geq 1)$ be the sequence of jump times of N. Then, the process L, the solution of*

$$dL_t = L_{t-}\mu_t dM_t, \quad L_0 = 1\,, \tag{8.4.1}$$

is a local martingale defined by

$$L_t = \begin{cases} \exp(-\int_0^t \mu_s\lambda_s\, ds) & \text{if } t < T_1 \\ \prod_{n, T_n \leq t}(1+\mu_{T_n})\exp(-\int_0^t \mu_s\lambda_s\, ds) & \text{if } t \geq T_1\,. \end{cases} \tag{8.4.2}$$

Moreover, if μ is such that $\mu_s > -1\, a.s. \forall s$, then

$$L_t = \exp\left[-\int_0^t \mu_s\lambda_s ds + \int_0^t \ln(1+\mu_s)\, dN_s\right].$$

Later, we shall simply write the equalities (8.4.2) as

$$L_t = \prod_{n, T_n \leq t}(1+\mu_{T_n})\exp\left(-\int_0^t \mu_s\lambda_s\, ds\right)$$

with the understanding that $\prod_\emptyset = 1$.

PROOF: From general results on SDE, the linear equation (8.4.1) admits a unique solution (see also Exercise 8.3.4.4). Between two consecutive jumps, the solution of the equation (8.4.1) satisfies

$$dL_t = -L_{t-}\mu_t\lambda_t dt$$

therefore, for $t \in [T_n, T_{n+1}[$, we obtain

$$L_t = L_{T_n}\exp\left(-\int_{T_n}^t \mu_s\lambda_s ds\right).$$

The jumps of L occur at the same times as the jumps of N and the size of the jumps is $\Delta L_t = L_{t-}\mu_t\Delta N_t$, therefore $L_{T_n} = L_{T_n-}(1+\mu_{T_n})$. By backward recurrence on n, we get (8.4.2). □

The local martingale L is denoted by $\mathcal{E}(\mu\star M)$ and called the **Doléans-Dade exponential** of the process $\mu\star M$. The process L can also be written

$$L_t = \prod_{0<s\leq t}(1+\mu_s\Delta N_s)\exp\left[-\int_0^t \mu_s\,\lambda_s\, ds\right]. \tag{8.4.3}$$

Moreover, if for every t, $\mu_t > -1$, then L is a positive local martingale, therefore it is a supermartingale and

$$\begin{aligned} L_t &= \exp\left[-\int_0^t \mu_s \lambda_s ds + \sum_{s \le t} \ln(1+\mu_s) \Delta N_s\right] \\ &= \exp\left[-\int_0^t \mu_s \lambda_s ds + \int_0^t \ln(1+\mu_s)\, dN_s\right] \\ &= \exp\left[\int_0^t [\ln(1+\mu_s) - \mu_s]\, \lambda_s\, ds + \int_0^t \ln(1+\mu_s)\, dM_s\right] . \end{aligned}$$

The process L is a martingale if $\forall t$, $\mathbb{E}(L_t) = 1$. This is the case if μ is bounded. We shall see a more general criterion in $\rightarrowtail$ Subsection 9.4.3.

If μ is not greater than -1, then the process L defined in (8.4.2) is still a local martingale which satisfies $dL_t = L_{t-} \mu_t dM_t$. However it may be negative.

Example 8.4.4.2 A useful example is the case where $\mu \equiv -1$. In this case, we obtain that $\mathbb{1}_{\{t<T_1\}} \exp\left(\int_0^t \lambda_s ds\right)$ is a local martingale. Note that we have obtained similar results in Chapter 7 for processes with a single jump.

8.4.5 Change of Probability Measure

We establish now a particular case of the general Girsanov theorem (see $\rightarrowtail$ Section 9.4 for a general case).

Proposition 8.4.5.1 *Let μ be a predictable process such that $\mu > -1$ and $\int_0^t \lambda_s |\mu_s| ds < \infty$ a.s.. Let L be the positive exponential local martingale solution of $dL_t = L_{t-} \mu_t dM_t$. Assume that L is a martingale and let $\mathbb{Q}$ be the probability measure (locally equivalent to $\mathbb{P}$) defined on $\mathcal{F}_t$ by $\mathbb{Q}|_{\mathcal{F}_t} = L_t \mathbb{P}|_{\mathcal{F}_t}$. Then, under $\mathbb{Q}$, the process M^μ defined as*

$$M_t^\mu := M_t - \int_0^t \mu_s \lambda_s ds = N_t - \int_0^t (\mu_s + 1) \lambda_s\, ds\ , t \ge 0$$

is a local martingale.

PROOF: From the integration by parts formula, we get

$$\begin{aligned} d(M^\mu L)_t &= M_{t-}^\mu dL_t + L_{t-} dM_t^\mu + d[L, M^\mu]_t \\ &= M_{t-}^\mu dL_t + L_{t-} dM_t^\mu + L_{t-} \mu_t dN_t \\ &= M_{t-}^\mu dL_t + L_{t-} dM_t + L_{t-} \mu_t dM_t = (M_{t-}^\mu \mu_t + 1 + \mu_t) L_{t-} dM_t\, , \end{aligned}$$

hence, the process $M^\mu L$ is a $\mathbb{P}$-local martingale and M^μ is a $\mathbb{Q}$-local martingale. If μ and λ are deterministic, the process N is a $\mathbb{Q}$-inhomogeneous Poisson process with deterministic intensity $(\mu(t)+1)\lambda(t)$. □

Comment 8.4.5.2 We have seen that a Poisson process with stochastic intensity can be viewed as a time-changed of a standard Poisson process. Here, we interpret a Poisson process with stochastic intensity as a Poisson process with constant intensity under a change of probability. Indeed, a Poisson process with intensity 1 under $\mathbb{P}$ is a Poisson process with stochastic intensity $(\lambda_t, t \geq 0)$ under $\mathbb{Q}^\lambda$, where $\mathbb{Q}^\lambda|_{\mathcal{F}_t} = L_t^\lambda \, \mathbb{P}|_{\mathcal{F}_t}$ and where $dL_t^\lambda = L_{t-}^\lambda(\lambda_t - 1) dM_t$.

8.4.6 An Elementary Model of Prices Involving Jumps

Suppose that S is a stochastic process with dynamics given by

$$dS_t = S_{t-}(b(t)dt + \phi(t)dM_t), \tag{8.4.4}$$

where M is the compensated martingale associated with an inhomogeneous Poisson process N with strictly positive deterministic intensity λ and where b, ϕ are deterministic continuous functions. We assume that $\phi > -1$ so that the process S remains strictly positive. The solution of (8.4.4) is

$$\begin{aligned} S_t &= S_0 \exp\left[-\int_0^t \phi(s)\lambda(s)ds + \int_0^t b(s)ds\right] \prod_{s \leq t}(1 + \phi(s)\Delta N_s) \\ &= S_0 \exp\left[\int_0^t b(s)ds\right] \exp\left[\int_0^t \ln(1+\phi(s))dN_s - \int_0^t \phi(s)\lambda(s)ds\right]. \end{aligned}$$

Hence $S_t \exp\left(-\int_0^t b(s)ds\right)$ is a strictly positive local martingale.

We assume now that S is the dynamics of the price of a financial asset under the historical probability measure. We denote by r the deterministic interest rate and by $R_t = \exp(-\int_0^t r(s)ds)$ the discount factor. It is important to give a necessary and sufficient condition under which the financial market with the asset S and the riskless asset is complete and arbitrage free when ϕ does not vanish. Therefore, our aim is to give conditions such that there exists a probability measure $\mathbb{Q}$, equivalent to $\mathbb{P}$, under which the discounted process SR is a local martingale.

Any $\mathbf{F}^M$-martingale admits a representation as a stochastic integral with respect to M. Hence, any strictly positive $\mathbf{F}^M$-martingale L can be written as $dL_t = L_{t-}\mu_t dM_t$ where μ is an $\mathbf{F}^M$-predictable process such that $\mu > -1$ and, if $L_0 = 1$, the martingale L can be used as a Radon-Nikodým density. We are looking for conditions on μ such that the process RS is a $\mathbb{Q}$-local martingale where $d\mathbb{Q}|_{\mathcal{F}_t} = L_t d\mathbb{P}|_{\mathcal{F}_t}$; or equivalently, the process $(Y_t = R_t S_t L_t, t \geq 0)$ is a $\mathbb{P}$-local martingale. Integration by parts yields

$$\begin{aligned} dY_t &\overset{\text{mart}}{=} Y_{t-}\left((b(t) - r(t))dt + \phi(t)\mu_t d[M]_t\right) \\ &\overset{\text{mart}}{=} Y_{t-}\left(b(t) - r(t) + \phi(t)\mu_t\lambda(t)\right)dt. \end{aligned}$$

Hence, Y is a $\mathbb{P}$-local martingale if and only if $\mu_t = -\dfrac{b(t) - r(t)}{\phi(t)\lambda(t)}$.

Assume that $\mu > -1$ and define $\mathbb{Q}|_{\mathcal{F}_t} = L_t \mathbb{P}|_{\mathcal{F}_t}$. The process N is an inhomogeneous $\mathbb{Q}$-Poisson process with intensity $((\mu(s)+1)\lambda(s), s \geq 0)$ and

$$dS_t = S_{t-}(r(t)dt + \phi(t)dM_t^\mu)$$

where $(M^\mu(t) = N_t - \int_0^t (\mu(s)+1)\lambda(s)\,ds\,,\, t \geq 0)$ is the compensated $\mathbb{Q}$-martingale. Hence, the discounted price SR is a $\mathbb{Q}$-local martingale. In this setting, $\mathbb{Q}$ is the unique equivalent martingale measure.

The condition $\mu > -1$ is needed in order to obtain at least one e.m.m. and, from the fundamental theorem of asset pricing, to deduce the absence of arbitrage property.

If μ fails to be greater than -1, there does not exist an e.m.m. and there are arbitrages in the market. We now make explicit an arbitrage opportunity in the particular case when the coefficients are constant with $\phi > 0$ and $\dfrac{b-r}{\phi\lambda} > 1$, hence $\mu < -1$. The inequality

$$S_t = S_0 \exp[(b-\phi\lambda)t] \prod_{s \leq t}(1+\phi\Delta N_s) > S_0 e^{rt} \prod_{s \leq t}(1+\phi\Delta N_s) > S_0 e^{rt}$$

proves that an agent who borrows S_0 and invests in a long position in the underlying has an arbitrage opportunity, since his terminal wealth at time T $S_T - S_0 e^{rT}$ is strictly positive with probability one. Note that, in this example, the process $(S_t e^{-rt}, t \geq 0)$ is increasing.

Comment 8.4.6.1 We have required that ϕ and b are continuous functions in order to avoid integrability conditions. Obviously, we can generalize, to some extent, to the case of Borel functions. Note that, since we have assumed that $\phi(t)$ does not vanish, there is the equality of σ-fields

$$\sigma(S_s, s \leq t) = \sigma(N_s, s \leq t) = \sigma(M_s, s \leq t)\,.$$

8.5 Poisson Bridges

Let N be a Poisson process with constant intensity λ, $\mathcal{F}_t^N = \sigma(N_s, s \leq t)$ its natural filtration and $T > 0$ a fixed time. Let $\mathcal{G}_t = \sigma(N_s, s \leq t; N_T)$ be the natural filtration of N enlarged with the terminal value N_T of the process N.

8.5.1 Definition of the Poisson Bridge

Proposition 8.5.1.1 *The process*

$$\eta_t = N_t - \int_0^t \frac{N_T - N_s}{T-s} ds,\ t \leq T$$

is a **G***-martingale with predictable bracket*

$$\Lambda_t = \int_0^t \frac{N_T - N_s}{T-s} ds\,.$$

PROOF: For $0 < s < t < T$,

$$\mathbb{E}(N_t - N_s|\mathcal{G}_s) = \mathbb{E}(N_t - N_s|N_T - N_s) = \frac{t-s}{T-s}(N_T - N_s)$$

where the last equality follows from the fact that, if X and Y are independent with Poisson laws with parameters μ and ν respectively, then

$$\mathbb{P}(X = k|X + Y = n) = \frac{n!}{k!(n-k)!}\alpha^k(1-\alpha)^{n-k}$$

where $\alpha = \dfrac{\mu}{\mu+\nu}$. Hence,

$$\begin{aligned}\mathbb{E}\left(\int_s^t du \frac{N_T - N_u}{T-u}|\mathcal{G}_s\right) &= \int_s^t \frac{du}{T-u}\left(N_T - N_s - \mathbb{E}(N_u - N_s|\mathcal{G}_s)\right) \\ &= \int_s^t \frac{du}{T-u}\left(N_T - N_s - \frac{u-s}{T-s}(N_T - N_s)\right) \\ &= \int_s^t \frac{du}{T-s}(N_T - N_s) = \frac{t-s}{T-s}(N_T - N_s)\,.\end{aligned}$$

Therefore,

$$\mathbb{E}\left(N_t - N_s - \int_s^t \frac{N_T - N_u}{T-u}du|\mathcal{G}_s\right) = \frac{t-s}{T-s}(N_T - N_s) - \frac{t-s}{T-s}(N_T - N_s) = 0$$

and the result follows.

Therefore, η is a compensated $\mathbf{G}$-Poisson process, time-changed by $\int_0^t \frac{N_T - N_s}{T-s}ds$, i.e., $\eta_t = \widetilde{M}(\int_0^t \frac{N_T - N_s}{T-s}ds)$ where $(\widetilde{M}(t), t \geq 0)$ is a compensated Poisson process. □

Comment 8.5.1.2 Poisson bridges are studied in Jeulin and Yor [496]. This kind of enlargement of filtration is used for modelling insider trading in Elliott and Jeanblanc [314], Grorud and Pontier [410] and Kohatsu-Higa and Øksendal [534].

8.5.2 Harness Property

The previous result may be extended in terms of the harness property.

Definition 8.5.2.1 *A process X fulfills the* **harness property** *if*

$$\mathbb{E}\left(\frac{X_t - X_s}{t-s}\middle|\,\mathcal{F}_{s_0],\,[T}\right) = \frac{X_T - X_{s_0}}{T - s_0}$$

for $s_0 \leq s < t \leq T$ where $\mathcal{F}_{s_0],\,[T} = \sigma(X_u, u \leq s_0, u \geq T)$.

A process with the harness property satisfies

$$\mathbb{E}\left(X_t \middle| \mathcal{F}_{s],[T}\right) = \frac{T-t}{T-s}X_s + \frac{t-s}{T-s}X_T ,$$

and conversely.

Proposition 8.5.2.2 *If X satisfies the harness property, then, for any fixed T,*

$$M_t^T = X_t - \int_0^t du \frac{X_T - X_u}{T-u}, \; t < T$$

is an $\mathcal{F}_{t],[T}$-martingale and conversely.

PROOF: If X satisfies the harness property, it is easy to check that M^T is an $\mathcal{F}_{t],[T}$-martingale. Conversely, assume that M^T is an $\mathcal{F}_{t],[T}$-martingale. Let us prove that the harness property holds, i.e.,

$$\mathbb{E}\left(\frac{X_t - X_s}{t-s} \middle| \mathcal{F}_{s],[T}\right) = \frac{X_T - X_s}{T-s} .$$

From the hypothesis

$$\begin{aligned}\mathbb{E}(X_t - X_s | \mathcal{F}_{s],[T}) &= \int_s^t du \, \mathbb{E}\left(\frac{X_T - X_u}{T-u} \middle| \mathcal{F}_{s],[T}\right) \\ &= (X_T - X_s)\int_s^t \frac{du}{T-u} - \int_s^t \frac{du}{T-u}\mathbb{E}(X_u - X_s | \mathcal{F}_{s],[T}) .\end{aligned}$$

Therefore, for fixed s, T, the process $\varphi(u) = \mathbb{E}(X_u - X_s | \mathcal{F}_{s],[T})$ defined for $u \geq s$, satisfies

$$\varphi(t) = (X_T - X_s)\int_s^t \frac{du}{T-u} - \int_s^t \frac{du}{T-u}\varphi(u) .$$

It follows that φ is a solution of the ODE

$$\varphi'(t) = \frac{X_T - X_s}{T-t} - \varphi(t)\frac{1}{T-t}$$

with initial condition $\varphi(s) = 0$. This ODE has a unique solution given by $\varphi(t) = (t-s)\frac{X_T - X_s}{T-s}$. □

Comment 8.5.2.3 See Exercise 6.19 in Chaumont and Yor [161] for other properties. See also Jacod and Protter [470] and Exercise 12.3 in Yor [868]. We shall prove in ↣ Subsection 11.2.7 that any integrable Lévy process enjoys the harness property (see also Mansuy and Yor [621]). This property is used in Corcuera et al. [194] for studying insider trading.

8.6 Compound Poisson Processes

8.6.1 Definition and Properties

Definition 8.6.1.1 *Let $\lambda > 0$ and let F be a cumulative distribution function on $\mathbb{R}$. A (λ, F)-***compound Poisson process** *is a process $X = (X_t, t \geq 0)$ of the form*

$$X_t = \sum_{k=1}^{N_t} Y_k, \; X_0 = 0$$

where N is a Poisson process with intensity λ and the $(Y_k, k \geq 1)$ are i.i.d. random variables with law $F(y) = \mathbb{P}(Y_1 \leq y)$, independent of N (we use the convention that $\sum_{k=1}^{0} Y_k = 0$). We assume that $\mathbb{P}(Y_1 = 0) = 0$.

The process X differs from a Poisson process since the sizes of the jumps are random variables. We denote by $F(dy)$ the measure associated with F and by F^{*n} its n-th convolution, i.e.,

$$F^{*n}(y) = \mathbb{P}\left(\sum_{k=1}^{n} Y_k \leq y\right).$$

We use the convention $F^{*0}(y) = \mathbb{P}(0 \leq y) = \mathbb{1}_{[0,\infty[}(y)$.

Proposition 8.6.1.2 *A (λ, F)-compound Poisson process has stationary and independent increments (i.e., it is a Lévy process $\rightarrowtail$ Chapter 11); the cumulative distribution function of the r.v. X_t is*

$$\mathbb{P}(X_t \leq x) = e^{-\lambda t} \sum_{n=0}^{\infty} \frac{(\lambda t)^n}{n!} F^{*n}(x).$$

PROOF: Since the (Y_k) are i.i.d., one gets

$$\mathbb{E}\left(\exp(i\lambda \sum_{k=1}^{n} Y_k + i\mu \sum_{k=n+1}^{m} Y_k)\right) = (\mathbb{E}[\exp(i\lambda Y_1)]\,)^n\; (\mathbb{E}[\exp(i\mu Y_1)]\,)^{m-n}.$$

Then, setting $\psi(\lambda, n) = (\mathbb{E}[\exp(i\lambda Y_1)]\,)^n$, the independence and stationarity of the increments $(X_t - X_s)$ and X_s with $t > s$ follows from

$$\begin{aligned}\mathbb{E}(\,\exp(i\lambda X_s + i\mu(X_t - X_s))\,) &= \mathbb{E}(\,\psi(\lambda, N_s)\,\psi(\mu, N_t - N_s)\,) \\ &= \mathbb{E}(\,\psi(\lambda, N_s)\,)\,\mathbb{E}(\,\psi(\mu, N_{t-s})\,).\end{aligned}$$

The independence of a finite sequence of increments follows by induction.

From the independence of N and the random variables $(Y_k, k \geq 1)$ and using the Poisson law of N_t, we get

$$\mathbb{P}(X_t \leq x) = \sum_{n=0}^{\infty} \mathbb{P}\left(N_t = n, \sum_{k=1}^{n} Y_k \leq x\right)$$
$$= \sum_{n=0}^{\infty} \mathbb{P}(N_t = n)\mathbb{P}\left(\sum_{k=1}^{n} Y_k \leq x\right) = e^{-\lambda t} \sum_{n=0}^{\infty} \frac{(\lambda t)^n}{n!} F^{*n}(x).$$

□

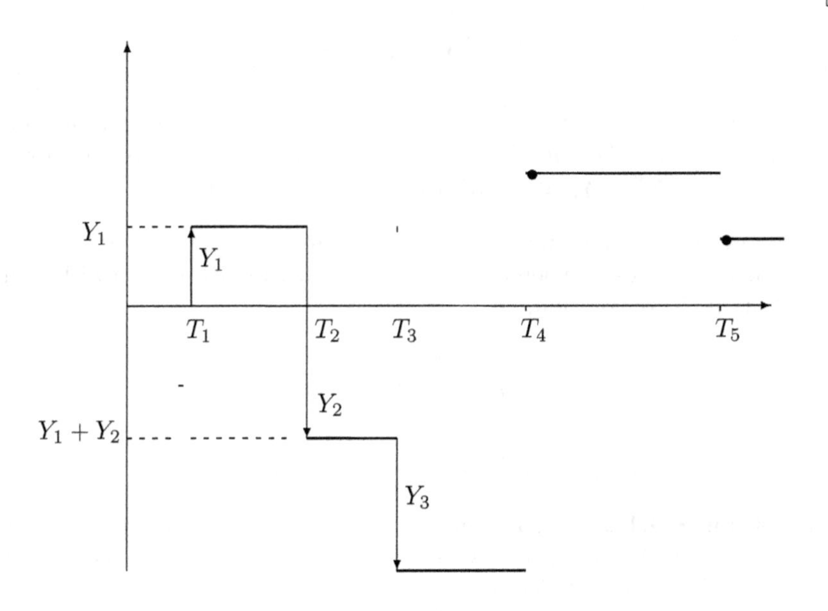

Fig. 8.2 Compound Poisson process

8.6.2 Integration Formula

If $Z_t = Z_0 + bt + X_t$ with X a (λ, F)-compound Poisson process, and if f is a C^1 function, the following obvious formula gives a representation of $f(Z_t)$ as a sum of integrals:

$$f(Z_t) = f(Z_0) + \int_0^t bf'(Z_s)ds + \sum_{s \leq t} f(Z_s) - f(Z_{s-})$$
$$= f(Z_0) + \int_0^t bf'(Z_s)ds + \sum_{s \leq t} (f(Z_s) - f(Z_{s-}))\Delta N_s$$
$$= f(Z_0) + \int_0^t bf'(Z_s)ds + \int_0^t (f(Z_s) - f(Z_{s-}))\, dN_s\,.$$

It is possible to write this formula as

$$f(Z_t) = f(Z_0)+\int_0^t (bf'(Z_s)+(f(Z_s)-f(Z_{s-})\lambda)ds+\int_0^t (f(Z_s) - f(Z_{s-}))\, dM_s$$

however this equality does not give immediately the canonical decomposition of the semi-martingale $f(Z_t)$. Indeed, the reader can notice that the process $\int_0^t (f(Z_s) - f(Z_{s-}))\, dM_s$ is not a martingale. See $\rightarrowtail$ Subsection 8.6.4 for the decomposition of this semi-martingale.

Exercise 8.6.2.1 Prove that the infinitesimal generator of Z is given, for C^1 functions f such that f and f' are bounded, by

$$\mathcal{L}f(x) = bf'(x) + \lambda \int_{-\infty}^{\infty} (f(x+y) - f(x))\, F(dy)\,.$$

$\triangleleft$

8.6.3 Martingales

Proposition 8.6.3.1 *Let X be a (λ, F)-compound Poisson process such that $\mathbb{E}(|Y_1|) < \infty$. Then, the process $(Z_t = X_t - t\lambda\mathbb{E}(Y_1), t \geq 0)$ is a martingale and in particular, $\mathbb{E}(X_t) = \lambda t\mathbb{E}(Y_1) = \lambda t \int_{-\infty}^{\infty} yF(dy)$.*

If $\mathbb{E}(Y_1^2) < \infty$, the process $(Z_t^2 - t\lambda\mathbb{E}(Y_1^2), t \geq 0)$ is a martingale and $Var\,(X_t) = \lambda t\mathbb{E}(Y_1^2)$.

PROOF: The martingale property of $(X_t - \mathbb{E}(X_t), t \geq 0)$ follows from the independence and stationarity of the increments of the process X. We leave the details to the reader. It remains to compute the expectation of the r.v. X_t as follows:

$$\begin{aligned}\mathbb{E}(X_t) &= \sum_{n=1}^{\infty} \mathbb{E}\left(\sum_{k=1}^{n} Y_k \mathbb{1}_{\{N_t=n\}}\right) = \sum_{n=1}^{\infty} n\mathbb{E}(Y_1)\mathbb{P}(N_t = n)\\ &= \mathbb{E}(Y_1)\sum_{n=1}^{\infty} n\mathbb{P}(N_t = n) = \lambda t\mathbb{E}(Y_1)\,.\end{aligned}$$

The proof of the second property can be done by the same method; however, it is more convenient to use the Laplace transform of X (See below, Proposition 8.6.3.4). □

Proposition 8.6.3.2 *Let $X_t = \sum_{i=1}^{N_t} Y_i$ be a (λ, F)-compound Poisson process, where the random variables Y_i are square integrable.*

Then $Z_t^2 - \sum_{i=1}^{N_t} Y_i^2$ is a martingale.

PROOF: It suffices to write

$$Z_t^2 - \sum_{i=1}^{N_t} Y_i^2 = Z_t^2 - \lambda t \mathbb{E}(Y_1^2) - \left(\sum_{i=1}^{N_t} Y_i^2 - \lambda t \mathbb{E}(Y_1^2) \right).$$

We have proved that $Z_t^2 - \lambda t \mathbb{E}(Y_1^2)$ is a martingale. Now, since $\sum_{i=1}^{N_t} Y_i^2$ is a compound Poisson process, $\sum_{i=1}^{N_t} Y_i^2 - \lambda t \mathbb{E}(Y_1^2)$ is a martingale. □

The process $A_t = \sum_{i=1}^{N_t} Y_i^2$ is an increasing process such that $X_t^2 - A_t$ is a martingale. Hence, as for a Poisson process, we have two (in fact an infinity of) increasing processes C_t such that $X_t^2 - C_t$ is a martingale. The particular process $C_t = t\lambda \mathbb{E}(Y_1^2)$ is predictable, whereas the process $A_t = \sum_{i=1}^{N_t} Y_i^2$ satisfies $\Delta A_t = (\Delta X_t)^2$. The predictable process $t\lambda \mathbb{E}(Y_1^2)$ is the predictable quadratic variation and is denoted $\langle X \rangle_t$, the process $\sum_{i=1}^{N_t} Y_i^2$ is the optional quadratic variation of X and is denoted $[X]_t$.

Proposition 8.6.3.3 *Let $X_t = \sum_{k=1}^{N_t} Y_k$ be a (λ, F)-compound Poisson process.*

(a) *Let $dS_t = S_{t-}(\mu dt + dX_t)$ (that is S is the Doléans-Dade exponential martingale $\mathcal{E}(U)$ of the process $U_t = \mu t + X_t$). Then,*

$$S_t = S_0 e^{\mu t} \prod_{k=1}^{N_t} (1 + Y_k).$$

In particular, if $1 + Y_1 > 0, \mathbb{P}.a.s.$, then

$$S_t = S_0 \exp\left(\mu t + \sum_{k=1}^{N_t} \ln(1 + Y_k) \right) = S_0 e^{\mu t + X_t^*} = S_0 e^{U_t^*}.$$

Here, X^* is the (λ, F^*)-compound Poisson process $X_t^* = \sum_{k=1}^{N_t} Y_k^*$, where $Y_k^* = \ln(1 + Y_k)$ (hence $F^*(y) = F(e^y - 1)$) and

$$U_t^* = U_t + \sum_{s \le t} (\ln(1 + \Delta X_s) - \Delta X_s) = U_t + \sum_{k=1}^{N_t} (\ln(1 + Y_k) - Y_k).$$

The process $(S_t e^{-rt}, t \ge 0)$ is a local martingale if and only if $\mu + \lambda \mathbb{E}(Y_1) = r$.

(b) *The process*

$$S_t = x \exp(bt + X_t) = x e^{V_t} \tag{8.6.1}$$

is a solution of

$$dS_t = S_{t-} dV_t^*,\ S_0 = x$$

(i.e., $S_t = x\, \mathcal{E}(V^*)_t$) where

$$V_t^* = V_t + \sum_{s \le t} (e^{\Delta X_s} - 1 - \Delta X_s) = bt + \sum_{s \le t} (e^{\Delta X_s} - 1) .$$

The process S is a martingale if and only if

$$\lambda \int_{-\infty}^{\infty} (1 - e^y) F(dy) = b .$$

PROOF: The solution of

$$dS_t = S_{t-}(\mu dt + dX_t), \; S_0 = x$$

is

$$S_t = x\mathcal{E}(U)_t = xe^{\mu t} \prod_{k=1}^{N_t} (1 + Y_k) = xe^{\mu t} e^{\sum_{k=1}^{N_t} \ln(1+Y_k)} = e^{\mu t + \sum_{k=1}^{N_t} Y_k^*}$$

where $Y_k^* = \ln(1 + Y_k)$. From

$$\mu t + \sum_{k=1}^{N_t} Y_k^* = \mu t + X_t + \sum_{k=1}^{N_t} Y_k^* - X_t = U_t + \sum_{s \le t} (\ln(1 + \Delta X_s) - \Delta X_s) ,$$

we obtain $S_t = xe^{U_t^*}$. Then,

$$\begin{aligned} d(e^{-rt} S_t) &= e^{-rt} S_{t-}((-r + \mu + \lambda \mathbb{E}(Y_1))dt + dX_t - \lambda \mathbb{E}(Y_1)dt) \\ &= e^{-rt} S_{t-}((-r + \mu + \lambda \mathbb{E}(Y_1))dt + dZ_t) , \end{aligned}$$

where $Z_t = X_t - \lambda \mathbb{E}(Y_1)t$ is a martingale. It follows that $e^{-rt}S_t$ is a local martingale if and only if $-r + \mu + \lambda \mathbb{E}(Y_1) = 0$.

The second assertion is the same as the first one, with a different choice of parametrization. Let

$$S_t = xe^{bt + X_t} = xe^{bt} \exp\left(\sum_1^{N_t} Y_k \right) = xe^{bt} \prod_{k=1}^{N_t} (1 + Y_k^*)$$

where $1 + Y_k^* = e^{Y_k}$. Hence, from part a), $dS_t = S_{t-}(bdt + dV_t^*)$ where $V_t^* = \sum_{k=1}^{N_t} Y_k^*$. It remains to note that

$$bt + V_t^* = V_t + V_t^* - X_t = V_t + \sum_{s \le t} (e^{\Delta X_s} - 1 - \Delta X_s) .$$

□

We now denote by ν the positive measure $\nu(dy) = \lambda F(dy)$. Using this notation, a (λ, F)-compound Poisson process will be called a **ν-compound Poisson** process. This notation, which is not standard, will make the various formulae more concise and will be of constant use in ↣ Chapter 11 when dealing with Lévy 's processes which are a generalization of compound Poisson processes. Conversely, to any positive finite measure ν on $\mathbb{R}$, we can associate a cumulative distribution function by setting $\lambda = \nu(\mathbb{R})$ and $F(dy) = \nu(dy)/\lambda$ and construct a ν-compound Poisson process.

Proposition 8.6.3.4 *If X is a ν-compound Poisson process, let*

$$\mathcal{J}(\nu) = \left\{\alpha : \int_{-\infty}^{\infty} e^{\alpha x}\nu(dx) < \infty\right\}.$$

The Laplace transform of the r.v. X_t is

$$\mathbb{E}(e^{\alpha X_t}) = \exp\left(-t\int_{-\infty}^{\infty}(1-e^{\alpha x})\nu(dx)\right) \; for \alpha \in \mathcal{J}(\nu).$$

The process

$$Z_t^{(\alpha)} = \exp\left(\alpha X_t + t\int_{-\infty}^{\infty}(1-e^{\alpha x})\nu(dx)\right)$$

is a martingale.

The characteristic function of the r.v. X_t is

$$\mathbb{E}(e^{iuX_t}) = \exp\left(-t\int_{-\infty}^{\infty}(1-e^{iux})\nu(dx)\right).$$

PROOF: From the independence between the random variables $(Y_k, k \geq 1)$ and the process N,

$$\mathbb{E}(e^{\alpha X_t}) = \mathbb{E}\left(\exp\left(\alpha\sum_{k=1}^{N_t} Y_k\right)\right) = \mathbb{E}(\Phi(N_t))$$

where $\Phi(n) = \mathbb{E}\left(\exp\left(\alpha\sum_{k=1}^{n} Y_k\right)\right) = [\Psi_Y(\alpha)]^n$, with $\Psi_Y(\alpha) = \mathbb{E}\left(\exp(\alpha Y_1)\right)$.

Now, $\mathbb{E}(\Phi(N_t)) = \sum_n [\Psi_Y(\alpha)]^n e^{-\lambda t}\frac{\lambda^n t^n}{n!} = \exp\left(-\lambda t(1-\Psi_Y(\alpha)\right)$. The martingale property follows from the independence and stationarity of the increments of X. □

Taking the derivative w.r.t. α of $Z^{(\alpha)}$ and evaluating it at $\alpha = 0$, we obtain that the process Z of Proposition 8.6.3.1 is a martingale, and using the second derivative of $Z^{(\alpha)}$ evaluated at $\alpha = 0$, one obtains that $Z_t^2 - \lambda t\mathbb{E}(Y_1^2)$ is a martingale.

Proposition 8.6.3.5 *Let X be a ν-compound Poisson process, and f a bounded Borel function. Then, the process*

$$\exp\left(\sum_{k=1}^{N_t} f(Y_k) + t\int_{-\infty}^{\infty}(1-e^{f(x)})\nu(dx)\right)$$

is a martingale. In particular

$$\mathbb{E}\left(\exp\left(\sum_{k=1}^{N_t} f(Y_k)\right)\right) = \exp\left(-t\int_{-\infty}^{\infty}(1-e^{f(x)})\nu(dx)\right).$$

PROOF: The proof is left as an exercise. □

For any bounded Borel function f, we denote by $\nu(f) = \int_{-\infty}^{\infty} f(x)\nu(dx)$ the product $\lambda\mathbb{E}(f(Y_1))$. Then, one has the following proposition:

Proposition 8.6.3.6 (i) *Let X be a ν-compound Poisson process and f a bounded Borel function. The process*

$$M_t^f = \sum_{s\leq t} f(\Delta X_s)\mathbb{1}_{\{\Delta X_s\neq 0\}} - t\nu(f)$$

is a martingale.

(ii) *Conversely, suppose that X is a pure jump process and that there exists a finite positive measure σ such that*

$$\sum_{s\leq t} f(\Delta X_s)\mathbb{1}_{\{\Delta X_s\neq 0\}} - t\sigma(f)$$

is a martingale for any bounded Borel function f, then X is a σ-compound Poisson process.

PROOF: (i) From the definition of M^f,

$$\begin{aligned}\mathbb{E}(M_t^f) &= \sum_n \mathbb{E}(f(Y_n))\mathbb{P}(T_n < t) - t\nu(f) = \mathbb{E}(f(Y_1))\sum_n \mathbb{P}(T_n < t) - t\nu(f)\\ &= \mathbb{E}(f(Y_1))\mathbb{E}(N_t) - t\nu(f) = 0\,.\end{aligned}$$

The proof of the proposition is now standard and results from the computation of conditional expectations which leads to, for $s > 0$

$$\mathbb{E}(M_{t+s}^f - M_t^f|\mathcal{F}_t) = \mathbb{E}\left(\sum_{t<u\leq t+s} f(\Delta X_u)\mathbb{1}_{\{\Delta X_u\neq 0\}} - s\nu(f)|\mathcal{F}_t\right) = 0\,.$$

Another proof relies on the fact that the process

$$\sum_{s\leq t} f(\Delta X_s)\mathbb{1}_{\{\Delta X_s\neq 0\}} = \sum_{k=1}^{N_t} f(Y_k) = \sum_{k=1}^{N_t} Z_k$$

is a compound Poisson process, hence

$$\sum_{k=1}^{N_t} f(Y_k) - t\lambda\mathbb{E}(Z_1) = \sum_{k=1}^{N_t} f(Y_k) - t\lambda\mathbb{E}(f(Y_1))$$

is a martingale.

(ii) For the converse, we write

$$\begin{aligned} e^{iuX_t} &= 1 + \sum_{s\leq t} e^{iuX_{s-}}(e^{iu\Delta X_s} - 1) \\ &= 1 + \int_0^t e^{iuX_{s-}} dM_s^f + \sigma(f)\int_0^t e^{iuX_s} ds \end{aligned}$$

where $f(x) = e^{iux} - 1$. Hence,

$$\mathbb{E}(e^{iuX_{t+s}}|\mathcal{F}_t) = e^{iuX_t} + \sigma(f)\int_0^s dr\,\mathbb{E}(e^{iuX_{t+r}}|\mathcal{F}_t)\,.$$

Setting $\varphi(s) = \mathbb{E}(e^{iuX_{t+s}}|\mathcal{F}_t)$, one gets $\varphi(s) = \varphi(0) + \sigma(f)\int_0^s \varphi(r)dr$, hence

$$\mathbb{E}(e^{iuX_{t+s}}|\mathcal{F}_t) = e^{iuX_t}\exp\left(s\int_{\mathbb{R}} \sigma(dx)(e^{iux} - 1)\right).$$

The remainder of the proof is standard and left to the reader. □

Introducing the random measure $\mu = \sum_n \delta_{T_n,Y_n}$ on $\mathbb{R}^+ \times \mathbb{R}$ and denoting by $(f * \mu)_t$ the integral[1]

$$\int_0^t \int_{\mathbb{R}} f(x)\mu(\omega; ds, dx) = \sum_{k=1}^{N_t} f(Y_k)\,,$$

we obtain that

$$M_t^f = (f * \mu)_t - t\nu(f) = \int_0^t \int_{\mathbb{R}} f(x)(\mu(\omega; ds, dx) - ds\,\nu(dx))$$

is a martingale. (We shall generalize this fact when studying marked point processes in ↣ Section 8.8 and Lévy processes in ↣ Chapter 11.)

Example 8.6.3.7 Let $U_s = \alpha s + \sigma W_s$ where W is a standard Brownian motion and let N be a Poisson process with intensity 1, independent of W. Define the process Z as $Z_t = U_{N_t}$ (that is a time change of the drifted Brownian motion U). Conditionally on $N_1 = n$, the r.v. Z_1 has a $\mathcal{N}(\alpha n, \sigma^2 n)$ law. The process Z is a compound Poisson process

$$(Z_t, t \geq 0) \stackrel{\text{law}}{=} \left(\sum_{k=1}^{N_t} Y_k,\, t \geq 0\right) \text{ where } Y_k \stackrel{\text{law}}{=} \mathcal{N}(\alpha, \sigma^2)\,.$$

[1] Later, in Chapter 11, we shall often use $\mathbf{N}(ds, dx)$ instead of $\mu(ds, dx)$

Example 8.6.3.8 Let $X^{(i)}, i = 1, 2$ be two compound Poisson processes

$$X_t^{(i)} = \sum_{k=1}^{N_t^{(i)}} Y_k^{(i)}$$

where $Y^{(i)}, N^{(i)}, i = 1, 2$ are independent and $Y_1^{(i)}$ is a reflected normal r.v. (i.e., with density $f(y)\mathbb{1}_{\{y>0\}}$ where $f(y) = \frac{\sqrt{2}}{\sigma\sqrt{\pi}} e^{-y^2/(2\sigma^2)}$). The characteristic function of the r.v. $X_t^{(1)} - X_t^{(2)}$ is

$$\Psi(u) = e^{-2\lambda t} e^{\lambda t(\Phi(u)+\Phi(-u))}$$

with $\Phi(u) = \mathbb{E}(e^{iuY_1})$. From

$$\begin{aligned}\Phi(u) + \Phi(-u) = \mathbb{E}(e^{iuY_1} + e^{-iuY_1}) &= \int_0^\infty e^{iuy} f(y)dy + \int_0^\infty e^{-iuy} f(y)dy \\ &= \int_{-\infty}^\infty e^{iuy} f(y)dy = 2e^{-\sigma^2 u^2/2}\end{aligned}$$

we obtain

$$\Psi(u) = \exp(2\lambda t(e^{-\sigma^2 u^2/2} - 1)) .$$

This is the characteristic function of $\sigma W(N_t^{(1)} + N_t^{(2)})$ where W is a BM, evaluated at time $N_t^{(1)} + N_t^{(2)}$.

Exercise 8.6.3.9 Let X be a (λ, F)-compound Poisson process. Compute $\mathbb{E}(e^{iuX_t})$ in the following two cases:

(a) Merton's case [643]: The law F is a Gaussian law, with mean c and variance δ,

(b) Kou's case [540] (double exponential model): The law F of Y_1 is

$$F(dx) = \left(p\theta_1 e^{-\theta_1 x}\mathbb{1}_{\{x>0\}} + (1-p)\theta_2 e^{\theta_2 x}\mathbb{1}_{\{x<0\}}\right) dx ,$$

where $p \in]0,1[$ and $\theta_i, i = 1, 2$ are positive numbers. See $\rightarrowtail$ Example 10.4.4.5 for a generalization and an answer. ◁

Exercise 8.6.3.10 (Example of a Compound Poisson Process.) (See Sato [761], p. 21.) Let X be a ν-compound Poisson process with ν a probability measure of the form $\nu(dx) = p\delta_1(dx) + q\delta_{-1}(dx)$ where $\delta_a(dx)$ denotes the Dirac measure at a and where $q = 1 - p, p \in]0, 1[$. Prove that

$$\mathbb{P}(X_t = k) = e^{-t}\left(\frac{p}{q}\right)^{k/2} I_k(2(pq)^{1/2}t)$$

where I_k is the Bessel function (see $\rightarrowtail$ A.5.2). ◁

Exercise 8.6.3.11 (Extension of Compound Poisson Process.) (See Sato [761], p. 143) Let X be a ν-compound Poisson process and h a bounded function. The sequence (T_k) is the sequence of jumps of the Poisson process N. Let $Z_t = \sum_{k=1}^{N_t} h(T_k, Y_k)$. Prove that

$$\mathbb{E}(e^{iuZ_t}) = \exp\left(\int_0^t ds \int (e^{iuh(s,y)} - 1)\nu(dy)\right).$$

The process Z has independent non-homogeneous increments; it is called an additive process. $\triangleleft$

8.6.4 Itô's Formula

Let X be a ν-compound Poisson process, and $Z_t = Z_0 + bt + X_t$. Then, Itô's formula

$$\begin{aligned} f(Z_t) - f(Z_0) &= b\int_0^t f'(Z_s)ds + \sum_{k,\, T_k \le t} f(Z_{T_k}) - f(Z_{T_k -}) \\ &= b\int_0^t f'(Z_s)ds + \int_0^t \int_{\mathbb{R}} [f(Z_{s-} + y) - f(Z_{s-})]\, \mu(ds, dy) \end{aligned}$$

(where $\mu = \sum_{n=1}^{\infty} \delta_{T_n, Y_n}$) can be written as

$$f(Z_t) - f(Z_0) = \int_0^t ds\, (\mathcal{L}f)(Z_s) + M(f)_t$$

where $\mathcal{L}f(x) = bf'(x) + \int_{\mathbb{R}} (f(x+y) - f(x))\, \nu(dy)$ is the infinitesimal generator of Z and

$$M(f)_t = \int_0^t \int_{\mathbb{R}} [f(Z_{s-} + y) - f(Z_{s-})]\, (\mu(ds, dy) - ds\, \nu(dy))$$

is a local martingale.

8.6.5 Hitting Times

Let $Z_t = ct - \sum_{k=1}^{N_t} Y_k$ be a (λ, F)-compound Poisson process with a drift term $c > 0$ and $T(x) = \inf\{t : x + Z_t \le 0\}$ where $x > 0$. The random variables Y can be interpreted as losses for insurance companies. The process Z is called the Cramer-Lundberg risk process.

▶ If $c = 0$ and if the support of the cumulative distribution function F is included in $[0, \infty[$, then the process Z is decreasing and

$$\{T(x) \ge t\} = \{Z_t + x \ge 0\} = \left\{x \ge \sum_{k=1}^{N_t} Y_k\right\},$$

hence,

$$\mathbb{P}(T(x) \geq t) = \mathbb{P}\left(x \geq \sum_{k=1}^{N_t} Y_k\right) = \sum_n \mathbb{P}(N_t = n) F^{*n}(x).$$

For a cumulative distribution function F with support in $\mathbb{R}$,

$$\mathbb{P}(T(x) \geq t) = \sum_n \mathbb{P}(N_t = n)\mathbb{P}(Y_1 \leq x, Y_1 + Y_2 \leq x, \ldots, Y_1 + \cdots + Y_n \leq x).$$

▶ Assume now that $c \neq 0$, that the support of F is included in $[0, \infty[$ and that, for every u, $\mathbb{E}(e^{uY_1}) < \infty$. Setting $\psi(u) = cu + \int_0^\infty (e^{uy} - 1)\nu(dy)$, the process $(\exp(uZ_t - t\psi(u)), t \geq 0)$ is a martingale (Corollary 8.6.3.3). Since the process Z has no negative jumps, the level cannot be crossed with a jump and therefore $Z_{T(x)} = -x$. From Doob's optional sampling theorem, $\mathbb{E}(e^{uZ_{t\wedge T(x)} - (t\wedge T(x))\psi(u)}) = 1$ and when t goes to infinity, one obtains

$$\mathbb{E}(e^{-ux - T(x)\psi(u)} \mathbb{1}_{\{T(x)<\infty\}}) = 1.$$

Hence one gets the Laplace transform of $T(x)$

$$\mathbb{E}(e^{-\theta T(x)} \mathbb{1}_{\{T(x)<\infty\}}) = e^{x\psi^\sharp(\theta)},$$

where $\psi^\sharp$ is the negative inverse of ψ (i.e., $\psi^\sharp(\theta)$ is the solution y of $\psi(y) = \theta$ for $\theta > 0$ which satisfies $y < 0$).

Example 8.6.5.1 (One-sided Exponential Law.)
If $F(dy) = \kappa e^{-\kappa y}\mathbb{1}_{\{y>0\}}dy$, one obtains $\psi(u) = cu - \frac{\lambda u}{\kappa + u}$, hence inverting ψ,

$$\mathbb{E}(e^{-\theta T(x)} \mathbb{1}_{\{T(x)<\infty\}}) = e^{x\psi^\sharp(\theta)},$$

with

$$\psi^\sharp(\theta) = \frac{\lambda + \theta - \kappa c - \sqrt{(\lambda + \theta - \kappa c)^2 + 4\theta\kappa c}}{2c}.$$

Exercise 8.6.5.2 Let $X_t = \sum_{i=1}^{N_t} Y_i$ and $X_t^* = \sum_{i=1}^{N_t^*} Y_i^*$ be two compound Poisson processes, where N, N^* are independent Poisson processes with respective intensities λ and λ^*. We assume that the four random objects N, N^*, Y, Y^* are independent and that the law of Y_1 (resp. the law of Y_1^*) has support in $[0, \infty[$. Prove that $e^{-\rho(X_t - X_t^*)}$ is a martingale for ρ a root of $\lambda\mathbb{E}(e^{-\rho Y_1} - 1) + \lambda^*\mathbb{E}(e^{\rho Y_1^*} - 1) = 0$. ◁

8.6.6 Change of Probability Measure

Two questions may be asked:

(a) Starting from a ν-compound Poisson process X under $\mathbb{P}$, find some changes of measures $\mathbb{Q} \ll \mathbb{P}$ such that, under $\mathbb{Q}$, X is still a compound Poisson process.

(b) Given two compound Poisson processes, when are their distributions locally equivalent?

We treat point (a) and leave (b) to the reader (see $\rightarrowtail$ Exercise 8.6.6.3).

Let X be a ν-compound Poisson process, $\widetilde{\nu}$ a positive finite measure on $\mathbb{R}$ absolutely continuous w.r.t. ν, and $\widetilde{\lambda} = \widetilde{\nu}(\mathbb{R}) > 0$. Let

$$L_t = \exp\left(t(\lambda - \widetilde{\lambda}) + \sum_{s\leq t} \ln\left(\frac{d\widetilde{\nu}}{d\nu}\right)(\Delta X_s)\right). \tag{8.6.2}$$

Proposition 8.6.3.4 proves that, if $\sum_{k=1}^{N_t} Z_k$ is a compound Poisson process, then

$$\exp\left(\sum_{k=1}^{N_t} Z_k + t\lambda\mathbb{E}(1 - e^{Z_1})\right)$$

is a martingale. It follows that

$$\exp\left(\sum_{k=1}^{N_t} f(Y_k) + t\int_{-\infty}^{\infty} (1 - e^{f(x)})\nu(dx)\right) \tag{8.6.3}$$

is a martingale, hence for $f = \ln\left(\frac{d\widetilde{\nu}}{d\nu}\right)$, the process L is a martingale. Set $\mathbb{Q}|_{\mathcal{F}_t} = L_t\mathbb{P}|_{\mathcal{F}_t}$.

Proposition 8.6.6.1 *Let X be a ν-compound Poisson process under $\mathbb{P}$. Define $d\mathbb{Q}|_{\mathcal{F}_t} = L_t d\mathbb{P}|_{\mathcal{F}_t}$ where L is given in (8.6.2). Then, the process X is a $\widetilde{\nu}$-compound Poisson process under $\mathbb{Q}$.*

PROOF: First we find the law of the random variable $X_t = \sum_{k=1}^{N_t} Y_k$ under $\mathbb{Q}$. Let $f = \ln\left(\frac{d\widetilde{\nu}}{d\nu}\right)$. Then

$$\begin{aligned}
\mathbb{E}_{\mathbb{Q}}(e^{iuX_t}) &= \mathbb{E}_{\mathbb{P}}\left(e^{iuX_t}\exp\left(t(\lambda - \widehat{\lambda}) + \sum_1^{N_t} f(Y_k)\right)\right) \\
&= \sum_{n=0}^{\infty} e^{-\lambda t}\frac{(\lambda t)^n}{n!} e^{t(\lambda-\widehat{\lambda})}\left(\mathbb{E}_{\mathbb{P}}(e^{iuY_1 + f(Y_1)})\right)^n \\
&= \sum_{n=0}^{\infty} e^{-\lambda t}\frac{(\lambda t)^n}{n!} e^{t(\lambda-\widehat{\lambda})}\left(\mathbb{E}_{\mathbb{P}}\left(\frac{d\widehat{\nu}}{d\nu}(Y_1)\, e^{iuY_1}\right)\right)^n \\
&= \sum_{n=0}^{\infty} \frac{(\lambda t)^n}{n!} e^{-t\widehat{\lambda}}\left(\frac{1}{\lambda}\int e^{iuy}\, d\widehat{\nu}(y)\right)^n = \exp\left(t\int (e^{iuy} - 1)\, d\widehat{\nu}(y)\right).
\end{aligned}$$

It remains to check that X has independent and stationary increments under $\mathbb{Q}$. Using Proposition 8.6.3.5, one gets, for $t > s$,

$$\mathbb{E}_{\mathbb{Q}}(e^{iu(X_t - X_s)}|\mathcal{F}_s) = \frac{1}{L_s}\mathbb{E}_{\mathbb{P}}(L_t e^{iu(X_t-X_s)}|\mathcal{F}_s)$$
$$= \exp\left((t-s)\int (e^{iux}-1)\widetilde{\nu}(dx)\right).$$

□

In that case, the change of measure changes the intensity (equivalently, the law of N_t) and the law of the jumps, but the independence of the Y^i is preserved and N remains a Poisson process. It is possible to change the measure using more general Radon-Nikodým densities, so that the process X does not remain a compound Poisson process.

Exercise 8.6.6.2 Prove that the process L defined in (8.6.3) satisfies

$$dL_t = L_{t-}\left(\int_{\mathbb{R}} (e^{f(y)}-1)(\mu(dt,dy) - \nu(dy)dt)\right).$$

◁

Exercise 8.6.6.3 Prove that two compound Poisson processes with measures ν and $\widetilde{\nu}$ are locally absolutely continuous, only if ν and $\widetilde{\nu}$ are equivalent.
`Hint:` Use $\mathbb{E}\left(\left(\sum_{s\leq t} f(\Delta X_s)\right)\right) = t\nu(f)$. ◁

8.6.7 Price Process

We consider, as in Mordecki [658], the stochastic differential equation

$$dS_t = (\alpha S_{t-} + \beta)\, dt + (\gamma S_{t-} + \delta)dX_t \tag{8.6.4}$$

where X is a ν-compound Poisson process.

Proposition 8.6.7.1 *The solution of* (8.6.4) *is a Markov process with infinitesimal generator*

$$\mathcal{L}(f)(x) = (\alpha x + \beta)f'(x) + \int_{-\infty}^{+\infty} [f(x+\gamma xy + \delta y) - f(x)]\,\nu(dy),$$

for suitable f (in particular for $f \in C^1$ with compact support).

PROOF: We use Stieltjes integration to write, path by path,

$$f(S_t) - f(x) = \int_0^t f'(S_{s-})(\alpha S_{s-} + \beta)\, ds + \sum_{0\leq s\leq t} \Delta(f(S_s)).$$

Hence,

$$\mathbb{E}(f(S_t)) - f(x) = \mathbb{E}\left(\int_0^t f'(S_s)(\alpha S_s + \beta)ds\right) + \mathbb{E}\left(\sum_{0\leq s\leq t} \Delta(f(S_s))\right).$$

From

$$\begin{aligned}\mathbb{E}\left(\sum_{0\leq s\leq t} \Delta(f(S_s))\right) &= \mathbb{E}\left(\sum_{0\leq s\leq t} f(S_{s-} + \Delta S_s) - f(S_{s-})\right)\\ &= \mathbb{E}\left(\int_0^t \int_{\mathbb{R}} d\nu(y)\left[f(S_{s-} + (\gamma S_{s-} + \delta)y) - f(S_{s-})\right]\right),\end{aligned}$$

we obtain the infinitesimal generator. □

Proposition 8.6.7.2 *The process $(e^{-rt}S_t, t \geq 0)$ where S is a solution of (8.6.4) is a local martingale if and only if*

$$\alpha + \gamma \int_{\mathbb{R}} y\nu(dy) = r, \quad \beta + \delta \int_{\mathbb{R}} y\nu(dy) = 0\,.$$

PROOF: Left as an exercise. □

Let $\widetilde{\nu}$ be a positive finite measure which is absolutely continuous with respect to ν and

$$L_t = \exp\left((\lambda - \widetilde{\lambda}) + \sum_{s\leq t} \ln\left(\frac{d\widetilde{\nu}}{d\nu}\right)(\Delta X_s)\right).$$

Let $\mathbb{Q}|_{\mathcal{F}_t} = L_t \mathbb{P}|_{\mathcal{F}_t}$. Under $\mathbb{Q}$,

$$dS_t = (\alpha S_{t-} + \beta)\,dt + (\gamma S_{t-} + \delta)dX_t$$

where X is a $\widetilde{\nu}$-compound Poisson process. The process $(S_t e^{-rt}, t \geq 0)$ is a $\mathbb{Q}$-martingale if and only if

$$\alpha + \gamma \int_{\mathbb{R}} y\widetilde{\nu}(dy) = r, \quad \beta + \delta \int_{\mathbb{R}} y\widetilde{\nu}(dy) = 0\,.$$

Hence, there is an infinite number of e.m.m's: one can change the intensity of the Poisson process, or the law of the jumps, while preserving the compound process setting. Of course, one can also change the probability so as to break the independence assumptions.

8.6.8 Martingale Representation Theorem

The martingale representation theorem will be presented in the following Section 8.8 on marked point processes.

8.6.9 Option Pricing

The valuation of perpetual American options will be presented in ↣ Subsection 11.9.1 in the chapter on Lévy processes, using tools related to Lévy processes. The reader can refer to the papers of Gerber and Shiu [388, 389] and Gerber and Landry [386] for a direct approach. The case of double-barrier options is presented in Sepp [781] for double exponential jump diffusions, the case of lookback options is studied in Nahum [664]. Asian options are studied in Bellamy [69].

8.7 Ruin Process

We present briefly some basic facts about the problem of ruin, where compound Poisson processes play an essential rôle.

8.7.1 Ruin Probability

In the **Cramer-Lundberg model** the surplus process of an insurance company is $x + Z_t$, with $Z_t = ct - X_t$, where $X_t = \sum_{k=1}^{N_t} Y_k$ is a compound Poisson process. Here, c is assumed to be positive, the Y_k are $\mathbb{R}^+$-valued and we denote by F the cumulative distribution function of Y_1. Let $T(x)$ be the first time when the surplus process falls below 0:

$$T(x) = \inf\{t > 0 \, : \, x + Z_t \leq 0\} .$$

The probability of ruin is $\Phi(x) = \mathbb{P}(T(x) < \infty)$. Note that $\Phi(x) = 1$ for $x < 0$.

Lemma 8.7.1.1 *If* $\infty > \mathbb{E}(Y_1) \geq \frac{c}{\lambda}$, *then for every* x, *ruin occurs with probability 1.*

PROOF: Denoting by T_k the jump times of the process N, and setting

$$S_n = \sum_1^n [Y_k - c(T_k - T_{k-1})] ,$$

the probability of ruin is

$$\Phi(x) = \mathbb{P}(\inf_n(-S_n) < -x) = \mathbb{P}(\sup_n S_n > x) .$$

The strong law of large numbers implies

$$\lim_{n\to\infty} \frac{1}{n} S_n = \lim_{n\to\infty} \frac{1}{n} \sum_1^n [Y_k - c(T_k - T_{k-1})] = \mathbb{E}(Y_1) - \frac{c}{\lambda} .$$

□

8.7.2 Integral Equation

Let $\Psi(x) = 1 - \Phi(x) = \mathbb{P}(T(x) = \infty)$ where $T(x) = \inf\{t > 0 \,:\, x + Z_t \leq 0\}$. Obviously $\Psi(x) = 0$ for $x < 0$. From the Markov property, for $x \geq 0$

$$\Psi(x) = \mathbb{E}(\Psi(x + cT_1 - Y_1))$$

where T_1 is the first jump time of the Poisson process N. Thus

$$\Psi(x) = \int_0^\infty dt \lambda e^{-\lambda t} \mathbb{E}(\Psi(x + ct - Y_1))\,.$$

With the change of variable $y = x + ct$ we get

$$\Psi(x) = e^{\lambda x/c} \frac{\lambda}{c} \int_x^\infty dy e^{-\lambda y/c} \mathbb{E}(\Psi(y - Y_1))\,.$$

Differentiating w.r.t. x, we obtain

$$\begin{aligned} c\Psi'(x) &= \lambda \Psi(x) - \lambda \mathbb{E}(\Psi(x - Y_1)) = \lambda \Psi(x) - \lambda \int_0^\infty \Psi(x-y) dF(y) \\ &= \lambda \Psi(x) - \lambda \int_0^x \Psi(x-y) dF(y)\,. \end{aligned}$$

In the case where the Y_k's are exponential with parameter μ,

$$c\Psi'(x) = \lambda \Psi(x) - \lambda \int_0^x \Psi(x-y) \mu e^{-\mu y} dy\,.$$

Differentiating w.r.t. x and using the integration by parts formula leads to

$$c\Psi''(x) = (\lambda - c\mu)\Psi'(x)\,.$$

▶ For $\beta = \frac{1}{c}(\lambda - c\mu) < 0$, the solution of this differential equation is

$$\Psi(x) = c_1 \int_x^\infty e^{\beta t} dt + c_2$$

where c_1 and c_2 are two constants such that $\Psi(\infty) = 1$ and $\lambda\Psi(0) = c\Psi'(0)$. Therefore $c_2 = 1$, $c_1 = \frac{\lambda}{c}\frac{\lambda - \mu c}{c\mu} < 0$ and $\Psi(x) = 1 - \frac{\lambda}{c\mu} e^{\beta x}$. It follows that $\mathbb{P}(T(x) < \infty) = \frac{\lambda}{c\mu} e^{\beta x}$.
▶ If $\beta > 0$, then $\Psi(x) = 0$. Note that the condition $\beta > 0$ is equivalent to $\mathbb{E}(Y_1) \geq \frac{c}{\lambda}$.

8.7.3 An Example

Let $Z_t = ct - X_t$ where $X_t = \sum_{k=1}^{N_t} Y_k$ is a compound Poisson process. We denote by F the cumulative distribution function of Y_1 and we assume that

$F(0) = 0$, i.e., that the random variable Y_1 is $\mathbb{R}^+$-valued. Yuen et al. [871] assume that the insurer is allowed to invest in a portfolio, with stochastic return $R_t = rt + \sigma W_t + X_t^*$ where W is a Brownian motion and $X_t^* = \sum_{k=1}^{N_t^*} Y_k^*$ is a compound Poisson process. We assume that $(Y_k, Y_k^*, k \geq 1, N, N^*, W)$ are independent. We denote by F^* the cumulative distribution function of Y_1^*.

The risk process S associated with this model is defined as the solution $S_t(x)$ of the stochastic differential equation

$$S_t = x + Z_t + \int_0^t S_{s-} dR_s \,, \tag{8.7.1}$$

i.e.,

$$S_t(x) = U_t \left(x + \int_0^t U_{s-}^{-1} dZ_s \right)$$

where $U_t = e^{rt}\mathcal{E}(\sigma W)_t \prod_{k=1}^{N_t^*}(1 + Y_k^*)$. Note that the process S jumps at the time when the processes N or N^* jump and that

$$\Delta S_t = \Delta Z_t + S_{t-} \Delta R_t \,.$$

Let $T(x) = \inf\{t : S_t(x) < 0\}$ and $\Psi(x) = \mathbb{P}(T(x) = \infty) = \mathbb{P}(\inf_t S_t(x) \geq 0)$, the survival probability.

Proposition 8.7.3.1 *For $x \geq 0$, the function Ψ is the solution of the implicit equation*

$$\Psi(x) = \int_0^\infty \int_0^\infty \frac{\gamma}{2y^{2+\alpha+a}} p_u^\alpha(1, y)(D(y, u) + D^*(y, u))\, dy du$$

where

$$\begin{aligned}
p_u^\alpha(z, y) &= \left(\frac{y}{z}\right)^\alpha \frac{y}{u} e^{-(z^2+y^2)/(2u)} I_\alpha\left(\frac{zy}{u}\right),\\
D^*(y, u) &= \frac{\lambda^*}{\lambda + \lambda^*} \int_{-1}^\infty \Psi((1+z)y^{-2}(x + 4c\sigma^{-2}u))\, dF^*(z),\\
D(y, u) &= \frac{\lambda}{\lambda + \lambda^*} \int_0^{y^{-2}(x+4c\sigma^{-2}u)} \Psi(y^{-2}(x + 4c\sigma^{-2}u) - z)\, dF(z),
\end{aligned}$$

$$a = \sigma^{-2}(2r - \sigma^2),\ \gamma = \frac{8(\lambda + \lambda^*)}{\sigma^2},\ \alpha = (a^2 + \gamma^2)^{1/2} \,.$$

PROOF: Let τ (resp. τ^*) be the first time when the process N (resp. N^*) jumps, $T = \tau \wedge \tau^*$ and $m = \inf_{t \geq 0} S_t$. Note that, from the independence between N and N^*, we have $\mathbb{P}(\tau = \tau^*) = 0$. On the set $\{t < T\}$, one has $S_t = e^{rt}\mathcal{E}(\sigma W)_t \left(x + c\int_0^t e^{-rs}[\mathcal{E}(\sigma W)_s]^{-1} ds\right)$. We denote by V the process

$V_t = e^{rt}\mathcal{E}(\sigma W)_t(x + c\int_0^t e^{-rs}[\mathcal{E}(\sigma W)_s]^{-1}ds)$. The optional stopping theorem applied to the bounded martingale

$$M_t = \mathbb{E}(\mathbb{1}_{m\geq 0}|\mathcal{F}_t)$$

and the strong Markov property lead to

$$\Psi(x) = \mathbb{P}(m \geq 0) = M_0 = \mathbb{E}(M_T) = \mathbb{E}(\Psi(S_T))\,.$$

Hence,

$$\begin{aligned}
\Psi(x) &= \mathbb{E}(\Psi(S_\tau)\mathbb{1}_{\tau<\tau^*}) + \mathbb{E}(\Psi(S_{\tau^*})\mathbb{1}_{\tau^*<\tau}) \\
&= \mathbb{E}(\Psi(V_\tau - Y_1)\mathbb{1}_{\tau<\tau^*}) + \mathbb{E}(\Psi(V_{\tau^*}(1+Y_1^*))\mathbb{1}_{\tau^*<\tau}) \\
&= \int_0^\infty dt\lambda e^{-\lambda t}\mathbb{E}(\Psi(V_t - Y_1))\,\mathbb{P}(t<\tau^*) \\
&\qquad + \int_0^\infty dt\lambda^* e^{-\lambda^* t}\mathbb{E}(\Psi(V_t(1+Y_1^*)))\,\mathbb{P}(t<\tau) \\
&= \int_0^\infty e^{-(\lambda+\lambda^*)t}\left(\lambda\mathbb{E}[\Psi(V_t - Y_1)] + \lambda^*\mathbb{E}[\Psi(V_t(1+Y_1^*))]\right)dt\,.
\end{aligned}$$

Employing the change of variable $t = 4\sigma^{-2}s$,

$$\Psi(x) = \frac{4}{\sigma^2}\int_0^\infty e^{-4\sigma^{-2}(\lambda+\lambda^*)s}\left(\lambda\Upsilon(s) + \lambda^*\Upsilon^*(s)\right)ds\,,$$

where

$$\Upsilon(s) = \mathbb{E}[\Psi(X_s - Y_1)], \Upsilon^*(s) = \mathbb{E}[\Psi(Z_s(1+Y_1^*))]$$

and

$$X_s = e^{2(as+B_s)}\left(x + \frac{4c}{\sigma^2}\int_0^s e^{-2(at+B_t)}dt\right),\ B_s = \frac{\sigma}{2}W_{4\sigma^{-2}s}\,,$$

where $a = \frac{2r}{\sigma^2} - 1$. Hence, using the symmetry of BM,

$$\Upsilon(s) = \mathbb{E}\left[\Psi\left(e^{-2(B_s - as)}\left(x + \frac{4c}{\sigma^2}\int_0^s e^{2(B_t - at)}dt\right) - Y_1\right)\right].$$

Therefore

$$\begin{aligned}
&\frac{4}{\sigma^2}\int_0^\infty e^{-4\sigma^{-2}(\lambda+\lambda^*)s}\lambda\Upsilon(s)ds \\
&= \frac{\lambda}{\lambda+\lambda^*}\mathbb{E}\left[\Psi\left(e^{-2(B_\Theta - a\Theta)}\left(x + \frac{4c}{\sigma^2}\int_0^\Theta e^{2(B_t - at)}dt\right) - Y_1\right)\right]
\end{aligned}$$

where Θ is an exponential random variable, independent of B, with parameter $4(\lambda+\lambda^*)\sigma^{-2}$. The law of the pair

$$\left(e^{-2(B_\Theta - a\Theta)}, \int_0^\Theta e^{2(B_t - at)} dt\right)$$

was presented in Corollary 6.6.2.2. It follows that

$$\frac{4}{\sigma^2}\int_0^\infty e^{-4\sigma^{-2}(\lambda+\lambda^*)s}\lambda\Upsilon(s)ds = \int_0^\infty \int_0^\infty \frac{\gamma}{2y^{2+\alpha+a}} p_u^\alpha(1,y)D(y,u)dydu\,.$$

The study of the second term can be carried out by the same method. □

Comment 8.7.3.2 See the main papers of Klüppelberg [526], Paulsen [700], Paulsen and Gjessing [701], Yuen et al [871], the books of Asmussen [23], Embrechts et al. [322], Mikosch [650] and Mel'nikov [639] and the thesis of Loisel [602]. Many applications to ruin theory can be found in Gerber and his co-authors, e.g., in [387].

8.8 Marked Point Processes

We now generalize compound Poisson processes, introducing briefly a class of processes which are no longer Lévy processes: we introduce a spatial dimension for the size of jumps which are no longer i.i.d. random variables; moreover, the time intervals between two consecutive jumps are no longer independent. Let $(E, \mathcal{E})$ be a measurable space and $(\Omega, \mathcal{F}, \mathbb{P})$ a probability space.

8.8.1 Random Measure

Definition 8.8.1.1 *A* ***random measure*** ϑ *on the space* $\mathbb{R}^+ \times E$ *is a family of positive measures* $(\vartheta(\omega; dt, dx); \omega \in \Omega)$ *defined on* $\mathbb{R}^+ \times E$ *such that, for* $[0,t] \times A \in \mathcal{B} \otimes \mathcal{E}$*, the map* $\omega \to \vartheta(\omega; [0,t], A)$ *is* $\mathcal{F}$*-measurable, and satisfying* $\vartheta(\omega; \{0\} \times E) = 0$.

8.8.2 Definition

Let (Z_n) be a sequence of random variables taking values in the measurable space $(E, \mathcal{E})$, and (T_n) an increasing sequence of positive random variables, with - to avoid explosion - $\lim_n T_n = +\infty$. We define the **marked point process** $\mathbf{N} = \{(T_n, Z_n)\}$ by: for each Borel set $A \subset E$,

$$N_t(A) = \sum_n \mathbb{1}_{\{T_n \leq t\}} \mathbb{1}_{\{Z_n \in A\}}\,.$$

We associate with $\mathbf{N}$ a random measure μ by

$$\mu(\cdot; [0,t], A) = N_t(A)\,.$$

The natural filtration of $\mathbf{N}$ is

$$\mathcal{F}_t^{\mathbf{N}} = \sigma(N_s(A), s \le t, A \in \mathcal{E}).$$

Let H be a map

$$(t, \omega, z) \in (\mathbb{R}^+, \Omega, E) \to H(t, \omega, z) \in \mathbb{R}.$$

The map H is predictable if it is $\mathcal{P} \otimes \mathcal{E}$ measurable. The random counting measure $\mu(\omega; ds, dz)$ acts on the set of predictable processes H as

$$\begin{aligned}(H\star\mu)_t &= \int_{]0,t]} \int_E H(s,z)\mu(ds,dz) = \sum_n H(T_n, Z_n)\mathbb{1}_{\{T_n \le t\}} \\ &= \sum_{n=1}^{N_t(E)} H(T_n, Z_n),\end{aligned}$$

where we have dropped ω in the notation.

Definition 8.8.2.1 *The* ***compensator*** *of μ is the (up to a null set) unique random measure ν such that, for every predictable process H,*

(i) the process $H\star\nu$ is predictable,
(ii) if moreover, the process $|H|\star\mu$ is increasing and locally integrable, the process $(H\star\mu - H\star\nu)$ is a local martingale.

The existence of a compensator is established in Brémaud and Jacod [125], Jacod and Shiryaev [471] and Kallenberg [505].

We now assume that $E = \mathbb{R}^d$. The compensator admits an explicit representation: let $G_n(dt, dz)$ be a regular version of the conditional distribution of (T_{n+1}, Z_{n+1}) with respect to $\mathcal{F}_{T_n}^N = \sigma\{((T_1, Z_1), \dots (T_n, Z_n)\}$. Then,

$$\nu(dt, dz) = \sum_n \mathbb{1}_{\{T_n < t \le T_{n+1}\}} \frac{G_n(dt, dz)}{G_n([t, \infty[\times \mathbb{R}^d)}. \tag{8.8.1}$$

A proof can be found in Prigent [725], Chapter 1 Proposition 1.1.30.

Comment 8.8.2.2 See Brémaud and Jacod [125], Brémaud [124], Prigent [725], Jacod [467] and Jacod and Shiryaev [471] for more details on marked point processes.

Warning 8.8.2.3 The notation in various papers in the literature can be very different from the above: authors may use $\mathbf{N}$ or N for various quantities.

8.8.3 An Integration Formula

Let $dX_t = \beta_t dt + \int_E \gamma(t,z)\mu(dt,dz)$, where β and γ are predictable and let F be a $C^{1,1}$ function. Then

$$\begin{aligned} dF(t,X_t) &= \partial_t F dt + \beta_t\, \partial_x F dt \\ &+ \int_E (F(t,X_{t-}+\gamma(t,z)) - F(t,X_{t-}))\,\mu(dt,dz) \end{aligned}$$

or, in an integrated form

$$\begin{aligned} F(t,X_t) = F(0,X_0) &+ \int_0^t \partial_t F(s,X_s)ds + \int_0^t \beta_s\, \partial_x F(s,X_s)ds \\ &+ \sum_{n=1}^{N_t(E)} [F(T_n,X_{T_n}) - F(T_n,X_{T_n^-})]\,. \end{aligned}$$

8.8.4 Marked Point Processes with Intensity and Associated Martingales

In what follows, we assume that, for every $A \in \mathcal{E}$, the process $N_t(A)$ admits the **F**-predictable intensity $\lambda_t(A)$, i.e., there exists a predictable process $(\lambda_t(A), t \geq 0)$ such that

$$N_t(A) - \int_0^t \lambda_s(A)ds$$

is a martingale. (The most common form of intensity is $\lambda_t(A) = \alpha_t m_t(A)$ where α_t is a positive predictable process and m_t a deterministic probability measure on $(E,\mathcal{E})$. In that case, $\nu(dt,dz) = \alpha_t m_t(A)dt$. We shall say that the marked point process admits $(\alpha_t, m_t(dz))$ as $\mathbb{P}$-local characteristics.)

If $X_t := \sum_{n=1}^{N_t(E)} H(T_n,Z_n)$ where H is an **F**-predictable process that satisfies

$$\mathbb{E}\left(\int_{]0,t]}\int_E |H(s,z)|\lambda_s(dz)ds\right) < \infty$$

the process

$$X_t - \int_0^t \int_E H(s,z)\lambda_s(dz)ds = \int_{]0,t]}\int_E H(s,z)\,[\mu(ds,dz) - \lambda_s(dz)ds]$$

is a martingale and in particular

$$\mathbb{E}\left(\int_{]0,t]}\int_E H(s,z)\mu(ds,dz)\right) = \mathbb{E}\left(\int_{]0,t]}\int_E H(s,z)\lambda_s(dz)ds\right)\,.$$

8.8.5 Girsanov's Theorem

Let μ be the random measure of a marked point process with intensity of the form $\lambda_t(A) = \alpha_t m_t(A)$ where m_t is, as above, a deterministic probability measure on $(E, \mathcal{E})$. Let $(\psi_t, h(t,z))$ be two predictable positive processes such that

$$\int_0^t \psi_s \alpha_s ds < \infty, \int_E h(t,z) m_t(dz) = 1 .$$

Let L be the local martingale solution of

$$dL_t = L_{t-} \int_E (\psi_t h_t(z) - 1)(\mu(dt, dz) - \alpha_t m_t(dz)dt) .$$

If $\mathbb{E}(L_t) = 1$, setting $\mathbb{Q}|_{\mathcal{F}_t} = L_t \mathbb{P}|_{\mathcal{F}_t}$, the marked point process has the $\mathbb{Q}$-local characteristics $(\psi_t \alpha_t, h(t,z) m_t(dz))$.

Exercise 8.8.5.1 Prove Proposition 8.6.6.1 using the above result. $\triangleleft$

8.8.6 Predictable Representation Theorem

Let $(\Omega, \mathcal{F}, \mathbf{F}, \mathbb{P})$ be a probability space where $\mathbf{F}$ is the filtration generated by the marked point process $\mathbf{N}$. Then, any $(\mathbb{P}, \mathbf{F})$-square integrable martingale M admits the representation

$$M_t = M_0 + \int_0^t \int_E H(s,x)(\mu(ds, dx) - \lambda_s(dx)ds)$$

where H is a predictable process such that

$$\mathbb{E}\left(\int_0^t \int_E |H(s,x)|^2 \lambda_s(dx)ds\right) < \infty .$$

See Brémaud [124] for a proof. More generally

Proposition 8.8.6.1 *Let W be a Brownian motion M, $\mathbf{N}$ a marked point process and $\mathcal{F}_t = \sigma(W_s, \mathcal{F}_s^{\mathbf{N}}; s \leq t)$ completed.*

Let $\widetilde{\mu}(ds, dz) = \mu(ds, dx) - \lambda_s(dx)ds$. Then, any $(\mathbb{P}, \mathbf{F})$-local martingale has the representation

$$M_t = M_0 + \int_0^t \varphi_s dW_s + \int_0^t \int_E H(s,z) \widetilde{\mu}(ds, dz) \quad (8.8.2)$$

where φ is a predictable process such that $\int_0^t \varphi_s^2 ds < \infty$ and H is a predictable process such that $\int_0^t \int_E |H(s,x)| \lambda_s(dx)ds < \infty$. If M is a square integrable martingale, each term on the right-hand side of the representation (8.8.2) is square integrable, and

$$\mathbb{E}\left(\left(\int_0^t \varphi_s dW_s\right)^2\right) = \mathbb{E}\left(\int_0^t \varphi_s^2 ds\right)$$

$$\mathbb{E}\left(\left(\int_0^t \int_E H(s,z)\widetilde{\mu}(ds,dz)\right)^2\right) = \mathbb{E}\left(\int_0^t \int_E H^2(s,z)\lambda_s(dz)ds\right)$$

PROOF: We refer to Kunita and Watanabe [550], Kunita [549], and to Chapter III in the book of Jacod and Shiryaev [471]. □

Comment 8.8.6.2 Björk et al. [103] and Prigent [725, 724] gave the first applications to finance of Marked point processes, which are now studied by many authors, especially in a BSDE framework.

Exercise 8.8.6.3 Check that the process

$$S_t = \exp\left(\int_0^t \left(\beta_s - \frac{1}{2}\sigma_s^2\right) ds + \int_0^t \sigma_s dW_s\right) \prod_{n, T_n \le t} (1 + \gamma(T_n, Z_n))$$

is a solution of

$$dS_t = S_{t-}\left(\beta_t dt + \sigma_t dW_t + \int_E \gamma(t,y)\mu(dt,dy)\right),$$

where μ is the random measure associated with the marked point process $\mathbf{N} = \{(T_n, Z_n)\}$. ◁

8.9 Poisson Point Processes

We end this chapter with a brief section on Poisson point processes, which are of major importance in the study of Brownian excursions.

8.9.1 Poisson Measures

Let $(E, \mathcal{E})$ be a measurable space. A random measure μ on $(E, \mathcal{E})$ is a Poisson measure with intensity ν, where ν is a σ-finite measure on $(E, \mathcal{E})$, if

(i) for every set $B \in \mathcal{E}$ with $\nu(B) < \infty$, $\mu(B)$ follows a Poisson distribution with parameter $\nu(B)$, and
(ii) for disjoint sets $B_i, i \le n$, the variables $\mu(B_i), i \le n$ are independent.

Example 8.9.1.1 Let π be a probability measure, $(Y_k, k \in \mathbb{N})$ i.i.d. random variables with law π and N a Poisson variable with mean m, independent of the Y_k's. The random measure $\sum_{k=1}^N \delta_{Y_k}$ is a Poisson measure with intensity $\nu = m\pi$. Here, δ_y is the Dirac measure at point y.

8.9.2 Point Processes

Let $(E, \mathcal{E})$ be a measurable space and δ an additional point. We introduce $E_\delta = E \cup \delta, \mathcal{E}_\delta = \sigma(\mathcal{E}, \{\delta\})$.

Definition 8.9.2.1 *Let* $\mathbf{e}$ *be a stochastic process defined on a probability space* $(\Omega, \mathcal{F}, \mathbb{P})$*, taking values in* $(E_\delta, \mathcal{E}_\delta)$*. The process* $\mathbf{e}$ *is a point process if:*

(i) *the map* $(t, \omega) \to \mathbf{e}_t(\omega)$ *is* $\mathcal{B}(]0, \infty[) \otimes \mathcal{F}$*-measurable,*
(ii) *the set* $D_\omega = \{t : \mathbf{e}_t(\omega) \neq \delta\}$ *is a.s. countable.*

For every measurable set B of $]0, \infty[\times E$, we set

$$N^B(\omega) := \sum_{s \geq 0} \mathbb{1}_B(s, \mathbf{e}_s(\omega)) .$$

In particular, if $B =]0, t] \times \Gamma$, we write

$$N_t^\Gamma = N^B = \text{Card}\{s \leq t : \mathbf{e}(s) \in \Gamma\} .$$

Let the space $(\Omega, \mathbb{P})$ be endowed with a filtration $\mathbf{F}$. A point process is $\mathbf{F}$-adapted if, for any $\Gamma \in \mathcal{E}$, the process N^Γ is $\mathbf{F}$-adapted. For any $\Gamma \in \mathcal{E}_\delta$, we define a point process $\mathbf{e}^\Gamma$ by

$$\begin{aligned} \mathbf{e}_t^\Gamma(\omega) &= \mathbf{e}_t(\omega) \quad \text{if } \mathbf{e}_t(\omega) \in \Gamma \\ \mathbf{e}_t^\Gamma(\omega) &= \delta \quad \text{otherwise} \end{aligned}$$

Definition 8.9.2.2 *A point process* $\mathbf{e}$ *is discrete if* $N_t^E < \infty$ *a.s. for every* t*. It is said to be* σ*-discrete if there is a sequence* E_n *of sets with* $E = \cup E_n$ *such that each* $\mathbf{e}^{E_n}$ *is discrete.*

8.9.3 Poisson Point Processes

Definition 8.9.3.1 *An* $\mathbf{F}$*-Poisson point process* $\mathbf{e}$ *is a* σ*-discrete point process such that:*

(i) *the process* $\mathbf{e}$ *is* $\mathbf{F}$*-adapted,*
(ii) *for any* s *and* t *and any* $\Gamma \in \mathcal{E}$, $N_{s+t}^\Gamma - N_t^\Gamma$ *is independent from* $\mathcal{F}_t$ *and distributed as* N_s^Γ.

In particular, for any disjoint family $(\Gamma_i, i = 1, \ldots, d)$, the d-dimensional process $(N_t^{\Gamma_i}, i = 1, \cdots, d)$ is a Poisson process. Moreover, if N^Γ is finite almost surely, then $\mathbb{E}(N_t^\Gamma) < \infty$ and the quantity $\frac{1}{t}\mathbb{E}(N_t^\Gamma)$ does not depend on t.

Definition 8.9.3.2 *The* σ*-finite measure on* $\mathcal{E}$ *defined by*

$$\mathbf{n}(\Gamma) = \frac{1}{t}\mathbb{E}(N_t^\Gamma)$$

is called the characteristic measure of $\mathbf{e}$.

If $\mathbf{n}(\Gamma) < \infty$, the process $N_t^\Gamma - t\mathbf{n}(\Gamma)$ is an $\mathbf{F}$-martingale.

Proposition 8.9.3.3 (Compensation Formula.) *Let H be a predictable positive process (i.e., measurable with respect to $\mathcal{P} \otimes \mathcal{E}_\delta$) vanishing at δ. Then*

$$\mathbb{E}\left[\sum_{s\geq 0} H(s,\omega,\mathbf{e}_s(\omega))\right] = \mathbb{E}\left[\int_0^\infty ds \int_E H(s,\omega,u)\mathbf{n}(du)\right].$$

If, for any t, $\mathbb{E}\left[\int_0^t ds \int_E H(s,\omega,u)\mathbf{n}(du)\right] < \infty$, the compensated process

$$\sum_{s\leq t} H(s,\omega,\mathbf{e}_s(\omega)) - \int_0^t ds \int_E H(s,\omega,u)\mathbf{n}(du)$$

is a martingale.

PROOF: By the Monotone Class Theorem, it is enough to prove this formula for $H(s,\omega,u) = K(s,\omega)\mathbb{1}_\Gamma(u)$. In that case, $N_t^\Gamma - t\mathbf{n}(\Gamma)$ is an $\mathbf{F}$-martingale.
□

Proposition 8.9.3.4 (Exponential Formula.) *If f is a $\mathcal{B}\otimes\mathcal{E}$-measurable function such that $\int_0^t ds \int_E |f(s,u)|\mathbf{n}(du) < \infty$ for every t, then,*

$$\mathbb{E}\left[\exp\left(i\sum_{0<s\leq t} f(s,\mathbf{e}_s)\right)\right] = \exp\left(\int_0^t ds \int_E (e^{if(s,u)} - 1)\mathbf{n}(du)\right).$$

Moreover, if $f \geq 0$,

$$\mathbb{E}\left[\exp\left(-\sum_{0<s\leq t} f(s,\mathbf{e}_s)\right)\right] = \exp\left(-\int_0^t ds \int_E (1 - e^{-f(s,u)})\mathbf{n}(du)\right).$$

8.9.4 The Itô Measure of Brownian Excursions

Let $(B_t, t \geq 0)$ be a Brownian motion and (τ_s) be the inverse of the local time (L_t) at level 0. The set $\cup_{s\geq 0}]\tau_{s-}(\omega), \tau_s(\omega)[$ is (almost surely) equal to the complement of the zeros set $\{u : B_u(\omega) = 0\}$. The excursion process $(\mathbf{e}_s, s \geq 0)$ is defined by

$$\mathbf{e}_s(\omega)(t) = \mathbb{1}_{\{t \leq \tau_s(\omega) - \tau_{s-}(\omega)\}} B_{(\tau_{s-}(\omega)} + t)(\omega), t \geq 0.$$

This is a path-valued process $\mathbf{e} : \mathbb{R}^+ \times \Omega \to \Omega_*$, where

$$\Omega_* = \{\varepsilon : \mathbb{R}^+ \to \mathbb{R} : \exists V(\varepsilon) < \infty, \text{ with } \varepsilon(V(\varepsilon) + t) = 0, \forall t \geq 0$$

$$\varepsilon(u) \neq 0, \forall 0 < u < V(\varepsilon),\ \varepsilon(0) = 0,\ \varepsilon \text{ is continuous}\}.$$

Hence, $V(\varepsilon)$ is the lifetime of ε.

The starting point of Itô's excursion theory is that the excursion process is a Poisson Point Process; its **characteristic measure n**, called Itô's measure, evaluated on the set Γ, i.e., $\mathbf{n}(\Gamma)$, is the intensity of the Poisson process

$$N_t^\Gamma := \sum_{s \leq t} \mathbb{1}_{\mathbf{e}_s \in \Gamma} .$$

The quantity $\mathbf{n}(\Gamma)$ is the positive real γ such that $N_t^\Gamma - t\gamma$ is an $(\mathcal{F}_{\tau_t})$-martingale.

Here, are some very useful descriptions of $\mathbf{n}$:

- **Itô**: Conditionally on $V = v$, the process

$$(|\epsilon_u|, u \leq v)$$

  is a BES^3 bridge of length v. The law of the lifetime V under $\mathbf{n}$ is

$$\mathbf{n}_V(dv) = \frac{dv}{\sqrt{2\pi v^3}} .$$

  Thanks to the symmetry of Brownian motion, a full description of $\mathbf{n}$ is

$$\mathbf{n}(d\epsilon) = \int_0^\infty \mathbf{n}_V(dv) \, \frac{1}{2}(\Pi_+^v + \Pi_-^v) \, (d\epsilon)$$

  where Π_+^v (resp. Π_-^v) is the law of the standard Bessel Bridge (resp. the law of its negative) with dimension 3 and length v.

- **Williams**: Let $M(\epsilon) = \sup_{u \leq v} |\epsilon_u|$. Then, conditionally on $M = m$, the two processes $(\epsilon_u, u \leq T_m)$ and $(\epsilon_{V-u}, u \leq V - T_m)$ are two independent BES^3 processes considered up to their first hitting time of m, and

$$\mathbf{n}_M(dm) = \mathbf{n}(M(\epsilon) \in dm) = \frac{dm}{m^2} .$$

  We leave to the reader the task of writing a disintegration formula for $\mathbf{n}$ with respect to $\mathbf{n}_M$.

Comment 8.9.4.1 See Jeanblanc et al. [483] for applications to decomposition of Brownian paths and Feynman-Kac formula. At the moment, there are very few applications to finance of excursion theory. One can cite Gauthier [376] for a study of Parisian options, and Chesney et al. [173] for Asian-Parisian options.

9

General Processes: Mathematical Facts

In this chapter, we consider studies involving càdlàg processes. We pay particular attention to semi-martingales with respect to a given filtration; these processes will always be taken with càdlàg paths. We present the definition of stochastic integrals with respect to a square integrable martingale, and we extend the definition to stochastic integrals with respect to a local martingale. Then, we introduce semi-martingales, quadratic covariation processes for semi-martingales and some general versions of Itô's formula and Girsanov's theorem. We give necessary and sufficient conditions for the existence of an equivalent martingale measure. We end the chapter with a brief survey of valuation in an incomplete market.

The reader may refer to the lectures of Itô [464], Meyer [647] and the books of Kallenberg [505] and Protter [727]. The general theory of stochastic processes is presented in Dellacherie [240], Dellacherie and Meyer [242, 244], Dellacherie, Maisonneuve and Meyer [241] and He et al. [427]. More advanced results are given in Jacod and Shiryaev [471] and Bichteler [88]. The survey papers of Kunita [549] and Runggaldier [750] are excellent introductions to this subject with finance in view. Applications to finance can be found in Prigent [725] and Shiryaev [791].

We assume that a filtered probability space $(\Omega, \mathcal{F}, \mathbf{F}, \mathbb{P})$ is given, with some filtration $\mathbf{F}$, satisfying the usual conditions. We recall that the notation $X \stackrel{\text{mart}}{=} Y$ means that $X - Y$ is a local martingale.

9.1 Some Basic Facts about càdlàg Processes

9.1.1 An Illustrative Lemma

To justify why càdlàg processes are considered throughout the development of semi-martingales, we note the very elementary lemma:

M. Jeanblanc, M. Yor, M. Chesney, *Mathematical Methods for Financial Markets*, Springer Finance, DOI 10.1007/978-1-84628-737-4_9,

Lemma 9.1.1.1 *let X be a càd process, such that $X_q = 0$, $\mathbb{P}$-a.s., $\forall q \in \mathbb{Q}_+$. Then, the process $(X_t, t \geq 0)$ is indistinguishable from 0.*

In contrast, here is a non-càdlàg process, with restriction to $\mathbb{Q}_+$ equal to 0, which is not indistinguishable from 0: take $X_t = N_t - N_{t-}$, where N is a Poisson process.

9.1.2 Finite Variation Processes, Pure Jump Processes

When a process X is càdlàg, (i.e., continuous on the right and with limits on the left), we denote by X_{t-} the left limit of X_s when $s \to t, s < t$, as previously. The jump of X at time t is denoted by $\Delta X_t = X_t - X_{t-}$, and we shall call $(\Delta X_t, t \geq 0)$ the jump process of X. We recall (see Definition 1.1.10.2) that an increasing process A is a càdlàg process $(A_t, t \geq 0)$ equal to 0 at time 0, such that $A_s \leq A_t$ for $s \leq t$. If A is an increasing process, there exists a continuous increasing process A^c and a purely discontinuous process A^d (i.e., A^d is equal to the sum of its jumps: $A^d_t = \sum_{0<s\leq t} \Delta A_s$) such that $A = A^c + A^d$. This decomposition is obviously unique.

If U is a càdlàg function, the set $\{t : |U(t) - U(t-)| > a\}$ is discrete for each $a > 0$ and the set $\{t : U(t) \neq U(t-)\}$ is at most countable. Obviously, the same property holds (almost surely) for càdlàg processes.

If moreover the càdlàg process U has finite variation on finite intervals (see Definition 1.1.10.8), then for every $t > 0$ the series $U^d_t = \sum_{0<s\leq t} \Delta U_s$ is absolutely convergent, U^d is càdlàg, it has finite variation on finite intervals and $U^c = U - U^d$ is continuous.

Definition 9.1.2.1 *If $U^c = 0$, that is $U_t = \sum_{0<s\leq t} \Delta U_s$, the process U is said to be a pure jump process. In general a finite variation U admits a unique decomposition as $U_t = U^c_t + \sum_{s\leq t} \Delta U_s$ with U^c continuous with finite variation.*

Let U be a càdlàg process with integrable variation. The **Stieltjes integral** $\int_0^\infty \theta_s dU_s$ is defined for elementary processes θ, i.e., processes of the form $\theta_s = \vartheta_a \mathbb{1}_{]a,b]}(s)$ with ϑ_a a r.v., by $U^\theta_t := \int_0^\infty \theta_s dU_s = \vartheta_a (U_b - U_a)$ and for θ such that $\int_0^\infty |\theta_s||dU_s| < \infty$ by linearity and passage to the limit. (Hence, the integral is defined path-by-path.) Then, one defines the integral

$$U^\theta_t = \int_0^t \theta_s dU_s = \int_{]0,t]} \theta_s dU_s = \int_0^\infty \mathbb{1}_{\{]0,t]\}} \theta_s dU_s \,.$$

Note that the jump processes of U^θ and U are related by:

$$\Delta U^\theta_t = \theta_t \, \Delta U_t \,.$$

If U and V are two finite variation processes, the Stieltjes' integration by parts formula can be written as follows:

$$\begin{aligned} U_tV_t &= U_0V_0 + \int_{]0,t]} V_s dU_s + \int_{]0,t]} U_{s-} dV_s \\ &= U_0V_0 + \int_{]0,t]} V_{s-} dU_s + \int_{]0,t]} U_{s-} dV_s + \sum_{s\leq t} \Delta U_s \, \Delta V_s \,. \end{aligned} \tag{9.1.1}$$

As a partial check, one can verify that the jump process of the left-hand side, i.e., $U_tV_t - U_{t-}V_{t-}$, is equal to the jump process of the right-hand side, i.e., $V_{t-}\Delta U_t + U_{t-}\Delta V_t + \Delta U_t \, \Delta V_t$.

We shall often write $\int_0^t V_s dU_s$ for $\int_{]0,t]} V_s dU_s$.

Proposition 9.1.2.2 *Let F be a C^2 function and U a càdlàg process with finite variation. Then, the process $F(U)$ is also càdlàg with finite variation and*

$$\begin{aligned} F(U_t) &= F(U_0) + \int_0^t F'(U_{s-})dU_s + \sum_{s\leq t} [F(U_s) - F(U_{s-}) - F'(U_{s-})\Delta U_s] \\ &= F(U_0) + \int_0^t F'(U_{s-})dU_s^c + \sum_{s\leq t} [F(U_s) - F(U_{s-})] \\ &= F(U_0) + \int_0^t F'(U_s)dU_s^c + \sum_{s\leq t} [F(U_s) - F(U_{s-})] \,, \end{aligned} \tag{9.1.2}$$

where $U = U^c + U^d$ is the decomposition of U into its continuous and discontinuous parts.

PROOF: The result is true for $F(x) = x$ and if the result holds for F, by integration by parts, it holds for xF. Hence, the result is true for polynomials and for C^1 functions by Weierstrass approximation. The C^2 assumption on F (a little bit too strong) is needed to ensure the convergence of the series. □

As for continuous path semi-martingales, we introduce, besides the ordinary exponential, the notion of the stochastic exponential in the form of the solution to a linear stochastic equation.

Lemma 9.1.2.3 *Let U be a càdlàg process with finite variation. The unique solution of*

$$dY_t = Y_{t-}dU_t, \; Y_0 = y$$

is the stochastic exponential of U (the Doléans-Dade exponential of U) equal to

$$Y_t = y \exp(U_t^c - U_0^c) \prod_{s \le t} (1 + \Delta U_s) \tag{9.1.3}$$

$$= y \exp(U_t - U_0) \prod_{s \le t} (1 + \Delta U_s) e^{-\Delta U_s} . \tag{9.1.4}$$

PROOF: Applying the integration by parts formula to the right-hand side of (9.1.3) shows that it is a solution to the equation $dY_t = Y_{t-} dU_t$. As for the uniqueness, if $Y^i, i = 1, 2$ are two solutions, then, setting $Z = Y^1 - Y^2$ we get $Z_t = \int_0^t Z_{s-} dU_s$. Let us denote by M the running maximum of Z, i.e.,$M_t := \sup_{s \le t} |Z_s|$, then, if V_t is the variation process of U_t

$$|Z_t| \le M_t V_t$$

which implies that

$$|Z_t| \le M_t \int_0^t V_{s-} dV_s = M_t \frac{V_t^2}{2} .$$

Iterating, we obtain $|Z_t| \le M_t \frac{V_t^n}{n!}$ and the uniqueness follows by letting $n \to \infty$. □

We will generalize later (see, e.g., ↣ Exercise 9.4.3.5 and ↣ Proposition 10.2.4.2) the stochastic exponential to semi-martingales.

Comment 9.1.2.4 At this point, for this class of processes, the second formula (9.1.4) may not be so useful, but later on, in ↣ Subsection 9.4.3, when U is a semi-martingale, we shall need this expression.

9.1.3 Some σ-algebras

Definition 9.1.3.1 *Let* $\mathbf{F}$ *be a given filtration (called the reference filtration).*

- *The* **optional** *σ-algebra $\mathcal{O}$ is the σ-algebra on $\mathbb{R}^+ \times \Omega$ generated by càdlàg* $\mathbf{F}$*-adapted processes (considered as mappings on $\mathbb{R}^+ \times \Omega$).*
- *The* **predictable** *σ-algebra $\mathcal{P}$ is generated by the* $\mathbf{F}$*-adapted càg (or continuous) processes. The inclusion $\mathcal{P} \subset \mathcal{O}$ holds.*

These two σ-algebras $\mathcal{P}$ and $\mathcal{O}$ are equal if all $\mathbf{F}$-martingales are continuous, in particular if $\mathbf{F}$ is a strong (or even weak) Brownian filtration. In general

$$\mathcal{O} = \mathcal{P} \vee \sigma(\Delta M, M \text{ describing the set of martingales}) .$$

The optional and predictable σ-algebras were defined with the help of stopping times in Subsection 1.2.3.

A process is said to be **predictable** (resp. **optional**) if it is measurable with respect to the predictable (resp. optional) σ-field. If X is a predictable (resp. optional) process and T a stopping time, then the stopped process X^T is also predictable (resp. optional). Every process which is càg and adapted is predictable, every process which is càd and adapted is optional. If X is a càdlàg adapted process, then $(X_{t^-}, t \geq 0)$ is a predictable process.

A stopping time T is **predictable** (see Definition 1.2.3.1) if and only if the process $(\mathbb{1}_{\{t<T\}} = 1 - \mathbb{1}_{\{T \leq t\}}, t \geq 0)$ is predictable, that is if and only if the stochastic interval $[\![0, T[\![= \{(\omega, t) : 0 \leq t < T(\omega)\}$ is predictable. Note that $\mathcal{O} = \mathcal{P}$ if and only if any stopping time is predictable (this is the case if the reference filtration is a weakly Brownian filtration; see Subsection 5.8.1). A stopping time T is **totally inaccessible** if $\mathbb{P}(T = S < \infty) = 0$ for all predictable stopping times S. See Dellacherie [240], Dellacherie and Meyer [242] and Elliott [313] for related results.

Often, when dealing with point processes, there is a space of marks $(E, \mathcal{E})$ besides the filtered probability space. In such a setting, the following definition makes sense: a **predictable function** is a map $H : \Omega \times \mathbb{R}^+ \times E \to \mathbb{R}$ which is $\mathcal{P} \times \mathcal{E}$ measurable.

9.2 Stochastic Integration for Square Integrable Martingales

We present, in this section the notion of stochastic integration with respect to a square integrable martingale. We follow closely the presentation of Meyer [647]. We shall extend this notion to local martingales and semi-martingales in the next section.

9.2.1 Square Integrable Martingales

We recall that $\mathbf{H}^2$ is the set of **square integrable** martingales, i.e., martingales such that $\sup_{t<\infty} E(M_t^2) < \infty$ (see Subsection 1.2.2). If M is a square integrable martingale, then $M_\tau = E(M_\infty | \mathcal{F}_\tau)$ for any stopping time τ where $M_\infty = \lim_{t\to\infty} M_t$ and $\|M\|_2^2 := E(M_\infty^2) = \sup_{t<\infty} E(M_t^2)$. Let $M_\infty^* = \sup_t |M_t|$. Then, (Doob's inequality)

$$\|M_\infty^*\|_2^2 \leq 4\|M_\infty\|_2^2 .$$

Definition 9.2.1.1 *Two martingales in* $\mathbf{H}^2$ *are* ***orthogonal*** *if their product is a martingale.*

Definition 9.2.1.2 *We denote by* $\mathbf{H}^{2,c}$ *the space of continuous square integrable martingales and by* $\mathbf{H}^{2,d}$ *the set of square integrable martingales orthogonal to* $\mathbf{H}^{2,c}$*. A martingale in* $\mathbf{H}^{2,d}$ *is called a purely discontinuous martingale.*

Warning 9.2.1.3 We stress that a purely discontinuous martingale is not necessarily of bounded variation (see Azéma's martingale in Example 9.3.3.6).

Very often, a martingale in $\mathbf{H}^{2,d}$ is said to be a **compensated sum of jumps**. We explain here the meaning of this definition. If $M \in \mathbf{H}^{2,d}$ there exists a sequence of stopping times $(T_n, n \geq 1)$ such that

$$\{(t,\omega) : \Delta M_t(\omega) \neq 0\} \subset \cup_n \{(t,\omega) : t = T_n(\omega)\},$$

i.e., the sequence $(T_n, n \geq 1)$ exhausts the set of jump times of M. Define $A_t^n = \Delta M_{T_n} \mathbb{1}_{\{T_n \leq t\}}$ and let $M^n = A^n - A^{n,(p)}$ be the compensated martingale associated with A^n, where $A^{n,(p)}$ is the dual predictable projection of A^n (defined in Section 5.2)). It can be proved that, as $k \to \infty$, $\sum_{n=1}^k M_t^n$ converges in L^2 to M_t.

Example 9.2.1.4 If M is the martingale associated with a Poisson process, i.e., $M_t = N_t - \lambda t$, then M is purely discontinuous and $A_t^n = \mathbb{1}_{\{T_n \leq t\}}$. It follows that $\sum_{n=1}^k M_t^n = N_{t \wedge T_k} - \lambda(t \wedge T_k)$.

Theorem 9.2.1.5 *Any martingale M in $\mathbf{H}^{2,d}$ is orthogonal to any square integrable martingale N with no jumps in common with M. For any square integrable martingale M,*

$$E\left(\sum_{s \leq t} (\Delta M_s)^2\right) \leq E(M_t^2).$$

For a purely discontinuous martingale $E(\sum_{s \leq t} (\Delta M_s)^2) = E(M_t^2)$.

PROOF: We refer the reader to Meyer [647]. □

Comment 9.2.1.6 Note that the sum $\sum_{s \leq t} (\Delta M_s)^2$ is convergent and that, in general, the sum $\sum_{s \leq t} |\Delta M_s|$ is not convergent (see ↣ Example of Azéma's martingale in Example 9.3.3.6).

Proposition 9.2.1.7 (a) *Let $M \in \mathbf{H}^{2,d}$. Then the process $M_t^2 - \sum_{s \leq t} (\Delta M_s)^2$ is a martingale.*

(b) *Let $M \in \mathbf{H}^{2,d}$ and $N \in \mathbf{H}^2$. Then, the process $M_t N_t - \sum_{s \leq t} \Delta M_s \, \Delta N_s$ is a martingale and $E(M_\infty N_\infty) = E(\sum_s \Delta M_s \, \Delta N_s)$.*

PROOF: The proof of a) is an application of Proposition 1.2.3.7. First, if τ is a stopping time, the stopped process M^τ is a martingale, hence $E(M_\tau^2 - \sum_{s \leq \tau} (\Delta M_s)^2) = 0$. This implies that the process $M_t^2 - \sum_{s \leq t} (\Delta M_s)^2$ is a martingale. □

Definition 9.2.1.8 *For any martingale $M \in \mathbf{H}^2$, we denote by M^c its projection on $\mathbf{H}^{2,c}$ and by M^d its projection on $\mathbf{H}^{2,d}$. Then, $M = M^c + M^d$ is the decomposition of any martingale in $\mathbf{H}^2$ into its continuous and purely discontinuous parts.*

The process M^2 is a submartingale bounded by $(M^*)^2$, and the process $E(M_\infty^2|\mathcal{F}_t) - M_t^2$ is a supermartingale of class (D), hence the Doob-Meyer supermartingale decomposition Theorem 1.2.1.6 applies: there exists a unique predictable increasing process $\langle M \rangle$ (called the **predictable bracket**) such that $M^2 - \langle M \rangle$ is a martingale equal to 0 at time 0.

The predictable bracket satisfies

$$\langle M \rangle_t = \mathbb{P} - \lim \sum_i E((M_{t_i^{(n)}} - M_{t_{i-1}^{(n)}})^2 | \mathcal{F}_{t_{i-1}^{(n)}})$$

where $0 = t_0^{(n)} < t_1^{(n)} < \cdots < t_{p(n)}^{(n)} = t$ and $\sup_{i=1,\ldots,p(n)}(t_i^{(n)} - t_{i-1}^{(n)})$ goes to 0. Furthermore, if M is continuous, then

$$\langle M \rangle_t = \mathbb{P} - \lim \sum_i (M_{t_i^{(n)}} - M_{t_{i-1}^{(n)}})^2 .$$

Definition 9.2.1.9 *Let M^c denote the continuous part of M. We define the* ***quadratic variation*** *process of M as*

$$[M]_t = \langle M^c \rangle_t + \sum_{0 \le s \le t} (\Delta M_s)^2 .$$

For $M \in \mathbf{H}^2$, the process $[M]$ is an increasing integrable process, and the process $M^2 - [M]$ is a martingale. Indeed, $M = M^c + M^d$, hence

$$M_t^2 - [M]_t = (M_t^c)^2 - \langle M^c \rangle_t + (M_t^d)^2 - \sum_{0 \le s \le t} (\Delta M_s)^2 + 2M_t^c M_t^d ,$$

is the sum of the 3 martingales

$$(M_t^c)^2 - \langle M^c \rangle_t, \quad (M_t^d)^2 - \sum_{0 \le s \le t} (\Delta M_s)^2, \quad 2M_t^c M_t^d .$$

By polarisation, one defines for M and N in $\mathbf{H}^2$, the quadratic covariation process:

$$\langle M, N \rangle = \frac{1}{2} (\langle M + N \rangle - \langle M \rangle - \langle N \rangle)$$

which is the unique predictable process with integrable variation such that $MN - \langle M, N \rangle$ is a martingale and

$$[M, N] = \frac{1}{2} ([M + N] - [M] - [N])$$

which is the unique optional process with integrable variation such that the process $MN - [M, N]$ is a martingale and $\Delta[M, N]_t = \Delta M_t \, \Delta N_t$.

The processes $MN - [M, N]$ and $[M, N] - \langle M, N \rangle$ are martingales, and the martingales M and N are orthogonal if and only if $\langle M, N \rangle$ is null.

Kunita-Watanabe Inequalities

Let H and K be two measurable processes on $\Omega \times \mathbb{R}^+$. Then

$$\int_0^\infty |H_s||K_s|\,|d\langle M,N\rangle_s| \leq \left(\int_0^\infty H_s^2 d\langle M\rangle_s\right)^{1/2} \left(\int_0^\infty K_s^2 d\langle N\rangle_s\right)^{1/2}$$

$$\int_0^\infty |H_s||K_s|\,|d[M,N_s]| \leq \left(\int_0^\infty H_s^2 d[M]_s\right)^{1/2} \left(\int_0^\infty K_s^2 d[N]_s\right)^{1/2}.$$

The idea of the proof is to establish the result for elementary processes, using the Cauchy-Schwarz inequality, and to pass to the limit using the monotone class theorem.

9.2.2 Stochastic Integral

Now let H be a predictable elementary bounded process, i.e., a process of the form

$$H_t = H_0 + \sum_{i=1}^n h_i \mathbb{1}_{]t_i,t_{i+1}]}(t)$$

for a finite sequence of real numbers $t_0 = 0 < t_1 < \cdots < t_n < t_{n+1} = \infty$, where the random variables $h_i \in \mathcal{F}_{t_i}$ are bounded. We call Λ the vector space of these elementary processes. We define, for $H \in \Lambda$:

$$\int_0^\infty H_s dM_s = H_0 M_0 + \sum_{i=1}^n h_i (M_{t_{i+1}} - M_{t_i})$$

and $(H\star M)_t := \int_0^t H_s dM_s = \int_0^\infty \mathbb{1}_{[0,t]}(s) H_s dM_s$. It is easy to check that the process $H\star M$ is a square integrable martingale and that

$$E((H\star M)_\infty^2) = E\left(\int_{[0,\infty[} H_s^2 d\langle M\rangle_s\right) = E\left(\int_{[0,\infty[} H_s^2 d[M]_s\right).$$

If H is a predictable process such that $E\left(\int_{[0,\infty[} H_s^2 d\langle M\rangle_s\right) < \infty$, then, for any $N \in \mathbf{H}^2$, $E\left(\int_{[0,\infty[} |H_s|\,|d\langle M,N\rangle_s|\right) < \infty$.

Theorem 9.2.2.1 *Let $L^2(M)$ be the set of predictable processes H such that*

$$\|H\|_{L^2(M)} := \left[E\left(\int_{[0,\infty[} H_s^2 d\langle M\rangle_s\right)\right]^{1/2} < \infty.$$

The linear mapping $H \to H\star M$ defined on Λ admits a unique extension as a continuous linear mapping from $L^2(M)$ to $\mathbf{H}^2$. Furthermore, the processes $\Delta(H\star M)_t$ and $H_t \Delta M_t$ are indistinguishable.

The stochastic integral $I = H\star M$ is the unique martingale in $\mathbf{H}^2$ such that, for any $N \in \mathbf{H}^2$

$$E(I_\infty N_\infty) = E\left(\int_{[0,\infty[} H_s d\langle M, N\rangle_s\right).$$

Moreover, one has

$$\langle I, N\rangle_t = \int_0^t H_s d\langle M, N\rangle_s\,, \qquad [I, N]_t = \int_0^t H_s\, d[M, N]_s\,.$$

If $H \in L^2(M)$, the continuous and the discontinuous parts of $H\star M$ are $H\star M^c$ and $H\star M^d$. It is important to note that if M is a martingale with integrable variation, and H a predictable process such that $E\left(\int_0^\infty H_s^2 d[M]_s\right) < \infty$ and $E\left(\int_0^\infty |H_s|\, |dM_s|\right) < \infty$, then the stochastic integral and the Stieltjes integral are equal.

9.3 Stochastic Integration for Semi-martingales

We now present the extension of stochastic integrals for a local martingale, then for a semi-martingale.

9.3.1 Local Martingales

In order to extend the definition of stochastic integrals to local martingales (see Subsection 1.2.4), we recall a result that will reduce the problem to the case of $\mathbf{H}^2$ martingales and integrable quadratic variation processes.

Lemma 9.3.1.1 *Let M be a local martingale. There exists a sequence T_n of stopping times such that, for any n the stopped martingale M^{T_n} is of the form $N^n + U^n$, where $N^n \in \mathbf{H}^2$ and U^n is a process with integrable variation.*

Sketch of the Proof: In a first step, one establishes that there exists a sequence of stopping times T_n such that the stopped processes $M^n = M^{T_n}$ are martingales and $E(|M_{T_n}|\,|\mathcal{F}_s)\mathbb{1}_{\{s<T_n\}}$ is bounded. Then, one denotes $C_t = M_t^n \mathbb{1}_{\{T_n \leq t\}} = M_{T_n}^n \mathbb{1}_{\{T_n \leq t\}}$ and $X_t = M_t^n \mathbb{1}_{\{T_n \leq t\}} = M_t^n - C_t$. Let $C^{(p)}$ denote the dual predictable projection of the integrable variation process C, and set $U = C - C^{(p)}$ and $N = X + C^{(p)} = M - U$. It remains to prove that the local martingale N is in $\mathbf{H}^2$ (see [647]). □

One starts by a definition of the quadratic variation of a local martingale. This is extremely important, as it is the main tool for defining stochastic integrals with respect to local martingales.

If M is a local martingale, its continuous part, defined in Definition 9.2.1.8 admits a predictable (in fact a continuous) bracket, denoted $\langle M^c, M^c\rangle$.

Definition 9.3.1.2 *The* ***quadratic variation*** *of a local martingale M is*

$$[M,M]_t = \langle M^c, M^c \rangle_t + \sum_{s \le t} (\Delta M_s)^2 \,.$$

It is important to emphasize that, for any local martingale M, the quantity $\sum_{0 \le s \le t} (\Delta M_s)^2$ is well defined. Note that in general, $[M,M]$ is optional and not predictable.

The process $[M,M]$ is an increasing process such that, by definition $\Delta[M,M]_t = (\Delta M_t)^2$. The quadratic variation of M is the limit in probability of the sum of the square of the increments, i.e.,

$$[M,M]_t = \mathbb{P} - \lim \sum_{i=1}^{p(n)} (M_{t_i^n} - M_{t_{i-1}^n})^2 \,, \tag{9.3.1}$$

where $0 = t_0 \le t_1^n \cdots \le t_{p(n)}^n = t$ and $\sup_i (t_i^n - t_{i-1}^n)$ goes to 0.

The continuous part of the increasing process $[M,M]$ is denoted by $[M,M]^c$, hence

$$[M,M]_t = [M,M]_t^c + \sum_{0 \le s \le t} (\Delta M_s)^2$$

and it follows that $[M,M]_t^c = \langle M^c, M^c \rangle_t$.

We now define the stochastic integral of a locally bounded predictable process with respect to a local martingale. Recall that a process is locally bounded if there exists a sequence of stopping times T_n and constants K_n such that $|H_t| \mathbb{1}_{\{t \le T_n\}} \le K_n$.

Theorem 9.3.1.3 *Let M be a local martingale and H a locally bounded predictable process. There exists a (unique) local martingale $H \star M = \int_0^{\cdot} H_s dM_s$ such that*

$$[H \star M, N] = H \star [M,N]$$

for any bounded martingale N.

Sketch of the Proof: One defines the martingale $H \star M^n$ for $M^n = M^{T_n}$ where $(T_n; n \ge 1)$ is a sequence of stopping times such that the decomposition given in Lemma 9.3.1.1 of the local martingale holds true, and H is bounded. It remains to check that there exists a process denoted $H \star M$ such that $H \star M^n = (H \star M)^{T_n}$ (see [647]). □

Example 9.3.1.4 Let $M_t = \mathcal{N}_t - \lambda t$ where $\mathcal{N}$ is a Poisson process with parameter λ. From the equality (8.2.2), the predictable quadratic variation $\langle M,M \rangle$ (the predictable bracket) of M is λt. The quadratic variation $[M,M]$ of M is $\mathcal{N}$. Indeed, the process $M^2 - \mathcal{N}$ is a martingale, hence, the quadratic variation of M is $\mathcal{N}$, since $\Delta \mathcal{N} = (\Delta \mathcal{N})^2$.

9.3.2 Quadratic Covariation and Predictable Bracket of Two Local Martingales

Proposition 9.3.2.1 *Let M be a local martingale. The* ***quadratic variation*** *process $([M,M]_t, t \geq 0)$ satisfies*

$$[M,M]_t = M_t^2 - M_0^2 - 2\int_0^t M_{s-} dM_s \,. \tag{9.3.2}$$

The process $M_t^2 - [M,M]_t$ is a local martingale.

PROOF: We assume that $M_0 = 0$. Let $t_i^n, i \leq p(n)$ be an increasing finite sequence of real numbers with $t_0^n = 0$ and $t_{p(n)+1}^n = t$. For fixed n, write (with $t_i = t_i^n$)

$$\begin{aligned} M_t^2 &= \sum_{i=0}^{p(n)} M_{t_{i+1}}^2 - M_{t_i}^2 = \sum_{i=0}^{p(n)} (M_{t_i} + \Delta M_{t_i,t_{i+1}})^2 - M_{t_i}^2 \\ &= 2\sum_{i=0}^{p(n)} M_{t_i} \Delta M_{t_i,t_{i+1}} + \sum_{i=0}^{p(n)} (\Delta M_{t_i,t_{i+1}})^2 \end{aligned}$$

where $\Delta M_{t_i,t_{i+1}} = M_{t_{i+1}} - M_{t_i}$ and pass to the limit when n goes to infinity and $\sup(t_i^n - t_{i-1}^n)$ goes to 0. □

Let M and N be two local martingales. The quadratic covariation of two local martingales is defined via the polarization identity

$$[M+N, M+N] = [M,M] + [N,N] + 2[M,N] \,.$$

By polarisation of formula (9.3.2) the quadratic covariation of M and N satisfies

$$[M,N]_t = M_t N_t - M_0 N_0 - \int_0^t M_{s-} dN_s - \int_0^t N_{s-} dM_s \,. \tag{9.3.3}$$

It follows that

$$\begin{aligned} \Delta[M,N]_t &= \Delta(MN)_t - M_{t-}\Delta N_t - N_{t-}\Delta M_t \\ &= M_t N_t - M_{t-}N_{t-} - M_{t-}(N_t - N_{t-}) - N_{t-}(M_t - M_{t-}) \\ &= \Delta M_t \Delta N_t \,. \end{aligned}$$

Proposition 9.3.2.2 *The quadratic covariation of the local martingales M and N is the unique càdlàg process $[M,N]$ with finite variation, such that*

(i) *$MN - [M,N]$ is a local martingale,*

(ii) *$\Delta[M,N]_t = \Delta M_t \Delta N_t$.*

The quadratic covariation $[M, N]$ is the limit in probability of

$$\sum_{i=0}^{p(n)} (M_{t_i^n} - M_{t_{i-1}^n})(N_{t_i^n} - N_{t_{i-1}^n}) \tag{9.3.4}$$

when n goes to infinity and $\sup(t_i^n - t_{i-1}^n)$ goes to zero, where the sequence t_i^n satisfies $0 = t_0^n < t_1^n < \cdots < t_{p(n)+1}^n = t$.

Let M and N be local martingales equal to 0 at time 0, such that the product MN is a special semi-martingale (see $\rightarrowtail$ Subsection 9.3.4). We denote by $\langle M, N\rangle$ (**mixed predictable bracket**) the unique predictable process with finite variation such that $MN - \langle M, N\rangle$ is a local martingale equal to 0 at time 0. If M is locally square integrable, we denote by $\langle M\rangle$ its predictable bracket, that is $\langle M\rangle = \langle M, M\rangle$.

We shall see later under which sufficient condition does the mixed predictable bracket exist. In particular, if M and N are continuous, the mixed predictable bracket exists and is the (continuous) quadratic covariation process defined in Subsection 1.3.1.

If M and N are two local martingales and if the predictable bracket $\langle M, N\rangle$ exists, then the process $[M, N] - \langle M, N\rangle$ is a local martingale.

Example 9.3.2.3 If M is the compensated martingale associated with a Poisson process $\mathcal{N}$ and if W is a Brownian motion with respect to the same filtration

$$[W, W]_t = t, \quad [M, M]_t = \mathcal{N}_t, \quad [W, M]_t = 0\,.$$

Example 9.3.2.4 Let $(\mathcal{N}_t, t \geq 0)$ be a Poisson process with constant intensity λ, and M the associated compensated martingale.

(a) Let f and g be two square integrable functions (or predictable square integrable processes) and

$$X_t = \int_0^t f(s)dM_s, \quad Y_t = \int_0^t g(s)dM_s\,.$$

Then,

$$[X, Y]_t = \sum_{s \leq t} f(s)g(s)\Delta\mathcal{N}_s, \quad \langle X, Y\rangle_t = \int_0^t f(s)g(s)\lambda ds\,.$$

(b) Let $(Y_i, i \geq 1, Z_i, i \geq 1)$ be i.i.d. random variables, independent of $\mathcal{N}$, and define the compound Poisson processes

$$U_t = \sum_{i=1}^{\mathcal{N}_t} Y_i, \quad V_t = \sum_{i=1}^{\mathcal{N}_t} Z_i\,.$$

Then

$$[U, V]_t = \sum_{i=1}^{\mathcal{N}_t} Y_i Z_i, \quad \langle U, V\rangle_t = t\lambda E(Y_1 Z_1)\,.$$

9.3.3 Orthogonality

Definition 9.3.3.1 *Two local martingales are* ***orthogonal*** *if their product is a local martingale.*

In particular, if M and N are independent, they are orthogonal in the filtration generated by M and N. The converse does not hold (see Exercise 1.5.2.2). We illustrate this fact with a new example.

Let $Z = X + iY$ a complex Brownian motion (see Subsection 5.1.3). Then, the two processes $\mathcal{A}_t = \int_0^t (X_s dY_s - Y_s dX_s)$ and $\rho_t = \int_0^t (X_s dX_s + Y_s dY_s) = (|Z_t|^2 - 2t)/2$ are orthogonal, but not independent: indeed they have the same predictable variation process which is not deterministic, as would be the case if the martingales $\mathcal{A}$ and ρ were independent.

Example 9.3.3.2 We present further examples of orthogonal martingales.

- The two local martingales X and Y in Example 9.3.2.4 are orthogonal if $fg = 0$, $ds \times \mathbb{P}$ a.s..
- If M^U and M^V are the compensated martingales associated with the compound Poisson process defined in Example 9.3.2.4, they are orthogonal whenever $E(Y_1 Z_1) = 0$. Note that the covariation process $[M^U, M^V]$ is not equal to 0, it is a local martingale.
- If X is a Lévy process (see ↣ Chapter 11) without drift and continuous part, with Lévy measure ν, the martingales $\sum_{s \leq t} f(\Delta X_s) - t \int \nu(dx) f(x)$ and $\sum_{s \leq t} g(\Delta X_s) - t \int \nu(dx) g(x)$ (see ↣ Exercise 11.2.3.13) are orthogonal iff $\int \nu(dx) f(x) g(x) = 0$ and are independent iff $fg = 0$.

Definition 9.3.3.3 *A local martingale X is* ***purely discontinuous*** *(a compensated sum of jumps) if $X_0 = 0$ and if it is orthogonal to any continuous local martingale.*

The preceding definition is justified by the following: from Corollary 9.3.7.2, any martingale with bounded variation is orthogonal to any continuous martingale. We emphasize again, as we did in Warning 9.2.1.3, that the expression *purely discontinuous martingale* comes as a whole, and one should not confuse this notion with that of a purely discontinuous bounded variation process.

Example 9.3.3.4 If $\mathcal{N}$ is a Poisson process of intensity λ, the compensated martingale $(M_t = \mathcal{N}_t - \lambda t, t \geq 0)$ is a purely discontinuous martingale. In that case, $\sum_{s \leq t} \Delta M_s = \mathcal{N}_t$.

Theorem 9.3.3.5 (Canonical Decomposition.) *Every local martingale M can be uniquely decomposed as follows: $M = m + M^c + M^d$ where M^c is a continuous local martingale, $M_0^c = 0$, M^d is a purely discontinuous local martingale $M_0^d = 0$ and $m = M_0$ is an $\mathcal{F}_0$-measurable random variable.*

Example 9.3.3.6 The Azéma martingale $\mu_t = (\text{sgn} B_t) \sqrt{t - g_t}$ (see Proposition 4.3.8.1) is a purely discontinuous martingale in its own filtration. From Exercise 4.3.8.3, its predictable bracket is $t/2$. Its continuous martingale part equals 0 and its quadratic variation process is

$$[\mu]_t = \sum_{s \leq t} (\Delta \mu_s)^2 = g_t .$$

It is important to note that $\sum_{s \leq t} |\Delta \mu_s| = \infty$, $\mathbb{P}$ a.s. (see Protter [727] for a proof). The Azéma martingale satisfies the equation

$$d[\mu]_t = \frac{dt}{2} - \mu_{t-} d\mu_t .$$

Indeed,

$$\mu_t^2 = 2 \int_0^t \mu_{s-} d\mu_s + [\mu]_t$$

that is $t - g_t = 2 \int_0^t \mu_{s-} d\mu_s + g_t$ or $g_t = - \int_0^t \mu_{s-} d\mu_s + \frac{t}{2}$ which leads immediately to $[\mu]_t = \frac{t}{2} - \int_0^t \mu_{s-} d\mu_s$. This is an example of the so-called **structure equations**, which are equations of the form

$$d[M, M]_t = \alpha_t dt + \Phi_t dM_t ,$$

where M is required to be a local martingale and α and Φ are predictable functionals of M. The particular case

$$d[M, M]_t = dt + \beta M_{t-} dM_t ,$$

admits a unique weak solution. For $-2 \leq \beta < 0$, $[M, M]^c = 0$ and M is bounded.

Comment 9.3.3.7 See $\rightarrowtail$ Subsection 10.6.2, Dellacherie et al. [241], Emery [325, 326], Protter [727] and Chapter 15 in Yor [868] for a general study of structure equations.

9.3.4 Semi-martingales

We give several results on semi-martingales. We refer the reader to Meyer [647] or Protter [727] for the proofs of these results.

Definition 9.3.4.1 *An* **F***-**semi-martingale** is a càdlàg process X which can be written as $X = M + A$ where M is an* **F***-local martingale and where A is an* **F***-adapted càdlàg process with finite variation with value 0 at time 0.*

In general, this decomposition is not unique, and we shall speak about decompositions of semi-martingales. It is necessary to add some conditions on the finite variation process to get the uniqueness.

Definition 9.3.4.2 *A* ***special semi-martingale*** *is a semi-martingale with a predictable finite variation part. Such a decomposition $X = M + A$ with A* ***predictable****, is unique. We call it the* **canonical decomposition** *of X, if it exists.*

The class of semi-martingales is stable under time-changes (see Section 5.1) and mutual absolute continuity changes of probability measures. Moreover, if X is a semi-martingale, then $f(X)$ is a semi-martingale if and only if f'' is a Radon measure (see Çinlar et al. [189]).

Example 9.3.4.3 Local martingales, super-martingales, finite variation processes, càdlàg adapted processes with independent and stationary increments, Itô and Lévy processes are semi-martingales.

Examples of non-semi-martingales: If B is a Brownian motion, the process $|B_t|^\alpha$ is not a semi-martingale for $0 < \alpha < 1$ (the second derivative of the function $f(x) = |x|^\alpha$ is not a Radon measure).

The process $S_t = \int_0^t \phi(B_s)\, ds$ where ϕ does not belong to L^1_{loc} (and if the integral $\int_0^t \phi(B_s)\, ds$ is defined) has zero quadratic variation but infinite variation, hence is not an $\mathbf{F}$-semi-martingale.

As an example, let $S_t = \int_0^t \frac{ds}{B_s} = \lim_{\varepsilon \to 0} \int \frac{da}{a} L_t^a \mathbb{1}_{\{|a| \geq \varepsilon\}}$. See Subsection 6.1.2.

Example 9.3.4.4 The Poisson process N with parameter λ is a special semi-martingale with *canonical* decomposition $N_t = M_t + \lambda t$. Note that we can also write a decomposition as $N_t = 0 + N_t$, where on the right-hand side, N is considered as an increasing process. Hence, the semi-martingale N admits at least two (optional) decompositions. (See Chapter 8.)

If $|\Delta X| \leq C$ where C is a constant, then the semi-martingale X is special and its canonical decomposition $X = M + A$ satisfies $|\Delta A| \leq C$ and $|\Delta M| \leq 2C$. In particular, if X is a continuous semi-martingale, it is special and the processes M and A in its canonical decomposition $X = M + A$ are continuous.

Note that, if $\mathcal{F}_t \subset \mathcal{G}_t$ for every t, i.e., if $\mathbf{F}$ is a subfiltration of $\mathbf{G}$, then an $\mathbf{F}$-semi-martingale is a $\mathbf{G}$-semi-martingale if and only if any $\mathbf{F}$-local martingale is a $\mathbf{G}$-semi-martingale (see Section 5.9). We recall the following result of Stricker [807]:

Proposition 9.3.4.5 *Let $\mathbf{F}$ and $\mathbf{G}$ be two filtrations such that for all $t \geq 0$, $\mathcal{F}_t \subset \mathcal{G}_t$. If X is a $\mathbf{G}$-semi-martingale which is $\mathbf{F}$-adapted, then it is also an $\mathbf{F}$-semi-martingale.*

Let X be a semi-martingale such that $\forall t \geq 0, \sum_{s \leq t} |\Delta X_s| < \infty$. The process $(X_t - \sum_{s \leq t} \Delta X_s, t \geq 0)$ is a continuous semi-martingale with unique decomposition $M + A$ where M is a continuous local martingale and A a continuous process with bounded variation. The continuous martingale M is called the continuous martingale part of X, it is denoted by X^c, and $[X,X]^c = [X^c, X^c] = \langle X^c \rangle$. Note however that not every semi-martingale satisfies $\sum_{s \leq t} |\Delta X_s| < \infty$ and looking for the continuous martingale part of X may be more complicated (see ↣ Chapter 11). If X is a martingale such that $\sum_{s \leq t} |\Delta X_s| < \infty$, it does not imply in general that the processes X and $(X_t^c + \sum_{s \leq t} \Delta X_s, t \geq 0)$ are equal. In fact, $X_t = X_t^c + \sum_{s \leq t} \Delta X_s$ if and only if $\sum_{s \leq t} \Delta X_s$ is a local martingale. This is a strong condition: for example, under the assumption $\Delta X_s = H_s \mathbb{1}_{\{\Delta X_s \neq 0\}}$ with H predictable, this condition is satisfied only for continuous martingales. Indeed, in that case, the process $\sum_{s \leq t} (\Delta X_s)^2 = \sum_{s \leq t} H_s \Delta X_s$ is a positive local martingale, hence a super-martingale, its expectation equals 0, hence the continuity of X follows.

Proposition 9.3.4.6 *If X and Y are semi-martingales and if X^c, Y^c are their continuous martingale parts, their quadratic covariation is*

$$[X,Y]_t = \langle X^c, Y^c \rangle_t + \sum_{s \leq t} (\Delta X_s)(\Delta Y_s) .$$

PROOF: In a first step, we note that $[X^c] = [X]^c$, and we extend this equality by polarization. Then, we note that the jumps of $[X,Y]_t$ and $\langle X^c, Y^c \rangle_t + \sum_{s \leq t} \Delta X_s \Delta Y_s$ are the same. □

9.3.5 Stochastic Integration for Semi-martingales

If X is a semi-martingale with decomposition $M+A$, then for any predictable, locally bounded process H, we can define the process

$$(H \star X)_t := \int_0^t H_s \, dX_s = \int_0^t H_s \, dM_s + \int_0^t H_s \, dA_s ,$$

where $\int_0^t H_s \, dA_s$ is a Stieltjes integral.

The process $(H \star X)_t$ does not depend on the decomposition of the semi-martingale X. It is a semi-martingale, in particular it is a càdlàg adapted process. The map $H \to H \star X$ is linear and the map $X \to H \star X$ is linear, the process $H \star X$ is a semi-martingale.

If X is a semi-martingale and H a predictable process, the jump process of the stochastic integral of H with respect to X is equal to H times the jump process of X: $(\Delta(H \star X))_t = H_t (\Delta X)_t$. It can also be checked that $H \star X^c = (H \star X)^c$.

Comment 9.3.5.1 See Dellacherie and Meyer [244] Chapter 8, Jacod [468], Jacod and Shiryaev [471] Chapter 1, Kallenberg [505] and Protter [727] Chapter 2, for more information on stochastic integration.

9.3.6 Quadratic Covariation of Two Semi-martingales

We examine which of the previous constructions and properties of $[X,Y]$ may be extended to pairs of semi-martingales. The **quadratic covariation** of two semi-martingales is the finite variation process defined by the integration by parts formula

$$[X,Y] = XY - X_0Y_0 - X_- \star Y - Y_- \star X\,.$$

For every predictable bounded process H, the quadratic covariation of Y and of $H\star X$ (the stochastic integral of H with respect to X), is the stochastic integral of H with respect to the quadratic covariation of X and Y: $[H\star X, Y] = H\star[X,Y]$.

If either X or Y has locally finite variation, then the sum $\sum_{0<s\leq t} |\Delta X_s|\,|\Delta Y_s|$ is almost surely finite and the quadratic covariation of the pair (X,Y) is given by $[X,Y]_t = X_0Y_0 + \sum_{0<s\leq t} \Delta X_s \Delta Y_s$. For general semi-martingales X and Y, we have

$$\sum_{0<s\leq t} |\Delta X_s|\,|\Delta Y_s| \leq \left(\sum_{0<s\leq t} (\Delta X_s)^2\right)^{1/2} \left(\sum_{0<s\leq t} (\Delta Y_s)^2\right)^{1/2}$$

hence the sum $\sum_{0<s\leq t} |\Delta X_s|\,|\Delta Y_s|$ is almost surely finite.

Note that, if $\mathbb{P}$ and $\mathbb{Q}$ are equivalent, the quadratic covariations of the semi-martingales X and Y under $\mathbb{P}$ and under $\mathbb{Q}$ are the same.

An adapted increasing process A is said to be a compensator for the semi-martingale Y if $Y - A$ is a local martingale. For example if X is a local martingale, the process $[X,X]$ is a compensator for X^2. In general, a semi-martingale admits many compensators. If there exists a predictable compensator, then it is unique (among predictable compensators). Once more, we note that in general, $[X,X]$ is optional and not predictable.

Remark 9.3.6.1 If M and N are two pure jump martingales with the same jumps, they are equal. Indeed, the difference $M-N$ is a continuous martingale which is orthogonal to itself, hence is equal to 0.

9.3.7 Particular Cases

The integration by parts formula

$$X_tY_t = X_0Y_0 + \int_0^t X_{s-}dY_s + \int_0^t Y_{s-}dX_s + [X,Y]_t\,,$$

can be simplified, as presented in Yoeurp [857], under some additional hypotheses.

Proposition 9.3.7.1 *Let X be a semi-martingale.*

(a) *If A is a bounded variation process*

$$X_tA_t = X_0A_0 + \int_0^t X_s dA_s + \int_0^t A_{s-} dX_s \tag{9.3.5}$$

and $[X,A] = \Delta X \star A$.

(b) *If A is a predictable process with bounded variation*

$$X_tA_t = X_0A_0 + \int_0^t X_{s-} dA_s + \int_0^t A_s dX_s \tag{9.3.6}$$

and $[X,A] = \Delta A \star X$.

PROOF: **Part a:** If A is a bounded variation process

$$\begin{aligned}\int_0^t X_{s-} dA_s + [X,A]_t &= \int_0^t X_{s-} dA_s + \sum_{s\leq t} \Delta X_s \Delta A_s \\ &= \int_0^t (X_{s-} + \Delta X_s) dA_s = \int_0^t X_s dA_s\,.\end{aligned}$$

Part b: If M is a martingale and A a predictable process with bounded variation, Yoeurp [857] established that

$$\begin{aligned}[M,A]_t &= \sum_{s\leq t} \Delta M_s \, \Delta A_s = \sum_{n,T_n\leq t} \Delta A_{T_n} \, \Delta M_{T_n} \\ &= \int_0^t \sum_n \Delta A_{T_n} \mathbb{1}_{\{T_n = s\}} dM_s = \int_0^t \Delta A_s \, dM_s\end{aligned}$$

where (T_n) is a sequence of stopping times which exhaust the jumps, therefore, the process $[M,A]_t$ is a local martingale. If A is a predictable process with bounded variation

$$\begin{aligned}\int_0^t A_{s-} dX_s + [X,A]_t &= \int_0^t A_{s-} dX_s + \sum_{s\leq t} \Delta X_s \Delta A_s \\ &= \int_0^t A_{s-} dX_s + \int_0^t (A_s - A_{s-}) dX_s = \int_0^t A_s dX_s\,.\end{aligned}$$

□

We recall that the notation $X \overset{\text{mart}}{=} Y$ means that $X - Y$ is a local martingale.

Corollary 9.3.7.2 *Assume that M is a local martingale and A a bounded variation process. Then*

$$M_tA_t \overset{\text{mart}}{=} \int_0^t M_s dA_s \,.$$

Consequently, if M is continuous and A is a local martingale with bounded variation, the process MA is also a local martingale.

If A is a predictable bounded variation process,

$$M_tA_t \overset{\text{mart}}{=} \int_0^t M_{s-} dA_s \,.$$

Then, here, we recover partly the orthogonality between continuous local martingales and purely discontinuous martingales (Definition 9.3.3.3).

9.3.8 Predictable Bracket of Two Semi-martingales

We first discuss the predictable bracket of two martingales, which we have defined earlier. We now give some conditions for the existence of this bracket. If the quadratic covariation process is locally integrable, then the predictable bracket exists. In particular this is the case if X and Y are locally square integrable and $\langle X, Y\rangle$ may be defined as the dual predictable projection of $[X, Y]$.

If X is a semi-martingale such that $X = M + A$ where M is locally square integrable and A predictable, one can define its predictable bracket as follows. Note that

$$[M + A]_t = [M]_t + 2[M, A]_t + [A]_t \,.$$

We have seen in Proposition 9.3.7.1 that the process $[M, A]_t = \int_0^t \Delta A_s dM_s$ is a local martingale, hence $[M+A]_t \overset{\text{mart}}{=} \langle M\rangle_t + [A]_t$. Since $[A]$ is predictable, the process $\langle X\rangle := \langle M\rangle + [A]$ is predictable too and is called the **predictable bracket** of the semi-martingale X. If X is a continuous semi-martingale, $\langle X\rangle_t = \langle M\rangle_t$.

In the general case, one defines $\langle X\rangle_t$ as the dual predictable projection of $[X]_t$. The polarization formula is used to define the mixed predictable bracket $\langle X, Y\rangle$.

Warning 9.3.8.1 In general, if $X = M + A$ is a semi-martingale with predictable bracket $\langle X\rangle$, the process $X_t^2 - \langle X\rangle_t$ is not a martingale. Indeed,

$$X_t^2 - \langle X\rangle_t = M_t^2 + A_t^2 + 2M_tA_t - \langle M\rangle_t - [A]_t \,.$$

Using

$$A_t^2 + 2M_tA_t - [A]_t = 2M_tA_t + 2\int_0^t A_{s-}dA_s \overset{\text{mart}}{=} 2\int_0^t X_{s-}dA_s$$

one obtains

$$X_t^2 \overset{\text{mart}}{=} \langle X\rangle_t + 2\int_0^t X_{s-}dA_s \,.$$

If $\mathbb{P}$ and $\mathbb{Q}$ are equivalent probabilities, the mixed predictable bracket of the *continuous* semi-martingales X and Y is the same under these two probabilities. This is no longer true if discontinuous martingales are considered. (See the case of Poisson processes in Comment 8.2.4.2.)

9.4 Itô's Formula and Girsanov's Theorem

We give here, without proof, a general Itô formula (see Protter [727] for a proof).

9.4.1 Itô's Formula: Optional and Predictable Forms

Itô's formula is an extremely powerful tool. It is, therefore, worthwhile writing it in different ways. We recall that we have discussed Itô's formula for continuous semi-martingales in Subsection 1.5.3.

If X is a semi-martingale, X^c its continuous martingale part and f a $C^2(\mathbb{R})$ function, then the sum $\sum_{s\leq t}|f(X_s)-f(X_{s-})-f'(X_{s-})\Delta X_s|$ is almost surely finite for any t and one has the **optional Itô formula**

$$\begin{aligned} f(X_t) = \; & f(X_0) + \textstyle\int_0^t f'(X_{s-})dX_s + \frac{1}{2}\displaystyle\int_0^t f''(X_s)d\langle X^c\rangle_s \\ & + \sum_{0<s\leq t}[f(X_s) - f(X_{s-}) - f'(X_{s-})\Delta X_s]\,. \end{aligned}$$

Note that in the particular case where X is continuous, we recover Itô's formula established in Theorem 1.5.3.1:

$$f(X_t) = f(X_0) + \int_0^t f'(X_s)dX_s + \frac{1}{2}\int_0^t f''(X_s)d\langle X\rangle_s \,.$$

Coming back to the general case, we recall that

$$\langle X^c\rangle_t = [X^c]_t = [X]_t^c = [X]_t - \sum_{s\leq t}(\Delta X_s)^2 \,.$$

If X is a special semi-martingale, and $[X]$ is locally integrable, then, for any $f \in C^2$, $f(X_t)$ is a special semi-martingale.

More generally, let $X = (X^1, \ldots, X^d)$ be a semi-martingale and f a $C^2(\mathbb{R}^d)$ function. Let us denote by $M^{i,c}$ the continuous martingale part of X^i. Then,

$$f(X_t) = f(X_0) + \sum_{i=1}^{d} \int_0^t \partial_i f(X_{s-}) dX_s^i + \frac{1}{2} \int_0^t \sum_{i,j=1}^{d} \partial_{ij} f(X_{s-})\, d\langle M^{i,c}, M^{j,c} \rangle_s$$
$$+ \sum_{s \leq t} \left[f(X_s) - f(X_{s-}) - \sum_{i=1}^{d} \partial_i f(X_{s-}) \Delta X_s^i \right]. \tag{9.4.1}$$

Obviously, this formula entails the form of Itô's formula for $f(M_t, A_t)$ where M is a multidimensional martingale, and A a multidimensional process with bounded variation.

Definition 9.4.1.1 *A process X is a* ***pseudo-continuous semi-martingale*** *(PCSM) if its additive decomposition is $X = M + V$ where M is a local martingale and V is an adapted process with finite variation such that:*

(i) *V is continuous (hence, predictable),*
(ii) *$\Delta X_s = H_s \mathbb{1}_{\{\Delta X_s \neq 0\}}$ where H is predictable,*
(iii) *the predictable compensator A of*
$\sum_{s \leq t} (\Delta X_s)^2 = \sum_{s \leq t} (\Delta M_s)^2 = \sum_{s \leq t} H_s^2 \mathbb{1}_{\{\Delta X_s \neq 0\}}$ *is continuous.*

In particular, PCSM's are special semi-martingales. Note that, if X is a PCSM, the predictable compensator of

$$\sum_{s \leq t} \Phi(\Delta X_s) \mathbb{1}_{\{\Delta X_s \neq 0\}} = \sum_{s \leq t} \Psi(H_s)(\Delta X_s)^2$$

is $\int_0^t \Psi(H_s) dA_s$, where $\Psi(x) = \frac{\Phi(x)}{x^2} \mathbb{1}_{\{x \neq 0\}}$.

If X is a PCSM with decomposition $X_t = M_t^c + M_t^d + V_t$, and f a smooth function, then $f(X)$ is a PCSM. We present the canonical decomposition of the special semi-martingale $f(X)$: from Itô's formula

$$\begin{aligned} f(X_t) &= f(X_0) + \int_0^t f'(X_{s-}) dX_s + \frac{1}{2} \int_0^t f''(X_{s-}) d\langle M^c, M^c \rangle_s \\ &\quad + \sum_{s \leq t} [f(X_{s-} + H_s) - f(X_{s-}) - f'(X_{s-}) H_s] \mathbb{1}_{\{\Delta X_s \neq 0\}}, \\ &= f(X_0) + \int_0^t f'(X_{s-}) dX_s + \frac{1}{2} \int_0^t f''(X_{s-}) d\langle M^c, M^c \rangle_s \\ &\quad + \sum_{s \leq t} f^{(T)}(X_{s-}, H_s)(\Delta X_s)^2, \end{aligned}$$

where $f^{(T)}(x,h) = \frac{f(x+h)-f(x)-f'(x)h}{h^2}\mathbb{1}_{\{h\neq 0\}}$ (here, T stands for Taylor). In particular,

$$f^{(T)}(X_{s-},H_s) = \int_0^1 dx(1-x)f''(X_{s-}+xH_s)\,.$$

We note that the jumps of

$$\int_0^t \frac{f(X_{s-}+H_s)-f(X_{s-})-f'(X_{s-})H_s}{H_s}\mathbb{1}_{\{H_s\neq 0\}}\,dM_s^d$$

are

$$\frac{f(X_{s-}+H_s)-f(X_{s-})-f'(X_{s-})H_s}{H_s}(\Delta X_s) = f^{(T)}(X_{s-},H_s)(\Delta X_s)^2\,.$$

From the hypothesis, the process

$$\sum_{s\leq t} f^{(T)}(X_{s-},H_s)(\Delta X_s)^2 - \int_0^t f^{(T)}(X_{s-},H_s)dA_s$$

is a martingale, and it has the same jumps as

$$\int_0^t \frac{f(X_{s-}+H_s)-f(X_{s-})-f'(X_{s-})H_s}{H_s}dM_s^d\,,$$

hence, these two purely discontinuous martingales are equal.

Proposition 9.4.1.2 *Let X be a PCSM with decomposition $M+V$ and let f be a function in C_b^2. Then,*

$$f(X_t) = M_t^f + V_t^f = M_t^{f,c} + M_t^{f,d} + V_t^f$$

where

$$\begin{aligned}
M_t^{f,c} &= \int_0^t f'(X_{s-})dM_s^c = \int_0^t f'(X_s)dM_s^c\,,\\
V_t^f &= \int_0^t f'(X_{s-})dV_s + \frac{1}{2}\int_0^t f''(X_s)d\langle M^c\rangle_s + \int_0^t f^{(T)}(X_{s-},H_s)dA_s\,,\\
M_t^{f,d} &= \int_0^t f^{(T)}(X_{s-},H_s)(d[M^d,M^d]_s - dA_s)\,.
\end{aligned}$$

Exercise 9.4.1.3 Let N be an inhomogeneous Poisson process with intensity $\lambda(t)$, M its compensated martingale, W a Brownian motion and

$$dX_t = h_t dt + f_t dW_t + g_t dM_t\,,$$

where f,g and h are (bounded) predictable processes. Using the identity $\sum_{s\leq t}\phi_s\Delta N_s = \int_0^t \phi_s dN_s$, prove that

$$\begin{aligned}F(t,X_t) = F(0,X_0) &+ \int_0^t \partial_s F(s,X_s)\,ds + \int_0^t \partial_x F(s,X_s)h_s ds \\ &+\tfrac{1}{2}\int_0^t \partial_{xx}F(s,X_s)f_s^2\,ds \\ &+\int_0^t [F(s,X_s+g_s)-F(s,X_s)-\partial_x F(s,X_s)g_s]\lambda_s\,ds \\ &+\int_0^t \partial_x F(s,X_s)f_s dW_s + \int_0^t [F(s,X_{s-}+g_s)-F(s,X_{s-})]\,dM_s\,.\end{aligned}$$

◁

9.4.2 Semi-martingale Local Times

Let X be a semi-martingale.

Proposition 9.4.2.1 *The measure $\varphi \to \int_0^t \varphi(X_s)d\langle X^c\rangle_s$ is absolutely continuous with respect to the Lebesgue measure. Its Radon-Nikodym density may be defined as L_t^x the local time at x which satisfies the Itô-Tanaka formula:*

$$\begin{aligned}(X_t-x)^+ = (X_0-x)^+ &+ \int_{]0,t]} \mathbb{1}_{\{X_{s-}>x\}}dX_s + \sum_{0<s\le t} \mathbb{1}_{\{X_{s-}>x\}}(X_s-x)^- \\ &+ \sum_{0<s\le t} \mathbb{1}_{\{X_{s-}\le x\}}(X_s-x)^+ + \frac{1}{2}L_t^x\,.\end{aligned}$$

This result is generalized to:

Theorem 9.4.2.2 *Let f be a convex function and X be a real-valued semi-martingale. Then, $f(X)$ is a semi-martingale and*

$$f(X_t) = f(X_0) + \int_{]0,t]} f'(X_{s-})\,dX_s + \frac{1}{2}\int_{-\infty}^{\infty} f''(dx)L_t^x + A_t$$

where f' is the left derivative of f and A is an adapted purely discontinuous increasing process. Moreover

$$\Delta A_t = f(X_t) - f(X_{t-}) - f'(X_{t-})\Delta X_t\,.$$

Proof: See Meyer [647], Protter [727] Chapter IV, Section 7. □

Example 9.4.2.3 Azéma Martingale: It can be proved (see Protter [727]) that the local times at all levels except 0 of Azéma's martingale (see Subsection 4.3.8 for the definition of this martingale) are 0 and that the local time at 0 for Brownian motion coincides with the local time at 0 of the Azéma martingale. An important consequence is that Azéma's martingale is not a semi-martingale for the Brownian filtration: Indeed, if it were, then its (continuous) martingale part would have zero quadratic variation, hence would be zero. Consequently, Azéma's martingale would have bounded variation, which is false.

9.4.3 Exponential Semi-martingales

We generalize the definition of stochastic exponential introduced for bounded variation processes in Lemma 9.1.2.3.

Definition 9.4.3.1 *Let $(X_t, t \geq 0)$ be a real-valued* $\mathbf{F}$*-semi-martingale. The stochastic differential equation*

$$Y_t = 1 + \int_0^t Y_{s-} dX_s$$

has a unique càdlàg $\mathbf{F}$*–adapted solution denoted by $(\mathcal{E}(X)_t, t \geq 0)$, and is called the* ***Doléans-Dade exponential*** *of X:*

$$\mathcal{E}(X)_t = \exp\left(X_t - X_0 - \frac{1}{2}\langle X^c, X^c\rangle_t \right) \prod_{s \leq t} (1 + \Delta X_s) e^{-\Delta X_s} .$$

The process $\mathcal{E}(X)$ has strictly positive values if and only if $\Delta X_s > -1$ for all s. If $\Delta X_s \geq -1, \forall s$ and $\tau = \inf\{t : \Delta X_t = -1\}$, the exponential is equal to zero after τ, and is strictly positive before τ.

If X is a local martingale and $\Delta X_s > -1$, the process $\mathcal{E}(X)$ is a positive local martingale and, therefore, a super-martingale and converges when t goes to infinity, to some random variable $\mathcal{E}(X)_\infty$ with $E(\mathcal{E}(X)_\infty) \leq 1$. It is a martingale if and only if its expectation is 1, i.e., $E(\mathcal{E}(X)_t) = 1$ for all t.

The process $\mathcal{E}(X)$ is a uniformly integrable martingale if and only if $E(\mathcal{E}(X)_\infty) = 1$.

The pure jump process $\prod_{s \leq t} (1 + \Delta X_s) e^{-\Delta X_s}$ has finite variation.

A condition which ensures the uniform integrability of $\mathcal{E}(X)$ is $\Delta X > -1$ and

$$E\left(\exp\left(\frac{1}{2} \langle X^c \rangle_\infty \right) \prod_t (1 + \Delta X_t) \exp\left(- \frac{\Delta X_t}{1 + \Delta X_t} \right) \right) < \infty .$$

(see Lépingle and Mémin [580]). If the martingale X is continuous, the usual formula is obtained: the process

$$Y_t = \exp\left(X_t - X_0 - \frac{1}{2} \langle X \rangle_t \right)$$

is the solution of $dY_t = Y_t dX_t$, $Y_0 = 1$. If X and Y are semi-martingales, then

$$\mathcal{E}(X)\mathcal{E}(Y) = \mathcal{E}(X + Y + [X, Y]) .$$

If the martingale X is BMO (i.e., if $\sup_{\tau \in \mathcal{T}} E([X]_\infty - [X]_{\tau-}) < m$ where $\mathcal{T}$ is the set of stopping times) and if $\Delta X > 1 - \delta$, with $0 < \delta < \infty$, then $\mathcal{E}(X)$ is a martingale. (See Doléans-Dade and Meyer [257]).

The mapping $X \to \mathcal{E}(X) = Z$ can be inverted under some assumptions on Z. As Choulli et al. [181], Kallsen and Shiryaev [508, 509] and Jamshidian

[475], we call its reciprocal function **the stochastic logarithm**: if Z is a semi-martingale such that Z and Z_- are $\mathbb{R}\setminus\{0\}$-valued, there exists a unique semi-martingale X, denoted $\mathcal{L}(Z)$ such that $X_0 = 0$ and $Z = Z_0\mathcal{E}(X)$. Indeed, from

$$Z_t = Z_0 + \int_0^t Z_{s-} dX_s \,,$$

we get $X_t = \int_0^t \frac{1}{Z_{s-}} dZ_s$.

In particular, any semi-martingale of the form $S = e^X$ where X is a semi-martingale can be written as a stochastic exponential $S = \mathcal{E}(\tilde{X})$ where

$$\tilde{X}_t = \mathcal{L}(S)_t = X_t + \frac{1}{2}\langle X\rangle_t^c + \sum_{s\le t}(e^{\Delta X_s} - 1 - \Delta X_s)\,.$$

Remark 9.4.3.2 Symbolic Calculus on Semi-martingales. If f is a smooth function, we denote $F(x) = \int_0^x dy f(y)$. If X is a given semi martingale, one may define, in a similar way, the process $\mathcal{F}(X)_t$ as

$$\mathcal{F}(X)_t = \int_0^t f(X_{s-})dX_s \,.$$

Setting $Y_t = \mathcal{F}(X)_t$, if $f(X_s)$ and $f(X_{s-})$ do not vanish, we can invert the mapping as

$$X_t = x + \int_0^t \frac{dY_s}{f(X_{s-})}$$

this last equality being a SDE, i.e., $dX_t = \varphi(X_{t^-})dY_t$, with $\varphi = 1/f$. The case $\varphi(x) = x$ leads to $X_t = \mathcal{E}(Y)_t$ and, if X_s and X_{s-} do not vanish, $Y_t = \int_0^t \frac{1}{X_{s-}} dX_s = \mathcal{L}(X)_t$.

Exercise 9.4.3.3 Let Z be an $\mathbb{R}$-valued semi-martingale such that Z and Z_- do not vanish. Prove that

$$\mathcal{L}(Z)_t = \ln\left(\left|\frac{Z_t}{Z_0}\right|\right) + \int_0^t \frac{1}{2Z_{s-}^2} d\langle Z^c\rangle_s - \sum_{s\le t}\left(\ln\left|\frac{Z_s}{Z_{s-}}\right| + 1 - \frac{Z_s}{Z_{s-}}\right)\,.$$

◁

Exercise 9.4.3.4 Prove that if X, Y are two semi-martingales such that their stochastic logarithms are well defined, then

$$\mathcal{L}(XY) = \mathcal{L}(X) + \mathcal{L}(Y) + [\mathcal{L}(X), \mathcal{L}(Y)]\,.$$

◁

Exercise 9.4.3.5 Let X be a semi-martingale. Check that the solution of the SDE $dS_t = S_{t-}(b(t)dt + \sigma(t)dX_t)$ is

$$S_t = S_0 \exp(U_t) \Pi_t$$

where

$$\begin{aligned} U_t &= \int_0^t \sigma(s)dX_s + \int_0^t b(s)ds - \frac{1}{2}\int_0^t \sigma^2(s)d\langle X^c \rangle_s \,, \\ \Pi_t &= \prod_{0<s\leq t} (1+\sigma(s)\Delta X_s) \exp(-\sigma(s)\Delta X_s) \,. \end{aligned}$$

◁

9.4.4 Change of Probability, Girsanov's Theorem

Let $\mathbb{Q}$ be equivalent to $\mathbb{P}$ on $\mathcal{F}_t$, for all t and $\mathbb{Q}|_{\mathcal{F}_t} = L_t \, \mathbb{P}|_{\mathcal{F}_t}$ where L is a strictly positive $\mathbb{P}$-martingale. Any $\mathbb{P}$-local martingale X is a $\mathbb{Q}$ semi-martingale and the semi-martingale decompositions are given by the following theorem:

Theorem 9.4.4.1 *Let X be a local martingale with respect to $\mathbb{P}$. Then,*

(i)

$$X_t - \int_0^t \frac{d[X,L]_s}{L_s} \quad \textit{is a } \mathbb{Q}\textit{-local martingale.} \tag{9.4.2}$$

(ii) *If $[X,L]$ is $\mathbb{P}$-locally integrable (which implies that $\langle X,L \rangle$ exists) the process*

$$X_t - \int_0^t \frac{d\langle X,L \rangle_s}{L_{s-}} \quad \textit{is a } \mathbb{Q}\textit{-local martingale.} \tag{9.4.3}$$

We may call the process in (9.4.2) *the optional Girsanov's transform of X and the process in* (9.4.3) *the predictable Girsanov's transform of X.*

PROOF: (i) Using Corollary 9.3.7.2 and setting $Z_t = X_t - \int_0^t \frac{d[X,L]_s}{L_s}$, the integration by parts formula yields

$$Z_t L_t \stackrel{\text{mart}}{=} L_t X_t - \int_0^t d[X,L]_s \stackrel{\text{mart}}{=} 0 \,,$$

where $\stackrel{\text{mart}}{=}$ refers to $\mathbb{P}$.

(ii) Note that $A_t := \int_0^t \frac{d\langle X,L\rangle_s}{L_{s-}}$ is a predictable process with bounded variation. The process $Y_t = X_t - \int_0^t \frac{d\langle X,L\rangle_s}{L_{s-}}$ is a $\mathbb{Q}$-local martingale if and only if the process YL is a $\mathbb{P}$-local martingale. The integration by parts formula leads to

$$\begin{aligned} Y_t L_t &= Y_0 L_0 + \int_0^t Y_{s-} dL_s + \int_0^t L_{s-} dY_s + [Y,L]_t \\ &\overset{\text{mart}}{=} -\int_0^t L_{s-} dA_s + [Y,L]_t = -\int_0^t L_{s-} dA_s + [X,L]_t - [A,L]_t \\ &\overset{\text{mart}}{=} -\langle X,L\rangle_t + [X,L]_t - \Sigma_{s\leq t} \Delta A_s \Delta L_s \,. \end{aligned}$$

The difference $-\langle X,L\rangle_t + [X,L]_t$ and the sum $\sum_{s\leq t} \Delta A_s \Delta L_s = \int_0^t \Delta A_s dL_s$ are local martingales. (Note that ΔA is predictable.) □

As a check, we prove directly the following consequence of (i) and (ii), i.e., the process

$$\int_0^t \frac{d[X,L]_s}{L_s} - \int_0^t \frac{d\langle X,L\rangle_s}{L_{s-}}$$

is a $\mathbb{Q}$-local martingale (in other terms, under $\mathbb{Q}$, the predictable compensator of $\displaystyle\int_0^t \frac{d[X,L]_s}{L_s}$ is $\displaystyle\int_0^t \frac{d\langle X,L\rangle_s}{L_{s-}}$). Let h be any predictable bounded process. Let us verify that

$$\mathbb{E}_{\mathbb{Q}}\left(\int_0^t h_s \frac{d[X,L]_s}{L_s}\right) = \mathbb{E}_{\mathbb{Q}}\left(\int_0^t h_s \frac{d\langle X,L\rangle_s}{L_{s-}}\right).$$

The left-hand side is equal to:

$$\mathbb{E}_{\mathbb{P}}\left(L_t \int_0^t h_s \frac{d[X,L]_s}{L_s}\right) = \mathbb{E}_{\mathbb{P}}\left(\int_0^t h_s d[X,L]_s\right).$$

From (9.3.5), the right-hand side is

$$\mathbb{E}_{\mathbb{P}}\left(L_t \int_0^t h_s \frac{d\langle X,L\rangle_s}{L_{s-}}\right) = \mathbb{E}_{\mathbb{P}}\left(\int_0^t h_s d\langle X,L\rangle_s\right),$$

hence, the equality between the left- and the right-hand sides. The following exercise amplifies the previous argument:

Exercise 9.4.4.2 Let A be an increasing process, with $\mathbb{Q}$ and $\mathbb{P}$ as above. Prove that the $\mathbb{Q}$-compensator of $\int_0^t \frac{dA_s}{L_s}$ is $\int_0^t \frac{dA_s^{p,P}}{L_{s-}}$, where $A^{p,P}$ is the $\mathbb{P}$-compensator of A.
Hint: Let H be a positive predictable process. Then,

$$\mathbb{E}_{\mathbb{Q}}\left(\int_0^t H_s \frac{dA_s}{L_s}\right) = \mathbb{E}_{\mathbb{P}}\left(L_t \int_0^t H_s \frac{dA_s}{L_s}\right)$$
$$= \mathbb{E}_{\mathbb{P}}\left(\int_0^t H_s dA_s\right) = \mathbb{E}_{\mathbb{P}}\left(\int_0^t H_s dA_s^{p,P}\right).$$

◁

Comment 9.4.4.3 It is worth remarking that any $\mathbb{Q}$-local martingale may be obtained as a Girsanov transform of a $\mathbb{P}$-local martingale. Indeed, let M be a $\mathbb{Q}$-local martingale such that $M_0 = 0$. Writing $M = N/L$ where $N = ML$ is a $\mathbb{P}$-local martingale, we obtain

$$\frac{N_t}{L_t} = \int_0^t \frac{1}{L_{s-}} dN_s + \int_0^t N_{s-} d(1/L_s) + [N, L^{-1}]_t .$$

We now assume for simplicity that L and N are continuous and leave the general case to the reader. From the three equalities

$$d\left(\frac{1}{L_t}\right) = -\frac{dL_t}{L_t^2} + \frac{d\langle L\rangle_t}{L_t^3},$$

$$\int_0^t N_s d(1/L_s) = -\int_0^t \frac{N_s}{L_s^2} dL_s + \int_0^t \frac{N_s}{L_s^3} d\langle L\rangle_s ,$$

$$\langle N, L^{-1}\rangle_t = -\int_0^t \frac{1}{L_s^2} d\langle N, L\rangle_s ,$$

we obtain

$$M_t = \int_0^t \frac{1}{L_s}\left(dN_s - \frac{d\langle N, L\rangle_s}{L_s}\right) - \int_0^t \frac{N_s}{L_s^2}\left(dL_s - \frac{d\langle L\rangle_s}{L_s}\right).$$

This last formula can be written as

$$M_t = X_t - \int_0^t \frac{d\langle X, L\rangle_s}{L_s}$$

where X is the $\mathbb{P}$-local martingale

$$X_t = \int_0^t \frac{L_s dN_s - N_s dL_s}{L_s^2} .$$

It follows that M is obtained from N through a Girsanov's transformation.

Example 9.4.4.4 Let $\mathbb{Q}$ be locally equivalent to the Wiener measure (on $\mathcal{F}_t$) and let X be the coordinate process. The process $\tilde{X}_t := X_t - \int_0^t \frac{d\langle X,L\rangle_s}{L_s}$ is a Brownian motion which generates the $(\mathbb{Q}, \mathbf{F})$-local martingales: any $(\mathbb{Q}, \mathbf{F})$-local martingale can be represented as a stochastic integral w.r.t. $\widetilde{X}$. This fact is widely used in finance: the model is often written under the historical probability measure, and the representation theorem is written under the risk-neutral probability measure. This discussion is related to that in Subsection 5.8.1 about strong and weak Brownian filtrations.

Corollary 9.4.4.5 *For ζ a $\mathbb{P}$-local martingale, if $\mathbb{Q} = \mathcal{E}(\zeta)\,\mathbb{P}$, and if X is a $\mathbb{P}$-local martingale such that $\langle X, \zeta \rangle$ exists, then $\widetilde{X} = X - \langle X, \zeta \rangle$ is a $\mathbb{Q}$-local martingale.*

PROOF: Indeed, in that case, $L_t = \mathcal{E}(\zeta)_t$ satisfies $dL_t = L_{t-} d\zeta_t$, hence $d\langle X, L \rangle_t = L_{t-} d\langle X, \zeta \rangle_t$. □

Example 9.4.4.6 Let W be a Brownian motion and let $M_t = N_t - \int_0^t \lambda(s)ds$ be the compensated martingale associated with an inhomogeneous Poisson process. Let L be the Radon-Nikodým density defined as the solution to

$$dL_t = L_{t-}\left[\psi_t dW_t + \gamma_t dM_t\right], L_0 = 1$$

where (ψ, γ) are predictable processes, and $\mathbb{Q}|_{\mathcal{F}_t} = L_t\,\mathbb{P}|_{\mathcal{F}_t}$. We assume here that L is a strictly positive martingale. The process W^ψ defined as

$$W_t^\psi := W_t - \int_0^t \psi_s \, ds$$

is a $\mathbb{Q}$-Brownian motion and

$$M_t^\gamma := M_t - \int_0^t \lambda(s)\gamma_s ds = N_t - \int_0^t \lambda(s)(1+\gamma_s)ds$$

is a $\mathbb{Q}$-local martingale.

9.5 Existence and Uniqueness of the e.m.m.

Let S be a càdlàg adapted process on a filtered probability space $(\Omega, \mathcal{F}, \mathbf{F}, \mathbb{P})$. This process represents the price of some financial asset. We restrict our study to the case of a finite horizon T, and we assume that there exists another asset, the savings account with constant value (in other words, the interest rate is null).

9.5.1 Predictable Representation Property

We recall the definition given in Chapter 1:

Definition 9.5.1.1 *The process $(S_t, 0 \leq t \leq T)$ admits an $\mathbf{F}$-equivalent martingale measure (e.m.m.) if there exists a probability measure $\mathbb{Q}$ on $\mathcal{F}_T$, equivalent to $\mathbb{P}$, such that the process $(S_t, t \leq T)$ is a $\mathbb{Q}$-$\mathbf{F}$-local martingale.*

We denote by $\mathcal{M}_P(S)$ the set of e.m.m's, i.e., the set of probabilities equivalent to $\mathbb{P}$ such that S is a $\mathbb{Q}$-local martingale. Of particular importance, is the case when $\mathcal{M}_P(S)$ is a singleton.

Definition 9.5.1.2 *Let S be an $(\mathbf{F},\mathbb{P})$-semi-martingale and $\mathbb{Q} \in \mathcal{M}_P(S)$. The process S enjoys the $\mathbf{F}$-predictable representation property (PRP) under $\mathbb{Q}$ if any $(\mathbb{Q},\mathbf{F})$-local martingale M admits a representation of the form*

$$M_t = m + \int_0^t m_s dS_s,\ t \le T$$

where m is a constant and $(m_s, s \le T)$ is $\mathbf{F}$-predictable.

We do not assume that $\mathbf{F}$ is the natural filtration of S.

Comment 9.5.1.3 In this section, our aim is to give some important results on e.m.m's. In finance, the existence of an e.m.m. is linked with the property of absence of arbitrage. We do not give a full discussion of this link for which the reader can refer to the papers of Stricker [809, 808] and Kabanov [500]. The books of Björk [102] and Steele [806] contain useful comments and the book of Delbaen and Schachermayer [236] is an exhaustive presentation of arbitrage theory.

9.5.2 Necessary Conditions for Existence

For $\mathbb{Q} \in \mathcal{M}_P(S)$ we denote by $(\Lambda_t, t \le T)$ the right-continuous version of the restriction to the σ-algebra $\mathcal{F}_t$ of the Radon-Nikodým density of $\mathbb{P}|_{\mathcal{F}_t}$ with respect to $\mathbb{Q}|_{\mathcal{F}_t}$ defined as $\mathbb{P}|_{\mathcal{F}_t} = \Lambda_t\, \mathbb{Q}|_{\mathcal{F}_t}$. We prove that, if S is a continuous process and if $\mathcal{M}_P(S)$ is non-empty, then S is a semi-martingale and that its finite variation part is absolutely continuous with respect to the bracket of the semi-martingale S, this last property being called the **structure condition**.

Theorem 9.5.2.1 ***Structure condition:*** *If S is a continuous process and if the set $\mathcal{M}_P(S)$ is non-empty, then S is a $\mathbb{P}$-semi-martingale with decomposition $S = M + A$ such that the finite variation process A is absolutely continuous with respect to the bracket $\langle S \rangle$, i.e., there exists a process H such that $A_t = \int_0^t H_s d\langle S\rangle_s$. Moreover, the integrability condition*

$$\int_0^t H_s^2 d\langle S\rangle_s < \infty \tag{9.5.1}$$

holds.

PROOF: Let S be a continuous process and $\mathbb{Q} \in \mathcal{M}_P(S)$. The process Λ, defined as $\mathbb{P}|_{\mathcal{F}_t} = \Lambda_t\, \mathbb{Q}|_{\mathcal{F}_t}$, is a strictly positive $\mathbb{Q}$-martingale. The process S is a $\mathbb{Q}$-local martingale (by definition of $\mathbb{Q}$), hence, from Girsanov's theorem, the process $\widetilde{S}_t := S_t - \displaystyle\int_0^t \frac{d\langle S, \Lambda\rangle_s}{\Lambda_{s-}}$ is a $(\mathbb{P},\mathbf{F})$-local martingale. The continuity of S implies that $\langle S, \Lambda\rangle = \langle S, \Lambda^c\rangle$. Therefore, $S_t = \widetilde{S}_t + A_t$ where the processes

$\widetilde{S}$ and A are continuous ($A_t = \int_0^t \frac{d\langle S, \Lambda^c\rangle_s}{\Lambda_{s-}}$), and $\widetilde{S}$ is a $\mathbb{P}$-local martingale. Hence, S is a $\mathbb{P}$-semi-martingale. Now, from the Kunita-Watanabe inequality, for any process φ such that $\int_0^T {\varphi_s}^2 d\langle S\rangle_s < \infty$,

$$\int_0^T |\varphi_s|\,|d\langle S, \Lambda^c\rangle_s| \leq \left(\int_0^T {\varphi_s}^2 d\langle S\rangle_s\right)^{1/2} \langle \Lambda^c\rangle_T^{1/2}$$

hence the absolute continuity of dA_s with respect to $d\langle S\rangle_s$ and the existence of H such that $dA_s = H_s d\langle S\rangle_s$. From a similar argument,

$$\int_0^T |\varphi_s|\,|H_s||d\langle S, \Lambda^c\rangle_s| \leq C\left(\int_0^T {\varphi_s}^2 d\langle S\rangle_s\right)^{1/2}$$

which implies that $\int_0^T H_s^2 d\langle S, \Lambda^c\rangle_s < \infty$. □

Comments 9.5.2.2 (a) In particular, if the continuous process S is not a semi-martingale, there does not exist an e.m.m. $\mathbb{Q}$ such that S is a $\mathbb{Q}$-martingale. This remark is important in finance: it is well known that a fractional Brownian motion (except for a BM) is not a semi-martingale (however, see Cheridito [165] where it is proved that a sum of a BM and a fBM may be a semi-martingale). Hence, the important problem: are there arbitrage opportunities in a fractional Brownian motion framework? The difficulty relies on a definition of a stochastic integral w.r.t. fractional Brownian motion. We do not give a complete list of references, and we mention only the paper of Coutin [201]. In [737], Rogers constructs an arbitrage, and in [202], Coviello and Russo avoid arbitrages using forward integrals. The full discussion is still open.

(b) The absolute continuity condition for the bounded variation part with respect to the predictable bracket of S is a cornerstone in the papers of Delbaen et al. [231, 232] and Mania and Tevzadze [619] while studying the closedness in L^2 of the set of stochastic integrals, and the existence of a minimal entropy martingale measure.

Example 9.5.2.3 Example of a Semi-martingale S of the Form $M+A$ with $A_t = \int_0^t \lambda_s d\langle S\rangle_s$ Such That $\mathcal{M}_P(S)$ is Empty. Let us consider the process:

$$S_t := B_t + \lambda \int_0^t \frac{ds}{s} B_s\,.$$

The process $H_s = \frac{B_s}{s}$ is not square integrable (see Jeulin [493], Jeulin and Yor [496]): $\int_0^t \frac{B_s^2}{s^2} ds = \infty$, hence the set $\mathcal{M}_P(S)$ is empty. Delbaen and Schachermayer [234] define and obtain some examples of immediate arbitrages that is arbitrages which occur immediately after 0 (in the above example, the lack of integrability of H_s is in the neighborhood of 0).

We now assume that S is locally square integrable and that $\mathcal{M}_P(S)$ is not empty, and we decompose the $\mathbb{Q}$-local martingale S, where $\mathbb{Q} \in \mathcal{M}_P(S)$, into the sum of its continuous and purely discontinuous parts:

$$S_t = S_t^c + S_t^d .$$

Theorem 9.5.2.4 *Suppose that S enjoys the PRP under $\mathbb{Q}$ and that $\langle S \rangle$ is continuous. Then,*

(i) $\Delta S_t = m_t \mathbb{1}_{\{\Delta S_t \neq 0\}}$, *where m is predictable,*
(ii) *the random measures $\langle S^c \rangle$ and $\langle S^d \rangle$ are mutually singular.*

PROOF: The process $[S^d]_t - \langle S^d \rangle_t$ is a $\mathbb{Q}$-local martingale. From the PRP of S under $\mathbb{Q}$, we have

$$[S^d]_t - \langle S^d \rangle_t = \int_0^t m_s dS_s \,, \tag{9.5.2}$$

for some predictable process m. Since $\langle S \rangle$ is continuous, $\langle S^d \rangle$ is continuous; recall that $[S^d]_t = \sum_{s \leq t} (\Delta S_s)^2$, hence, we get $\Delta S_t = m_t \mathbb{1}_{\{\Delta S_t \neq 0\}}$.

Moreover, from the PRP assumption,

$$S_t^c = \int_0^t \varphi_s^{(c)} dS_s, \; S_t^d = \int_0^t \varphi_s^{(d)} dS_s$$

for some predictable processes $\varphi^{(c)}$ and $\varphi^{(d)}$. Then, since $\langle S^c, S^d \rangle = 0$, we obtain

$$\int_0^t \varphi_s^{(c)} \varphi_s^{(d)} d\langle S \rangle_s = 0$$

Therefore,

$$0 = \varphi_s^{(c)} \varphi_s^{(d)} \, d\langle S \rangle_s \,,$$

or, equivalently

$$(\varphi_s^{(c)})^2 (\varphi_s^{(d)})^2 = 0, \quad d\langle S \rangle \, a.s..$$

Then, using the fact that

$$d\langle S^c \rangle_s = (\varphi_s^{(c)})^2 \, d\langle S \rangle_s, \; d\langle S^d \rangle_s = (\varphi_s^{(d)})^2 d\langle S \rangle_s$$

the property (ii) follows. □

Note that under the hypotheses of the above theorem, S is a PCSM (see Definition 9.4.1.1).

Sufficient Condition for Existence

Theorem 9.5.2.5 *Let S be an $(\mathbf{F}, \mathbb{Q})$-local martingale and suppose that:*

(i) *every continuous* $\mathbf{F}$*-martingale is a stochastic integral with respect to* S^c,
(ii) *every discontinuous* $\mathbf{F}$*-martingale is a stochastic integral with respect to* S^d,
(iii) *the random measures* $\langle S^c \rangle$ *and* $\langle S^d \rangle$ *have disjoint supports.*

Then S enjoys the PRP.

PROOF: Let $M = M^c + M^d$ be a martingale. It admits a representation of the form

$$M_t = m + \int_0^t m_s^{(c)} dS_s^c + \int_0^t m_s^{(d)} dS_s^d \,.$$

Let Γ_c (resp. Γ_d) be the support of the measure $d\langle S^c \rangle_t$ (resp. $d\langle S^d \rangle_t$). Then

$$S_t^c = \int_0^t \mathbb{1}_{\Gamma_c}(s) dS_s, \; S_t^d = \int_0^t \mathbb{1}_{\Gamma_d}(s) dS_s \,.$$

Indeed

$$\begin{aligned}
&\mathbb{E}\left(S_t^c - \int_0^t \mathbb{1}_{\Gamma_c}(s) dS_s \right)^2 \\
&= \mathbb{E}\left(\langle S^c \rangle_t - 2\int_0^t \mathbb{1}_{\Gamma_c}(s) d\langle S^c \rangle_s + \int_0^t \mathbb{1}_{\Gamma_c}(s) d\langle S \rangle_s \right) \\
&= \mathbb{E}\left(\int_0^t \mathbb{1}_{\Gamma_c}(s) (d\langle S \rangle_s - \langle S^c \rangle_s) \right) = \mathbb{E}\left(\int_0^t \mathbb{1}_{\Gamma_c}(s) (d\langle S^d \rangle_s \right) = 0 \,.
\end{aligned}$$

Hence

$$M_t = M_t^c + M_t^d = m + \int_0^t \left[m_s^{(c)} \mathbb{1}_{\Gamma_c}(s) + m_s^{(d)} \mathbb{1}_{\Gamma_d}(s) \right] dS_s \,.$$

□

We first present a case where the PRP property is not satisfied, and an example where it is.

Let B be a Brownian motion and Π the compensated martingale associated with a Poisson process N of intensity $\lambda = 1$ in the same filtration (hence B and Π are independent). We leave it to the reader to write the result for a general fixed λ or even for a deterministic function $\lambda(t)$.

(1) Let $S_t = B_t + \Pi_t$. This process does not satisfy the PRP. Note that (i) in Theorem 9.5.2.4 is satisfied, but (ii) is not: indeed, $\langle S^c \rangle_t = \langle S^d \rangle_t = t$.

(2) Let $dX_t = f(t)dB_t + g(t)d\Pi_t$ where f and g are deterministic functions such that $fg = 0$. The filtration $\mathbf{F}^X$ is the filtration generated by the processes

$(\int_0^t f(s)dB_s, t \geq 0)$ and $(\int_0^t g(s)d\Pi_s, t \geq 0)$ and is included in $\mathbf{F}^B \vee \mathbf{F}^N$. Let us set $d\widetilde{X}_t = f(t)dB_t + g(t)dN_t$. The set $\mathcal{S}$ which contains all the r.v.'s

$$\Phi_\infty = \exp\left(\int_0^\infty \varphi_s d\widetilde{X}_s - \frac{1}{2}\int_0^\infty \varphi_s^2 f^2(s)ds - \int_0^\infty ds\left(e^{\varphi_s g(s)} - 1\right)\right)$$

where φ belongs to the set Δ of deterministic functions, is total in $L^2(\mathcal{F}_\infty^X, \mathbb{P})$. Indeed, if $Y \in L^2(\mathcal{F}_\infty^X, \mathbb{P})$ is orthogonal to Φ_∞, then Y is orthogonal to $\exp(\int_0^\infty \varphi_s d\widetilde{X}_s)$ for every $\varphi \in \Delta$, hence to Y itself.

The process X enjoys the PRP w.r.t. $\mathbf{F}^X$. Indeed,

$$\begin{aligned}
&\exp\left(\int_0^t \varphi_s d\widetilde{X}_s - \frac{1}{2}\int_0^t \varphi_s^2 f^2(s)ds - \int_0^t ds\left(e^{\varphi_s g(s)} - 1\right)\right)\\
&= \exp\left(\int_0^t \varphi_s f(s)dB_s - \frac{1}{2}\int_0^t \varphi_s^2 f^2(s)ds\right)\\
&\quad \times \exp\left(\int_0^t \varphi_s g(s)dN_s - \lambda\int_0^t ds\left(e^{\varphi_s g(s)} - 1\right)\right) = \mathcal{E}_t^{(1)}\mathcal{E}_t^{(2)}
\end{aligned}$$

and the two martingales $\mathcal{E}^{(i)}$ are orthogonal. It follows that

$$\begin{aligned}
\mathcal{E}_t^{(1)}\mathcal{E}_t^{(2)} &= 1 + \int_0^t \mathcal{E}_s^{(1)}\mathcal{E}_{s-}^{(2)}\,\varphi_s f(s)dB_s + \int_0^t \mathcal{E}_s^{(1)}\mathcal{E}_{s-}^{(2)}\,\varphi_s g(s)d\Pi_s\\
&= 1 + \int_0^t \mathcal{E}_s^{(1)}\mathcal{E}_{s-}^{(2)}\,\varphi_s dX_s\,,
\end{aligned}$$

since $fg = 0$.

9.5.3 Uniqueness Property

Theorem 9.5.3.1 *Let S be a continuous $\mathbf{F}$-adapted process. Suppose that $\mathcal{M}_P(S)$ is not empty and let $\mathbb{Q} \in \mathcal{M}_P(S)$. The following two properties are equivalent:*

(i) *PRP holds under $\mathbb{Q}$ with respect to S,*
(ii) *The set $\mathcal{M}_P(S)$ is a singleton.*

PROOF: (i) implies (ii): Let $\mathbb{Q}$ and $\mathbb{Q}^*$ belong to $\mathcal{M}_P(S)$ and let q be the Radon-Nikodým density of $\mathbb{Q}^*$ w.r.t. $\mathbb{Q}$: $\mathbb{Q}^*|_{\mathcal{F}_t} = q_t\mathbb{Q}|_{\mathcal{F}_t}$. The canonical decomposition of S as a $\mathbb{Q}^*$-semi-martingale is

$$S_t = \widetilde{S}_t + \int_0^t \frac{d\langle S, q\rangle_s}{q_{s-}}$$

where $\widetilde{S}$ is a $\mathbb{Q}^*$-local martingale. Since S is a $\mathbb{Q}^*$-local martingale, $\langle S, q\rangle = 0$, hence the $\mathbb{Q}$-martingale q is orthogonal to S under $\mathbb{Q}$, which is not possible unless $q = 1$, because S enjoys the PRP under $\mathbb{Q}$, hence $\mathbb{Q}^* = \mathbb{Q}$.

(ii) implies (i). If (i) does not hold, there exists a locally bounded non-trivial martingale orthogonal to S, say q^0 (see Jacod and Yor [472] and Protter [727] for details). By stopping we may assume that q^0 is uniformly bounded, say by $1/2$, and $q^0_0 = 0$ and we consider $q_t = 1 + q^0_t$. The probability $\mathbb{Q}^*$ defined by $\mathbb{Q}^*|_{\mathcal{F}_t} = q_t \mathbb{Q}|_{\mathcal{F}_t}$ belongs to $\mathcal{M}_P(S)$ and is different from $\mathbb{Q}$. □

A Particular Case

We recall a result in a simple case, when the coefficient of the Brownian motion, i.e., the volatility, may vanish (see Steele [806]).

Proposition 9.5.3.2 *Let* $\mathbf{W}$ *be the Wiener measure,* X *the canonical process* $X_t(\omega) = \omega(t)$ *and* $Z_t = \int_0^t \Phi(s,\omega) dX_s$ *where* $\Phi \in L^2_{loc}(X)$. *There exists a unique probability* $\mathbb{Q}$ *equivalent to* $\mathbf{W}$ *such that* Z *is a* $(\mathbb{Q}, \mathbf{F})$*-local martingale if and only if*

$$\{(s,\omega) : \Phi(s,\omega) = 0\} \tag{9.5.3}$$

is $ds \times d\mathbf{W}$ *negligible. (Of course, then, this unique probability coincides with* $\mathbf{W}$*.)*

PROOF: ▶ Suppose that (9.5.3) does not hold. Then, for any $\lambda \neq 0$ the process

$$L^\lambda_t = \exp\left(\lambda \int_0^t \mathbb{1}_{\{\Phi(s,\omega)=0\}} dX_s - \frac{1}{2}\lambda^2 \int_0^t \mathbb{1}_{\{\Phi(s,\omega)=0\}} ds\right)$$

is a martingale which is not identically equal to 1. Let $\mathbb{Q}^\lambda|_{\mathcal{F}_t} = L^\lambda_t \mathbf{W}|_{\mathcal{F}_t}$; the process $X^\lambda_t = X_t - \lambda \int_0^t \mathbb{1}_{\{\Phi(s,\omega)=0\}} ds$ is a $(\mathbb{Q}^\lambda, \mathbf{F})$-local martingale and $Z_t = \int_0^t \Phi(s,\omega) dX^\lambda_s$, hence there is an infinity of e.m.m's.
▶ The converse study is easy. □

9.5.4 Examples

Lemma 9.5.4.1 *Let* M *and* N *be two martingales in their own respective filtrations.*

(i) *If* M *has the* $\mathbf{F}^M$*-PRP, if* N *has the* $\mathbf{F}^N$*-PRP, and if* M *and* N *are* $\mathbf{F}^M \vee \mathbf{F}^N$*-martingales, then they are orthogonal in* $\mathbf{F}^M \vee \mathbf{F}^N$ *if and only if they are independent.*

(ii) *If* M *has the* $\mathbf{F}^M$*-PRP and* N *the* $\mathbf{F}^N$*-PRP, and if* M *and* N *are independent, then the pair* (M,N) *enjoys the* $\mathbf{F}^M \vee \mathbf{F}^N$*-PRP.*

PROOF: (i) Take $\Phi \in L^2(\mathcal{F}^M_\infty)$ and $\Psi \in L^2(\mathcal{F}^N_\infty)$. Then,

$$\Phi = \mathbb{E}(\Phi) + \int_0^\infty \varphi_s dM_s, \quad \Psi = \mathbb{E}(\Psi) + \int_0^\infty \psi_s dN_s .$$

Now, we consider these stochastic integrals as written in $\mathbf{F}^M \vee \mathbf{F}^N$ and from the orthogonality hypothesis, we deduce

$$\mathbb{E}(\Phi\Psi) = \mathbb{E}(\Phi)\mathbb{E}(\Psi)\,.$$

(ii) Under the hypotheses, we have, with the previous notation

$$\Phi\Psi = \mathbb{E}(\Phi\Psi) + \int_0^\infty \Psi_{s-}\varphi_s dM_s + \int_0^\infty \Phi_{s-}\psi_s dN_s$$

where

$$\begin{aligned}\Psi_s &= \mathbb{E}(\Psi|\mathcal{F}_s^N) = \mathbb{E}(\Psi|\mathcal{F}_s^N \vee \mathcal{F}_\infty^M)\,,\\ \Phi_s &= \mathbb{E}(\Phi|\mathcal{F}_s^M) = \mathbb{E}(\Phi|\mathcal{F}_s^M \vee \mathcal{F}_\infty^N)\,.\end{aligned}$$

The general representation result follows from the totality of the products $\Phi\Psi$ in $L^2(\mathcal{F}_\infty^M \mathcal{F}_\infty^N)$. □

While using this result, it is important to prove that M and N are martingales in the "large" filtration $\mathbf{F}^M \vee \mathbf{F}^N$. One may recall that it is possible to construct a Poisson process adapted to a Brownian filtration $\mathbf{F}^B$, i.e., a Poisson process N such that $\mathbf{F}^N \subset \mathbf{F}^B$ (see Jeulin [494]), so that a fortiori it is not true that a Poisson process and a Brownian motion constructed on the same probability space are independent.

Theorem 9.5.4.2 *Suppose that the SDE*

$$dS_t = b(S_t)dt + \sigma(S_t)dB_t$$

admits a unique weak solution. Then, any $\mathbf{F}^S$-local martingale can be written as $M_t = m + \int_0^t m_s dS_s^{\mathrm{mar}}$ where S^{mar} is the martingale part of S, i.e.,

$$S_t^{\mathrm{mar}} = S_t - S_0 - \int_0^t b(S_u)du = \int_0^t \sigma(S_u)dB_u\,.$$

PROOF: See Hunt and Kennedy [455] and Jacod and Yor [472]. □

9.6 Self-financing Strategies and Integration by Parts

We present a simple, but useful application of the integration by parts formula for the characterization of self-financing strategies in finance.

9.6.1 The Model

Assume that there is a financial market where $S_t^1, S_t^2, \ldots, S_t^k$ represent values at time t of k assets. Here, S^1, $S^2, \ldots, S^k$ are semi-martingales on some probability space $(\Omega, \mathcal{F}, \mathbf{F}, \mathbb{P})$, satisfying the usual conditions. We assume that S^1 is stricly positive. We emphasize that a priori, there is no riskless asset being traded in the market. Let $\pi = (\pi^1, \pi^2, \ldots, \pi^k)$ be a trading strategy; in particular, the processes π^i are $\mathbf{F}$-predictable. The component π_t^i represents the number of units of the i-th asset held in the portfolio at time t. Then the wealth $V_t(\pi)$ of the trading strategy $\pi = (\pi^1, \pi^2, \ldots, \pi^k)$ equals

$$V_t(\pi) = \sum_{i=1}^{k} \pi_t^i S_t^i \tag{9.6.1}$$

and we say that π is a *self-financing strategy* if

$$V_t(\pi) = V_0(\pi) + \sum_{i=1}^{k} \int_0^t \pi_u^i \, dS_u^i \,. \tag{9.6.2}$$

By combining the last two formulae, we obtain

$$dV_t(\pi) = \left(V_t(\pi) - \sum_{i=2}^{k} \pi_t^i S_t^i \right) (S_t^1)^{-1} \, dS_t^1 + \sum_{i=2}^{k} \pi_t^i \, dS_t^i .$$

The latter representation shows that the wealth process only depends on $k-1$ components of π.

9.6.2 Self-financing Strategies and Change of Numéraire

Choosing S^1 as a numéraire, and denoting

$$V_t^1(\pi) = V_t(\pi)(S_t^1)^{-1}, \; S_t^{i,1} = S_t^i (S_t^1)^{-1},$$

we get the following well-known result which proves that the self-financing property does not depend on the choice of the numéraire,

Lemma 9.6.2.1 (i) *Let $\pi = (\pi^1, \pi^2, \ldots, \pi^k)$ be a self-financing strategy. Then we have*

$$V_t^1(\pi) = V_0^1(\pi) + \sum_{i=2}^{k} \int_0^t \pi_u^i \, dS_u^{i,1}, \quad \forall \, t \in [0, T]. \tag{9.6.3}$$

(ii) *Conversely, let X be an $\mathcal{F}_T$-measurable random variable, and assume that there exist $x \in \mathbb{R}$ and predictable processes π^i, $i = 2, \ldots, k$ such that*

$$X = S_T^1 \left(x + \sum_{i=2}^{k} \int_0^t \pi_u^i \, dS_u^{i,1} \right) .$$

Then there exists a predictable process π^1 such that the strategy π is self-financing and replicates X.

PROOF: We give the proof for $k = 2$.

(i) Let $V = (V_t(\pi), t \geq 0)$ be the value of a self-financing strategy π, and $V^1 = V/S^1$ the value of this strategy in the numéraire S^1. From the integration by parts formula

$$d(V_t^1) = V_{t-} d(S_t^1)^{-1} + (S_{t-}^1)^{-1} dV_t + d[(S^1)^{-1}, V]_t \,.$$

From the self-financing condition

$$\begin{aligned} d(V_t^1) &= \pi_t^1 S_{t-}^1 d(S_t^1)^{-1} + \pi_t^2 S_{t-}^2 d(S_t^1)^{-1} + (S_{t-}^1)^{-1} \pi_t^1 dS_t^1 \\ &\quad + \pi_t^2 (S_{t-}^1)^{-1} dS_t^2 + \pi_t^1 d[(S^1)^{-1}, S^1]_t + \pi_t^2 d[(S^1)^{-1}, S^2]_t \\ &= \pi_t^1 \left(S_{t-}^1 d(S^1)_t^{-1} + (S^1)_{t-}^{-1} dS_t^1 + d[(S^1)^{-1}, S^1]_t \right) \\ &\quad + \pi_t^2 \left(S_{t-}^2 d(S_t^1)^{-1} + (S_{t-}^1)^{-1} dS_{t-}^2 + d[(S^1)^{-1}, S^2]_t \right) . \end{aligned}$$

We now note that

$$S_{t-}^1 d(S_t^1)^{-1} + (S_{t-}^1)^{-1} dS_t^1 + d[(S^1)^{-1}, (S^1)]_t = d(S^1 (S^1)^{-1})_t = 0$$

and

$$S_{t-}^2 d(S_t^1)^{-1} + (S_{t-}^1)^{-1} dS_t^2 + d[(S^1)^{-1}, S^2]_t = d((S_t^1)^{-1} S_t^2)$$

hence,

$$dV_t^1 = \pi_t^2 dS_t^{2,1} \,.$$

We now prove part (ii). We define

$$V_t^1 = x + \sum_{i=2}^{k} \int_0^t \pi_u^i \, dS_u^{i,1}, \tag{9.6.4}$$

and we set

$$\pi_t^1 = V_t^1 - \sum_{i=2}^{k} \pi_t^i S_t^{i,1} = (S_t^1)^{-1} \left(V_t - \sum_{i=2}^{k} \pi_t^i S_t^i \right),$$

where $V_t = V_t^1 S_t^1$. From the definition of V, we have $dV_t^1 = \sum_{i=2}^k \pi_t^i \, dS_t^{i,1}$ and thus

$$\begin{aligned} dV_t &= d(V_t^1 S_t^1) = V_{t-}^1 dS_t^1 + S_{t-}^1 dV_t^1 + d[S^1, V^1]_t \\ &= V_{t-}^1 dS_t^1 + \sum_{i=2}^{k} \pi_t^i (S_{t-}^1 \, dS_t^{i,1} + d[S^1, S^{i,1}]_t). \end{aligned}$$

From the obvious equality

$$dS_t^i = d(S_t^{i,1} S_t^1) = S_{t-}^{i,1} dS_t^1 + S_{t-}^1 dS_t^{i,1} + d[S^1, S^{i,1}]_t,$$

it follows that

$$dV_t = V^1_{t-} dS^1_t + \sum_{i=2}^{k} \pi^i_t \left(dS^i_t - S^{i,1}_{t-} dS^1_t \right)$$

$$= \left(V^1_{t-} - \sum_{i=2}^{k} \pi^i_t S^{i,1}_{t-} \right) dS^1_t + \sum_{i=2}^{k} \pi^i_t \, dS^i_t \, .$$

The needed equality $dV_t = \sum_{i=1}^k \pi^i_t \, dS^i_t$ holds if

$$\pi^1_t = V^1_t - \sum_{i=2}^{k} \pi^i_t S^{i,1}_t = V^1_{t-} - \sum_{i=2}^{k} \pi^i_t S^{i,1}_{t-} ,$$

i.e., if $\Delta V^1_t = \sum_{i=2}^k \pi^i_t \Delta S^{i,1}_t$, which is the case from the definition (9.6.4) of V^1_t. Note also that the process π^1 is indeed predictable. □

For more examples, in particular for a study of constrained strategies, see Bielecki et al [92, 93].

9.7 Valuation in an Incomplete Market

We study a market in which both a savings account with constant interest rate r, and d risky assets $(S^i, i = 1, \dots, d)$ are traded. We shall denote for simplicity by $\pi_t S_t$ the amount of money invested in the risky assets (that is, we set $\pi_t S_t = \sum_{i=1}^d \pi^i_t S^i_t$) and $\widetilde{S}_t = S_t e^{-rt}$ is the discounted price. We assume here that the market is incomplete and arbitrage free. More precisely, we assume that the set $\mathcal{M}_{\mathbb{P}}(\widetilde{S})$ of e.m.m's is not empty and not reduced to a singleton. By definition of an incomplete market, if H is a contingent claim, it is in general not possible to construct a hedging strategy for H, i.e., a self-financing portfolio with terminal value equal to H (note that contingent claims of the form $H = e^{rT}\left(x + \int_0^T \pi_s d\widetilde{S}_s\right)$ such as for example $c + \kappa S_T$, are hedgeable). We recall that some extra conditions are required to avoid doubling strategies, the most common one being that the value of the strategy is bounded below.

The set

$$\{\mathbb{E}_{\mathbb{Q}}(He^{-rT}), \mathbb{Q} \in \mathcal{M}_{\mathbb{P}}(\widetilde{S})\}$$

is called the set of **viable** prices. If the asset H is traded at price $\mathbb{E}_{\mathbb{Q}}(He^{-rT})$ for $\mathbb{Q} \in \mathcal{M}_{\mathbb{P}}(\widetilde{S})$, this does not induce arbitrage opportunities. If the asset H is traded at a price which is not in the set of viable prices, this induces an arbitrage opportunity. We present briefly some methods of evaluating contingent claims in an incomplete market.

One way is to find a hedging strategy which replicates as best as possible the contingent claim. The dual approach is to choose a particular equivalent martingale measure, e.g., the Föllmer-Schweizer minimal probability measure or the minimal entropy measure. In that setting, one gives an arbitrary value to the market price of risk.

Another method is to relate the price of the contingent claim to a utility function approach. We now present briefly these three approaches.

9.7.1 Replication Criteria

Super-replication

The **super-replication price** of H is the smallest initial wealth v such that there exists a self-financing strategy π which super-hedges the contingent claim, i.e., in the case $r = 0$, v satisfies

$$v = \inf\{w \,:\, \exists \pi,\ w + \int_0^T \pi_s dS_s \geq H\,\}.$$

(See $\rightarrowtail$ Subsection 10.5.3 for an example). The super-replication price v is also called the **selling price**. It is proved in El Karoui and Quenez [307] that this super-replication price is the supremum of the viable prices, i.e.,

$$\sup_{\mathbb{Q}\in\mathcal{M}_{\mathbb{P}}(\widetilde{S})} \mathbb{E}_{\mathbb{Q}}(e^{-rT}H)\,.$$

See also Kramkov [543]. However, from the practitioner's viewpoint, this price is too large (see $\rightarrowtail$ Subsection 10.5.2 for an example). In the case of stochastic volatility (see Frey and Sin [360]) the selling price of a European option is often infinite, except if the volatility is bounded (see El Karoui et al. [301]). This method can be applied in models where transaction costs are taken into account, with the same drawback: for example, the super-replication price for a call is the price of the underlying asset (see Soner et al. [796], Hubalek and Schachermayer [449]).

Quadratic Hedging

Föllmer and Sonderman [352] suggest minimizing the quadratic error under the historical probability measure, i.e., finding v and π which minimize

$$\mathbb{E}_{\mathbb{P}}((H - V_T^{v,\pi})^2)$$

over initial wealth v and self-financing strategies π, where $V_T^{v,\pi} = x + \int_0^T \pi_s dS_s$ is the terminal wealth associated to the strategy π (we assumed that the interest rate is null). The solution is the L^2-projection of H on the vector space $x + \int_0^T \pi_s dS_s$. (One needs to find some conditions which ensure that this space is closed, see Delbaen et al. [232].)

It can be proved that there exists, at least in markets with continuous asset prices, a Radon-Nikodým density L^{qh} which does not depend on the choice of H such that $v = \mathbb{E}_{\mathbb{P}}(e^{-rT} H L^{qh})$. Despite this explicit solution to the original question, again practitioners dislike this criterion, which gives the same weight for losses and gains of V relative to H. There are also asymmetric criteria, but then the mathematical theory is more involved and results are less explicit. We refer to the book of Föllmer and Schied [350] for a complete study.

In the case of weather derivatives, or in a default risk setting, an interesting question is the measurability criteria for the strategies. Indeed, even if the market is incomplete, it is possible to include the information about the weather or on the default in the choice of the portfolio (see Bielecki et al. [89] in the case of credit risk).

9.7.2 Choice of an Equivalent Martingale Measure

This method consists in the choice of an equivalent martingale measure (or a state price density) in an appropriate way. One possibility is to minimize $\mathbb{E}_{\mathbb{P}}(f(L_T))$ over the set of Radon-Nikodým densities for a given convex function f. However, the solution depends on the choice of numéraire. We now present two different, but classical, choices of convex functions.

Minimal Entropy Measure: $f(x) = x \ln x$

Let $\mathbb{P}$ and $\mathbb{Q}$ be two equivalent probability measures. The relative entropy of $\mathbb{Q}$ w.r.t. $\mathbb{P}$ corresponds to the choice $f(x) = x \ln x$ and is

$$H(\mathbb{Q}|\mathbb{P}) = \mathbb{E}_{\mathbb{Q}}\left(\ln\left(\frac{d\mathbb{Q}}{d\mathbb{P}}\right)\right) = \mathbb{E}_{\mathbb{P}}\left(\frac{d\mathbb{Q}}{d\mathbb{P}} \ln\left(\frac{d\mathbb{Q}}{d\mathbb{P}}\right)\right) . \tag{9.7.1}$$

Let S denote the value of the asset and $\mathcal{M}_{\mathbb{P}}\left(\widetilde{S}\right)$ the set of e.m.m's. Any probability $\mathbb{Q}^*$ such that

$$i) \ \mathbb{Q}^* \in \mathcal{M}_{\mathbb{P}}\left(\widetilde{S}\right)$$
$$ii) \ \forall \mathbb{Q} \in \mathcal{M}_{\mathbb{P}}\left(\widetilde{S}\right), \ H(\mathbb{Q}^*|\mathbb{P}) \leq H(\mathbb{Q}|\mathbb{P})$$

is called a minimal entropy measure. See Choulli and Stricker [182], Delbaen et al. [231], Frittelli [362], Frittelli et al. [364], Hobson [441] and Miyahara [654] for a complete study. In Rouge and El Karoui [310], the authors provide a general framework for pricing contingent claims, using a BSDE approach.

Mean Variance Hedging: $f(x) = x^2$

Any probability $\mathbb{Q}^*$ such that

(i) $\mathbb{Q}^* \in \mathcal{M}_{\mathbb{P}}(\widetilde{S})$
(ii) $\mathbb{E}_{\mathbb{P}}((d\mathbb{Q}^*/d\mathbb{P})^2) = \inf\{\mathbb{E}_{\mathbb{P}}((d\mathbb{Q}/d\mathbb{P})^2), \mathbb{Q} \in \mathcal{M}_{\mathbb{P}}(\widetilde{S})\}$

is called a minimal measure. The existence of $\mathbb{Q}^*$ is established in Mania and Schweizer [618] for continuous processes. The existence of $\mathbb{Q}^*$ in the case of discontinuous processes can fail. In the case $r = 0$, this probability is related to the existence of a strategy π and an initial wealth v which minimize, for a given H

$$\mathbb{E}_{\mathbb{P}}\left(\left(v + \int_0^T \pi_s dS_s - H\right)^2\right).$$

See also Föllmer and Schweizer [351].

9.7.3 Indifference Prices

Another method, studied by Davis [220], is to value contingent claims for an agent endowed with a particular utility function. Related results have been obtained by a number of authors in various contexts.

A different approach was initiated by Hodges and Neuberger [444]. We briefly explain the framework of this approach. Let x be the initial endowment of an agent and U a utility function. The reservation price of the contingent claim H is defined as the infimum of h's such that

$$\sup_\pi \mathbb{E}[U(V_T^{x+h,\pi} - H)] \geq \sup_\pi \mathbb{E}[U(V_T^{x,\pi})],$$

where the supremum is taken over the admissible strategies. The agent selling the contingent claim starts with an initial endowment $x+h$. Using the strategy π, he obtains a portfolio with terminal value $V_T^{x+h,\pi}$ and he has to deliver the contingent claim H; hence, his terminal wealth is $V_T^{x+h,\pi} - H$. He agrees to sell the claim if his utility $\sup_\pi \mathbb{E}[U(V_T^{x+h,\pi} - H)]$ is greater than his utility when he does not sell the claim. The particular case where U is an exponential function is studied in detail in Rouge and El Karoui [310], whereas Delbaen et al. [231] make precise the link between this approach and the one based on entropy.

We do not discuss here these interesting approaches which are based on optimization portfolio theory (see Karatzas and Shreve [514]), but we refer the reader to the papers of Hugonnier [452], Bouchard-Denize [111], Henderson [430] and Musiela and Zariphopoulou [662], to the book of Pham [710] and to the collective book [142].

10

Mixed Processes

In this chapter, we present stochastic calculus for mixed processes (also often called jump-diffusions), i.e., loosely speaking they are processes whose dynamics are driven by a pair of processes consisting of a Brownian motion and a compound Poisson process. The jump-diffusion approach models the large number of small movements from the diffusion process part, while the jump process part captures rare large moves. It is worthwhile mentioning that a jump-diffusion process is not a diffusion, in the usual acceptance of this terminology, i.e., a Markov process with continuous paths; however, we keep this commonly used terminology of jump-diffusion. We draw the reader's attention to the fact that, in general, mixed processes in our sense are not Markov processes.

10.1 Definition

For the moment, up to Subsection 10.4.4, we only consider the restricted class of **mixed processes** of the form

$$X_t = X_0 + \int_0^t h_s \, ds + \int_0^t f_s dW_s + \int_0^t g_s dM_s \,.$$

Here, W is a Brownian motion and $(M_t = N_t - \int_0^t \lambda(s)ds, t \geq 0)$ is the compensated martingale associated with an inhomogeneous Poisson process N with deterministic intensity $\lambda(t)$. The processes M and W are independent (a justification of this independence assumption is presented in $\rightarrowtail$ Proposition 10.2.6.2) and f, g, h are predictable processes (with respect to the filtration generated by the pair (W, M)) such that, for any t

$$\int_0^t |h_s| ds < \infty, \int_0^t f_s^2 ds < \infty, \int_0^t |g_s| \lambda(s) ds < \infty \quad a.s..$$

In what follows, we write $X = (f, g, h)$ for such processes. In this general setting, a mixed process is not a Markov process. The continuous martingale

M. Jeanblanc, M. Yor, M. Chesney, *Mathematical Methods for Financial Markets*, Springer Finance, DOI 10.1007/978-1-84628-737-4_10,

part of the semi-martingale X is $\int_0^t f_s dW_s$ and the purely discontinuous martingale part is $\int_0^t g_s dM_s$.

In the particular case where the coefficients f, g, h are constant, or deterministic functions of time, the filtration generated by the process X is equal to the filtration generated by the pair (W, M) and the process X is an inhomogeneous Markov process.

We shall also consider, in ↣ Subsection 10.2.4, the case of stochastic differential equations where the coefficients f, g, h are defined in terms of a given function of time and space, i.e., $f_s = \widehat{f}(s, X_{s-})$.

The jump times of the process X are those of N, the jump size of X is $\Delta X_t = X_t - X_{t-} = g_t \Delta N_t$.

Clearly, the model could be extended to the case where W and N are multidimensional processes (see Shirakawa [789]) or to the case where M is the compensated martingale of a compound Poisson process. To keep the notation simple, we shall consider only one-dimensional processes and the case where M is the compensated martingale of a Poisson process (except in Subsection 10.4.4).

Some authors prefer to write the dynamics of a jump-diffusion process using the Poisson process N instead of the compensated martingale. We have chosen to write the dynamics in terms of martingales in order to recognize easily the martingale part. However, writing the dynamics

$$X_t = X_0 + \int_0^t \widehat{h}_s \, ds + \int_0^t f_s dW_s + \int_0^t g_s dN_s \, ,$$

with $\widehat{h}_s = h_s - g_s \lambda(s)$ is more convenient for recognizing the jump part.

10.2 Itô's Formula

Let $X = (f, g, h)$ and $Y = (\tilde{f}, \tilde{g}, \tilde{h})$ be two mixed processes driven by the same pair consisting of a Brownian motion W and an inhomogeneous Poisson process N:

$$\begin{aligned} dX_t &= h_t dt + f_t dW_t + g_t dM_t = (h_t - g_t \lambda(t))dt + f_t dW_t + g_t dN_t \, , \\ dY_t &= \tilde{h}_t dt + \tilde{f}_t dW_t + \tilde{g}_t dM_t = (\tilde{h}_t - \tilde{g}_t \lambda(t))dt + \tilde{f}_t dW_t + \tilde{g}_t dN_t \, . \end{aligned}$$

We study the stability of this class of mixed processes under multiplication and composition with a regular function.

10.2.1 Integration by Parts

The integration by parts formula (9.3.3) reads

$$d(XY) = X_- \, dY + Y_- \, dX + d[X, Y]$$

with

$$d[X,Y]_t = f_t\tilde{f}_t dt + g_t\tilde{g}_t dN_t = (f_t\tilde{f}_t + g_t\tilde{g}_t\lambda(t))dt + g_t\tilde{g}_t dM_t .$$

Hence,

$$\begin{aligned} d(X_tY_t) = &(\tilde{h}_tX_{t-} + h_tY_{t-} + f_t\tilde{f}_t + g_t\tilde{g}_t\lambda(t))\,dt \\ &+ (\tilde{f}_tX_{t-} + f_tY_{t-})dW_t + (\tilde{g}_tX_{t-} + g_tY_{t-} + g_t\tilde{g}_t)dM_t , \end{aligned}$$

from which we conclude that XY is a mixed process. Recall that, with respect to the dt and dW_t differentials, X_{t-} and Y_{t-} may be replaced by X_t and Y_t, whereas the presence of left limits with respect to dM_t is important.

10.2.2 Itô's Formula: One-dimensional Case

The following result is obtained from the general Itô formula (9.4.1). We shall write Itô's formula in two forms, the "optional" one and the "predictable" one.

Proposition 10.2.2.1 (Optional Itô Formula.) *Let W be a Brownian motion and M the compensated martingale associated with a Poisson process N with deterministic intensity λ, independent of W. Let F be a $C^{1,2}$ function defined on $\mathbb{R}^+ \times \mathbb{R}$, and X a mixed process with dynamics*

$$dX_t = h_t dt + f_t dW_t + g_t dM_t .$$

Then,

$$\boxed{\begin{aligned} F(t,X_t) = F(0,X_0) &+ \int_0^t \partial_s F(s,X_s)\,ds + \int_0^t \partial_x F(s,X_{s-})dX_s \\ &+ \frac{1}{2}\int_0^t \partial_{xx}F(s,X_s)f_s^2\,ds \\ &+ \sum_{s\le t}[F(s,X_s) - F(s,X_{s-}) - \partial_x F(s,X_{s-})\Delta X_s] . \end{aligned}}$$

Here, the sum is taken over the almost surely finite number of jump times which occur prior to t and is equal to

$$\int_0^t [F(s,X_{s-}+g_s) - F(s,X_{s-}) - \partial_x F(s,X_{s-})g_s]\,dN_s .$$

We may justify our terminology *optional Itô formula* from the fact that, in general, the last term is only optional, in contrast with the last term in the *predictable Itô formula* (10.2.1).

In the integral with respect to ds, X_{s-} or X_s can be used. Hence, writing Itô's formula in a differential form, one obtains

$$dF(t,X_t) = \partial_t F(t,X_t)\,dt + \partial_x F(t,X_{t-})dX_t + \cdots$$

or

$$dF(t,X_t) = \partial_t F(t,X_{t-})\,dt + \partial_x F(t,X_{t-})dX_t + \cdots .$$

We like to emphasize that these formulae are easy to memorize. Let us write

$$dX_t = h_t dt + f_t dW_t + g_t dM_t .$$

In order to obtain Itô's formula, we begin with "standard terms"

$$\partial_t F\,dt + \partial_x F\,dX_t + \tfrac{1}{2} f_t^2 \partial_{xx} F\,dt . \tag{S}$$

Next, we concentrate on the jumps terms both in (S) and in $F(t,X_t)$: the process $F(t,X_t)$, has a jump of size $F(t,X_t) - F(t,X_{t-})$ when the process X jumps; the standard term (S) has a jump of size $\partial_x F(t,X_{t-})\Delta X_t$ when X jumps. It suffices to add to (S) the jumps of $F(t,X_t)$ and to subtract its own jumps to obtain Itô's formula.

We now present the **predictable Itô Formula.**

Proposition 10.2.2.2 (Predictable Itô Formula.) *Let W be a Brownian motion and M the compensated martingale associated with an inhomogeneous Poisson process N with deterministic intensity λ, independent of W. Let F be a $C^{1,2}$ function defined on $\mathbb{R}^+ \times \mathbb{R}$, and*

$$dX_t = h_t dt + f_t dW_t + g_t dM_t .$$

Then,

$$\boxed{\begin{aligned} F(t,X_t) = F(0,X_0) &+ \int_0^t \partial_x F(s,X_s) f_s dW_s \\ &+ \int_0^t [F(s,X_{s-}+g_s) - F(s,X_{s-})]\,dM_s \\ &+ \int_0^t \Big[\partial_t F(s,X_s) + h_s \partial_x F(s,X_s) + \frac{1}{2} f_s^2 \partial_{xx} F(s,X_s) \\ &\qquad + \lambda(s)[F(s,X_s+g_s) - F(s,X_s) - \partial_x F(s,X_s) g_s] \Big]\,ds . \end{aligned}} \tag{10.2.1}$$

It follows that the process $(F(t,X_t), t \geq 0)$ is a local martingale if and only if

$$\begin{aligned} &\partial_t F(t,X_t) + h_t \partial_x F(t,X_t) + \frac{1}{2} f_t^2 \partial_{xx} F(t,X_t) \\ &\qquad + \lambda(t)[F(t,X_t+g_t) - F(t,X_t) - g_t \partial_x F(t,X_t)] = 0,\ dt \times d\mathbb{P}\,a.s.. \end{aligned}$$

The use of the predictable Itô formula was important in obtaining this last result. In the case where the coefficients f, g, h, λ are constant (or deterministic, or even functions of time and state), we are led to solve the PDEI (Partial Differential Equation with Integral term)

$$\partial_t F(t,x) + h\partial_x F(t,x) + \frac{1}{2} f^2 \partial_{xx} F(t,x) \\ + \lambda [F(t, x+g) - F(t,x) - g\,\partial_x F(t,x)] = 0\,, \ \forall x \in \mathbb{R}, \forall t \geq 0.$$

Exercise 10.2.2.3 Let $dX_t = h_t dt + \sigma_t dW_t + \varphi_t dM_t$ be a mixed process and $S_t = e^{X_t}$. Prove that the dynamics of S are

$$dS_t = S_{t-} \left(\left(h_t + \frac{1}{2}\sigma_t^2 + (e^{\varphi_t} - 1 - \varphi_t)\lambda(t) \right) dt + \sigma_t dW_t + (e^{\varphi_t} - 1) dM_t \right).$$

Conversely, if

$$dS_t = S_{t-}(\mu_t dt + \sigma_t dW_t + \psi_t dM_t)$$

with $\psi_t > -1$, prove that $S_t = e^{Y_t}$, where Y is a mixed process. ◁

Exercise 10.2.2.4 Let $dS_t = S_{t-}(b dt + \sigma dW_t + \phi dM_t)$ where b, σ and ϕ are constant coefficients and $\phi > -1$. Let $Y_t = (S_t)^{-1}$. Prove that

$$dY_t = -Y_{t-} \left\{ \left(b - \sigma^2 + \lambda \left(\frac{\phi}{1+\phi} - \phi \right) \right) dt + \sigma dW_t + \frac{\phi}{1+\phi} dM_t \right\}.$$

Hint: The jumps of S occur when the Poisson process N jumps, and the sizes of the jumps are $\Delta S_t = \phi S_{t-} \Delta N_t$, hence $S_t = S_{t-}(1 + \phi \Delta N_t)$. The coefficient of dM (or of dN) may also be obtained by looking at the size of the jumps of the process S^{-1}. ◁

10.2.3 Multidimensional Case

Let F be a $C^{1,2}$ function defined on $\mathbb{R}^+ \times \mathbb{R}^d$ and let $X_t = (X_i(t), i \leq d)$ be a d-dimensional mixed process with components having dynamics

$$X_i(t) = X_i(0) + \int_0^t h_i(s)\, ds + \int_0^t f_i(s) dW_s + \int_0^t g_i(s) dM_s\,,$$

where for simplicity, we have taken W and M uni-dimensional and where g_i are predictable processes and f_i, h_i are optional. Then, the optional Itô formula is

$$F(t, X_t) = F(0, X_0) + \int_0^t \partial_s F(s, X_s)\, ds + \sum_{i=1}^d \int_0^t \partial_{x_i} F(s, X_{s-}) dX_i(s) \\ + \frac{1}{2} \sum_{i,j=1}^d \int_0^t \partial_{x_i x_j} F(s, X_s) f_i(s) f_j(s)\, ds \\ + \sum_{s \leq t} [F(s, X_s) - F(s, X_{s-}) - \sum_{i=1}^d \partial_{x_i} F(s, X_{s-}) g_i(s) \Delta N_s]\,,$$

A further multidimensional generalization (with a d-dimensional Brownian motion and a k-dimensional Poisson process) could be developed.

Exercise 10.2.3.1 Prove that the predictable Itô's formula is

$$\begin{aligned} F(t,X_t) &= F(0,X_0) + \int_0^t \partial_s F(s,X_s)\,ds \\ &+ \sum_{i=1}^d \int_0^t \partial_i F(s,X_s) f_i(s) dW_s + \sum_{i=1}^d \int_0^t \partial_i F(s,X_s) h_i(s) ds \\ &+ \frac{1}{2}\sum_{i,j=1}^d \int_0^t \partial_{ij} F(s,X_s) f_i(s) f_j(s)\,ds \\ &+ \int_0^t [F(s,X_{s-}+g_s) - F(s,X_{s-})]\,dM_s \\ &+ \int_0^t [F(s,X_s+g_s) - F(s,X_s) - \sum_{i=1}^d \partial_i F(s,X(s)) g_i(s)] \lambda(s)\,ds\,, \end{aligned}$$

where $X_s = (X_s^i;\, i=1,\dots,d)$. ◁

10.2.4 Stochastic Differential Equations

Proposition 10.2.4.1 *Assume that λ is bounded and μ, σ, ϕ are functions: $\mathbb{R}^+ \times \mathbb{R} \to \mathbb{R}$ which satisfy*

$$\begin{aligned} |\mu(t,x) - \mu(t,y)| + |\sigma(t,x) - \sigma(t,y)| + |\phi(t,x) - \phi(t,y)| &\le C|x-y|, \forall t,x,y \\ |\mu(t,0)| + |\sigma(t,0)| + |\phi(t,0)| &\le C, \quad \forall t\,. \end{aligned}$$

Then the SDE

$$dX_t = \mu(t,X_t)dt + \sigma(t,X_t)dW_t + \phi(t,X_{t-})dM_t, \; X_0 = x_0 \text{ given,}$$

admits a unique (pathwise) solution. The solution is an inhomogeneous Markov process.

PROOF: See Bass [58, 57], Doléans-Dade and Meyer [256], Ikeda and Watanabe [456] or Jacod and Shiryaev [471]. □

Proposition 10.2.4.2 *Let W be a Brownian motion, $\mathbf{N}$ the random measure associated with a ν-compound Poisson process, and $\beta_i, \gamma_i, \delta_i, i=0,1$ continuous functions, with $\delta_1 > -1$. The solution of the equation*

$$\begin{aligned} dX_t = (\beta_0(t) + \beta_1(t)X_t)\,dt + (\gamma_0(t) + \gamma_1(t)X_t)\,dW_t \\ + \int (\delta_0(t,x) + \delta_1(t,x)X_{t-})\,\mathbf{N}(dt,dx) \end{aligned}$$

is

$$X_t = Z_t^{-1}\beta(t)\left(1 + \int_0^t \frac{Z_s}{\beta(s)}(\beta_0(s) - \gamma_0(s)\gamma_1(s))ds \right.$$
$$\left. + \int_0^t \frac{Z_s}{\beta(s)}\gamma_0(s)dW_s + \int_0^t \frac{Z_{s-}}{\beta(s)}\int \delta_0(s,x)\,\mathbf{N}(ds,dx)\right)$$

where

$$Z_t = \mathcal{E}(-\gamma_1 \star W)_t \exp\left(-\int_0^t \int \ln(1+\delta_1(s,x))\,\mathbf{N}(ds,dx)\right)$$

and $\beta(t) = \exp \int_0^t \beta_1(u)du$.

PROOF: The proof is obtained from Itô's calculus using the method of variation of constants. This is a particular case of affine equations of the form $dX_t = X_{t-}dY_t + dH_t$. See Gapeev [372] for a study of the solutions in this particular case. □

Comment 10.2.4.3 Despite the affine property of the coefficients, this model is not what is called in mathematical finance an affine model. We shall present the later in ↣ Subsection 10.4.4.

10.2.5 Feynman-Kac Formula

As in the Brownian motion case (see Subsection 1.5.6), we can interpret the expectations of certain functionals (see below) as the values of solutions of integro-differential equations. Consider the case where

$$dX_t = \mu(t, X_t)dt + \sigma(t, X_t)dW_t + \phi(t, X_{t-})dM_t\,.$$

The Markov property implies that, if h is a real-valued Borel function and T a fixed time, the conditional expectation $\mathbb{E}(h(X_T)|\mathcal{F}_t)$ is of the form $H(t, X_t)$. Let us assume that H is a $C^{1,2}$ function. Since $H(t, X_t)$ is a martingale, its predictable finite variation part is equal to zero. Therefore, from (10.2.1), we are led to consider the PDEI

$$\begin{cases} \dfrac{\partial H}{\partial t}(t,x) + \mu(t,x)\dfrac{\partial H}{\partial x}(t,x) + \dfrac{\sigma^2(t,x)}{2}\dfrac{\partial^2 H}{\partial x^2}(t,x) \\ \qquad + \lambda(t)\left[H(t, x+\phi(t,x)\,) - H(t,x) - \phi(t,x)\dfrac{\partial H}{\partial x}(t,x)\right] = 0\,, \\ H(T,x) = h(x)\,. \end{cases} \tag{10.2.2}$$

If H solves (10.2.2), then $H(t, X_t)$ is a local martingale. If h is bounded, we obtain

$$\mathbb{E}(h(X_T)|\mathcal{F}_t) = H(t, X_t)\,.$$

10.2.6 Predictable Representation Theorem

Let M be the compensated martingale associated with an inhomogeneous Poisson process N with *deterministic* intensity and W a Brownian motion defined on the same space. We assume that W and M are independent and we denote by $\mathbf{F}$ the canonical filtration generated by the pair (W,N) i.e., $\mathcal{F}_t = \sigma(W_s, N_s\,, s \le t) = \sigma(W_s, M_s\,, s \le t)$.

Theorem 10.2.6.1 *Let Z be a square integrable $\mathbf{F}$-martingale. Then, there exist two predictable processes (φ, ψ) such that $Z = z + \varphi\star W + \psi\star M$, with*

$$\mathbb{E}\left(\int_0^t \varphi_s^2 ds\right) < \infty\,, \quad \mathbb{E}\left(\int_0^t \psi_s^2 \lambda(s) ds\right) < \infty, \,\forall t\,.$$

If Z is a local martingale, there exist two predictable processes (φ, ψ) such that $Z = z + \varphi\star W + \psi\star M$.

PROOF: Let $F \in L^2(\mathcal{F}_\infty^W)$ and $G \in L^2(\mathcal{F}_\infty^N)$. Then, from the predictable representation theorem for a Brownian filtration (resp. for a Poisson filtration) there exist two predictable processes φ and ψ such that

$$F = \mathbb{E}(F) + \int_0^\infty \varphi_s dW_s,\; G = \mathbb{E}(G) + \int_0^\infty \psi_s dM_s,$$

with

$$\mathbb{E}\left(\int_0^\infty \varphi_s^2 ds\right) < \infty,\, \mathbb{E}\left(\int_0^\infty \psi_s^2 \lambda(s) ds\right) < \infty\,.$$

Define

$$F_t = \mathbb{E}(F|\mathcal{F}_t^W) = \mathbb{E}(F) + \int_0^t \varphi_s dW_s,\; G_t = \mathbb{E}(G|\mathcal{F}_t^N) = \mathbb{E}(G) + \int_0^t \psi_s dM_s\,.$$

These processes are $\mathbf{F}$-martingales and, since W and M are independent (see Lemma 9.5.4.1), the integration by parts formula yields

$$\begin{aligned} F_t G_t &= F_0 G_0 + \int_0^t F_{s-} dG_s + \int_0^t G_{s-} dF_s \\ &= F_0 G_0 + \int_0^t F_{s-} \psi_s dM_s + \int_0^t G_{s-} \varphi_s dW_s\,. \end{aligned}$$

The result for square integrable martingales now follows from the totality in $L^2(\mathcal{F}_\infty)$ of these products FG. The result for local martingales can be easily deduced. □

We now discuss the independence assumption:

Proposition 10.2.6.2 *Let $\mathbf{G}$ be a given filtration. If W is a $\mathbf{G}$-Brownian motion and N a $\mathbf{G}$-Poisson process, then W and N are independent.*

PROOF: Let $F \in L^2(\mathcal{F}^W_\infty)$ and $G \in L^2(\mathcal{F}^N_\infty)$. With the same notation as in the proof of the previous theorem, we obtain

$$FG = \mathbb{E}(F)\mathbb{E}(G) + \int_0^\infty F_s dG_s + \int_0^\infty G_{s-} dF_s$$

and $\mathbb{E}(FG) = \mathbb{E}(F)\mathbb{E}(G)$. Note that we have used the fact that $(F_s, s \geq 0)$ and $(G_s, s \geq 0)$ are **G**-martingales. □

Comment 10.2.6.3 See Chou and Meyer [180], Davis [219, 222], Jacod [468] and Watanabe [837] for complements. Note again that the hypothesis that W is a Brownian motion and N a Poisson process in the same filtration is very important.

10.3 Change of Probability

Throughout this section, we always assume that W is a Brownian motion and M is the compensated martingale of an inhomogeneous Poisson process with deterministic intensity $\lambda(t)$, with W and M independent. The filtration generated by W and M is denoted by **F**.

10.3.1 Exponential Local Martingales

Let γ and ψ be two **F**-predictable processes such that $\gamma_t > -1$. The solution of

$$dL_t = L_{t^-}(\psi_t dW_t + \gamma_t dM_t), \; L_0 > 0 \tag{10.3.1}$$

is the strictly positive exponential local martingale

$$\begin{aligned}
L_t &= L_0 \prod_{s \leq t} (1 + \gamma_s \Delta N_s) \, e^{-\int_0^t \gamma_s \lambda(s) ds} \; \exp\left(\int_0^t \psi_s dW_s - \frac{1}{2} \int_0^t \psi_s^2 ds \right) \\
&= L_0 \exp\left(\int_0^t \ln(1+\gamma_s) dN_s - \int_0^t \lambda(s)\gamma_s ds + \int_0^t \psi_s dW_s - \frac{1}{2} \int_0^t \psi_s^2 ds \right) \\
&= L_0 \exp\left(\int_0^t \ln(1+\gamma_s) dM_s + \int_0^t [\ln(1+\gamma_s) - \gamma_s]\lambda(s) ds \right) \\
&\qquad\qquad \times \exp\left(\int_0^t \psi_s dW_s - \frac{1}{2} \int_0^t \psi_s^2 ds \right).
\end{aligned}$$

In the last expression, the exponential term is the Doléans-Dade exponential martingale of $\psi \star W$, whereas the first one is the Doléans-Dade exponential martingale of $\gamma \star M$, therefore one can write

$$L_t = L_0 \, \mathcal{E}(\gamma \star M)_t \, \mathcal{E}(\psi \star W)_t \, .$$

Since the process L is a positive local martingale, it is a martingale if and only if $\mathbb{E}(L_t) = L_0, \forall t$. It suffices that

$$\mathbb{E}\left(\exp\left(\sup_{t\leq T}\int_0^t \ln(1+\gamma_s)dN_s\right)\right) < \infty\,,$$
$$\mathbb{E}\left(\exp\left(\sup_{t\leq T}\int_0^t \psi_s dW_s\right)\right) < \infty\,.$$

Comment 10.3.1.1 The study of uniformly integrable exponential martingales in that setting can be found in many papers, including Cherny and Shiryaev [169] and Lépingle and Mémin [580]. See also the list at the end of this book in Appendix B.

10.3.2 Girsanov's Theorem

From the representation theorem 10.2.6.1, if $\mathbb{P}$ and $\mathbb{Q}$ are equivalent probabilities, there exist two predictable processes ψ and γ, with $\gamma > -1$ such that the Radon-Nikodým density L of $\mathbb{Q}$ with respect to $\mathbb{P}$ is of the form

$$dL_t = L_{t^-}\left(\psi_t dW_t + \gamma_t dM_t\right).$$

Then, from Theorem 9.4.4.1, $\widetilde{W}$ and $\widetilde{M}$ are $\mathbb{Q}$-local martingales where

$$\widetilde{W}_t = W_t - \int_0^t \psi_s ds,\ \widetilde{M}_t = M_t - \int_0^t \lambda(s)\gamma_s ds\,.$$

If γ is deterministic, the process N is a $\mathbb{Q}$-inhomogeneous Poisson process with deterministic intensity $(\lambda(t)(1+\gamma(t)), t \geq 0)$ and $\widetilde{W}$ and $\widetilde{M}$ are $\mathbb{Q}$-independent. In the general case, $\widetilde{W}$ and $\widetilde{M}$ can fail to be independent as shown in the following example.

Example 10.3.2.1 Suppose that the intensity of the Poisson process N is equal to 1 and let $dL_t = L_{t^-}\gamma_t dM_t$, $L_0 = 1$ where γ is a non-deterministic $\mathbf{F}^W$-predictable process. Denote by $\mathbb{Q}^\gamma$ the probability $\mathbb{Q}^\gamma|_{\mathcal{F}_t} = L_t\,\mathbb{P}|_{\mathcal{F}_t}$. The filtration of M^γ is that of both M and $\int_0^t \gamma_s ds$, hence, the processes W and M^γ are not independent.

Comments 10.3.2.2 (a) Define $\mathbb{Q}|_{\mathcal{F}_T} = L_T\,\mathbb{P}|_{\mathcal{F}_T}$, where L follows (10.3.1) with $\gamma > -1$. If $\mathbb{E}(L_T) < 1$ (this implies that L is a strict local martingale), then $\mathbb{Q}$ is a positive finite measure on $\mathcal{F}_T$, but this measure is not a probability measure.

(b) If L satisfies (10.3.1) and is a martingale without the condition $\gamma > -1$, then the measure $\mathbb{Q}$ is no longer a positive measure. Nevertheless, one can define a $\mathbb{Q}$-martingale as a process Z such that ZL is a $\mathbb{P}$-martingale. See Ruiz de Chavez [748] and Begdhadi-Sakrani [65] for an extended study and Gaussel [375] for an application to finance.

Exercise 10.3.2.3 Let $(\mu, \sigma, a_i, i = 1, 2)$ be given constants with $a_i > 0$ and $X_t = \mu t + \sigma B_t + a_1 N_t^{(1)} - a_2 N_t^{(2)}$ where B is a BM and $(N^{(i)}, i = 1, 2)$ are two independent Poisson processes with respective intensities λ_i. Let Ψ be defined by $\mathbb{E}(e^{\theta X_t}) = e^{t\Psi(\theta)}$ and $\mathbb{Q}|_{\mathcal{F}_t} = e^{\theta X_t - t\Psi(\theta)}\, \mathbb{P}|_{\mathcal{F}_t}$ be a change of probability. Let r be a given number.

Characterize θ such that $(e^{-rt+\theta X_t}, t \geq 0)$ is a $\mathbb{Q}$-martingale.

This is a particular Esscher transform; see $\rightarrowtail$ Subsection 11.3.1 for a more general setting. $\triangleleft$

10.4 Mixed Processes in Finance

Bachelier [39] assumed that the dynamics of asset prices are Brownian motion with drift, and Samuelson, in order to preserve positivity of prices worked with the ordinary (or stochastic) exponential of drifted Brownian motion. Following this idea, we introduce dynamics of prices as (stochastic) exponentials of mixed processes (see Exercise 10.2.2.3). Let M be the compensated martingale associated with a Poisson process N with deterministic intensity $(\lambda(t), t \geq 0)$, and W an independent Brownian motion. The dynamics of the price are supposed to be given by

$$dS_t = S_{t^-}(b_t dt + \sigma_t dW_t + \phi_t dM_t) \tag{10.4.1}$$

where b, σ and ϕ are predictable processes. In a closed form

$$S_t = S_0 \exp\left(\int_0^t b_s ds\right) \mathcal{E}(\sigma \star W)_t\, \mathcal{E}(\phi \star M)_t\,.$$

The jumps of S occur when the process N jumps, and $\Delta S_t = S_{t^-}\phi_t \Delta N_t$, hence $S_t = S_{t^-}(1 + \phi_t \Delta N_t)$. Therefore, in order that the price S remains positive, we assume that $\phi > -1$.

Note that the process

$$S_t \exp\left(-\int_0^t b_s ds\right), t \geq 0$$

is a local martingale.

10.4.1 Computation of the Moments

In the constant coefficients case, with $\phi > -1$, the solution of (10.4.1) may be written as

$$S_t = S_0 \exp\left[\left(b - \phi\lambda - \frac{\sigma^2}{2}\right)t + \sigma W_t + [\ln(1+\phi)]N_t\right] = S_0 e^{X_t} \tag{10.4.2}$$

where X is the Lévy process (see $\rightarrowtail$ Chapter 11 for information on Lévy processes)

$$X_t = \left(b - \phi\lambda - \frac{\sigma^2}{2}\right)t + \sigma W_t + [\ln(1+\phi)]N_t\,. \qquad (10.4.3)$$

The moments of the r.v. S_t can be computed as follows: first, we have

$$\begin{aligned}\left(e^{-bt}S_t\right)^k &= S_0^k \exp\left(k\ln(1+\phi)\,N_t - \lambda k\phi t + \sigma k W_t - \frac{tk\sigma^2}{2}\right)\\ &= S_0^k\,\mathcal{E}(\phi_k M)_t\,\mathcal{E}(\sigma k W)_t \exp\left[t\left(\lambda[\phi_k - k\phi] + \frac{1}{2}\sigma^2 k(k-1)\right)\right]\end{aligned}$$

where $\phi_k = (1+\phi)^k - 1$. Therefore,

$$S_t^k = Z_t^{(k)} \exp(tg(k))$$

where we have denoted by $Z^{(k)}$ the martingale

$$Z_t^{(k)} = S_0^k \mathcal{E}(\phi_k M)_t\,\mathcal{E}(\sigma k W)_t = S_0^k \exp(kX_t - tg(k)) \qquad (10.4.4)$$

and $g(k)$ is the Laplace exponent (see $\rightarrowtail$ Subsection 11.2.3 for a generalization to Lévy processes)

$$g(k) = bk + \frac{1}{2}\sigma^2 k(k-1) + \lambda[(1+\phi)^k - 1 - k\phi]\,. \qquad (10.4.5)$$

Therefore, $\mathbb{E}(S_t^k) = S_0^k \exp(tg(k))$. In particular,

$$\mathbb{E}(S_t^2) = S_0^2 \exp((2b + \sigma^2 + \lambda\phi^2)t)\,.$$

Note that, for every θ, the process $\exp(\theta X_t - tg(\theta))$ is a martingale.

Exercise 10.4.1.1 Let S be given by (10.4.2). Give the dynamics of S^k. $\triangleleft$

10.4.2 Symmetry

We present a put-call symmetry in the case where, under a domestic risk-neutral probability $\mathbb{Q}$, the dynamics of the process S are assumed to be,

$$dS_t = S_{t-}((r-\delta)dt + \sigma dW_t + \phi dM_t)$$

where $M_t = N_t - \lambda t$ is a $\mathbb{Q}$-martingale and r, δ, σ, ϕ are constants with $\phi > -1$. Setting $X_t = (-\phi\lambda - \frac{\sigma^2}{2})t + \sigma W_t + [\ln(1+\phi)]N_t$ and $Z_t = \exp(X_t - tg(1))$, where g is defined in (10.4.5) and setting $b = r - \delta$, one obtains

$$S_t = S_0 e^{(r-\delta)t} Z_t = S_0 e^{(r-\delta)t} \exp(X_t - tg(1))$$

where Z is a $\mathbb{Q}$-martingale (see equality (10.4.4)).

We can write

$$\mathbb{E}_{\mathbb{Q}}(e^{-rt}(K-S_t)^+) = \mathbb{E}_{\mathbb{Q}}\left(e^{-\delta t}Z_t\left(\frac{KS_0}{S_t}-S_0\right)^+\right)$$
$$= \mathbb{E}_{\widehat{\mathbb{Q}}}\left(e^{-\delta t}\left(\frac{KS_0}{S_t}-S_0\right)^+\right),$$

where $\widehat{\mathbb{Q}}|_{\mathcal{F}_t} = Z_t\,\mathbb{Q}|_{\mathcal{F}_t}$. Under the foreign risk-neutral probability $\widehat{\mathbb{Q}}$, the process $Y = 1/S$ follows (see Exercise 10.2.2.4):

$$dY_t = Y_{t-}((\delta - r)dt - \sigma d\widehat{W}_t - \frac{\phi}{1+\phi}d\widehat{M}_t)$$

where $\widehat{W}_t = W_t - \sigma t$ is a $\widehat{\mathbb{Q}}$-BM and $\widehat{M}_t = N_t - \lambda(1+\phi)t$ is a $\widehat{\mathbb{Q}}$-martingale. Hence, denoting by C_E (resp. P_E) the price of a European call (resp. put) on the underlying S with strike K,

$$P_E(x,K,r,\delta;\sigma,\phi,\lambda) = C_E\left(K,x,\delta,r;\sigma,-\frac{\phi}{1+\phi},\lambda(1+\phi)\right).$$

The same method establishes that American call and put prices satisfy

$$P_A(x,K,r,\delta;\sigma,\phi,\lambda) = KxC_A\left(\frac{1}{x},\frac{1}{K},\delta,r,;\sigma,\frac{-\phi}{1+\phi},\lambda(1+\phi)\right).$$

Therefore, if $b_p(r,\delta;\phi,\lambda)$ (resp. $b_c(\delta,r;\frac{-\phi}{1+\phi},\lambda(1+\phi))$) is the put (resp. call) exercise boundary, then

$$b_p(r,\delta;\phi,\lambda)\,b_c\left(\delta,r;\frac{-\phi}{1+\phi},\lambda(1+\phi)\right) = K^2.$$

Comment 10.4.2.1 The put-call symmetry formulae for currency are well known in the case of continuous processes (see e.g., Detemple [251]). Fajardo and Mordecki [339] and Mordecki [659] establish, from the Wiener-Hopf decomposition, a general symmetry relationship for Lévy processes. See also Eberlein and Papapantoleon [293] and Eberlein et al. [294].

10.4.3 Hitting Times

We assume that the dynamics of the underlying process $S_t = S_0e^{X_t}$ are given as in (10.4.2) by

$$S_t = S_0\exp\left[\left(b - \phi\lambda - \frac{\sigma^2}{2}\right)t + \sigma W_t + [\ln(1+\phi)]N_t\right].$$

Let us denote by $T_L(S)$ the first passage time of the process S at level L, for $L > S_0$, i.e.,

$$T_L(S) = \inf\{t \geq 0 \,:\, S_t \geq L\}$$

and by $T_\ell(X) = T_L(S)$ the companion first passage time of the process X at level $\ell = \ln(L/S_0)$, for $\ell > 0$, i.e., $T_\ell(X) = \inf\{t \geq 0 \,:\, X_t \geq \ell\}$.

Assuming that the process S has no positive jumps, i.e., in the case where $\phi \in]-1, 0[$, we apply the optional sampling theorem to the bounded martingale $(Z^{(k)}_{t\wedge T_\ell}, t \geq 0)$ where $Z^{(k)}_t = \exp(kX_t - tg(k))$ and $k > 0$. Then, relying on the continuity of the process S (hence of X) at the stopping time T_ℓ, we obtain the following formula:

$$\mathbb{E}[\exp(-g(k)T_\ell)] = \exp(-k\ell)\,.$$

Inverting the Laplace exponent $g(k)$ we obtain the Laplace transform

$$\mathbb{E}(\exp(-uT_\ell)) = \begin{cases} \exp(-g^{-1}(u)\,\ell), & \text{for } \ell > 0 \\ 1 & \text{otherwise}\,. \end{cases} \tag{10.4.6}$$

Here, $g^{-1}(u)$ is the positive root of $g(k) = u$. Indeed, the function $k \to g(k)$ is strictly convex, and, therefore, the equation $g(k) = u$ admits no more than two solutions; a straightforward computation proves that for $u \geq 0$, there are two solutions, one of them is greater than 1 and the other one negative. Therefore, by solving numerically the latter equation, the positive root $g^{-1}(u)$ can be obtained, and the Laplace transform of T_ℓ is known.

If the jump size is positive, there is a non-zero probability that X_{T_ℓ} is strictly greater than ℓ. In this case, we introduce the **overshoot** O_ℓ)

$$O_\ell = X_{T_\ell} - \ell\,. \tag{10.4.7}$$

The difficulty is to obtain the law of the overshoot. See $\rightarrowtail$ Subsection 10.6.2 and references therein for more information on overshoots in the general case of Lévy processes.

Exercise 10.4.3.1 (See Volpi [832]) Let $X_t = bt + W_t + Z_t$ where W is a Brownian motion and $Z_t = \sum_{k=1}^{N_t} Y_k$ a (λ, F)-compound Poisson process independent of W. The first passage time above the level x is

$$T_x = \inf\{t \,:\, X_t \geq x\}$$

and the overshoot is $O_x = X_{T_x} - x$. Let Φ_x be the Laplace transform of the pair (T_x, O_x), i.e.,

$$\Phi_x(\theta, \mu, x) = \mathbb{E}(e^{-\theta T_x - \mu O_x} \mathbb{1}_{\{T_x < \infty\}})\,.$$

Let τ_1 be the first jump time of N. We wish to establish an integral equation for Φ_x using the following computation:

(1) Prove that

$$\mathbb{E}(e^{-\theta T_x - \mu O_x} \mathbb{1}_{\{T_x < \tau_1\}}) = e^{(b-\alpha)x}$$

where $\alpha = \sqrt{b^2 + 2(\theta + \lambda)}$.

(2) Prove that

$$\begin{aligned}\mathbb{E}(e^{-\theta T_x - \mu O_x} \mathbb{1}_{\{T_x = \tau_1\}}) &= \frac{e^{(b-\alpha)x}}{\alpha(\mu - b + \alpha)} \int_{[0,x[} (e^{(\alpha-b)y} - e^{-\mu y}) F(dy) \\ &+ \frac{1}{\alpha(\mu - b - \alpha)} \int_{[x,\infty[} (e^{(b-\alpha)(y-x)} - e^{-\mu(y-x)}) F(dy) \\ &+ \frac{e^{\mu x} - e^{(b-\alpha)x}}{\alpha(\mu - b + \alpha)} \int_{[x,\infty[} e^{-\mu y} F(dy) \\ &+ \frac{e^{(b-\alpha)x}}{\alpha(\mu - b - \alpha)} \int_{[0,\infty[} (e^{-(\alpha+b)y} - e^{-\mu y}) F(dy).\end{aligned}$$

(3) Prove that

$$\begin{aligned}&\mathbb{E}(e^{-\theta T_x - \mu O_x} \mathbb{1}_{\{\tau_1 < T_x < \infty\}}) \\ &= \frac{1}{\alpha} \int_{\mathbb{R}} F(dy) \int_{-\infty}^{(x-y)\wedge x} e^{bz} (e^{-\alpha|z|} - e^{(2x-z)\alpha}) \Phi_{x-z-y}(\theta, \mu) dz.\end{aligned}$$

(4) Deduce an equation for Φ by adding the three components above. ◁

10.4.4 Affine Jump-Diffusion Model

These affine processes are defined as the solutions of the following equation (10.4.8):

Proposition 10.4.4.1 *Suppose that*

$$dX_t = \mu(X_t)dt + \sigma(X_t)dW_t + dZ_t \tag{10.4.8}$$

where μ and σ^2 are affine functions $\mu(x) = \mu_0 + \mu_1 x$; $\sigma^2(x) = \sigma_0 + \sigma_1 x$ and Z is a ν-compound Poisson process such that $\int e^{zy}\nu(dy) < \infty, \forall z$. Then, for any affine function $\psi(x) = \psi_0 + \psi_1 x$, for all θ, there exist two functions α and β such that,

$$\mathbb{E}\left(e^{\theta X_T} \exp\left(-\int_t^T \psi(X_s)ds\right) \Big| \mathcal{F}_t\right) = e^{\alpha(t)+\beta(t)X_t}.$$

PROOF: It suffices to prove the existence of α and β such that the process

$$e^{\alpha(t)+\beta(t)X_t} \exp\left(-\int_0^t \psi(X_s)ds\right)$$

is a martingale, and $\alpha(T) = 0, \beta(T) = \theta$. From Itô's calculus, this leads to a Riccati ODE and to a linear ODE:

$$\beta'(t) = \psi_1 - \mu_1\beta(t) - \frac{1}{2}\sigma_1\beta^2(t)\,,$$
$$\alpha'(t) = \psi_0 - \mu_0\beta(t) - \frac{1}{2}\sigma_0\beta^2(t) - \lambda(\widehat{\nu}(\beta(t)) - 1)$$

where $\widehat{\nu}(z) = \int e^{zy}\nu(dy)$. It remains to check that there exists a solution satisfying the boundary conditions. □

Comment 10.4.4.2 See Duffie et al. [274, 272] for a generalization of this model.

Exercise 10.4.4.3 Let $\lambda(t,x) = \lambda_1 + \lambda_2 x$. Prove that

$$dX_t = \mu(X_t)dt + \sigma(X_t)dW_t + dM_t$$

where M is the compensated martingale of an inhomogeneous Poisson process with intensity $\lambda(t, X_t)$ has a solution.
`Hint:` Start from the previous process (10.4.8) and make a change of probability. ◁

A Particular Case

As an example, let us study the case where S is an exponential of an affine process with constant coefficients (see Exercise 10.2.2.3).

Proposition 10.4.4.4 *Let W be a Brownian motion and Z a ν-compound Poisson process independent of W of the form $Z_t = \sum_{n=1}^{N_t} Y_n$. Let*

$$dS_t = S_{t^-}(\mu dt + \sigma dW_t + dZ_t)\,, \tag{10.4.9}$$

where μ and σ are constants. The infinitesimal generator of S is given by

$$\mathcal{L}f = \partial_t f + x\mu\partial_x f + \frac{1}{2}\sigma^2 x^2\partial_{xx} f + \int_{\mathbb{R}} (f(x(1+y),t) - f(x,t))\,d\nu(y)$$

for f sufficiently regular. The process $(S_t e^{-rt}, t \geq 0)$ is a martingale if and only if $\mathbb{E}(|Y_1|) < \infty$ and $\mu + \lambda\mathbb{E}(Y_1) = r$.

If $Y_1 \geq -1$ a.s., the process S can be written in an exponential form as

$$S_t = S_0 e^{X_t},\; X_t = bt + \sigma W_t + V_t$$

where $b = \mu - \frac{1}{2}\sigma^2$ and V is the $(\lambda, \widetilde{F})$-compound Poisson process

$$V_t = \sum_{n=1}^{N_t} \ln(1+Y_n) = \sum_{n=1}^{N_t} U_n \,,$$

with $\widetilde{F}(u) = F(e^u - 1)$.

If moreover $\mathbb{E}(e^{\theta U_1}) < \infty$ *for all* θ,

$$\mathbb{E}(S_t^\theta) = S_0^\theta e^{t\Psi(\theta)}$$

where

$$\Psi(\theta) = \theta b + \frac{1}{2}\sigma^2\theta^2 - \lambda \int_{-\infty}^{\infty} (1 - e^{\theta u})\, \widetilde{F}(du)\,.$$

PROOF: The expression of the infinitesimal generator is straightforward, the details are left to the reader. In order to give conditions which imply that $(S_t e^{-rt}, t \geq 0)$ is a local martingale, it suffices to write its dynamics as

$$d(e^{-rt}S_t) = e^{-rt}S_{t-}((-r + \mu + \lambda\mathbb{E}(Y_1))dt + \sigma dW_t + dZ_t - \lambda\mathbb{E}(Y_1)dt)\,,$$

and to use the martingale property of the process $(Z_t - \lambda\mathbb{E}(Y_1)t, t \geq 0)$.
If $Y_1 \geq -1$, the solution of (10.4.9) is

$$\begin{aligned} S_t &= S_0 e^{\mu t} e^{\sigma W_t - \frac{1}{2}\sigma^2 t} \exp\left(\sum_{n=1}^{N_t} \ln(1+Y_n)\right) \\ &= S_0 e^{bt + \sigma W_t + V_t} = S_0 e^{X_t} \end{aligned}$$

with

$$X_t = bt + \sigma W_t + V_t, \quad V_t = \sum_{n=1}^{N_t} \ln(1+Y_n)$$

and $b = \mu - \frac{1}{2}\sigma^2$. Then, denoting by $\widetilde{F}$ the cumulative distribution function of $\ln(1+Y_1)$ (i.e., $\widetilde{F}(y) = F(e^y - 1)$), if $\mathbb{E}(e^{\theta \ln(1+Y_1)}) < \infty$,

$$\begin{aligned} \mathbb{E}(S_t^\theta) &= S_0^\theta \mathbb{E}(e^{\theta X_t}) = S_0^\theta e^{\theta b t} e^{\frac{1}{2}\sigma^2\theta^2 t} \mathbb{E}(e^{\theta V_t}) \\ &= S_0^\theta e^{\theta b t} e^{\frac{1}{2}\sigma^2\theta^2 t} \exp\left(-t\lambda \int_{-\infty}^{\infty} (1 - e^{\theta u})\widetilde{F}(du)\right)\,, \end{aligned}$$

where the last equation follows from Proposition 8.6.3.4. □

Example 10.4.4.5 We present two examples of jump-diffusion processes. We assume that

$$X_t = \mu t + \sigma W_t + \sum_{i=1}^{N_t} Y_i$$

where $\sum_{k=1}^{N_t} Y_k$ is a compound Poisson process.

- **Merton's Model.** Merton [643] chooses a model where prices are of the form $S_t := xe^{X_t}$ with $X_t = bt + \sigma W_t + \sum_{k=1}^{N_t} Y_k$ where the law of Y_1 is Gaussian, with mean μ and variance α^2. In differential form

$$dS_t = S_{t-}\left(\left(b + \frac{1}{2}\sigma^2\right) dt + \sigma dW_t + dZ_t\right)$$

with $Z_t = \sum_{k=1}^{N_t} \widetilde{Y}_k$ where the r.v's $\widetilde{Y}_n$ are log-normal. In that case, the density of the r.v. X_t is easily obtained by conditioning on the number of jumps, i.e., by $N_t = n$, and

$$\mathbb{P}(X_t \in dx) = e^{-\lambda t} \sum_{n=0}^{\infty} \frac{(\lambda t)^n}{n!} \frac{1}{\sqrt{2\pi(\sigma^2 t + n\alpha^2)}} e^{-(x-bt-n\mu)^2/2(\sigma^2 t+n\alpha^2)} dx\,,$$

$$\Psi(\theta) = \frac{1}{2}\sigma^2\theta^2 + b\theta + \lambda(1 - e^{-\mu\theta+\alpha^2\theta^2/2})\,.$$

- **Kou's Double-exponential Jumps Model.** A particular jump-diffusion model is the double exponential jumps model, introduced by Kou [540] and Kou and Wang [541, 542]. In this model

$$X_t = \mu t + \sigma W_t + \sum_{i=1}^{N_t} Y_i\,,$$

where the density of the law of Y_1 is

$$\nu(dx) = \left(p\eta_1 e^{-\eta_1 x} \mathbb{1}_{\{x>0\}} + (1-p)\eta_2 e^{\eta_2 x} \mathbb{1}_{\{x<0\}}\right) dx\,.$$

Here, η_i are positive real numbers, and $p \in [0,1]$. With probability p (resp. $(1-p)$), the jump size is positive (resp. negative) with exponential law with parameter η_1 (resp. η_2). Then,

$$\mathbb{E}(e^{iuX_t}) = \exp\left(t\left\{-\frac{1}{2}\sigma^2 u^2 + ibu + \lambda\left(\frac{p\eta_1}{\eta_1 - iu} + \frac{(1-p)\eta_2}{\eta_2 + iu} - 1\right)\right\}\right),$$

and $\mathbb{E}(e^{\beta X_t}) = \exp(\Psi(\beta)t)$ is defined for $-\eta_2 < \beta < \eta_1$ where

$$\Psi(\beta) = \beta\mu + \frac{1}{2}\beta^2\sigma^2 + \lambda\left(\frac{p\eta_1}{\eta_1 - \beta} + \frac{(1-p)\eta_2}{\beta + \eta_2} - 1\right)$$

(in terms of Lévy processes (see ↣ Subsection 11.2.3), Ψ is the Laplace exponent). Let $T_x = \inf\{t : X_t \geq x\}$. Then, Kou and Wang [541] establish that, for $r > 0$, $y > 0$, and $x > 0$,

$$\mathbb{E}(e^{-rT_x}) = \frac{\eta_1 - \beta_1}{\eta_1} \frac{\beta_2}{\beta_2 - \beta_1} e^{-x\beta_1} + \frac{\beta_2 - \eta_1}{\eta_1} \frac{\beta_1}{\beta_2 - \beta_1} e^{-x\beta_2}\,,$$

$$\mathbb{E}(e^{-rT_x}\mathbb{1}_{\{X_{T_x}-x>y\}}\mathbb{1}_{\{T_x<\infty\}}) = e^{\eta_1 y}\frac{\eta_1-\beta_1}{\eta_1}\frac{\beta_2-\eta_1}{\beta_2-\beta_1}\left(e^{-x\beta_1}-e^{-x\beta_2}\right),$$

$$\mathbb{E}(e^{-rT_x}\mathbb{1}_{\{X_{T_x}=x\}}) = \frac{\eta_1-\beta_1}{\beta_2-\beta_1}e^{-x\beta_1}+\frac{\beta_2-\eta_1}{\beta_2-\beta_1}e^{-x\beta_2}$$

and, for $-\eta_2 < \theta < \eta_1$

$$\mathbb{E}(e^{\theta X_{T_x}-rT_x}) = e^{\theta x}\left(\frac{\eta_1-\beta_1}{\beta_2-\beta_1}\frac{\beta_2-\theta}{\eta_1-\theta}e^{-x\beta_1}+\frac{\beta_2-\eta_1}{\beta_2-\beta_1}\frac{\beta_1-\theta}{\eta_1-\theta}e^{-x\beta_2}\right)$$

where $0 < \beta_1 < \eta_1 < \beta_2$ are roots of $G(\beta) = r$. The method is based on finding an explicit solution of $\mathcal{L}u = ru$ where $\mathcal{L}$ is the infinitesimal generator of the process X.

10.4.5 General Jump-Diffusion Processes

Let W be a Brownian motion and $p(ds,dz)$ the random measure associated with a marked point process. Let $\mathcal{F}_t = \sigma(W_s,\, p([0,s],A), A \in \mathcal{E};\ s \leq t)$. The solution of

$$dS_t = S_{t^-}\left(\mu_t dt + \sigma_t dW_t + \int_{\mathbb{R}}\varphi(t,x)p(dt,dx)\right)$$

can be written in an exponential form as

$$S_t = S_0 \exp\left(\int_0^t \left[\mu_s - \frac{1}{2}\sigma_s^2\right]ds + \int_0^t \sigma_s dW_s\right)\prod_{n=1}^{N_t}(1+\varphi(T_n,Z_n))$$

where $N_t = p((0,t],\mathbb{R})$ is the total number of jumps on the time interval $[0,t]$. See Björk et al. [103] and Mercurio and Runggaldier [640] for applications to finance.

10.5 Incompleteness

We consider a financial market with finite horizon T, where a risky asset with price S is traded as well as the savings account, with deterministic interest rate $(r(t), t \geq 0)$. We denote by $R(t)$ the discount factor $R(t) = \exp\left(-\int_0^t r(s)ds\right)$. The natural filtration of S is denoted by $\mathbf{F}$. In an incomplete arbitrage free market, the set $\mathcal{Q}$ of e.m.m's contains several equivalent probability measures and perfect hedging is not possible. More precisely, pricing a contingent claim H using an e.m.m. $\mathbb{Q}$ as $\mathbb{E}_{\mathbb{Q}}(R(T)H)$ does not correspond to the initial price of a hedging strategy. However, trading the contingent claim H at price $\mathbb{E}_{\mathbb{Q}}(R(T)H)$ for some $\mathbb{Q} \in \mathcal{Q}$, does not induce an arbitrage opportunity. Due to the convexity of the set $\mathcal{Q}$ of e.m.m's, the set $\{\mathbb{E}_{\mathbb{Q}}(R(T)H), \mathbb{Q} \in \mathcal{Q}\}$ is an interval, and any choice of initial price outside this interval would induce an arbitrage.

We assume here that the dynamics of S under the historical probability follow a particular case of (10.4.1), i.e.,

$$dS_t = S_{t^-}(b(t)dt + \sigma(t)dW_t + \phi(t)dM_t), \; S_0 = x. \tag{10.5.1}$$

Here, b, σ and ϕ are deterministic bounded Borel functions assumed to satisfy $|\sigma(t)| > \epsilon, \phi(t) > -1, \epsilon < |\phi(t)| < c$ where ϵ and c are strictly positive constants. The process W is a Brownian motion and M is the compensated martingale associated with an inhomogeneous Poisson process having deterministic intensity λ.

In this section, we address the problem of the range of viable prices and we give a dual formulation of the problem in terms of super-strategies for mixed diffusion dynamics.

10.5.1 The Set of Risk-neutral Probability Measures

In a first step, we determine the set of equivalent martingale measures. Note that

$$d(RS)_t = R(t)S_{t^-}([b(t) - r(t)]dt + \sigma(t)dW_t + \phi(t)dM_t)\,. \tag{10.5.2}$$

Proposition 10.5.1.1 *The set $\mathcal{Q}$ of e.m.m's is the set of probability measures $\mathbb{P}^{\psi,\gamma}$ such that $\mathbb{P}^{\psi,\gamma}|_{\mathcal{F}_t} = L_t^{\psi,\gamma}\ \mathbb{P}|_{\mathcal{F}_t}$ where $L_t^{\psi,\gamma} := L_t^{\psi}(W)\,L_t^{\gamma}(M)$ is the product of the Doléans-Dade martingales*

$$\begin{cases} L_t^{\psi}(W) = \mathcal{E}(\psi\star W)_t = \exp\left[\int_0^t \psi_s dW_s - \frac{1}{2}\int_0^t \psi_s^2 ds\right], \\ \\ L_t^{\gamma}(M) = \mathcal{E}(\gamma\star M)_t = \exp\left[\int_0^t \ln(1+\gamma_s)dN_s - \int_0^t \lambda(s)\gamma_s ds\right]. \end{cases}$$

In these formulae, the predictable processes ψ and γ satisfy the following constraint

$$b(t) - r(t) + \sigma(t)\psi_t + \lambda(t)\phi(t)\gamma_t = 0 \;\;,\;\; d\mathbb{P}\otimes dt \; a.s.. \tag{10.5.3}$$

Here, $L^{\gamma}(M)$ is assumed to be a strictly positive $\mathbb{P}$-martingale. In particular, the process γ satisfies $\gamma_t > -1$.

PROOF: Let $\mathbb{Q}$ be an e.m.m. with Radon-Nikodým density equal to L. Using the predictable representation theorem for the pair (W, M), the strictly positive $\mathbb{P}$-martingale L can be written in the form

$$dL_t = L_{t^-}[\psi_t dW_t + \gamma_t dM_t]$$

where (ψ, γ) are predictable processes. It remains to choose this pair such that the process RSL is a $\mathbb{P}$-martingale. Itô's lemma gives formula (10.5.3). Indeed,

$$\begin{aligned} d(RSL)_t &= R_t S_{t^-} dL_t + L_{t^-} d(RS)_t + d[RS, L]_t \\ &\overset{\text{mart}}{=} (LRS)_{t^-} (b(t) - r(t) + \sigma(t)\psi_t + \lambda(t)\phi(t)\gamma_t) dt \,. \end{aligned}$$

□

The terms $-\psi$ and $-\gamma$ are respectively the risk premium associated with the Brownian risk and the jump risk.

Definition 10.5.1.2 *Let us denote by Γ the set of predictable processes γ such that $L^{\psi,\gamma}$ is a strictly positive $\mathbb{P}$-martingale.*

As recalled in Subsection 10.3.2, the process W^ψ defined by

$$W_t^\psi := W_t - \int_0^t \psi_s \, ds$$

is a $\mathbb{P}^{\psi,\gamma}$-Brownian motion and the process M^γ with $M_t^\gamma := M_t - \int_0^t \lambda(s)\gamma_s ds$ is a $\mathbb{P}^{\psi,\gamma}$-martingale. In terms of these $\mathbb{P}^{\psi,\gamma}$-martingales, the price process follows

$$dS_t = S_{t^-} \left[r(t)dt + \sigma(t) dW_t^\psi + \phi(t) dM_t^\gamma \right]$$

and satisfies

$$R(t) S_t = x \, \mathcal{E}(\sigma \star W^\psi)_t \, \mathcal{E}(\phi \star M^\gamma)_t \,.$$

We shall use the decomposition

$$W_t^\psi = W_t + \int_0^t \theta(s) ds + \int_0^t \lambda(s)\phi(s) \frac{\gamma_s}{\sigma(s)} ds$$

where $\theta(s) = \frac{b(s)-r(s)}{\sigma(s)}$. Note that, in the particular case $\gamma = 0$, under $\mathbb{P}^{-\theta,0}$, the risk premium of the jump part is equal to 0, the intensity of the Poisson process N is equal to λ and the process $W_t + \int_0^t \theta(s) ds$ is a Brownian motion independent of N.

Warning 10.5.1.3 In the case where γ is a deterministic function, the inhomogeneous Poisson process N has a $\mathbb{P}^{\psi,\gamma}$ deterministic intensity equal to $\lambda(t)(1+\gamma(t))$. The martingale M^γ has the predictable representation property and is independent of W^ψ. This is not the case when γ depends on W and the pair (W^ψ, M^γ) can fail to be independent under $\mathbb{P}^{\psi,\gamma}$ (see Example 10.3.2.1).

Comment 10.5.1.4 For a general study of changes of measures for jump-diffusion processes, the reader can refer to Cheridito et al. [166].

10.5.2 The Range of Prices for European Call Options

As the market is incomplete, it is not possible to give a hedging price for each contingent claim $B \in L^2(\mathcal{F}_T)$. In this section, we shall write $\mathbb{P}^\gamma = \mathbb{P}^{\psi,\gamma}$ where ψ and γ satisfy the relation (10.5.3); thus ψ is given in terms of γ and $\gamma > -1$. At time t, we define a viable price V_t^γ for the contingent claim B using the conditional expectation (with respect to the σ-field $\mathcal{F}_t$) of the discounted contingent claim under the martingale-measure $\mathbb{P}^\gamma$, that is, $R(t)V_t^\gamma := \mathbb{E}^\gamma(R(T)\,B|\mathcal{F}_t)$.

We now study the range of viable prices associated with a European call option, that is, the interval $]\inf_{\gamma\in\Gamma} V_t^\gamma\,,\,\sup_{\gamma\in\Gamma} V_t^\gamma[$, for $B = (S_T - K)^+$. We denote by $\mathcal{BS}$ the Black and Scholes function, that is, the function such that

$$R(t)\mathcal{BS}(x,t) = \mathbb{E}(R(T)(X_T - K)^+\,|X_t = x)\;,\;\; \mathcal{BS}(x,T) = (x-K)^+$$

when

$$dX_t = X_t(r(t)dt + \sigma(t)\,dW_t)\,. \tag{10.5.4}$$

In other words,

$$\mathcal{BS}(x,t) = x\mathcal{N}(d_1) - K(R_T/R_t)\mathcal{N}(d_2)$$

where

$$d_1 = \frac{1}{\Sigma(t,T)}\left(\ln\left(\frac{x}{K}\right) + \int_t^T r(u)du + \frac{1}{2}\Sigma^2(t,T)\right)\,,$$

and $\Sigma^2(t,T) = \int_t^T \sigma^2(s)ds$. We recall (see end of Subsection 2.3.2) that $\mathcal{BS}$ is a convex function of x which satisfies

$$\mathcal{L}(\mathcal{BS})(x,t) = r(t)\mathcal{BS}(x,t) \tag{10.5.5}$$

where

$$\mathcal{L}(f)(x,t) = \frac{\partial f}{\partial t}(x,t) + r(t)x\frac{\partial f}{\partial x}(x,t) + \frac{1}{2}x^2\sigma^2(t)\frac{\partial^2 f}{\partial x^2}(x,t)\,.$$

Furthermore, $|\partial_x \mathcal{BS}(x,t)| \leq 1$.

Theorem 10.5.2.1 *Let $\mathbb{P}^\gamma \in \mathcal{Q}$. Then, any associated viable price of a European call is bounded below by the Black and Scholes function, evaluated at the underlying asset value, and bounded above by the underlying asset value, i.e.,*

$$R(t)\mathcal{BS}(S_t,t) \leq \mathbb{E}^\gamma(R(T)\,(S_T-K)^+|\mathcal{F}_t) \leq R(t)\,S_t\,.$$

The range of viable prices $V_t^\gamma = \frac{R(T)}{R(t)}\mathbb{E}^\gamma((S_T-K)^+|\mathcal{F}_t)$, is exactly the interval $]\mathcal{BS}(S_t,t),\,S_t[$.

PROOF: We give the proof in the case $t = 0$, the general case follows from the Markov property. Setting

$$\Lambda(f)(x,t) = f((1+\phi(t))\,x,\,t) - f(x,t) - \phi(t)x\,\frac{\partial f}{\partial x}(x,t)\,,$$

Itô's formula (10.2.1) for mixed processes leads to

$$\begin{aligned}
R(T)\mathcal{BS}(S_T,T) &= \mathcal{BS}(S_0,0) \\
&\quad + \int_0^T \left[\mathcal{L}(R\,\mathcal{BS})(S_s,s) + R(s)\lambda(s)(\gamma_s+1)\Lambda(\mathcal{BS})(S_s,s)\right] ds \\
&\quad + \int_0^T R(s)S_{s_-}\,\frac{\partial \mathcal{BS}}{\partial x}(S_{s_-},s)\,(\sigma(s)dW_s^\gamma + \phi(s)dM_s^\gamma) \\
&\quad + \int_0^T R(s)\Lambda(\mathcal{BS})(S_{s_-},s)\,dM_s^\gamma\,.
\end{aligned}$$

The convexity of $\mathcal{BS}(\cdot,t)$ implies that $\Lambda(\mathcal{BS})(x,t) \geq 0$ and the Black-Scholes equation (10.5.5) implies $\mathcal{L}[R\,\mathcal{BS}] = 0$. The stochastic integrals are martingales; indeed $\left|\frac{\partial \mathcal{BS}}{\partial x}(x,t)\right| \leq 1$ implies that $|\Lambda\mathcal{BS}(x,t)| \leq 2xc$ where c is the bound for the size of the jumps ϕ. Taking expectation with respect to $\mathbb{P}^\gamma$ gives

$$\begin{aligned}
\mathbb{E}^\gamma(R(T)\mathcal{BS}(S_T,T)) &= \mathbb{E}^\gamma(R(T)(S_T-K)^+) \\
&= \mathcal{BS}(S_0,0) + \mathbb{E}^\gamma\left(\int_0^T R(s)\lambda(s)(\gamma_s+1)\Lambda\mathcal{BS}(S_s,s)\,ds\right).
\end{aligned}$$

The lower bound follows. The upper bound is a trivial one.

To establish that the range is the whole interval, we can restrict our attention to the case of constant parameters γ. In that case, W^ψ and M^γ are independent, and the convexity of the Black-Scholes price and Jensen's inequality would lead us easily to a comparison between the $\mathbb{P}^\gamma$ price and the Black-Scholes price, since

$$V^\gamma(0,x) = \mathbb{E}(\mathcal{BS}(x\mathcal{E}(\phi M^\gamma)_T,T)) \geq \mathcal{BS}(x\mathbb{E}(\mathcal{E}(\phi M^\gamma)_T),\,T) = \mathcal{BS}(x,T)\,.$$

We establish the following lemma.

Lemma 10.5.2.2 *We have*

$$\lim_{\gamma\to -1} \mathbb{E}^\gamma((S_T-K)^+) = \mathcal{BS}(x,0)$$
$$\lim_{\gamma\to +\infty} \mathbb{E}^\gamma((S_T-K)^+) = x\,.$$

PROOF: From the inequality $|\Lambda\mathcal{BS}(x,t)| \leq 2xc$, it follows that

$$
\begin{aligned}
0 \leq \mathbb{E}\left(\int_0^T R(s)\lambda(s)(\gamma+1)\Lambda\mathcal{BS}(S_s,s)\,ds\right) \\
&\leq 2(\gamma+1)c\int_0^T \lambda(s)\mathbb{E}^\gamma(R(s)S_s)\,ds \\
&= 2(\gamma+1)cS_0\int_0^T \lambda(s)\,ds
\end{aligned}
$$

where the right-hand side converges to 0 when γ goes to -1. It can be noted that, when γ goes to -1, the risk-neutral intensity of the Poisson process goes to 0, that is there are no more jumps in the limit.

The equality for the upper bound relies on the convergence (in law) of S_T towards 0 as γ goes to infinity. The convergence of $\mathbb{E}^\gamma((K-S_T)^+)$ towards K then follows from the boundedness character and the continuity of the pay-off. The put-call parity gives the result for a call option. See Bellamy and Jeanblanc [70] for details. □

The range of prices in the case of American options is studied in Bellamy and Jeanblanc [70] and in the case of Asian options in Bellamy [69]. The results take the following form:

- Let
$$R(t)\,\mathcal{P}^\gamma(S_t,t) = \text{esssup}_{\tau\in\mathcal{T}(t,T)}\mathbb{E}^\gamma(R(\tau)(K-S_\tau)^+\,|\mathcal{F}_t)$$
be a discounted American viable price, evaluated under the risk-neutral probability $\mathbb{P}^\gamma$ (see Subsection 1.1.1 for the definition of esssup). Here, $\mathcal{T}(t,T)$ is the class of stopping times with values in the interval $[t,T]$. Let $\mathcal{P}^{Am}$ be defined as the American-Black-Scholes function for an underlying asset following (10.5.4), that is,
$$\mathcal{P}^{Am}(X_t,t) := \text{esssup}_{\tau\in\mathcal{T}(t,T)}\mathbb{E}^\gamma(R(\tau)(K-X_\tau)^+\,|X_t)\,.$$
Then
$$\mathcal{P}^{Am}(S_t,t) \leq \mathcal{P}^\gamma(S_t,t) \leq K\,.$$
- The range of Asian-option prices is the whole interval
$$]x\,\mathcal{A}(x,0), \frac{xR(T)}{T}\int_0^T \frac{1}{R(u)}du[$$
where $\mathcal{A}$ is the function solution of the PDE equation (6.6.8) for the evaluation of Asian options in a Black-Scholes framework.

Comments 10.5.2.3 (a) As we shall establish in the next section 10.5.3, $R(t)\,\mathcal{BS}(S_t,t)$ is the greatest sub-martingale with terminal value equal to the terminal pay-off of the option, i.e., $(K-S_T)^+$.

(b) El Karoui and Quenez [307] is the main paper on super-replication prices. It establishes that when the dynamics of the stock are driven by a Wiener process, then the supremum of the viable prices is equal to the minimal initial value of an admissible self-financing strategy that super-replicates the contingent claim. This result is generalized by Kramkov [543]. See Mania [617] and Hugonnier [191] for applications.

(c) Eberlein and Jacod [290] establish the absence of non-trivial bounds on European option prices in a model where prices are driven by a purely discontinuous Lévy process with unbounded jumps. The results can be extended to a more general case, where $S_t = e^{X_t}$ where X is a Lévy process.(See Jakubenas [473].) These results can also be extended to the case where the pay-off is of the form $h(S_T)$ as long as the convexity of the Black and Scholes function [which is defined, with the notation of (10.5.4), as $\mathbb{E}(h(X_T)|X_t = x)$], is established. Bergman et al. [75], El Karoui et al. [302, 301], Hobson [440] and Martini [625] among others have studied the convexity property. See also Ekström et al. [296] for a generalization of this convexity property to a multi-dimensional underlying asset. The papers of Mordecki [413] and Bergenthum and Rüschendorf [74] give bounds for option prices in a general setting.

10.5.3 General Contingent Claims

More generally, let B be any contingent claim, i.e., $B \in L^2(\mathcal{F}_T)$. This contingent claim is said to be hedgeable if there exists a process π and a constant b such that $R(T)\,B = b + \int_0^T \pi_s\, d[RS]_s$.

Let $X^{y,\pi,C}$ be the solution of

$$\begin{aligned} dX_t &= r(t)X_t dt + \pi_t X_{t-}\left[\sigma(t)dW_t^0 + \phi(t)dM_t\right] - dC_t \\ X_0 &= y\,. \end{aligned}$$

Here, (π, C) belongs to the class $\mathcal{V}(y)$ consisting of pairs of adapted processes (π, C) such that $X_t^{y,\pi,C} \geq 0$, $\forall t \geq 0$, π being a predictable process and C an increasing process. The minimal value

$$\inf\{y \,:\, \exists(\pi, C) \in \mathcal{V}(y)\,, X_T^{y,\pi,C} \geq B\}$$

which represents the minimal price that allows the seller to hedge his position, is the **selling price** of B or the **super-replication price**.

The non-negative assumption on the wealth process precludes arbitrage opportunities.

Proposition 10.5.3.1 *Here, γ is a generic element of Γ (see Definition 10.5.1.2). We assume that* $\sup_\gamma \mathbb{E}^\gamma(B) < \infty$. *Let*

$$]\inf_\gamma \mathbb{E}^\gamma(R(T)B), \sup_\gamma \mathbb{E}^\gamma(R(T)B)[$$

be the range of prices. The upper bound $\sup_\gamma \mathbb{E}^\gamma(R(T)B)$ *is the selling price of* B*.*

The contingent claim is hedgeable if and only if there exists $\gamma^* \in \Gamma$ *such that* $\sup_\gamma \mathbb{E}^\gamma(R(T)B) = \mathbb{E}^{\gamma^*}(R(T)B)$*. In this case* $\mathbb{E}^\gamma(R(T)B) = \mathbb{E}^{\gamma^*}(R(T)B)$ *for any* γ*.*

PROOF: Let us introduce the random variable

$$R(\tau)\, V_\tau := \text{ess sup}_{\gamma\in\Gamma}\mathbb{E}^\gamma[BR(T)\,|\mathcal{F}_\tau]$$

where τ is a stopping time. This defines a process $(R(t)V_t, t \geq 0)$ which is a $\mathbb{P}^\gamma$-super-martingale for any γ. This super-martingale can be decomposed as

$$R(t)V_t = V_0 + \int_0^t \mu_s dW_s^\gamma + \int_0^t \nu_s dM_s^\gamma - A_t^\gamma \tag{10.5.6}$$

where A^γ is an increasing process. It is easy to check that μ and ν do not depend on γ and that

$$A_t^\gamma = A_t^0 + \int_0^t \left(\frac{\mu_s\,\phi_s}{\sigma(s)} - \nu_s\right)\lambda(s)\gamma_s ds \tag{10.5.7}$$

where A^0 is the increasing process obtained for $\gamma = 0$. It is useful to write the decomposition of the process RV under $\mathbb{P}^0$ as:

$$R(t)V_t = V_0 + \int_0^t \frac{\mu_s}{\sigma(s)}[\sigma(s)dW_s^0 + \phi(s)dM_s] - \int_0^t R(s)dC_s$$

where the process C is defined via

$$R(t)\,dC_t := dA_t^0 + \left(\frac{\mu_t\,\phi(t)}{\sigma(t)} - \nu_t\right)(dN_t - \lambda(t)dt)\,. \tag{10.5.8}$$

Note that W^0 is the Brownian motion W^γ for $\gamma = 0$ and that $M^0 = M$.

Lemma 10.5.3.2 *The processes* μ *and* ν *defined in* (10.5.6) *satisfy:*

(a) $\dfrac{\mu_t\,\phi(t)}{\sigma(t)} - \nu_t \geq 0$*, a.s.* $\forall t \in [0,T]$.
(b) The process C*, defined in* (10.5.8)*, is an increasing process.*

Proof of the Lemma

▶ **Part a:** Suppose that the positivity condition is not satisfied and introduce $F_t := \{\omega : \left(\frac{\mu_t\,\phi(t)}{\sigma(t)} - \nu_t\right)(\omega) < 0\}$. Let $n \in \mathbb{N}$. The process γ^n defined by $\gamma_t^n = n\mathbb{1}_{F_t}$ belongs to Γ (see Definition 10.5.1.2) and for this process γ^n the r.v.

$$A_t^{\gamma^n} = A_t^0 + \int_0^t \left(\frac{\mu_s \phi(s)}{\sigma(s)} - \nu_s\right) \lambda(s)\gamma_s^n ds = A_t^0 - n \int_0^t \left(\frac{\mu_s \phi(s)}{\sigma(s)} - \nu_s\right)^- \lambda(s) ds$$

fails to be positive for large values of n.

▶ **Part b:** The process C defined as

$$\begin{aligned} R(t)dC_t &= dA_t^0 + \left(\frac{\mu_t \phi(t)}{\sigma(t)} - \nu_t\right)(dN_t - \lambda(t)dt) \\ &= dA_t^0 - \left(\frac{\mu_t \phi(t)}{\sigma(t)} - \nu_t\right)\lambda(t)dt + \left(\frac{\mu_t \phi(t)}{\sigma(t)} - \nu_t\right) dN_t \quad (10.5.9) \end{aligned}$$

will be shown to be the sum of two increasing processes. To this end, notice that, on the one hand

$$\frac{\mu_t \phi(t)}{\sigma(t)} - \nu_t \geq 0$$

which establishes the positivity of $\left(\frac{\mu_t \phi(t)}{\sigma(t)} - \nu_t\right) dN_t$. On the other hand, passing to the limit when γ goes to -1 on the right-hand side of (10.5.7) establishes that the remaining part in (10.5.9),

$$A_t^0 - \int_0^t \left(\frac{\mu_s \phi(s)}{\sigma(s)} - \nu_s\right) \lambda(s)\, ds\,,$$

is an increasing (optional) process. □

We now complete the proof of Proposition 10.5.3.1. It is easy to check that the triple (V, π, C), with $V_t R_t \pi_t = \dfrac{\mu_t}{\sigma(t)}$ and C being the increasing process defined via (10.5.8) satisfies

$$dV_t = r(t)V_t dt + \pi_t V_{t-}[\sigma(t)dW_t^0 + \phi(t)dM_t] - dC_t\,.$$

As in El Karoui and Quenez [307], it can be established that a bounded contingent claim B is hedgeable if there exists γ^* such that

$$\mathbb{E}^{\gamma^*}[R(T)B] = \sup_{\gamma \in \Gamma} \mathbb{E}^{\gamma}[R(T)B]\,.$$

In this case, the expectation of the discounted value does not depend on the choice of the e.m.m.: $\mathbb{E}^{\gamma}[R(T)B] = \mathbb{E}^{\gamma^*}[R(T)B]$ for any γ. In our framework, this is equivalent to $A_t^0 = 0$, $dt \times d\mathbb{P}\, a.s.$, which implies, from the second part of the lemma, that $\frac{\mu_t \phi(t)}{\sigma(t)} - \nu_t = 0$. In this case B is obviously hedgeable. □

Comment 10.5.3.3 The proof goes back to El Karoui and Quenez [306, 307] and is used again in Cvitanić and Karatzas [208], for price processes driven by continuous processes. Nevertheless, in El Karoui and Quenez, the reference filtration may be larger than the Brownian filtration. In particular, there may be a Poisson subfiltration in the reference filtration.

10.6 Complete Markets with Jumps

In this section, we present some models involving jump-diffusion processes for which the market is complete.

Our first model consists in a simple jump-diffusion model whereas our second model is more sophisticated.

10.6.1 A Three Assets Model

Assume that the market contains a riskless asset with interest rate r and two risky assets (S^1, S^2) with dynamics

$$\begin{aligned} dS_t^1 &= S_{t-}^1 (b_1(t)dt + \sigma_1(t)dW_t + \phi_1(t)dM_t) \\ dS_t^2 &= S_{t-}^2 (b_2(t)dt + \sigma_2(t)dW_t + \phi_2(t)dM_t) , \end{aligned}$$

where W is a Brownian motion and M the compensated martingale associated with an inhomogeneous Poisson process with deterministic intensity $\lambda(t)$. We assume that W and M are independent. Here, the coefficients b_i, σ_i, ϕ_i and λ are deterministic functions and $\phi_i > -1$. The unique risk-neutral probability $\mathbb{Q}$ is defined by (see Subsection 10.5.1) $\mathbb{Q}|_{\mathcal{F}_t} = L_t^{\psi,\gamma}\mathbb{P}|_{\mathcal{F}_t}$, where

$$dL_t = L_{t-} [\psi_t dW_t + \gamma_t dM_t] .$$

Here, the processes ψ and γ are $\mathbf{F}^{W,M}$-predictable, defined as a solution of

$$b_i(t) - r(t) + \sigma_i(t)\psi_t + \lambda(t)\phi_i(t)\gamma_t = 0, i = 1, 2 .$$

It is easy to check that, under the conditions

$$\begin{aligned} |\sigma_1(t)\phi_2(t) - \sigma_2(t)\phi_1(t)| \geq \epsilon > 0 , \\ \gamma(t) = \frac{[b_2(t) - r(t)]\,\sigma_1(t) - [b_1(t) - r(t)]\,\sigma_2(t)}{\lambda\,(\sigma_2(t)\phi_1(t) - \sigma_1(t)\phi_2(t))} > -1 , \end{aligned}$$

there exists a unique solution such that L is a strictly positive local martingale. Hence, we obtain an arbitrage free complete market.

Comments 10.6.1.1 (a) See Jeanblanc and Pontier [490] and Shirakawa [789] for applications to option pricing. Using this setting makes it possible to complete a financial market where the only risky asset follows the dynamics (10.4.1), i.e.,

$$dS_t = S_{t-} (b_t dt + \sigma_t dW_t + \phi_t dM_t) ,$$

with a second asset, for example a derivative product.

(b) The same methodology applies in a default setting. In that case if

$$\begin{aligned} dS_t^0 &= rS_t^0 dt \\ dS_t^1 &= S_t^1(\mu_1 dt + \sigma_1 dW_t) \\ dS_t^2 &= S_{t-}^2 (\mu_2 dt + \sigma_2 dW_t + \varphi_2 dM_t) \end{aligned}$$

are three assets traded in the market, where S^0 is riskless, S^1 is default free and S^2 is a defaultable asset, the market is complete (see Subsection 7.5.6). Here, W is a standard Brownian motion and M is the compensated martingale of the default process, i.e., $M_t = \mathbb{1}_{\{\tau \leq t\}} - \int_0^{t\wedge\tau} \lambda_s ds$ (see Chapter 7). Vulnerable claims can be hedged (see Subsection 7.5.6). See Bielecki et al. [90] and Ayache et al. [33, 34] for an approach using PDEs.

10.6.2 Structure Equations

It is known from Emery [325] that, if X is a martingale and β a bounded Borel function, then the equation

$$d[X,X]_t = dt + \beta(t)dX_t \tag{10.6.1}$$

has a unique solution. This equation is called a **structure equation** (see also Example 9.3.3.6) and its solution enjoys the predictable representation property (see Protter [727], Chapter IV). Relation (10.6.1) implies that the process X has predictable quadratic variation $d\langle X,X\rangle_t = dt$. If $\beta(t)$ is a constant β, the martingale solution of

$$d[X,X]_t = dt + \beta dX_t \tag{10.6.2}$$

is called Azéma-Emery martingale with parameter β.

Dritschel and Protter's Model

In [266], Dritschel and Protter studied the case where the dynamics of the risky asset are

$$dS_t = S_{t^-}\sigma dZ_t$$

where Z is a semi-martingale whose martingale part satisfies (10.6.2) with $-2 \leq \beta < 0$ and proved that, under some condition on the drift of Z, the market is complete and arbitrage free.

Privault's Model

In [485], the authors consider a model where the asset price is driven by a Brownian motion on some time interval, and by a Poisson process on the remaining time intervals. More precisely, let ϕ and α be two bounded deterministic Borel functions, with $\alpha > 0$, defined on $\mathbb{R}^+$ and

$$\lambda(t) = \begin{cases} \alpha^2(t)/\phi^2(t) & \text{if } \phi(t) \neq 0, \\ 0 & \text{if } \phi(t) = 0\,. \end{cases}$$

We assume, to avoid trivial results, that the Lebesgue measure of the set $\{t : \phi(t) = 0\}$ is neither 0 nor ∞. Let B be a standard Brownian motion, and N an inhomogeneous Poisson process with intensity λ. We denote by i the indicator function

$$i(t) = \mathbb{1}_{\{\phi(t)=0\}}$$

We assume that B and N are independent and that $\lim_{t\to\infty} \lambda(t) = \infty$ and $\lambda(t) < \infty$, $\forall t$. The process $(X_t, t \geq 0)$ defined by

$$dX_t = i(t)dB_t + \frac{\phi(t)}{\alpha(t)} (dN_t - \lambda(t)dt), X_0 = 0 \tag{10.6.3}$$

satisfies the structure equation

$$d[X,X]_t = dt + \frac{\phi(t)}{\alpha(t)} dX_t .$$

From X, we construct a martingale Z with predictable quadratic variation $d\langle Z,Z\rangle_t = \alpha^2(t)dt$, by setting

$$dZ_t = \alpha(t)dX_t, \; Z_0 = 0,$$

that is,

$$dZ_t = i(t)\alpha(t)dB_t + \phi(t)(dN_t - \lambda(t)dt), \; Z_0 = 0.$$

Proposition 10.6.2.1 *The martingale Z satisfies*

$$d[Z,Z]_t = \alpha^2(t)dt + \phi(t)dZ_t, \tag{10.6.4}$$

and $d\langle Z,Z\rangle_t = \alpha^2(t)dt$.

PROOF: Using the relations $d[B,N]_t = 0$ and $i(t)\phi(t) = 0$, we have

$$\begin{aligned} d[Z,Z]_t &= i(t)\alpha^2(t)dt + \phi^2(t)dN_t \\ &= i(t)\alpha^2(t)dt + \phi(t)\left(dZ_t - i(t)\alpha(t)dB_t + (1-i(t))\frac{\alpha^2(t)}{\phi(t)}dt\right) \\ &= \alpha^2(t)dt + \phi(t)dZ_t . \end{aligned}$$

□

From the general results on structure equations which we have already invoked, the martingale Z enjoys the predictable representation property.

Let S denote the solution of the equation

$$dS_t = S_{t-}(\mu(t)dt + \sigma(t)dZ_t),$$

with initial condition S_0 where the coefficients μ and σ are assumed to be deterministic. Then,

$$\begin{aligned} S_t = S_0 \exp&\left(\int_0^t \sigma(s)\alpha(s)i(s)dB_s - \frac{1}{2}\int_0^t i(s)\sigma^2(s)\alpha^2(s)ds\right) \\ \times \exp&\left(\int_0^t (\mu(s) - \phi(s)\lambda(s)\sigma(s))ds\right) \prod_{k=1}^{N_t} (1 + \sigma(T_k)\phi(T_k)), \end{aligned}$$

where $(T_k)_{k\geq 1}$ denotes the sequence of jump times of N.

Proposition 10.6.2.2 *Let us assume that* $\phi(t)\,\frac{r(t)-\mu(t)}{\sigma(t)\alpha^2(t)} > -1$, *and let*

$$\psi(t) := \phi(t)\,\frac{r(t)-\mu(t)}{\sigma(t)\alpha^2(t)}\,.$$

Then, the unique e.m.m. is the probability $\mathbb{Q}$ *such that* $\mathbb{Q}|_{\mathcal{F}_t} = L_t\mathbb{P}|_{\mathcal{F}_t}$, *where* $dL_t = L_{t^-}\psi(t)dZ_t$, $L_0 = 1$.

PROOF: It is easy to check that

$$d(S_tL_tR(t)) = S_{t^-}L_{t^-}R(t)\sigma(t)dZ_t\,,$$

where $R(t) = e^{rt}$, hence SRL is a $\mathbb{P}$-local martingale. □

We now compute the price of a European call written on the underlying asset S. Note that the two processes $(\widetilde{B}_t = B_t - \int_0^t \psi(s)i(s)\alpha(s)ds, t \geq 0)$ and $(N_t - \int_0^t \lambda(s)(1+\phi(s))ds, t \geq 0)$ are $\mathbb{Q}$-martingales. Furthermore,

$$S_t = S_0 \exp\left(\int_0^t \sigma(s)\alpha(s)i(s)d\widetilde{B}_s - \frac{1}{2}\int_0^t i(s)\sigma^2(s)\alpha^2(s)ds\right)$$
$$\times \exp\left(\int_0^t \left(r(s) - \phi(s)\lambda(s)\sigma(s)(1+\psi(s))\right)ds\right)\prod_{k=1}^{N_t}\left(1+\sigma(T_k)\phi(T_k)\right).$$

In order to price a European option, we compute $\mathbb{E}_{\mathbb{Q}}\left[R(T)(S_T-K)^+\right]$. Let

$$\mathcal{BS}(x,T;r,\sigma^2;K) = \mathbb{E}[e^{-rT}(xe^{rT-\sigma^2T/2+\sigma W_T} - K)^+]$$

denote the classical Black-Scholes function, where W_T is a Gaussian centered random variable with variance T. In the case of deterministic volatility $(\sigma(s), s \geq 0)$ and interest rate $(r(s), s \geq 0)$, the price of a call in the Black-Scholes model is

$$\mathcal{BS}(x,T;R,\Sigma(T);K), \quad \text{with } R = \frac{1}{T}\int_0^T r(s)ds \text{ and } \Sigma(T) = \frac{1}{T}\int_0^T \sigma^2(s)ds.$$

Let $\Gamma^\sigma(t) = \int_0^t i(s)\alpha^2(s)\sigma^2(s)ds$ denote the variance of $\int_0^t i(s)\alpha(s)\sigma(s)d\widetilde{B}_s$, and $\Gamma(t) = \int_0^t \gamma(s)ds$, where $\gamma(t) = \lambda(t)(1+\phi(t)\psi(t))$ denote the compensator of N under $\mathbb{Q}$.

Proposition 10.6.2.3 *In this model, the price of a European option is*

$$\mathbb{E}_{\mathbb{Q}}\left[\exp\left(-\int_0^T r(s)ds\right)(S_T-K)^+\right]$$
$$= \exp\left(-\Gamma_T\right)\sum_{k=0}^{\infty}\frac{1}{k!}\int_0^T\cdots\int_0^T dt_1\cdots dt_k\,\gamma_{t_1}\cdots\gamma_{t_k}$$
$$\mathcal{BS}\left(S_0\exp\left(-\int_0^T(\phi\gamma\sigma)(s)ds\right)\prod_{i=1}^{k}\left(1+\sigma(t_i)\phi(t_i)\right), T; R, \frac{\Gamma_T^\sigma}{T}; K\right).$$

PROOF: We have

$$\mathbb{E}_{\mathbb{Q}}\left[R(T)(S_T - K)^+\right] = \sum_{k=0}^{\infty} \mathbb{E}_{\mathbb{Q}}\left[R(T)(S_T - K)^+ \mid N_T = k\right] \mathbb{Q}(N_T = k),$$

with $\mathbb{Q}(N_T = k) = \exp(-\Gamma_T)(\Gamma_T)^k/k!$. Conditionally on $\{N_T = k\}$, the jump times $(T_1, \ldots, T_n)$ have the law

$$\frac{1}{\Gamma_T^k} 1_{\{0<t_1<\cdots<t_k<T\}} \gamma_{t_1} \cdots \gamma_{t_k} dt_1 \cdots dt_k,$$

since the process $(N_{\Gamma_t^{-1}}, t \geq 0)$ is a standard Poisson process. Hence, conditionally on $\{N(\Gamma^{-1}(\Gamma_T)) = k\} = \{N_T = k\}$, its jump times $(\Gamma_{T_1}, \ldots, \Gamma_{T_k})$ have a uniform law on $[0, \Gamma_T]^k$ (see Subsection 8.2.1). We then use the fact that $\tilde{B}$ and N are also independent under $\mathbb{Q}$, since $(\mu(t), t \geq 0)$ is deterministic, and the identity in law

$$S_T \stackrel{\text{law}}{=} S_0 X_T e^{-\int_0^T \phi(s)\lambda(s)\sigma(s)ds} \prod_{k=1}^{N_T} (1 + \sigma(T_k)\phi(T_k)),$$

where

$$X_T = \exp\left(\int_0^T r(s)ds - \Gamma_T^\sigma/2 + \left(\frac{\Gamma_T^\sigma}{T}\right)^{1/2} W_T\right),$$

with W_T is independent of N.

Comment 10.6.2.4 Within the framework of Privault, the valuation and hedging of European options is presented in Jeanblanc and Privault [485], the valuation of exotic options is studied in El-Khatib [312].

The reader can refer to Hobson and Rogers [442] for a different model of complete market with jumps.

10.7 Valuation of Options

Several articles have focused on the valuation of European options when the underlying value follows a jump-diffusion process. Merton [643] was the first one to obtain a closed form solution assuming that the market price of jump is null (see Subsection 10.5.1). Kou and Wang [542] have studied the case of double exponential jumps. Jump-diffusion models with stochastic volatility and interest rate have also been developed (see for example Scott [776]). The problem of American option valuation is more complex. It was investigated by several authors. Bates [59] derived the early exercise premium by relying on an extension of the MacMillan [608] and Barone-Adesi and Whaley [56] approaches in a jump-diffusion setting.

As shown by Chesney and Jeanblanc [174], this extension generates good results only if the underlying process is continuous at the exercise boundary with probability one (e.g., in the case of perpetual currency calls, when jumps are negative). Otherwise, if the overshoot is strictly positive at the exercise boundary (positive jumps for the currency in the call case) the pricing problem is more difficult to tackle and one should be very cautious when applying a MacMillan approximation. Therefore, a new approach is developed in the paper [174].

Pham [709] considers the American put option valuation in a jump-diffusion model (Merton's assumptions), and relates this problem (which is indeed an optimal stopping problem) to a parabolic integro-differential free boundary problem. By extending the Riesz decomposition obtained by Carr et al. [154] for a diffusion model, Pham derives a decomposition of the American put price as the sum of a European price and an early exercise premium. The latter term requires the identification of the exercise boundary. In the same context, Zhang [873, 875] relies on variational inequalities and shows how to use numerical methods, (finite difference methods), to price the American put. Zhang [874] describes this problem as a free boundary problem, and by using the MacMillan approximation obtains a price for the perpetual put, and an approximation of the finite maturity put price. These results are obtained only when jumps are positive. Mastroeni and Matzeu [628, 627] obtain an extension of Zhang results in a multidimensional state space.

Boyarchenko and Levendorskii [119], Mordecki [659, 658] and Gerber and Shiu [388] also consider the American option pricing problem. They obtain solutions which are explicit only if the distribution of the jump size is exponential or if the jump size is negative for a call (resp. positive for a put). Mordecki [659] establishes the value of the boundary for a perpetual option in terms of the law of the extrema of the underlying Lévy process.

The structure of this section is as follows. We first present the valuation of European calls. We give the explicit solution when the jumps are log-normally distributed. In the case of constant jump size, the PDE for option pricing is then obtained. In Subsection 10.7.2, the perpetual American currency call option and its exercise boundary are considered. Exact analytical solutions are derived when the jump size is negative, i.e., when the overshoot of the exercise boundary is equal to 0. They are based on the computation of the Laplace transform of the first passage time of the process at the exercise boundary, which is obtained by use of the optimal sampling theorem. As in Zhang [874] or Bates [59], we then show, in Subsection 10.7.2, how an accurate approximation for the valuation of finite maturity options can be obtained by relying on the perpetual maturity case. See also $\rightarrowtail$ Chapter 11 for a study if the jumps can take positive values.

10.7.1 The Valuation of European Options

In this subsection, we suppose that the price of the risky asset has dynamics

$$dS_t = S_{t^-}(\mu dt + \sigma dW_t + dM_t)$$

where W is a Brownian motion and M is the compensated martingale associated with a compound Poisson process, i.e., $M_t = \sum_{k=1}^{N_t} Y_k - t\lambda\mathbb{E}(Y_1)$. The coefficients μ, σ, λ are constant. We assume that the law of the jumps Y_k has support in $]-1, \infty[$, so that S remains strictly positive. In a closed form,

$$S_t = S_0 e^{\mu t} \exp\left(\sigma W_t - \frac{1}{2}\sigma^2 t\right) \exp(-t\lambda\mathbb{E}(Y_1)) \prod_{k=1}^{N_t}(1+Y_k)\,.$$

When S is an exchange rate, Merton [643], Lamberton and Lapeyre [559] and Nahum [664] choose to evaluate the derivatives under the e.m.m. $\mathbb{Q}$ under which the underlying spot foreign exchange rate $(S_t, t \geq 0)$ follows the dynamics

$$dS_t = S_{t^-}((r-\delta)dt + \sigma d\widehat{W}_t + dM_t)$$

where $\widehat{W}$ is a $\mathbb{Q}$-Brownian motion and $M_t = \sum_{k=1}^{N_t} Y_k - t\lambda\mathbb{E}(Y_1)$ is a $\mathbb{Q}$-martingale. In an explicit form,

$$S_t = S_0 \exp\left(\left(r - m - \frac{1}{2}\sigma^2\right)t + \sigma\widehat{W}_t\right) \prod_{k=1}^{N_t}(1+Y_k)\,, \tag{10.7.1}$$

where $m = \delta + \lambda\mathbb{E}(Y_1)$. In particular, the Poisson process N has the same intensity under $\mathbb{P}$ and $\mathbb{Q}$, the law of Y_1 is the same under both probabilities (note that, in particular, $\mathbb{E}_{\mathbb{Q}}(Y_1) = \mathbb{E}_{\mathbb{P}}(Y_1)$) and $\sum_{k=1}^{N_t} Y_k$ is a $\mathbb{Q}$-compound Poisson process. As in Merton [643], there is no risk premium for jumps.

The price of a European call can be evaluated as follows: from (10.7.1), we deduce

$$\begin{aligned}
&e^{-rT}\mathbb{E}_{\mathbb{Q}}\left((S_T - K)^+)\right) \\
&= \sum_{n=0}^{\infty} \frac{e^{-(r+\lambda)T}(\lambda T)^n}{n!} \mathbb{E}_{\mathbb{Q}}\left[\left(S_0 e^{(r-m)T}\mathcal{E}(\sigma\widehat{W})_T \prod_{k=1}^{n}(1+Y_k) - K\right)^+\right] \qquad (10.7.2) \\
&= \sum_{n=0}^{\infty} \frac{e^{-\lambda T}(\lambda T)^n}{n!} \int_{-1}^{\infty} \int_{-1}^{\infty} F(dy_1)\cdots F(dy_n) \\
&\qquad \times e^{-rT}\mathbb{E}_{\mathbb{Q}}\left[\left(S_0 e^{(r-m)T}\mathcal{E}(\sigma\widehat{W})_T \prod_{k=1}^{n}(1+y_k) - K\right)^+\right] \\
&= \sum_{n=0}^{\infty} \frac{e^{-\lambda T}(\lambda T)^n}{n!} \int_{-1}^{\infty} \int_{-1}^{\infty} F(dy_1)\cdots F(dy_n) \mathcal{BS}\left(S_0 e^{-mT} \prod_{k=1}^{n}(1+y_k), r, \sigma, T\right)
\end{aligned}$$

where $\mathcal{BS}$ is the value of a plain vanilla call given by

$$\mathcal{BS}(S_0, \theta, \Sigma, T) = S_0 \mathcal{N}(d_1) - K e^{-\theta T} \mathcal{N}(d_2), \qquad (10.7.3)$$

$$d_1 = \frac{\ln(S_0/K) + (\theta + \Sigma^2/2)T}{\Sigma\sqrt{T}}, \quad d_2 = d_1 - \Sigma\sqrt{T}.$$

In the case where $1 + Y_1 \stackrel{\text{law}}{=} e^Z$, where Z is a Gaussian r.v., we obtain, as in Merton, a more pleasant form, using the stability of independent Gaussian random variables under addition in the equality (10.7.2).

Proposition 10.7.1.1 *When* $\ln(1+Y) \stackrel{\text{law}}{=} \mathcal{N}(\mu, \alpha^2)$, *the price of a European call with maturity* T *is given by*

$$C(S_0, T) = \sum_{n=0}^{\infty} \frac{e^{-\lambda T}(\lambda T)^n}{n!} e^{(\theta_n - r)T} \mathcal{BS}(S_0, \theta_n, \Sigma_n, T) \qquad (10.7.4)$$

where $\mathcal{BS}$ *is the value of a plain vanilla call defined in* (10.7.3), *with*

$$\begin{cases} \theta_n = r - \delta - \lambda \mathbb{E}(Y_1) + \frac{n \ln(1+\mathbb{E}(Y_1))}{T} \\ \Sigma_n^2 = \sigma^2 + n\alpha^2/T, \end{cases} . \qquad (10.7.5)$$

PROOF: Assume that the $1 + Y_j$ are independent log-normally distributed r.v.'s, i.e., $\ln(1 + Y_1) \stackrel{\text{law}}{=} \mathcal{N}(\mu, \alpha^2)$. The expectation

$$\mathbb{E}_{\mathbb{Q}}\left[\left(e^{(r-m)T}\mathcal{E}(\sigma\widehat{W})_t \prod_{j=1}^{n}(1 + Y_j) - K\right)^+\right]$$

in equation (10.7.2) is equal to

$$\mathbb{E}_{\mathbb{Q}}\left[\left(\exp\left(\left(r - m - \frac{\sigma^2}{2}\right)T + \sum_{k=1}^{n}\ln(1 + Y_k) + \sigma\widehat{W}_T\right) - K\right)^+\right].$$

The r.v. $\sum_{k=1}^n \ln(1+Y_k) + \sigma\widehat{W}_T$ is normally distributed with expectation and variance respectively equal to $n\mu$ and $n\alpha^2 + \sigma^2 T = T\Sigma_n^2$. Therefore,

$$\begin{aligned} &\left(r - m - \frac{\sigma^2}{2}\right)T + \sum_{k=1}^{n}\ln(1 + Y_k) + \sigma\widehat{W}_T \\ &= \left(r - \delta - \lambda\mathbb{E}(Y_1) - \frac{\sigma^2}{2}\right)T + \sum_{k=1}^{n}\ln(1 + Y_k) + \sigma\widehat{W}_T \\ &\stackrel{\text{law}}{=} \theta_n T + \Sigma_n\sqrt{T}G - \frac{\Sigma_n^2 T}{2} \end{aligned}$$

where Σ_n and θ_n are given by equations (10.7.5) and G is a standard Gaussian variable.

It is straightforward to obtain the result given by formula (10.7.4). Indeed, the latter expectation corresponds to the payoff of a standard European call, with underlying adjusted drift and volatility respectively equal to θ_n and Σ_n.

The term $e^{(\theta_n - r)T} \mathcal{BS}(S_0, \theta_n, \Sigma_n, T)$ in formula (10.7.4) is, therefore, the value of the option conditional on knowing that exactly n Poisson jumps will occur during the life of the option. The call price is a weighted average of these prices. The weights are the probabilities that a Poisson random variable takes the value n, for $n \in \mathbb{N}$. □

Comments 10.7.1.2 (a) See Lipton [597] for a general discussion.

(b) The valuation of lookback options is studied in Nahum [664].

(c) Sepp [780] studies the barrier options by means of the Laplace transform in time of the price.

(d) Asian options are presented in Bayraktar and Xing [62] and in Boughamoura et al. [113] in a Lévy setting.

PDE for Option Prices

Suppose now that the dynamics of the risky asset (the currency) are

$$dS_t = S_{t^-}((r - \delta)dt + \sigma dW_t + \phi\, dM_t) \tag{10.7.6}$$

under the chosen risk-neutral probability. Here, W is a BM and M is the compensated martingale of a Poisson process with constant intensity λ and the coefficients r, δ, σ and $\phi > -1$ are supposed to be constant. Our aim is to compute $C(S_t, T - t) = \mathbb{E}(e^{-r(T-t)} h(S_T) | \mathcal{F}_t)$ where h is a smooth function. Applying the Feynman-Kac formula we obtain that, if C satisfies

$$\begin{aligned} \frac{\sigma^2}{2} x^2 \frac{\partial^2 C}{\partial x^2}(x,\tau) + (r - \delta - \phi\lambda) x \frac{\partial C}{\partial x}(x,\tau) - rC(x,\tau) - \frac{\partial C}{\partial \tau}(x,\tau) \\ + \lambda[C((1+\phi)x, \tau) - C(x,\tau)] = 0 \\ C(x,0) = h(x) \end{aligned} \tag{10.7.7}$$

where $\tau = T - t$, then $C(S_t, T-t)$ is the value of the European call. However this kind of PDE is difficult to solve.

10.7.2 American Option

We apply the previous computation to the case of an American option. We suppose that the dynamics of the underlying asset are given by (10.7.6) and that the jump size is constant and equal to ϕ. The discounted American call price is a martingale in the continuation region. This means that if the spot price x is lower than the exercise boundary value at time t, the American call value C_A satisfies (10.7.7).

If the jump is positive, and if x belongs to the interval $[\frac{\bar{x}}{1+\phi}, \bar{x}]$, where $\bar{x}$ is the exercise boundary level at time t, the American call value satisfies

$$\frac{\sigma^2}{2}x^2\frac{\partial^2 C_A}{\partial x^2}(x,\tau) + (r - \delta - \phi\lambda)x\frac{\partial C_A}{\partial x}(x,\tau) - rC_A(x,\tau)$$
$$- \frac{\partial C_A}{\partial \tau}(x,\tau) + \lambda((1+\phi)x - K - C_A(x,\tau)) = 0\,,$$

because in this case, after the jump, the American call value is equal to its intrinsic value.

The Valuation of the Perpetual American Option: Negative Jumps

We now assume that the jump is negative $-1 < \phi \leq 0$. Zhang's results [874] concerning the perpetual option value and the exercise boundary will be rederived in this context.

The exercise boundary is defined as follows: for a given time $t \in [0, T]$

$$b_c(T-t) = \inf\{x \geq 0 \,:\, x - K = C_A(x, T-t)\}\,. \tag{10.7.8}$$

It is worth mentioning that, with a strictly positive foreign interest rate, American and European call options have different prices. When the maturity tends to infinity, the exercise boundary admits a limit b_c^* (see Mordecki [659]). As shown below, this limit b_c^* is finite (for $\delta > 0$). The option value is given by:

$$C_A(S_0, +\infty) = \sup_\tau \mathbb{E}((S_\tau - K)e^{-r\tau}) := C_A(S_0) \tag{10.7.9}$$

where τ runs over stopping times. Here, the $+\infty$ indicates that the maturity is infinite. It is proved in Mordecki [659] that one can restrict attention to the case of first passage time stopping times. Since the jump size is negative, the process is continuous on the boundary, therefore,

$$C_A(S_0) = \sup_{L \geq S_0} [(L-K)\mathbb{E}(e^{-rT_L})] = (b_c^* - K)\mathbb{E}(e^{-rT_{b_c^*}}) \tag{10.7.10}$$

where T_L is the first passage time of the process S out of the continuation region :

$$T_L = \inf\{t \geq 0 \,:\, S_t \geq L\}\,.$$

From (10.4.6), with $\ell = \ln(L/S_0)$, the Laplace transform of the hitting time is

$$\mathbb{E}_{\mathbb{Q}}(e^{-rT_L}) = \left(\frac{S_0}{L}\right)^\rho \wedge 1$$

where ρ is equal to $g^{-1}(r)$ (defined in 10.4.6) and is strictly greater than 1. Indeed, in the case of the call, we use the positive root k of $g(k) = r$ because we want the call value to be an increasing function of the underlying asset value. (See (10.4.5) for the definition of the Laplace exponent $g(k)$.) We can thus derive the value of the exercise boundary as the value where the supremum in equation (10.7.10) is attained:

$$b_c^* = \frac{\rho K}{\rho - 1}. \tag{10.7.11}$$

In the continuation region, (i.e., if $S_0 < \frac{\rho K}{\rho - 1}$), the option value is:

$$C_A(S_0) = \left(\frac{S_0}{b_c^*}\right)^\rho (b_c^* - K). \tag{10.7.12}$$

Without jumps ($\lambda = 0$), we obtain the known formula (3.11.13). Indeed, in this case $g(k) = (r - \delta)k + \frac{1}{2}\sigma^2 k(k-1)$, hence $\rho = \frac{-\nu + \sqrt{\nu^2 + 2r}}{\sigma}$ with $\nu = \left(r - \delta - \sigma^2/2\right)/\sigma$.

Put prices can be obtained by the symmetry formula (see Subsection 10.4.2), used in the case $\phi > 0$:

$$P_A(S_0, K, r, \delta; \sigma, \phi, \lambda) = K S_0 C_A(1/S_0, 1/K, \delta, r; \sigma, -\frac{\phi}{1+\phi}, \lambda(1+\phi)).$$

Do not forget, when applying this formula, that the price of the put on the left-hand side is evaluated under the risk-neutral probability in the domestic economy, whereas on the right-hand side, the call should be evaluated under the foreign risk-neutral probability. By relying on Subsection 10.4.2, the exercise boundary of the put can be obtained:

$$b_p(K, T-t, r, \delta; \phi, \lambda)\, b_c(K, T-t, \delta; r, \frac{-\phi}{1+\phi}, \lambda(1+\phi)) = K^2, \tag{10.7.13}$$

where now $\phi > 0$.

An Approximation of the American Option Value

Let us now rely on the Barone-Adesi and Whaley approach [56] and on Bates's article [59]. If the American and European option values satisfy the same linear PDE (10.7.7) in the continuation region, their difference DC (the American premium) must also satisfy this PDE in the same region. Write:

$$DC(S_0, T) = y f(S_0, y)$$

where $y = 1 - e^{-rT}$, and where f is a function of two arguments that has to be determined. In the continuation region, f satisfies the following PDE which is obtained by a change of variables:

$$\frac{\sigma^2}{2}x^2\frac{\partial^2 f}{\partial x^2}+(r-\delta)x\frac{\partial f}{\partial x}-\frac{rf}{y}-(1-y)r\frac{\partial f}{\partial y}-\lambda[\phi x\frac{\partial f}{\partial x}-f((1+\phi)x,y)+f(x,y)]=0\,.$$

Let us now assume that the derivative of f with respect to y may be neglected. Whether or not this is a good approximation is an empirical issue that we do not discuss here.

The equation now becomes an ODE

$$\frac{\sigma^2}{2}x^2\frac{\partial^2 f}{\partial x^2}+(r-\delta)x\frac{\partial f}{\partial x}-\frac{rf}{y}-\lambda[\phi x\frac{\partial f}{\partial x}-f((1+\phi)x,y)+f(x,y)]=0\,.$$

The value of the perpetual option satisfies almost the same ODE. The only difference is that y is equal to 1 in the perpetual case, and therefore we have r/y instead of r in the third term of the left-hand side. The form of the solution is known (see (10.7.12)):

$$f(S_0,y)=zS_0^\rho\,.$$

Here, z is still unknown, and ρ is the positive solution of equation $g(k)=r/y$.

When S_0 tends to the exercise boundary $b_c(T)$, by the continuity of the option value, the following equation is satisfied from the definition of the American premium $DC(x,T)$:

$$b_c(T)-K=C_E(b_c(T),T)+yz\,b_c(T)^\rho\,, \tag{10.7.14}$$

and by use of the smooth-fit condition[1], the following equation is obtained:

$$1=\frac{\partial C_E}{\partial x}(b_c(T)),T)+yz\rho(b_c(T))^{\rho-1}\,. \tag{10.7.15}$$

In a jump-diffusion model this condition was derived by Zhang [873] in the context of variational inequalities and by Pham [709] with a free boundary formulation. We thus have a system of two equations (10.7.14) and (10.7.15) and two unknowns z and $b_c(T)$. This system can be solved: $b_c(T)$ is the implicit solution of

$$b_c(T)=K-C_E(b_c(T),T)+\left(1-\frac{\partial C_E}{\partial x}(b_c(T),T)\right)\frac{b_c(T)}{\rho}\,.$$

If $S_0>b_c(T)$, $C_A(S_0,T)=S_0-K$. Otherwise, if $S_0<b_c(T)$, the approximate formula is

$$C_A(S_0,T)=C_E(S_0,T)+A(S_0/b_c(T))^\rho \tag{10.7.16}$$

with

$$A=\left(1-\frac{\partial C_E}{\partial x}(b_c(T),T)\right)\frac{b_c(T)}{\rho}\,.$$

[1] The smooth-fit condition ensures that the solution of the PDE is C^1 at the boundary. See Villeneuve [830].

Here, (see (10.7.4)),

$$C_E(S_0,T) = \sum_{n=0}^{\infty} \frac{e^{-\lambda T}(\lambda T)^n}{n!} e^{\Gamma_n T} \mathcal{BS}(S_0, r + \Gamma_n, \sigma, T)$$

and $\mathcal{BS}$ is the Black and Scholes function:

$$\mathcal{BS}(S_0,\theta,\sigma,T) = S_0 \mathcal{N}(d_1) - K e^{-\theta T} \mathcal{N}(d_2)\,,$$

$$\Gamma_n = -\delta - \phi\lambda + \frac{n \ln(1+\phi)}{T}\,,$$

$$d_1 = \frac{\ln(S_0/K) + (\theta + \frac{\sigma^2}{2})T}{\sigma\sqrt{T}}\,,\; d_2 = d_1 - \sigma\sqrt{T}\,.$$

This approximation was obtained by Bates [59], for the put, in a more general context in which $1+\phi$ is a log-normal random variable. This means that his results could even be used with positive jumps for an American call. However, as shown in Subsection 10.7.2, in this case the differential equation whose solution is the American option approximation value, takes a specific form just below the exercise boundary. Unfortunately, there is no known solution to this differential equation. Positive jumps generate discontinuities in the process on the exercise boundary, and therefore the problem is more difficult to solve (see $\rightarrowtail$ Chapter 11).

11

Lévy Processes

In this chapter, we present briefly Lévy processes and some of their applications to finance. Lévy processes provide a class of models with jumps which is sufficiently rich to reproduce empirical data and allow for some explicit computations.

In a first part, we are concerned with infinitely divisible laws and in particular, the stable laws and self-decomposable laws. We give, without proof, the Lévy-Khintchine representation for the characteristic function of an infinitely divisible random variable.

In a second part, we study Lévy processes and we present some martingales in that setting. We present stochastic calculus for Lévy processes and changes of probability. We study more carefully the case of exponentials and stochastic exponentials of Lévy processes.

In a third part, we develop briefly the fluctuation theory and we proceed with the study of Lévy processes without positive jumps and increasing Lévy processes.

We end the chapter with the introduction of some classes of Lévy processes used in finance such as the CGMY processes and we give an application of fluctuation theory to perpetual American option pricing.

The main books on Lévy processes are Bertoin [78], Doney [260], Itô [463], Kyprianou [553], Sato [761], Skorokhod [801, 802] and Zolotarev [878]. Each of these books has a particular emphasis: Bertoin's has a general Markov processes flavor, mixing deeply analysis and study of trajectories, Sato's concentrates more on the infinite divisibility properties of the one-dimensional marginals involved and Skorokhod's describes in a more general setting processes with independent increments. Kyprianou [553] presents a study of global and local path properties and local time. The tutorial papers of Bertoin [81] and Sato [762] provide a concise introduction to the subject from a mathematical point of view.

Various applications to finance can be found in the books by Barndorff-Nielsen and Shephard [55], Boyarchenko and Levendorskii [120], Cont and Tankov [192], Overhaus et al. [690], Schoutens [766]. Barndorff-Nielsen and

M. Jeanblanc, M. Yor, M. Chesney, *Mathematical Methods for Financial Markets*, Springer Finance, DOI 10.1007/978-1-84628-737-4_11,

Shephard deal with simulation of Lévy processes and stochastic volatility, Cont and Tankov with simulation, estimation, option pricing and integro-differential equations and Overhaus et al. (the quantitative research team of Deutsche Bank) with various applications of Lévy processes in quantitative research, (as equity linked structures and volatility modeling); Schoutens presents the mathematical tools in a concise manner.

The books [554] and [53] contain many interesting papers with application to finance. Control theory for Lévy processes can be found in Øksendal and Sulem [685].

11.1 Infinitely Divisible Random Variables

11.1.1 Definition

In what follows, we denote by $x \cdot y$ the scalar product of x and y, two elements of $\mathbb{R}^d$, and by $|x|$ the euclidean norm of x.

Definition 11.1.1.1 *A random variable X taking values in $\mathbb{R}^d$ with distribution μ is said to be* ***infinitely divisible*** *if its characteristic function $\hat{\mu}(u) = \mathbb{E}(e^{iu \cdot X})$ where $u \in \mathbb{R}^d$, may be written for any integer n as the nth-power of a characteristic function $\widehat{\mu}_n$, that is if*

$$\widehat{\mu}(u) = (\widehat{\mu}_n(u))^n .$$

By a slight abuse of language, we shall also say that such a characteristic function (or distribution function) is infinitely divisible. Equivalently, X is infinitely divisible if

$$\forall n, \ \exists (X_i^{(n)}, i \leq n, \text{ i.i.d.}) \text{ such that } \ X \stackrel{\text{law}}{=} X_1^{(n)} + X_2^{(n)} + \cdots + X_n^{(n)} .$$

Example 11.1.1.2 A Gaussian variable, a Cauchy variable, a Poisson variable and the hitting time of the level a for a Brownian motion are examples of infinitely divisible random variables. Gamma, Inverse Gaussian, Normal Inverse Gaussian and Variance Gamma variables are also infinitely divisible (see ↣ Examples 11.1.1.9 for details and ↣ Appendix A.4.4 and A.4.5 for definitions of these laws). A uniformly distributed random variable, and more generally any bounded random variable is not infinitely divisible. The next Proposition 11.1.1.4 will play a crucial rôle in the description of infinitely divisible laws.

Definition 11.1.1.3 *A* ***Lévy measure*** *on $\mathbb{R}^d$ is a positive measure ν on $\mathbb{R}^d \setminus \{0\}$ such that*

$$\int_{\mathbb{R}^d \setminus \{0\}} (1 \wedge |x|^2)\, \nu(dx) < \infty ,$$

i.e.,

$$\int_{|x|>1} \nu(dx) < \infty \text{ and } \int_{0<|x|\leq 1} |x|^2 \nu(dx) < \infty\,.$$

In Feller [343], Lukacs [605], Sato [761] and Zolotarev [878], the reader will find more properties of infinitely divisible random variables, as well as the proofs of the following results.

Proposition 11.1.1.4 (Lévy-Khintchine Representation.) *If X is an infinitely divisible random variable with characteristic function $\widehat{\mu}$, there exists a unique triple (m, A, ν) where $m \in \mathbb{R}^d$, A is a positive matrix and ν is a Lévy measure such that*

$$\boxed{\widehat{\mu}(u) = \exp\left(iu \cdot m - \tfrac{1}{2} u \cdot Au + \int_{\mathbb{R}^d} (e^{iu \cdot x} - 1 - iu \cdot x \mathbb{1}_{\{|x|\leq 1\}})\nu(dx)\right)} \tag{11.1.1}$$

SKETCH OF THE PROOF: The Lévy-Khintchine representation formula may be obtained by proving that an infinitely divisible random variable is the limit in law of random variables of the form

$$Z^{(n)} + \sum_{k=1}^{N_1^{(n)}} Y_k^{(n)}$$

where $Z^{(n)}$ is a Gaussian r.v. and $\sum_{k=1}^{N_1^{(n)}} Y_k^{(n)}$ a compound Poisson process (see Section 8.6) evaluated at time $t = 1$. The characteristic function for a compound Poisson process is given in Proposition 8.6.3.4. See Bertoin [78] or Sato [761] for a complete proof of this representation formula. □

▶ When $\int_{\mathbb{R}^d \setminus \{0\}} \mathbb{1}_{\{|x|\leq 1\}} |x| \nu(dx) < \infty$, we can write (11.1.1) in the **reduced form**

$$\widehat{\mu}(u) = \exp\left(iu \cdot m_0 - \frac{1}{2} u \cdot Au + \int_{\mathbb{R}^d} (e^{iu \cdot x} - 1)\nu(dx)\right), \tag{11.1.2}$$

where $m_0 = m - \int_{\mathbb{R}^d} x \mathbb{1}_{\{|x|\leq 1\}} \nu(dx)$. We shall use this reduced form representation whenever possible.

▶ When $\int_{\mathbb{R}^d} \mathbb{1}_{\{|x|>1\}} |x| \nu(dx) < \infty$, it is possible to write (11.1.1) in the form

$$\widehat{\mu}(u) = \exp\left(iu \cdot \tilde{m} - \frac{1}{2} u \cdot Au + \int_{\mathbb{R}^d} (e^{iu \cdot x} - 1 - iu \cdot x)\nu(dx)\right),$$

where $\tilde{m} = m + \int_{|x|>1} x\, \nu(dx)$. In that case, by differentiation of the Fourier transform $\widehat{\mu}(u)$ with respect to u,

$$\mathbb{E}(X) = -i\widehat{\mu}'(0) = \tilde{m} = m + \int_{|x|>1} x\,\nu(dx)\,.$$

If the r.v. X is positive, the Lévy measure ν is a measure on $]0,\infty[$ with $\int_{]0,\infty[}(1 \wedge x)\nu(dx) < \infty$ (see $\rightarrowtail$ Proposition 11.2.3.11). It is more natural, in this case, to consider, for $\lambda > 0$, $\mathbb{E}(e^{-\lambda X})$ the Laplace transform of X, and now, the Lévy-Khintchine representation takes the form

$$\mathbb{E}(e^{-\lambda X}) = \exp\left(-\left(\lambda m_0 + \int_{]0,\infty[} \nu(dx)(1-e^{-\lambda x})\right)\right).$$

Remark 11.1.1.5 The following converse of the Lévy-Khintchine representation holds true: any function ϑ of the form

$$\vartheta(u) = \exp\left(iu\centerdot m - u\centerdot Au + \int_{\mathbb{R}^d}(e^{iu\centerdot x} - 1 - iu\centerdot x\mathbb{1}_{\{|x|\leq 1\}})\nu(dx)\right)$$

where ν is a Lévy measure and A a positive matrix, is a characteristic function which is obviously infinitely divisible (take $m_n = m/n, A_n = A/n, \nu_n = \nu/n$). Hence, there exists μ, infinitely divisible, such that $\vartheta = \widehat{\mu}$.

Warning 11.1.1.6 Some authors use a slightly different representation for the Lévy-Khintchine formula. They define a **centering function** (also called a truncation function) as an $\mathbb{R}^d$-valued measurable bounded function h such that $(h(x)-x)/|x|^2$ is bounded. Then they prove that, if X is an infinitely divisible random variable, there exists a triple (m_h, A, ν), where ν is a Lévy measure, such that

$$\widehat{\mu}(u) = \exp\left(iu\centerdot m_h - \frac{1}{2}u\centerdot Au + \int_{\mathbb{R}^d}(e^{iu\centerdot x} - 1 - iu\centerdot h(x))\nu(dx)\right).$$

Common choices of centering functions on $\mathbb{R}$ are $h(x) = x\mathbb{1}_{\{|x|\leq 1\}}$, as in Proposition 11.1.1.4 or $h(x) = \frac{x}{1+x^2}$ (Kolmogorov centering). The triple (m_h, A, ν) is called a characteristic triple. The parameters A and ν do not depend on h; when h is the centering function of Proposition 11.1.1.4, we do not indicate the dependence on h for the parameter m.

Remark 11.1.1.7 The choice of the level 1 in the common centering function $h(x) = \mathbb{1}_{\{|x|\leq 1\}}$ is not essential and, up to a change of the constant m, any centering function $h_r(x) = \mathbb{1}_{\{|x|\leq r\}}$ can be considered.

Comment 11.1.1.8 In the one-dimensional case, when the law of X admits a second order moment, the Lévy-Khintchine representation was obtained by Kolmogorov [536]. Kolmogorov's measure $\widetilde{\nu}$ corresponds to the representation

$$\exp\left(ium + \int_{\mathbb{R}} \frac{e^{iux} - 1 - iux}{x^2}\widetilde{\nu}(dx)\right).$$

Hence, $\nu(dx) = \mathbb{1}_{\{x\neq 0\}}\widetilde{\nu}(dx)/x^2$ and the mass of $\widetilde{\nu}$ at 0 corresponds to the Gaussian term.

Example 11.1.1.9 We present here some examples of infinitely divisible laws and give their characteristics in reduced form whenever possible (see equation (11.1.2)).

- **Gaussian Laws.** The Gaussian law $\mathcal{N}(a,\sigma^2)$ has characteristic function $\exp(iua - u^2\sigma^2/2)$. Its characteristic triple in reduced form is $(a,\sigma^2,0)$.
- **Cauchy Laws.** The Cauchy law with parameter $c > 0$ has the characteristic function

$$\exp(-c|u|) = \exp\left(\frac{c}{\pi}\int_{-\infty}^{\infty}(e^{iux}-1)x^{-2}dx\right).$$

Here, we make the convention

$$\int_{-\infty}^{\infty}(e^{iux}-1)x^{-2}dx = \lim_{\epsilon\to 0}\int_{-\infty}^{\infty}(e^{iux}-1)x^{-2}\mathbb{1}_{\{|x|\geq\epsilon\}}dx.$$

The reduced form of the characteristic triple for a Cauchy law is

$$(0,0,c\pi^{-1}x^{-2}dx).$$

- **Gamma Laws.** If X follows a $\Gamma(a,\nu)$ law, its Laplace transform is, for $\lambda > 0$,

$$\mathbb{E}(e^{-\lambda X}) = \left(1+\frac{\lambda}{\nu}\right)^{-a} = \exp\left(-a\int_0^{\infty}(1-e^{-\lambda x})e^{-\nu x}\frac{dx}{x}\right).$$

Hence, the reduced form of the characteristic triple for a Gamma law is $(0,0,\mathbb{1}_{\{x>0\}}\,ax^{-1}e^{-\nu x}dx)$.
- **Brownian Hitting Times.** The first hitting time of $a > 0$ for a Brownian motion has characteristic triple (in reduced form)

$$\left(0,0,\frac{a}{\sqrt{2\pi}}\,x^{-3/2}\mathbb{1}_{\{x>0\}}dx\right).$$

Indeed, we have seen in Proposition 3.1.6.1 that $\mathbb{E}(e^{-\lambda T_a}) = e^{-a\sqrt{2\lambda}}$. Moreover, from ↣ Appendix A.5.8

$$\sqrt{2\lambda} = \frac{1}{\sqrt{2}\Gamma(1/2)}\int_0^{\infty}(1-e^{-\lambda x})x^{-3/2}dx,$$

hence, using that $\Gamma(1/2) = \sqrt{\pi}$

$$\mathbb{E}(e^{-\lambda T_a}) = \exp\left(-\frac{a}{\sqrt{2\pi}}\int_0^{\infty}(1-e^{-\lambda x})\,x^{-3/2}dx\right).$$

- **Inverse Gaussian Laws.** The Inverse Gaussian laws (see ↣ Appendix A.4.4) have characteristic triple (in reduced form)

$$\left(0, 0, \frac{a}{\sqrt{2\pi x^3}} \exp\left(-\frac{1}{2}\nu^2 x\right) \mathbb{1}_{\{x>0\}} dx\right).$$

Indeed

$$\begin{aligned}
&\exp\left(-\frac{a}{\sqrt{2\pi}} \int_0^\infty \frac{dx}{x^{3/2}} (1 - e^{-\lambda x})\, e^{-\nu^2 x/2}\right) \\
&= \exp\left(-\frac{a}{\sqrt{2\pi}} \int_0^\infty \frac{dx}{x^{3/2}} \left((e^{-\nu^2 x/2} - 1) + (1 - e^{-(\lambda+\nu^2/2)x})\right)\right) \\
&= \exp(-a(-\nu + \sqrt{\nu^2 + 2\lambda})
\end{aligned}$$

is the Laplace transform of the first hitting time of a for a Brownian motion with drift ν.

11.1.2 Self-decomposable Random Variables

We now focus on a particular class of infinitely divisible laws.

Definition 11.1.2.1 *A random variable is **self-decomposable** (or of class L) if*

$$\forall c \in]0, 1[, \ \forall u \in \mathbb{R}, \ \widehat{\mu}(u) = \widehat{\mu}(cu)\widehat{\mu}_c(u),$$

where $\widehat{\mu}_c$ is a characteristic function.

In other words, X is self-decomposable if

$$\text{for } 0 < c < 1, \ \exists X_c \ \text{ such that } \ X \overset{\text{law}}{=} cX + X_c$$

where on the right-hand side the r.v's X and X_c are independent. Intuitively, we "compare" X with its multiple cX, and need to add a "residual" variable X_c to recover X.

We recall some properties of self-decomposable variables (see Sato [761] for proofs). Self-decomposable variables are infinitely divisible. The Lévy measure of a self-decomposable r.v. is of the form

$$\nu(dx) = \frac{h(x)}{|x|} dx$$

where h is increasing for $x > 0$ and decreasing for $x < 0$.

Proposition 11.1.2.2 (Sato's Theorem.) *If the r.v. X is self-decomposable, then $X \overset{\text{law}}{=} Z_1$ where the process Z satisfies $Z_{ct} \overset{\text{law}}{=} cZ_t$ and has independent increments.*

Processes with independent increments are called additive processes in Sato [761], Chapter 2. Self-decomposable random variables are linked with Lévy processes in a number of ways, in particular the following Jurek-Vervaat representation [499] of self-decomposable variables X, which, for simplicity, we assume to take values in $\mathbb{R}^+$. For the definitions of Lévy processes and subordinators, see $\rightarrowtail$ Subsection 11.2.1.

Proposition 11.1.2.3 (Jurek-Vervaat Representation.) *A random variable $X \geq 0$ is self-decomposable if and only if it satisfies $X \stackrel{\text{law}}{=} \int_0^\infty e^{-s} dY_s$, where $(Y_s, s \geq 0)$ denotes a subordinator, called the* ***background driving Lévy process*** *(BDLP) of X.*

The Laplace transforms of the random variables X and Y_1 are related by the following

$$\mathbb{E}(\exp(-\lambda Y_1)) = \exp\left(\lambda \frac{d}{d\lambda} \ln \mathbb{E}(\exp(-\lambda X))\right).$$

Example 11.1.2.4 We present some examples of self-decomposable random variables:

- **First Hitting Times for Brownian Motion.** Let W be a Brownian motion. The random variable $T_r = \inf\{t : W_t = r\}$ is self-decomposable. Indeed, for $0 < \lambda < 1$,

$$\begin{aligned} T_r &= T_{\lambda r} + (T_r - T_{\lambda r}) \\ &= \lambda^2 \widehat{T}_r + (T_r - T_{\lambda r}) \end{aligned}$$

  where $\widehat{T}_r = \frac{1}{\lambda^2} T_{\lambda r} \stackrel{\text{law}}{=} T_r$, and $\widehat{T}_r$ and $(T_r - T_{\lambda r})$ are independent, as a consequence of both the scaling property of the process $(T_a, a \geq 0)$ and the strong Markov property of the Brownian motion process at $T_{\lambda r}$ (see Subsection 3.1.2).
- **Last Passage Times for Transient Bessel Processes.** Let R be a transient Bessel process (with index $\nu > 0$) and

$$\Lambda_r = \sup\{t : R_t = r\}.$$

  The random variable Λ_r is self-decomposable. To prove the self-decomposability, we use an argument similar to the previous one, i.e., independence for pre- and post-Λ_r processes, although Λ_r is not a stopping time.

Comment 11.1.2.5 See Sato [760], Jeanblanc et al. [484] and Shanbhag and Sreehari [783] for more information on self-decomposable r.v's and BDLP's. In Madan and Yor [613] the self-decomposability property is used to construct martingales with given marginals. In Carr et al. [152] the risk-neutral process is modeled by a self-decomposable process.

11.1.3 Stable Random Variables

Definition 11.1.3.1 *A real-valued r. v. is* **stable** *if for any $a > 0$, there exist $b > 0$ and $c \in \mathbb{R}$ such that $[\widehat{\mu}(u)]^a = \widehat{\mu}(bu)\, e^{icu}$. A random variable is* **strictly stable** *if for any $a > 0$, there exists $b > 0$ such that $[\widehat{\mu}(u)]^a = \widehat{\mu}(bu)$.*

In terms of r.v's, X is stable if

$$\forall n, \ \exists(\beta_n, \gamma_n), \ \text{ such that } \ X_1^{(n)} + \cdots + X_n^{(n)} \stackrel{\text{law}}{=} \beta_n X + \gamma_n$$

where $(X_i^{(n)}, i \leq n)$ are i.i.d. random variables with the same law as X.

For a strictly stable r.v., it can be proved, with the notation of the definition, that b (which depends on a) is of the form $b(a) = ka^{1/\alpha}$ with $0 < \alpha \leq 2$. The r.v. X is then said to be **α-stable.** For $0 < \alpha < 2$, an α-stable random variable satisfies $\mathbb{E}(|X|^\gamma) < \infty$ if and only if $\gamma < \alpha$. The second order moment exists if and only if $\alpha = 2$, and in that case X is a Gaussian random variable, hence has all moments.

A stable random variable is self-decomposable and hence is infinitely divisible.

Proposition 11.1.3.2 *The characteristic function μ of an α-stable law can be written*

$$\widehat{\mu}(u) = \begin{cases} \exp(imu - \frac{1}{2}\sigma^2 u^2), & \text{for } \alpha = 2 \\ \exp\left(imu - \gamma|u|^\alpha[1 - i\beta \,\text{sgn}(u) \tan(\pi\alpha/2)]\right), & \text{for } \alpha \neq 1, \neq 2 \\ \exp\left(imu - \gamma|u|(1 + i\beta \ln|u|)\right), & \alpha = 1 \end{cases}$$

where $\beta \in [-1, 1]$ and $m, \gamma, \sigma \in \mathbb{R}$. For $\alpha \neq 2$, the Lévy measure of an α-stable law is absolutely continuous with respect to the Lebesgue measure, with density

$$\nu(dx) = \begin{cases} c^+ x^{-\alpha-1} dx & \text{if } x > 0 \\ c^- |x|^{-\alpha-1} dx & \text{if } x < 0\,. \end{cases} \tag{11.1.3}$$

Here, $c^\pm$ are positive real numbers given by

$$\begin{aligned} c^+ &= \frac{1}{2}(1+\beta)\frac{\alpha\gamma}{\Gamma(1-\alpha)\cos(\alpha\pi/2)}\,, \\ c^- &= \frac{1}{2}(1-\beta)\frac{\alpha\gamma}{\Gamma(1-\alpha)\cos(\alpha\pi/2)}\,. \end{aligned}$$

Conversely, if ν is a Lévy measure of the form (11.1.3), we obtain the characteristic function of the law on setting $\beta = (c^+ - c^-)/(c^+ + c^-)$.

For $\alpha = 1$, the definition of $c^{\pm}$ can be given by passing to the limit: $c^{\pm} = 1 \pm \beta$. For $\beta = 0, m = 0$, X is said to have a symmetric stable law. In that case

$$\widehat{\mu}(u) = \exp(-\gamma |u|^{\alpha}) .$$

Example 11.1.3.3 A Gaussian variable is α-stable with $\alpha = 2$. The Cauchy law is stable with $\alpha = 1$. The hitting time $T_1 = \inf\{t : W_t = 1\}$ where W is a Brownian motion is a $(1/2)$-stable variable.

Comment 11.1.3.4 The reader can refer to Bondesson [108] for a particular class of infinitely divisible random variables and to Samorodnitsky and Taqqu [756] and Zolotarev [878] for an extensive study of stable laws and stable processes. See also Lévy-Véhel and Walter [586] for applications to finance.

11.2 Lévy Processes

11.2.1 Definition and Main Properties

Definition 11.2.1.1 *Let $(\Omega, \mathcal{F}, \mathbb{P})$ be a probability space. An $\mathbb{R}^d$-valued process X such that $X_0 = 0$ is a* ***Lévy process*** *if*

(a) *for every $s, t \geq 0, X_{t+s} - X_s$ is independent of $\mathcal{F}_s^X$,*
(b) *for every $s, t \geq 0$ the r.v's $X_{t+s} - X_s$ and X_t have the same law,*
(c) *X is continuous in probability, i.e., for fixed t, $\mathbb{P}(|X_t - X_u| > \epsilon) \to 0$ when $u \to t$ for every $\epsilon > 0$.*

It can be shown that up to a modification, a Lévy process is càdlàg (it is in fact a semi-martingale, see $\rightarrowtail$ Corollary 11.2.3.8).

Example 11.2.1.2 Brownian motion, Poisson processes and compound Poisson processes are examples of Lévy processes (see Section 8.6). If X is a Lévy process, C a matrix and D a vector, $CX_t + Dt$ is also a Lévy process. More generally, the sum of two independent Lévy processes is a Lévy process.

One can easily generalize this definition to **F**-Lévy processes, where **F** is a given filtration, by changing (a) into:

(a') for every s, t, $X_{t+s} - X_s$ is independent of $\mathcal{F}_s$.

Another generalization consists of the class of additive processes that satisfy (a) and often (c), but not (b). Natural examples of additive processes are the processes $(T_a, a \geq 0)$ of first hitting times of levels by a diffusion Y, a consequence of the strong Markov property.

A particular class of Lévy processes is that of subordinators:

Definition 11.2.1.3 *A Lévy process that takes values in* $[0, \infty[$ *(equivalently, which has increasing paths) is called a* ***subordinator***.

In this case, the parameters in the Lévy-Khintchine formula are $m \geq 0, \sigma = 0$ and the Lévy measure ν is a measure on $]0, \infty[$ with $\int_{]0,\infty[}(1 \wedge x)\nu(dx) < \infty$. This last property is a consequence of $\rightarrowtail$ Proposition 11.2.3.11.

Proposition 11.2.1.4 (Strong Markov Property.) *Let* X *be an* $\mathbf{F}$*-Lévy process and* τ *an* $\mathbf{F}$*-stopping time. Then, on the set* $\{\tau < \infty\}$ *the process* $Y_t = X_{\tau+t} - X_\tau$ *is an* $(\mathcal{F}_{\tau+t}, t \geq 0)$*-Lévy process; in particular,* Y *is independent of* $\mathcal{F}_\tau$ *and has the same law as* X.

PROOF: Let us set $\varphi(t; u) = \mathbb{E}(e^{iuX_t})$. Let us assume that the stopping time τ is bounded and let $A \in \mathcal{F}_\tau$. Let $u_j, j = 1, \ldots, n$ be a sequence of real numbers and $0 \leq t_0 < \cdots < t_n$ an increasing sequence of positive numbers. Then, applying the optional sampling theorem several times, for the martingale $Z_t(u) = e^{iuY_t}/\mathbb{E}(e^{iuY_t})$,

$$\mathbb{E}\left(\mathbb{1}_A e^{i\sum_{j=1}^n u_j(Y_{t_j} - Y_{t_{j-1}})}\right) = \mathbb{E}\left(\mathbb{1}_A \prod_j \frac{Z_{\tau+t_j}(u_j)}{Z_{\tau+t_{j-1}}(u_j)} \varphi(t_j - t_{j-1}, u_j)\right)$$
$$= \mathbb{P}(A) \prod_j \varphi(t_j - t_{j-1}, u_j)\,.$$

The general case is obtained by passing to the limit. □

Proposition 11.2.1.5 *Let* X *be a one-dimensional Lévy process. Then, for any fixed* t,

$$(X_u, u \leq t) \overset{\text{law}}{=} (X_t - X_{(t-u)-}, u \leq t)\,. \tag{11.2.1}$$

Consequently, $(X_t, \inf_{u \leq t} X_u) \overset{\text{law}}{=} (X_t, X_t - \sup_{u \leq t} X_u)$*. Moreover, for any* $\alpha \in \mathbb{R}$*, the process* $(e^{\alpha X_t} \int_0^t du\, e^{-\alpha X_u}, t \geq 0)$ *is a Markov process and for any fixed* t,

$$\left(e^{X_t}, e^{X_t} \int_0^t e^{-X_{s-}} ds\right) \overset{\text{law}}{=} \left(e^{X_t}, \int_0^t e^{X_{s-}} ds\right).$$

PROOF: It is straightforward to prove (11.2.1), since the right-hand side has independent increments which are distributed as those of the left-hand side. The proof of the remaining part can be found in Carmona et al. [141] and Donati-Martin et al. [258]. □

Proposition 11.2.1.6 *If* X *is a Lévy process, for any* $t > 0$*, the r.v.* X_t *is infinitely divisible.*

PROOF: Using the decomposition $X_t = \sum_{k=1}^n (X_{kt/n} - X_{(k-1)t/n})$ we observe the infinitely divisible character of any variable X_t. □

We shall prove later (see ↣ Corollary 11.2.3.8) that a Lévy process is a semi-martingale. Hence, the integral of a locally bounded predictable process with respect to a Lévy process is well defined.

Proposition 11.2.1.7 *Let (X, Y) be a two-dimensional Lévy process. Then*

$$U_t = e^{-X_t}\left(u + \int_0^t e^{X_{s-}} dY_s\right), \; t \geq 0$$

is a Markov process, whose semigroup is given by

$$Q_t(u, f) = \mathbb{E}\left(f(ue^{-X_t} + \int_0^t e^{-X_{s-}} dY_s)\right).$$

Comment 11.2.1.8 The last property in Proposition 11.2.1.5 is a particular case of Proposition 11.2.1.7.

Exercise 11.2.1.9 Prove that $(e^{\alpha X_t} \int_0^t du\, e^{-\alpha X_u}, t \geq 0)$ is a Markov process whereas $\int_0^t du\, e^{\alpha X_u}$ is not. Give an explanation of this difference.
`Hint:` For $s < t$, write

$$Y_t = e^{\alpha X_t} \int_0^t du\, e^{-\alpha X_u} = e^{\alpha(X_t - X_s)} e^{\alpha X_s} Y_s + e^{\alpha(X_t - X_s)} \int_s^t e^{-\alpha(X_u - X_s)} du$$

and use the independence property of the increments of X. ◁

11.2.2 Poisson Point Processes, Lévy Measures

Let X be a Lévy process. For every bounded Borel set $\Lambda \in \mathbb{R}^d$, such that $0 \notin \bar{\Lambda}$, where $\bar{\Lambda}$ is the closure of Λ, we define

$$N_t^\Lambda = \sum_{0<s\leq t} \mathbb{1}_\Lambda(\Delta X_s)$$

to be the number of jumps up to time t which take values in Λ. The map $\Lambda \to N_t^\Lambda(\omega)$ defines a σ-finite measure on $\mathbb{R}^d \setminus 0$ denoted by $N_t(\omega, dx)$.

Definition 11.2.2.1 *The σ-additive measure ν defined on $\mathbb{R}^d \setminus 0$ by*

$$\nu(\Lambda) = \mathbb{E}(N_1^\Lambda)$$

is called the ***Lévy measure*** *of the process X.*

We shall soon show (see ↣ Remark 11.2.3.3) that the Lévy measure is indeed the same measure which occurs in the Lévy-Khintchine representation of the r.v. X_1.

Proposition 11.2.2.2 *Let X be a Lévy process with Lévy measure ν.*

(i) *Assume $\nu(\Lambda) < \infty$. Then, the process*

$$N_t^\Lambda = \sum_{0<s\leq t} \mathbb{1}_\Lambda(\Delta X_s),\ t \geq 0$$

is a Poisson process with intensity $\nu(\Lambda)$.

(ii) *Let Λ_i be a finite collection of sets and assume that, $\forall i, \nu(\Lambda_i) < \infty$. The processes N^{Λ_i} are independent if and only if $\nu(\Lambda_i \cap \Lambda_j) = 0, i \neq j$, in particular if the sets Λ_i are disjoint. If $\Lambda = \cup_i \Lambda_i$, with disjoint Λ_i, then $N^\Lambda = \sum_i N^{\Lambda_i}$ is a Poisson process with intensity $\sum_i \nu(\Lambda_i)$.*

PROOF: (i) The process N^Λ has independent and stationary increments, its paths are increasing and right continuous and increase only by jumps of size 1, hence it is a Poisson process (Watanabe's characterization, Proposition 8.3.3.1).

(ii) We give the proof for two sets Λ and Γ.

(a) We first assume that Λ and Γ are disjoint. The processes N^Λ and N^Γ are Poisson processes in the same filtration and they never jump at the same time. Hence, they are independent (see Proposition 8.3.6.2).

(b) Conversely, assume that N^Λ and N^Γ are independent. From the independence property, for any pair (λ, μ) of positive real numbers

$$\mathbb{E}\left(\exp\left(-(\lambda N_t^\Gamma + \mu N_t^\Lambda)\right)\right) = \mathbb{E}\left(\exp\left(-\lambda N_t^\Gamma\right)\right) \mathbb{E}\left(\exp\left(-\mu N_t^\Lambda\right)\right) ,$$

hence

$$\begin{aligned}
&\exp\left(-t\int \nu(dx)\left(1 - e^{-(\lambda \mathbb{1}_\Gamma(x) + \mu \mathbb{1}_\Lambda(x))}\right)\right) \\
&= \exp\left(-t\int \nu(dx)\left(1 - e^{-\lambda \mathbb{1}_\Gamma(x)}\right)\right) \exp\left(-t\int \nu(dx)\left(1 - e^{-\mu \mathbb{1}_\Lambda(x)}\right)\right) .
\end{aligned}$$

It follows that

$$\begin{aligned}
\int_{\Lambda\cap\Gamma} \nu(dx)(1 - e^{-(\lambda+\mu)}) + \int_{\Lambda\setminus\Gamma} \nu(dx)(1 - e^{-\lambda}) + \int_{\Gamma\setminus\Lambda} \nu(dx)(1 - e^{-\mu}) \\
= \int_\Gamma \nu(dx)(1 - e^{-\lambda}) + \int_\Lambda \nu(dx)(1 - e^{-\mu}) ,
\end{aligned}$$

therefore

$$\int_{\Lambda\cap\Gamma} \nu(dx)(1 - e^{-(\lambda+\mu)}) = \int_{\Lambda\cap\Gamma} \nu(dx)(1 - e^{-\lambda}) + \int_{\Lambda\cap\Gamma} \nu(dx)(1 - e^{-\mu})$$

which implies that $\nu(\Gamma \cap \Lambda) = 0$. □

It follows that, if $\nu(\Gamma) < \infty$, the process $(N_t^\Gamma - \nu(\Gamma)t,\ t \geq 0)$ is a martingale. Note that if Γ and Λ are disjoint, N^Γ and N^Λ are independent, and the processes $N_t^\Gamma - \nu(\Gamma)t$ and $N_t^\Lambda - \nu(\Lambda)t$ are orthogonal martingales. The jump process of a Lévy process is a Poisson point process (see Section 8.9).

Let Λ be a Borel set of $\mathbb{R}^d$ with $0 \notin \bar{\Lambda}$, and f a positive Borel function defined on Λ. We have, by definition of N_t,

$$\int_\Lambda f(x) N_t(\omega, dx) = \sum_{0<s\leq t} f(\Delta X_s(\omega)) \mathbb{1}_\Lambda(\Delta X_s(\omega)) .$$

Proposition 11.2.2.3 (Compensation Formula.) *Let f be a positive function such that $f(0) = 0$. Then,*

$$\mathbb{E}\left(\sum_{0<s\leq t} f(\Delta X_s) \right) = t \int_{\mathbb{R}^d} f(x) \nu(dx) .$$

(i) *If $\int_{\mathbb{R}^d} f(x)\nu(dx) < \infty$, then $\sum_{0<s\leq t} f(\Delta X_s)$ is a Lévy process, with Lévy measure $\nu \circ f^{-1}$, the image of ν by f. The process*

$$M_t^f := \sum_{0<s\leq t} f(\Delta X_s) - t \int_{\mathbb{R}^d} f(x) \nu(dx)$$

is a martingale. More generally, if H is a positive predictable function (i.e., $H : \Omega \times \mathbb{R}^+ \times \mathbb{R}^d \to \mathbb{R}^+$ is $\mathcal{P} \times \mathcal{B}$ measurable) such that $H_s(\omega, 0) = 0$

$$\mathbb{E}\left[\sum_{s\leq t} H_s(\omega, \Delta X_s) \right] = \mathbb{E}\left[\int_0^t ds \int d\nu(x) H_s(\omega, x) \right] .$$

(ii) *Let Λ be a Borel set in $\mathbb{R}^d$ such that $0 \notin \bar{\Lambda}$ and $g := f \mathbb{1}_\Lambda$. Then if $g \in L^1(d\nu) \cap L^2(d\nu)$, the process $(M_t^g)^2 - t \int_\Lambda f^2(x) \nu(dx)$ is a martingale. In particular,*

$$\mathbb{E}\left(\int_\Lambda f(x) N_t(\cdot, dx) - t \int_\Lambda f(x) \nu(dx) \right)^2 = t \int_\Lambda f^2(x) \nu(dx) .$$

PROOF: In a first step, these results are established for functions f of the form $f(x) = \sum \alpha_i \mathbb{1}_{A_i}(x)$ where the sets A_i are disjoint, using properties of compound Poisson processes. Then, it suffices to use the dominated convergence theorem. □

Example 11.2.2.4 The Gamma process is a Lévy process with Lévy measure $dx\, x^{-1} e^{-x}$, and for $\alpha > 0$, the process $\sum_{0<s\leq t} (\Delta\gamma_s)^\alpha$ is a subordinator with Lévy measure $\nu_\alpha(dy) = \alpha^{-1}\, dy\, y^{-1} e^{-y^{1/\alpha}}$.

In the particular cases of Poisson processes and compound Poisson processes, the compensation formula in Proposition 11.2.2.3 was already obtained respectively as a direct consequence of (8.2.4) for Poisson processes and as a consequence of Proposition 8.6.3.6 for compound Poisson processes.

One can extend the exponential formula (8.6.2) to Lévy processes:

Proposition 11.2.2.5 (Exponential Formula.) *Let X be a Lévy process and ν its Lévy measure.*

(i) *For any $t \in \mathbb{R}^+$, $u_i \geq 0$, and f satisfying*

$$\int_\Lambda (1 - e^{-u \centerdot f(x)})\nu(dx) < \infty \,,$$

one has:

$$\mathbb{E}\left(\exp\left(-u \centerdot \int_\Lambda f(x) N_t(\cdot, dx)\right)\right) = \exp\left(-t \int_\Lambda (1 - e^{-u \centerdot f(x)})\nu(dx)\right).$$

(ii) *For every t and every Borel function f defined on $\mathbb{R}^+ \times \mathbb{R}^d$ such that $\int_0^t ds \int |1 - e^{f(s,x)}|\nu(dx) < \infty$, one has*

$$\mathbb{E}\left[\exp\left(\sum_{s \leq t} f(s, \Delta X_s)\mathbb{1}_{\{\Delta X_s \neq 0\}}\right)\right] = \exp\left(-\int_0^t ds \int_\mathbb{R} (1 - e^{f(s,x)})\nu(dx)\right).$$

Exercise 11.2.2.6 Let X be a Lévy process with finite variance. Check that $\mathbb{E}(X_t) = t\mathbb{E}(X_1)$ and $\operatorname{Var}(X_t) = t \operatorname{Var}(X_1)$. ◁

Exercise 11.2.2.7 Prove that, if $f \in L^2(\nu \mathbb{1}_\Lambda)$,

$$\operatorname{Var} \int_\Lambda f(x) N(t, dx) = t \int_\Lambda f^2(x)\nu(dx) \,.$$

Prove that $\mathbb{E}((N_t^\Lambda)^2) = t^2(\nu(\Lambda))^2 + t\,\nu(\Lambda)$. ◁

Exercise 11.2.2.8 This exercise will provide another proof of Proposition 11.2.2.2. Assuming that N^Γ and N^Λ are independent, prove that $\nu(\Gamma \cap \Lambda) = 0$.
`Hint:` The integration by parts formula leads to

$$\begin{aligned} N_t^\Gamma N_t^\Lambda &= \int_0^t N_{s-}^\Gamma dN_s^\Lambda + \int_0^t N_{s-}^\Lambda dN_s^\Gamma + \sum_{s \leq t} \Delta N_s^\Gamma \, \Delta N_s^\Lambda \\ &= \int_0^t N_{s-}^\Gamma dN_s^\Lambda + \int_0^t N_{s-}^\Lambda dN_s^\Gamma + \sum_{s \leq t} \Delta N_s^{\Gamma \cap \Lambda} . \end{aligned}$$

Hence

$$\begin{aligned} \mathbb{E}(N_t^\Gamma N_t^\Lambda) &= \nu(\Lambda)\mathbb{E}\left(\int_0^t N_s^\Gamma ds\right) + \nu(\Gamma)\mathbb{E}\left(\int_0^t N_s^\Lambda ds\right) + t\nu(\Gamma \cap \Lambda) \\ &= t^2 \nu(\Gamma)\nu(\Lambda) + t\nu(\Gamma \cap \Lambda) . \end{aligned}$$

It remains to note that $\mathbb{E}(N_t^\Gamma N_t^\Lambda) = t^2\nu(\Lambda)\nu(\Gamma)$, because of our hypothesis of independence, thus $\nu(\Lambda \cap \Gamma) = 0$. We might also have obtained the result by noting that $\sum_{s \leq t} \Delta N_s^\Gamma \, \Delta N_s^\Lambda = 0$, since the independent processes N^Γ and N^Λ do not have common jumps. ◁

Exercise 11.2.2.9 Let X be a real-valued Lévy process. Prove the following assertions:

1. If $\mathbb{E}(|X_1|) < \infty$, then $(X_t - t\mathbb{E}(X_1), t \geq 0)$ is a martingale.
2. If $\mathbb{E}(\exp(\lambda X_t)) < \infty$, then $(\exp(\lambda X_t)[\mathbb{E}(\exp(\lambda X_t))]^{-1}, t \geq 0)$ is a martingale. (See also $\rightarrowtail$ Subsection 11.3.1.)
3. If $\widehat{\mu}_t(u) = \mathbb{E}(e^{iuX_t})$, the process $(e^{iuX_t}/\widehat{\mu}_t(u), t \geq 0)$ is a martingale.
4. For $t, s \geq 0$, $\widehat{\mu}_{t+s} = \widehat{\mu}_t \widehat{\mu}_s$ and $\widehat{\mu}_t$ does not vanish.

$\triangleleft$

Exercise 11.2.2.10 Let Λ be such that $0 \notin \bar{\Lambda}$ and $f \in L^1(\nu \mathbb{1}_\Lambda)$. Prove that the process $\int_\Lambda f(x) N_t(\cdot, dx)$ is a compound Poisson process. $\triangleleft$

Exercise 11.2.2.11 The Ornstein-Uhlenbeck Process Driven by a Lévy Process. The OU process driven by the Lévy process $(X_t, t \geq 0)$ with initial state U_0 and parameter c is the solution of

$$U_t = U_0 + X_t - c \int_0^t U_s ds \,.$$

Check that

$$U_t = e^{-ct} \left(U_0 + \int_0^t e^{cs} dX_s \right) .$$

`Hint:` This is a particular case of Proposition 11.2.1.7. See Novikov [678] and Hadjiev [416] for a study of these OU processes. $\triangleleft$

Exercise 11.2.2.12 The Structure of Compound Poisson Processes. Let (Y_n) be a sequence of random variables, and (T_n) an increasing sequence of random times such that the series $\sum_n |Y_n| \mathbb{1}_{\{T_n \leq t\}}$ converges and that the process $X_t = \sum_n Y_n \mathbb{1}_{\{T_n \leq t\}}$ is a Lévy process. Prove that the random variables $(S_n = T_n - T_{n-1}, n \geq 1)$ are i.i.d. with exponential law and that the $(Y_n, n \geq 0)$ are i.i.d. and independent of $(S_n, n \geq 1)$. $\triangleleft$

Exercise 11.2.2.13 Let X be a Lévy process and $Q_t = \sum_{s \leq t} (\Delta X_s)^2$. Prove that

$$\mathbb{E}(e^{-\lambda Q_t}) = \exp \left(-t \int_{-\infty}^{\infty} (1 - e^{-\lambda x^2}) \nu(dx) \right) .$$

See Carr et al. [151] for some applications to finance. $\triangleleft$

Exercise 11.2.2.14 (a) Prove that, if $(X_t, t \geq 0)$ is a Lévy process, then for every $n \in \mathbb{N}$, one has

$$(X_{nt}, t \geq 0) \stackrel{\text{law}}{=} (X_t^{(1)} + X_t^{(2)} + \cdots + X_t^{(n)}, t \geq 0) \tag{11.2.2}$$

where the $X_t^{(i)}$ are n independent copies of X.

(b) Let $(X_t, t \geq 0)$ be a Lévy process starting from 0, and let $a, b > 0$. Define

$$Y_t = \int_{at}^{bt} \frac{du}{u} X_u, \; t \geq 0.$$

Prove that Y also satisfies (11.2.2), although the process $(Y_t, t \geq 0)$ is not a Lévy process, unless $X_u = cu, u \geq 0$.

(c) Prove that, if the process $(X_t, t \geq 0)$ satisfies (11.2.2), then the process X considered as a r.v. with values in $D([0, \infty[)$ is infinitely divisible.

Let us call processes that satisfy (11.2.2) IDT processes (infinitely divisible in time). This exercise shows that there are many IDT processes which are not Lévy processes. See Mansuy [620] and Es-Sebaiy and Ouknine [334]. $\triangleleft$

11.2.3 Lévy-Khintchine Formula for a Lévy Process

Proposition 11.2.3.1 *Let X be a Lévy process taking values in $\mathbb{R}^d$. Then, for each t, the r.v. X_t is infinitely divisible and its characteristic function is given by the Lévy-Khintchine formula: for $u \in \mathbb{R}^d$,*

$$\begin{aligned} &\mathbb{E}(\exp(iu \cdot X_t)) \\ &= \exp\left(t\left(iu \cdot m - \frac{u \cdot Au}{2} + \int_{\mathbb{R}^d} (e^{iu \cdot x} - 1 - iu \cdot x \mathbb{1}_{|x| \leq 1}) \nu(dx)\right)\right), \end{aligned}$$

where $m \in \mathbb{R}^d$, A is a positive semi-definite matrix, and ν is a Lévy measure on $\mathbb{R}^d \setminus \{0\}$. The r.v. X_t admits the characteristic triple $(tm, tA, t\nu)$. We shall say in short that X is a (m, A, ν) Lévy process.

PROOF: Let X be a Lévy process. We have seen that X_t is an infinitely divisible random variable. In particular the distribution of X_1 is infinitely divisible and admits a Lévy-Khintchine representation (see Proposition 11.1.1.4). The Lévy-Khintchine representation for X_t is then obtained from that of X_1, first for $t \in \mathbb{N}$, then for $t \in \mathbb{Q}^+$ and for any t, using the continuity in law w.r.t. time for Lévy processes. $\square$

Exercise 11.2.3.2 Express the Lévy-Khintchine representation of the r.v. $\sum_{i=1}^n \lambda_i (X_{t_{i+1}} - X_{t_i})$ in terms of the Lévy-Khintchine representation of the process X, or of the r.v. X_1. $\triangleleft$

Remark 11.2.3.3 Let us check in a particular case that the Lévy measure which appears in the Lévy-Khintchine formula is the same as the one which appears in Definition 11.2.2.1 when X is a real-valued Lévy process with bounded variation (i.e., the difference of two subordinators) and without drift term (see Bertoin [78] for the general case). From

$$e^{iuX_t} = 1 + \sum_{s \leq t} (e^{iuX_s} - e^{iuX_{s-}}) = 1 + \sum_{s \leq t} e^{iuX_{s-}} (e^{iu\Delta X_s} - 1),$$

we obtain

$$\mathbb{E}(e^{iuX_t}) = 1 + \mathbb{E}\left(\sum_{s\leq t} e^{iuX_{s-}}(e^{iu\Delta X_s} - 1)\right)$$
$$= 1 + \mathbb{E}\left(\int_0^t ds\, e^{iuX_{s-}} \int \nu(dx)(e^{iux} - 1)\right),$$

where, in the second equality, we have used the compensation formula in Proposition 11.2.2.3. Therefore, setting $\widehat{\mu}_t(u) = \mathbb{E}(e^{iuX_t})$, we get the linear integral equation

$$\widehat{\mu}_t(u) = 1 + \int_0^t ds\, \widehat{\mu}_s(u) \int \nu(dx)(e^{iux} - 1).$$

(The integral $\int \nu(dx)(1 \wedge |x|)$ is finite.) Solving this linear equation leads to the Lévy-Khintchine formula.

Definition 11.2.3.4 *The continuous function* $\Phi : \mathbb{R}^d \to \mathbb{C}$ *such that*

$$\mathbb{E}\left[\exp(iu \centerdot X_1)\right] = \exp(-\Phi(u))$$

is called the ***characteristic exponent*** *(sometimes the Lévy exponent) of the Lévy process X.*

If $\mathbb{E}\left[e^{\lambda \centerdot X_1}\right] < \infty$ *for any* λ *with positive components, the function* Ψ *defined on* $[0,\infty[^d$*, such that*

$$\mathbb{E}\left[\exp(\lambda \centerdot X_1)\right] = \exp(\Psi(\lambda))$$

is called the ***Laplace exponent*** *of the Lévy process X.*

It follows that

$$\mathbb{E}\left[\exp(iu \centerdot X_t)\right] = \exp(-t\Phi(u))$$

and, if $\Psi(\lambda)$ *exists,*

$$\mathbb{E}\left[\exp(\lambda \centerdot X_t)\right] = \exp(t\Psi(\lambda))$$

and

$$\Psi(\lambda) = -\Phi(-i\lambda).$$

From Proposition 11.2.3.1,

$$\begin{aligned}\Phi(u) &= -iu \centerdot m + \tfrac{1}{2}u \centerdot Au - \int (e^{iu \centerdot x} - 1 - iu \centerdot x\mathbb{1}_{|x|\leq 1})\nu(dx)\\ \Psi(\lambda) &= \lambda \centerdot m + \tfrac{1}{2}\lambda \centerdot A\lambda + \int (e^{\lambda \centerdot x} - 1 - \lambda \centerdot x\mathbb{1}_{|x|\leq 1})\nu(dx)\end{aligned} \qquad (11.2.3)$$

▶ If $\nu(\mathbb{R}^d \setminus 0) < \infty$, the process X has a finite number of jumps in any finite time interval. In finance, when $\nu(\mathbb{R}^d) < \infty$, one refers to **finite activity**.

If moreover $A = 0$, the process X is a compound Poisson process with "drift."

▶ If $\nu(\mathbb{R}^d \setminus 0) = \infty$, the process corresponds to **infinite activity**. Assume moreover that $A = 0$. Then

- If $\int_{|x|\leq 1} |x|\nu(dx) < \infty$, the paths of X are of bounded variation on any finite time interval.
- If $\int_{|x|\leq 1} |x|\nu(dx) = \infty$, the paths of X are no longer of bounded variation on any finite time interval.

Example 11.2.3.5 We present examples of Lévy processes and their characteristics.

- **Drifted Brownian Motion.** The process $(mt + \sigma B_t,\, t \geq 0)$ where B is a BM is a Lévy process with characteristic exponent $-ium + u^2\sigma^2/2$, hence, its Lévy measure is zero. We shall see that the family of Lévy processes with zero Lévy measure consists precisely of all the Lévy processes with continuous paths, which are exactly $(mt + \sigma B_t, t \geq 0)$ for any $m, \sigma \in \mathbb{R}$.
- **Poisson Process.** The Poisson process with intensity λ is a Lévy process with characteristic exponent (see (8.2.1))
$$\lambda(1 - e^{iu}) = \int (1 - e^{iux})\lambda\, \delta_1(dx)$$
hence its Lévy measure is $\lambda\, \delta_1$, where δ is the Dirac measure.
- **Compound Poisson Process.** Let $X_t = \sum_{k=1}^{N_t} Y_k$ be a ν-compound Poisson process. Its characteristic exponent is (see Proposition 8.6.3.4)
$$\Phi(u) = \int (1 - e^{iux})\, \nu(dx)\,.$$
Its Lévy measure is $\nu(dx) = \lambda F(dx)$, where F is the common law of all the Y_k's, and λ the intensity of the Poisson process N.
- **Process of Brownian Hitting Times.** Let W be a BM. The process $(T_r, r \geq 0)$ where $T_r = \inf\{t : W_t \geq r\}$ is an increasing Lévy process. The process $(-T_r, r \geq 0)$ admits as Laplace exponent $-\sqrt{2\lambda}$. Hence, the Lévy measure of the Lévy process T_r is $\nu(dx) = dx \mathbb{1}_{\{x>0\}}/\sqrt{2\pi x^3}$. We have recalled above that if a Lévy process is continuous, then, it is a Brownian motion with drift. Hence, the process $(T_r, r \geq 0)$ is not continuous.
- **Stable Processes.** With any stable r.v. of index α, we may associate a Lévy process which will be called a stable process of index α. The process T_r is a stable process (in fact a subordinator) of index $1/2$. The linear Brownian motion (resp. the Cauchy process) is symmetric stable of index 2 (resp. 1).

Comment 11.2.3.6 The Laplace exponent is also called the cumulant function. When considering X as a semi-martingale, the triple (m, σ^2, ν)

is the triple of predictable characteristics of X (see Jacod and Shiryaev [471]). Obviously, the characteristics of the **dual process** $\widehat{X} = -X$ are $(-m, \sigma^2, \widehat{\nu})$ where $\widehat{\nu}(dx) = \nu(-dx)$. The Laplace exponent of the dual process is $\widehat{\Psi}(\lambda) = \Psi(-\lambda)$.

Proposition 11.2.3.7 (Lévy-Itô's Decomposition.) *If X is an $\mathbb{R}^d$-valued Lévy process, it can be decomposed into*

$$X = Y^{(0)} + Y^{(1)} + Y^{(2)} + Y^{(3)}$$

where $Y^{(0)}$ is a constant drift, $Y^{(1)}$ is a linear transform of a Brownian motion, $Y^{(2)}$ is a compound Poisson process with jump sizes greater than or equal to 1 and $Y^{(3)}$ is a Lévy process with jump sizes smaller than 1. The processes $Y^{(i)}$ are independent.

SKETCH OF THE PROOF: The characteristic exponent of the Lévy process can be written

$$\Phi = \Phi^{(0)} + \Phi^{(1)} + \Phi^{(2)} + \Phi^{(3)}$$

with

$$\begin{aligned}
\Phi^{(0)}(\lambda) &= -im \cdot \lambda, \ \Phi^{(1)}(\lambda) = \frac{1}{2}\lambda \cdot A\lambda\,, \\
\Phi^{(2)}(\lambda) &= \int (1 - e^{i\lambda \cdot x}) \mathbb{1}_{\{|x|>1\}}\, \nu(dx)\,, \\
\Phi^{(3)}(\lambda) &= \int (1 - e^{i\lambda \cdot x} + i\lambda \cdot x) \mathbb{1}_{\{|x|\leq 1\}} \nu(dx)\,.
\end{aligned}$$

Each $\Phi^{(i)}$ is a characteristic exponent. The function $\Phi^{(2)}$ can be viewed as the characteristic exponent of a compound Poisson process with Lévy measure $\mathbb{1}_{\{|x|\geq 1\}}\nu(dx)$ (see Proposition 8.6.3.4). In the case $\int_{\{|x|\leq 1\}} |x|\nu(dx) < \infty$, and in particular if ν is finite, one can write

$$\Phi^{(3)}(\lambda) = -i\lambda \cdot \tilde{m}_0 + \int (1 - e^{i\lambda \cdot x}) \mathbb{1}_{\{|x|\leq 1\}}) \nu(dx)$$

and one can construct a Lévy process with characteristic exponent $\Phi^{(3)}$. This approach fails if $\int_{\{|x|<1\}} |x|\nu(dx) = \infty$. In that case, one constructs a Lévy process with Lévy measure ν by approximating $\Phi^{(3)}$ by

$$\Phi^{(3,\epsilon)}(\lambda) = \int (1 - e^{i\lambda \cdot x} - i\lambda \cdot x)\, \mathbb{1}_{\{\epsilon<|x|\leq 1\}} \nu(dx)\,,$$

and letting $\epsilon \to 0$. See Chapter 4 in Sato [761] or Chapter 1 in Bertoin [79] for details. □

This result is often used to yield the following representation:

$$X_t = mt + Z_t + \int_0^t \int x \mathbb{1}_{\{|x| \le 1\}} \widetilde{\mathbf{N}}(ds, dx) + \int_0^t \int x \mathbb{1}_{\{|x|>1\}} \mathbf{N}(ds, dx) \tag{11.2.4}$$

where Z is an $\mathbb{R}^d$-valued Brownian motion with correlation matrix A, $\mathbf{N}$ the random measure[1] of the jumps of X, and $\widetilde{\mathbf{N}}(dt, dx) = \mathbf{N}(dt, dx) - dt\, \nu(dx)$ is the compensated martingale measure. In other words,

$$X_t = mt + Z_t + \int_0^t \int x \mathbb{1}_{\{|x| \le 1\}} \widetilde{\mathbf{N}}(ds, dx) + \sum_{s \le t} \Delta X_s \mathbb{1}_{\{|\Delta X_s| > 1\}} \,.$$

▶ If $\int_{\{|x| \le 1\}} |x| \nu(dx) < \infty$, the process X can be represented as

$$X_t = m_0 t + Z_t + \int_0^t \int x \mathbf{N}(ds, dx)$$

where $m_0 = m - \int x \mathbb{1}_{\{|x| \le 1\}} \nu(dx)$. If ν is a finite measure, the process $\int_0^t \int x \mathbf{N}(ds, dx)$ is a compound Poisson process: indeed, let $T_k, k \ge 1$ be the jump times of X and $N_t = \sum \mathbb{1}_{\{T_k \le t\}}$ the associated counting process. The process N is a Poisson process with intensity $\lambda = \nu(\mathbb{R}^d \setminus 0)$. Furthermore, the random variables $Y_k = \int x \mathbf{N}(\{T_k\}, dx)$ are i.i.d. with law $\nu(dx)/\lambda$ and $\sum_1^{N_t} Y_k = \int_0^t \int x \mathbf{N}(ds, dx)$.

▶ If $\int_{\{|x| \le 1\}} |x| \nu(dx) = \infty$, it is not possible to separate the integral

$$\int_{|x| \le 1} x (N_t(\cdot, dx) - t \nu(dx))$$

into the two parts $\int_{|x| \le 1} x N_t(\cdot, dx)$ and $t \int_{|x| \le 1} x \nu(dx)$ which may both be ill-defined.

Corollary 11.2.3.8 *A Lévy process is a semi-martingale.*

PROOF: This is a consequence of the Lévy-Itô decomposition. □

Comment 11.2.3.9 As a direct consequence, the series $\sum_{s \le t} (\Delta X_s)^2$ converges, and for any constant a the process

$$\sum_{s \le t} ((\Delta X_s)^2 \wedge a) - t \int (x^2 \wedge a)\, \nu(dx)$$

[1] We indicate these random measures in boldface to avoid possible confusion with Poisson processes which are always denoted by N.

is a martingale. It follows that $\int (x^2 \wedge a)\, \nu(dx) < \infty$, which justifies the integrability condition on the Lévy measures.

Lemma 11.2.3.10 *Let F be a C^2 function and X a Lévy process. Then, the series*

$$\sum_{s \leq t} |F(\Delta X_s) - F(0) - F'(0)\Delta X_s|$$

converges.

PROOF: The series $\sum_{0<s\leq t} \mathbb{1}_{\{|\Delta X_s|>1\}}(F(\Delta X_s) - F(0) - F'(0)\Delta X_s)$ is obviously convergent. Furthermore, the inequality

$$0 \leq |F(x) - F(0) - xF'(0)| \leq cx^2$$

for $|x| \leq 1$ and the convergence of the series $\sum_{s\leq t}(\Delta X_s)^2$ imply that the series $\sum_{0<s\leq t} \mathbb{1}_{\{|\Delta X_s|\leq 1\}}|F(\Delta X_s) - F(0) - F'(0)\Delta X_s|$ is also convergent. □

Proposition 11.2.3.11 *Let X be a (m, σ^2, ν)-Lévy process taking values in $\mathbb{R}$. Then, X is increasing if and only if*

$$\sigma = 0,\ \nu(-\infty, 0) = 0,\ \int \mathbb{1}_{\{x\leq 1\}} x\, \nu(dx) < \infty,\ m_0 = m - \int \mathbb{1}_{\{x\leq 1\}} x\, \nu(dx) \geq 0\,.$$

PROOF: If $\nu(-\infty, 0) = 0$, the process X has no negative jumps. Hence, assuming moreover that $\sigma = 0$, and the convergence of $\int x\nu(dx)$, we obtain $X_t = \int_0^t \int x\mathbf{N}(ds, dx) + tm_0$ and, if $m_0 \geq 0$, X is increasing.

For the converse, since X has no negative jumps, $\nu(-\infty, 0) = 0$. The proof of $\int \mathbb{1}_{\{x\leq 1\}} x\, \nu(dx) < \infty$ relies on the remark that an increasing function remains increasing after deleting a finite number of its jumps. We shall delete jumps of size greater than ϵ. Then, setting $X_t^\epsilon = \int_0^t \int_{]\epsilon,\infty[} x\mathbf{N}(dx, ds)$, the process $X_t - X_t^\epsilon$ is increasing, hence $X_t - X_t^\epsilon \geq 0$. It follows that

$$\lim_{\epsilon\to 0} X_t^\epsilon = \int_0^t \int_{\mathbb{R}^+} x\mathbf{N}(dx, ds) =: \widehat{X}_t$$

exists and is bounded by X_t. Now,

$$\begin{aligned}
\mathbb{E}(e^{-\lambda X_t^\epsilon}) &= \exp\left(t \int_{]\epsilon,\infty[} (e^{-\lambda x} - 1)\nu(dx)\right) \\
&= \exp\left(t \int_{]\epsilon,\infty[} (e^{-\lambda x} - 1 + \lambda x \mathbb{1}_{\{x\leq 1\}})\nu(dx) - t\lambda \int_{]\epsilon,1]} x\nu(dx)\right).
\end{aligned}$$

As ϵ tends to 0, the quantity $\mathbb{E}(e^{-\lambda X_t^\epsilon})$ tends to $\mathbb{E}(e^{-\lambda \widehat{X}_t})$ and the quantity $\int_{]\epsilon,\infty[}(e^{-\lambda x} - 1 + \lambda x\mathbb{1}_{\{x\leq 1\}})\nu(dx)$ tends to $\int_{\mathbb{R}^+}(e^{-\lambda x} - 1 + \lambda x\mathbb{1}_{\{x\leq 1\}})\nu(dx)$ which is finite. It follows that $\int_{]0,1]} x\nu(dx) < \infty$. The conclusion of the proof can be found in Sato [761], Theorem 21.5 page 137. □

Exercise 11.2.3.12 Let X be a (m, σ^2, ν) real-valued Lévy process. Check that the quadratic variation of the semi-martingale X is

$$[X]_t = \sigma^2 t + \int_0^t \int_{\mathbb{R}} x^2 \mathbf{N}(ds, dx).$$

The predictable quadratic variation of X is

$$\langle X \rangle_t = \sigma^2 t + t \int_{\mathbb{R}} x^2 \nu(dx),$$

as long as $\int_{\mathbb{R}} x^2 \nu(dx) < \infty$. ◁

Exercise 11.2.3.13 Let X be a Lévy process without drift or continuous part, and $f \in L^1(\nu)$. Prove that $M_t^f := \sum_{s \leq t} f(\Delta X_s) - t \int \nu(dx) f(x)$ is a martingale. Prove that, for f, g satisfying some integrability conditions, $\langle M^f, M^g \rangle_t = t \int \nu(dx)\, f(x)\, g(x)$. ◁

11.2.4 Itô's Formulae for a One-dimensional Lévy Process

Let X be a (m, σ^2, ν) real-valued Lévy process and $f \in C^{1,2}(\mathbb{R}^+ \times \mathbb{R}, \mathbb{R})$. Then, since X is a semi-martingale, we can apply Itô's formula in the optional or predictable form:

▶ The optional Itô's formula is

$$\begin{aligned} f(t, X_t) &= f(0, X_0) + \int_0^t \partial_t f(s, X_s) ds \\ &+ \frac{\sigma^2}{2} \int_0^t \partial_{xx} f(s, X_s) ds + \int_0^t \partial_x f(s, X_{s-}) dX_s \\ &+ \sum_{s \leq t} f(s, X_{s-} + \Delta X_s) - f(s, X_{s-}) - (\Delta X_s)\, \partial_x f(s, X_{s-}). \end{aligned}$$

▶ If f and its first and second derivatives w.r.t. x are bounded, the predictable Itô's formula is

$$\begin{aligned} df(t, X_t) &= \partial_x f(t, X_t) \sigma dW_t + \int_{\mathbb{R}} (f(t, X_{t-} + x) - f(t, X_{t-}))\, \widetilde{\mathbf{N}}(dt, dx) \\ + & \left[\frac{1}{2} \sigma^2 \partial_{xx} f(t, X_t) + m\, \partial_x f(t, X_t) + \partial_t f(t, X_t) \right. \\ + & \left. \int_{\mathbb{R}} \left(f(t, X_{t-} + x) - f(t, X_{t-}) - x\, \partial_x f(t, X_{t-}) \mathbb{1}_{|x| \leq 1} \right) \nu(dx) \right] dt. \end{aligned}$$

The quantity $\partial_x f(t, X_t) \sigma dW_t + \int_{x \in \mathbb{R}} (f(t, X_{t-} + x) - f(t, X_{t-}))\, \widetilde{\mathbf{N}}(dt, dx)$ is the local martingale part of the semi-martingale $f(t, X_t)$, written in "differential " form, and

$$\left[\frac{1}{2}\sigma^2\partial_{xx}f(t,X_t) + m\,\partial_x f(t,X_t) + \partial_t f(t,X_t)\right.$$
$$\left. + \int_{\mathbb{R}} \left[f(t,X_{t-}+x) - f(t,X_{t-}) - x\,\partial_x f(t,X_{t-})\mathbb{1}_{|x|\le 1}\right]\nu(dx)\right] dt$$

is the continuous finite variation part, again written in differential form. One may note that this formula is a generalization of the formula obtained for compound Poisson processes (see Subsection 8.6.4).

11.2.5 Itô's Formula for Lévy-Itô Processes

In this section, we step out of the framework of Lévy processes, by considering the more general class of **Lévy-Itô processes**

$$dX_t = a_t dt + \sigma_t dW_t + \int \mathbb{1}_{\{|x|\le 1\}}\gamma_t(x)\widetilde{\mathbf{N}}(dt,dx) + \int \mathbb{1}_{\{|x|>1\}}\gamma_t(x)\mathbf{N}(dt,dx)\,, \tag{11.2.5}$$

where we assume that a and σ are adapted processes, $\gamma(x)$ is predictable and that the integrals are well defined. These processes are semi-martingales. Itô's formula can be generalized as follows

Proposition 11.2.5.1 (Itô's Formula for Lévy-Itô Processes.) *Let X be defined as in (11.2.5). If f is bounded and $f \in C_b^{1,2}$, then the process $Y_t := f(t,X_t)$ is a semi-martingale:*

$$\begin{aligned}
dY_t &= \partial_x f(t,X_t)\sigma_t dW_t + \int_{|x|\le 1} \left(f(t,X_{t-}+\gamma_t(x)) - f(t,X_{t-})\right)\widetilde{\mathbf{N}}(dt,dx) \\
&+ \left(\partial_t f(t,X_t) + \partial_x f(t,X_t)a_t + \frac{1}{2}\sigma_t^2\partial_{xx}f(t,X_t)\right) dt \\
&+ \int_{|x|>1} \left(f(t,X_{t-}+\gamma_t(x)) - f(t,X_{t-})\right)\mathbf{N}(dt,dx) \\
&+ \left(\int_{|x|\le 1} \left(f(t,X_{t-}+\gamma_t(x)) - f(t,X_{t-}) - \gamma_t(x)\partial_x f(t,X_{t-})\right)\nu(dx)\right) dt\,.
\end{aligned}$$

Note that in the last integral, we do not need to write (X_{t-}); (X_t) would also do, but the minus sign may help to understand the origin of this quantity. More generally, let $X^i, i = 1,\dots,n$ be n processes with dynamics

$$\begin{aligned}
X_t^i &= X_0^i + V_t^i + \sum_{k=1}^m \int_0^t f_{i,k}(s)dW_s^k + \int_0^t \int_{|x|>1} g_i(s,x)\mathbf{N}(ds,dx) \\
&+ \int_0^t \int_{|x|\le 1} h_i(s,x)(\mathbf{N}(ds,dx) - \nu(dx)\,ds)
\end{aligned}$$

where V is a continuous bounded variation process. Then, for Θ a C_b^2 function

$$\Theta(X_t) = \Theta(X_0) + \sum_{i=1}^{n} \int_0^t \partial_{x_i}\Theta(X_s)dV_s^i + \sum_{i=1}^{n}\sum_{k=1}^{m} \int_0^t f_{i,k}(s)\partial_{x_i}\Theta(X_s)dW_s^k$$
$$+ \frac{1}{2}\sum_{i,j=1}^{n} \int_0^t \sum_{k=1}^{m} f_{i,k}(s)f_{j,k}(s)\partial_{x_i x_j}\Theta(X_s)ds$$
$$+ \int_0^t \int_{|x|>1} (\Theta(X_{s^-} + g(s,x)) - \Theta(X_{s^-}))\mathbf{N}(ds,dx)$$
$$+ \int_0^t \int_{|x|\leq 1} (\Theta(X_{s^-} + h(s,x)) - \Theta(X_{s^-}))\widetilde{\mathbf{N}}(ds,dx)$$
$$+ \int_0^t \int_{|x|\leq 1} (\Theta(X_{s^-} + h(s,x)) - \Theta(X_{s^-}) - \sum_{i=1}^{n} h_i(s,x)\partial_{x_i}\Theta(X_{s^-}))ds\,\nu(dx)\,.$$

Exercise 11.2.5.2 Let W be a Brownian motion and $\mathbf{N}$ a random Poisson measure. Let $X_t^i = \int_0^t \varphi_s^i dW_s + \int_0^t \int \psi^i(s,x)\widetilde{\mathbf{N}}(ds,dx), i = 1,2$ be two real-valued martingales.

Prove that $[X^1,X^2]_t = \int_0^t \varphi_s^1\varphi_s^2 ds + \int_0^t \int \psi^1(s,x)\psi^2(s,x)\mathbf{N}(ds,dx)$ and that, under suitable conditions

$$\langle X^1, X^2\rangle_t = \int_0^t \varphi_s^1\varphi_s^2 ds + \int_0^t \int \psi^1(s,x)\psi^2(s,x)\,\nu(dx)\,ds\,.$$

◁

Exercise 11.2.5.3 Prove that, if

$$S_t = S_0 \exp\left(bt + \sigma W_t + \int_0^t \int_{|x|\leq 1} x\widetilde{\mathbf{N}}(ds,dx) + \int_0^t \int_{|x|>1} x\mathbf{N}(ds,dx)\right)$$

then

$$dS_t = S_{t_-}\left(\left(b + \frac{1}{2}\sigma^2\right)dt + \sigma dW_t + \int_{|x|\leq 1} (e^x - 1 - x)\nu(dx)\,dt\right.$$
$$\left. + \int_{|x|\leq 1} (e^x - 1)\widetilde{\mathbf{N}}(dt,dx) + \int_{|x|>1} (e^x - 1)\mathbf{N}(dt,dx)\right)\,.$$

◁

Exercise 11.2.5.4 Geometric Itô-Lévy process This is a generalization of the previous exercise. Prove that the solution of

$$dX_t = X_{t^-}\left(adt + \sigma dW_t + \int_{|x|\leq 1} \gamma(t,x)\widetilde{\mathbf{N}}(dt,dx) + \int_{|x|>1} \gamma(t,x)\mathbf{N}(dt,dx)\right)$$

and $X_0 = 1$ where $\gamma(t,x) \geq -1$ and a, σ are constant, is

$$X_t = \exp\Bigg(\left(a - \frac{1}{2}\sigma^2\right)t + \sigma W_t + \int_0^t ds \int_{|x|\leq 1} (\ln(1+\gamma(s,x)) - \gamma(s,x))\,\nu(dx)$$

$$+ \int_0^t \int_{|x|\leq 1} \ln(1+\gamma(s,x))\widetilde{\mathbf{N}}(ds,dx) + \int_0^t \int_{|x|>1} \ln(1+\gamma(s,x))\mathbf{N}(ds,dx)\Bigg).$$

$\triangleleft$

Exercise 11.2.5.5 Let f be a predictable (bounded) process and $g(t,x)$ a predictable function. Let $g_1 = g\mathbb{1}_{|x|>1}$ and $g_2 = g\mathbb{1}_{|x|\leq 1}$. We assume that $e^{g_1} - 1$ is integrable w.r.t. $\widetilde{\mathbf{N}}$ and that g_2 is square integrable w.r.t. $\widetilde{\mathbf{N}}$. Prove that the solution of

$$dX_t = X_{t-}\left(f_t dW_t + \int (e^{g(t,x)} - 1)\widetilde{\mathbf{N}}(dt,dx)\right)$$

is

$$X_t = \mathcal{E}(f\star W)_t \exp\left(\mathbf{N}_t(g_1) - \int_0^t \int (e^{g_1(s,x)} - 1)\nu(dx)\,ds\right)$$
$$\exp\left(\widetilde{\mathbf{N}}_t(g_2) - \int_0^t \int (e^{g_2(s,x)} - 1 - g_2(s,x))\nu(dx)\,ds\right)$$

where $\mathbf{N}_t(g) = \int_0^t \int g(s,x)\mathbf{N}(ds,dx)$ and $\widetilde{\mathbf{N}}_t(g) = \int_0^t \int g(s,x)\widetilde{\mathbf{N}}(ds,dx)$. $\triangleleft$

Exercise 11.2.5.6 Let $S_t = e^{X_t}$ where X is a (m,σ^2,ν)-one dimensional Lévy process and λ a constant such that $\mathbb{E}(e^{\lambda X_T}) = \mathbb{E}(S_T^\lambda) < \infty$. Using the fact that

$$(e^{-t\Psi(\lambda)}S_t^\lambda = e^{\lambda X_t - t\Psi(\lambda)}, t \geq 0)$$

is a martingale, prove that S^λ is a special-semimartingale with canonical decomposition

$$S_t^\lambda = S_0^\lambda + M_t^{(\lambda)} + A_t^{(\lambda)}$$

where $A_t^{(\lambda)} = \Psi(\lambda)\int_0^t S_s^\lambda ds$. Prove that

$$\langle M^{(\lambda)}, M^{(\mu)}\rangle_t = (\Psi(\lambda+\mu) - \Psi(\lambda) - \Psi(\mu))\int_0^t S_u^{\lambda+\mu} du\,.$$

$\triangleleft$

11.2.6 Martingales

We now come back to the Lévy processes framework. Let X be a real-valued (m,σ^2,ν) Lévy process.

Proposition 11.2.6.1 (a) *An* $\mathbf{F}$*-Lévy process is a martingale if and only if it is an* $\mathbf{F}$*-local martingale.*

(b) *Let* X *be a Lévy process such that* X *is a martingale. Then, the process* $\mathcal{E}(X)$ *is a martingale.*

PROOF: See He et al. [427], Theorem 11.4.6. for part (a) and Cont and Tankov [192] for part (b). □

Proposition 11.2.6.2 *We assume that, for any* t, $\mathbb{E}(|X_t|) < \infty$, *which is equivalent to* $\int \mathbb{1}_{\{|x|\geq 1\}}|x|\nu(dx) < \infty$. *Then, the process* $(X_t - \mathbb{E}(X_t), t \geq 0)$ *is a martingale; hence, the process* $(X_t, t \geq 0)$ *is a martingale if and only if* $\mathbb{E}(X_t) = 0$, *i.e.,* $m + \int \mathbb{1}_{\{|x|\geq 1\}} x\nu(dx) = 0$.

PROOF: The first part is obvious. The second part of the proposition follows from the computation of $\mathbb{E}(X_t)$ which is obtained by differentiation of the characteristic function $\mathbb{E}(e^{iuX_t})$ at $u = 0$. The condition $\int \mathbb{1}_{\{|x|\geq 1\}}|x|\nu(dx) < \infty$ is needed for X_t to be integrable. □

Proposition 11.2.6.3 (Wald Martingale.)
For any λ *such that* $\Psi(\lambda) = \ln \mathbb{E}(e^{\lambda X_1}) < \infty$, *the process* $(e^{\lambda X_t - t\Psi(\lambda)}, t \geq 0)$ *is a martingale .*

PROOF: Obvious from the independence of increments. □

Note that $\Psi(\lambda)$ is well defined for every $\lambda > 0$ in the case where the Lévy measure has support in $(-\infty, 0[$. In that case, the Lévy process is said to be **spectrally negative** (see ↣ Section 11.5).

Corollary 11.2.6.4 *The process* $(e^{X_t}, t \geq 0)$ *is a martingale if and only if* $\int_{|x|\geq 1} e^x \nu(dx) < \infty$ *and*

$$\frac{1}{2}\sigma^2 + m + \int (e^x - 1 - x\mathbb{1}_{\{|x|\leq 1\}})\nu(dx) = 0\,.$$

PROOF: This follows from the above proposition and the expression of $\Psi(1)$. □

Proposition 11.2.6.5 (Doléans-Dade Exponential.) *Let* X *be a real-valued* (m, σ^2, ν)*-Lévy process and* Z *the Doléans-Dade exponential of* X, *i.e., the solution of* $dZ_t = Z_{t-}dX_t$, $Z_0 = 1$. *Then*

$$Z_t = e^{X_t - \sigma^2 t/2} \prod_{0<s\leq t} (1 + \Delta X_s)e^{-\Delta X_s} := \mathcal{E}(X)_t\,.$$

It is important to note that the product

$$\prod_{0<s\leq t} (1 + |\Delta X_s|)e^{-\Delta X_s} = \prod_{0<s\leq t} e^{-\Delta X_s + \ln(1+|\Delta X_s|)}$$

is convergent: indeed, the product

$$\prod_{0<s\leq t} (1 + |\Delta X_s| \mathbb{1}_{\{|\Delta X_s|>1/2\}})\, e^{-\Delta X_s}$$

is obviously convergent, and the inequality $0 < x - \ln(1+x) < x^2$ for $|x| < 1/2$ together with the convergence of the series $\sum_{s\leq t}(\Delta X_s)^2$ prove that the product $\prod_{s\leq t}(1 + |\Delta X_s| \mathbb{1}_{\{|\Delta X_s|>1/2\}}) e^{-\Delta X_s}$ is convergent too. Note that, with obvious definition, $(\mathcal{E}(X)_t, t \geq 0)$ is a multiplicative Lévy process.

We now investigate, as in Goll and Kallsen [400] the link between Doléans-Dade exponentials and ordinary exponentials (this is very close to the previous Exercises 11.2.5.3 and 11.2.5.4).

Proposition 11.2.6.6 *Let X be a real-valued (m, σ^2, ν)-Lévy process.*

(i) *Let $S_t = e^{X_t}$ be the ordinary exponential of the process X. The stochastic logarithm of S (i.e., the process Y which satisfies $S_t = \mathcal{E}(Y)_t$) is a Lévy process and is given by*

$$Y_t := \mathcal{L}(S)_t = X_t + \frac{1}{2}\sigma^2 t - \sum_{0<s\leq t} \left(1 + \Delta X_s - e^{\Delta X_s}\right) .$$

The Lévy characteristics of Y are

$$\begin{aligned} m_Y &= m + \frac{1}{2}\sigma^2 + \int \left((e^x - 1)\mathbb{1}_{\{|e^x-1|\leq 1\}} - x\mathbb{1}_{\{|x|\leq 1\}}\right) \nu(dx) , \\ \sigma_Y^2 &= \sigma^2 , \\ \nu_Y(A) &= \nu(\{x \,:\, e^x - 1 \in A\}) = \int \mathbb{1}_A(e^x - 1)\, \nu(dx) . \end{aligned}$$

(ii) *Let $Z_t = \mathcal{E}(X)_t$ the Doléans-Dade exponential of X. If $Z > 0$, the ordinary logarithm of Z is a Lévy process L given by*

$$L_t := \ln(Z_t) = X_t - \frac{1}{2}\sigma^2 t + \sum_{0<s\leq t} \left(\ln(1 + \Delta X_s) - \Delta X_s\right) .$$

Its Lévy characteristics are

$$\begin{aligned} m_L &= m - \frac{1}{2}\sigma^2 + \int \left(\ln(1+x)\mathbb{1}_{\{|\ln(1+x)|\leq 1\}} - x\mathbb{1}_{\{|x|\leq 1\}}\right) \nu(dx) , \\ \sigma_L^2 &= \sigma^2 , \\ \nu_L(A) &= \nu(\{x \,:\, \ln(1+x) \in A\}) = \int \mathbb{1}_A(\ln(1+x))\, \nu(dx) . \end{aligned}$$

PROOF: We only prove part (i) and leave part (ii) to the reader. Note that the series $\sum_{0<s\leq t}(1 + \Delta X_s - e^{\Delta X_s})$ is absolutely convergent by application of Lemma 11.2.3.10.

The process $Y_t = X_t + \frac{1}{2}\sigma^2 t - \sum_{0<s\leq t}\left(1 + \Delta X_s - e^{\Delta X_s}\right)$ is a Lévy process, $\sigma_Y^2 = \sigma^2$, and $\Delta Y_t = e^{\Delta X_t} - 1$. This implies $\nu_Y(dx) = (e^x - 1)\,\nu(dx)$. Using the equality

$$\sum_{s\leq t}\left(1 + \Delta X_s - e^{\Delta X_s}\right) = \int_0^t \int (1 + x - e^x)\mathbf{N}(ds, dx)$$

we obtain that the Lévy-Itô decomposition of Y is (where m_Y is defined in Proposition 11.2.6.6)

$$\begin{aligned}
Y_t &= mt + \sigma B_t + \int_0^t \int_{\{|x|\leq 1\}} x\widetilde{\mathbf{N}}(ds, dx) + \int_0^t \int_{\{|x|>1\}} x\mathbf{N}(ds, dx) + \frac{1}{2}\sigma^2 t \\
&\quad - \int_0^t \int (1 + x - e^x)\mathbf{N}(ds, dx) \\
&= m_Y t + \sigma B_t + \int_0^t \int (e^x - 1)\mathbb{1}_{\{|e^x-1|\leq 1\}}\widetilde{\mathbf{N}}(ds, dx) \\
&\quad + \int_0^t \int (e^x - 1)\mathbb{1}_{\{|e^x-1|>1\}}\mathbf{N}(ds, dx) \\
&= m_Y t + \sigma B_t + \int_0^t \int y\mathbb{1}_{\{|y|\leq 1\}}\widetilde{\mathbf{N}}_Y(ds, dy) + \int_0^t \int y\mathbb{1}_{\{|y|>1\}}\mathbf{N}_Y(ds, dy)\,.
\end{aligned}$$

The result follows. □

The following proposition may help the reader to become more familiar with another class of martingales for Lévy processes.

Proposition 11.2.6.7 (Asmussen-Kelly-Whitt Martingale.) *Let X be a real-valued (m, σ^2, ν)-Lévy process:*

$$X_t = mt + \sigma B_t + \int_0^t \int_{\{|x|\leq 1\}} x\widetilde{\mathbf{N}}(ds, dx) + \int_0^t \int_{\{|x|>1\}} x\mathbf{N}(ds, dx)\,.$$

Let $Z_t = \Sigma_t - X_t$, where $\Sigma_t = \sup_{s\leq t} X_s$ and Σ^c is the continuous part of the increasing process Σ. Let f be a C^2 function. Then, the process

$$\begin{aligned}
&f(Z_t) - f(Z_0) - f'(0)\Sigma_t^c - \frac{\sigma^2}{2}\int_0^t f''(Z_s)ds + m\int_0^t f'(Z_s)ds \\
&\quad - \int_0^t ds \int \nu(dx)[f(Z_{s-} + h_s(x)) - f(Z_{s-}) + x\mathbb{1}_{\{|x|\leq 1\}}f'(Z_{s-})]
\end{aligned}$$

is a local martingale, where $h_t(x) = -(x \wedge Z_{t-})$ is a predictable function.

PROOF: Note that if, at time t, the process X has a jump of size x smaller than Z_{t-} then $Z_t = Z_{t-} - x$; if the process X has a jump of size x greater than Z_{t-} then $\Sigma_t = X_t$ and $Z_t = 0$. In other words, the jumps of the process

Z are $\Delta Z_t = h_t(\Delta X_t)$ with $h_t(x) = -(x \wedge Z_{t-})$ a predictable function. From Itô's formula, the process

$$\begin{aligned} Y_t = f(Z_t) - f(Z_0) - \int_0^t f'(Z_s)d\Sigma_s - \frac{\sigma^2}{2}\int_0^t f''(Z_s)ds + m\int_0^t f'(Z_s)ds \\ - \int_0^t ds \int \nu(dx)[f(Z_{s-} + h_s(x)) - f(Z_{s-}) - h_s(x)\mathbb{1}_{\{|x|\leq 1\}}f'(Z_{s-})] \end{aligned}$$

is a local martingale.

We split the last integral into two parts

$$\begin{aligned} \int_0^t ds \int_{-\infty}^{Z_{s-}} \nu(dx)[f(Z_{s-} + h_s(x)) - f(Z_{s-}) + x\mathbb{1}_{\{|x|\leq 1\}}f'(Z_{s-})] \\ + \int_0^t ds \int_{Z_{s-}}^{\infty} \nu(dx)[f(Z_{s-} + h_s(x)) - f(Z_{s-}) + Z_{s-}\mathbb{1}_{\{|x|\leq 1\}}f'(Z_{s-})] \,. \end{aligned}$$

We denote by Σ^c the continuous part of the increasing process Σ and by $\Sigma_t^d = \sum_{s\leq t} \Delta\Sigma_s$ its discontinuous part, then

$$\int_0^t f'(Z_{s-})d\Sigma_s = \int_0^t f'(Z_s)d\Sigma_s^c + \int_0^t f'(Z_{s-})d\Sigma_s^d \,.$$

The support of the measure $d\Sigma_t^c$ is

$$\{t \,:\, X_{t-} = X_t = \Sigma_t = \Sigma_{t-}\} = \{t \,:\, Z_t = Z_{t-} = 0\}$$

hence $\int_0^t f'(Z_{s-})d\Sigma_s^c = f'(0)\Sigma_t^c$. It remains to use

$$\begin{aligned} -\int_0^t f'(Z_s)d\Sigma_s^d + \int_0^t ds\, f'(Z_s)\int_{Z_s}^{\infty} \nu(dx)Z_s\mathbb{1}_{\{|x|\leq 1\}} \\ = \int_0^t ds\, f'(Z_s)\int_{Z_s}^{\infty} \nu(dx)x\mathbb{1}_{\{|x|\leq 1\}} \,. \end{aligned}$$

□

We recall that a spectrally negative Lévy process is a Lévy process whose Lévy measure ν has its support in $]-\infty, 0[$.

Corollary 11.2.6.8 *Let X be a spectrally negative Lévy process. Then, the one-sided maximum of X, i.e., $\Sigma_t = \sup_{s\leq t} X_s$ is continuous. For $a > 0$, the process*

$$e^{-a(\Sigma_t - X_t)} + a\Sigma_t - \Psi(a)\int_0^t e^{-a(\Sigma_s - X_s)}ds$$

is a local martingale, where

$$\Psi(a) = \frac{\sigma^2}{2}a^2 + am + \int_{-\infty}^{0} \nu(dx)\left(e^{ax} - 1 - ax\mathbb{1}_{\{|x|\le 1\}}\right)$$

is the Laplace exponent of X.

PROOF: We apply the previous proposition for $f(x) = e^{-ax}$. Since X is a spectrally negative Lévy process, the process Σ is continuous. We obtain that

$$e^{-a(\Sigma_t - X_t)} + a\Sigma_t - \frac{\sigma^2}{2}\int_0^t a^2 e^{-a(\Sigma_s - X_s)}ds - m\int_0^t a e^{-a(\Sigma_s - X_s)}ds$$
$$-\int_0^t ds \int \nu(dx)\left(e^{-a(\Sigma_s - X_s - x)} - e^{-a(\Sigma_s - X_s)} - a e^{-a(\Sigma_s - X_s)} x \mathbb{1}_{\{|x|\le 1\}}\right)$$

is a local martingale. This last expression is equal to

$$e^{-a(\Sigma_t - X_t)} + a\Sigma_t - \int_0^t ds\, e^{-a(\Sigma_s - X_s)}\Psi(a)\,.$$

□

In the case where $X = B$ is a Brownian motion, we obtain that

$$e^{-a(\Sigma_t - B_t)} + a\Sigma_t - \frac{a^2}{2}\int_0^t ds\, e^{-a(\Sigma_s - B_s)}$$

is a martingale, or

$$e^{-a|B_t|} + aL_t - \frac{a^2}{2}\int_0^t ds\, e^{-a|B_s|}$$

is a martingale, where L is the local time at 0 for B.

11.2.7 Harness Property

Proposition 11.2.7.1 *Any Lévy process such that* $\mathbb{E}(|X_1|) < \infty$ *enjoys the harness property given in Definition* 8.5.2.1.

PROOF: Let X be a Lévy process. Then, for $s < t$

$$\mathbb{E}[\exp(i(\lambda X_s + \mu X_t))] = \exp(-s\Phi(\lambda + \mu) - (t - s)\Phi(\mu))\,.$$

Therefore, by differentiation with respect to λ, and taking $\lambda = 0$,

$$i\mathbb{E}[X_s \exp(i\mu X_t)] = -s\Phi'(\mu)\exp(-t\Phi(\mu))$$

which implies

$$t\mathbb{E}[X_s \exp(i\mu X_t)] = s\mathbb{E}[X_t \exp(i\mu X_t)]\,.$$

It follows that

$$\mathbb{E}\left[\frac{X_s}{s}\Big|\sigma(X_u, u \geq t)\right] = \frac{X_t}{t}.$$

Then, using the homogeneity of the increments and recalling the notation $\mathcal{F}_{s],[T} = \sigma(X_u, u \leq s, u \geq T)$ already given in Definition 8.5.2.1

$$\mathbb{E}\left(\frac{X_t - X_s}{t-s}|\mathcal{F}_{s],[T}\right) = \frac{X_T - X_s}{T-s}.$$

□

See Proposition 8.5.2.2, and Mansuy and Yor [621] for more comments on the harness property.

Exercise 11.2.7.2 The following amplification of Proposition 11.2.7.1 is due to Pal. Let X be a Lévy process and $\tau = (t_k, k \in \mathbb{N})$ a sequence of subdivisions of $\mathbb{R}^+$ and define $\mathcal{F}^{(\tau)} = \sigma(X_{t_k}, k \in \mathbb{N})$.

Prove that $\mathbb{E}(X_t|\mathcal{F}^{(\tau)}) = X_t^{(\tau)}$ where

$$X_t^{(\tau)} = \sum_k \left(X_{t_k} + \frac{t-t_k}{t_{k+1}-t_k}\left(X_{t_{k+1}} - X_{t_k}\right)\right)\mathbb{1}_{t_k < t \leq t_{k+1}}$$

is the linear interpolation of X along the subdivision τ. If $\tau \subset \tau'$, prove that

$$\mathbb{E}(X_t^{(\tau)}|\mathcal{F}^{(\tau')}) = X_t^{(\tau')}.$$

`Hint:` Write, for $t_k < t < t_{k+1}$

$$\mathbb{E}\left(\frac{X_t - X_{t_k}}{t-t_k}|X_{t_k}, X_{t_{k+1}}\right) = \frac{1}{t_{k+1}-t_k}\left(X_{t_{k+1}} - X_{t_k}\right).$$

◁

11.2.8 Representation Theorem of Martingales in a Lévy Setting

Proposition 11.2.8.1 *Let X be an $\mathbb{R}^d$-valued Lévy process and $\mathbf{F}^X$ its natural filtration. Let M be an $\mathbf{F}^X$-local martingale. Then, there exist an $\mathbb{R}^d$-valued predictable process $\varphi = (\varphi^i)$ and $\psi : \mathbb{R}^+ \times \Omega \times \mathbb{R}^d \to \mathbb{R}$ a predictable function such that*

$$\int_0^t (\varphi_s^i)^2 ds < \infty, \text{ a.s.},$$

$$\int_0^t \int_{|x|\leq 1} |\psi(s,x)| ds\, \nu(dx) < \infty, \text{ a.s.}, \quad \int_0^t \int_{|x|>1} \psi^2(s,x) ds\, \nu(dx) < \infty, \text{ a.s.}$$

and

$$M_t = M_0 + \sum_{i=1}^d \int_0^t \varphi_s^i dW_s^i + \int_0^t \int_{\mathbb{R}^d} \psi(s,x)\widetilde{\mathbf{N}}(ds,dx).$$

Moreover, if $(M_t, t \le T)$ is a square integrable martingale, then

$$\mathbb{E}\left[\left(\int_0^T \varphi^i_s dW^i_s\right)^2\right] = \mathbb{E}\left[\int_0^T (\varphi^i_s)^2 ds\right] < \infty\,,$$

$$\mathbb{E}\left[\left(\int_0^T \int \psi(s,x)\widetilde{\mathbf{N}}(ds,dx)\right)^2\right] = \mathbb{E}\left[\int_0^T ds \int \psi^2(s,x)\nu(dx)\right] < \infty\,.$$

The processes φ^i, ψ are essentially unique.

PROOF: See Kunita and Watanabe [550] and Kunita [549]. □

Proposition 11.2.8.2 *If L is a strictly positive local martingale such that $L_0 = 1$, then there exist*

a predictable process $f = (f_1, \dots, f_d)$ such that $\int_0^T |f_s|^2 ds < \infty$,
a predictable function g where

$$\int_0^T \int_{|x|\le 1} |e^{g(s,x)} - 1| ds\, \nu(dx) < \infty, \; \int_0^T \int_{|x|>1} g^2(s,x)\, ds\, \nu(dx) < \infty\,,$$

such that

$$dL_t = L_{t-}\left(\sum_{i=1}^d f_i(t) dW^i_t + \int (e^{g(t,x)} - 1)\widetilde{\mathbf{N}}(dt,dx)\right).$$

In a closed form, we obtain

$$\begin{aligned} L_t = &\exp\left(\sum_{i=1}^d \int_0^t f_i(s) dW^i_s - \frac{1}{2}\int_0^t |f_s|^2 ds\right)\\ &\times \exp\left(\mathbf{N}_t(g_0) - \int_0^t ds \int_{|x|\le 1} (e^{g_0(s,x)} - 1)\, \nu(dx)\right)\\ &\times \exp\left(\widetilde{\mathbf{N}}_t(g_1) - \int_0^t ds \int_{|x|>1} (e^{g_1(s,x)} - 1 - g_1(s,x))\, \nu(dx)\right), \end{aligned}$$

with $g_0(s,x) = \mathbb{1}_{\{|x|\le 1\}} g(s,x)$, $g_1(s,x) = \mathbb{1}_{\{|x|>1\}} g(s,x)$.

Comment 11.2.8.3 Nualart and Schoutens [682] have established the following predictable representation theorem for Lévy processes which satisfy $\int \mathbb{1}_{\{|x|\ge 1\}} e^{\delta |x|} \nu(dx) < \infty$ for some $\delta > 0$. These processes have moments of all orders. The processes

$$\begin{aligned} X^{(1)}_s &= X_s\\ X^{(i)}_s &= \sum_{0<u\le s} (\Delta X_u)^i\,,\; i \ge 2 \end{aligned}$$

are Lévy processes and, from Proposition 11.2.2.3, $\mathbb{E}(X^{(i)}_s) = s\int x^i \nu(dx)$.

Let $Y^{(i)}$ be martingales defined as $Y_s^{(1)} = X_s - \mathbb{E}(X_s)$ and, for $i \geq 2$,

$$Y_s^{(i)} = \sum_{0<u\leq s} (\Delta X_u)^i - \mathbb{E}\left(\sum_{0<u\leq s} (\Delta X_u)^i\right).$$

Let $H^{(i)}$ be a set of pairwise strongly orthogonal martingales, obtained by the Gram-Schmidt orthogonalization procedure of $Y^{(i)}$. Then, for $F \in L^2(\mathcal{F}_T)$, there exist predictable processes $(\varphi^{(i)})$ such that

$$F = \mathbb{E}(F) + \sum_{i=1}^{\infty} \int_0^T \varphi_s^{(i)} dH_s^{(i)}.$$

See Corcuera et al. [195] and Leon et al. [577] for applications.

11.3 Absolutely Continuous Changes of Measures

11.3.1 Esscher Transform

Let X be a Lévy process, and assume that $\mathbb{E}(e^{\theta \cdot X_t}) < \infty$ for some $\theta \in \mathbb{R}^d$. We define a probability $\mathbb{P}^{(\theta)}$, locally equivalent to $\mathbb{P}$ by the formula

$$\mathbb{P}^{(\theta)}|_{\mathcal{F}_t} = \frac{e^{\theta \cdot X_t}}{\mathbb{E}(e^{\theta \cdot X_t})} \mathbb{P}|_{\mathcal{F}_t}. \tag{11.3.1}$$

Note that, from Proposition 11.2.6.3, the process

$$L_t = \frac{e^{\theta \cdot X_t}}{\mathbb{E}(e^{\theta \cdot X_t})} = e^{\theta \cdot X_t - t\Psi(\theta)}$$

is a martingale. This particular choice of measure transformation, (called a θ-Esscher transform, or exponential tilting) preserves the Lévy process property as we now prove.

Proposition 11.3.1.1 *Let X be a $\mathbb{P}$-Lévy process with parameters (m, A, ν). Let θ be such that $\mathbb{E}(e^{\theta \cdot X_t}) < \infty$ and suppose that $\mathbb{P}^{(\theta)}$ is defined by (11.3.1). Then X is a $\mathbb{P}^{(\theta)}$-Lévy process and the Lévy-Khintchine representation of X under $\mathbb{P}^{(\theta)}$ is*

$$\mathbb{E}_{\mathbb{P}^{(\theta)}}(e^{iu \cdot X_t}) = \exp\left(iu \cdot m^{(\theta)} - \frac{u \cdot Au}{2} + \int_{\mathbb{R}^d} (e^{iu \cdot x} - 1 - iu \cdot x \mathbb{1}_{|x|\leq 1}) \nu^{(\theta)}(dx)\right)$$

with

$$m^{(\theta)} = m + \frac{1}{2}(A + A^T)\theta + \int_{|x|\leq 1} x(e^{\theta \cdot x} - 1)\nu(dx),$$

$$\nu^{(\theta)}(dx) = e^{\theta \cdot x}\nu(dx).$$

The characteristic exponent of X *under* $\mathbb{P}^{(\theta)}$ *is*

$$\Phi^{(\theta)}(u) = \Phi(u - i\theta) - \Phi(-i\theta)\,,$$

and the Laplace exponent is $\Psi^{(\theta)}(u) = \Psi(u+\theta) - \Psi(\theta)$ *for* $u \geq \min(-\theta, 0)$.

PROOF: It is not difficult to prove that X has independent and stationary increments under $\mathbb{P}^{(\theta)}$. The characteristic exponent of X under $\mathbb{P}^{(\theta)}$ is $\Phi^{(\theta)}$ such that

$$\begin{aligned} e^{-t\Phi^{(\theta)}(u)} &= \mathbb{E}_{\mathbb{P}^{(\theta)}}(e^{iu\,\bullet\, X_t}) = \mathbb{E}(e^{iu\,\bullet\, X_t + \theta\,\bullet\, X_t})e^{t\Phi(-i\theta)} \\ &= e^{-t(\Phi(u-i\theta)-\Phi(-i\theta))}\,. \end{aligned}$$

A simple computation leads to

$$\begin{aligned} \Phi(u-i\theta) - \Phi(-i\theta) &= -iu\,\bullet\, m + \frac{1}{2}u\,\bullet\, Au - \frac{1}{2}iu\,\bullet\, A\theta - \frac{1}{2}i\theta\,\bullet\, Au \\ &\quad - \int \left(e^{\theta\,\bullet\, x}(e^{iu\,\bullet\, x}-1) - iu\,\bullet\, x\mathbb{1}_{\{|x|\leq 1\}}\right)\nu(dx) \\ &= -iu\,\bullet \left(m + \frac{1}{2}(A + A^T)\theta + \int (e^{\theta\,\bullet\, x}-1)x\mathbb{1}_{\{|x|\leq 1\}}\nu(dx)\right) \\ &\quad + \frac{1}{2}u\,\bullet\, Au + \int e^{\theta\,\bullet\, x}(e^{iu\,\bullet\, x} - 1 - iu\,\bullet\, x\mathbb{1}_{\{|x|\leq 1\}})\nu(dx)\,. \end{aligned}$$

Hence, X_1 has the required Lévy-Khintchine representation under $\mathbb{P}^{(\theta)}$. □

Corollary 11.3.1.2 *Let* X *be a* (m, σ^2, ν) *real-valued Lévy process and* β *such that*

$$\int \left|e^{\beta x}(e^x - 1) - x\mathbb{1}_{\{|x|\leq 1\}}\right|\nu(dx) < \infty$$

and

$$m + \left(\beta + \frac{1}{2}\right)\sigma^2 + \int \left(e^{\beta x}(e^x-1) - x\mathbb{1}_{\{|x|\leq 1\}}\right)\nu(dx) = 0\,.$$

Let $\mathbb{P}^{(\beta)}|_{\mathcal{F}_t} = \dfrac{e^{\beta X_t}}{\mathbb{E}(e^{\beta X_t})}\mathbb{P}|_{\mathcal{F}_t}$. *Then, the process* $(e^{X_t}, t \geq 0)$ *is a* $\mathbb{P}^{(\beta)}$-*martingale.*

PROOF: This is a consequence of Proposition 11.3.1.1 and Corollary 11.2.6.4. □

Comment 11.3.1.3 The Esscher transform was introduced by Esscher [335] and again by Gerber and Shiu [388]. See Bühlmann et al. [136] for a discussion on Esscher transforms in finance and Fujiwara and Miyahara [369] for some relations between the Esscher transform and minimal entropy measure.

Exercise 11.3.1.4 Let X be a real-valued Lévy process such that, for any λ, $\mathbb{E}(e^{\lambda X_t}) < \infty$ and define

$$\mathbb{P}^{(\lambda)}|_{\mathcal{F}_t} = \frac{e^{\lambda X_t}}{\mathbb{E}(e^{\lambda X_t})}\mathbb{P}|_{\mathcal{F}_t} .$$

Prove that $(\mathbb{P}^{(\lambda)})^{(\mu)} = \mathbb{P}^{(\lambda+\mu)}$ and that if $S_t = S_0 e^{X_t}$, then, for any μ, for any positive Borel function g,

$$\mathbb{E}^{(\lambda)}[S_t^\mu g(S_t)] = \mathbb{E}^{(\lambda)}[S_t^\mu]\, \mathbb{E}^{(\lambda+\mu)}[g(S_t)] .$$

Extend this formula to $\mathbb{R}^d$-valued Lévy processes. ◁

Exercise 11.3.1.5 Prove that the Esscher transforms of the 1/2-stable process are the inverse Gaussian processes. (See ↣ Example 11.6.1.2 for the definition of Inverse Gaussian processes.) ◁

Exercise 11.3.1.6 Prove that the Esscher transforms of a double exponential process are double exponential processes. ◁

Exercise 11.3.1.7 Let γ be a $\Gamma(1,1)$ process:

$$\mathbb{E}(e^{-\lambda\gamma_t}) = \exp\left(-t\int_0^\infty \frac{dx}{x} e^{-x}(1-e^{-\lambda x})\right) = \frac{1}{(1+\lambda)^t} .$$

(a) Prove the double identity, for $1+\mu > 0$

$$\begin{aligned}\mathbb{E}(e^{-\lambda\gamma_t} e^{-\mu\gamma_t}(1+\mu)^t) &= \frac{1}{(1+\frac{\lambda}{1+\mu})^t} \\ &= \mathbb{E}(e^{-\frac{\lambda}{1+\mu}\gamma_t}) .\end{aligned}$$

(b) Conclude that the $(-\mu)$-Esscher transform of γ_t is $\frac{1}{1+\mu}\gamma_t$, in other terms, any multiple $(a\gamma_t, t \geq 0)$ is an Esscher transform of $(\gamma_t, t \geq 0)$.

(c) Prove that, for $a > 0$, the process $(a\frac{\gamma_t}{\gamma_T}, t \leq T)$ is distributed as the γ bridge process on $[0,T]$, starting from 0 and ending at a, at time T. ◁

11.3.2 Preserving the Lévy Property with Absolute Continuity

Given a Lévy process X under $\mathbb{P}$ with generating triple (m, A, ν), we define, following Sato [761], its compensated jump part as

$$X_t^\nu = \lim_{\epsilon\to 0} \sum_{(s,\Delta X_s)\in(0,t]\times\{|x|>\epsilon\}} \Delta X_s - t\int_{\epsilon<|x|\leq 1} x\,\nu(dx) .$$

We now examine which other Girsanov-type transformations than Esscher transforms preserve the Lévy property of a Lévy process.

Proposition 11.3.2.1 *Let* $(X, \mathbb{P})$ *(resp.* $(X, \mathbb{P}^*)$*) be Lévy processes with generating triples* (m, A, ν) *(resp.* (m^*, A^*, ν^*)*). The following statements are equivalent:*

(i) $\mathbb{P}$ *and* $\mathbb{P}^*$ *are locally equivalent,*
(ii) $A = A^*$, $d\nu^*(x) = e^{\varphi(x)} d\nu(x)$ *with the function* φ *satisfying*

$$\int_{\mathbb{R}^d} (e^{\varphi(x)/2} - 1)^2 \nu(dx) < \infty$$

and

$$m^* - m - \int_{|x| \le 1} x \, (\nu^* - \nu)(dx) \in \mathcal{R}(A)$$

where $\mathcal{R}(A) = \{Ax, x \in \mathbb{R}^d\}$.

Furthermore, if the previous conditions (ii) *are satisfied*

$$\mathbb{P}^*|_{\mathcal{F}_t} = e^{U_t} \, \mathbb{P}|_{\mathcal{F}_t} ,$$

where

$$\begin{aligned} U_t &= \eta \centerdot (X_t - X_t^\nu) - \frac{t}{2} \eta \centerdot A\eta - tm \centerdot \eta \\ &+ \lim_{\epsilon \to 0} \sum_{s \le t, \, |\Delta X_s| > \epsilon} \varphi(\Delta X_s) - t \int_{\epsilon < |x| \le 1} (e^{\varphi(x)} - 1) \nu(dx) , \end{aligned}$$

and η *is such that* $m^* - m - \int_{|x| \le 1} x \, (\nu^* - \nu)(dx) = A\eta$.

PROOF: See Sato [761], Theorem 33.1. Note that $(e^{U_t}, t \ge 0)$ is a martingale and that U is a Lévy process with generating triple

$$\begin{aligned} \sigma_U^2 &= \eta \centerdot A\eta, \\ m_U &= -\frac{1}{2} \eta \centerdot A\eta - \int_{\mathbb{R}} (e^y - 1 - y \mathbb{1}_{\{|y| \le 1\}}(y)) \, \nu_U(dy) , \\ \nu_U &= (\varphi(\nu)) \,|_{\mathbb{R} \setminus \{0\}} \end{aligned}$$

i.e., ν_U is the image of ν by φ. □

Example 11.3.2.2 Let X_1 and X_2 be CGMY Lévy processes (see ↣ Section 11.8) with parameters $(C, G_i, M_i, Y), i = 1, 2$. Then, it is easy to check that the hypotheses of Proposition 11.3.2.1 are satisfied. See Cont and Tankov [192] Chapter 9, Example 9.2 for the case of tempered stable processes.

11.3.3 General Case

To complete this discussion, we study the Girsanov transformation of Lévy processes using a general positive martingale density L. In this generality, the Lévy property will be lost under the new probability. From the previous martingale representation Theorem 11.2.8.2, we can and do associate with the martingale L a pair (f, g) such that

$$dL_t = L_{t-}\left(\sum_{i=1}^{d} f_t^i dW_t^i + \int (e^{g(t,x)} - 1)[\mathbf{N}(dt, dx) - \nu(dx)\,dt]\right). \qquad (11.3.2)$$

Sometimes, we shall denote this process L by $L(f, g)$.

Proposition 11.3.3.1 *Let* $\mathbb{Q}|_{\mathcal{F}_t} = L_t(f,g)\,\mathbb{P}|_{\mathcal{F}_t}$ *where* $L(f,g)$ *is defined in* (11.3.2). *Then, with respect to* $\mathbb{Q}$:

(i) *The process* W^f *defined by* $W_t^f := W_t - \int_0^t f_s ds$ *is a Brownian motion.*

(ii) *The random measure* $\mathbf{N}$ *is compensated by* $e^{g(s,x)} ds\,\nu(dx)$ *meaning that for any Borel function* h *such that*

$$\int_0^T ds \int_{\mathbb{R}} |h(s,x)| e^{g(s,x)} \nu(dx) < \infty\,,$$

the process

$$M_t^h := \int_0^t \int_{\mathbb{R}} h(s,x) \left(\mathbf{N}(ds, dx) - e^{g(s,x)} \nu(dx)\,ds\right)$$

is a local martingale.

PROOF: Using Itô's calculus, it is easy to check that $W^f L$ and $M^h L$ are $\mathbb{P}$-martingales. □

The $\mathbb{P}$-Lévy process X is a $\mathbb{Q}$-semi-martingale; in general, it is not a Lévy process, nor an additive process. One should note the important gap between the framework of Propositions 11.3.2.1 and 11.3.3.1: with the latter, the Markovian property may be lost.

Comment 11.3.3.2 A frequently asked question is to find a condition for a local exponential martingale to be a martingale. In order that $L(f, g)$ is a martingale, Kunita [549] gives the following condition: if (f, g) satisfies

$$\mathbb{E}\left[\exp\left(\int_0^t \left(af^2 + \int_{|g|>\delta} e^{2ag^+} d\nu + 2ae^{2a\delta} \int_{|g|\le\delta} g^2 d\nu\right) ds\right)\right] < \infty$$

for some $a > 1$ and some $\delta > 0$, then $L(f, g)$ is a martingale.

11.4 Fluctuation Theory

Here, X is a real-valued Lévy process. We are interested in the law of the running maximum.

11.4.1 Maximum and Minimum

We start with a simple lemma.

Lemma 11.4.1.1 *Let X be a Lévy process and Θ be an exponential variable with parameter q, independent of X. Then, the pair (Θ, X_Θ) has an infinitely divisible law, and its Lévy measure is $\mu(dt, dx) = t^{-1}e^{-qt}\mathbb{P}(X_t \in dx)dt$.*

PROOF: For any pair α, β, with $\alpha > 0$, we have

$$\begin{aligned}\mathbb{E}(e^{-\alpha\Theta+i\beta X_\Theta}) &= \int_0^\infty qe^{-qt}e^{-t\alpha-t\Phi(\beta)}dt \\ &= \exp\left(\int_0^\infty (e^{-t\alpha-t\Phi(\beta)} - 1)t^{-1}e^{-qt}dt\right)\end{aligned}$$

where in the second equality, we have used the Frullani integral

$$\int_0^\infty (1 - e^{-\lambda t})t^{-1}e^{-bt}dt = \ln\left(1 + \frac{\lambda}{b}\right).$$

It remains to write

$$\int_0^\infty (e^{-t\alpha-t\Phi(\beta)} - 1)t^{-1}e^{-qt}dt = \int_0^\infty \int_{\mathbb{R}} (e^{-t\alpha+i\beta x} - 1)\,\mathbb{P}(X_t \in dx)\,t^{-1}e^{-qt}dt.$$

□

Let $\Sigma_t = \sup_{s \leq t} X_s$ be the running maximum of the Lévy process X. The reflected process $\Sigma - X$ enjoys the strong Markov property.

Proposition 11.4.1.2 (Wiener-Hopf Factorization.) *Let Θ be an exponential variable with parameter q, independent of X. Then, the random variables Σ_Θ and $X_\Theta - \Sigma_\Theta$ are independent and*

$$\mathbb{E}(e^{iu\Sigma_\Theta})\mathbb{E}(e^{iu(X_\Theta-\Sigma_\Theta)}) = \frac{q}{q + \Phi(u)}. \tag{11.4.1}$$

SKETCH OF THE PROOF: Note that

$$\mathbb{E}(e^{iuX_\Theta}) = q\int_0^\infty \mathbb{E}(e^{iuX_t})e^{-qt}dt = q\int_0^\infty e^{-t\Phi(u)}e^{-qt}dt = \frac{q}{q + \Phi(u)}.$$

Using excursion theory, the random variables Σ_Θ and $X_\Theta - \Sigma_\Theta$ are shown to be independent (see Bertoin [78]), hence

$$\mathbb{E}(e^{iu\Sigma_\Theta})\mathbb{E}(e^{iu(X_\Theta-\Sigma_\Theta)}) = \mathbb{E}(e^{iuX_\Theta}) = \frac{q}{q+\Phi(u)}\,.$$

□

The equality (11.4.1) is known as the **Wiener-Hopf factorization**, the factors being the characteristic functions $\mathbb{E}(e^{iu\Sigma_\Theta})$ and $\mathbb{E}(e^{iu(X_\Theta-\Sigma_\Theta)})$. We now prepare for the computation of these factors.

There exists a family (L_t^x) of local times of the reflected process $\Sigma - X$ which satisfies

$$\int_0^t ds f(\Sigma_s - X_s) = \int_0^\infty dx f(x) L_t^x$$

for all positive functions f. We then consider $L = L^0$ and τ its right continuous inverse $\tau_\ell = \inf\{u > 0 \,:\, L_u > \ell\}$. Introduce

$$H_\ell = \Sigma_{\tau_\ell} \quad \text{for } \tau_\ell < \infty, \qquad H_\ell = \infty \text{ otherwise}\,. \tag{11.4.2}$$

The two-dimensional process (τ, H) is called the **ladder process** and is a Lévy process.

Proposition 11.4.1.3 *Let κ be the Laplace exponent of the ladder process defined as*

$$e^{-\ell\kappa(\alpha,\beta)} = \mathbb{E}(\exp(-\alpha\tau_\ell - \beta H_\ell))\,.$$

There exists a constant $k > 0$ such that

$$\kappa(\alpha,\beta) = k \exp\left(\int_0^\infty dt \int_0^\infty t^{-1}(e^{-t} - e^{-\alpha t-\beta x})\,\mathbb{P}(X_t \in dx)\right)\,.$$

SKETCH OF THE PROOF:
Using excursion theory, and setting $G_\Theta = \sup\{t < \Theta \,:\, X_t = \Sigma_t\}$, it can be proved that

$$\mathbb{E}(e^{-\alpha G_\Theta - \beta\Sigma_\Theta}) = \frac{\kappa(q,0)}{\kappa(\alpha+q,\beta)}\,. \tag{11.4.3}$$

The pair of random variables (Θ, X_Θ) can be decomposed as the sum of $(G_\Theta, \Sigma_\Theta)$ and $(\Theta - G_\Theta, X_\Theta - \Sigma_\Theta)$, which are shown to be independent infinitely divisible random variables. Let μ, μ^+ and μ^- denote the respective Lévy measures of these three two-dimensional variables. The Lévy measure μ^+ (resp μ^-) has support in $[0,\infty[\times[0,\infty[$ (resp. $[0,\infty[\times]-\infty,0]$) and $\mu = \mu^+ + \mu^-$. From Lemma 11.4.1.1, the Lévy measure of (Θ, X_Θ) is $t^{-1}e^{-qt}\,\mathbb{P}(X_t \in dx)\,dt$ and noting that this quantity is

$$t^{-1}e^{-qt}\,\mathbb{P}(X_t \in dx)\,dt\mathbb{1}_{\{x>0\}} + t^{-1}e^{-qt}\,\mathbb{P}(X_t \in dx)\,dt\mathbb{1}_{\{x<0\}}$$

one establishes that $\mu^+(dt,dx) = t^{-1}e^{-qt}\,\mathbb{P}(X_t \in dx)\,dt\,\mathbb{1}_{\{x>0\}}$.

It follows from (11.4.3) that

$$\frac{\kappa(q,0)}{\kappa(\alpha+q,\beta)} = \exp\left(-\int_0^\infty \int_0^\infty (1 - e^{-\alpha t-\beta x})\,\mu^+(dt,dx)\right)\,.$$

Hence,

$$\kappa(\alpha+q,\beta)=\kappa(q,0)\exp\left(\int_0^\infty dt\int_0^\infty t^{-1}(1-e^{-\alpha t-\beta x})e^{-qt}\mathbb{P}(X_t\in dx)\right).$$

In particular, for $q=1$,

$$\kappa(\alpha+1,\beta)=\kappa(1,0)\exp\left(\int_0^\infty dt\int_0^\infty t^{-1}(e^{-t}-e^{-(\alpha+1)t-\beta x})\mathbb{P}(X_t\in dx)\right).$$

□

Note that, for $\alpha=0$ and $\beta=-iu$, one obtains from equality (11.4.3)

$$\mathbb{E}(e^{iu\Sigma_\Theta})=\frac{\kappa(q,0)}{\kappa(q,-iu)}.$$

From Proposition 11.2.1.5, setting $\underline{m}_t=\inf_{s\le t}X_s$, we have

$$\underline{m}_\Theta\overset{\text{law}}{=}X_\Theta-\Sigma_\Theta.$$

Let $\widehat{X}:=-X$ be the dual process of X. The Laplace exponent of the dual ladder process is, for some constant $\widehat{k}$,

$$\begin{aligned}\widehat{\kappa}(\alpha,\beta)&=\widehat{k}\exp\left(\int_0^\infty dt\int_0^\infty t^{-1}(e^{-t}-e^{-\alpha t-\beta x})\mathbb{P}(-X_t\in dx)\right)\\&=\widehat{k}\exp\left(\int_0^\infty dt\int_{-\infty}^0 t^{-1}(e^{-t}-e^{-\alpha t-\beta x})\mathbb{P}(X_t\in dx)\right).\end{aligned}$$

From the Wiener-Hopf factorization and duality, one deduces

$$\mathbb{E}(e^{iu\underline{m}_\Theta})=\frac{\widehat{\kappa}(q,0)}{\widehat{\kappa}(q,iu)}$$

and

$$\mathbb{E}(e^{iu(X_\Theta-\Sigma_\Theta)})=\frac{\widehat{\kappa}(q,0)}{\widehat{\kappa}(q,iu)}=\mathbb{E}(e^{iu\underline{m}_\Theta}).$$

Note that, by definition

$$\begin{aligned}\kappa(q,0)&=k\exp\left(\int_0^\infty dt\,t^{-1}(e^{-t}-e^{-qt})\mathbb{P}(X_t\ge 0)\right),\\\widehat{\kappa}(q,0)&=\widehat{k}\exp\left(\int_0^\infty dt\,t^{-1}(e^{-t}-e^{-qt})\mathbb{P}(X_t\le 0)\right),\end{aligned}$$

hence,

$$\kappa(q,0)\widehat{\kappa}(q,0)=k\widehat{k}\exp\left(\int_0^\infty dt\,t^{-1}(e^{-t}-e^{-qt})\right)=k\widehat{k}q,$$

where the last equality follows from Frullani's integral, and we deduce the following relationship between the ladder exponents

$$\kappa(q, -iu)\,\widehat{\kappa}(q, -iu) = k\widehat{k}\,(q + \Phi(u))\,. \tag{11.4.4}$$

Example 11.4.1.4 In the case of the Lévy process $X_t = \sigma W_t + \mu t$, one has $\Phi(u) = -i\mu u + \frac{1}{2}\sigma^2 u^2$, hence the quantity $\frac{q}{q+\Phi(u)}$ is

$$\frac{q}{q - i\mu u + \frac{1}{2}\sigma^2 u^2} = \frac{a}{a + iu}\frac{b}{b - iu}$$

with

$$a = \frac{\mu + \sqrt{\mu^2 + 2\sigma^2 q}}{\sigma^2}, \quad b = \frac{-\mu + \sqrt{\mu^2 + 2\sigma^2 q}}{\sigma^2}.$$

The first factor of the Wiener-Hopf factorization $\mathbb{E}(e^{iu\Sigma_\Theta})$ is the characteristic function of an exponential distribution with parameter a, and the second factor $\mathbb{E}(e^{iu(X_\Theta - \Sigma_\Theta)})$ is the characteristic function of an exponential on the negative half axis with parameter b.

Comment 11.4.1.5 Since the publication of the paper of Bingham [100], many results have been obtained about fluctuation theory. See, e.g., Doney [260], Greenwood and Pitman [407], Kyprianou [553] and Nguyen-Ngoc and Yor [670].

11.4.2 Pecherskii-Rogozin Identity

For $x > 0$, denote by T_x the first passage time above x defined as

$$T_x = \inf\{t > 0 \,:\, X_t > x\}$$

and by $O_x = X_{T_x} - x$ the overshoot which can be written

$$O_x = \Sigma_{T_x} - x = H_{\eta_x} - x$$

where $\eta_x = \inf\{t \,:\, H_t > x\}$ and where H is defined in (11.4.2). Indeed, $T_x = \tau(\eta_x)$.

Proposition 11.4.2.1 (Pecherskii-Rogozin Identity.) *For every triple of positive numbers* (α, β, q),

$$\int_0^\infty e^{-qx}\mathbb{E}(e^{-\alpha T_x - \beta O_x})dx = \frac{\kappa(\alpha, q) - \kappa(\alpha, \beta)}{(q - \beta)\kappa(\alpha, q)}\,. \tag{11.4.5}$$

PROOF: See Pecherskii and Rogozin [702], Bertoin [78] or Nguyen-Ngoc and Yor [670] for different proofs of the Pecherskii-Rogozin identity. □

Comment 11.4.2.2 Roynette et al. [745] present a general study of overshoot and asymptotic behavior. See Hilberink and Rogers [435] for application to endogeneous default, Klüppelberg et al. [526] for applications to insurance.

11.5 Spectrally Negative Lévy Processes

A **spectrally negative Lévy process** X is a real-valued Lévy process with no positive jumps, equivalently its Lévy measure is supported by $(-\infty, 0)$. Then, X admits positive exponential moments

$$\mathbb{E}(\exp(\lambda X_t)) = \exp(t\Psi(\lambda)) < \infty, \ \forall \lambda > 0$$

where

$$\Psi(\lambda) = \lambda m + \frac{1}{2}\sigma^2\lambda^2 + \int_{-\infty}^{0} (e^{\lambda x} - 1 - \lambda x \mathbb{1}_{\{-1<x<0\}})\nu(dx).$$

We recall that the process $e^{\lambda X_t - t\Psi(\lambda)}$ is a martingale. Introduce the inverse of Ψ,

$$\Psi^{\sharp}(q) = \inf\{\lambda \geq 0 : \Psi(\lambda) \geq q\}.$$

Then, the process

$$(\exp(\Psi^{\sharp}(q)X_t - tq), t \geq 0)$$

is a martingale.

11.5.1 Two-sided Exit Times

We present here some results on two-sided exit times. We look for formulae for two-sided exit, i.e., for $0 < x < a$,

$$\begin{aligned} T_0^- &= \inf\{t \geq 0 \, : \, X_t \leq 0\}, \\ T_a^+ &= \inf\{t \geq 0 \, : \, X_t \geq a\}. \end{aligned}$$

Assume now that X is a spectrally negative $\mathbf{F}$-Lévy process with $X_0 = x$. Then, the process $(\exp \lambda(X_t - x) - t\psi(\lambda))$ is a $(\mathbb{P}_x, \mathbf{F})$ martingale and, for $a > x$, the optional stopping theorem (with a careful check of integrability conditions) implies

$$\mathbb{E}_x(e^{-qT_a^+}\mathbb{1}_{\{T_a^+<\infty\}}) = e^{-\Psi^{\sharp}(q)(a-x)}.$$

We can express the Laplace transforms of the two-sided exit times in terms of a family of two scale functions $z^{(q)}$ and $w^{(q)}$, $q > 0$

$$\begin{aligned} \mathbb{E}_x\left[\exp(-qT_0^-)\right] &= z^{(q)}(x) - \frac{q}{\Psi^{\sharp}(q)}w^{(q)}(x), \\ \mathbb{E}_x\left[\exp(-qT_a^+)\mathbb{1}_{\{T_a^+<T_0^-\}}\right] &= w^{(q)}(x)/w^{(q)}(a), \\ \mathbb{E}_x\left[\exp(-qT_0^-)\mathbb{1}_{\{T_0^-<T_a^+\}}\right] &= z^{(q)}(x) - z^{(q)}(a)w^{(q)}(x)/w^{(q)}(a). \end{aligned}$$

The functions $z^{(q)}$ and $w^{(q)}$ are characterized via their Laplace transforms

$$\int_0^\infty e^{-\beta x} w^{(q)}(x)dx = \frac{1}{\Psi(\beta) - q}, \text{ for } \beta > \Psi^\sharp(q), \tag{11.5.1}$$

$$z^{(q)}(x) = 1 + q \int_0^x w^{(q)}(y)dy .$$

As a consequence, one obtains

$$\int_0^\infty e^{-\beta x} z^{(q)}(x)dx = \frac{\Psi(\beta)}{\beta(\Psi(\beta) - q)}, \quad \beta > \Psi^\sharp(q) .$$

11.5.2 Laplace Exponent of the Ladder Process

Proposition 11.5.2.1 *Let X be a spectrally negative Lévy process and Θ be an exponential variable with parameter q, independent of X. Then, Σ_Θ has an exponential law with parameter $\Psi^\sharp(q)$.*

PROOF: For any $x > 0$,

$$\mathbb{P}(\Sigma_\Theta \geq x) = \mathbb{P}(T_x \leq \Theta) = \mathbb{E}\left(\int_{T_x}^\infty q e^{-qt} dt\right) = \mathbb{E}(e^{-qT_x}) = e^{-x\Psi^\sharp(q)} ,$$

where we have used the fact that, since X is spectrally negative, $X_{T_x} = x$. □

Proposition 11.5.2.2 *The Laplace exponent of the ladder process is*

$$\kappa(\alpha, \beta) = \Psi^\sharp(\alpha) + \beta .$$

The Laplace exponent of the dual ladder process is

$$\widehat{\kappa}(\alpha, \beta) = k\widehat{k} \frac{\alpha - \Psi(\beta)}{\Psi^\sharp(\alpha) - \beta} .$$

PROOF: The absence of positive jumps ensures that $H_t = \Sigma_{\tau_t} = t$, and

$$\kappa(\alpha, \beta) = \Psi^\sharp(\alpha) + \beta, \quad \widehat{\kappa}(\alpha, \beta) = k\widehat{k} \frac{\alpha - \Psi(\beta)}{\Psi^\sharp(\alpha) - \beta} .$$

(See formula (11.4.4)). □

11.5.3 D. Kendall's Identity

Kendall's identity states that for a Lévy process with no positive jumps, if $x > 0$ and $T_x = \inf\{t : X_t \geq x\}$, then

$$\boxed{t\,\mathbb{P}(T_x \in dt)\,dx = x\,\mathbb{P}(X_t \in dx)\,dt}$$

In particular, if X_t admits a density $p_X(t,x)$, then T_x has a density $p_T(t,x)$ and

$$tp_T(t,x) = xp_X(t,x) .$$

(See Borovkov and Burq [110] or Dozzi and Vallois [265] for a proof.) Note that, in the case of Brownian motion, this equality was obtained in 3.1.4.1.

Example 11.5.3.1 Let $X_t = \lambda t - \gamma_t$ where γ is a Gamma process. Then

$$\begin{aligned}\mathbb{E}(f(\lambda t - \gamma_t)) &= \frac{1}{\Gamma(t)} \int_0^\infty dy f(\lambda t - y)\, y^{t-1} e^{-y} \\ &= \frac{1}{\Gamma(t)} \int_{-\infty}^{\lambda t} dx f(x) (\lambda t - x)^{t-1} e^{-(\lambda t - x)}\end{aligned}$$

hence,

$$p_T(t,x) = \frac{x}{t} p_X(t,x) = \frac{x}{\Gamma(1+t)} (\lambda t - x)^{t-1} e^{-(\lambda t - x)} \mathbb{1}_{\{x < \lambda t\}} .$$

11.6 Subordinators

11.6.1 Definition and Examples

Recall (see Definition 11.2.1.3) that a Lévy process Z which takes values in $[0,\infty[$ is called a subordinator. The Laplace transform $\mathbb{E}(e^{-\theta Z_t})$ of Z_t exists for $\theta \geq 0$ and is

$$\mathbb{E}(e^{-\theta Z_t}) = \exp\left(-t \left(m\theta + \int_{]0,\infty[} (1 - e^{-\theta x}) \nu(dx) \right) \right) .$$

The constant m is called the drift of Z. If Z has no drift and f is a positive function, the exponential formula leads to

$$\mathbb{E}\left(\exp - \int_0^\infty f(s) dZ_s \right) = \exp\left(- \int_0^\infty ds \int (1 - e^{-x f(s)}) \nu(dx) \right) .$$

The term subordinator originates from the following operation:

Definition 11.6.1.1 *Let Z be a subordinator and X an independent Lévy process. The process $\widetilde{X}_t = X_{Z_t}$ is a Lévy process, called the subordinated Lévy process, i.e., it is the X process subordinated by Z.*

Example 11.6.1.2 We present some examples of subordinators (S) and subordinated processes (SP), in various degrees of generality:

- **Compound Poisson Process (S).** A ν-compound Poisson process X where the support of ν is included in $]0, \infty[$, is a subordinator. If X is a ν-compound Poisson process, its quadratic variation
$$[X]_t = \sum_{s \leq t} (\Delta X_s)^2 = \sum_{k=1}^{N_t} Y_k^2$$
is a compound Poisson process and a subordinator.
- **(S)** Let B be a BM, and
$$\tau_r = \inf\{t \geq 0 \, : \, B_t \geq r\}\,.$$
The process $(\tau_r, r \geq 0)$ is a $1/2$-stable subordinator, whose Lévy measure is $\frac{1}{\sqrt{2\pi}\, x^{3/2}} \mathbb{1}_{\{x>0\}} dx$ (see Example 11.2.3.5).
- **(SP)** Let X be a Brownian motion and Z an independent subordinator. Then, if $\widetilde{X}_t = X_{Z_t}$, and $\widetilde{\nu}$ denotes the Lévy measure of $\widetilde{X}$, we have
$$\begin{aligned} \mathbb{E}(e^{i\theta \widetilde{X}_t}) &= \mathbb{E}\left(\exp -\frac{\theta^2}{2} Z_t\right) = \exp\left(-t \int_{\mathbb{R}^+} \nu(dz)(1 - e^{-\theta^2 z/2})\right) \\ &= \exp\left(-t \int_{\mathbb{R}^+} \widetilde{\nu}(dx)(1 - e^{i\theta x})\right)\,. \end{aligned}$$
Using
$$1 - e^{-\theta^2 z/2} = \frac{1}{\sqrt{2\pi z}} \int_{-\infty}^{\infty} dx\, e^{-\frac{x^2}{2z}} (1 - e^{i\theta x})\,,$$
we obtain
$$\boxed{\widetilde{\nu}(dx) = dx \int_{\mathbb{R}^+} \nu(dz) \frac{1}{\sqrt{2\pi z}}\, e^{-\frac{x^2}{2z}}}$$
a formula which goes back at least to Huff [450].
- **Cauchy Process (SP).** We present the most well-known example of the subordination operation. Let B be a Brownian motion and, for $t \geq 0$ define $\tau_t = \inf\{s \geq 0 \, : \, B_s \geq t\}$. Let W be a BM, independent of B. The process $\widetilde{X}_t = W_{\tau_t}$ is a Cauchy process (i.e., W_{τ_t} has the Cauchy law with parameter t) whose Lévy measure is $dx/(\pi x^2)$. The proof relies on the equality
$$\mathbb{E}(e^{iu\widetilde{X}_t}) = \mathbb{E}\left(\exp -\frac{u^2}{2} \tau_t\right) = e^{-|u|t}\,.$$
This result will be vastly generalized in the following Proposition 11.6.2.1.
- **Gamma Process (S).** The Gamma process $(G(t; a, \nu), t \geq 0)$ is a Lévy process with $G(1; a, \nu)$ having a Gamma $\Gamma(a, \nu)$ distribution (See

Example 11.1.1.9 or Appendix.) The Gamma process is an increasing Lévy process, hence a subordinator, with Lévy measure

$$\frac{a}{x}\exp(-x\nu)\mathbb{1}_{\{x>0\}}dx\,.$$

11.6.2 Lévy Characteristics of a Subordinated Process

Proposition 11.6.2.1 (Changes of Lévy Characteristics Under Subordination.) *Let X be a (m^X, A^X, ν^X) Lévy process and Z a subordinator with drift β and Lévy measure ν^Z, independent of X. The process $\widetilde{X}_t = X_{Z_t}$ is a Lévy process with characteristic exponent*

$$\Phi(u) = i\widetilde{m}\centerdot u + \frac{1}{2}u\centerdot\widetilde{A}u - \int(e^{iu\centerdot x} - 1 - iu\centerdot x\mathbb{1}_{\{|x|\leq 1\}})\widetilde{\nu}(dx)$$

with

$$\begin{aligned}\widetilde{m} &= \beta m^X + \int\int \nu^Z(ds)\mathbb{1}_{\{|x|\leq 1\}}x\,\mathbb{P}(X_s\in dx)\,,\\ \widetilde{A} &= \beta A^X\,,\\ \widetilde{\nu}(dx) &= \beta\nu^X(dx) + \int_0^\infty \nu^Z(ds)\mathbb{P}(X_s\in dx)\,.\end{aligned}$$

PROOF: The proof may be performed in two parts. First, one establishes that $\widetilde{X}$ is a Lévy process, and second, its Lévy characteristics may be computed. We refer to Sato [761], Theorem 30.1 for the details. □

Comment 11.6.2.2 A precise study of subordinators with many examples may be found in Bertoin [79, 80].

Exercise 11.6.2.3 Let B be a Brownian motion, $X_t = \mu t + \sigma B_t$ and N a Poisson process with intensity λ independent of B. Prove that the process $Z_t = X_{N_t}$ is a compound Poisson process. Give the law of Z_1. (See also Exercise 8.6.3.7.) ◁

11.7 Exponential Lévy Processes as Stock Price Processes

We now present, in the following sections, some applications to finance.

11.7.1 Option Pricing with Esscher Transform

Let $S_t = S_0 e^{rt+X_t}$ where under the historical probability $\mathbb{P}$ the process X is a real-valued Lévy process with characteristic triple (m, σ^2, ν). The process S is called an **exponential Lévy process**.

Let us assume that $\mathbb{E}(e^{\alpha X_1}) < \infty$, for $\alpha \in [-\epsilon, \epsilon]$. In terms of the Lévy measure ν, this condition can be written

$$\int \mathbb{1}_{\{|x|\geq 1\}} e^{\alpha x} \nu(dx) < \infty .$$

This implies that X has finite moments of all orders.

Proposition 11.7.1.1 *We assume that $\mathbb{E}(e^{\alpha X_1}) < \infty$ for any α in an open interval $]a, b[$ with $b - a > 1$ and that there exists a real number θ such that $\Psi(\theta) = \Psi(\theta + 1)$. Then, the process $e^{-rt} S_t = S_0 e^{X_t}$ is a martingale under the Esscher transform $\mathbb{P}^{(\theta)}$ defined as $\mathbb{P}^{(\theta)}|_{\mathcal{F}_t} = Z_t \mathbb{P}|_{\mathcal{F}_t}$ where $Z_t = \frac{e^{\theta X_t}}{\mathbb{E}(e^{\theta X_t})}$.*

PROOF: In order to prove that $(e^{X_t}, t \geq 0)$ is a martingale, since X is a Lévy process under the Esscher transform Z, one has to check that $\mathbb{E}_{\mathbb{P}^{(\theta)}}(e^{X_t}) = 1$, which follows from the choice of θ and

$$\mathbb{E}_{\mathbb{P}^{(\theta)}}(e^{X_t}) = \frac{1}{\mathbb{E}_{\mathbb{P}}(e^{\theta X_t})} \mathbb{E}_{\mathbb{P}}(e^{(\theta+1)X_t}) = e^{t(\Psi(\theta+1) - \Psi(\theta))} .$$

□

We assume that the e.m.m. chosen by the market is the probability $\mathbb{P}^{(\theta)}$ defined in Proposition 11.7.1.1. The value of a contingent claim $h(S_T)$ is

$$\begin{aligned} V_t &= e^{-r(T-t)} \mathbb{E}_{\mathbb{P}^{(\theta)}}(h(S_T)|\mathcal{F}_t) \\ &= e^{-r(T-t)} \frac{1}{\mathbb{E}_{\mathbb{P}}(e^{\theta X_T})} \mathbb{E}_{\mathbb{P}}(h(y e^{r(T-t) + X_{T-t}}) e^{\theta X_{T-t}}) \big|_{y=S_t} . \end{aligned}$$

11.7.2 A Differential Equation for Option Pricing

Let $S_t = S_0 e^{X_t}$ where X is a (m, σ^2, ν)-Lévy process under the risk-neutral probability $\mathbb{Q}$. We assume that $\mathbb{E}_{\mathbb{Q}}(e^{X_t}) < \infty$.

By definition of a risk-neutral probability, the discounted price process $(Z_t = e^{-rt} S_t / S_0, t \geq 0)$ is a $\mathbb{Q}$-strictly positive martingale with initial value equal to 1. We know that $e^{X_t - t\Psi(1)}$ is a martingale, hence Z is a martingale if $\Psi(1) = r$. In other terms, we assume that the $\mathbb{Q}$-characteristic triple (m, σ^2, ν) of X is such that

$$m = r - \sigma^2/2 - \int (e^y - 1 - y \mathbb{1}_{\{|y|\leq 1\}}) \nu(dy) .$$

Assume that H is a function which is regular enough so that

$$V(t, S) = e^{-r(T-t)} \mathbb{E}_{\mathbb{Q}}(H(S_T)|S_t = S)$$

belongs to $C^{1,2}$. Then, since $e^{-rt} V(t, S_t)$ is a $\mathbb{Q}$-martingale, Itô's formula yields

$$\begin{aligned} rV &= \frac{1}{2}\sigma^2 S^2 \partial_{SS} V + \partial_t V + rS\partial_S V \\ &+ \int_{\mathbb{R}} (V(t, Se^y) - V(t,S) - S(e^y - 1)\partial_S V(t,S))\,\nu(dy)\,. \end{aligned}$$

Introducing the function $u(t,y) = e^{r(T-t)}V(t, S_0 e^y)$ where $y = \ln(S/S_0)$, we see that u satisfies

$$\begin{aligned} &\partial_t u + \left(r - \frac{\sigma^2}{2}\right)\partial_y u + \frac{1}{2}\sigma^2 \partial_{yy} u \\ &+ \int (u(t, y+z) - u(t,z) - y(e^z - 1)\partial_y u(\tau, x))\,\nu(dz) = 0\,. \end{aligned}$$

See Cont and Tankov [192] Chapter 12, for regularity conditions.

11.7.3 Put-call Symmetry

Let us study a financial market with a riskless asset with constant interest rate r, and a price process (a currency) $S_t = S_0 e^{X_t}$ where X is a $\mathbb{Q}$-Lévy process such that $\mathbb{E}(e^{X_t}) < \infty$. The choice of $\mathbb{Q}$ as an e.m.m. implies that the process $(Z_t = e^{-(r-\delta)t} S_t/S_0, t \geq 0)$ is a $\mathbb{Q}$-strictly positive martingale with initial value equal to 1. We know that $e^{X_t - t\Psi(1)}$ is a martingale, hence Z is a martingale if $\Psi(1) = r - \delta$. In other words, we assume that the $\mathbb{Q}$-characteristic triple (m, σ^2, ν) of X is such that

$$m = r - \delta - \sigma^2/2 - \int (e^y - 1 - y\mathbb{1}_{\{|y|\leq 1\}})\nu(dy)\,.$$

Then,

$$\begin{aligned} \mathbb{E}_{\mathbb{Q}}(e^{-rT}(S_T - K)^+) &= \mathbb{E}_{\mathbb{Q}}(e^{-\delta T} Z_T (S_0 - KS_0/S_T)^+) \\ &= \mathbb{E}_{\widehat{\mathbb{Q}}}(e^{-\delta T}(S_0 - KS_0/S_T)^+) \end{aligned}$$

with $\widehat{\mathbb{Q}}|_{\mathcal{F}_t} = Z_t \mathbb{Q}|_{\mathcal{F}_t}$. The process X is a $\widehat{\mathbb{Q}}$-Lévy process, with characteristic exponent $\Psi(\lambda+1) - \Psi(1)$ (see Proposition 11.3.1.1). The process $S_0/S_t = e^{-X_t}$ is the exponential of the Lévy process, $Y = -X$ which is the dual of the Lévy process X, and the characteristic exponent of Y is $\Psi(1-\lambda) - \Psi(1)$. Hence, the following symmetry between call and put prices holds:

$$C_E(S_0, K, r, \delta, T, \Psi) = P_E(K, S_0, \delta, r, T, \widetilde{\Psi})\,,$$

where $\widetilde{\Psi}(\lambda) = \Psi(1-\lambda) - \Psi(1)$.

Comment 11.7.3.1 This result was proved by Fajardo and Mordecki [339] and generalized by Eberlein and Papapantoleon [293].

11.7.4 Arbitrage and Completeness

We give here some results related to arbitrage opportunities. The reader can refer to Cherny and Shiryaev [170] and Selivanov [778] for proofs. Let X be a (m, σ^2, ν) Lévy process, not identically equal to 0 and $S_t = e^{X_t}$ for $t < T$. The set

$$\mathcal{M} = \{\mathbb{Q} \sim \mathbb{P} : S \text{ is an } (\mathbf{F}, \mathbb{Q})\text{-martingale}\}$$

is empty if and only if X is increasing or if X is decreasing. If $\mathcal{M}$ is not empty, then $\mathcal{M}$ is a singleton if and only if one of the following two conditions is satisfied: $\sigma = 0,\ \nu = \lambda\delta_a$, or $\sigma \neq 0,\ \nu = 0$. Hence, there is no arbitrage in a Lévy model (except if the Lévy process is increasing or decreasing) and the market is incomplete, except in the basic cases of a Brownian motion and of a Poisson process.

The same kind of result holds for a time-changed exponential model: if $S_t = e^{X(\tau_t)}$ where τ is an increasing process, independent of X, such that $\mathbb{P}(\tau_T > \tau_0) > 0$, then the set $\mathcal{M} = \{\mathbb{Q} \sim \mathbb{P} : S \text{ is an } (\mathcal{F}_t, \mathbb{Q})\text{-martingale}\}$ is empty if and only if X is increasing or decreasing. If $\mathcal{M}$ is not empty, then $\mathcal{M}$ is a singleton if and only if τ is deterministic and continuous and one of the following two conditions is satisfied: $\sigma = 0,\ \nu = \lambda\delta_a$, or $\sigma \neq 0,\ \nu = 0$.

Comment 11.7.4.1 Esscher transforms appear while looking for specific changes of measures, which minimize some criteria, such as variance minimal martingale measure, f^q-minimal martingale measure or minimal entropy martingale measure. See Fujiwara and Miyahara [369] and Klöppel et al.[525].

11.8 Variance-Gamma Model

In a series of papers, Madan and several co-authors [155, 609, 610, 612] introduce and exploit the Variance-Gamma Model (see also Seneta [779]). The **Variance-Gamma process** is a Lévy process where X_t has a Variance-Gamma law (see $\rightarrowtail$ Subsection A.4.6) $\mathrm{VG}(\sigma, \nu, \theta)$. Its characteristic function is

$$\mathbb{E}(\exp(iuX_t)) = \left(1 - iu\theta\nu + \frac{1}{2}\sigma^2\nu u^2\right)^{-t/\nu}.$$

The Variance-Gamma process may be characterized as a time-changed BM with drift as follows: let W be a BM, and $\gamma(t)$ a $G(t; 1/\nu, 1/\nu)$ process independent of W. Then

$$X_t = \theta\gamma(t) + \sigma W_{\gamma(t)}$$

is a $\mathrm{VG}(\sigma, \nu, \theta)$ process. The Variance-Gamma process is a *finite variation process* and is the difference of two increasing Lévy processes. More precisely, it is the difference of two independent Gamma processes

$$X_t = G(t;\nu^{-1},\nu_1) - G(t;\nu^{-1},\nu_2),$$

where $\nu_i, i = 1,2$ are given below. Indeed, the characteristic function can be factorized:

$$\mathbb{E}(\exp(iuX_t)) = \left(1 - \frac{iu}{\nu_1}\right)^{-t/\nu}\left(1 + \frac{iu}{\nu_2}\right)^{-t/\nu}$$

with

$$\nu_1^{-1} = \frac{1}{2}\left(\theta\nu + \sqrt{\theta^2\nu^2 + 2\nu\sigma^2}\right) > 0,$$
$$\nu_2^{-1} = \frac{1}{2}\left(-\theta\nu + \sqrt{\theta^2\nu^2 + 2\nu\sigma^2}\right) > 0.$$

The Lévy density of the process X is

$$\frac{1}{\nu}\frac{1}{|x|}\exp(-\nu_2|x|), \text{ for } x < 0,$$
$$\frac{1}{\nu}\frac{1}{x}\exp(-\nu_1 x), \text{ for } x > 0.$$

The Variance-Gamma process has infinite activity. The density of the r.v. X_1 is

$$\frac{2e^{\theta x/\sigma^2}}{\nu^{1/\nu}\sqrt{2\pi}\sigma\Gamma(1/2)}\left(\frac{x^2}{\theta^2 + 2\sigma^2/\nu}\right)^{\frac{1}{2\nu}-\frac{1}{4}} K_{\frac{1}{\nu}-\frac{1}{2}}\left(\frac{1}{\sigma^2}\sqrt{x^2(\theta^2 + 2\sigma^2/\nu)}\right)$$

where K_α is the modified Bessel function.

In the risk-neutral world, stock prices driven by a Variance-Gamma process have dynamics

$$S_t = S_0 \exp\left(rt + X_t + \frac{t}{\nu}\ln\left(1 - \theta\nu - \frac{\sigma^2\nu}{2}\right)\right).$$

Indeed, from

$$\mathbb{E}(e^{X_t}) = \exp\left(-\frac{t}{\nu}\ln\left(1 - \theta\nu - \frac{\sigma^2\nu}{2}\right)\right),$$

we get that the process $(S_t e^{-rt}, t \geq 0)$ is a martingale. The parameters ν and θ give control on skewness and kurtosis.

The **tempered stable process** is a Lévy process without Gaussian component and with Lévy density

$$\frac{C_+}{x^{1+Y_+}}e^{-xM_+}\mathbb{1}_{\{x>0\}} + \frac{C_-}{|x|^{1+Y_-}}e^{xM_-}\mathbb{1}_{\{x<0\}}$$

where $C_\pm > 0, M_\pm \geq 0$, and $Y_\pm < 2$. It was introduced by Koponen [537] and used by Bouchaud and Potters [112] for financial modelling.

The **CGMY model**, introduced by Carr et al. [150] is a particular case of the tempered stable process (i.e., the case where $C_+ = C_- = C$ and $Y_+ = Y_- = Y$). The Lévy density is

$$\frac{C}{x^{Y+1}} e^{-Mx} \mathbb{1}_{\{x>0\}} + \frac{C}{|x|^{Y+1}} e^{Gx} \mathbb{1}_{\{x<0\}}$$

with $C > 0, M \geq 0, G \geq 0$ and $Y < 2$.

If $Y < 0$, there is a finite number of jumps in any finite interval; if not, the process has infinite activity. If $Y \in [1, 2[$, the process has infinite variation. This process is also called KoBol (see Schoutens [766]).

11.9 Valuation of Contingent Claims

11.9.1 Perpetual American Options

In this section, we present the model derived in Chesney and Jeanblanc [174]. Suppose that the dynamics of the risky asset (a currency or a paying dividend asset) are

$$S_t = S_0 e^{X_t} \tag{11.9.1}$$

under the chosen risk-neutral probability $\mathbb{Q}$ where X is a Lévy process with parameters (m, σ^2, ν). We assume that $\mathbb{E}(e^{\lambda X_1}) < \infty$ for every $\lambda \in \mathbb{R}$. As seen in Subsection 11.7.3, one has

$$-(r - \delta) + m + \frac{1}{2}\sigma^2 + \int (e^x - 1 - x\mathbb{1}_{|x|\leq 1})\nu(dx) = 0\,. \tag{11.9.2}$$

By definition, the value of a perpetual American call with strike K is $C_A(S_0) = \sup_{\tau \in \mathcal{T}} \mathbb{E}((S_\tau - K)e^{-r\tau})$ where $\mathcal{T}$ is the set of stopping times. Mordecki [659] has proved that one can reduce attention to first hitting times, i.e., to stopping times τ of the form

$$T(L) = \inf\{t \geq 0 \,:\, S_t \geq L\}\,,$$

where L is a fixed boundary, with $L \geq S_0$. Therefore,

$$C_A(S_0) = \sup_L \mathbb{E}\left((S_{T(L)} - K)e^{-rT(L)}\right)\,.$$

Let us define f as

$$f(x, L) = \mathbb{E}\left(e^{-rT(L)}(x \exp(X_{T(L)}) - K)\right)\,.$$

Then, the value of the American call is

$$C_A(x) = \sup_{L \geq x} f(x, L)\,.$$

Let $\ell = \ln(L/x)$, $T^X(\ell) = \inf\{t \geq 0, X_t \geq \ell\}$ the hitting time of ℓ for the process X, and O_ℓ the overshoot defined in terms of X by $X_{T(\ell)} = \ell + O_\ell$. (Obviously, $T^X(\ell) = T^S(L) := T_\ell$.) We introduce the function

$$\begin{aligned} g(x, \ell) &= f(x, L) = f(x, xe^\ell) = x\mathbb{E}(e^{-rT_\ell + X_{T_\ell}}) - K\mathbb{E}(e^{-rT_\ell}) \\ &= xe^\ell \mathbb{E}(e^{-rT_\ell + 0_\ell}) - K\mathbb{E}(e^{-rT_\ell}). \end{aligned}$$

Then, $C_A(x) = \sup_{\ell \geq 0} g(x, \ell)$.

General Formula

On the one hand, we define

$$\varphi(q, x) = x\alpha(q, r) - K\beta(q, r)$$

with

$$\alpha(q, r) = \int_0^{+\infty} e^{-q\ell} \mathbb{E}(e^{-rT_\ell + O_\ell}) d\ell, \tag{11.9.3}$$

$$\beta(q, r) = \int_0^{+\infty} e^{-q\ell} \mathbb{E}(e^{-rT_\ell}) d\ell. \tag{11.9.4}$$

It is then not difficult to check that

$$\varphi(q, x) = \int_0^\infty e^{-q\ell} \bar{g}(x, \ell) d\ell$$

where

$$\bar{g}(x, \ell) = g(xe^{-\ell}, \ell) = x\mathbb{E}(e^{-rT_\ell + O_\ell}) - K\mathbb{E}(e^{-rT_\ell}) = f(xe^{-\ell}, x).$$

Thus

$$\begin{aligned} \varphi(q, x) &= \int_0^\infty e^{-q\ell} \bar{g}(x, \ell) d\ell = \int_0^x e^{-q\ln(x/y)} \bar{g}(x, \ln(x/y)) \frac{1}{y} dy \\ &= \int_0^x \frac{1}{y} f(y, x) e^{-q\ln(x/y)} dy. \end{aligned} \tag{11.9.5}$$

On the other hand, as in Gerber and Landry [386], by definition of the perpetual call exercise boundary b_c:

$$\text{for } x < b_c, \quad f(x, b_c) = \sup_{L \geq x} f(x, L)$$

hence, assuming that f is differentiable,

$$\frac{\partial f}{\partial L}(x, b_c) = 0, x < b_c. \tag{11.9.6}$$

Therefore, by differentiation of (11.9.5) with respect to x at point b_c

$$\frac{\partial \varphi}{\partial x}(q, b_c) = \frac{f(b_c, b_c)}{b_c} - \frac{q}{b_c}\varphi(q, b_c)$$

hence

$$\alpha(q,r) = \frac{b_c - K}{b_c} - \frac{q}{b_c}(b_c\alpha(q,r) - K\beta(q,r))\,.$$

Now, due to the Pecherski-Rogozin identity (see Subsection 11.4.2), the functions α and β are known in terms of the ladder exponent κ:

$$\alpha(q,r) = \frac{\kappa(r,q) - \kappa(r,-1)}{(q+1)\kappa(r,q)},\ \beta(q,r) = \frac{\kappa(r,q) - \kappa(r,0)}{q\kappa(r,q)} \tag{11.9.7}$$

and an easy computation leads to

$$b_c = \frac{\kappa(r,0)}{\kappa(r,-1)}K\,. \tag{11.9.8}$$

Proposition 11.9.1.1 *The boundary of a perpetual American call is given by*

$$b_c = \frac{\kappa(r,0)}{\kappa(r,-1)}K$$

where κ is the ladder exponent of X.

Comment 11.9.1.2 If $\mathbb{E}(e^{X_1}) < \infty$, using the Wiener-Hopf factorization, Mordecki [659] proves that the boundaries for perpetual American options are given by

$$b_p = K\mathbb{E}(e^{\underline{m}_\Theta}), b_c = K\mathbb{E}(e^{\Sigma_\Theta})$$

where $\underline{m}_t = \inf_{s\leq t} X_s$ and Θ is an exponential r.v. independent of X with parameter r, hence $b_c b_p = \dfrac{rK^2}{1-\ln\mathbb{E}(e^{X_1})}$. This last equality follows from

$$\mathbb{E}(e^{iu\underline{m}_\Theta})\mathbb{E}(e^{iu\Sigma_\Theta}) = \frac{\kappa(r,0)\widehat{\kappa}(r,0)}{\kappa(r,-iu)\widehat{\kappa}(r,iu)} = \frac{r}{r+\Phi(u)}$$

with $\Phi(u) = -\ln\mathbb{E}(e^{iuX_1})$. The equality between $\mathbb{E}(e^{\Sigma_\theta})$ and $\frac{\kappa(r,0)}{\kappa(r,-1)}$ is obtained in Nguyen-Ngoc and Yor [671] using Wiener-Hopf decomposition.

A Particular Case: X without Positive Jumps

If the process X has no positive jumps, then, we obtain the well-known result derived e.g. in Zhang [874] (see Subsection 10.7.2). Indeed, in that case, if Ψ is the Laplace exponent of X,

$$\kappa(\alpha, \beta) = \Psi^\sharp(\alpha) + \beta$$

where $\Psi^\sharp(z)$ is the positive number y such that $\Psi(y) = z$. Hence

$$b_c = K \frac{\Psi^\sharp(r)}{\Psi^\sharp(r) - 1}.$$

Note that

$$\Psi(\lambda) = \lambda m + \lambda^2 \frac{\sigma^2}{2} + \int (e^{\lambda x} - 1 - \lambda x \mathbb{1}_{|x|\leq 1})\nu(dx). \tag{11.9.9}$$

The function Ψ is convex, $\Psi(0) = 0$ and $\Psi(1) = r - \delta$. (This last property is a consequence of the martingale criterion (11.9.2).)

A Second Particular Case: X without Negative Jumps

We now assume that the jumps of X are positive hence the dual process $\widehat{X} = -X$ has no positive jumps. Then

$$\kappa(u, k) = \frac{u - \widehat{\Psi}(k)}{\widehat{\Psi}^\sharp(u) - k} \tag{11.9.10}$$

where $\widehat{\Psi}$ is the Lévy exponent of $-X$, given by $\widehat{\Psi}(k) = \Psi(-k)$ and $\widehat{\Psi}^\sharp(z)$ is the positive root of $\widehat{\Psi}(y) = z$. Relying on equations (11.9.8) and (11.9.10), the exercise boundary is given by:

$$b_c = \frac{r - \widehat{\Psi}(0)}{\widehat{\Psi}^\sharp(r)} \frac{\widehat{\Psi}^\sharp(r) + 1}{r - \widehat{\Psi}(-1)} K.$$

Using that $\widehat{\Psi}(0) = 0$, $\widehat{\Psi}(-1) = \Psi(1) = r - \delta$, and $\widehat{\Psi}^\sharp(r) = -\Psi^{\sharp,n}(r)$, where $\Psi^{\sharp,n}(r)$ is the negative root of $\Psi(k) = r$, we obtain the following proposition:

Proposition 11.9.1.3 *The exercise boundary for a perpetual call written on the exponential of a Lévy process without negative jumps is*

$$b_c = \frac{r}{\delta} \frac{\Psi^{\sharp,n}(r) - 1}{\Psi^{\sharp,n}(r)} K. \tag{11.9.11}$$

Comment 11.9.1.4 It is straightforward to check that in the case $\nu = 0$, the formula coincides with the usual formula (3.11.12). Indeed, in this case,

$$\frac{r}{\delta} \frac{\Psi_0^{\sharp,n}(r) - 1}{\Psi_0^{\sharp,n}(r)} = \frac{\Psi_0^\sharp(r)}{\Psi_0^\sharp(r) - 1} \tag{11.9.12}$$

where Ψ_0 is the Lévy exponent in the case $\phi = 0$, so that $\Psi_0^\sharp(r)$ is the positive root of $bk + \frac{1}{2}\sigma^2 k(k-1) = r$ and $\Psi_0^{\sharp,n}(r)$ is the negative root. Usual relations between the sum and the product of roots and the coefficients for a second order polynomial, lead to the result, $b_c = \frac{\gamma_1}{\gamma_1 - 1} K$ with $\gamma_1 = \frac{-\nu + \sqrt{\nu^2 + 2r}}{\sigma}$ (see (3.11.10)).

The Perpetual American Currency Put

The put case can be solved by using the symmetrical relationship (see Fajardo and Mordecki [339]) between the American call and put boundaries:

$$b_p(K, r, \delta, \Psi) b_c(K, \delta, r, \widetilde{\Psi}) = K^2$$

where

$$\widetilde{\Psi}(u) = \Psi(1-u) - \Psi(1). \tag{11.9.13}$$

And by relying on (11.9.11), the exercise boundary b_p of the perpetual put in a jump-diffusion setting with negative jumps can be obtained:

$$b_p = \frac{rK}{\delta} \frac{\widetilde{\Psi}^{\sharp,n}(\delta)}{\widetilde{\Psi}^{\sharp,n}(\delta) - 1}. \tag{11.9.14}$$

The case where the size of the jumps of X is a strictly positive constant φ corresponds to $\nu(dx) = \lambda \delta_\varphi(dx)$, with $\lambda > 0$. Set $\phi = e^\varphi - 1$. In that case, the Laplace exponent of the Lévy process X is

$$\begin{aligned}
\Psi(u) &= u\mu + u^2 \frac{\sigma^2}{2} + \lambda(e^{u\varphi} - 1) \\
&= u\left(r - \delta - \frac{\sigma^2}{2}\right) + u^2 \frac{\sigma^2}{2} + \lambda((1+\phi)^u - 1 - u\phi)
\end{aligned}$$

with $\mu = r - \delta - \lambda\phi - \frac{\sigma^2}{2}$. Recall that in the case of jumps of constant size

$$b_p(K, r, \delta, \lambda, \varphi) b_c(K, \delta, r, \lambda(1+\varphi), -\frac{\varphi}{1+\varphi}) = K^2.$$

The price of a perpetual American call option can be decomposed as

$$\begin{aligned}
C_A(x) = {} & \delta x \sum_{n=0}^{+\infty} \frac{1}{n!} \int_0^{+\infty} e^{-(\delta+\lambda(1+\phi))s} (\lambda(1+\phi)s)^n \mathcal{N}(d_1(b_c, n; s)) ds \\
& - rK \sum_{n=0}^{+\infty} \frac{1}{n!} \int_0^{+\infty} e^{-(r+\lambda)s} (\lambda s)^n \mathcal{N}(d_2(b_c, n; s)) ds
\end{aligned} \tag{11.9.15}$$

with

$$\begin{aligned}
d_1(z, n; s) &= \frac{\ln(x/z) + (r - \delta - \lambda\phi + \sigma^2/2)s + n\ln(1+\phi)}{\sigma\sqrt{s}}, \\
d_2(z, n; s) &= d_1(z, n; s) - \sigma\sqrt{s}.
\end{aligned}$$

Comment 11.9.1.5 The perpetual exercise boundary for the put can also be obtained by relying on the procedure used for the call. It can also be checked that, when there is no dividend and when the Lévy measure corresponds to a compound Poisson process, this formula is the same as in Gerber and Landry [386], i.e.,

$$b_c = rK \left(m + \sigma^2 + \int (xe^x - x\mathbb{1}_{|x|\leq 1})\nu(dx) \right)^{-1} .$$

See Chesney and Jeanblanc [174] for details.

Comment 11.9.1.6 The problem of American options is studied in Alili and Kyprianou [8], Asmussen et al. [24], Boyarchenko and Levendorskii [117], Chan [160, 159], Kyprianou and Pistorius [555] and Mordecki [658]. Russian options are presented in Avram et al. [32], Asmussen et al. [24] and Mordecki and Moreira [660]. Barrier and lookback options are studied in Avram et al. [31] and Nguyen-Ngoc and Yor [670, 671].

Il faut imaginer Sisyphe heureux.
A. Camus, Le Mythe de Sisyphe

A

List of Special Features, Probability Laws, and Functions

A.1 Main Formulae

For the reader's convenience, we give here a list of some main formulae and the page numbers where they appeared. They are presented, for each topic, in alphabetical order. We have not given details of notation, which should be clear from the context.

A.1.1 Absolute Continuity Relationships

Cameron-Martin's Formula. (Page 73)

$$\mathbf{W}^{(\nu)}[F(X_t, t \le T)] = \mathbf{W}^{(0)}[e^{\nu X_T - \nu^2 T/2} F(X_t, t \le T)]$$

Girsanov's Theorem. (Page 73)

$$\mathbf{W}^{(f)}[F(X_t, t \le T)]$$
$$= \mathbf{W}^{(0)}\left[\exp\left(\int_0^T f(X_s)dX_s - \frac{1}{2}\int_0^T f^2(X_s)ds\right) F(X_t, t \le T)\right]$$

Squared Radial Ornstein-Uhlenbeck Processes and Squared Bessel Processes. (Page 356)

$${}^b\mathbb{Q}_x^a|_{\mathcal{F}_t} = \exp\left(-\frac{b}{4}(X_t - x - at) - \frac{b^2}{8}\int_0^t X_s ds\right)\mathbb{Q}_x^a|_{\mathcal{F}_t}$$

M. Jeanblanc, M. Yor, M. Chesney, *Mathematical Methods for Financial Markets*, Springer Finance, DOI 10.1007/978-1-84628-737-4,

Poisson Processes of Intensities λ and $(1+\beta)\lambda$. (Page 466)

$$\Pi^{(1+\beta)\lambda}|_{\mathcal{F}_t} = \left((1+\beta)^{N_t} e^{-\lambda\beta t}\right) \Pi^{\lambda}|_{\mathcal{F}_t}$$

A.1.2 Bessel Processes

The CIR Process r is a space-time changed BESQ process ρ. (Page 357)

$$r_t = e^{-kt}\rho\left(\frac{\sigma^2}{4k}(e^{kt}-1)\right)$$

Laplace Transform for the BESQ. (Page 343)

$$\mathbb{Q}_x^{\delta}[\exp(-\lambda\rho_t)] = \frac{1}{(1+2\lambda t)^{\delta/2}} \exp\left(-\frac{\lambda x}{1+2\lambda t}\right)$$

Scale Functions.
For a squared Bessel process (Page 336)

$$-x^{1-(\delta/2)} \text{ for } \delta > 2;\ \ln x \text{ for } \delta = 2;\ x^{1-(\delta/2)} \text{ for } \delta < 2$$

Transition Densities. (Page 343)
For a squared Bessel process of index ν

$$q_t^{(\nu)}(x,y) = \frac{1}{2t}\left(\frac{y}{x}\right)^{\nu/2} \exp\left(-\frac{x+y}{2t}\right) I_\nu\left(\frac{\sqrt{xy}}{t}\right)$$

For a Bessel process of index ν

$$p_t^{(\nu)}(x,y) = \frac{y}{t}\left(\frac{y}{x}\right)^{\nu} \exp\left(-\frac{x^2+y^2}{2t}\right) I_\nu\left(\frac{xy}{t}\right)$$

A.1.3 Brownian Motion

Let W be a Brownian motion, and $M_t = \sup_{s \le t} W_s$.

Brownian Bridges.
The following processes are Brownian bridges on $[0, T]$ (Page 237)

$$X_t = (T-t) \int_0^t \frac{dW_s}{T-s} \,;\, 0 \le t \le T$$
$$Y_t = (T-t)\, W\left(\frac{t}{T-t}\right), \, 0 \le t \le T$$

Hitting Times.

Law of the hitting time of $y > 0$ for a drifted BM: $X_t = W_t + \nu t$, $\nu > 0$ (Page 148)

$$\mathbb{P}(T_y(X) \in dt) = \frac{y}{\sqrt{2\pi t^3}} \exp\left(-\frac{1}{2t}(y - \nu t)^2\right) \mathbb{1}_{\{t \ge 0\}}\, dt$$

Joint law of W_t and first hitting time of 0. (Page 142)

$$\mathbb{P}_z(W_t \in dx, T_0 \ge t) = \frac{\mathbb{1}_{\{x \ge 0\}}}{\sqrt{2\pi t}} \left[\exp\left(-\frac{(z-x)^2}{2t}\right) - \exp\left(-\frac{(z+x)^2}{2t}\right)\right] dx$$

Joint Law of B and its Local Time. (Page 222)

$$\mathbb{P}(|B_t| \in dx,\, L_t^0 \in d\ell) = \mathbb{1}_{\{x \ge 0\}} \mathbb{1}_{\{\ell \ge 0\}} \frac{2(x+\ell)}{\sqrt{2\pi t^3}} \exp\left(-\frac{(x+\ell)^2}{2t}\right) dx\, d\ell$$

Lévy's Equivalence Theorem. (Page 218)

$$(|W_t|, L_t \,;\, t \ge 0) \overset{\text{law}}{=} (M_t - W_t, M_t \,;\, t \ge 0)$$

Occupation Time Formula. (Page 212)

$$\int_0^t f(W_s)\, ds = \int_{-\infty}^{+\infty} L_t^x f(x)\, dx$$

Reflection Principle.(Page 136) for $y \geq 0,\, x \leq y$

$$\mathbb{P}(W_t \leq x\,, M_t \geq y) = \mathbb{P}(W_t \geq 2y - x)$$

Joint law of W_t and M_t (Page 137)

$$\mathbb{P}(W_t \in dx, M_t \in dy) = \mathbb{1}_{\{y \geq 0\}} \mathbb{1}_{\{x \leq y\}} \frac{2(2y-x)}{\sqrt{2\pi t^3}} \exp\left(-\frac{(2y-x)^2}{2t}\right) dx\, dy$$

Tanaka's Formulae. (Pages 214 and 215)

$$(W_t - x)^+ = (W_0 - x)^+ + \int_0^t \mathbb{1}_{\{W_s > x\}}\, dW_s + \frac{1}{2} L_t^x$$

$$|W_t - x| = |W_0 - x| + \int_0^t \operatorname{sgn}(W_s - x)\, dW_s + L_t^x$$

$$f(W_t) = f(W_0) + \int_0^t (D_- f)(W_s)\, dW_s + \frac{1}{2} \int_{\mathbb{R}} L_t^a f''(da)$$

A.1.4 Diffusions

Scale Function. (Page 271)

$$s(x) = \int_c^x \exp\left(-2 \int_c^u b(v)/\sigma^2(v)\, dv\right) du$$

Speed Measure Density. (Page 272)

$$m(x) = \frac{2}{\sigma^2(x) s'(x)}$$

A.1.5 Finance

Dupire's Formula. (Page 228)

$$\frac{1}{2} K^2 \sigma^2(T, K) = \frac{\partial_T C(K,T) + rK \partial_K C(K,T)}{\partial^2_{KK} C(K,T)}$$

Let $dS_t = S_t((r - \delta)dt + \sigma dW_t)$, and let $C_E\,(x, K; r, \delta; T - t)$ be the price of a call option at time t, with strike K.

Black and Scholes' Formula. (Pages 97 and 160)

$$C_E(x, K; r, \delta; T-t) = xe^{-\delta(T-t)}\mathcal{N}\left[d_1\left(\frac{xe^{-\delta(T-t)}}{Ke^{-r(T-t)}}, T-t\right)\right] - Ke^{-r(T-t)}\mathcal{N}\left[d_2\left(\frac{xe^{-\delta(T-t)}}{Ke^{-r(T-t)}}, T-t\right)\right]$$

where

$$d_1(y,u) := \frac{1}{\sqrt{\sigma^2 u}}\ln(y) + \frac{1}{2}\sqrt{\sigma^2 u}$$

$$d_2(y,u) := d_1(y,u) - \sqrt{\sigma^2 u}$$

A.1.6 Girsanov's Theorem

(Page 534) Let X be a local martingale with respect to $\mathbb{P}$ and

$$\mathbb{Q}|_{\mathcal{F}_t} = L_t \mathbb{P}|_{\mathcal{F}_t}$$

Then, $X_t - \displaystyle\int_0^t \frac{d[X,L]_s}{L_s}$ is a $\mathbb{Q}$-local martingale.

If $[X,L]$ is $\mathbb{P}$-locally integrable, $X_t - \int_0^t \frac{d\langle X,L\rangle_s}{L_{s-}}$ is a $\mathbb{Q}$-local martingale.

A.1.7 Hitting Times

Laplace transform of the first hitting times for diffusions. (Page 278)

$$\mathbb{E}_x\left(e^{-\lambda T_y}\right) = \begin{cases} \Phi_{\lambda\uparrow}(x)/\Phi_{\lambda\uparrow}(y) \text{ if } x < y \\ \Phi_{\lambda\downarrow}(x)/\Phi_{\lambda\downarrow}(y) \text{ if } x > y \end{cases}$$

A.1.8 Itô's Formulae

Integration by parts formula for general semi-martingales. (Page 469)

$$X_t Y_t = X_0 Y_0 + \int_0^t Y_{s-}\,dX_s + \int_0^t X_{s-}\,dY_s + [X,Y]_t$$

Optional Itô's Formula for General Semi-martingales. (Page 528)

$$f(X_t) = f(X_0) + \int_0^t f'(X_{s-})dX_s + \frac{1}{2}\int_0^t f''(X_s)d\langle X^c\rangle_s + \sum_{0<s\leq t}[f(X_s) - f(X_{s-}) - f'(X_{s-})\Delta X_s]$$

Itô–Kunita-Ventzel's Formula. (Page 40)

$$dF_t(x) = \sum_{j=1}^n f_t^j(x)dM_t^j$$

Let $X = (X^1, \dots, X^d)$ be a continuous semi-martingale. Then

$$F_t(X_t) = F_0(X_0) + \sum_{j=1}^n \int_0^t f_s^j(X_s)dM_s^j + \sum_{i=1}^d \int_0^t \frac{\partial F_s}{\partial x_i}(X_s)dX_s^i + \sum_{i=1}^d \sum_{j=1}^n \int_0^t \frac{\partial f_s}{\partial x_i}(X_s)d\langle M^j, X^i\rangle_s + \frac{1}{2}\sum_{i,k=1}^d \int_0^t \frac{\partial^2 F_s}{\partial x_i \partial x_k}d\langle X^k, X^i\rangle_s .$$

Mixed Processes.
Let $dX_t = h_t dt + f_t dW_t + g_t dM_t$ where $dM_t = dN_t - \lambda(t)dt$.

Optional Itô's Formula for Mixed Processes. (Page 553)

$$F(t, X_t) = F(0, X_0) + \int_0^t \partial_s F(s, X_s)\,ds + \int_0^t \partial_x F(s, X_{s-})dX_s + \frac{1}{2}\int_0^t \partial_{xx}F(s, X_s)f_s^2\,ds + \sum_{s\leq t}[F(s, X_s) - F(s, X_{s-}) - \partial_x F(s, X_{s-})\Delta X_s]$$

Predictable Itô's Formula for Mixed Processes. (Page 612)

$$\begin{aligned}F(t,X_t) = F(0,X_0) &+ \int_0^t \partial_x F(s,X_s) f_s dW_s \\ &+ \int_0^t \left[F(s,X_{s-}+g_s) - F(s,X_{s-})\right] dM_s \\ &+ \int_0^t \Big[\partial_t F(s,X_s) + h_s \partial_x F(s,X_s) + \frac{1}{2} f_s^2 \partial_{xx} F(s,X_s) \\ &+ \lambda(s)[F(s,X_s+g_s) - F(s,X_s) - \partial_x F(s,X_s) g_s] \Big] ds\end{aligned}$$

Optional Itô formula for Lévy Processes. (Page 612)

$$\begin{aligned}f(t,X_t) = f(0,X_0) &+ \int_0^t \partial_t f(s,X_s) ds \\ &+ \frac{\sigma^2}{2} \int_0^t \partial_{xx} f(s,X_s) ds + \int_0^t \partial_x f(s,X_{s-}) dX_s \\ &+ \sum_{s \le t} \left(f(s,X_{s-} + \Delta X_s) - f(s,X_{s-}) - (\Delta X_s)\, \partial_x f(s,X_{s-}) \right)\end{aligned}$$

Predictable Itô Formula for Lévy Processes.(Page 612)

$$\begin{aligned}df(t,X_t) = \partial_x f(t,X_t) \sigma dW_t &+ \int_{\mathbb{R}} \left(f(t,X_{t-}+x) - f(t,X_{t-})\right) \widetilde{\mathbf{N}}(dt,dx) \\ + &\quad \Big[\frac{1}{2}\sigma^2 \partial_{xx} f(t,X_t) + m\, \partial_x f(t,X_t) + \partial_t f(t,X_t) \\ + &\quad \int_{\mathbb{R}} \left(f(t,X_{t-}+x) - f(t,X_{t-}) - x\, \partial_x f(t,X_{t-}) \mathbb{1}_{|x|\le 1} \right) \nu(dx) \Big] dt\end{aligned}$$

A.1.9 Lévy Processes

Lévy-Khintchine Representation. (Page 593)

$$\widehat{\mu}(u) = \exp\left(iu \cdot m - \frac{1}{2} u \cdot Au + \int_{\mathbb{R}^d} (e^{iu \cdot x} - 1 - iu \cdot x \mathbb{1}_{|x|\le 1}) \nu(dx) \right)$$

Laplace and Characteristic Exponents. (Page 607)

$$\Phi(u) = -iu \cdot m + \frac{1}{2} u \cdot Au - \int (e^{iu \cdot x} - 1 - iu \cdot x \mathbb{1}_{|x| \leq 1}) \nu(dx)$$

$$\Psi(\lambda) = \lambda \cdot m + \frac{1}{2} \lambda \cdot \sigma^2 \lambda + \int (e^{\lambda \cdot x} - 1 - \lambda \cdot x \mathbb{1}_{|x| \leq 1}) \nu(dx)$$

Subordination. (Page 636)

The characteristics of $\widetilde{X}_t = X_{Z_t}$ are

$$\begin{aligned} \widetilde{a} &= \beta a^X + \int \int \nu^Z(ds) \mathbb{1}_{\{|x| \leq 1\}} x \, \mathbb{P}(X_s \in dx) \\ \widetilde{A} &= \beta A^X \\ \widetilde{\nu}(dx) &= \beta \nu^X(dx) + \int_0^\infty \nu^Z(ds) \mathbb{P}(X_s \in dx) \end{aligned}$$

A.1.10 Semi-martingales

Tanaka-Meyer Formula. (Page 224)

$$|X_t - x| = |X_0 - x| + \int_0^t \operatorname{sgn}(X_s - x) \, dX_s + L_t^x(X)$$

$$(X_t - x)^+ = (X_0 - x)^+ + \int_0^t \mathbb{1}_{\{X_s > x\}} \, dX_s + \frac{1}{2} L_t^x(X)$$

A.2 Processes

Process	Law		Equation	Page
Drifted BM	$\mathbf{W}^{(\mu)}$	drift μ	$dX_t = \mu dt + dW_t$	32
Squared Bessel BESQ^δ	$\mathbb{Q}^\delta$	δ	$d\rho_t = \delta dt + 2\sqrt{\rho_t}\, dW_t$	334
Bessel BES^δ	$\mathbb{P}^\delta$	dimension $\delta > 1$	$dR_t = dW_t + \frac{\delta - 1}{2} \frac{1}{R_t} dt$	334
CIR	${}^a\mathbb{Q}^{b,\sigma}$		$dr_t = (a - br_t)dt + \sigma\sqrt{r_t}dW_t$	357
CEV			$dS_t = S_t(\mu dt + \sigma S_t^\beta dW_t)$	365

A.3 Some Main Models

Model	Parameters	Dynamics	Page
Black and Scholes	r, σ	$dS_t = S_t(rdt + \sigma dW_t)$	94
Garman and Kohlhagen	r, δ, σ	$dS_t = S_t((r - \delta)dt + \sigma dW_t)$	129
Hull and White (interest rate)	k, θ, σ Borel funct.	$dr_t = k(t)(\theta(t) - r_t)dt + \sigma(t)dW_t$	120
Vasicek	k, θ, σ	$dr_t = k(\theta - r_t)dt + \sigma dW_t$	120
CIR or square root	k, θ, σ	$dr_t = k(\theta - r_t)\, dt + \sigma\sqrt{r_t}dW_t$	357
CEV	μ, σ, β	$dS_t = S_t(\mu dt + \sigma S_t^\beta dW_t)$	365
Hull and White (stochastic vol.)	μ, a, λ	$dS_t = S_t(\mu dt + \sigma_t dW_t)$ $d\sigma_t = \sigma_t(adt + \lambda dB_t)$	393

A.4 Some Important Probability Distributions

A.4.1 Laws with Density

Law	Density	Characteristic function $\Phi(u) = \mathbb{E}(e^{iuX})$ or Mellin function $M(\lambda) = \mathbb{E}(X^\lambda)$
Exponential	$\lambda e^{-\lambda x} \mathbb{1}_{\{x>0\}}$	$\Phi(u) = \dfrac{\lambda}{\lambda - iu}$
Gaussian	$\dfrac{1}{\sigma\sqrt{2\pi}} \exp\left(-\dfrac{(x-m)^2}{2\sigma^2}\right)$	$\Phi(u) = \exp(ium - \frac{1}{2}\sigma^2 u^2)$
Cauchy	$\dfrac{1}{\pi} \dfrac{a}{a^2 + x^2}$	$\Phi(u) = \exp(-a\|u\|)$
Arcsine	$\dfrac{1}{\pi} \dfrac{1}{\sqrt{s(1-s)}} \mathbb{1}_{0 \le s \le 1}$	$M(\lambda) = \dfrac{B(\lambda + 1/2, 1/2)}{B(1/2, 1/2)}$
Beta (a,b)	$\dfrac{t^{a-1}(1-t)^{b-1}}{B(a,b)} \mathbb{1}_{0 \le t \le 1}$	$M(\lambda) = \dfrac{B(a+\lambda, b)}{B(a,b)}$
Gamma(α)	$\dfrac{t^{\alpha-1}e^{-t}}{\Gamma(\alpha)} \mathbb{1}_{0 \le t}$	$M(\lambda) = \dfrac{\Gamma(\alpha+\lambda)}{\Gamma(\alpha)}$

A.4.2 Some Algebraic Properties for Special r.v.'s

- If C follows a Cauchy law with parameter $a = 1$, the random variable $\dfrac{1}{1+C^2}$ follows the Arcsine law.
- If G and $\widehat{G}$ are two independent standard Gaussian r.v's., then $\dfrac{G}{\widehat{G}}$ follows a Cauchy law with parameter $a = 1$.

- "Beta-Gamma" algebra: If γ_a is a gamma r.v. with parameter a, then

$$\gamma_a \stackrel{\text{law}}{=} \beta_{a,b}\gamma_{a+b}$$

where $\beta_{a,b}$ is a beta(a, b) r.v. independent of γ_{a+b}.

A particular case of the beta-gamma algebra is

$$G^2 \stackrel{\text{law}}{=} 2\gamma_{1/2} \stackrel{\text{law}}{=} 2\beta_{1/2,1/2}\mathbf{e}$$

where G is a gaussian variable, $\mathbf{e} \stackrel{\text{law}}{=} \gamma_1$ is a standard exponential variable, independent of the arc-sine variable $\beta_{1/2,1/2}$.

See Chaumont and Yor [161] and Dufresne [278] for other algebraic properties.

A.4.3 Poisson Law

Poisson law, $\lambda > 0$	$\mathbb{P}(X = n)$	$e^{-\lambda}\dfrac{\lambda^n}{n!}$
	Characteristic function	$\exp\left(\lambda(e^{iu} - 1)\right)$
	Mean	λ
	Variance	λ

A.4.4 Gamma and Inverse Gaussian Law

$\Gamma(a,\nu)$ $a,\nu>0$	Density	$\frac{\nu^a}{\Gamma(a)} x^{a-1} e^{-x\nu} \mathbb{1}_{\{x>0\}}$
	Characteristic function	$(1-iu/\nu)^{-a}$
	Mean	a/ν
	Variance	a/ν^2
	Lévy density	$ax^{-1}e^{-\nu x}\mathbb{1}_{x>0}$
$\mathrm{IG}(a,\nu)$	Density	$\frac{a}{\sqrt{2\pi}} e^{a\nu} x^{-3/2}$ $\times \exp\left(-\frac{1}{2}(a^2x^{-1}+\nu^2 x)\right) \mathbb{1}_{\{x>0\}}$
	Characteristic function	$\exp\left(-a(\sqrt{\nu^2-2iu}-\nu)\right)$
	Mean	a/ν
	Variance	a/ν^3
	Lévy density	$\frac{a}{\sqrt{2\pi x^3}} \exp\left(-\frac{1}{2}\nu^2 x\right) \mathbb{1}_{\{x>0\}}$

If X follows a $\Gamma(a,\nu)$ law, then cX follows a $\Gamma(a,\nu/c)$ law.

A.4.5 Generalized Inverse Gaussian and Normal Inverse Gaussian

$\text{GIG}(\theta, a, \nu), a > 0, \nu > 0$

Density	$\frac{(\nu/a)^\theta}{2K_\theta(a\nu)} x^{\theta-1} \exp\left(-\frac{1}{2}(a^2x^{-1} + \nu^2 x)\right) \mathbb{1}_{\{x>0\}}$
Characteristic f.	$\frac{1}{K_\theta(a\nu)} \left(1 - 2iu/\nu^2\right)^{\theta/2} K_\theta(a\nu\sqrt{1 - 2iu\nu^{-2}})$
Mean	$aK_{\theta+1}(a\nu)/(\nu K_\theta(a\nu))$
Variance	$a^2\nu^2 K_\theta^{-2}(a\nu)\left(K_{\theta+2}(a\nu)K_\theta(a\nu) + K_{\theta+1}^2(a\nu)\right)$
Lévy density	$x^{-1}e^{-\nu^2 x/2}\left(a^2 \int_0^\infty e^{-xy} g(y)dy + \max(0,\theta)\right)$ $g(y) = \left(\pi^2 a^2 y \left[J_{\lvert\theta\rvert}^2(a\sqrt{2y}) + N_{\lvert\theta\rvert}^2(a\sqrt{2y}))\right]\right)^{-1}$

$\mathrm{NIG}(\alpha, \beta, \mu, \delta), |\beta| < \alpha, \delta > 0$

Density $q(x) = \sqrt{1+x^2}$	$a\left[q\left(\frac{x-\mu}{\delta}\right)\right]^{-1} K_1\left(\delta\,\alpha\,q\left(\frac{x-\mu}{\delta}\right)\right) e^{\beta(x-\mu)}$ $a = \pi^{-1}\alpha \exp\left(\delta\sqrt{\alpha^2-\beta^2}\right)$
Characteristic f.	$\exp\left(-\delta\left(\sqrt{\alpha^2-(\beta+iu)^2}-\sqrt{\alpha^2-\beta^2}\right)-iu\mu\right)$
Mean	$\frac{\delta\beta}{\sqrt{\alpha^2-\beta^2}}+\mu$
Variance	$\alpha^2\delta(\alpha^2-\beta^2)^{3/2}$
Lévy density	$\frac{\delta\alpha}{\pi}\frac{e^{\beta x}}{K_1(\alpha\|x\|)}\|x\|$

In particular $IG(a, \nu) = \mathrm{GIG}(-\frac{1}{2}, a, \nu)$.

A.4.6 Variance Gamma $\mathrm{VG}(\sigma,\nu,\theta)$

$\mathrm{VG}(\sigma,\nu,\theta)$	Density	$c\lvert x\rvert^{\frac{1}{\nu}-\frac{1}{2}}e^{\theta/\sigma^2}K_{\frac{1}{\nu}-\frac{1}{2}}\left(\frac{\sqrt{\theta^2+2\sigma^2/\nu}}{\sigma^2}\lvert x\rvert\right)$ $c=\sigma\sqrt{\frac{\nu}{2\pi}}\frac{(\theta^2\nu+2\sigma^2)^{\frac{1}{4}-\frac{\theta}{2\nu}}}{\Gamma(1/\nu)}$
	Characteristic function	$\left(1-iu\theta\nu+\frac{1}{2}\sigma^2\nu u^2\right)^{-1/\nu}$
	Mean	θ
	Variance	$\sigma^2+\nu\theta^2$
	Lévy density	$\frac{1}{\nu\lvert x\rvert}e^{Ax-B\lvert x\rvert}$ $A=\frac{\theta}{\sigma^2},\ B=\frac{\sqrt{\theta^2+2\sigma^2/\nu}}{\sigma^2}$

A.4.7 Tempered Stable $\mathrm{TS}(Y^{\pm},C^{\pm},M_{\pm})$

Lévy density	$\nu(x)=\frac{C^+}{x^{Y^++1}}e^{-M_+x}\mathbb{1}_{x>0}+\frac{C^-}{\lvert x\rvert^{Y^-+1}}e^{M_-x}\mathbb{1}_{x<0}$
Characteristic function (for $Y^{\pm}\neq 1,0$)	$\Gamma(-Y^+)M_+^{Y^+}C^+\left((1-\frac{iu}{M_+})^{Y^+}-1+\frac{iuY^+}{M_+}\right)$ $+\Gamma(-Y^-)M_-^{Y^-}C^-\left((1+\frac{iu}{M_-})^{Y^-}-1-\frac{iuY^-}{M_-}\right)$
Mean	0
Variance	$\Gamma(2-Y^+)C^+M_+^{Y^+-2}$ $+\Gamma(2-Y^-)C^-M_-^{Y^--2}$

A.5 Special Functions

A.5.1 Gamma and Beta Functions

Let $a, b \in \mathbb{R}^+$

$$\Gamma(a) = \int_0^{+\infty} x^{a-1} \exp(-x) dx \,,$$

$$B(a,b) = \frac{\Gamma(a)\Gamma(b)}{\Gamma(a+b)}$$

A.5.2 Bessel Functions

The Bessel functions of the first kind J_ν and of the second kind N_ν are solutions to the Bessel equation with parameter ν

$$z^2 u'' + z u' + (z^2 - \nu^2) u = 0 \,.$$

$$J_\nu(z) = (z/2)^\nu \sum_{k=0}^{\infty} \frac{(-z^2/4)^k}{k!\, \Gamma(\nu + k + 1)}$$

$$N_\nu(z) = \frac{J_\nu(z)\cos(\nu\pi) - J_{-\nu}(z)}{\sin(\nu\pi)} \,.$$

The modified Bessel functions I_ν and K_ν are solutions to the Bessel equation with parameter ν

$$z^2 u''(z) + z u'(z) - (z^2 + \nu^2) u(z) = 0$$

and are given by:

$$I_\nu(z) = \left(\frac{z}{2}\right)^\nu \sum_{n=0}^{\infty} \frac{z^{2n}}{2^{2n}\, n!\, \Gamma(\nu + n + 1)}$$

$$K_\nu(z) = \frac{\pi(I_{-\nu}(z) - I_\nu(z))}{2 \sin \pi\nu} \,.$$

Some recurrence relations:

$$\frac{d}{dz}(z^\nu I_\nu(z)) = z^\nu I_{\nu-1}(z) \qquad \frac{d}{dz}(z^\nu K_\nu(z)) = -z^\nu K_{\nu-1}(z)$$

$$\frac{d}{dz}(z^{-\nu} I_\nu(z)) = z^{-\nu} I_{\nu+1}(z) \qquad \frac{d}{dz}(z^{-\nu} K_\nu(z)) = -z^{-\nu} K_{\nu+1}(z)$$

$$I_{\nu-1}(z) - I_{\nu+1}(z) = \frac{2\nu}{z} I_\nu(z) \qquad K_{\nu-1}(z) - K_{\nu+1}(z) = -\frac{2\nu}{z} K_\nu(z) \tag{A.5.1}$$

The Bessel function K_ν can be written in integral form

$$K_\nu(x) = \frac{1}{2}\int_0^\infty y^{\nu-1}\exp\left(-\frac{1}{2}x\left(y+\frac{1}{y}\right)\right)dy,\ x>0\,. \tag{A.5.2}$$

Explicit formulae for some values of ν:

$$I_{1/2}(z) = \sqrt{2/\pi z}\,\sinh z, I_{3/2}(z) = \sqrt{2/\pi z}\,(\cosh z - z^{-1}\sinh z)$$
$$I_{5/2}(z) = \sqrt{2/\pi z}\,((1+3z^{-2})\sinh z - 3z^{-1}\sinh z)$$

$$K_{1/2}(z) = \sqrt{\frac{2}{\pi z}}e^{-z}, \quad K_{3/2}(z) = \sqrt{\frac{2}{\pi z}}\left(1+\frac{1}{z}\right)e^{-z}\,. \tag{A.5.3}$$

A.5.3 Hermite Functions

The equation

$$u'' - 2zu' + 2\nu u = 0$$

admits fundamental solutions of the form $\mathcal{H}_\nu(\pm z)$ where

$$\mathcal{H}_\nu(z) = \frac{1}{2\Gamma(-\nu)}\sum_{k=0}^{\infty}\frac{(-1)^k}{k!}\Gamma\left(\frac{k-\nu}{2}\right)(2z)^k\,.$$

A.5.4 Parabolic Cylinder Functions

The Weber equation

$$u'' + \left(\nu + \frac{1}{2} - \frac{z^2}{4}\right)u = 0$$

has as a particular solution the parabolic cylinder function $D_\nu(z)$ where

$$D_\nu(z) = \exp\left(-\frac{z^2}{4}\right)2^{-\nu/2}\sqrt{\pi}\,\mathcal{H}_\nu(z/\sqrt{2})$$

and $\mathcal{H}_\nu$ is the Hermite function.

$$D'_{-\nu}(z) = -\frac{z}{2}D_{-\nu}(z) - \nu D_{-\nu-1}(z)\,.$$

A.5.5 Airy Function

The Airy function $(\mathrm{Ai})(x)$ is a solution of the Sturm-Liouville equation

$$u''(x) = x\,u(x),\ u(0) = 1\,,$$

and is given by

$$(\mathrm{Ai})(x) = \frac{1}{\pi}\left(\frac{x}{3}\right)^{1/2}K_{1/3}\left(\frac{2}{3}x^{3/2}\right)\,.$$

A.5.6 Kummer Functions

The Kummer functions $M(\alpha,\gamma;x)$ and $U(\alpha,\gamma;x)$ are solutions of

$$xu'' + (\gamma - x)u' - \alpha u = 0$$

and are given by

$$M(\alpha,\gamma;x) = \sum_k \frac{(\alpha)_k}{(\gamma)_k}\frac{x^k}{k!} \tag{A.5.4}$$

$$U(\alpha,\gamma;x) = \frac{\pi}{\sin\pi\gamma}\left(\frac{M(\alpha,\gamma;x)}{\Gamma(1+\alpha-\gamma)\Gamma(\gamma)} + x^{1-\gamma}\frac{M(1+\alpha-\gamma,2-\gamma;x)}{\Gamma(\alpha)\Gamma(2-\gamma)}\right)$$

where $(\alpha)_k = \alpha(\alpha+1)\cdots(\alpha+k-1)$ (See Lebedev [570] and Slater [803].)

The Kummer function M is also called the confluent hypergeometric function, and is denoted ${}_1F_1$. When $0<\alpha<\gamma$, one has

$$M(\alpha,\gamma;x) = \frac{\Gamma(\gamma)}{\Gamma(\gamma-\alpha)\Gamma(\alpha)}\int_0^1 e^{xt}t^{\alpha-1}(1-t)^{\gamma-\alpha-1}dt$$

$$U(\alpha,\gamma;x) = \frac{1}{\Gamma(\alpha)}\int_0^\infty e^{-xt}t^{\alpha-1}(1+t)^{\gamma-\alpha-1}dt\,,$$

The solutions of

$$x^2u'' + (\alpha x + 1)u' = \lambda u$$

are $\Phi_{\lambda\uparrow}(x)$ and $\Phi_{\lambda\downarrow}(x)$ defined as

$$\Phi_{\lambda\uparrow}(x) = \left(\frac{1}{x}\right)^{(\nu+\mu)/2} M\left(\frac{\nu+\mu}{2}, 1+\mu, \frac{1}{x}\right)$$

$$\Phi_{\lambda\downarrow}(x) = \left(\frac{1}{x}\right)^{(\nu+\mu)/2} U\left(\frac{\nu+\mu}{2}, 1+\mu, \frac{1}{x}\right)$$

where M and U denote the Kummer functions and $\mu = \sqrt{\nu^2+4\lambda}$, $1+\nu=\alpha$. It follows that the solutions of

$$x^2u'' + (\alpha x + \beta)u' = \lambda u$$

are $\Phi_{\lambda\uparrow}(x/\beta)$ and $\Phi_{\lambda\downarrow}(x/\beta)$.

A.5.7 Whittaker Functions

The Whittaker functions $W_{k,m}$ and $M_{k,m}$ are solutions of the second order differential equation

$$u'' + \left(-\frac{1}{4} + \frac{k}{x} - \frac{m^2-1/4}{x^2}\right)u = 0\,.$$

Whittaker functions are related to the Kummer functions:

$$W_{k,m}(x) = e^{-x/2}x^{m+1/2}\,U(m-k+1/2; 1+2m; x)$$

$$M_{k,m}(x) = e^{-x/2}x^{m+1/2}\,M(m-k+1/2; 1+2m; x)$$

A.5.8 Some Laplace Transforms

$$\int_0^\infty \exp(-\lambda x) f(x) dx = \Psi(\lambda) . \tag{A.5.5}$$

Function	Laplace transform
$x^{\alpha-1}$	$\frac{\Gamma(\alpha)}{\lambda^\alpha}$
$\frac{1}{\sqrt{2\pi x^{3/2}}} \sum_{n\in\mathbb{Z}} (v-u+2nv) e^{-(v-u+2nv)^2/2x}$	$\frac{\sinh(u\sqrt{2\lambda})}{\lambda \sinh(v\sqrt{2\lambda})}$
$\frac{\pi^2}{b^2} \sum_{n\geq 1} (-1)^{n+1} n^2 e^{-n^2\pi^2 x/(2b^2)}$	$\frac{b\sqrt{2\lambda}}{\sinh(b\sqrt{2\lambda})}$
$\frac{\|a\|}{\sqrt{2\pi x^3}} \exp(-\frac{a^2}{2x})$	$e^{-\|a\|\sqrt{2\lambda}}$

References

1. J. Akahori. Some formulae for a new type of path-dependent option. *The Annals of Applied Prob.*, 5:383–388, 1995.
2. J. Akahori, S Ogawa, and Watanabe S., editors. *Proceedings of 3th Ritsumeikan Conference.* World Scientific, 2003.
3. J. Akahori, S. Ogawa, and Watanabe S., editors. *Proceedings of 5th Ritsumeikan Conference.* World Scientific, 2005.
4. J. Akahori, S. Ogawa, and Watanabe S., editors. *Proceedings of 6th Ritsumeikan Conference.* World Scientific, 2006.
5. J. Akahori, S. Ogawa, and Watanabe S., editors. *Proceedings of 7th Ritsumeikan Conference.* World Scientific, 2007.
6. L. Alili. *Fonctionnelles exponentielles et certaines valeurs principales des temps locaux browniens.* Thèse, Paris VI, 1995.
7. L. Alili, D. Dufresne, and M. Yor. Sur l'identité de Bougerol pour les fonctionnelles exponentielles du mouvement brownien avec drift. In M. Yor, editor, *Exponential Functionals and Principal Values Related to Brownian Motion, a Collection of Research Papers*, pages 3–14. Biblioteca de la Revista mathemática Iberoamericana, Madrid, 1997.
8. L. Alili and A.E. Kyprianou. Some remarks on first passage of Lévy processes, the American put and pasting principle. *The Annals of Applied Prob.*, 15:2062–2080, 2005.
9. L. Alili and P. Patie. On the first crossing times of a Brownian motion and a family of continuous curves. *C.R.A.S.*, 340:225–228, 2005.
10. L. Alili, P. Patie, and J.L. Pedersen. Representation of the first hitting time density of an Ornstein-Uhlenbeck process. *Stochastics models*, 21:967–980, 2005.
11. B. Alziary, J.P. Decamps, and P.F. Koehl. A P.D.E. approach to Asian options: analytical and numerical evidence. *Journal of Banking and Finance*, 21:613–640, 1997.
12. J. Amendinger. Martingale representation theorems for initially enlarged filtrations. *Stochastic Processes and their Appl.*, 89:101–116, 2000.
13. J. Amendinger, P. Imkeller, and M. Schweizer. Additional logarithmic utility of an insider. *Stochastic Processes and their Appl.*, 75:263–286, 1998.
14. J.H.M. Anderluh and J.A.M. Van der Weide. Parisian options: The implied barrier concept. In M. Bubak, G.D. van Albada, P.M.A. Sloot, and J.J.

M. Jeanblanc, M. Yor, M. Chesney, *Mathematical Methods for Financial Markets*, Springer Finance, DOI 10.1007/978-1-84628-737-4,

Dongarra, editors, *Lecture Notes in Computer Science*, Computational Science, ICCS 2004, pages 851–858. Springer, 2004.

15. J.H.M. Anderluh and J.A.M. van der Weide. Double sided Parisian option pricing. *Finance and Stochastics*, 13:205–238, 2009.
16. L.G.B. Andersen, J. Andreasen, and D. Eliezer. Static replication of barrier options: some general results. *Journal of Computational Finance*, 5:1–25, 2002.
17. J. Andreasen. Dynamite dynamics. In J. Gregory, editor, *Credit Derivatives, the Definite Guide*, Application networks, pages 371–385. Risk Books, 2003.
18. J. Andreasen, B. Jensen, and R. Poulsen. Eight different derivations of the Black-Scholes formula. *Mathematical Scientist*, 23:18–40, 1998.
19. S. Ankirchner, S. Dereich, and P. Imkeller. Enlargement of filtrations and continuous Girsanov-type embeddings. In C. Donati-Martin, M. Emery, A. Rouault, and Ch. Stricker, editors, *Séminaire de Probabilités XL*, volume 1899 of *Lecture Notes in Mathematics*. Springer-Verlag, 2007.
20. J-P. Ansel and Ch. Stricker. Quelques remarques sur un théorème de Yan. In J. Azéma and M. Yor, editors, *Séminaire de Probabilités XXIV*, volume 1426 of *Lecture Notes in Mathematics*, pages 266–274. Springer-Verlag, 1990.
21. D. Applebaum. *Lévy Processes and Stochastic Calculus.* Cambridge University Press, 2004.
22. J. Aquilina and L.C.G. Rogers. The squared Ornstein-Uhlenbeck market. *Math. Finance*, 14:487–514, 2004.
23. S. Asmussen. *Ruin Probabilities.* World Scientific, 2000.
24. S. Asmussen, F. Avram, and M.R. Pistorius. Russian and American put options under exponential phase-type Lévy models. *Stochastic Processes and their Appl.*, 109:79–111, 2003.
25. M. Atlan. Localizing volatilities. *Preprint*, arXiv:math.PR/0604316, 2005.
26. M. Atlan, H. Geman, D. Madan, and M. Yor. Correlation and the pricing of risks. *Annals of Finance*, 3:411–453, 2007.
27. M. Atlan and B. Leblanc. Time-changed Bessel processes and credit risk. *Preprint*, 2006.
28. M. Avellaneda and P. Laurence. *Quantitative Modeling of Derivative Securities.* Chapman & Hall/CRD, 2000.
29. M. Avellaneda, A. Levy, and A. Paras. Pricing and hedging derivative securities in markets with uncertain volatilities. *Applied Mathematical Finance*, 2:73–88, 1995.
30. M. Avellaneda and L. Wu. Pricing Parisian-style options with a lattice method. *International Journal of Theoretical and Applied Finance*, 2:1–16, 1995.
31. F. Avram, T. Chan, and M. Usabel. On the valuation of constant barrier options under spectrally negative exponential Lévy models. *Stochastic Processes and their Appl.*, 100:75–107, 2002.
32. F. Avram, A.E. Kyprianou, and M.R. Pistorius. Exit problems for spectrally negative Lévy processes and applications to Russian, American and Canadized options. *The Annals of Applied Prob.*, 14:215–238, 2004.
33. E. Ayache, P.A. Forsyth, and K.R. Vetzal. Next generation models for convertible bonds with credit risk. *Wilmott Magazine*, December:68–77, 2002.
34. E. Ayache, P.A. Forsyth, and K.R. Vetzal. The valuation of convertible bonds with credit risk. *J. of Derivatives*, 11 (Fall):9–29, 2003.
35. J. Azéma. Quelques applications de la théorie générale des processus, I. *Invent. Math.*, 18:293–336, 1972.

36. J. Azéma, R.F. Gundy, and M. Yor. Sur l'intégrabilité uniforme des martingales continues. In J. Azéma and M. Yor, editors, *Séminaire de Probabilités XIV*, volume 784 of *Lecture Notes in Mathematics*, pages 53–61. Springer-Verlag, 1980.
37. J. Azéma and M. Yor, editors. *Temps locaux, volume 52-53.* Astérisque, Paris, 1978.
38. J. Azéma and M. Yor. Etude d'une martingale remarquable. In J. Azéma and M. Yor, editors, *Séminaire de Probabilités XXIII*, volume 1557 of *Lecture Notes in Mathematics*, pages 88–130. Springer-Verlag, 1989.
39. L. Bachelier. Théorie de la spéculation, thèse. *Annales Scientifiques de l'Ecole Normale Supérieure*, III-17:21–86, 1900. Reprinted by Jacques Gabay, Paris (1995), English Translation: Cootner (ed.), (1964) *Random Character of Stock Market Prices*, Massachusetts Institute of Technology, pages 17-78 or S. Haberman and T.A. Sibett (1995) (eds.), History of Actuarial Science, VII, 15-78. London.
40. L. Bachelier. Probabilités des oscillations maxima. *C.R.A.S.*, 212:836–838, 1941. Erratum au volume 213, 1941, p.220.
41. L. Bachelier. *Louis Bachelier's Theory of Speculation: The Origins of Modern Finance.* Princeton University Press, 2006. M. Davis, and A. Etheridge, Translators.
42. P. Baldi, L. Caramellino, and M.G. Iovino. Pricing single and double barrier options via sharp large deviations techniques. *Math. Finance*, 9:293–322, 1999.
43. G. Barles, R. Buckdhan, and E. Pardoux. Backward stochastic differential equations and integral-partial differential equations. *Stochastics and Stochastics Reports*, 60:57–83, 1997.
44. G. Barles, J. Burdeau, M. Romano, and N. Samsoen. Critical stock price near expiration. *Math. Finance*, 5:77–95, 1995.
45. M.T. Barlow. Study of filtration expanded to include an honest time. *Z. Wahr. Verw. Gebiete*, 44:307–323, 1978.
46. M.T. Barlow. One-dimensional stochastic differential equation with no strong solution. *J. London Math. Soc.*, 26:335–347, 1982.
47. M.T. Barlow. Skew Brownian motion and a one-dimensional stochastic differential equation. *Stochastics*, 25:1–2, 1988.
48. M.T. Barlow, M. Emery, F.B. Knight, S. Song, and M. Yor. Autour d'un théorème de Tsirel'son sur des filtrations browniennes et non-browniennes. In J. Azéma and M. Yor, editors, *Séminaire de Probabilités XXXII*, volume 1686 of *Lecture Notes in Mathematics*, pages 264–305. Springer-Verlag, 1998.
49. M.T. Barlow and E. Perkins. Strong existence, uniqueness and non-uniqueness in an equation involving local time. In J. Azéma and M. Yor, editors, *Séminaire de Probabilités XVII*, volume 986 of *Lecture Notes in Mathematics*, pages 32–61. Springer-Verlag, 1983.
50. M.T. Barlow, J.W. Pitman, and M. Yor. On Walsh Brownian motions. In J. Azéma and M. Yor, editors, *Séminaire de Probabilités XXIII*, volume 1372 of *Lecture Notes in Mathematics*, pages 275–293. Springer-Verlag, 1989.
51. O.E. Barndorff-Nielsen. Processes of normal inverse Gaussian type. *Finance and Stochastics*, 2:41–61, 1998.
52. O.E. Barndorff-Nielsen, P. Blaesild, and C. Halgreen. First hitting times models for the generalized inverse Gaussian distribution. *Stochastic Processes and their Appl.*, 7:49–54, 1978.

53. O.E. Barndorff-Nielsen, T. Mikosch, and S.I. Resnick. *Lévy Processes. Theory and Applications.* Birkhäuser, 2001.
54. O.E. Barndorff-Nielsen and N. Shephard. Modelling by Lévy processes for financial econometrics. In O.E. Barndorff-Nielsen, T. Mikosch, and S.I. Resnick, editors, *Lévy Processes. Theory and Applications*, pages 283–318. Birkhäuser, 2001.
55. O.E. Barndorff-Nielsen and N. Shephard. *Continuous Time Approach to Financial Volatility.* Forthcoming Cambridge University Press, Cambridge, 2009.
56. G. Barone-Adesi and R. Whaley. Efficient analytic approximation of American option values. *J. of Finance*, 42:301–320, 1987.
57. R.F. Bass. General theory of processes. http://www.math.uconn. edu/~bass, 1998.
58. R.F. Bass. Stochastic calculus for discontinuous processes. http://www. math.uconn.edu/~bass, 1998.
59. D.S. Bates. The crash of '87: Was it expected? the evidence from options markets. *J. of Finance*, 46:1009–1044, 1991.
60. F. Baudoin. Conditioned stochastic differential equations: theory and applications. *Stochastic Processes and their Appl.*, 100:109–145, 2002.
61. F. Baudoin. Modeling anticipations on financial markets. In *Paris-Princeton Lecture on mathematical Finance 2002*, volume 1814 of *Lecture Notes in Mathematics*, pages 43–92. Springer-Verlag, 2003.
62. E. Bayraktar and H. Xing. Pricing asian options for jump diffusions. *Preprint*, 2008.
63. D. Becherer. The numéraire portfolio for unbounded semimartingales. *Finance and Stochastics*, 5:327–341, 2001.
64. S. Beckers. The constant elasticity of variance model and its implications for option pricing. *J. of Finance*, 35:661–673, 1980.
65. S. Begdhadi-Sakrani. *Martingales continues, filtrations faiblement browniennes et mesures signées.* PhD thesis, Paris VI, July 2000.
66. S. Beghdadi-Sakrani. Calcul stochastique pour les mesures signées. In J. Azéma, M. Emery, M. Ledoux, and M. Yor, editors, *Séminaire de Probabilités XXXVI*, volume 1801 of *Lecture Notes in Mathematics*, pages 366–382. Springer-Verlag, 2002.
67. A. Bélanger, S.E. Shreve, and D. Wong. A unified model for credit derivatives. *Math. Finance*, 14:317–350, 2004.
68. M. Bellalah. Analysis and valuation of exotic and real options: a survey of important results. *Finance*, 20, 1999.
69. N. Bellamy. Asian options in a market driven by a discontinuous process. *Finance*, 20:69–93, 1999.
70. N. Bellamy and M. Jeanblanc. Incomplete markets with jumps. *Finance and Stochastics*, 4:209–222, 1999.
71. S. Benninga, T. Björk, and Z. Wiener. On the use of numéraires in option pricing. *J. of Derivatives*, 10:43–58, Winter 2002.
72. A. Bentata and M. Yor. From Black-Scholes and Dupire formulae to last passage times of local martingales. *Preprint*, 2008.
73. H. Berestycki, J. Busca, and I. Florent. An inverse parabolic problem arising in finance. *CRAS*, 331:965–969, 2000.
74. J. Bergenthum and L. Rüschendorf. Comparison of option prices in semimartingale models. *Finance and Stochastics*, 10:222–249, 2006.

75. Y.Z. Bergman, D.B. Grundy, and Z. Wiener. General properties of option prices. *J. of Finance*, 51:1573–1610, 1996.
76. C. Bernard, O. Le Courtois, and F. Quittard-Pinon. A new procedure for pricing Parisian options. *J. of Derivatives*, 12:45–54, 2005.
77. C. Bernard, O. Le Courtois, and F. Quittard Pinon. A new procedure for pricing Parisian options. *J. of Derivatives*, 12:45–54, 2005.
78. J. Bertoin. *Lévy Processes*. Cambridge University Press, Cambridge, 1996.
79. J. Bertoin. Subordinators: Examples and applications. In P. Bernard, editor, *Ecole d'été de Saint Flour, XXVII*, volume 1717 of *Lecture Notes in Mathematics*, pages 1–91. Springer-Verlag, 1997.
80. J. Bertoin. Subordinators, Lévy processes with no negative jumps, and branching processes. *Lecture Notes, MaPhySto Aarhus*, 2000.
81. J. Bertoin. Some elements on Lévy processes. In D.N. Shanbhag and C.R. Rao, editors, *Handbook of Statistics*, pages 117–144. Elsevier, 2001.
82. J. Bertoin and J.W. Pitman. Path transformations connecting Brownian bridge, excursion and meander. *Bull. Sci. Math.*, 118:147–166, 1994.
83. J. Bertoin and M. Yor. Exponential functionals of Lévy processes. *Probability Surveys*, 2:191–212, 2005.
84. V. Bhansali. *Pricing and Managing Exotic and Hybrid Options*. McGraw Hill, 1998.
85. Ph. Biane, J.W. Pitman, and M. Yor. Probability laws related to Riemann zeta and theta functions. *Bull. AMS*, 38:435–469, 2001.
86. Ph. Biane and M. Yor. Valeurs principales associées aux temps locaux Browniens. *Bull. Sci. Mathematics*, 111:23–101, 1987.
87. Ph. Biane and M. Yor. Quelques précisions sur le méandre Brownien. *Bull. Sci. Mathematics*, 112:101–109, 1988.
88. K. Bichteler. *Stochastic Integration with Jumps*. Cambridge University Press, Cambridge, 2002.
89. T.R. Bielecki, M. Jeanblanc, and M. Rutkowski. Hedging of defaultable claims. In R. Carmona, editor, *Paris-Princeton on Mathematical Finance 2003*, volume 1847 of *Lecture Notes in Mathematics*. Springer, 2004.
90. T.R. Bielecki, M. Jeanblanc, and M. Rutkowski. PDE approach to valuation and hedging of credit derivatives. *Quantitative Finance*, 5:257–270, 2004.
91. T.R. Bielecki, M. Jeanblanc, and M. Rutkowski. Stochastic methods in credit risk modelling, valuation and hedging. In M. Frittelli and W. Rungaldier, editors, *CIME-EMS Summer School on Stochastic Methods in Finance, Bressanone*, volume 1856 of *Lecture Notes in Mathematics*. Springer, 2004.
92. T.R. Bielecki, M. Jeanblanc, and M. Rutkowski. Market completeness under constrained trading. In A.N. Shiryaev, M.R. Grossinho, P.E. Oliveira, and M.L. Esquivel, editors, *Stochastic Finance, Proceedings of International Lisbon Conference*, pages 83–107. Springer, 2005.
93. T.R. Bielecki, M. Jeanblanc, and M. Rutkowski. Completeness of a reduced-form credit risk model with discontinuous asset prices. *Stochastic Models*, 22:661–687, 2006.
94. T.R. Bielecki, M. Jeanblanc, and M. Rutkowski. Hedging of basket credit derivatives in credit default swap market. *Journal of Credit Risk*, 3:91–132, 2007.
95. T.R. Bielecki, M. Jeanblanc, and M. Rutkowski. *Introduction to Mathematics of Credit Risk Modeling, CIMPA School, Marrakech 2007*. Forthcoming, Hermann, Paris, 2007.

96. T.R. Bielecki, M. Jeanblanc, and M. Rutkowski. Hedging of credit default swaptions and first-to-default swaps in cds markets. *Preprint*, 2008.
97. T.R. Bielecki, M. Jeanblanc, and M. Rutkowski. Pricing and trading credit default swaps in a hazard process model. *Advances in Applied Prob.*, 2008.
98. T.R. Bielecki, H. Jin, S.R. Pliska, and X.Y. Zhou. Continuous-time mean-variance portfolio selection with bankruptcy prohibition. *Math. Finance*, 15:213–244, 2005.
99. T.R. Bielecki and M. Rutkowski. *Credit risk: Modelling Valuation and Hedging.* Springer Verlag, Berlin, 2001.
100. N.H. Bingham. Fluctuation theory in continuous time. *Adv. Appl. Prob.*, 7:705–766, 1975.
101. N.H. Bingham and R. Kiesel. *Risk-neutral Valuation.* Springer-Finance, Berlin, second edition, 2004.
102. T. Björk. *Arbitrage Theory in Continuous Time.* Oxford University Press, Oxford, second edition, 2004.
103. T. Björk, Yu. Kabanov, and W.J. Runggaldier. Bond market structure in the presence of marked point processes. *Math. Finance*, 40:211–239, 1997.
104. F. Black and J.C. Cox. Valuing corporate securities: Some effects of bond indenture provisions. *J. Finance*, 31:351–367, 1976.
105. F. Black and M. Scholes. The pricing of options and corporate liabilities. *Journal of Political Economy*, 81:637–654, 1973.
106. Ch. Blanchet-Scalliet and F. Patras. Counterparty risk valuation for cds. *Preprint*, 2009.
107. R.M. Blumenthal and R.K. Getoor. *Markov Processes and Potential Theory.* Academic Press, 1968.
108. L. Bondesson. Classes of infinitely divisible distributions and densities. *Z. Wahr. Verw. Gebiete*, 57:39–71, 1981. Correction and addendum, Ib. 59, p.277.
109. A. Borodin and P. Salminen. *Handbook of Brownian Motion: Facts and Formulae.* Birkhäuser, second edition, 2002.
110. K. Borovkov and Z. Burq. Kendall's identity for the first crossing time revisited. *Elec. Comm. in Probab.*, 6:91–94, 2001.
111. B. Bouchard-Denize. *Contrôle Stochastique Appliqué à la Finance.* Thèse, Paris 1, 2000.
112. J-P. Bouchaud and M. Potters. *Théorie des risques financiers.* Aléa-Saclay, CEA, 1997.
113. W. Boughamoura, A.N. Pandey, and F. Trabelsi. Variance reduction with control variate for pricing asian options in a geometric lévy model. *Preprint*, 2008.
114. N. Bouleau. *Martingales et marchés financiers.* Odile Jacob, Paris, 2004.
115. E. Bouyé, V. Durrleman, A. Nikeghbali, G. Riboulet, and Th. Roncalli. Copulas for finance, a reading guide and some applications. *Unpublished manuscript*, www.creditlyonnais.fr, 2000.
116. J. Bowie and P. Carr. Static simplicity. *Risk*, 7:45–49, 1994.
117. S.I. Boyarchenko and S.Z. Levendorskii. Option pricing and hedging under regular Lévy processes of exponential type. In M. Kohlmann and S. Tang, editors, *Mathematical Finance*, Trends in Mathematics, pages 121–130, Basel, 2001. Birkhäuser.
118. S.I. Boyarchenko and S.Z. Levendorskii. Barrier options and touch-and-out options under regular Lévy processes of exponential type. *The Annals of Applied Prob.*, 12:1261–1298, 2002.

119. S.I. Boyarchenko and S.Z. Levendorskii. Perpetual American options under Lévy processes. *Siam J. on Control and Optimization*, 40:1663–1696, 2002.
120. S.I. Boyarchenko and S.Z. Levendorskii. *Non-Gaussian Merton-Black-Scholes Theory*. World scientific, Singapore, 2003.
121. P.P. Boyle and Y.S. Tian. Pricing lookback and barrier options under the CEV model. *J. of Financial and Quantitative Analysis*, 34:241–264, 1999.
122. L. Breiman. First exit times from a square root boundary. In L. Le Cam and J. Neyman, editors, *Proc. 5th Berkeley Symp. Math. Statist. Prob.*, 2, pages 9–16. U.C. Press, Berkeley, 1966.
123. L. Breiman. *Probability*. Addison-Wesley, Reading MA, 1968.
124. P. Brémaud. *Point Processes and Queues: Martingale Dynamics*. Springer Verlag, Berlin, 1981.
125. P. Brémaud and J. Jacod. Processus ponctuels et martingales: résultats récents sur la modélisation et le filtrage. *Adv. Appl. Prob.*, 12:362–416, 1977.
126. P. Brémaud and M. Yor. Changes of filtration and of probability measures. *Z. Wahr. Verw. Gebiete*, 45:269–295, 1978.
127. M. Brennan and E. Schwartz. Evaluating natural resources investments. *Journal of Business*, 58:135–157, 1985.
128. D. Brigo and A. Alfonsi. Credit default swaps calibration and option pricing with the SSRD stochastic intensity and interest-rate model. *Finance and Stochastics*, 9:29–42, 2005.
129. M. Broadie and J.B. Detemple. Option pricing: valuation models and applications. *Management Science*, 50:1145–1177, 2004.
130. M. Broadie, P. Glasserman, and S. Kou. A continuity correction for discrete barrier options. *Math. Finance*, 7:325–347, 1997.
131. O. Brockhaus, M. Farkas, A. Ferraris, D. Long, and M. Overhaus. *Equity Derivatives and Market Risk Models*. Risk books, London, 2000.
132. J. Brossard. Deux notions équivalentes d'unicité en loi pour les équations différentielles stochastiques. In J. Azéma, M. Emery, M. Ledoux, and M. Yor, editors, *Séminaire de Probabilités XXXVII*, volume 1832 of *Lecture Notes in Mathematics*, pages 246–250. Springer-Verlag, 2003.
133. G. Brunick. *A Weak Existence Result with Application to the Financial Engineers Calibration Problem*. PhD thesis, Carnegie Mellon, 2008.
134. R. Buckdahn. Backward stochastic differential equations and viscosity solutions of semilinear parabolic deterministic and stochastic PDE of second order. In R. Buckdahn, H-J. Engelbert, and M. Yor, editors, *Stochastic Processes and Related Topics*, pages 1–54, London, 2001. Taylor and Francis.
135. D.K. Buecker and D. Kelly-Lyth. The value of an Asian option as a double integral. *Unpublished manuscript*, 1999.
136. H. Bühlmann, F. Delbaen, P. Embrechts, and A.N. Shiryaev. No-arbitrage, change of measure and conditional Esscher transform in a semi-martingale model of stock price. *CWI Quarterly*, 9:291–317, 1996.
137. R.H. Cameron and W.T. Martin. Transformation of Wiener integrals under translations. *Ann. Math.*, 45:386–396, 1944.
138. L. Campi, S. Polbenikov, and A. Sbuelz. Systematic equity-based credit risk: A CEV model with jump to default. *Journal of Economic Dynamics and Control*, 33:93–108, 2009.
139. L. Campi and A. Sbuelz. Closed form pricing of benchmark equity default swaps under the CEV assumption. *Risk Letters*, 1, 2005.

140. Ph. Carmona. *Généralisation de la loi de l'arc sinus et entrelacements de processus de Markov.* Thèse, Paris 6, 1994.
141. Ph. Carmona, F. Petit, and M. Yor. On the distribution and asymptotic results for exponential functionals of Lévy processes. In M. Yor, editor, *Exponential Functionals and Principal Values Related to Brownian Motion, a Collection of Research Papers*, pages 73–126. Biblioteca de la Revista Matemática Iberoamericana, Madrid, 1997.
142. R. Carmona, editor. *Indifference Pricing, Theory and Applications.* Princeton University Press, Princeton, 2006.
143. R.A. Carmona, E. Çinlar, I. Ekeland, E. Jouini, J.A. Scheinkman, and N. Touzi, editors. *Paris-Princeton Lecture on Mathematical Finance 2002*, volume 1814 of *Lecture Notes in Mathematics.* Springer-Verlag, 2003.
144. R.A. Carmona, E. Çinlar, I. Ekeland, E. Jouini, J.A. Scheinkman, and N. Touzi, editors. *Paris-Princeton Lecture on Mathematical Finance 2003*, volume 1847 of *Lecture Notes in Mathematics.* Springer-Verlag, 2004.
145. R.A. Carmona, E. Çinlar, I. Ekeland, E. Jouini, J.A. Scheinkman, and N. Touzi, editors. *Paris-Princeton Lecture on Mathematical Finance, 2004*, volume 1919 of *Lecture Notes in Mathematics.* Springer-Verlag, 2007.
146. P. Carr. Randomization and the American put. *Review of Financial Studies*, 11:597–626, 1998.
147. P. Carr and M. Chesney. American put-call symmetry. *Unpublished Manuscript, HEC School*, 2000.
148. P. Carr and A. Chou. Hedging complex barrier options. *Unpublished manuscript*, 1997.
149. P. Carr, K. Ellis, and V. Gupta. Static hedging of path-dependent options. *J. of Finance*, 53:1165–1190, 1998.
150. P. Carr, H. Geman, D. Madan, and M. Yor. The fine structure of asset returns: an empirical investigation. *Journal of Business*, 75:305–332, 2002.
151. P. Carr, H. Geman, D. Madan, and M. Yor. Pricing options on realized variance. *Finance and Stochastics*, 9:453–475, 2005.
152. P. Carr, H. Geman, D. Madan, and M. Yor. Self-decomposability and option pricing. *Finance and Stochastics*, 17:31–57, 2007.
153. P. Carr and R. Jarrow. The stop-loss start-gain paradox and option valuation: A new decomposition into intrinsic and time value. *Review of Financial Studies*, 3:469–492, 1990.
154. P. Carr, R. Jarrow, and R. Myneni. Alternative characterizations of American put options. *Math. Finance*, 2:87–105, 1992.
155. P. Carr, D. Madan, and E.C. Chang. The variance Gamma process and option pricing. *European Finance Review*, 2:79–105, 1998.
156. P. Carr and M. Schröder. Bessel processes, the integral of geometric Brownian motion and Asian options. *Theory of Probability and Its Applications*, 48:400–425, 2004.
157. L. Carraro, N. El Karoui, A. Meziou, and J. Obłój. On Azéma-Yor martingales: further properties and applications. *Preprint*, 2008.
158. U. Çetin, R. Jarrow, Ph. Protter, and Y. Yildirim. Modeling credit risk with partial information. *The Annals of Applied Prob.*, 14:1167–1178, 2004.
159. T. Chan. Pricing contingent claims on stocks driven by Lévy processes. *The Annals of Applied Prob.*, 9:504–528, 1999.

160. T Chan. American options driven spectrally by one-sided Lévy process. In A.E. Kyprianou, W. Schoutens, and P. Wilmott, editors, *Exotic Option Pricing and Advanced Lévy Models*, pages 195–216. Wiley, 2005.
161. L. Chaumont and M. Yor. *Exercises in Probability: A Guided Tour from Measure Theory to Random Processes, via Conditioning.* Cambridge University Press, 2003.
162. A. Chen and M. Suchanecki. Default risk, bankruptcy and the market value of life insurance liabilities. *Insurance: Mathematics and Economics*, 40:231–255, 2007.
163. A. Chen and M. Suchanecki. Parisian exchange options. *Working paper*, 2008.
164. R. Chen and L. Scott. Pricing interest rate option in a two factor Cox-Ingersoll-Ross model of the term structure. *Review Financial Studies*, 5:613–636, 1992.
165. P. Cheridito. Arbitrage in fractional brownian motion models. *Finance and Stochastics*, 7:533–553, 2003.
166. P. Cheridito, D. Filipovic, and M. Yor. Equivalent and absolutely continuous measure changes for jump-diffusion processes. *The Annals of Applied Prob.*, 15:1713–1732, 2005.
167. A.S. Cherny. General arbitrage pricing model: probability and possibility approaches. In C. Donati-Martin, M. Emery, A. Rouault, and Ch. Stricker, editors, *Séminaire de Probabilités XL*, volume 1899 of *Lecture Notes in Mathematics*. Springer-Verlag, 2007.
168. A.S. Cherny and H-J. Engelbert. *Singular Stochastic Differential Equations*, volume 1858 of *Lecture Notes in Mathematics.* Springer, 2005.
169. A.S. Cherny and A.N. Shiryaev. On criteria for the uniform integrability of Brownian stochastic exponentials. In IOS Press, editor, *Optimal Control and Partial Differential Equations*, volume in Honor of Alain Bensoussan's 60th Birthday, pages 80–92, 2001.
170. A.S. Cherny and A.N. Shiryaev. Change of time and measures for Lévy processes. In *Lectures for the Summer School: From Lévy Processes to Semimartingales. Recent Theoretical Developments and Applications to Finance.* Aarhus, 2002.
171. M. Chesney, R.J. Elliott, and R. Gibson. Analytical solutions for the pricing of American bond and yield option. *Math. Finance*, 3:277–294, 1993.
172. M. Chesney and L. Gauthier. American Parisian options. *Finance and Stochastics*, 10:475–506, 2006.
173. M. Chesney, H. Geman, M. Jeanblanc-Picqué, and M. Yor. Some combinations of Asian, Parisian and barrier options. In M.A.H. Dempster and S. Pliska, editors, *Mathematics of Derivative Securities*, Publication of Newton Institute, pages 61–87. Cambridge University Press, 1997.
174. M. Chesney and M. Jeanblanc. Pricing American currency options in an exponential Lévy model. *Applied Math. Fin.*, 11:207–225, 2004.
175. M. Chesney, M. Jeanblanc-Picqué, and M. Yor. Brownian excursions and Parisian barrier options. *Adv. Appl. Prob.*, 29:165–184, 1997.
176. M. Chesney, B. Marois, and R. Wojakowski. *Les options de change.* Economica, Paris, 1995.
177. M. Chesney and L. Scott. Pricing European currency option: a comparison of the modified Black-Scholes model and a random variance model. *Journal of Financial and Quantitative Analysis*, 24:267–285, 1989.

178. R.J. Chitashvili and M.G. Mania. On the decomposition of a maximum of semimartingales and Itô's generalized formula. In *New Trends in Probability and Statistics*, Proceedings of Bakutiani Colloquium in honor of Acad. Ju. V. Prohorov, pages 301–360. VSP-Mokslas, 1990.
179. R.J. Chitashvili and M.G. Mania. On functions transforming a Wiener process into a semi-martingale. *Probability Theory and Related Fields*, 109:57–76, 1997.
180. C.S. Chou and P-A. Meyer. Sur la représentation des martingales comme intégrales stochastiques dans les processus ponctuels. In P-A. Meyer, editor, *Séminaire de Probabilités IX*, volume 1557 of *Lecture Notes in Maths.*, pages 226–236. Springer-Verlag, 1975.
181. T. Choulli, L. Krawczyk, and Ch. Stricker. $\mathcal{E}$-martingales and their applications in mathematical finance. *The Annals of Probability*, 26:853–876, 1998.
182. T. Choulli and Ch. Stricker. Minimal entropy Hellinger martingale measure in incomplete markets. *Math. Finance*, 15:465–490, 2005.
183. K.L. Chung. Excursions in Brownian motion. *Ark. für Math.*, 14:155–177, 1976.
184. K.L. Chung. *Lectures from Markov Processes to Brownian Motion*, volume 249 of *A Series of Comprehensive Studies in Mathematics.* Springer-Verlag, 1982.
185. K.L. Chung. *Green, Brown and Probability and Brownian motion on the line.* World Scientific, 2002.
186. K.L. Chung and R.J. Williams. *Introduction to Stochastic Integration.* Birkhäuser-Verlag, second edition, 1990.
187. K.L. Chung and Z. Zhao. *From Brownian Motion to Schrödinger's Equation.* Springer, Berlin, 1995.
188. E. Çinlar. *Introduction to Stochastic Processes.* Prentice Hall, 1975.
189. E. Çinlar, J. Jacod, Ph. Protter, and M. Sharpe. Semimartingales and Markov processes. *Z. Wahr. Verw. Gebiete*, 54:161–220, 1980.
190. C. Cocozza-Thivent. *Processus stochastiques et fiabilité des systèmes.* Springer-Smai, Paris, 1997.
191. P. Collin-Dufresne and J.N. Hugonnier. Pricing and hedging of contingent claims in the presence of extraneous risks. *Stochastic Processes and Appli.*, 117:742–765, 2007.
192. R. Cont and P. Tankov. *Financial Modeling with Jump Processes.* Chapman & Hall/CRC, 2004.
193. A. Conze and R. Viswanathan. Path dependent options: the case of lookback options. *J. of Finance*, 46:1893–1907, 1991.
194. J.M. Corcuera, P. Imkeller, A. Kohatsu-Higa, and D. Nualart. Additional utility of insiders with imperfect dynamical information. *Finance and Stochastics*, 8:437–450, 2004.
195. J.M. Corcuera, D. Nualart, and W. Schoutens. Completion of a Lévy market by power-jump assets. *Finance and Stochastics*, 9:109–127, 2005.
196. M.J. Cornwall, M. Chesney, M. Jeanblanc-Picqué, G.W. Kentwell, and M. Yor. Parisian barrier options : a discussion. *Risk Magazine*, 10:77–79, 1997.
197. D. Cossin and H. Pirotte. *Advanced Credit Risk Analysis.* Wiley, Chichester, 2001.
198. M. Costabile. A combinatorial approach for pricing Parisian options. *Decisions in Economics and Finance*, 25:111–125, 2002.
199. J.M. Courtault and Yu. Kabanov. *Louis Bachelier. Aux origines de la finance mathématique.* Presses Universitaires Franc-Comtoises, 2002.

200. S. Coutant, V. Durrleman, G. Rapuch, and T. Roncalli. Copulas, multivariate risk-neutral distributions and implied dependence functions. *Preprint*, www.creditlyonnais.fr, 2001.
201. L. Coutin. Fractional Brownian motion. In C. Donati-Martin, M. Emery, A. Rouault, and Ch. Stricker, editors, *Séminaire de Probabilités XL*, volume 1899 of *Lecture Notes in Mathematics*. Springer-Verlag, 2007.
202. R. Coviello and F. Russo. Modeling financial assets without semimartingale. *Preprint*, 2006.
203. A.M.G. Cox and D. Hobson. Local martingales, bubbles and option prices. *Finance and Stochastics*, 9:477–492, 2005.
204. J. Cox and M. Rubinstein. *Options Markets*. Prentice-Hall. Englewood Cliffs., 1991.
205. J.C. Cox. Notes on option pricing (I): Constant elasticity of variance diffusions. *Journal of Portfolio Management*, 22:15–17, 1996.
206. J.C. Cox, J.E. Ingersoll, and S.A. Ross. A theory of term structure of interest rates. *Econometrica*, 53:385–408, 1985.
207. S. Crépey. Calibration of the local volatility in a generalized Black-Scholes model using Tikhonov regularization. *SIAM Journal on Mathematical Analysis*, 2003.
208. J. Cvitanić and I. Karatzas. Hedging contingent claims with constrained portfolios. *The Annals of Applied Prob.*, 3:652–681, 1993.
209. R-A Dana and M. Jeanblanc. *Financial Markets in Continuous Time, Valuation and Equilibrium*. Springer-Finance, Berlin, second edition, 2007.
210. H.E. Daniels. The first crossing time density for Brownian motion with a perturbed linear boundary. *Bernoulli*, 6:571–580, 2000.
211. A. Dassios. The distribution of the quantile of a Brownian motion with drift and the pricing of related path dependent options. *The Annals of Applied Prob.*, 5:389–398, 1995.
212. A. Dassios. On the quantiles of Brownian motion and their hitting times. *Bernoulli*, 11:29–36, 2005.
213. A. Dassios. On Parisian type ruin probabilities. *Preprint*, 2006.
214. A. Dassios and J. Nagaradjasarma. The square root process and Asian options. *Quantitative Finance*, 6:337–347, 2006.
215. A. Dassios and S. Wu. Brownian excursions in a corridor and related Parisian options. *Preprint*, 2008.
216. A. Dassios and S. Wu. Brownian excursions outside a corridor and two-sided Parisian options. *Preprint*, 2008.
217. A. Dassios and S. Wu. Two-sided Parisian options with a single barrier. *Preprint*, 2008.
218. M.H. Davis, D. Duffie, W. Fleming, and S.E. Shreve, editors. *Math. Finance*. IMA, Springer-Verlag. Springer-Verlag, 1995.
219. M.H.A. Davis. The representation of martingales of jump processes. *Siam J. Control and Optim.*, 14:623–638, 1976.
220. M.H.A. Davis. Option Pricing in Incomplete Markets. In M.A.H. Dempster and S.R. Pliska, editors, *Mathematics of Derivative Securities*, Publication of the Newton Institute, pages 216–227. Cambridge University Press, 1997.
221. M.H.A. Davis. Mathematics of financial markets. In E. Bjorn and W. Schmid, editors, *Mathematics Unlimited: 2001 and Beyond*, Berlin, 2001. Springer-Verlag.

222. M.H.A. Davis. Martingale representation and all that. In E.H. Abed, editor, *Advances in Control, Communication Networks, and Transportation Systems: In Honor of Pravin Varaiya*, Systems and Control: Foundations and Applications Series, Boston, 2005. Birkhäuser.
223. M.H.A. Davis and J. Obłój. Market completion using options. *Preprint*, 2007.
224. M.H.A. Davis and P. Varaiya. On the multiplicity of an increasing family of sigma-fields. *The Annals of Probability.*, 2:958–963, 1974.
225. D. Davydov and V. Linetsky. Pricing and hedging path-dependent options under the CEV process. *Management Science*, 47:949–965, 2001.
226. D. Davydov and V. Linetsky. Structuring, pricing and hedging double barrier step options. *Journal of Computational Finance*, 45:55–87, 2002. Winter
227. D. Davydov and V. Linetsky. Pricing options on scalar diffusions: an eigenfunction expansion approach. *Oper. Res.*, 51:185–209, 2003.
228. R.D. DeBlassie. One dimensional scale invariant diffusions. *Stochastics and Stochastics Reports*, 70:131–151, 2000.
229. G. Deelstra and G. Parker. A covariance equivalent discretization of the CIR model. *Proceedings of the 5th AFIR International Colloquium*, 2:731–747, 1995.
230. F. Delbaen. Consols in the CIR model. *Math. Finance*, 53:125–134, 1993.
231. F. Delbaen, P. Grandits, Th. Rheinländer, D. Sampieri, M. Schweizer, and Ch. Stricker. Exponential hedging and entropic penalties. *Math. Finance*, 12:99–124, 2002.
232. F. Delbaen, P. Monat, W. Schachermayer, M. Schweizer, and Ch. Stricker. Weighted norm inequalities and hedging in incomplete markets. *Finance and Stochastics*, 1:181–227, 1997.
233. F. Delbaen and W. Schachermayer. A general version of the fundamental theorem of asset pricing. *Math. Annal*, 300:463–520, 1994.
234. F. Delbaen and W. Schachermayer. The existence of absolutely continuous local martingale measures. *The Annals of Applied Prob.*, 5:926–945, 1995.
235. F. Delbaen and W Schachermayer. No-arbitrage and the Fundamental Theorem of Asset Pricing: summary of main results. In D. Heath and G. Swindle, editors, *Introduction to Mathematical finance*, Proceedings of Symposia in Applied mathematics, pages 49–58, Providence, 1999. American Mathematical Society.
236. F. Delbaen and W. Schachermayer. *The Mathematics of Arbitrage*. Springer, Berlin, 2005.
237. F. Delbaen and H. Shirakawa. Arbitrage possibilities in Bessel processes and their relations to local martingales. *Probab. Theory Related Fields*, 3:357–366, 1995.
238. F. Delbaen and H. Shirakawa. A note on option pricing for constant elasticity of variance model. *Asia-Pacific Financial Markets*, 9:85–99, 2002.
239. F. Delbaen and M. Yor. Passport options. *Math. Finance*, 12:299–328, 2002.
240. C. Dellacherie. *Capacités et processus stochastiques*, volume 67 of *Ergebnisse*. Springer, 1972.
241. C. Dellacherie, B. Maisonneuve, and P-A. Meyer. *Probabilités et Potentiel, chapitres XVII-XXIV, Processus de Markov (fin). Compléments de calcul stochastique*. Hermann, Paris, 1992.
242. C. Dellacherie and P-A. Meyer. *Probabilités et Potentiel, chapitres I-IV*. Hermann, Paris, 1975. English translation: Probabilities and Potentiel, chapters I-IV, North-Holland, (1982).

243. C. Dellacherie and P-A. Meyer. A propos du travail de Yor sur les grossissements des tribus. In C. Dellacherie, P-A. Meyer, and M. Weil, editors, *Séminaire de Probabilités XII*, volume 649 of *Lecture Notes in Mathematics*, pages 69–78. Springer-Verlag, 1978.
244. C. Dellacherie and P-A. Meyer. *Probabilités et Potentiel, chapitres V-VIII.* Hermann, Paris, 1980. English translation : Probabilities and Potentiel, chapters V-VIII, North-Holland, (1982).
245. D.M. Delong. Crossing probabilities for a square root boundary for a Bessel process. *Comm. Statist A-Theory Methods*, 10:2197–2213, 1981.
246. B. DeMeyer, B. Roynette, P. Vallois, and M. Yor. On independent times and positions for Brownian motions. *CRAS Paris*, 233:1017–1022, 2001.
247. B. DeMeyer, B. Roynette, P. Vallois, and M. Yor. On independent times and positions for Brownian motions. *Revista Matemática Iberoamericana*, 18:541–586, 2002.
248. M.A.H. Dempster, editor. *Risk Management Value at Risk and Beyond.* Cambridge University Press, 2002.
249. M.A.H. Dempster and S.R. Pliska, editors. *Mathematics of Derivative Securities.* Cambridge University Press, Cambridge, 1997.
250. E. Derman and I. Kani. Riding on a smile. *Risk*, 6:18–20, 1994.
251. J. Detemple. American options symmetry properties. In E. Jouini, J. Cvitanić, and M. Musiela, editors, *Option Pricing, Interest Rates and Risk Management*, pages 67–104. Cambridge University Press, 2001.
252. J. Detemple. *American-style Derivatives Valuation and Computation*, volume 4 of *Financial Mathematics Series.* Chapman & Hall/CRC, 2005.
253. A.K. Dixit. Entry and exit decisions under uncertainty. *The Journal of Political Economy*, 97:620–638, 1989.
254. A.K. Dixit and R. Pindyck. *Investment under Uncertainty.* Princeton University Press, 1994.
255. W. Doeblin. Sur l'équation de Kolmogoroff. *CRAS, special issue, December*, 331:1059–1100, 2000.
256. C. Doléans-Dade and P-A. Meyer. Intégrales stochastiques par rapport aux martingales locales. In P-A. Meyer, editor, *Séminaire de Probabilités IV*, volume 1224 of *Lecture Notes in Mathematics*, pages 77–107. Springer-Verlag, 1970.
257. C. Doléans-Dade and P-A. Meyer. Une caractérisation de BMO. In P-A. Meyer, editor, *Séminaire de Probabilités XI*, volume 581 of *Lecture Notes in Mathematics*, pages 383–389. Springer-Verlag, 1977.
258. C. Donati-Martin, R. Ghomrasni, and M. Yor. On certain Markov processes attached to exponential functionals of Brownian motion: Applications to Asian options. *Revista Matemática Iberoamericana*, 17:179–193, 2001.
259. C. Donati-Martin, H. Matsumoto, and M. Yor. The law of geometric Brownian motion and its integral, revisited; Application to conditional moments. In H. Geman, D. Madan, S.R. Pliska, and T. Vorst, editors, *Mathematical Finance, Bachelier Congress 2000*, Springer Finance, pages 221–243. Springer-Verlag, 2002.
260. R.A. Doney. Fluctuation theory for Lévy processes. In P. Bernard, editor, *Ecole d'été de Saint Flour, XXXV*, volume 1897 of *Lecture Notes in Mathematics.* Springer-Verlag, 2005.
261. H. Doss. Liens entre équations différentielles stochastiques et ordinaires. *Ann. Inst. H. Poincaré*, 13:99–125, 1977.

262. R. Douady. A convolution method for option pricing European and barrier options. *Unpublished manuscript*, 1996.
263. R. Douady. Closed form formulas for exotic options and their lifetime distribution. *International Journal of Theoretical and Applied Finance*, 2:17–42, 1999.
264. R. Douady and M. Jeanblanc. A rating-based model for credit derivatives. *European Investment Review*, 1:17–29, 2002.
265. M. Dozzi and P. Vallois. Level crossing for certain processes without positive jumps. *Bull. Sci. math.*, 121:355–376, 1997.
266. M. Dritschel and Ph. Protter. Complete markets with discontinuous security price. *Finance and Stochastics*, 3:203–214, 1999.
267. L. Dubins, J. Feldman, M. Smorodinsky, and B. Tsirel'son. Decreasing sequences of σ-fields and a measure change for Brownian motion. *The Annals of Probability*, 24:882–904, 1996.
268. L. Dubins and G. Schwarz. On continuous martingales. *Proc. Nat. Acad. Sci. USA*, 53:913–916, 1965.
269. R.M. Dudley. Wiener functionals as Itô integrals. *The Annals of Probability*, 5:140–141, 1977.
270. D. Duffie. *Dynamic Asset Pricing Theory*. Princeton University Press, Princeton, third edition, 2001.
271. D. Duffie. *Credit Risk Modeling with Affine Processes*. Scuola Normale Superiore, Pisa, 2004.
272. D. Duffie, D. Filipović, and W. Schachermayer. Affine processes and applications in finance. *The Annals of Applied Prob.*, 13:984–1053, 2003.
273. D. Duffie and D. Lando. Term structure of credit spreads with incomplete accounting information. *Econometrica*, 69:633–664, 2000.
274. D. Duffie, J. Pan, and K. Singleton. Transform analysis and asset pricing for affine jump-diffusions. *Econometrica*, 68:1343–1376, 2000.
275. D. Duffie and K. Singleton. Modeling term structure of defaultable bonds. *Review of Financial Studies*, 12:687–720, 1998.
276. D. Duffie and K. Singleton. *Credit Risk: Pricing, Measurement and Management*. Princeton University Press, Princeton, 2002.
277. D. Dufresne. Weak convergence of random growth processes with applications to insurance. *Insurance: Mathematics and Economics*, 8:187–201, 1989.
278. D. Dufresne. Algebraic properties of beta and gamma distributions, and applications. *Advances in Applied Mathematics*, 20:285–299, 1998.
279. D. Dufresne. Laguerre series for Asian and other options. *Math. Finance*, 10:407–428, 2000.
280. D. Dufresne. The integral of geometric Brownian motion. *Adv. Appl. Prob.*, 33:223–241, 2001.
281. D. Dufresne. The integrated square root process. *Preprint*, 2002.
282. D. Dufresne. Bessel processes and Asian options. In H. Ben-Ameur and M. Breton, editors, *Numerical Methods in Finance*, pages 35–57. Springer, 2005.
283. B. Dupire. Pricing with a smile. *Risk Magazine*, 7:17–20, 1994.
284. B. Dupire. Pricing and hedging with smile. In M.A.H. Dempster and S.R. Pliska, editors, *Mathematics of Derivative Securities*, Isaac Newton Institute, pages 103–111. Cambridge University Press, 1997.

285. J. Durbin. The first passage density of the Brownian motion process to a curved boundary, (with an Appendix by D. Williams). *J. Appl. Prob.*, 29:291–304, 1992.
286. R. Durrett. *Brownian Motion and Martingales in Analysis.* Wadsworth, Belmont, California, 1984.
287. R. Durrett. *Stochastic Calculus: a Practical Introduction.* CRC Press, Boca Raton, 1996.
288. E.B. Dynkin. *Markov Processes.* Springer, Berlin, 1965.
289. E. Eberlein. Applications of generalized hyperbolic Lévy motions to finance. In O.E. Barndorff-Nielsen, T. Mikosch, and S.I. Resnick, editors, *Lévy Processes. Theory and Applications*, pages 319–336. Birkhäuser, 2001.
290. E. Eberlein and J. Jacod. On the range of option prices. *Finance and Stochastics*, 1:131–140, 1997.
291. E. Eberlein and W. Kluge. Valuation on floating range notes in Lévy structure term models. *Math. Finance*, 16:237–254, 2006.
292. E. Eberlein and A. Papapantoleon. Equivalence of floating and fixed strike Asian and lookback options. *Stochastic Processes and their Appl.*, 115:31–40, 2004.
293. E. Eberlein and A. Papapantoleon. Symmetries and pricing of exotic options in Lévy models. In A.E. Kyprianou, W. Schoutens, and P. Wilmott, editors, *Exotic Option Pricing and Advanced Lévy Models*, pages 88–124. Wiley, 2005.
294. E. Eberlein, A. Papapantoleon, and A.N. Shiryaev. On the duality principle in option pricing: semimartingale settings. *Finance and Stochastics*, 12:265–292, 2008.
295. E. Ekström. Perpetual American put options in a level-dependent volatility model. *J. Appl. Prob.*, 40:783–789, 2003.
296. E. Ekström, S. Janson, and J. Tysk. Superreplication of options on several underlying assets. *J. Appl. Prob.*, 42:27–38, 2005.
297. N. El Karoui. Exotic options. *Cours de l'Ecole Polytechnique*, 1999.
298. N. El Karoui and T. Cherif. Arbitrage entre deux marchés: Application aux options quanto. *Unpublished manuscript*, 1993.
299. N. El Karoui, H. Geman, and J-Ch. Rochet. Changes of numéraire, changes of probability measure and option pricing. *J. Appl. Prob.*, 32:443–458, 1995.
300. N. El Karoui and M. Jeanblanc. Options exotiques. *Finance*, 20:49–67, 1999.
301. N. El Karoui, M. Jeanblanc-Picqué, and S. Shreve. Robustness of Black and Scholes formula. *Math. Finance*, 8:93–126, 1997.
302. N. El Karoui, M. Jeanblanc-Picqué, and R. Viswanathan. Bounds for the price of options. *Proceedings of a US-French Workshop, Rutgers University. 1991. Lecture Notes in Control and Information Sciences, Springer Verlag.*, 177:234–237, 1992.
303. N. El Karoui and L. Mazliak. *Backward Stochastic Differential Equations.* Longman, Pitman Research Notes in Mathematics series, 364, Harlow, 1997.
304. N. El Karoui and A. Meziou. Constrained optimization with respect to stochastic dominance: Application to portfolio insurance. *Mathematical Finance*, 16:103–117, 2006.
305. N. El Karoui and A. Meziou. Max-plus decomposition of supermartingales and convex order. Application to American options and portfolio insurance. *Annals of Probability*, 36:647–697, 2008.
306. N. El Karoui and M-Cl. Quenez. Programmation dynamique et évaluation des actifs contingents en marchés incomplets. *CRAS, Paris*, 331:851–854, 1991.

307. N. El Karoui and M-Cl. Quenez. Dynamic programming and pricing of contingent claims in an incomplete market. *SIAM J. Control and Optim.*, 33:29–66, 1995.
308. N. El Karoui and M-Cl. Quenez. Non-linear pricing theory and backward stocjhastic differential equations. In W.J. Runggaldier, editor, *Financial Mathematics, Bressanone, 1996*, volume 1656 of *Lecture Notes in Mathematics.* Springer-Verlag, Berlin, 1997.
309. N. El Karoui, M-Cl. Quenez, and S. Peng. Backward stochastic differential equations in finance. *Math. Finance*, 7:1–71, 1997.
310. N. El Karoui and R. Rouge. Pricing via utility maximization and entropy. *Math. Finance*, 10:259–276, 2000.
311. N. El Karoui and G. Weidenfeld. Théorie générale et changement de temps. In P-A. Meyer, editor, *Séminaire de Probabilités XI*, volume 581 of *Lecture Notes in Mathematics*, pages 79–108. Springer-Verlag, 1977.
312. Y. El-Khatib. *Contributions à l'étude des marchés discontinus par le calcul de Malliavin.* Thèse, Université de La Rochelle, February 2003.
313. R.J. Elliott. *Stochastic Calculus and Applications.* Springer, Berlin, 1982.
314. R.J. Elliott and M. Jeanblanc. Incomplete markets and informed agents. *Mathematical Method of Operations Research*, 50:475–492, 1998.
315. R.J. Elliott, M. Jeanblanc, and M. Yor. On models of default risk. *Math. Finance*, 10:179–196, 2000.
316. R.J. Elliott and P.E. Kopp. *Mathematics of Financial Markets.* Springer Finance. Springer, Berlin, 1999.
317. R.J. Elliott and P.E. Kopp. Equivalent martingale measures for bridge processes. *Stochastic Analysis and Applications*, 9:429–444, 1991.
318. R.J. Elliott and J. Van der Hoek. *Binomial Models in Finance.* Springer Finance. Springer, Berlin, 2004.
319. K.D. Elworthy, X.M. Li, and M. Yor. The importance of strictly local martingales: applications to radial Ornstein-Uhlenbeck processes. *Probability Theory and Related Fields*, 115:325–355, 1999.
320. D. Emanuel and J. MacBeth. Further results on the constant elasticity of variance call option pricing model. *Journal of Financial and Quantitative Analysis*, 17:533–554, 1982.
321. P. Embrechts. The wizards of Wall street: did mathematics change finance? *Nieuw Archief voor Wiskunde*, 5(1):26–33, 2003. March
322. P. Embrechts, C. Klüppelberg, and T. Mikosch. *Modelling Extremal Events.* Springer, Berlin, 1997.
323. P. Embrechts, A.J. McNeil, and D. Strautman. Correlation and dependence in risk management: properties and pitfalls. In M.A.H. Dempster, editor, *Risk Management value at Risk and Beyond*, pages 196–223. Cambridge University Press, 2002.
324. P. Embrechts, L.C.G. Rogers, and M. Yor. A proof of Dassios's representation of the α-quantile of Brownian motion with drift. *The Annals of Applied Prob.*, 5:757–767, 1995.
325. M. Emery. On the Azéma martingales. In J. Azéma and M. Yor, editors, *Séminaire de Probabilités XXIII*, volume 1372 of *Lecture Notes in Mathematics*, pages 66–87. Springer-Verlag, 1989.
326. M. Emery. Sur les martingales d'Azéma (suite). In J. Azéma and M. Yor, editors, *Séminaire de Probabilités XXIV*, volume 1426 of *Lecture Notes in Mathematics*, pages 442–447. Springer-Verlag, 1990.

327. M. Emery. Espaces probabilisés filtrés : de la théorie de Vershik au mouvement Brownien, via des idées de Tsirelson. In *Séminaire Bourbaki, 53ième année*, volume 282, pages 63–83. Astérisque, 2002.
328. M. Emery and W. Schachermayer. Brownian filtrations are not stable under equivalent time-changes. In J. Azéma and M. Yor, editors, *Séminaire de Probabilités XXXIII*, volume 1709 of *Lecture Notes in Mathematics*, pages 267–276. Springer-Verlag, 1999.
329. M. Emery and W. Schachermayer. A remark on Tsirelson's stochastic differential equation. In J. Azéma and M. Yor, editors, *Séminaire de Probabilités XXXIII*, volume 1709 of *Lecture Notes in Mathematics*, pages 291–303. Springer-Verlag, 1999.
330. M. Emery, Ch. Stricker, and J-A. Yan. Valeurs prises par les martingales locales continues à un instant donné. *The Annals of Probability*, 11:635–641, 1983.
331. H.J. Engelbert and W. Schmidt. On the behaviour of certain functionals of the Wiener process and applications to stochastic differential equations. In *Stochastic Differential Systems*, volume 36 of *Lecture Notes Control Inf. Sci.*, pages 47–55. Springer-Verlag, 1981.
332. H.J. Engelbert and W. Schmidt. On one-dimensional stochastic differential equations with generalized drift. volume 69 of *Lecture Notes Control Inf. Sci.*, pages 143–155. Springer-Verlag, 1985.
333. H.J. Engelbert and W. Schmidt. On solutions of one-dimensional stochastic differential equations without drift. *Z. Wahr. Verw. Gebiete*, 68:287–314, 1985.
334. K. Es-Sebaiy and Y. Ouknine. How rich is the class of processes which are infinitely divisible with respect to time? *Stat. Prob. Letters*, 78:537–547, 2008.
335. F. Esscher. On the probability function in the collective theory of risk. *Skandinavisk Aktuarietidskrift*, 15:175–195, 1932.
336. S.N. Ethier and T.G. Kurtz. *Markov Processes: Characterization and Convergence*. Wiley, New York, 1986.
337. S.N. Evans. Multiplicities of a Wiener sausage. *Ann. Inst. H. Poincaré, Prob. et Stat.*, 30:501–518, 1994.
338. A. Eyraud-Loisel. BSDE with enlarged filtration. option hedging of an insider trader in a financial market with jumps. *Stochastic Processes and their Appl.*, 115:1745–1763, 2005.
339. J. Fajardo and E. Mordecki. A note on pricing duality and symmetry for two dimensional Lévy markets. In Yu. Kabanov, R. Lipster, and J. Stoyanov, editors, *From Stochastic Calculus to Mathematical Finance: The Shiryaev Festschrift*, Springer Finance, pages 249–256. Springer, 2006.
340. S. Fang and T. Zhang. Stochastic differential equations with non-Lipschitz coefficients: I. pathwise uniqueness and large deviations. *Preprint*, http://arxiv.org/abs/math.PR/0311032, 2003.
341. S. Fang and T. Zhang. Stochastic differential equations with non-Lipschitz coefficients: pathwise uniqueness and no explosion. *CRAS*, 337:737–740, 2003.
342. W. Feller. Two singular diffusion problems. *Annals of Mathematics*, 54:173–182, 1951.
343. W. Feller. *An Introduction to Probability Theory and its Applications, Volume 2*. Wiley, New York, second edition, 1971.
344. B. Ferebee. The tangent approximation to one-sided Brownian exit densities. *Z. Wahr. Verw. Gebiete*, 61:309–326, 1982.
345. S. Fischer. Call option pricing when the exercise price is uncertain and the valuation of index bonds. *J. of Finance*, 33, 1978.

346. P.J. Fitzsimmons, J.W. Pitman, and M. Yor. Markovian bridges: Construction, Palm interpolation, and splicing. In *Seminar on Stochastic Processes*, pages 101–134. Birkhäuser, 1993.
347. J-P. Florens and D. Fougere. Noncausality in continuous time. *Econometrica*, 64:1195–1212, 1996.
348. H. Föllmer and P. Imkeller. Anticipation cancelled by a Girsanov transformation: a paradox on Wiener space. *Ann. Inst. H. Poincaré, Prob. Stat.*, 26:569–586, 1993.
349. H. Föllmer, Ph. Protter, and A.N. Shiryaev. Quadratic variation and an extension of Itô's formula. *Bernoulli*, 1:149–169, 1995.
350. H. Föllmer and A. Schied. *Stochastic Finance: an Introduction in Discrete Time.* Studies in Mathematics, 27. De Gruyter, Berlin, second edition, 2004.
351. H. Föllmer and M. Schweizer. Hedging of contingent claims under incomplete information. In M.H.A. Davis and R.J. Elliott, editors, *Applied Stochastic Analysis*, pages 101–134, London, 1990. Gordon and Breach.
352. H. Föllmer and D. Sondermann. Hedging of contingent claims under incomplete information. In W. Hildenbrand and A. MasCollel, editors, *Contribution to Mathematical Economics*, pages 205–223, Amsterdam, 1986. North-Holland.
353. H. Föllmer, C-T Wu, and M. Yor. Canonical decomposition of linear transformations of two independent Brownian motions motivated by models of insider trading. *Stochastic Processes and their Appl.*, 84:137–164, 1999.
354. M. Forde. Semi model-independent computation of smile dynamics and greeks for barrier, under a CEV stochastic volatility hybrid model. *Journal of Risk*, 5:17–37, 2005.
355. R. Fortet. Les fonctions aléatoires du type Markoff associées à certaines équations linéaires aux dérivées partielles du type parabolique. *J. Math. Pures Appli.*, 22:177–243, 1943.
356. J-P. Fouque, G. Papanicolaou, and X. Sircar. *Derivatives in Financial Markets with Stochastic Volatilities.* Cambridge University Press, Cambridge, 2000.
357. D. Freedman. *Brownian Motion and Diffusions.* Holden-Day, San Francisco, 1971.
358. R. Frey and A.J. McNeil. Dependent defaults in models of portfolio credit risk. *Journal of Risk*, 6:59–62, 2003.
359. R. Frey, A.J. McNeil, and P. Embrechts. *Risk Management.* Cambridge University Press, Cambridge, 2006.
360. R. Frey and C. Sin. Bounds on European option prices under stochastic volatility. *Math. Finance*, 9:97–116, 1999.
361. A. Friedman. *Stochastic Differential Equations and Applications, Volume I.* Academic Press, New York, 1975.
362. M. Frittelli. The minimal entropy martingale measure and the valuation problem in incomplete markets. *Math. Finance*, 10:215–225, 2000.
363. M. Frittelli, editor. *CIME-EMS Summer School on Stochastic Methods in Finance, Bressanone*, volume 1856 of *Lecture Notes in Mathematics.* Springer, 2004.
364. M. Frittelli, S. Biagini, and G. Scandolo. *Duality in Mathematical Finance.* Springer Finance. Springer, Berlin, 2007.
365. M.C. Fu, R.A. Jarrow, J.-Y.J. Yen, and R.J. Elliott, editors. *Advances in Mathematical Finance, The Madan Festschrift.* Applied and Numerical Harmonic Analysis. Birkhäuser, 2007.

366. T. Fujita and R. Miura. Edokko options: a new framework of barrier options. *Asia-Pacific Financial Markets*, 9:141–151, 2002.
367. T. Fujita, F. Petit, and M. Yor. Pricing path-dependent options in some Black-Scholes market, from the distribution of homogeneous Brownian functionals. *Adv. Appl. Prob.*, 41:1–18, 2004.
368. T. Fujita and M. Yor. Perpetual Brownian and Bessel quantiles. *Preprint*, 2007.
369. T. Fujiwara and Y. Miyahara. The minimal entropy martingale measure for geometric Lévy processes. *Finance and Stochastics*, 27:509–531, 2003.
370. G. Fusai. Corridor options and Arc-sine law. *The Annals of Applied Prob.*, 10:634–663, 2000.
371. G. Fusai and A. Tagliani. Pricing of occupation time derivatives: continuous and discrete monitoring. *Journal of Computational Finance*, 5:1–37, 2001.
372. P.V. Gapeev. Solving stochastic jump differential equations. *Preprint*, 2003.
373. M.B. Garman and S.W. Kohlhagen. Foreign currency option values. *Journal of International Money and Finance*, 2:231–237, 1983.
374. D. Gasbarra, E. Valkeika, and L. Vostrikova. Enlargement of filtration and additional information in pricing models: a Bayesian approach. In Yu. Kabanov, R. Lipster, and J. Stoyanov, editors, *From Stochastic Calculus to Mathematical Finance: The Shiryaev Festschrift*, pages 257–286, 2006.
375. N. Gaussel. Selected problems on integration and transmission of information by financial markets. *Thesis*, 2001.
376. L. Gauthier. *Options réelles et options exotiques, une approche probabiliste.* Thèse de doctorat, Univ. Paris I, 2002.
377. D. Geman and J. Horowitz. Occupation densities. *The Annals of Probability*, 8:1–67, 1980.
378. H. Geman, D. Madan, S.R. Pliska, and T. Vorst, editors. *Mathematical Finance, Bachelier Congress 2000.* Springer Finance. Springer-Verlag, 2002.
379. H. Geman, D. Madan, and M. Yor. Asset prices are Brownian motions: only in business time. *Journal of Business*, pages 103–146, 1998.
380. H. Geman, D. Madan, and M. Yor. Time changes for Lévy processes. *Math. Finance*, 11:79–96, 2001.
381. H. Geman, D. Madan, and M. Yor. Stochastic volatility, jumps and hidden time changes. *Finance and Stochastics*, 6:63–90, 2002.
382. H. Geman and M. Yor. Quelques relations entre processus de Bessel, options asiatiques et fonctions confluentes géométriques. *C.R.A.S.*, 314:471–474, 1992.
383. H. Geman and M. Yor. Bessel processes, Asian options and perpetuities. *Math. Finance*, 4:345–371, 1993.
384. H. Geman and M. Yor. Pricing and hedging double barrier options: a probabilistic approach. *Math. Finance*, 6:365–378, 1996.
385. H. Geman and M. Yor. Stochastic time changes in catastrophe option pricing. *Insurance: Mathematics and Economics*, 21:185–193, 1997.
386. H.U. Gerber and B. Landry. On the discounted penalty at ruin in a jump diffusion and the perpetual put option. *Insurance: Mathematics and Economics*, 22:263–276, 1998.
387. H.U. Gerber and E.S.W. Shiu. On the time value of ruin. *North American Actuarial Journal*, 2:48–78, 1998.
388. H.U. Gerber and E.S.W. Shiu. Pricing perpetual options for jump processes. *North American Actuarial Journal*, 2:101–112, 1998.

389. H.U. Gerber and E.S.W. Shiu. From ruin theory to pricing reset guarantees and perpetual put options. *Insurance: Mathematics and Economics*, 24:3–14, 1999.
390. R.K. Getoor and M.J. Sharpe. Conformal martingales. *Invent. Math.*, 16:271–308, 1972.
391. R. Gibson and E. Schwartz. Stochastic convenience yield and the pricing of contingent claims. *J. of Finance*, 45:959–976, 1990.
392. H-J. Girlich. Bachelier's predecessors and the situation in 1900. *Preprint*, 2003.
393. I.V. Girsanov. On transforming a certain class of stochastic processs by absolutely continuous substitution of measures. *Theory of Probability and Its Applications*, 5:285–301, 1960.
394. I.V. Girsanov. An example of non-uniqueness of a solution of Itô's stochastic equation. *Theory of Probability and Its Applications*, 7:336–342, 1962.
395. E. Gobet. Les mathématiques appliquées au cœur de la finance. *Images des Mathématiques, CNRS*, 2004.
396. E. Gobet, G. Pagès, and M. Yor. Mathématiques et finance. In M. Yor, editor, *Aspects des mathématiques financières, Journée organisée à l'Académie des sciences, 1 Février 2005*, pages 77–98. Lavoisier, Paris, 2006.
397. A. Göing-Jaeschke and M. Yor. A clarification about hitting times densities for Ornstein-Uhlenbeck processes. *Finance and Stochastics*, 7:413–415, 2003.
398. A. Göing-Jaeschke and M. Yor. A survey and some generalizations of Bessel processes. *Bernoulli*, 9:313–349, 2003.
399. B.M. Goldman, H.B. Sosin, and M.A. Gatto. Path-dependent options: buy at low, sell at high. *J. of Finance*, 34:111–127, 1979.
400. T. Goll and J. Kallsen. Optimal portfolio for logarithm utility. *Stochastic Processes and their Appl.*, 89:31–48, 2000.
401. C. Gourieroux, J-P. Laurent, and H. Pham. Quadratic hedging and numeraire. *Math. Finance*, 10:179–200, 2000.
402. M. Grasselli. *La gestion de portefeuille à long terme : une approche de finance mathématique*. Thèse, Paris 1, 2001.
403. A.J. Grau. Moving window. *Preprint, Working Paper, School of Computer Science, University of Waterloo*, 2003.
404. A.J. Grau and J. Kallsen. Speedy Monte Carlo pricing of path-dependant options. *Working Paper, HVB-Institute for Mathematical Finance, Technische Universität München, Germany*, 2004.
405. S. Graversen, A.N. Shiryaev, and M. Yor. On the problem of stochastic integral representations of functionals of the brownian motion. II. *Theory of Probability and its Applications*, 51(1):65–77, 2007.
406. Y.M. Greenfeld. *Hedging of the Credit Risk Embedded in Derivative Transactions*. Ph. D. thesis, Carnegie Mellon, May 2000.
407. P.E. Greenwood and J.W. Pitman. Fluctuation identities for Lévy processes and splitting at the maximum. *Adv. Appl. Prob.*, 12:839–902, 1977.
408. J. Gregory, editor. *Credit Derivatives, the Definite Guide*. Risk Books, 2003.
409. P. Groeneboom. Brownian motion with a parabolic drift and Airy functions. *Probability Theory and Related Fields*, 81:79–109, 1989.
410. A. Grorud and M. Pontier. Insider trading in a continuous time market model. *International Journal of Theoretical and Applied Finance*, 1:331–347, 1998.
411. A. Grorud and M. Pontier. Asymmetrical information and incomplete markets. *International Journal of Theoretical and Applied Finance*, 4:285–302, 2001.

412. X. Guo and L.A. Shepp. Some optimal stopping problems with non-trivial boundaries for pricing exotic options. *J. Appl. Prob.*, 38:1–12, 2001.
413. A. Gushin and E. Mordecki. Bounds on option prices for semi-martingale market models. *Proc. Steklov Inst. math.*, 237:73–113, 2002.
414. I. Gyöngy. One-dimensional marginal distributions of processes having an Ito differential. *Probability Theory and Related Fields*, 71:501–516, 1986.
415. R.J. Haber, P. Schönbucher, and P. Wilmott. Pricing Parisian options. *J. of Derivatives, Spring*, pages 71–79, 1999.
416. D.I. Hadjiev. The first passage problem for generalized Orstein-Uhlenbeck processes with non-positive jumps. In J. Azéma and M. Yor, editors, *Séminaire de Probabilités, XIX*, volume 1123 of *Lecture Notes in Mathematics*, pages 80–90. Springer-Verlag, 1985.
417. P.S. Hagan, D. Kumar, A.S. Lesniewski, and D.E. Woodward. Managing smile risk. *WilmottMagazine, September*, pages 84–108, 2003.
418. P.S. Hagan, A.S. Lesniewski, and D.E. Woodward. Probability distribution in the SABR model of stochastic volatility. *Preprint*, 2003.
419. S. Hamadéne. Mixed zero-sum differential game and American game options. *SIAM J. Control Oper.*, 2006.
420. J.M. Harrison. *Brownian Motion and Stochastic Flow Systems.* Wiley, New York, 1985.
421. J.M. Harrison and D. Kreps. Martingales and arbitrage in multiperiod securities markets. *Journal of Economic Theory*, 20:381–408, 1979.
422. J.M. Harrison and S.P. Pliska. Martingales and stochastic integrals in the theory of continuous trading. *Stochastic Processes and their Appl.*, 11:215–260, 1981.
423. J.M. Harrison and S.P. Pliska. A stochastic calculus model of continuous trading: Complete markets. *Stochastic Processes and their Appl.*, 15:313–316, 1983.
424. J.M. Harrison and L.A. Shepp. On skew Brownian motion. *The Annals of Probability*, 9:309–131, 1981.
425. E.G. Haug. *The Complete Guide to Option Pricing Formulas.* McGraw-Hill, New York, 1998.
426. H. He, W.P. Keirstead, and J. Rebholz. Double lookbacks. *Math. Finance*, 8:201–228, 1998.
427. S-W. He, J-G. Wang, and J-A. Yan. *Semimartingale Theory and Stochastic Calculus.* CRC Press Inc, 1992.
428. D. Heath and E. Platen. Consistent pricing and hedging for a modified constant elasticity of variance model. *Quantitative Finance*, 2:459–467, 2002.
429. D. Heath and E. Platen. *Introduction to Quantitative Finance: a Benchmark Approach.* Springer Finance. Springer, Berlin, 2006.
430. V. Henderson. Price comparison results and super-replication: an application to passport options. *Appl. Stoch. Models Bus. Ind.*, 16:297–310, 2000.
431. V. Henderson and D. Hobson. Local time, coupling and the passport option. *Finance and Stochastics*, 4:81–93, 2000.
432. V. Henderson and R. Wojakowski. On the equivalence of floating and fixed-strike Asian options. *J. Appl. Prob.*, 39:391–394, 2002.
433. S.I. Heston. A closed-form solution for options with stochastic volatility with applications to bond and currency options. *The Review of Financial Studies*, 6:327–343, 1993.

434. R. Heynen and H. Kat. Crossing barriers. *Risk Magazine*, 7:46–51, 1995.
435. B. Hilberink and L.C.G. Rogers. Optimal capital structure and endogenous default. *Finance and Stochastics*, 6:237–263, 2002.
436. C. Hillairet. Comparison of insiders' optimal strategies depending on the type of side-information. *Stochastic Processes and their Appl.*, 115:1603–1627, 2005.
437. F. Hirsch and S. Song. Two-parameter Bessel processes. *Stochastic Processes and their Appl.*, 83:187–203, 1999.
438. F. Hirsch and M. Yor. A construction of processes with one-dimensional martingale marginals, associated with a lévy process, via its lévy sheet. *Preprint, Université d'Evry*, 2009.
439. F. Hirsch and M. Yor. A construction of processes with one dimensional martingale marginals, based upon path-space ornstein-uhlenbeck processes and the brownian sheet. *Preprint, Université d'Evry*, 2009.
440. D. Hobson. Volatility misspecification, option pricing and superreplication via coupling. *The Annals of Applied Prob.*, 8:193–205, 1996.
441. D. Hobson. Stochastic volatility models, correlation; and the q-optimal measure superreplication via coupling. *Math. Finance*, 14:537–556, 2004.
442. D. Hobson and L.C.G. Rogers. Complete models with stochastic volatility. *Math. Finance*, 5:17–48, 1998.
443. D. Hobson, D. Williams, and A.T. Wood. Taylor expansions of curve-crossing probabilities. *Bernoulli*, 5:779–795, 1999.
444. S.D. Hodges and A. Neuberger. Optimal replication of contingent claims under transaction costs. *Rev. Future Markets*, 8:222–239, 1989.
445. T. Hoggart, A.E. Whalley, and P. Wilmott. Hedging option portfolio in the presence of transaction costs. *Advances in Futures and Options Research*, 7:222–239, 1994.
446. H. Hörfelt. The moment problem for some Wiener functionals: Corrections to previous proofs (with an appendix by H. L. Pedersen). *J. Appl Prob.*, 42:851–860, 2005.
447. Y. Hu and B. Øksendal. Optimal time to invest when the price processes are geometric Brownian motion. *Finance and Stochastics*, 2:295–310, 1998.
448. Y. Hu and X.Y. Zhou. Constrained stochastic LQ control with random coefficients and application to mean-variance portfolio selection. *SIAM J. Control Optim.*, 44:444–466, 2005.
449. F. Hubalek and W. Schachermayer. The limitations of no-arbitrage arguments for real options. *International Journal of Theoretical and Applied Finance*, 4:361–363, 2001.
450. W. Huf. The strict subordination of differential processes. *Sankhya: The Indian Journal of Statistics*, 4:403–412, 1969.
451. J.N. Hugonnier. The Feynman-Kac formula and pricing occupation time derivatives. *International Journal of Theoretical and Applied Finance*, 2:153–178, 1999.
452. J.N. Hugonnier, D. Kramkov, and W. Schachermayer. On utility based pricing of contingent claims in incomplete markets. *Math. Finance*, 15:203–212, 2005.
453. J. Hull and A. White. The pricing of options on assets with stochastic volatilities. *J. of Finance*, 42:281–300, 1987.
454. J. Hull and A. White. Valuing credit default swaps (i): no counterparty default risk. *J. of Derivatives*, 8:29–40, 2000.
455. P.J. Hunt and J.E. Kennedy. *Financial Derivatives in Theory and Practice.* Wiley Series in Probability and Statistics, Chichester, 2000.

456. N. Ikeda and S. Watanabe. *Stochastic Differential Equations and Diffusion Processes.* North Holland, second edition, 1989.
457. P. Imkeller. Random times at which insiders can have free lunches. *Stochastics and Stochastics Reports*, 74:465–487, 2002.
458. P. Imkeller, M. Pontier, and F. Weisz. Free lunch and arbitrage possibilities in a financial market model with an insider. *Stochastic Processes and their Appl.*, 92:103–130, 2001.
459. J.E. Ingersoll. Digital contracts: simple tools for pricing complex derivatives. *Journal of Business*, 73:67–88, 2000.
460. M. Ismail and D. Kelker. Special functions, Stieltjes transforms and infinite divisibility. *SIAM J. Math. Anal.*, 10:884–901, 1979.
461. K. Itô. Extension of stochastic integrals. *Proc. of Intern. Symp. SDE. Kyoto*, pages 95–109, 1976.
462. K. Itô. In D.W. Stroock and S.R.S. Varadhan, editors, *Selected Papers.* Springer, Berlin, 1987.
463. K. Itô. *Stochastic Processes. Lectures Given at Aarhus University.* Springer, 2004.
464. K. Itô. *Essentials of Stochastic Processes.* AMS, 2006.
465. K. Itô and H.P. McKean. *Diffusion Processes and their Sample Paths.* Springer, Berlin, 1974.
466. S. Iyengar. Hitting lines with two-dimensional Brownian motion. *SIAM J. of Applied Mathematics*, 45:983–989, 1985.
467. J. Jacod. Multivariate point process predictable projection: Radon-Nikodym derivatives, representation of martingales. *Z. Wahr. Verw. Gebiete*, 31:235–253, 1975.
468. J. Jacod. *Calcul stochastique et Problèmes de martingales*, volume 714 of *Lecture Notes in Mathematics.* Springer-Verlag, Berlin, 1979.
469. J. Jacod. Grossissement initial, hypothèse H' et théorème de Girsanov. In *Séminaire de Calcul Stochastique 1982-83*, volume 1118 of *Lecture Notes in Mathematics.* Springer-Verlag, 1987.
470. J. Jacod and Ph. Protter. Time reversal of Lévy processes. *The Annals of Probability*, 16:620–641, 1988.
471. J. Jacod and A.N. Shiryaev. *Limit Theorems for Stochastic Processes.* Springer Verlag, Berlin, second edition, 2003.
472. J. Jacod and M. Yor. Etude des solutions extrémales et représentation intégrale des solutions pour certains problèmes de martingales. *Z. Wahr. Verw. Gebiete*, 38:83–125, 1977.
473. P. Jakubenas. Range of prices. *Unpublished*, 1999.
474. F. Jamshidian. Bond and bond option evaluation in the Gaussian interest rate model. *Research in Finance*, 9:131–170, 1991.
475. F. Jamshidian. Libor market model with semi-martingales. *Unpublished*, 1999.
476. F. Jamshidian. H-hypothesis. *Private communication*, 2003.
477. F. Jamshidian. Valuation of credit default swap and swaptions. *Finance and Stochastics*, 8:343–371, 2004.
478. F. Jamshidian. Numeraire invariance and application to option pricing and hedging. *Working paper*, 2008.
479. S. Janson. *Gaussian Hilbert Spaces.* Cambridge University Press, 1997.
480. R. Jarrow and Ph. Protter. A short history of stochastic integration and mathematical finance. The early years, 1880-1970. In *The Herman Rubin Festschrift*, pages 75–91. IMS Lecture Notes 45, 2004.

481. R.A. Jarrow and F. Yu. Counterparty risk and the pricing of defaultable securities. *Journal of Finance*, 56:1756–1799, 2001.
482. M. Jeanblanc and Y. Le Cam. Progressive enlargement of filtration with initial times. *Stochastic Processes and their Applic.*, 2009.
483. M. Jeanblanc, J.W. Pitman, and M. Yor. Feynman-Kac formula and decompositions of Brownian paths. *Computational and applied Mathematics*, 16:27–52, 1997.
484. M. Jeanblanc, J.W. Pitman, and M. Yor. Self-similar processes with independent increments associated with Lévy and Bessel processes. *Stochastic Processes and their Appl.*, 100:223–232, 2002.
485. M. Jeanblanc and N. Privault. A complete market model with Poisson and Brownian components. In R.C. Dalang, M. Dozzi, and F. Russo, editors, *Proceedings of the Ascona 99 Seminar on Stochastic Analysis, Random Fields and Applications*, volume 52 of *Progress in Probability*, pages 189–204. Birkhäuser Verlag, 2002.
486. M. Jeanblanc and M. Rutkowski. Modeling default risk: an overview. In *Mathematical Finance: Theory and Practice, Fudan University*, pages 171–269. Modern Mathematics Series, High Education press. Beijing, 2000.
487. M. Jeanblanc and M. Rutkowski. Modeling default risk: Mathematical tools. *Fixed Income and Credit risk modeling and Management, New York University, Stern School of Business, Statistics and Operations Research Department, Workshop*, 2000.
488. M. Jeanblanc and M. Rutkowski. Hedging of credit derivatives. In J. Gregory, editor, *Credit Derivatives, the Definite Guide*, Application Networks, pages 385–416. Risk Books, 2003.
489. M. Jeanblanc and S. Valchev. Partial information, default hazard process, and default-risky bonds. *IJTAF*, 8:807–838, 2005.
490. M. Jeanblanc-Picqué and M. Pontier. Optimal portfolio for a small investor in a market with discontinuous prices. *Applied Mathematics and Optimization*, 22:287–310, 1990.
491. C. Jennen and H.R. Lerche. First exit densities of Brownian motion through one- sided moving boundaries. *Z. Wahr. Verw. Gebiete*, 55:133–148, 1981.
492. C. Jennen and H.R. Lerche. Asymptotic densities of stopping times associated with test of power one. *Z. Wahr. Verw. Gebiete*, 61:501–511, 1982.
493. Th. Jeulin. *Semi-martingales et grossissement de filtration*, volume 833 of *Lecture Notes in Mathematics*. Springer-Verlag, 1980.
494. Th. Jeulin. Filtrations, sous-filtrations : propriétés élémentaires. In *Hommage à P-A. Meyer et J. Neveu*, volume 236 of *Astérisque*, pages 163–170. SMF, 1996.
495. Th. Jeulin and M. Yor. Grossissement d'une filtration et semi-martingales : formules explicites. In C. Dellacherie, P-A. Meyer, and M. Weil, editors, *Séminaire de Probabilités XII*, volume 649 of *Lecture Notes in Mathematics*, pages 78–97. Springer-Verlag, 1978.
496. Th. Jeulin and M. Yor. Inégalité de Hardy, semimartingales et faux-amis. In P-A. Meyer, editor, *Séminaire de Probabilités XIII*, volume 721 of *Lecture Notes in Mathematics*, pages 332–359. Springer-Verlag, 1979.
497. Th. Jeulin and M Yor, editors. *Grossissements de filtrations: exemples et applications*, volume 1118 of *Lecture Notes in Mathematics*. Springer-Verlag, 1985.

498. E. Jouini, J. Cvitanić, and M. Musiela, editors. *Option Pricing, Interest Rates and Risk Management*. Cambridge University Press, 2001.
499. Z. Jurek and W. Vervaat. An integral representation for self-decomposable Banach space valued random variables. *Z. Wahr. Verw. Gebiete*, 62:247–262, 1983.
500. Yu. Kabanov. Arbitrage theory. In E. Jouini, J. Cvitanić, and M. Musiela, editors, *Option Pricing, Interest Rates and Risk Management*, pages 3–42. Cambridge University Press, 2001.
501. Yu. Kabanov, R. Lipster, and J. Stoyanov, editors. *From Stochastic Calculus to Mathematical Finance: The Shiryaev Festschrift*. Springer Finance. Springer, 2006.
502. M. Kac. On distributions of certain Wiener functionals. *Trans. Amer. Math. Soc.*, 65:1–13, 1949.
503. N. Kahalé. Analytic crossing probabilities for certain barriers. *To appear in Annals of Applied Probability*, 2008.
504. O. Kallenberg. *Random Measures*. Academic Press, 1984.
505. O. Kallenberg. *Foundations of Modern Probability*. Probability and its applications. Springer-Verlag, New-York, Second edition, 2002.
506. G. Kallianpur and R.L. Karandikar. *Introduction to Option Pricing Theory*. Birkhäuser, 1999.
507. G. Kallianpur and J. Xiong. Asset pricing with stochastic volatility. *Appl. Math. Optim.*, 43:47–62, 2001.
508. J. Kallsen and A.N. Shiryaev. The cumulant process and Esscher's change of measure. *Finance and Stochastics*, 6:397–428, 2002.
509. J. Kallsen and A.N. Shiryaev. Time change representation of stochastic integrals. *Theory of Probability and Its Applications*, 46:522–528, 2002.
510. I. Karatzas. *Lectures on the Mathematics of Finance*. American Mathematical Society, Providence, 1997.
511. I. Karatzas and C. Kardaras. The numéraire portfolio in semimartingale financial models. *Finance and Stochastics*, 11:447–493, 2007.
512. I. Karatzas and I. Pikovsky. Anticipative portfolio optimization. *Adv. Appl. Prob.*, 28:1095–1122, 1996.
513. I. Karatzas and S.E. Shreve. *Brownian Motion and Stochastic Calculus*. Springer-Verlag, Berlin, 1991.
514. I. Karatzas and S.E. Shreve. *Methods of Mathematical Finance*. Springer-Verlag, Berlin, 1998.
515. S. Karlin and H. Taylor. *A First Course in Stochastic Processes*. Academic Press, San Diego, 1975.
516. H.M. Kat. *Structured Equity Derivatives*. Wiley, Chichester, 2001.
517. N. Kazamaki. *Continuous Exponential Martingales and BMO*. Lecture Notes in Mathematics, 1579, Springer-Verlag, Berlin, 1994.
518. D.G. Kendall. The Mardia-Dryden shape distribution for triangles: a stochastic calculus approach. *J. Appl. Prob.*, 28:225–230, 1991.
519. J.T. Kent. Some probabilistic properties of Bessel functions. *The Annals of Probability*, 6:760–770, 1978.
520. J.T. Kent. Eigenvalue expansions for diffusion hitting times. *Z. Wahr. Verw. Gebiete*, 52:309–319, 1980.
521. J.T. Kent. The spectral decomposition of a diffusion hitting time. *The Annals of Probability*, 10:207–219, 1982.

522. R.Z. Khasminskii. Ergodic properties of recurrent processes and stabilisation of the solution of the Cauchy problem for parabolic equation. *Theory of Probability and Its Applications*, 5:179–196, 1960.
523. J.F.C. Kingman. *Poisson Processes.* Oxford University Press, Oxford, 1993.
524. P.E. Kloeden and E. Platen. *Numerical Solutions of Stochastic Differential Equations.* Springer, Berlin, 1992.
525. S. Klöppel, M. Jeanblanc, and Y. Miyahara. Minimal f^q-martingale measures for exponential Lévy processes. *The Annals of Applied Prob.*, 17:1615–1638, 2007.
526. Cl. Klüppelberg, A.E. Kyprianou, and R.A. Maller. Ruin probabilities and overshoots for general Lévy insurance risk processes. *The Annals of Applied Prob.*, 14:1766–1801, 2004.
527. F.B. Knight. A reduction of continuous square-integrable martingales to Brownian motion. In H. Dinges, editor, *Martingales, a Report on a Meeting at Oberwolfach*, volume 190 of *Lecture Notes in Mathematics*, pages 19–31. Springer-Verlag, 1970.
528. F.B. Knight. *Essentials of Brownian Motion and Diffusion.* American Math. Soc. Math. Surveys 18, 1981.
529. F.B. Knight. Calculating the compensator: method and example. In *Seminar Stoc. proc. 1990*, Lecture Notes in Mathematics, pages 241–252. Birkhäuser, Basel, 1991.
530. F.B. Knight and B. Maisonneuve. A characterization of stopping times. *The Annals of Probability*, 22:1600–1606, 1994.
531. P. Koch Medina and S. Merino. *Mathematical Finance and Probability (A discrete Introduction).* Birkhäuser, Basel, 2003.
532. A. Kohatsu-Higa. Enlargement of filtrations and models for insider trading. In J. Akahori, S. Ogawa, and S. Watanabe, editors, *Stochastic Processes and Applications to Mathematical Finance*, pages 151–166. World Scientific, 2004.
533. A. Kohatsu-Higa. Enlargement of filtration. In *Paris-Princeton Lecture on Mathematical Finance, 2004*, volume 1919 of *Lecture Notes in Mathematics*, pages 103–172. Springer-Verlag, 2007.
534. A. Kohatsu-Higa and B. Øksendal. Enlargement of filtration and insider trading. *Preprint*, 2004.
535. M. Kohlmann and S. Tang, editors. *Mathematical Finance.* Trends in Mathematics. Birkhäuser, Basel, 2001.
536. A.N. Kolmogorov. Sulla forma generale di un processo stocastico omogeneo. *Atti Accad. Naz. Lincei Rend*, 15:805–808, 1932.
537. I. Koponen. Analytic approach to the problem of convergence of truncated Lévy flights towards the Gaussian stochastic process. *Physical Review, E*, 52:1197–1199, 1995.
538. R. Korn. *Optimal Portfolio.* World Scientific, Singapore, 1997.
539. S. Kotani. On condition that one-dimensional diffusion processes are martingales. In J. Azéma, M. Emery, M. Ledoux, and M. Yor, editors, *Séminaire de Probabilités XXXIX*, volume 1874 of *Lecture Notes in Mathematics*, pages 149–156. Springer-Verlag, 2006.
540. S.G. Kou. A jump diffusion model for option pricing. *Management Science*, 48:1086–1101, 2002.
541. S.G. Kou and H. Wang. First passage times of a jump diffusion process. *Adv. Appl. Prob.*, 35:504–531, 2003.

542. S.G. Kou and H. Wang. Option pricing under a double exponential jump diffusion model. *Management Science*, 50:1178–1192, 2004.
543. D. Kramkov. Optional decomposition of supermartingales and hedging contingent claims in incomplete security markets. *Probability Theory and Related Fields*, 105:459–479, 1996.
544. D. Kramkov and A.N. Shiryaev. Sufficient conditions for the uniform integrability of exponential martingales. *Progress in Mathematics*, 8:289–295, 1998.
545. D. Kreps. Arbitrage and equilibrium in economics with infinitely many commodities. *J. of Mathematical Economics*, 9:15–35, 1981.
546. H. Kunita. Some extensions of Itô's formula. In J. Azéma and M. Yor, editors, *Séminaire de Probabilités XV*, volume 850 of *Lecture Notes in Mathematics*, pages 118–141. Springer-Verlag, 1981.
547. H. Kunita. Stochastic differential equations and stochastic flows of diffeomorphisms. In *Ecole d'été de Probabilité de Saint-Flour*, volume 1097 of *Lect. Notes Mat*, pages 143–303. Springer, 1982.
548. H. Kunita. *Stochastic Flows and Stochastic Differential Equations.* Cambridge University Press, Cambridge, 1990.
549. H. Kunita. Representation of martingales with jumps and applications to mathematical finance. In H. Kunita, S. Watanabe, and Y. Takahashi, editors, *Stochastic Analysis and Related Topics in Kyoto. In honour of Kiyosi Itô*, Advanced studies in Pure mathematics, pages 209–233. Oxford University Press, 2004.
550. H. Kunita and S. Watanabe. On square integrable martingales. *Nagoya J. Math.*, 30:209–245, 1967.
551. N. Kunitomo and N. Ikeda. Pricing options with curved boundaries. *Math. Finance*, 2:275–298, 1992.
552. S. Kusuoka. A remark on default risk models. *Adv. Math. Econ.*, 1:69–82, 1999.
553. A.E. Kyprianou. *Introductory Lectures on Fluctuations of Lévy Processes with Applications.* Universitext, Springer, Berlin, 2006.
554. A.E. Kyprianou and R. Loeffen. Lévy processes in finance distinguished by their coarse and fine path properties. In A.E. Kyprianou, W. Schoutens, and P. Wilmott, editors, *Exotic Option Pricing and Advances in Lévy Processes*, pages 1–26. Wiley, 2005.
555. A.E. Kyprianou and M. Pistorius. Perpetual option and canadization through fluctuation theory. *The Annals of Applied Prob.*, 13:1077–1098, 2003.
556. C. Labart and J. Lelong. Pricing Parisian options. Master thesis, TU, Delft, 2003.
557. C. Labart and J. Lelong. Pricing double barrier Parisian options using Laplace transforms. *International Journal of Theoretical and Applied Finance*, 2008.
558. P. Lakner and L.M. Nygren. Portfolio optimization with downside constraints. *Mathematical Finance*, 16:283–299, 2006.
559. D. Lamberton and B. Lapeyre. *Introduction to Stochastic Calculus Applied to Finance.* English edition, translated by N. Rabeau and F. Mantion, Chapman & Hall, London, second edition, 2007.
560. D. Lamberton, B. Lapeyre, and A. Sulem, editors. *A Special Volume Dedicated to Malliavin Calculus*, volume 13. Mathematical Finance, Blackwell, 2003.

561. B.M. Lambrecht and W.R.M. Perraudin. Real options and preemption under incomplete information. *Journal of Economics and Dynamic Control*, 27:619–643, 2003.
562. J. Lamperti. Semi-stable Markov processes. *Z. Wahr. Verw. Gebiete*, 22:205–255, 1972.
563. D. Lando. *Three Essays on Contingent Claims Pricing.* PhD. thesis, Cornell university, May 1994.
564. D. Lando. *Credit Risk Modeling.* Princeton University Press, Princeton, 2004.
565. G. Last and A. Brandt. *Marked Point Processes on the Real Line. The Dynamic Approach.* Springer, 1995.
566. J-F. Le Gall. Applications du temps local aux équations différentielles stochastiques unidimensionnelles. In J. Azéma and M. Yor, editors, *Séminaire de Probabilités XVII*, volume 986 of *Lecture Notes in Mathematics*, pages 15–31. Springer-Verlag, 1983.
567. J-F. Le Gall. One-dimensional stochastic differential equation involving the local time of the unknown process. In *Stochastic Analysis*, volume 1095 of *Lecture Notes in Mathematics*, pages 51–82. Springer-Verlag, 1985.
568. J-F. Le Gall and M. Yor. Sur l'équation stochastique de Tsirel'son. In J. Azéma and M. Yor, editors, *Séminaire de Probabilités XVII*, volume 986 of *Lecture Notes in Mathematics*, pages 81–88. Springer-Verlag, 1983.
569. Y. Le Jan. Martingales et changement de temps. In P-A. Meyer, editor, *Séminaire de Probabilités XIII*, volume 721 of *Lecture Notes in Mathematics*, pages 385–399. Springer-Verlag, 1979.
570. N. Lebedev. *Special Functions and their Applications.* Dover, 1972.
571. B. Leblanc. Une approche unifiée pour une forme exacte du prix d'une option dans les différents modèles à volatilité stochastique. *Stochastics and Stochastics Reports*, 57:1–35, 1996.
572. B. Leblanc. *Modélisations de la volatilité d'un actif financier et applications.* Thèse, Paris VII, Septembre 1997.
573. B. Leblanc, O. Renault, and O. Scaillet. A correction note on the first passage time of an Ornstein-Uhlenbeck process to a boundary. *Finance and Stochastics*, 4:109–111, 2000.
574. D. Lefebvre, B. Øksendal, and A. Sulem. An introduction to optimal consumption with partial observation. In M. Kohlmann and S. Tang, editors, *Mathematical Finance*, Trends in Mathematics, pages 239–249, Basel, 2001. Birkhäuser.
575. A. Lejay. On the constructions of the skew Brownian motion. *Probability Surveys*, 3:413–466, 2006.
576. E. Lenglart, D. Lépingle, and M. Pratelli. Une présentation unifiée des inégalités en théorie des martingales. In J. Azéma and M. Yor, editors, *Séminaire de Probabilités XIV*, volume 784 of *Lecture Notes in Mathematics*, pages 26–48. Springer-Verlag, 1980.
577. J.A. León, J.L. Solé, F. Utzet, and J. Vives. On Lévy processes, Malliavin calculus and market models with jumps. *Finance and Stochastics*, 6:197–225, 2002.
578. J-P. Lepeltier and J. San Martin. Backward SDE with continuous coefficient. *Statistics and Probability Letters*, 34:347–354, 1997.
579. J-P. Lepeltier and J. San Martin. Existence for BSDE with superlinear-quadratic coefficient. *Stochastics and Stochastics Reports*, 63:227–240, 1998.

580. D. Lépingle and J. Mémin. Sur l'intégrabilité uniforme des martingales exponentielles. *Z. Wahr. Verw. Gebiete*, 42:175–203, 1978.
581. H.R. Lerche. *Boundary crossing of Brownian motion*, volume 40 of *Lecture Notes in Statistics*. Springer-Verlag, 1986.
582. K.S. Leung and Y.K. Kwok. Distribution of occupation times for CEV diffusions and pricing of a α quantile option. *Quantitative Finance*, 7:87–94, 2007.
583. S. Levental and A.V. Skorokhod. A necessary and sufficient condition for absence of arbitrage with tame portfolio. *The Annals of Applied Prob.*, 5:906–925, 1995.
584. P. Lévy. Sur certains processus stochastiques homogènes. *Compositio Math.*, 7:283–359, 1939.
585. P. Lévy. *Processus stochastiques et mouvement Brownien*. Gauthier-Villars, 1948. Reprinted by Jacques Gabay, Paris, 1992.
586. J. Lévy-Véhel and Ch. Walter. *Les marchés fractals*. PUF Finance, 2002.
587. A. Lewis. *Option Valuation under Stochastic Volatility*. Finance Press, 2001.
588. D.X. Li. On default correlation: a copula approach. *The Journal of Fixed Income*, 9:43–54, 2000.
589. A. Lim. Mean-variance hedging when there are jumps. *SIAM Journal on Control and Optimization*, 44:1893–1922, 2005.
590. V. Linetsky. Step options. *Math. Finance*, 9:55–96, 1999.
591. V. Linetsky. Computing hitting time densities for CIR and OU diffusions: applications to mean-reverting models. *Journal of Computational Finance*, 7:1–22, 2004.
592. V. Linetsky. Lookback options and hitting times of one-dimensional diffusions: a spectral expansion approach. *Finance and Stochastics*, 8:373–398, 2004.
593. V. Linetsky. The spectral decomposition of the option value. *International Journal of Theoretical and Applied Finance*, 7:337–384, 2004.
594. V. Linetsky. Spectral expansions for Asian (average price) options. *Operations Research*, 52:856–867, 2004.
595. V. Linetsky. Spectral methods in derivatives pricing. In J.R. Birge and V. Linetsky, editors, *Handbooks in OR & MS*, volume 15, pages 213–289. Elsevier B.V., 2008.
596. A. Lipton. *Mathematical Methods for Foreign Exchange*. World Scientific, Singapore, 2001.
597. A. Lipton. Assets with jumps. *Risk magazine,September*, pages 149–153, 2002.
598. R.S.R. Liptser and A.N. Shiryaev. *Statistics of Random Processes*. Springer, 2nd printing 2001, 1977.
599. C.F. Lo, H.M. Tang, K.C. Ku, and C.H. Hui. Valuation of single barrier CEV options with time dependent model parameters. In *Proceedings of the 2nd IASTED International Conference on Financial Engineering and Applications*, Cambridge, 2004.
600. C.F. Lo, P.H. Yuen, and C.H. Hui. Comments on pricing double barrier options using Laplace transforms by Anton Pelsser. *Finance and Stochastics*, 4:105–107, 2000.
601. C.F. Lo, P.H. Yuen, and C.H. Hui. Constant elasticity of variance option pricing model with time dependent parameters. *International Journal of Theoretical and Applied Finance*, 4:661–674, 2000.
602. St. Loisel. *Contribution à l'étude de processus univariés et multivariés de la théorie de la ruine*. PhD. thesis, Université Lyon 1, December 2004.

603. J.B. Long. The numéraire portfolio. *Journal of Financial Economics*, 26:29–69, 1990.
604. H. Loubergé, S. Villeneuve, and M. Chesney. Long term risk management of nuclear waste: a real option approach. *Journal of Economic Dynamic and Control*, 27:157–180, 2002.
605. E. Lukacs. *Characteristic Functions.* Griffin, 1970.
606. A. Lyasoff. The integral of geometric Brownian motion revisited. *Preprint*, 2005.
607. J. Ma and J. Yong. *Forward-backward Stochastic Differential Equations*, volume 1702 of *Lecture Notes in Mathematics.* Springer-Verlag, Berlin, 1999.
608. L.W. MacMillan. Analytic approximation for the American put option. *Advances in Futures and Option Research*, 1:119–139, 1986.
609. D. Madan. Purely discontinuous asset price process. In E. Jouini, J. Cvitanić, and M. Musiela, editors, *Option pricing, Interest rates and Risk Management*, pages 67–104. Cambridge University Press, 2001.
610. D. Madan and F. Milne. Option pricing with V.G. martingale components. *Math. Finance*, 1:39–55, 1991.
611. D. Madan, B. Roynette, and M. Yor. An alternative expression for the Black-Scholes formula in terms of Brownian first and last passage times. *Preprint, Université Paris 6*, 2008.
612. D. Madan and E. Seneta. The V.G. model for share market returns. *Journal of Business*, 63:511–524, 1990.
613. D. Madan and M. Yor. Making Markov martingales meet marginals: with explicit construction. *Bernoulli*, 8:509–536, 2002.
614. D. Madan and M. Yor. Itô's integrated formula for strict local martingales. In M. Emery and M. Yor, editors, *Séminaire de Probabilités XXXIX*, volume 1874 of *Lecture Notes in Mathematics*, pages 157–170. Springer-Verlag, 2006.
615. Y. Maghsoodi. Solutions of the extended CIR term structure and bond option valuation. *Math. Finance*, 6:89–109, 1996.
616. P. Malliavin and A. Thalmaier. *Stochastic Calculus of Variation in Mathematical Finance.* Springer-Verlag, New York, 2005.
617. M.G. Mania. A general problem of an optimal equivalent measure and contingent claim pricing in an incomplete market. *Stochastic Processes and their Appl.*, 90:19–42, 2000.
618. M.G. Mania and M. Schweizer. Mean variance. *Preprint*, 2004.
619. M.G. Mania and R. Tevzadze. Backward stochastic PDE and imperfect hedging. *International Journal of Theoretical and Applied Finance*, 6:663–692, 2003.
620. R. Mansuy. *Infinitely Divisible in Time Processes.* PhD thesis, Paris VI, 2005.
621. R. Mansuy and M. Yor. Harnesses, Lévy bridges and Monsieur Jourdain. *Stochastic Processes and their Appl.*, 115:329–338, 2005.
622. R. Mansuy and M. Yor. *Random Times and (Enlargement of) Filtrations in a Brownian Setting*, volume 1873 of *Lectures Notes in Mathematics.* Springer, 2006.
623. W. Margrabe. The value of an option to exchange one asset for another. *J. of Finance*, 33:77–86, 1978.
624. L. Martellini, Ph. Priaulet, and S. Priaulet. *Fixed Income Securities.* Wiley Finance, New York, 2003.
625. C. Martini. Propagation of convexity by Markovian and martingalian semigroups. *Potential Analysis*, 10:133–175, 1999.

626. G. Maruyama. On the transition probability functions of the Markov process. *Nat. Sci. Rep. Ochanomizu Univ.*, 5:10–20, 1954.
627. L. Mastroeni and M. Matzeu. An integro-differential parabolic variational inequality connected with the American option pricing problem. *Zeitschrift fur Analysis und ihre Anwendungen*, 14:869–880, 1995.
628. L. Mastroeni and M. Matzeu. Stability for the integro-differential variational inequalities of the American option pricing problem. *Advances in Mathematical Sciences and Applications*, 7:651–666, 1997.
629. H. Matsumoto and M. Yor. A relationship between Brownian motions with opposite drifts via certain enlargements of the Brownian filtration. *Osaka J. Maths*, 38:383–398, 2001.
630. H. Matsumoto and M. Yor. Exponential functionals of Brownian motion, I: probability laws at fixed time. *Probability Surveys*, 2:312–347, 2005.
631. H. Matsumoto and M. Yor. Exponential functionals of Brownian motion, II: some related diffusion processes. *Probability Surveys*, 2:348–384, 2005.
632. G. Mazziotto and J. Szpirglas. Modèle général de filtrage non linéaire et équations différentielles stochastiques associées. *Ann. Inst. H. Poincaré*, 15:147–173, 1979.
633. R.L. McDonald and M.D. Schroder. A parity result for American options. *Journal of Computational Finance*, 1:5–13, 1998.
634. R.L. McDonald and R. Siegel. The value of waiting to invest. *Quarterly Journal of Economics*, 101:707–728, 1986.
635. H.P. McKean. Appendix: A free boundary problem for the heat equation arising from a problem in mathematical economics. *Industr. Manage. Rev*, 16:32–39, 1965.
636. H.P. McKean. *Stochastic Integrals*. Academic Press, reprinted AMS Chelsea Publishing, 2005, New York, 1969.
637. H.P. McKean. Brownian Motion and the General Diffusion: Scale and Clock. In H. Geman, D. Madan, S.R. Pliska, and T. Vorst, editors, *Mathematical Finance, Bachelier Congress 2000*, Springer Finance, pages 75–92. Springer-Verlag, 2002.
638. A.V. Mel'nikov. *Financial Markets. Stochastic Analysis and the Pricing of Derivative Securities*. American Mathematical Society, Providence, 1999.
639. A.V. Mel'nikov. *Risk Analysis in Finance and Insurance*. Chapman & Hall/CRC, Boca Raton, 2004.
640. F. Mercurio and W.J. Runggaldier. Option pricing for jump diffusions. Approximation and their interpretation. *Math. Finance*, 3:191–200, 1993.
641. R. Merton. An intertemporal capital asset pricing model. *Econometrica*, 41:867–888, 1973.
642. R. Merton. On the pricing of corporate debt: the risk structure of interest rates. *J. of Finance*, 3:449–470, 1974.
643. R. Merton. Option pricing when underlying stock returns are discontinuous. *Journal of Financial Economics*, 3:125–144, 1976.
644. R. Merton. Application of option pricing theory, twenty-five years later. Stockholm, December 1997. Nobel lecture, http://nobelprize.org/nobel-prizes/economics/laureates/1997/merton-lecture.pdf.
645. M. Métivier. Pathwise differentiability with respect to a parameter of solutions of stochastic differential equations . In P-A. Meyer, editor, *Séminaire de Probabilités XVI*, volume 920 of *Lecture Notes in Mathematics*, pages 490–502. Springer-Verlag, 1982.

646. S.A.K. Metwally and A.F. Atiya. Using Brownian bridge for fast simulation of jump-diffusion processes and barrier options. *J. of Derivatives*, 10:43–54, 2002. Fall
647. P-A. Meyer. Un cours sur les intégrales stochastiques. In P-A. Meyer, editor, *Séminaire de Probabilités X*, volume 511 of *Lecture Notes in Mathematics*, pages 246–400. Springer-Verlag, 1976.
648. P-A. Meyer. Les processus stochastiques de 1950 à nos jours. In J-P. Pier, editor, *Developments of Mathematics 1950-2000*, volume 1899, pages 813–849. Birkhäuser, 2007.
649. P-A. Meyer and Ch. Yoeurp. Sur la décomposition multiplicative des sousmartingales positives. In *Séminaire de Probabilités X*, volume 511, pages 501–504. Springer, 1976.
650. T. Mikosch. *Non-life Insurance Mathematics, An introduction with Stochastic Processes.* Springer, Universitext, Berlin, 2004.
651. T. Mikosch. *A Point Process Approach to Collective Risk Theory.* Springer, Berlin, 2009.
652. T. Mikosch. *Non-Life Insurance Mathematics. An Introduction with the Poisson Process.* Springer, Berlin, second edition, 2009.
653. R. Miura. A note on a look-back option based on order statistics. *Hitosubashi Journal of Commerce and Management*, 27:15–28, 1992.
654. Y. Miyahara. Canonical martingale measures of incomplete assets markets. In S. Watanabe, Yu. V. Prohorov, M. Fukushima, and A.N. Shiryaev, editors, *Proceedings of the Seventh Japan-Russia Symposium*, Probability Theory and Mathematical Statistics, pages 343–352. World Scientific, 1997.
655. I. Monroe. Embedding right-continuous martingales into Brownian motion. *Ann. Math. Stat.*, 43:1293–1311, 1972.
656. I. Monroe. Processes that can be embedded in Brownian motion. *The Annals of Probability*, 6:42–56, 1978.
657. F. Moraux. On cumulative Parisian options. *Finance*, 23:127–132, 2002.
658. E. Mordecki. Optimal stopping for a diffusion with jumps. *Finance and Stochastics*, 3:227–236, 1999.
659. E. Mordecki. Optimal stopping and perpetual options for Lévy processes. *Finance and Stochastics*, 6:473–494, 2002.
660. E. Mordecki and W. Moreira. Russian options for a diffusion with negative jumps. *Publicationes Matemáticas del Uruguay*, 9:37–51, 2001.
661. M. Musiela and M. Rutkowski. *Martingale Methods in Financial Modelling.* Springer-Verlag, Heidelberg-Berlin-New York, second edition, 2005.
662. M. Musiela and T. Zariphopoulou. An example of indifference prices under exponential preferences. *Finance and Stochastics*, 8:229–240, 2004.
663. M. Nagasawa. *Schrödinger Equations and Diffusion Theory.* Birkhäuser, 1993.
664. E. Nahum. *On the Pricing of Lookback Options.* PhD. thesis, Berkeley University, Spring 1999.
665. H. Nakagawa. A filtering model on default risk. *J. Math. Sci. Univ. Tokyo*, 8:107–142, 2001.
666. S. Nakao. On the pathwise uniqueness of solutions of one–dimensional stochastic differential equations. *Osaka J. Math*, 9:513–518, 1972.
667. P. Navatte and F. Quittard-Pinon. The valuation of digital interest rate options and range notes revisited. *European Financial Management*, 14:79–97, 2004.
668. R.B. Nelsen. *An Introduction to Copulas*, volume 139 of *Lecture Notes in Stats.* Springer-Verlag, Berlin, 1999.

669. J. Neveu. Processus ponctuels. In *Ecole d'été de Saint Flour IV, 1976*, volume 598 of *Lecture Notes in Mathematics*. Springer-Verlag, 1977.
670. L. Nguyen-Ngoc and M. Yor. Exotic options and Lévy processes. *Preprint*, 2001.
671. L. Nguyen-Ngoc and M. Yor. Wiener-Hopf factorization and the pricing of barrier and lookback options under general Lévy processes. In Y. Ait-Sahalia and L-P. Hansen, editors, *Handbook of Financial Econometrics*. 2007.
672. J.A. Nielsen and K. Sandmann. The pricing of Asian options under stochastic interest rates. *Applied Math. Finance*, 3:209–26, 1996.
673. A. Nikeghbali. Moment problem for some convex functionals of Brownian motion and related processes. *Preprint 706, Laboratoire de probabilités et modèles aléatoires, Paris 6 et 7*, 2002.
674. A. Nikeghbali. An essay on the general theory of stochastic processes. *Probability Surveys*, 3:345–412, 2006.
675. A. Nikeghbali and M. Yor. A definition and some properties of pseudo-stopping times. *The Annals of Probability*, 33:1804–1824, 2005.
676. A. Nikeghbali and M. Yor. Doob's maximal identity, multiplicative decompositions and enlargements of filtrations. In D. Burkholder, editor, *Joseph Doob: A Collection of Mathematical Articles in his Memory*, volume 50, pages 791–814. Illinois Journal of Mathematics, 2007.
677. I. Norros. A compensator representation of multivariate life length distributions, with applications. *Scand. J. Statist.*, 13:99–112, 1986.
678. A. Novikov. Martingales and first passage times for Ornstein-Uhlenbeck processes with a jump component. *Teor. Veroyatnost. i Priminen*, 48:340–358, 2003.
679. A.A. Novikov. The stopping times of a Wiener process. *Theory of Probability and its Applications*, 16:449–456, 1971.
680. A.A. Novikov. Conditions for uniform integrability of continuous non-negative martingales. *Theory of Probability and its Applications*, 24:820–824, 1980.
681. D. Nualart. *The Malliavin Calculus and Related Topics*. Springer-Verlag, Heidelberg-Berlin-New York, 1995.
682. D. Nualart and W. Schoutens. Chaotic and predictable representations for Lévy processes. *Stochastic Processes and their Appl.*, 90:109–122, 2000.
683. D. Nualart and W. Schoutens. Backward stochastic differential equations and Feynman-Kac formula for Lévy processes with applications in finance. *Bernoulli*, 7:761–776, 2001.
684. B. Øksendal. *Stochastic Differential Equations*. Springer-Verlag, Berlin, sixth edition, 1998.
685. B. Øksendal and A. Sulem. *Control Theory for Jump Processes*. Springer-Verlag, Berlin, 2005.
686. E. Omberg. The valuation of American put options with exponential exercise policies. *Advances in Futures and Options Research*, 2:117–142, 1987.
687. Y. Ouknine. Skew-Brownian motion and associated processes. *Theory of Probability and Its Applications*, 35:163–169, 1991.
688. M. Overhaus, A. Bermudez, H. Buehler, A. Ferraris, Ch. Jordinson, and A. Lamnouar. *Equity Hybrid Derivative*. Wiley Finance, New York, 2006.
689. M. Overhaus, O. Brockhaus, A. Ferraris, C. Gallus, D. Long, and R. Martin. *Modelling and Hedging Equity Derivatives*. Risk books, London, 1999.

690. M. Overhaus, A. Ferraris, T. Knudsen, R. Milward, L. Nguyen-Ngoc, and G. Schindlmayr. *Equity Derivatives, Theory and Applications.* Wiley Finance, New York, 2002.
691. G. Pagès. *Introduction to Numerical Probability for Finance.* Lecture Notes, LPMA-Université Paris 6, 2008.
692. S. Pal and Ph. Protter. Strict local martingales, bubbles, and no early exercise. *Preprint*, 2008.
693. R. Panini and R.P. Srivastav. Option pricing with Mellin transforms. *Mathematical and Computer Modelling*, 40:43–56, 2004.
694. E. Pardoux and S. Peng. Adapted solution of a backward stochastic differential equation. *Systems and Control Letters*, 14:55–61, 1990.
695. C. Park and S.R. Paranjape. Probabilities of Wiener paths crossing differentiable curves. *Pacific Journal of Mathematics*, 53:579–583, 1974.
696. C. Park and F.J. Schuurmann. Evaluation of barrier crossing probabilities of Wiener paths. *J. Appl. Prob.*, 13:267–275, 1976.
697. P. Patie. *On some First Passage Time Problems Motivated by Financial Applications.* Ph.D. thesis, E.T.H. Zurich, 2005.
698. F. Patras. A reflection principle for correlated defaults. *Stochastic Processes and their Appl.*, 116:690–698, 2006.
699. S.J. Patterson. *An Introduction to the Theory of the Riemann Zeta-function.* Cambridge University Press, 1995.
700. J. Paulsen. Ruin theory in a stochastic environment. *Stochastic Processes and their Appl.*, 21:327–361, 1993.
701. J. Paulsen and H. Gjessing. Ruin theory with stochastic returns on investments. *Adv. Appl. Prob.*, 29:965–985, 1997.
702. E. Pecherskii and B.A. Rogozin. On joint distributions of random variables associated with fluctuations of a process with independent increments. *Theory of Probability and Its Applications*, 14:410–423, 1969.
703. A. Pechtl. Some applications of occupation times of Brownian motion with drift in mathematical finance. *Journal of Applied Mathematics and Decision Sciences*, 3(1):63–73, 1999.
704. A. Pelsser. Pricing double barrier options using Laplace transforms. *Finance and Stochastics*, 4:95–104, 2000.
705. S. Peng. Bsde and related g-expectation. In N. El Karoui and L. Mazliak, editors, *Backward Stochastic Differential Equations*, volume 364 of *Pitman Research Notes in Mathematics Series, 364*, pages 141–159. Longman, Harlow, 1997.
706. S. Peng. Nonlinear expectations, nonlinear evaluations and risk measures. In W.J. Runggaldier, editor, *CIME-EMS Summer School on Stochastic Methods in Finance, Bressanone*, volume 1856 of *Lecture Notes in Mathematics.* Springer, 2004.
707. G. Peskir. On integral equations arising in the first-passage problem for Brownian motion. *Journal of Integral Equations and Applications*, 14:397–424, 2002.
708. G. Peskir and A.N. Shiryaev. On the Brownian first-passage time over a one-sided stochastic boundary. *Theory of Probability and Its Applications*, 42:444–453, 1999.
709. H. Pham. Optimal stopping free boundary and American option in a jump diffusion model. *Applied Math. and Optim*, 35:145–164, 1997.

710. H. Pham. *Imperfections de marchés et méthodes d'évaluation et couverture d'options.* Corsi della Scuola Normale Superiore, Pisa, 1999.
711. H. Pham and M-Cl. Quenez-Kammerer. Optimal portfolio in partially observed stochastic volatility models. *The Annals of Applied Prob.*, 11:210–238, 2001.
712. J.W. Pitman. One-dimensional Brownian motion and the three-dimensional Bessel process. *Adv. Appl. Prob.*, 7:511–526, 1975.
713. J.W. Pitman. The distribution of the local times of a Brownian bridge. In J. Azéma and M. Yor, editors, *Séminaire de Probabilités XXXIII*, volume 1709 of *Lecture Notes in Mathematics*, pages 388–394. Springer-Verlag, 1999.
714. J.W. Pitman and L.C.G. Rogers. Markov functions. *The Annals of Probability*, 9:573–582, 1981.
715. J.W. Pitman and M. Yor. Bessel processes and infinitely divisible laws. In D. Williams, editor, *Stochastic integrals, LMS Durham Symposium, Lect. Notes 851*, pages 285–370. Springer, 1980.
716. J.W. Pitman and M. Yor. A decomposition of Bessel bridges. *Z. Wahr. Verw. Gebiete*, 59:425–457, 1982.
717. J.W. Pitman and M. Yor. Sur une décomposition des ponts de Bessel. In M. Fukushima, editor, *Functional analysis in Markov processes, Proceedings Katata and Kyoto 1981*, volume 923 of *Lecture Notes in Mathematics*, pages 276–285. Springer-Verlag, 1982.
718. J.W. Pitman and M. Yor. Laplace transforms related to excursions of a one-dimensional diffusion. *Bernoulli*, 5:249–255, 1999.
719. J.W. Pitman and M. Yor. Hitting, occupation and inverse local times of one-dimensional diffusions: martingale and excursion approaches. *Bernoulli*, 9:1–24, 2003.
720. J.W. Pitman and M. Yor. Infinitely divisible laws associated with hyperbolic functions. *Canadian Journal of Mathematics*, 55:292–330, 2003.
721. S.R. Pliska. *Introduction to mathematical finance.* Blackwell, Oxford, 1997.
722. P. Poncet and F. Quittard-Pinon. Pricing and hedging Asian options. *Banque et marchés*, 47:Juillet/Aout, 2000.
723. R. Portait and P. Poncet. *Finance de Marché.* Dalloz, Paris, 2008.
724. J-L. Prigent. Option pricing with a general marked point process. *Mathematics of Operations Research*, 26:50–66, 2001.
725. J-L. Prigent. *Weak Convergence of Financial markets.* Springer Finance. Springer, Berlin, 2003.
726. Ph. Protter. A partial introduction to financial asset pricing theory. *Stochastic Processes and their Appl.*, 91:169–204, 2001.
727. Ph. Protter. *Stochastic Integration and Differential Equations.* Springer, Berlin, Second edition, 2005.
728. A. Ramakrishnan. A stochastic model of a fluctuating density field. *Astrophys. J.*, 119:682–675, 1954.
729. M.M. Rao. Martingales and some applications. In D.N. Shanbhag and C.R. Rao, editors, *Handbook of Statistics, 19*, pages 765–815. Elsevier, 2001.
730. D. Revuz and M. Yor. *Continuous Martingales and Brownian Motion.* Springer Verlag, Berlin, third edition, 1999.
731. L. Ricciardi, L. Sacerdote, and S. Sato. On an integral equation for first passage time probability densities. *J. Appl. Prob.*, 21:302–314, 1984.
732. L. Ricciardi and S. Sato. First passage time density and moments of the Ornstein-Uhlenbeck process. *J. Appl. Prob.*, 25:43–57, 1988.

733. D.R. Rich. The mathematical foundations of barrier option-pricing theory. *Advances in Futures and Options Research*, 7:267–311, 1994.
734. H. Robbins and D. Siegmund. Boundary crossing probabilities for the Wiener process and sample sums. *Ann. Math. Stat*, 41:1410–1429, 1970.
735. G.O. Roberts and C.F. Shortland. Pricing barrier options with time-dependent coefficients. *Math. Finance*, 7:83–93, 1997.
736. L.C.G. Rogers. Which models for term structure of interest rates should one use. In M.H.A. Davis, D. Duffie, W. Fleming, and S.E. Shreve, editors, *Math. Finance*, IMA, Springer-Verlag, pages 93–115. Springer-Verlag, 1995.
737. L.C.G. Rogers. Arbitrage with fractional Brownian motion. *Math. Finance*, 7:95–105, 1997.
738. L.C.G. Rogers. The origins of risk-neutral pricing and the Black-Scholes formula. *Risk Management and Analysis*, 2:81–94, 1998.
739. L.C.G. Rogers and L. Shepp. The correlation of the maxima of correlated Brownian motions. *J. Appl. Prob.*, 43:880–883, 2006.
740. L.C.G. Rogers and Z. Shi. The value of an Asian option. *J. Appl. Prob.*, 32:1077–1088, 1995.
741. L.C.G. Rogers and D. Williams. *Diffusions, Markov processes and Martingales, Vol 1. Foundations.* Cambridge University Press, Cambridge, second edition, 2000.
742. L.C.G. Rogers and D. Williams. *Diffusions, Markov Processes and Martingales, Vol 2. Itô Calculus.* Cambridge University Press, Cambridge, second edition, 2000.
743. S. Rong. On solutions of backward stochastic differential equations with jumps and application. *Stochastic Processes and their Appl.*, 66:209–236, 1997.
744. M. Royer. *Equations différentielles stochastiques rétrogrades et martingales non-linéaires.* Thesis, Angers, 2003.
745. B. Roynette, P. Vallois, and A. Volpi. Asymptotic behavior of the hitting time, overshoot and undershoot for some Lévy processes. *ESAIM PS*, 12:58–97, 2008.
746. M. Rubinstein and E. Reiner. Breaking down the barriers. *Risk*, 9:28–35, 1991.
747. M. Rubinstein and E. Reiner. Unscrambling the binary code. *Risk Magazine*, 4:75–93, 1991.
748. J. Ruiz de Chavez. Le théorème de Paul Lévy pour les mesures signées. In J. Azéma and M. Yor, editors, *Séminaire de Probabilités XXVIII*, volume 1583 of *Lecture Notes in Mathematics*, pages 245–255. Springer-Verlag, 1994.
749. W.J. Runggaldier, editor. *Financial Mathematics, Bressanone, 1996*, volume 1656 of *Lecture Notes in Mathematics.* Springer-Verlag, Berlin, 1997.
750. W.J. Runggaldier. Jump-diffusion models. In S.T. Rachev, editor, *Handbook of Heavy Tailed Distributions in Finance.* North Holland, 2004.
751. M. Rutkowski. On solutions of stochastic differential equations with drift. *Probability Theory and Related Fields*, 85:387–402, 1990.
752. P. Salminen. On conditioned Ornstein-Uhlenbeck processes. *Adv. Appl. Prob.*, 16:920–922, 1984.
753. P. Salminen. On the first hitting time and the last exit for a Brownian motion to/from a moving boundary. *Adv. Appl. Prob.*, 20:411–426, 1988.
754. P. Salminen. On last exit decomposition of linear diffusions. *Studia Sci. Math. Hungar.*, 33:251–262, 1997.
755. P. Salminen, P. Vallois, and M. Yor. On the excursion theory for linear diffusions. *Japanese Journal of Mathematics*, 2:137–143, 2007.

756. G. Samorodnitsky and M. Taqqu. *Stable Non-Gaussian Random Processes.* Chapman & Hall, New-York, 1994.
757. P. Samuelson. Rational theory of warrant pricing. *Industrial Management Review*, 6:13–39, 1965.
758. P. Samuelson. Modern finance theory within one lifetime. In H. Geman, D. Madan, S.R. Pliska, and T. Vorst, editors, *Mathematical Finance, Bachelier Congress 2000*, Springer Finance, pages 41–46. Springer-Verlag, 2002.
759. K. Sandmann and Schönbucher P., editors. *Advances in Finance and Stochastics.* Essays in Honour of Dieter Sondermann. Springer-Verlag, Basel, 2002.
760. K. Sato. Self-similar processes with independent increments. *Probability Theory and Related Fields*, 89:285–300, 1991.
761. K. Sato. *Lévy Processes and Infinitely Divisible Distributions.* Cambridge University Press, Cambridge, 1999.
762. K. Sato. Basic results on Lévy processes. In O.E. Barndorff-Nielsen, T. Mikosch, and S.I. Resnick, editors, *Lévy Processes. Theory and Applications*, pages 3–37. Birkhäuser, 2001.
763. W. Schachermayer. Introduction to the mathematics of financial markets. In P. Bernard, editor, *Ecole d'été de Saint-Flour 2000*, volume 1816 of *Lecture Notes in Probability Theory and Statistics.*, pages 107–179. Springer-Verlag, Berlin, 2003.
764. M. Scholes. Derivatives in a dynamic environnement. Stockholm, December 1997. Nobel lecture, http://nobelprize.org/nobel-prizes/economics/laureates/1997/scholes-lecture.pdf.
765. Ph.J. Schönbucher. *Credit Derivatives Pricing Models.* Wiley Finance, Chichester, 2003.
766. W. Schoutens. *Lévy Processes in Finance, Pricing Financial Derivatives.* Wiley, Chichester, 2003.
767. M. Schröder. On the valuation of arithmetic-average Asian options: integral representations. *Unpublished manuscript. Mannheim University*, 2000.
768. M. Schröder. On the valuation of double barrier options: computational aspects. *J. Computational Finance*, 5:5–33, 2000.
769. M. Schröder. The Laplace approach to valuing exotic options: the case of the Asian option. In M. Kohlmann and S. Tang, editors, *Mathematical Finance*, Trends in Mathematics, pages 328–338, Basel, 2001. Birkhäuser.
770. M. Schröder. Brownian excursions and Parisian barrier options: a note. *Journal of Applied Prob.*, 40:855–864, 2003.
771. M. Schröder. Laguerre series in contingent claim valuation with application to Asian options. *Math. Finance*, 15:691–532, 2005.
772. M.D. Schroder. Computing the constant elasticity of variance option pricing formula. *J. of Finance*, 45:211–219, 1989.
773. M.D. Schroder. Changes of numeraire for pricing futures, forwards and options. *Review of Financial Studies*, 12:1143–1163, 1999.
774. K. Schürger. Laplace transform and supremum of stochastic processes. In *Advances in Finance and Stochastics*, Essays in Honour of Dieter Sondermann, pages 286–293, Basel, 2002. Springer-Verlag.
775. L.O. Scott. Option pricing when the variance changes randomly: theory, estimation, and an application. *Journal of Financial and Quantitative Analysis*, 22:419–438, 1987.

776. L.O. Scott. Pricing stock options in a jump diffusion model with stochastic volatility and interest rates: Applications of Fourier inversion methods. *Math. Finance*, 7:413–426, 1997.
777. L.O. Scott. Simulating a continuous time term structure model over discrete time periods: an application of Bessel bridges processes. *Unpublished manuscript*, 1998.
778. A.V. Selivanov. On the martingale measures in exponential Lévy models. *Theory of Probability and Its Applications*, 49:261–274, 2004.
779. E. Seneta. The early years of the variance-gamma process. In M.C. Fu, R.A. Jarrow, J.-Y.J. Yen, and R.J. Elliott, editors, *Advances in Mathematical Finance, The Madan Festschrift*, Applied and Numerical Harmonic Analysis. Birkhäuser, 2007.
780. A. Sepp. *Pricing Path-dependent Options under Jump-diffusion Process: Applications of Laplace transform* . Ph. D. Thesis, Tartu university, March 2003.
781. A. Sepp. Analytical pricing of double-barrier options under a double-exponential jump diffusion process: Applications of Laplace transform. *International Journal of Theoretical and Applied Finance*, 7:151–176, 2004.
782. M. Shaked and J.G. Shanthikumar. The multivariate hazard construction. *Stochastic Processes and their Appl.*, 24:241–258, 1987.
783. D.N. Shanbhag and A. Sreehari. On certain self decomposable distributions. *Z. Wahr. Verw. Gebiete*, 38:217–222, 1977.
784. M.J. Sharpe. Some transformations of diffusion by time reversal. *The Annals of Probability*, 8:1157–1162, 1980.
785. M.J. Sharpe. *General Theory of Markov Processes.* Academic Press, New-York, 1988.
786. L.A. Shepp. A first passage problem for the Wiener process. *Annals of Mathematics and Statistics*, 38:1912–1914, 1967.
787. L.A. Shepp and A.N. Shiryaev. The Russian option: reduced regret. *The Annals of Applied Prob.*, 3:631–640, 1993.
788. T. Shiga and S. Watanabe. Bessel diffusion as a one-parameter family of diffusion processes. *Z. Wahr. Verw. Gebiete*, 27:37–46, 1973.
789. H. Shirakawa. Security market model with Poisson and diffusion type return process. *Unpublished manuscript*, 1990.
790. H. Shirakawa. Squared Bessel processes and their applications to the square root interest rate model. *Asia-Pacific Financial markets*, 9:169–190, 2002.
791. A.N. Shiryaev. *Essentials of Stochastic Finance.* World Scientific, Singapore, 1999.
792. A.N. Shiryaev and A.S. Cherny. Some distributional properties of a Brownian motion with a drift and an extension of P. Lévy's theorem. *Theory of Probabability and Appl.*, 43:412–418, 2000.
793. A.N. Shiryaev and M. Yor. On the problem of stochastic integral representations of functionals of the Brownian motion. *Theory of Probability and Appl.*, 48:304–313, 2004.
794. S.E. Shreve. *Stochastic Calculus Models for Finance, Discrete Time.* Springer, 2004.
795. S.E. Shreve. *Stochastic Calculus Models for Finance, II: Continuous Time Models.* Springer, 2004.

796. S.E. Shreve, H.M. Soner, and J. Cvitanic. There is no nontrivial hedging portfolio for option pricing with transaction costs. *The Annals of Applied Prob.*, 5:327–355, 1995.
797. S.E. Shreve and Y. Večeř. Options on a traded account: vacation calls, vacation puts and passport options. *Finance and Stochastics*, 4:255–274, 2000.
798. D. Siegmund and Y-S. Yuh. Brownian approximations to first passage probabilities. *Z. Wahr. Verw. Gebiete*, 59:239–248, 1982.
799. C. Sin. *Strictly Local Martingales and Hedge Ratios on Stochastic Volatility Models.* Thesis, Cornell University, 1996.
800. C. Sin. Complications with stochastic volatility models. *Adv. Appl. Prob.*, 30:256–268, 1998.
801. A.V. Skorokhod. *Studies in the Theory of Random Processes.* Addison-Wesley, Reading, Mass., 1965.
802. A.V. Skorokhod. *Lectures on the Theory of Stochastic Processes.* VSP, Utrecht, 1996.
803. L.J. Slater. *Confluent Hypergeometric Functions.* Cambridge University Press, Cambridge, 1966.
804. D. Sondermann. *Introduction to Stochastic Calculus for Finance.* Lecture Notes in Economics and Mathematical Systems. Springer, Berlin, 2006.
805. R. Stanton. Path-dependent payoffs and contingent claims valuation: single premium deferred annuities. *Unpublished*, 1989.
806. J.M. Steele. *Stochastic Calculus and Financial Applications.* Springer Verlag, Berlin, second edition, 2002.
807. Ch. Stricker. Quasi-martingales, martingales locales, semimartingales et filtration naturelle. *Z. Wahr. Verw. Gebiete*, 39:55–63, 1977.
808. Ch. Stricker. Integral representation in the theory of continuous trading. *Stochastics*, 13:249–256, 1984.
809. Ch. Stricker. Arbitrage et lois de martingales. *Ann. Inst. H. Poincaré*, 26:451–460, 1990.
810. D.W. Stroock. *Probability Theory. An Analytical View.* Cambridge University Press, 1994.
811. D.W. Stroock. *Markov Processes from K. Ito's Perspective.* Annals of Mathematics Studies. 2003.
812. D.W. Stroock and S.R.S. Varadhan. *Multi-dimensional Diffusion Processes.* Springer Verlag, Berlin, 1979.
813. D.W. Stroock and M. Yor. Some remarkable martingales. In J. Azéma and M. Yor, editors, *Séminaire de Probabilités XV*, volume 850 of *Lecture Notes in Mathematics*, pages 590–603. Springer-Verlag, 1981.
814. M. Suchanecki. The lateral Chapman Kolmogorov relation and its application to barrier option pricing. *Working paper, University of Bonn*, 2004.
815. H.J. Sussmann. On the gap between deterministic and stochastic ordinary differential equations. *The Annals of Probability*, 6:19–41, 1976.
816. W. Szatzschneider. Comments about CIR model as a part of a financial market. *Unpublished manuscript*, 2001.
817. W. Szatzschneider. Extended Cox, Ingersoll and Ross model. *Unpublished manuscript*, 2001.
818. N. Taleb. *Dynamic Hedging.* Wiley, New York, 1997.
819. M.S. Taqqu. Bachelier and his times: a conversation with Bernard Bru. In H. Geman, D. Madan, S.R. Pliska, and T. Vorst, editors, *Mathematical*

Finance, Bachelier Congress 2000, Springer Finance, pages 1–40. Springer-Verlag, 2002.

820. L. Trigeorgis. *Real Options, Managerial Flexibility and Strategy in Resource Allocation.* 1996.
821. H.F. Trotter. A property of Brownian motion paths. *Illinois Journal of Mathematics*, 2:425–433, 1958.
822. B. Tsirel'son. An example of a stochastic differential equation having no strong solution. *Theory of Probability and Applications*, 20:427–430, 1975.
823. B. Tsirel'son. Triple points: from non-Brownian filtrations to harmonic measures. *GAFA, Geom. Funct. Ana.*, 7:1096–1142, 1997.
824. B. Tsirel'son. Within and beyond the reach of Brownian innovations. In *Proceedings of the International Congress of Mathematicians*, pages 311–320. Documenta Mathematica, 1998.
825. J.H. Van Schuppen and E. Wong. Transformations of local martingales under a change of law. *The Annals of Probability*, 2:879–888, 1974.
826. S.R.S. Varadhan. Diffusion Processes. In D.N. Shanbhag and C.R. Rao, editors, *Handbook of Statistics, 19*, pages 853–872. Elsevier, 2001.
827. Y. Večeř and M. Xu. Pricing Asian options in a semi-martingale model. *Quantitative Finance*, 4:170–175, 2004.
828. A.D. Ventzel. On equations of the theory of conditional Markov processes. *Theory of Probability and its Applications*, 10:357–361, 1965.
829. S. Villeneuve. Exercise regions of American options on several assets. *Finance and Stochastics*, 3:295–322, 1999.
830. S. Villeneuve. On threshold strategies and the smooth-fit principle for optimal stopping problems. *J. Appl. Probab.*, 44:181–198, 2007.
831. V.A. Volkonski. Random time changes in strong Markov processes. *Teoria*, 3:310–326, 1958.
832. A. Volpi. *Temps d'atteinte.* Thèse, Université Nancy, 2003.
833. J. Walsh. A diffusion with a discontinuous local time. In *Temps locaux*, volume 52-53, pages 37–45, Paris, 1978. Astérisque.
834. A.T. Wang. Generalized Itô's formula and additive functionals of Brownian path. *Z. Wahr. Verw. Gebiete*, 41:153–159, 1977.
835. J. Warren. Branching processes, the Ray-Knight theorem, and sticky Brownian motion. In M. Azéma, J. Emery and M. Yor, editors, *Séminaire de Probabilités XXXI*, volume 1655 of *Lecture Notes in Mathematics*, pages 1–15. Springer-Verlag, 1997.
836. S. Watanabe. Itô's calculus and its applications. In D.N. Shanbhag and C.R. Rao, editors, *Handbook ofStatistics, 19*, pages 873–931. Elsevier, 2001.
837. S. Watanabe. Martingale representation theorem and chaos expansion. In J. Akahori, S. Ogawa, and Watanabe S., editors, *Proceedings of 5th Ritsumeikan Conference*, pages 195–217, 2005.
838. S. Weinryb. Etude d'une équation différentielle stochastique avec temps local. In J. Azéma and M. Yor, editors, *Séminaire de Probabilités XVII*, volume 986 of *Lecture Notes in Mathematics*, pages 72–77. Springer-Verlag, 1983.
839. W. Werner. Girsanov's transformation for SLE(κ, ρ) processes, intersection, exponents and hitting exponents. *Ann. Sci. Toulouse*, 13:121–147, 2004.
840. D. Williams. Path decomposition and continuity of local time for one dimensionel diffusions. *Proc. London Math. Soc.*, 28:438–768, 1974.

841. D. Williams. Conditional excursion theory. In P-A. Meyer, editor, *Séminaire de Probabilités XIII*, volume 721 of *Lecture Notes in Mathematics*, pages 490–494. Springer, Berlin, 1979.
842. D. Williams. *Probability with Martingales.* Cambridge University Press, Cambridge, 1991.
843. D. Williams. *Weighing the Odds.* Cambridge University Press, Cambridge, 2001.
844. D. Williams. A non-stopping time with the optional-stopping property. *Bull. London Math. Soc.*, 34:610–612, 2002.
845. R.J. Williams. *Introduction to the Mathematics of Finance.* AMS, 2006.
846. P. Wilmott. *Derivatives; the Theory and Practice of Financial Engineering.* University Edition, Wiley, Chichester, 1998.
847. P. Wilmott, J. Dewynne, and S. Howison. *Options Pricing. Mathematical Models and Computation.* Oxford Financial Press, Oxford, 1994.
848. B. Wong and C.C. Heyde. On the martingale property of stochastic exponentials. *J. Appl. Probab.*, 41:654–664, 2004.
849. D. Wong. A unifying credit model. Technical report, Capital Markets Group, 1998.
850. E. Wong and B. Hajek. *Stochastic Processes in Engineering Systems.* Springer-Verlag, New-York, second edition, 1985.
851. J. Xia and J-A Yan. Some remarks on arbitrage pricing theory. In J. Yong, editor, *International Conference on Mathematical Finance: Recent Developments in Mathematical Finance*, pages 218–227. World Scientific, 2001.
852. J. Xia and J-A Yan. A new look at some basic concepts in arbitrage pricing theory. *Science in China (Series A)*, 46:764–774, 2003.
853. C. Xu and Y.K. Kwok. Integral price formulas for lookback options. *Journal of Applied Mathematics*, 2:117–125, 2005.
854. M. Yamazato. Hitting time distributions of single points for 1-dimensional generalized diffusion processes. *Nagoya Math. J.*, 119:143–172, 1990.
855. J-A Yan. A propos de l'intégrabilité uniforme des martingales exponentielles. In *Séminaire de Probabilités XVI*, volume 920 of *Lecture Notes in Mathematics*, pages 338–347. Springer, 1982.
856. J-Y. Yen and M. Yor. Call option prices based on Bessel processes. *Forthcoming in Methodology and Computing in Applied Probability.* Springer, 2009.
857. C. Yoeurp. *Contributions au calcul stochastique.* Thèse de doctorat d'état, Paris VI, 1982.
858. Ch. Yoeurp. Grossissement de filtration et théorème de Girsanov généralisé. In Th. Jeulin and M. Yor, editors, *Grossissements de filtrations: exemples et applications*, volume 1118. Springer, 1985.
859. M. Yor. Sur quelques approximations d'intégrales stochastiques. In *Séminaire de Probabilités XI*, volume 581 of *Lecture Notes in Mathematics*, pages 518–528. Springer-Verlag, 1977.
860. M. Yor. Grossissement d'une filtration et semi-martingales : théorèmes généraux. In C. Dellacherie, P-A. Meyer, and M. Weil, editors, *Séminaire de Probabilités XII*, volume 649 of *Lecture Notes in Mathematics*, pages 61–69. Springer-Verlag, 1978.
861. M. Yor. Sous-espaces denses de L^1 et H^1 et représentation des martingales. In P-A. Meyer, editor, *Séminaire de Probabilités XII*, volume 649 of *Lecture Notes in Mathematics*, pages 264–309. Springer-Verlag, 1978.

862. M. Yor. Sur l'étude des martingales continues extrémales. *Stochastics*, 2:191–196, 1979.
863. M. Yor. On some exponentials of Brownian motion. *Adv. Appl. Prob.*, 24:509–531, 1992.
864. M. Yor. *Some Aspects of Brownian Motion, Part I: Some Special Functionals.* Lectures in Mathematics. ETH Zürich. Birkhäuser, Basel, 1992.
865. M. Yor. Sur certaines fonctionnelles exponentielles du mouvement Brownien. *J. Appl. Prob.*, 29:202–208, 1992.
866. M. Yor. The distribution of Brownian quantiles. *J. Appl. Prob*, 32:405–416, 1995.
867. M. Yor. *Local Times and Excursions for Brownian Motion: a Concise Introduction.* Lecciones en Matemáticas. Facultad de Ciencias. Universidad central de Venezuela, Caracas, 1995.
868. M. Yor. *Some Aspects of Brownian Motion, Part II: Some Recent Martingale Problems.* Lectures in Mathematics. ETH Zürich. Birkhäuser, Basel, 1997.
869. M. Yor. Some remarks about the joint law of Brownian motion and its supremum. In J. Azéma, M. Emery, and M. Yor, editors, *Séminaire de Probabilités XXXI*, volume 1655 of *Lecture Notes in Mathematics*, pages 306–314. Springer-Verlag, 1997.
870. M. Yor, editor. *Aspects of Mathematical Finance, Académie des Sciences de Paris.* Springer, Berlin, 2008.
871. K.C. Yuen, G. Wang, and K.W. Ng. Ruin probabilities for a risk process with stochastic return on investments. *Stochastic Processes and their Appl.*, 110:259–274, 2004.
872. P.G. Zhang. *Exotic Options.* World Scientific, Singapore, 1997.
873. X. Zhang. *Analyse numérique des options américaines dans un modèle de diffusion des sauts.* Thèse, Ecole nationale des Ponts et Chaussées, Paris, 1994.
874. X. Zhang. Formules quasi-explicites pour les options américaines dans un modèle de diffusion avec sauts. *Mathematics and Computers in Simulation*, 38:151–161, 1995.
875. X. Zhang. Valuation of American options in jump-diffusion models. In L.C.G. Rogers and D. Talay, editors, *Numerical methods in finance*, Publication of Newton Institute, pages 93–114. Cambridge University Press, 1998.
876. C. Zhou. An analysis of default correlations and multiple defaults. *Review of Financial Studies, Summer*, 14:555–576, 2001.
877. C. Zhou. The term structure of credit spreads with jumps risk. *Journal of Banking and Finance*, 25:2015–2040, 2001.
878. V.M. Zolotarev. *One-dimensional Stable Distributions.* Amer. Math. Soc., Providence, 1986.

B

Some Papers and Books on Specific Subjects

We give here a list of references quoted in our book on some important subjects. We do not give any detail, the reader can refer to the author index and find the page number where the author appear to have some information. Of course, this list is far from being exhaustive, and we apologize for important papers or books which are not quoted.

B.1 Theory of Continuous Processes

B.1.1 Books

Chung [184, 185],Chung and Williams [186], Dellacherie and Meyer [242, 244], He et al. [427], Ikeda and Watanabe [456], Karatzas and Shreve [513], McKean [637], Øksendal [684], Protter [727], Revuz and Yor [730], Rogers and Williams [741], Stroock and Varadhan [812].

B.1.2 Stochastic Differential Equations

Barlow and Perkins [49], Barlow [47, 46], Engelbert and Schmidt [333, 331, 332], Fang and Zhang [340], Harrison and Shepp [424].

B.1.3 Backward SDE

Buckdhan [134], collective book [303], El Karoui and coauthors [308, 309, 310], Hu and Zhou [448], Lepeltier and San Martin [578, 579], Ma and Yong [607], Peng [706], Mania and Tevzadze [619].

Applications to Finance: Bielecki et al. [89, 98], El Karoui and Quenez [308], Lim [589].

BSDE with Jumps: Barles et al. [43], Royer [744], Nualart and Schoutens [683], Rong [743].

M. Jeanblanc, M. Yor, M. Chesney, *Mathematical Methods for Financial Markets*, Springer Finance, DOI 10.1007/978-1-84628-737-4,

B.1.4 Martingale Representation Theorems

Amendinger [12], Cherny and Shiryaev [170], Chou and Meyer [180], Davis [222, 219], Jacod [467, 468], Jacod and Yor [472], Kunita [549], Kunita and Watanabe [550], Nikeghbali [674], Nualart and Schoutens [682], Watanabe [837].

B.1.5 Enlargement of Filtrations

Theoritical Point of View: Amendinger et al. [13], Barlow [45], Brémaud and Yor [126], Dellacherie, Maisonneuve and Meyer [241], Föllmer and Imkeller [348], Jacod [468, 469], Jeulin [493, 494], Jeulin and Yor [495], Mansuy and Yor [622], Nikeghbali [674], Protter [727], Yor [868].

Application to Finance: Amendinger et al. [13], Bielecki et al. [91, 92, 93], Corcuera et al. [194], Elliott et al.[315], Eyraud-Loisel [338], Föllmer et al. [353], Gasbarra et al. [374], Grorud and Pontier [410], Hillairet [436], Imkeller [457], Imkeller et al. [458], Karatzas and Pikovsky [512], Kohatsu-Higa [533, 532], Kohatsu-Higa and Øksendal [534], Kusuoka [552].

B.1.6 Exponential Functionals

Alili et al [7], Bertoin and Yor [83], Matsumoto and Yor [630, 631], Yor [865].

B.1.7 Uniform Integrability of Martingales

Cherny and Shiryaev [169], Kramkov and Shiryaev [544], Heyde and Wang [848], Lépingle and Mémin [580], Novikov [680], Kazamaki [517], Yan [855].

B.2 Particular Processes

B.2.1 Ornstein-Uhlenbeck Processes

Alili et al. [10], Aquilina and Rogers [22], Breiman [122], Elworthy et al. [319], Going-Jaeschke and Yor [397], Linetsky [591].

B.2.2 CIR Processes

Chen and Scott [164], Cox et al. [206], Dassios and Nagaradjasarma [214], Deelstra and Parker [229], Delbaen [230], Dufresne [281], Linetsky [591], Maghsoodi [615], Rogers [736] and Shirakawa [790], Szatzschneider [816, 817].

B.2.3 CEV Processes

Boyle and Tian [121], Campi [138, 139], Cox [205], Davydov and Linetsky [225, 227], Delbaen and Shirakawa [238], Emanuel and MacBeth [320], Lewis [587], Hagan et al. [417, 418], Heath and Platen [428], Linetsky [592, 591], Lo et al. [601], Schroder [772].

B.2.4 Bessel Processes

Aquilina and Rogers [22], Carmona [140] Chen and Scott [164], Dassios and Nagaradjasarma [214], Deelstra and Parker [229], Delbaen [230], Davydov and Linetsky [225], Delbaen and Shirakawa [238], Dufresne [279], Duffie and Singleton [275], Grasselli [402], Heath and Platen [428], Geman and Yor [383], Going-Jaeschke and Yor [385], Hirsch and Song [437], Leblanc [572], Linetsky [594], Pitman and Yor [715, 717], Shirakawa [790].

B.3 Processes with Discontinuous Paths

B.3.1 Some Books

Bichteler [88], Brémaud [124], Dellacherie and Meyer [242, 244], He et al. [427], Jacod and Shiryaev [471], Kallenberg [505], Prigent [725], Protter, version 2.1. [727], Shiryaev [791].

B.3.2 Survey Papers

Bass [57, 58], Kunita [549], Runggaldier [750].

B.4 Hitting Times

Alili and Kyprianou [8], Alili et al. [10], Bachelier [39, 40], Barndorff-Nielsen et al. [52], Biane, Pitman and Yor [85], Borodin and Salminen [109], Borovkov and Burq [110], Breiman [122], Daniels [210], Dozzi and Vallois [265], Durbin [285], Ferebee [344], Freedman [357], Geman and Yor [384], Going-Jaeschke and Yor [397], Groeneboom [409], Hobson et al. [445], Iyengar [466], Jennen and Lerche [491, 492], Kent [519, 520, 521], Kou and Wang [541, 542], Kunitomo and Ikeda [551], Knight [528], Lerche [581], Linetsky [591, 592], Mordecki [659], Novikov [678], Patras [698], Peskir and Shiryaev [708], Pitman and Yor [715, 719], Ricciardi et al. [731, 732], Robbins and Siegmund [734], Roynette et al. [745], Salminen [753] Siegmund and Yuh [798], Yamazato [854].

B.5 Lévy Processes

B.5.1 Books

Applebaum [21], Bertoin [78], Boyarchenko and Levendorskii [120], Cont and Tankov [192], Doney [260], Jacod and Shiryaev [471], Kyprianou [553], Overhaus et al. [690], Sato [761], Schoutens [766],

B.5.2 Some Papers

Alili and Kyprianou [8], Asmussen et al. [24], Avram et al.[31, 32], Barndorff-Nielsen et al. [53, 54, 55], Bertoin [80, 81, 79], Bingham [100], Boyarchenko and Levendorskii [118, 119, 117], Carr et al. [155, 150], Chan [160, 159], Cherny and Shiryaev [170], Chesney and Jeanblanc [174], Corcuera et al. [195], Eberlein [289, 293, 292], Fajardo and Mordecki [339], Geman et al. [380, 381], Goll and Kallsen [400], Greenwood and Pitman [407], Hilberink and Rogers [435], Klüppelberg et al. [526], Kyprianou and coauthors [554, 555], Madan and coauthors [155, 609, 610, 612], Mordecki [659], Nguyen-Ngoc and Yor [670, 671], Nualart and Schoutens [682], Øksendal and Sulem [685], Sato [762], Skorokhod [801, 802], Zolotarev [878].

B.6 Some Books on Finance

B.6.1 Discrete Time

Avellaneda and Laurence [28], Elliott and Van der Hoek [318], Föllmer and Schied [350], Pliska [721], Koch Medina and Merino [531], Mel'nikov [638], Shreve [794].

B.6.2 Continuous Time

Avellaneda and Laurence [28], Bingham and Kiesel [101], Björk [102], Dana and Jeanblanc [209], Elliott and Kopp [316], Hunt and Kennedy [455], Kallianpur and Karandikar [506], Karatzas and Shreve [514], Lamberton and Lapeyre [559], Lewis [587], Musiela and Rutkowski [661], Overhaus et al. [688, 690], Portait and Poncet [723], Williams [845], Shreve [795].

B.6.3 Collective Books

The Shiryaev Festschrift [501], the Madan Festschrift [365], Paris-Princeton Lectures [143, 144, 145], Bressanone courses [363, 749], volume in honor of D. Sondermann, [759], Ritsumeikan conferences [2, 3, 4, 5], Option pricing, Interest Rates and Risk Management [498], Isaac Newton Institute Publications [248, 249], Trends in Mathematics [535], Bachelier Conference 2000 [378], IMA [218]

B.6.4 History

Bachelier [39, 41], Broadie and Detemple [129], Davis [221], Embrechts [321], Girlich [392], Gobet [395, 396], Jarrow and Protter [480], Merton [644], Rogers [738], Samuelson [758], Rogers [738], Scholes [764] Taqqu [819]

B.7 Arbitrage

Björk [102], Bühlmann et al. [136], Cherny [167], Cherny and Shiryaev [170], Delbaen and Schachermayer [236, 233, 235], Harrison and Kreps [421], Imkeller et al. [458], Kabanov [500], Levental and Skorokhod [583], Rogers [737] Selivanov [778], Schachermayer [763], Stricker [809], Xia and Yan [851, 852].

B.8 Exotic Options

B.8.1 Books

Bhansali [84], Brockaus et al. [689], Detemple [252], Haug [425], Lewis [587], Lipton [596], Musiela and Rutkowski [661], Taleb [818], Wilmott et al. [847, 846].

B.8.2 Articles

Asian Options: Bayraktar, E. and Xing, H. [62], Boughamoura et al. [113], Bellamy [69], Buecker and Kelly-Lyth [135], Chan [160], Dassios and Nagaradjasarma [214], Donati Martin et al. [259], Dufresne [279, 280, 281], Geman and Yor [383, 384], Henderson [430, 432], Rogers and Shi [740], Schröder [771, 767].

American Options: Asmussen et al. [24], Boyarchenko and Levendorski [119], Carr and coauthors [149, 155], Detemple [252] Villeneuve [829], Zhang [872].

Barrier Options: Andersen et al. [16], Avram and Chan [31], Baldi et al. [42], Boyarchenko and Levendorski [118], Broadie et al. [130], Carr and coauthors [148], Davydov and Linetsky [226], Kunitomo and Ikeda [551], Heynen and Kat [434], Pelsser [704], Rich [733], Roberts and Shortland [735], Rubinstein and Reiner [746], Sepp [781], Schröder [768], Suchanecki [814].

Parisian Options: Anderluh and Van der Weide [14, 15], Avellaneda and Wu [30], Bernard et al.[77], Chesney et al. [175], Cornwall et al. [196], Dassios [213], Dassios and Wu [217], Schröder [770].

Various: Boyle and Tian [121], Conze and Viswanathan [193], Dassios [211], Davydov and Linetsky [227], Delbaen and Yor [239], Douady [262, 263], El Karoui and coauthors [297, 300, 298], Fujita et al. [367], Fusai [371, 370], Garman and Kohlhagen [373], Goldman et al. [399], Guo and Shepp [412], He et al. [426], Henderson and Hobson [431], Lo et al. [600], Linetsky [593, 592], Nguyen-Ngoc and Yor[670].

Index of Authors

M. Jeanblanc, M. Yor, M. Chesney, *Mathematical Methods for Financial Markets*, Springer Finance, DOI 10.1007/978-1-84628-737-4,

Index of Symbols

M. Jeanblanc, M. Yor, M. Chesney, *Mathematical Methods for Financial Markets*, Springer Finance, DOI 10.1007/978-1-84628-737-4,

Subject Index

M. Jeanblanc, M. Yor, M. Chesney, *Mathematical Methods for Financial Markets*, Springer Finance, DOI 10.1007/978-1-84628-737-4,